D0204276

About Island Press

Island Press is the only nonprofit organization in the United States whose principal purpose is the publication of books on environmental issues and natural resource management. We provide solutions-oriented information to professionals, public officials, business and community leaders, and concerned citizens who are shaping responses to environmental problems.

Since 1984, Island Press has been the leading provider of timely and practical books that take a multidisciplinary approach to critical environmental concerns. Our growing list of titles reflects our commitment to bringing the best of an expanding body of literature to the environmental community throughout North America and the world.

Support for Island Press is provided by the Agua Fund, The Geraldine R. Dodge Foundation, Doris Duke Charitable Foundation, The Ford Foundation, The William and Flora Hewlett Foundation, The Joyce Foundation, Kendeda Sustainability Fund of the Tides Foundation, The Forrest & Frances Lattner Foundation, The Henry Luce Foundation, The John D. and Catherine T. MacArthur Foundation, The Marisla Foundation, The Andrew W. Mellon Foundation, Gordon and Betty Moore Foundation, The Curtis and Edith Munson Foundation, Oak Foundation, The Overbrook Foundation, The David and Lucile Packard Foundation, Wallace Global Fund, The Winslow Foundation, and other generous donors.

ENERGY FOR SUSTAINABILITY

ENERGY FOR SUSTAINABILITY
Technology, Planning, Policy

John Randolph
Virginia Tech

Gilbert M. Masters
Stanford University

ISLANDPRESS

Washington • Covelo • London

Island Press is a trademark of The Center for Resource Economics.

Randolph, John.
Energy for sustainability : technology, planning, policy / John Randolph and Gilbert Masters.
p. cm.
ISBN-13: 978-1-59726-103-6 (cloth : alk. paper)
ISBN-10: 1-59726-103-3 (cloth : alk. paper)
ISBN-13: 978-1-59726-104-3 (pbk. : alk. paper)
ISBN-10: 1-59726-104-1 (pbk. : alk. paper)
1. Renewable energy sources. I. Masters, Gilbert M. II. Title.
TJ808.R36 2008
333.79–dc22
2007040768

British Cataloguing-in-Publication data available.

Printed on recycled, acid-free paper.

Design by Black Dot Group

Manufactured in the United States of America

10 9 8 7 6 5 4 3 2 1

Keywords: natural resources, transportation, fossil fuels, oil, electricity, solar, wind, power, heating, land use, climate change, efficiency, renewable energy, green building, community planning

Dedication

*To Sandy and Mary, who are the sources of our inspiration,
and to our boys who inherit the world we leave them.*

Contents

Preface

Energy has emerged as one of the most significant and pervasive issues society will face in the twenty-first century. Our 40 percent dependence on oil raises questions of national security and economic stability. Our 85 percent dependence on fossil fuels is generating carbon emissions that are causing global warming, perhaps the most severe environmental problem of the century. Our exploding global demand for energy s inequitable and exacerbates both oil and carbon problems.

This book is based on the premise that our current patterns of global and U.S. energy use are not sustainable. Further, it contends that, although a mix of energy sources will continue to be necessary, the security and sustainability of our energy use depend on significant improvements in efficiency of use and increased development of renewable energy systems. The book provides the rationale for this premise and presents the technical background, system design fundamentals, economic analysis, and planning and policy approaches to advance the transition to more sustainable patterns of energy production and consumption.

As a critical public, economic, environmental, and social issue, energy has become an interdisciplinary field, yet few citizens and government and industry leaders fully appreciate or have sufficient understanding of its many dimensions, challenges, and uncertainties. As a result, this book is interdisciplinary to provide the multidimensional perspectives and knowledge necessary to achieve sustainable energy. To cross disciplinary boundaries necessary for people to understand and implement energy solutions, we see two audiences for this book: the engineers and scientists (the "techies") who must develop the energy systems needed to transform our patterns of energy use; and the social scientists, planners, and policy makers (the "fuzzies") who must develop the social, economic, and political case necessary to adopt the policies and public attitudes needed to transform our patterns of energy use. To achieve sustainable energy, each of these groups must understand the other. We need to make the "techies" more fuzzy and the "fuzzies" more techy, and this book attempts to address that challenge.

To do this, the book integrates energy data analysis, engineering design, life-cycle economic cost-effectiveness and environmental impact assessment, as well as planning and policy measures. The book is the result of a combined sixty-five years of teaching and researching energy patterns, efficient and renewable energy systems, and energy planning and policy. It has six sections:

I. Energy Patterns and Trends
II. Energy Fundamentals
III. Buildings and Energy

The book's first section provides "energy literacy" for the user by reviewing the importance energy plays in our economy, our environmental quality, and our quality of life; our current patterns of global and U.S. production and consumption; and future scenarios for energy. The second section provides a primer on energy physics, engineering, and economics as it relates to our production, conversion, and consumption of energy.

Sections III, IV, and V explore the energy technologies and opportunities in three of our most important energy sectors: buildings, electricity, and transportation. There are some policy issues discussed in the first five sections of the book, and policy and planning are the focus of section VI. It presents fundamentals of energy policy and the critical role of public policy and consumer choice in transforming energy markets to greater sustainability. Energy is one of the most complex problems of the new century, and the book argues that sustainable energy is within our grasp and uses current developments in technology, planning, and policy as hopeful signs.

Another premise of this book is that you cannot understand energy without understanding the numbers. Techies appreciate this; fuzzies may not. Throughout the book, analytical methods for energy and economic analysis aim to give users a quantitative appreciation for and understanding of energy systems. The emphasis is on simple, practical, mostly back-of-the-envelope and spreadsheet-based tools for design, sizing, and analysis of small-scale systems, including assessment of economic cost-effectiveness. In addition to analytical methods, the book uses case studies extensively to demonstrate current experience and illustrate the possibilities.

In late 2007, energy has become a fast moving field. While the intent of this book is to deal with basics and not fads, it also aims to articulate the prospects and possibilities we now face. Therefore, it is necessary to deal with recent developments in buildings (e.g., Green Buildings, "zero-energy buildings"), electricity (e.g., distributed energy, rooftop photovoltaics, wind farms, vehicles-to-grid), and transportation (e.g., plug-in hybrid, all-electric, and flex-fuel vehicles; biofuels; and transit-oriented land development), potential integrating notions such as Whole Community Energy, and energy policy (the significant energy and climate action by U.S. states and localities and the European Union).

But the energy world continues to change rapidly, and serious students of energy must work to keep up with new developments. As we finish this manuscript at the end of 2007, the global community has agreed in Bali to take the next steps to reduce GHG emissions beyond the Kyoto agreement, and the U.S. Congress has approved the latest federal energy act. Most analysts believe both of these initiatives fall well short of the policy actions needed for a timely transition to sustainable energy.

We encourage users of this book to consult the book Web site http://energyforsustainability.org for valuable links to updated and supplementary information. The Web site also contains an instructor's guide including problem sets and discussion questions to assist teaching and learning the book's practical analytical tools and planning and policy approaches.

Acknowledgments

This book is our own, but it also builds on a foundation of incredible work by countless institutions, organizations, and individuals, all are dedicated to the quest for sustainable energy. Their efforts have affected how we envision our energy situation and future, and many of the products of their work are reflected in this book. Their work is an inspiration not only for our work but also for the future of the world.

Among the individuals, we must list Donald Aitken, Peter Calthorpe, Ralph Cavanaugh, Howard Geller, Denis Hayes, John Holdren, Daniel Kammen, Skip Laitner, Amory Lovins, Stephen Nadel, Arthur Rosenfeld, and Robert Socolow, to name but a few.

Among the organizations and associations, we acknowledge the work of the American Council for an Energy Efficient Economy (ACEEE), the American Wind Energy Association (AWEA), the International Solar Energy Society (ISES), the Natural Resources Defense Council (NRDC), the Pew Center for Global Climate Change, the Renewable Fuels Association (RFA), the Rocky Mountain Institute (RMI), the Union of Concerned Scientists (UCS), and the U.S. Green Building Council (USGBC).

And among international and governmental agencies, the following continue to expand our understanding of the complex energy issues we face: the International Energy Agency (IEA), the Intergovernmental Panel on Climate Change (IPCC), the European Commission, the California Energy Commission, and the U.S. Department of Energy and its many arms, including the Energy Information Administration (EIA), the National Renewable Energy Laboratory (NREL), the Lawrence Berkeley Laboratory (LBL), the Oak Ridge National Laboratory (ORNL), and the other national labs.

We also wish to thank the Island Press staff for their insights and expertise in bringing this project to fruition, especially Todd Baldwin, Emily Davis, and Katherine Macdonald. In addition, the Black Dot Group for project management and copyediting.

Finally, we thank our many colleagues and students who have inspired and pushed us to think through and think anew about sustainable energy and its many dimensions in science, technology, economics, planning, policy, and society.

Energy Patterns and Trends

CHAPTER 1

The Energy Imperative and Patterns of Use

Energy is the keystone of nature and society. All life on Earth is made possible by incident solar energy captured and stored by plants and passed through ecosystems. Human civilization was spawned by innovation in acquiring and using diverse sources of energy, first by cultivating plants and domesticating animals and eventually by building machines that could use energy stored in fossil fuels. In fact, each phase of development of civilization was triggered by changes in energy use that provided opportunities for growth of human populations and economic systems.

Today, human society is in an unprecedented growth period. Since 1850 and the dawn of the Industrial Revolution, the population, the economy, and energy use have surged, fueled by oil, natural gas, and coal. This growth will soon be limited by diminishing availability of oil and gas and environmental constraints on fossil fuel use, probably sooner than most realize.

Some envision catastrophe ahead, characterized by abrupt climate change resulting from increasing carbon emissions from fossil fuel consumption, or constraints on oil and natural gas supplies, or political and military upheaval over access to energy resources, or economic depression triggered by increasingly volatile and rising energy prices, or all of the above.

Others see our beginning a period of transition to stabilized population and sustainable energy. **Sustainability** is defined as patterns of economic, environmental, and social progress that meet the needs of the present day without reducing the capacity to meet future needs. **Sustainable energy** refers to those patterns of energy production and use that can support society's present and future needs with the least life-cycle economic, environmental, and social costs. By **life cycle,** we mean the cost of a product from acquiring its original raw materials to manufacturing, transporting, and using it to its final demolition and disposal. Life-cycle analysis is fundamental to sustainability because it aims to capture full costs over an extended time period.

Global population is currently forecast to rise to about 9 billion by 2050 and stabilize by 2100. Although there is a huge appetite for more energy, especially among developing countries, ultimately the current growth in energy use may slow along with slower population growth. Population stabilization and slower energy growth do not mean that economic growth would also subside: as energy efficiency improves and structural changes in the economy continue to divorce it from growth of energy, labor, and materials, the economy may continue to grow despite slower population and energy growths.

The critical uncertainty is whether the transition in population and energy use will occur soon enough to avoid a potentially catastrophic situation. Already we are witnessing the symptoms of climate change, energy price volatility, and political turmoil.

In the years before his untimely death in 2005, Nobel Laureate Richard Smalley (2005) characterized the world's quest for sustainability in the following ten prioritized problems:

1. Energy
2. Water
3. Food
4. Environment
5. Poverty

6. Terrorism and war
7. Disease
8. Education
9. Democracy
10. Population

Smalley argued that energy tops the list because abundant, available, affordable, clean, efficient, and secure energy would enable the resolution of all the other problems. We need energy to reclaim and treat water, grow food, and manage the environment. If we can provide food, water, and a clean environment, we need energy to arrest poverty and disease and expand education and communication. By meeting these basic needs, we can control the root causes of terrorism and war, expand democracy, and stabilize population. Energy is the key for achieving a sustainable world system.

Our need for energy to create order in the world stems from the second law of thermodynamics, which states that matter and energy tend to degrade into an increased state of disorder, chaos, or randomness. Only through a flow of quality energy through the system (and a corresponding flow of lower-quality energy out) can order and structure be created. A constant flow of energy is required to maintain that order. Nature and human society on Earth are able to produce order and structure only through their ability to acquire energy. Chapter 4 will cover this fundamental principle in greater detail.

1.1 Our Energy Dilemma

Today we have an energy dilemma. Simply put, our energy problem has three components:

- *Oil*—37% of world energy still comes from petroleum. Reserves are concentrated in the politically volatile Middle East, and the date when conventional oil production will peak looms closer.
- *Carbon*—The global climate is already changing due to carbon emissions from fossil fuels, which still provide 86% of our energy.

- *Expanding global demand*—The developing world needs more energy to achieve basic needs. China's energy use is doubling every decade. Global energy usage grew by 2% per year from 1970 to 2002 and 4.1% per year from 2002 to 2005.

And there are three complicating factors:

- *Progress is slow* toward alternatives to oil, carbon, and growth in demand. We are nearly as dependent on fossil fuels now as we were in the 1970s. Although demand growth in developed countries has slowed, it has been offset by the increasing demand in the developing world. World energy usage nearly doubled from 1975 to 2005, and we remain dependent on fossil fuels, especially oil.
- *Change is hard* because of uncertainty, social norms, and vested interests. Transition to sustainable energy faces barriers to change, including uncertainty about supply options and their impacts, economic and political interests that fight to protect their status quo, and people resistant to changing their behavior. Consumers continue to desire bigger cars and houses and more energy-consuming products.
- *Time is short.* The time to act was yesterday. Over the past three decades, the economy and environment have provided clear signals that our energy patterns are not sustainable. Despite these warnings, we have done little to alter our patterns of use.

In this chapter we provide background necessary to understanding the importance of energy in history and the current global and U.S. energy situations. After giving a historical view of changing energy patterns that parallels the development of human civilization, we describe recent patterns and trends of energy production and consumption. Then in Chapters 2 and 3 we discuss the environmental, geologic, and geopolitical implications of these energy trends and a number of future energy scenarios based on different assumptions of energy demand, economic factors, and policy directions, including scenarios that may accelerate the transition to a sustainable energy future.

1.2 Historical Perspective: Energy and Civilization

It is interesting to trace the history of human society and see how major milestones in population growth, technology, living standards, and economy are linked to changes in our ability to acquire and convert energy for useful purposes.

The discovery of striking stones to ignite fires for thermal uses, perhaps 100,000 years ago, appears to be the first conscious human-engineered energy conversion. The invention of the wheel and stone tools and the domestication of work animals extended mechanical energy uses in the period 8000 B.C. to 4000 B.C. From 4000 B.C. to 1000 B.C., thermal energy from wood fire, then coal, not only provided warmth and cooking but also was essential in developing both ceramic and metal materials, such as pottery (4000 B.C.), bronze (2500 B.C.), and iron (1500 B.C.).

After the start of the first century A.D., devices to harness water and wind power extended human ability to use mechanical power for grist milling and water pumping. There were 10,000 windmills and 5,600 water mills in England by A.D. 1400. The first windsail-driven boats were used by Egyptians about 2000 B.C., but advanced sailing after A.D. 1250 ushered in the age of trade and exploration.

Coal and certain oils were used for heat and illumination since A.D. 100. Coal and oil were later put to a different use as fuel for the newly developed steam engine of the 1800s, and later other heat and mechanical engines, which revolutionized industry, transportation, and mechanized agriculture. The first commercial oil well (1859), invention of the internal combustion engine (1877), oil discoveries in Texas (1901) and Iran (1908), invention of the airplane (1903), and the Model T and assembly production (1908) ushered in the age of petroleum, the automobile, and air transport.

The electrical age had its founding in the invention of the generator and motor (1831) but waited for further inventions of the electric light (1879), refrigeration (1891), and air-conditioning (1902), and development of electric companies and transmission (first in 1891), before taking off after 1920 and revolutionizing living standards.

After 1950, further growth of fossil fuels, electricity (including nuclear power), and related technologies for electronics and telecommunications, agriculture, manufacturing, and transportation, set the stage for global expansion and unprecedented population and economic growth.

Human population had been constrained to 1 billion people by 1800 by limits on energy use and technology. Before 1850, society had to rely on human and animal labor to plow fields, harvest crops, chop firewood, mine and haul coal, and transport people and materials. This drove the market not only for draft animals, but also for human African slaves in the United States and elsewhere. But after 1850, advances in industry, agriculture, transportation, and communication brought about by new energy technologies, freed society from the constraints of slave and animal labor and expanded agricultural and industrial productivity. Human population ascended to 2 billion by 1927, 4 billion by 1974, 6 billion by 1999, and 6.6 billion by 2007.

The world economy grew (in constant dollars) from an estimated $100 billion in A.D. 1 to $700 billion in 1820. The Industrial Revolution spawned a fivefold increase by 1970 to $3,500 billion, then another tenfold increase from 1970 to 2000 to $36,000 billion (OECD, 2001). Prior to 1980, most energy analysts believed energy consumption and economic growth were inextricably correlated, but after 1980 economic product grew far faster than both energy consumption and population. Higher energy prices and new technologies led to greater energy efficiency, and the service and information sectors grew faster than the energy- and material-intensive manufacturing sectors of the economy.

Table 1.1 and Figure 1.1 show the huge growth in population, economy, and energy use since 1800. These exponential growth rates are impressive historically as they have changed the nature of the world in which we live; but they are equally impressive as we look to the future and consider how we can sustain them, modify them, and live with the consequences.

table 1.1   Global Energy through History and Its Relationship to Population and the Economy

Date	Population	Economy	Energy Age
< A.D. 1	<0.3 billion	<0.1 T$	Human, animal power; wood thermal
1–1800	<1 billion	<0.7 T$	Human, animal power; wood/coal thermal; wind/water
1800–1900	1.5 billion	0.9 T$	Coal/steam power, telegraph, railroad, Industrial Revolution
1900–1950	2 billion	2.0 T$	Petroleum and electrical ages begin; automobile, telephone, air travel
1950–1980	4 billion	4.0 T$	Nuclear power; intensive agriculture, computer, space exploration
1980–2000	6 billion	36.0 T$	Information Age; energy and economic growth rates begin to diverge

figure 1.1   Global Growth of Population, Energy, and Economy, 1800–2000

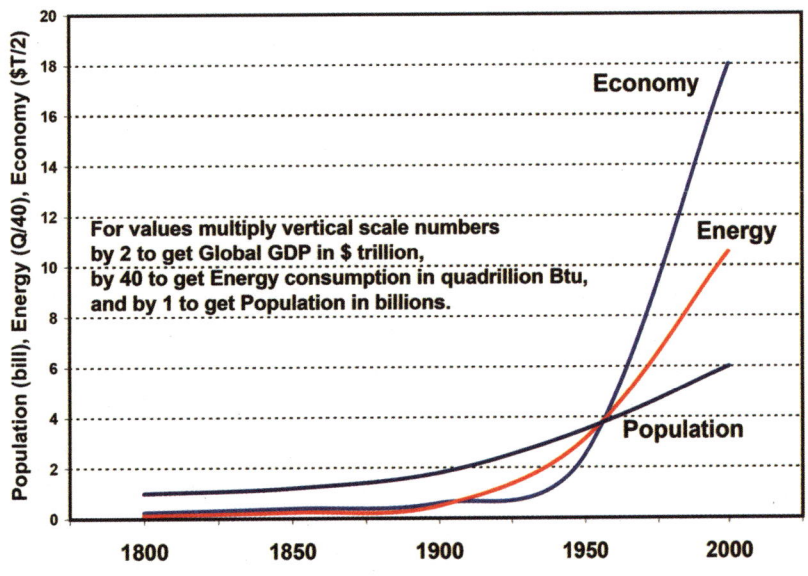

1.3 Global Energy Supply and Consumption

From this long-term perspective, let's zero in on the past several decades until today so that we can understand the current energy situation. We will look separately at global energy and U.S. energy. In Chapters 2 and 3, we will explore the implications of this situation and what it bodes for the future.

figure 1.2   **World Energy Consumption, 1860–2005**

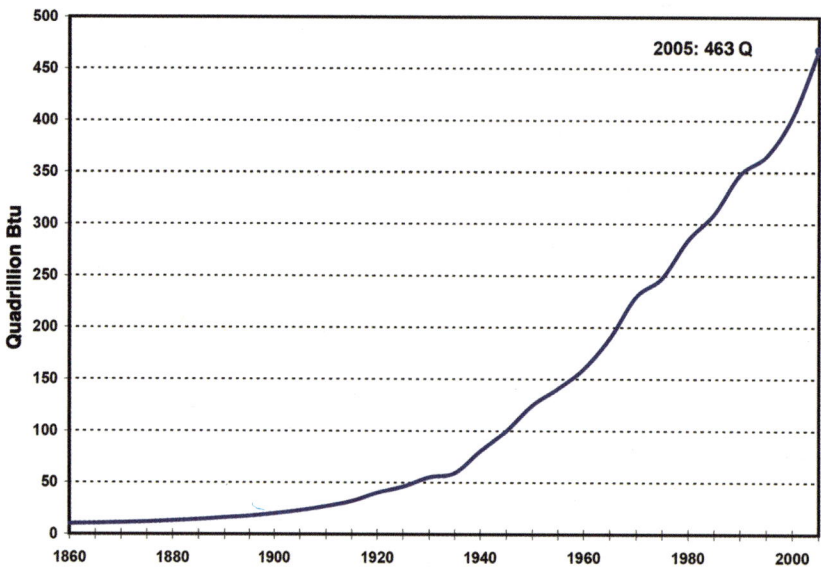

1.3.1 Explosive Growth of Energy, Inequitably Distributed

Energy supply and demand has exploded in the last century. As it has done so, our reliance on fossil fuels has grown. Not coincidentally, so has the disparity of use between industrialized and developing countries. Figure 1.2 shows the tremendous growth in commercial energy since 1860, especially since 1940. Consumption doubled from 1970 to 2005, growing at 2% per year until 2002 and 4.5% per year from 2003 to 2005, when consumption reached 463 quadrillion (10^{15}) British thermal units (Btu), or 463 quads. A **Btu** is a traditional English unit of energy equal to 1054 joules, the standard metric unit, and 0.293 watt-hour, the standard unit of electrical energy. A quad is one quadrillion (10^{15}) Btu and is a standard unit for national and global energy production and consumption used by U.S. energy agencies. Chapter 4 discusses energy units and conversion.

This rate of growth is likely to continue because of the expected demand by developing countries for energy. These countries have a big appetite for energy because of the current economic and energy disparity among rich and poor countries, shown in Table 1.2. Annual energy use worldwide averaged 72 million Btu per capita in 2005. While the developed countries, with 20% of the world population, consume energy at an annual rate of more than 150 million Btu per capita, the developing countries with 80% of the population, consume energy at a rate of less than 40 million Btu per capita. The average U.S. citizen consumed 340 million Btu in 2005; the average Japanese and Brit consumed about 170 million Btu; the average Chinese 51 million Btu (up from 33 in 2002); the average Indian 15 million Btu; the average Bengali 5 million Btu; and the average Ethiopian 1 million Btu per person.

table 1.2 Indicators of Energy, Economy, and Population for Selected Countries and the World, 2005

	Energy	Energy/GDP*	Energy/GDP**	% Pop	% Energy	% GDP*	% GDP**	% CO$_2$
Canada	**436**	17.4	13.8	0.5%	3.1%	2.3%	1.8%	2.2%
United States	340	9.1	9.1	4.6%	**21.8%**	**30.4%**	**19.2%**	**21.1%**
Australia	273	12.1	9.0	0.3%	1.2%	1.2%	1.1%	1.4%
Sweden	260	8.7	9.0	*0.1%*	0.5%	0.7%	0.4%	0.2%
Russia	212	**86.7**	**14.9**	2.2%	6.5%	1.0%	3.5%	6.0%
France	182	8.0	7.2	1.0%	2.5%	3.9%	2.7%	1.5%
Korea, South	191	14.5	12.5	0.8%	2.0%	1.8%	1.3%	1.8%
Germany	176	7.4	7.0	1.3%	3.1%	5.4%	3.6%	3.0%
Japan	177	*4.5*	6.5	2.0%	4.9%	13.8%	6.0%	4.4%
United Kingdom	166	6.1	6.0	0.9%	2.2%	4.5%	2.9%	2.0%
South Africa	114	31.5	10.0	0.7%	1.1%	0.4%	0.9%	1.5%
Mexico	65	10.8	6.6	1.6%	1.5%	1.8%	1.8%	1.4%
Brazil	50	13.9	6.3	2.9%	2.0%	1.8%	2.6%	1.3%
China	51	35.8	7.9	**20.3%**	14.5%	5.2%	14.7%	18.9%
Indonesia	23	25.3	5.8	3.6%	1.2%	0.6%	1.6%	1.3%
India	15	24.8	4.0	17.0%	3.5%	1.8%	7.0%	4.1%
Pakistan	14	30.0	5.3	2.5%	0.5%	0.2%	0.7%	0.4%
Nigeria	8	16.7	6.6	2.0%	0.2%	0.2%	0.3%	0.4%
Bangladesh	5	11.8	*1.1*	2.2%	0.1%	0.2%	1.1%	0.1%
Ethiopia	*1*	11.0	1.5	1.1%	*0.0%*	*0.0%*	*0.1%*	*0.0%*
World	72	12.7	8.0	6,445	463	43,920	55,500	28,193
units	million	1000Btu/$GDP	1000Btu/$GDP	million	quad Btu	billion $	billion $	million

Highest values in **bold**; lowest values in *italic bold*
* GDP data based on market exchange rates, a traditional measure of GDP.
** GDP data based on product purchasing power, which more accurately reflects strength of the national economy.
SOURCE: U.S. EIA (2007), *International Energy Annual 2005*

Table 1.2 gives other indicators of energy use disparity. The United States, with less than 5% of the world's population, accounted for 20 to 22% or about 4 to 5 times its share of the world's energy consumption, economic output, and carbon dioxide emissions in 2005. This disparity in energy use is important. It is an indicator of the economic and social disparity in the world, and we will never achieve a sustainable world system until basic human and societal needs for food, education, employment, transportation, and other necessities are met. As developing countries advance, they will require a huge increase in their energy use to fuel the industry, transportation, electrification, telecommunications, and human services to provide these basic needs.

One hopeful sign is that the United States and other countries have reduced the energy intensity of their economies (see Equation 1.1). **Energy intensity** indicates how much a national economy is dependent on energy per unit of economic output, or gross domestic product (GDP). It is measured in energy/$GDP. If energy intensity is low, then energy efficiency is high in that economy. Since 1980, global energy intensity has decreased by 25% (Figure 1.3); the United States has improved by 35%. Although the U.S. economy is considerably more energy intensive than those of Japan, Germany, France, the United Kingdom, and Sweden, it is ten times more energy efficient than Russia's economy based on energy/$GDP.

Eq. 1.1
$$\text{Energy intensity} = \frac{\text{energy used}}{\$\text{GDP}}$$

Figure 1.3 plots indicators of change from 1980 to 2004 on a scale normalized to 1980 values. The graph shows the large increase in the global economy, which has outpaced population by 80%, and the 45% increase in carbon dioxide emissions since 1980. Energy consumption growth has tracked population growth very closely, and this is shown in energy/capita that has been relatively constant.

Increase in developed countries

Eq. 1.2 $$\text{Average energy per capita} = \frac{\text{energy used}}{\text{Person}} = \frac{\text{Total energy used}}{\text{Total population}}$$

Increase in developing countries

figure 1.3   **Global Indicators of Change, 1980–2004**

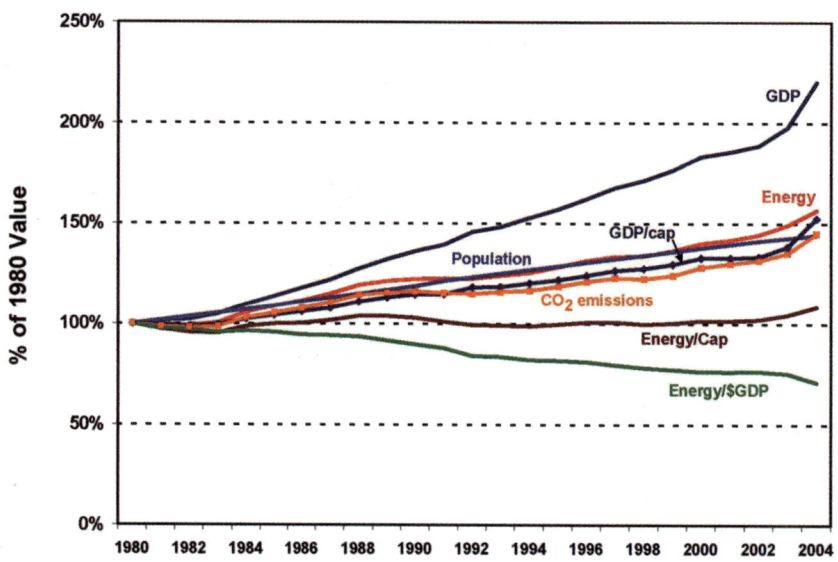

Energy consumption, population, GDP, energy/capita, GDP/capita, energy intensity, carbon dioxide emissions.

Source: data from U.S. EIA, 2007b

Average energy/capita is calculated by dividing total energy consumption by total population. This average disguises the high variance among developed countries (e.g., 436 for Canada) and less developed countries (e.g., 1 for Ethiopia). The average has remained constant because both energy used (the numerator in Equation 1.2) and population (the denominator) grew. The energy increased primarily in developed countries and population grew primarily in the less developed countries, so they offset each other to produce a near constant average energy per capita worldwide between 1980 and 2000, even though the disparity increased.

1.3.2 Continuing Dependency on Oil and Fossil Fuels

In 1973, the world got a wake-up call on the geopolitics of oil. The Organization of Petroleum Exporting Countries (OPEC) increased its influence on oil markets and the Arab oil embargo of the United States sent oil prices skyrocketing. Oil shocks in 1980 and 1991 gave a similar message about the volatility of oil supply and price, but the sources of world energy have changed little.

As Figure 1.4 shows, the world relied on fossil fuels for 90% and on oil for 46% of its commercial energy in 1980. However, those percentages dropped only slightly by 2005, to 86% and 37%, respectively. And in quantity, world oil consumption increased 33% and total fossil fuels increased 40% in that time, with a corresponding increase in carbon dioxide emissions.

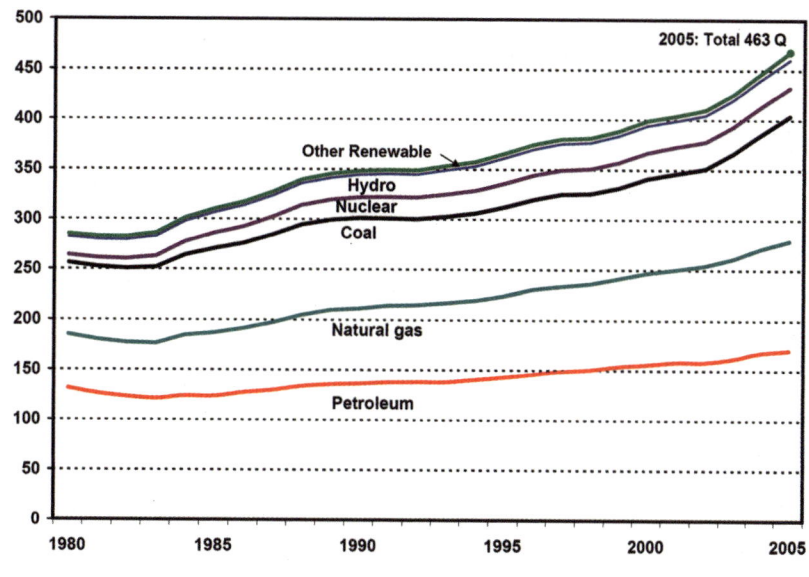

figure 1.4 World Energy Consumption by Source, 1980–2005

SOURCE: data from U.S. EIA, 2007b

Other sources of energy have not made a difference. Nuclear power grew to 6.5% by 1990, but has slipped to 5.9% in 2005. Hydroelectric power has not grown, and renewable power from wind, solar, geothermal, and biomass electricity, despite fast growth since 1999, has yet to make an impact.

The bottom line is that, despite fossil fuel price volatility and supply disruptions, political conflicts affected by access to energy resources, and increasing recognition of the dangers of continued fuel combustion emissions on global climate, the world has done little to change its fossil-fueled patterns of energy use. Despite skyrocketing prices pushing $100/barrel, world oil demand in 2007 is rising twice as fast as in 2006 and is expected to hit 88 million barrels per day by the end of 2007.

With 20% of the world population consuming 75% of the world energy and controlling 75% of the global economy, while the poorest 80% struggle toward development, we are far from an equitable energy and economic system. Perhaps the best news in these trends is that our global economy has become less energy intensive through efficiency improvements and structural change. If this can be translated to the developing world, perhaps their economic advancement can occur without the large energy requirements experienced by the industrialized world.

Still, the material and energy requirements needed to support the physical infrastructure to provide for basic needs of very large populations will be enormous, and global energy consumption will have to grow considerably. Chapters 2 and 3 explore some of consequences of that growth and our future options.

1.4 U.S. Energy Supply and Consumption

We should investigate U.S. energy patterns in greater detail. The United States is the world's largest energy consumer, and its behavior has a significant influence on other developed and developing nations. In addition, if we are to avoid or delay looming impacts of these patterns as the rest of the world develops, there is no better place to start our investigation than by examining the United States.

There are volumes of data and analyses of U.S. consumption and production available from various sources, especially the U.S. Energy Information Administration (EIA), an arm of the Department of Energy (DOE). Using EIA data, this section highlights three dramatic trends that characterize the U.S. energy situation:

1. *Energy intensity of the economy has steadily declined while energy use per capita has remained constant.* This is good news. We are getting far more economic output from our energy, and we are not increasing our average use of energy per person despite driving more, occupying bigger houses, and using more energy gadgets, such as cell phones and computers. In other words, we are getting more out of the energy we use.

2. *Electricity has increased significantly, and 92% of it is generated by steam power that is inherently inefficient.* Electricity is increasingly the source of choice because of its multiple uses

and convenience. But fossil-fuel and nuclear steam power requires three units of source energy for every one unit of electricity generated, and these electricity losses amount to more than 27% of our total energy use.

3. *Oil imports have increased dramatically due to rising consumption for transportation and declining domestic oil production.* Two-thirds of United States oil is now imported from other countries, and this is likely to increase as the United States runs out of conventional oil reserves.

1.4.1 U.S. Patterns of Energy Consumption and Production

Before looking more closely at each of these issues, it is important to understand some basic patterns of energy production and use. If we are to achieve more sustainable energy use patterns, we need to know where our energy comes from and how we use it. We need to be able to access energy data and analyze them to answer questions about past trends and current uses. Solution Box 1.1 gives a short primer on accessing and interpreting energy data.

Figure 1.5(a) indicates the historic growth of energy consumption relative to domestic production and the sources of energy used. Total consumption doubled from 1950 to 1973, but the oil shocks of 1973 and 1980 led to higher prices, economic recession, and temporary declines in energy consumption. Since the early 1980s, energy use increased to about 100 quads per year in 2003–2007.

Consumption has outpaced domestic energy production, and the growing gap (30% in 2005) must be met with net imports, almost entirely petroleum. Figure 1.5(b) shows the sources of energy use in the United States from 1950 to 2006. The United States is nearly as dependent on fossil fuels today (85% in 2007) as it was in 1973 (93%). We have nearly doubled our use of coal in that time.

Oil and natural gas use declined with higher prices from 1975–1985, but use has increased since 1985. Oil is no longer used as much to generate electricity or heat buildings, and its growth is attributed to increased use in vehicles for transportation. Petroleum use reached a record high of 20.8 million barrels per day in 2005. In 2006, despite higher crude oil and gasoline prices, oil still contributed 40% of total U.S. energy.

Other sources are still small compared to the fossil fuels. Figure 1.5(b) shows that nuclear power has grown steadily since 1970, but still amounts to only 8% of total energy in 2006. No new nuclear plants have been added since the 1980s. Renewable energy still amounts to only 7% of total energy consumption in 2006. This includes hydroelectric production; wood, waste, ethanol, and other biomass energy; and commercial wind and solar electricity. Noncommercial biomass, such as residential wood heat, and solar heating are not included in these figures.

Figure 1.6 shows the sources and distribution of consumed energy to various sectors in 2005, and Figure 1.5(c) gives the trends in consumption in each sector. Industry is still the largest user (32%), but total industrial use has remained about the same since the 1970s,

SOLUTION

Accessing and Interpreting Energy and Related Economic Data

Most of the energy data presented in this chapter come from the databases of the U.S. Department of Energy's Energy Information Administration (EIA) (http://www.eia.doe.gov/) and the International Energy Agency (IEA) (http://www.iea.org/). Some additional economic data come from the U.S. Bureau of Economic Analysis (BEA) (http://www.bea.doc.gov/). To answer simple questions about energy production and consumption patterns and trends, energy analysts must know how to access data and conduct simple statistical calculations.

EIA data are well organized through two annual reports. The Annual Energy Review (AER) outlines domestic consumption by fuel and use sector, production by fuel, imports by source country, and energy resource data from 1949 to the most current year. The AER is produced in August each year with the previous year's data. More recent monthly data are provided in EIA's Monthly Energy Review (MER) and several other fuel-specific monthly reports. Data are given in tabular and graphical HTML and PDF formats, as well as downloadable spreadsheet format. EIA's International Energy Annual and the International Energy Agency provide online energy data for countries and regions, and a variety of other fuel- and country-specific information.

Accessing and downloading data is one thing but presenting and interpreting data is another. Most interpretation requires simple analysis including indicators such as averages, quantity per capita, energy per $GDP, and so on. Trend analysis incorporates time in the calculations and often uses percent change and average annual percent change as important indicators. The list below gives some simple equations for such analysis.

1. Average of a number of values of A:

$$\text{Average (mean)} = \frac{\text{Sum of } A}{\#\text{Values}} = \frac{(\Sigma A)}{N}$$

where A = value

N = number of values

What is the average energy consumption of the top five energy consuming countries in Table 1.2?

Solution: The top five countries (from Table 1.2, column 5): United States, 21.8%; China, 14.5%; Russia, 6.5%; Japan, 4.9%; Germany, 3.1%.

$$\text{Country Consumption} = \%\text{ Energy}/100 \times \text{Total World Energy}$$

$$\text{Average} = \frac{\Sigma\text{Country Consumption}}{\#\text{ Countries}} = \frac{(\Sigma\text{Country\%Energy}/100) \times \text{TWE}}{\#\text{ Countries}}$$

$$\text{Average} = \frac{(0.218 + 0.145 + 0.065 + 0.049 + 0.031) \times 463Q}{5} = \frac{(0.508) \times 463Q}{5} = 47 \text{ quads}$$

2. % change in values between time 1 and time 2:

$$\% \text{ change} = 100\left(\frac{V_2}{V_1} - 1\right)$$

where V_1 = value at time 1
V_2 = value at time 2

The EIA data show that the United States consumed 84.60 quads in 1991 and 98.16 quads in 2003. What is the percent change in consumption during those 12 years?

Solution: % change = $100\left(\dfrac{98.16}{84.60} - 1\right)$ = 100 (1.160 – 1) = 100 (0.160) = **16.0%**

3. Value at time 2 with constant or average periodic (e.g., annual) growth rate from time 1:

$$V_2 = V_1(1 + r)^n$$

where r = periodic (e.g., annual) growth rate
n = number of periods (e.g., years)

4. Average periodic (e.g., annual) growth rate from value at time 1 to value at time 2:

$$r = \left(\frac{V_2}{V_1}\right)^{1/n} - 1$$

where r is given as a decimal rate and the %rate = $R = 100r$

What is the average annual rate of change in U.S. consumption between 1991 and 2003?

Solution: $r = \left(\dfrac{98.16}{84.60}\right)^{1/12} - 1$ = $(1.160)^{1/12} - 1$ = 1.0125 – 1 = .0125 or **R = 1.25% per year**

If the U.S. consumption were to continue at the average annual growth rate it had from 1991 to 2003, what would its consumption be in 2010?

Solution: $V_{2010} = V_{2003}(1 + .0125)^7 = 98.16Q(1.0125)^7$ = **107.1 quads**

5. Doubling time for constant or average periodic rate of growth:

$$DT = \frac{70}{R}$$

where DT = doubling time
R = %rate = $100r$

If the U.S. consumption were to continue at the average annual growth rate it had from 1991 to 2003 (R = 1.25%), in what year would it be double the 2003 consumption, or 196 quads?

Solution: $DT = \dfrac{70}{1.25}$ = **56 years or in 2059**

figure 1.5a U.S. Energy: Consumption, Production, and Net Imports, 1950–2006

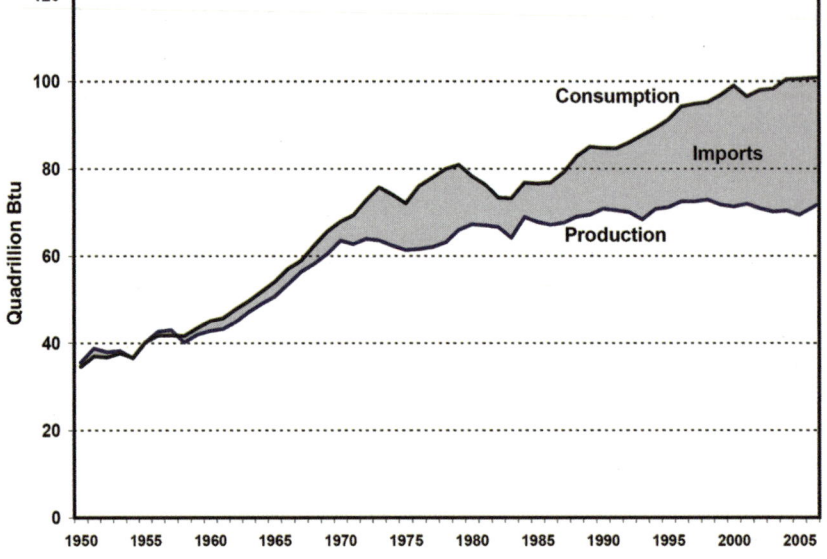

Note the widening gap between rising consumption and flattening domestic production. The gap (30% of consumption in 2005) is filled with net imports, especially petroleum.

figure 1.5b U.S. Energy Consumption by Source, 1950–2006

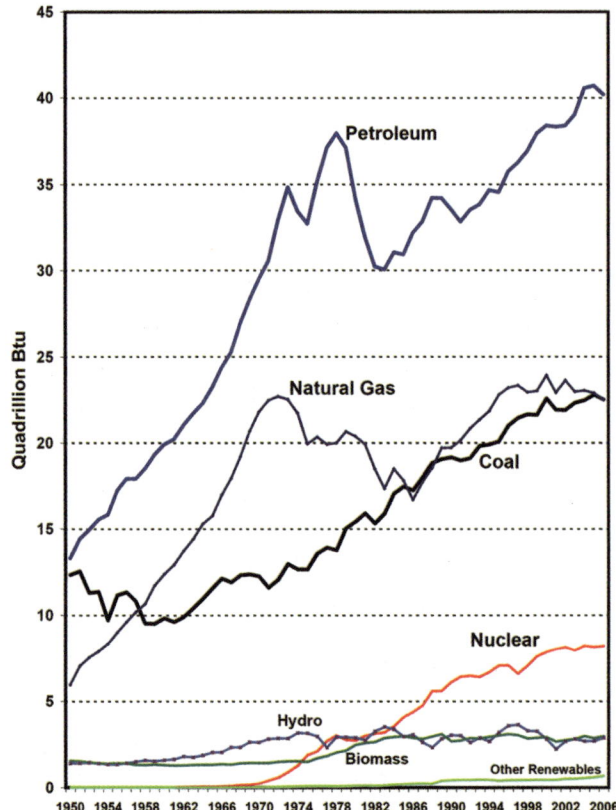

Note the continued heavy reliance on fossil fuels (86%), oil (40%), natural gas (23%), and coal (23%) in 2006. Nuclear power contributed 8% and renewable energy 6%.

SOURCE: data from U.S. EIA, 2007a, 2007c

figure 1.5c U.S. Energy Consumption by Sector, 1950–2006

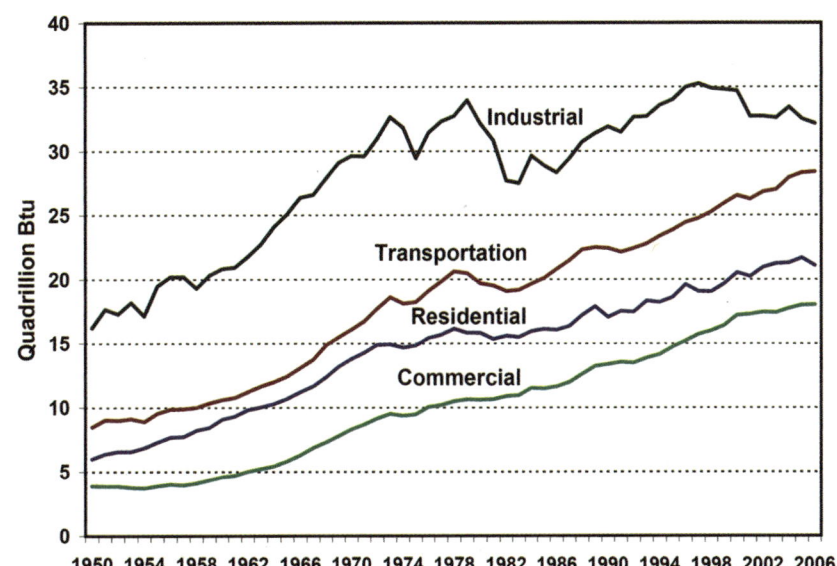

Note that industry is still the top energy consumer, but energy for transportation and residential and commercial buildings is rising at a faster rate. Taken together, buildings are the biggest consuming sector.

SOURCE: data from U.S. EIA, 2007a, 2007c

figure 1.6 Energy Flowchart for United States, 2005

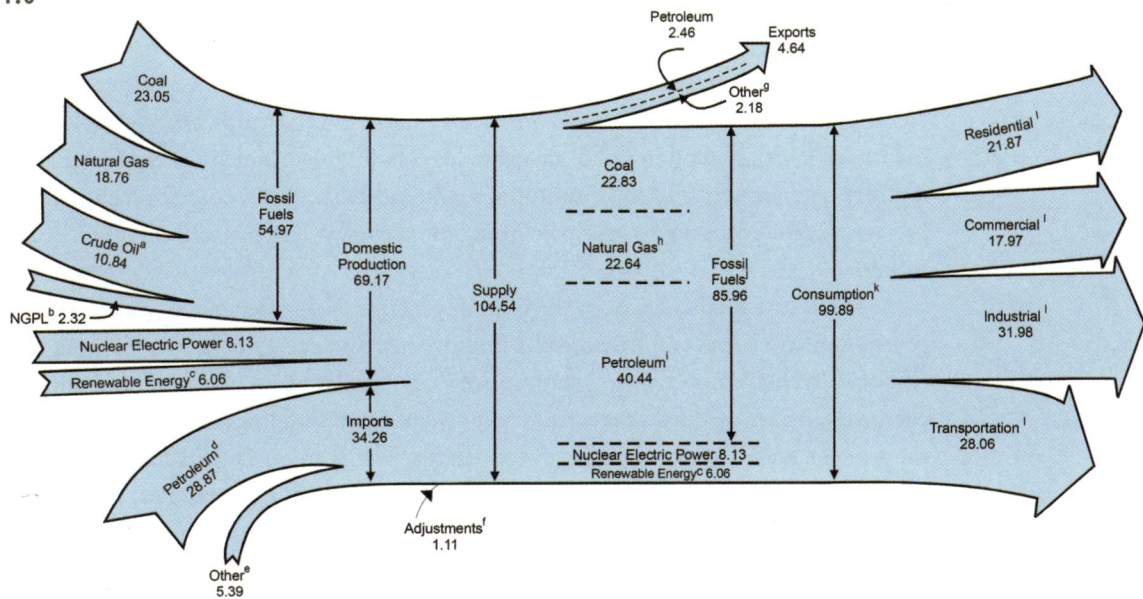

Energy sources on left and end uses on right: industrial (32%), transportation (28%), residential (22%), commercial (18%). Imports of petroleum are 28% of total consumption.

SOURCE: U.S. EIA, 2007b

despite some ups and downs. About 20% of industrial "energy" is actually used for nonenergy material feedstocks (e.g., asphalt, road oil, petrochemicals, lubricants, solvents). But energy use has grown steadily in all other sectors, especially transportation (now 28% of total use and 70% of oil) and residential and commercial buildings (40% of total use). Any effort to reduce energy demands, oil imports, and carbon emissions must focus on transportation and buildings.

1.4.2 U.S. Energy and Economy: Efficiency and Structural Changes

Amid concerns about stagnant energy production, shrinking domestic reserves of oil, and increasing oil imports, one bit of good news in U.S. energy patterns is that the energy intensity of the economy (see Equation 1.1, p. 10) has improved significantly. The main drivers of this improvement have been energy efficiency and structural changes in the economy.

From 1949 to 1973, energy steadily increased from 215 to 358 million Btu per capita, and energy intensity or the energy required per dollar GDP remained constant at 18 to 20 thousand Btu per $GDP. While most analysts in the early 1970s thought energy per capita might level off, many thought energy and economic growth were inextricably tied. Indeed, this theory was supported when both energy use and the economy declined in the mid-1970s and early 1980s. However, writers such as Daniel Bell foretold of structural changes in the economy. His book *The Coming of Post-Industrial Society* (1973) argued that we were moving toward post-industrialism that would be dominated by information, science-based industries, and services rather than material manufacturing.

Much of what Bell suggested has turned out to be true. The information-based economy began in 1980 and took off in the mid-1990s; science- and technology-based industries, such as computer and biotechnology, have grown much faster than traditional industry; and services (financial institutions, commercial services, entertainment, etc.) have become a mainstay of the economy, eclipsing more energy- and material-intensive manufacturing.

These structural changes in the economy to less energy-dependent modes of income generation have been complemented by improvements in energy efficiency. Spurred by new efficiency technologies, higher energy prices, and government mandates and incentives, energy users have invested in efficiency improvements in buildings and equipment that get the same or greater performance with less energy. These improvements have been made to vehicles, motors, furnaces, appliances, electronics, and building envelopes. We will see in later chapters that major opportunities for efficiency improvement remain untapped.

Figure 1.7 gives U.S. economic and energy consumption indicators since 1973. U.S. GDP has grown considerably. Despite four recessionary dips, the economy grew by 160% from 1973 to 2006, while energy use and population grew by about 40%. Energy per capita has remained relatively constant, actually dipping from 358 million Btu in 1973 to 334 million Btu in 2006. However, in constant 2000 $, energy use per $GDP has dropped by half from 17.4 thousand in 1973 to 8.7 thousand Btu per $GDP in 2006.

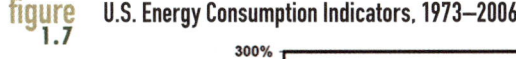 U.S. Energy Consumption Indicators, 1973–2006

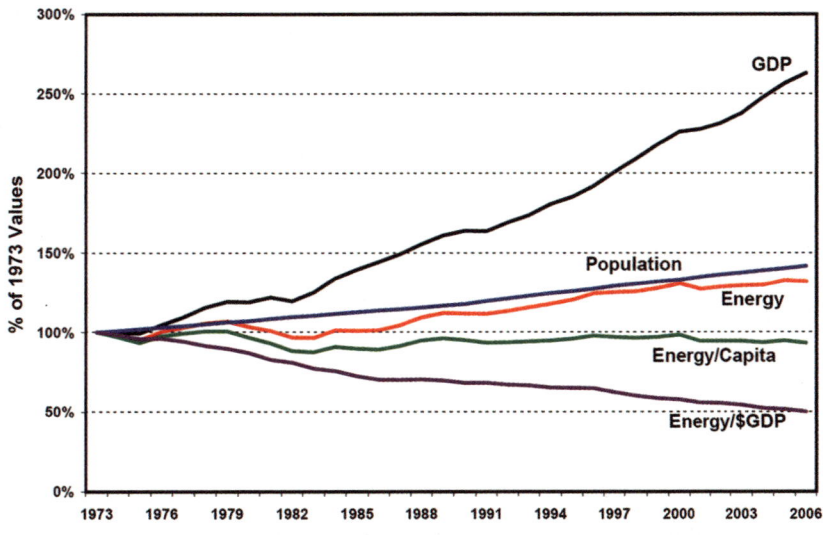

SOURCE: data from U.S. EIA, 2007a, 2007c

As Figure 1.7 shows, the most dramatic drop in energy intensity came in the "energy crisis" period of 1973–1986 and in the period of structural change in the economy from 1996–2000. Between 1949 and 1973, energy and economic growth were closely linked and energy intensity declined by only 0.4% per year. Between 1973 and 1986, however, higher energy prices prompted efficiency investments primarily in industry, vehicles, and buildings, and energy intensity dropped by 2.7% per year. The period from 1986 to 1996 saw minor improvement of 0.7% per year as investment in efficiency declined.

Between 1996 and 2000, the information economy driven by the "dot-coms" surged the economy forward with little increase in energy use, dropping energy intensity at an unprecedented rate of 2.8% per year. Even with the "bust" of the dot-coms and the sluggish economy during 2000–2005, energy intensity continued to drop at a rate of 2.1%. In addition to the lasting effects of the changing economic structure, higher energy prices in this period stimulated investment in efficiency and conservation.

Arthur Rosenfeld of Lawrence Berkeley Laboratory and the California Energy Commission developed Figures 1.8 and 1.9 to illustrate the economic effect of this trend. Figure 1.8 plots the actual drop in energy intensity from 1949 to 2005, and also continues the trend line from 1949 to 1973 (0.4% decline) to 2005. Figure 1.9 shows the resulting energy consumption if the trend to 1973 had continued. The improved energy intensity resulted in saving 70 quads of energy at a value of $0.7 trillion. Rosenfeld attributes one-third of this improvement to energy efficiency improvements in buildings, one-third to vehicle efficiency, and one-third to structural changes in the economy.

figure
1.8   **U.S. Energy Intensity, 1949–2005**

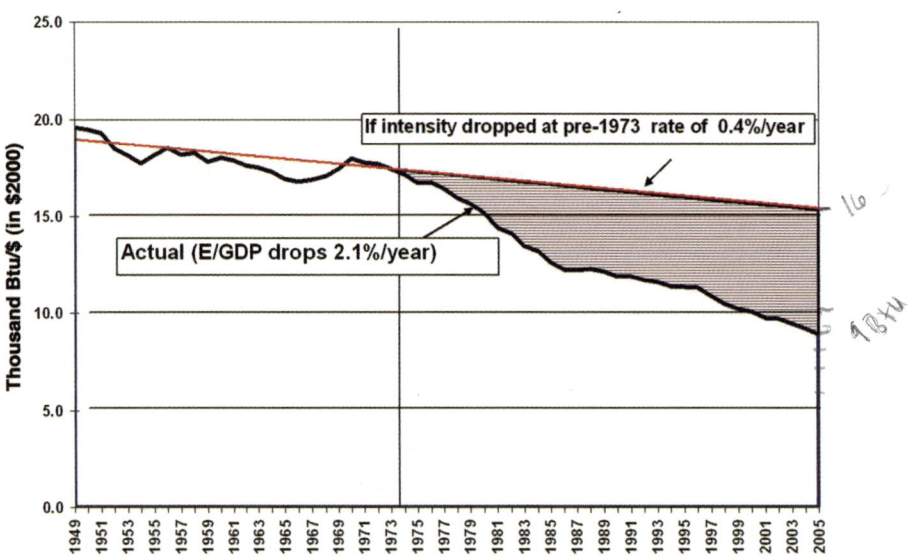

Actual U.S. energy intensity from 1949–2005 and continued trend line of 1949–1973. In year 2005, E/GDP would have been 16 Btu/$.

Source: Rosenfeld, 2006, used with permission

figure
1.9   **Effect of Improved Energy Intensity on U.S. Energy Consumption, 1973–2005**

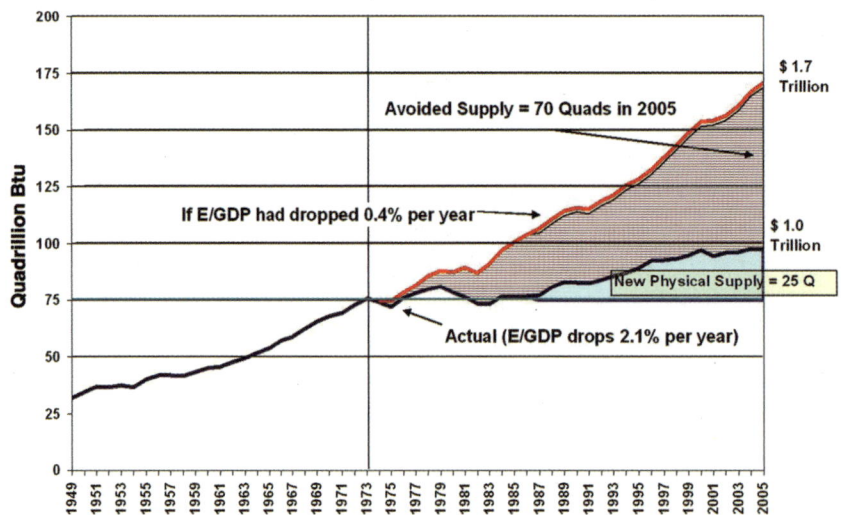

Actual U.S energy consumption and resulting consumption if pre-1973 energy intensity trend had continued.

Source: Rosenfeld, 2006, used with permission

figure 1.10   U.S. Energy Used for Generation of Electricity, 1950–2006

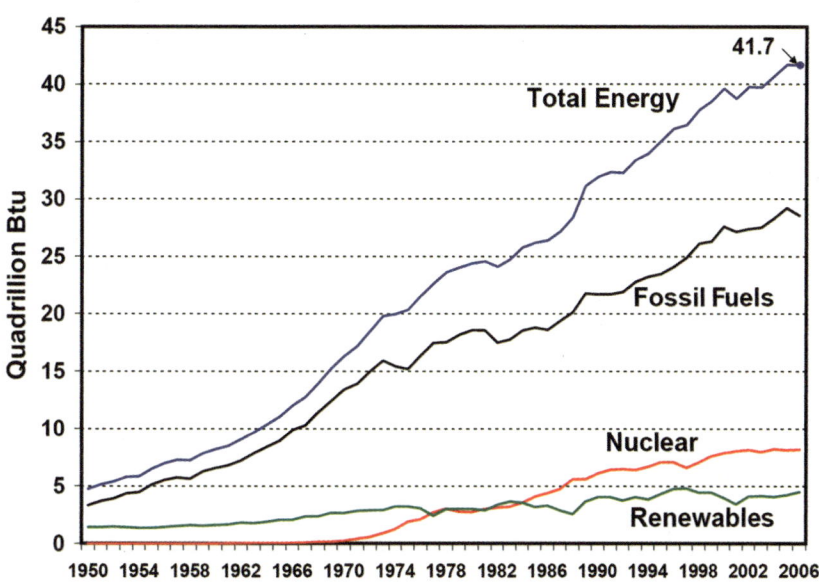

Continued high growth in electricity fueled mostly by coal (see Figure 1.11).

SOURCE: data from U.S. EIA, 2007a, 2007c

1.4.3 U.S. Electricity: Increasing Energy of Choice and Increasing Energy Conversion Losses

Another important trend in U.S. energy is the increasing reliance on electricity. Electricity is a high-quality form of energy that can be applied to a wide range of uses, including motors and electronics, lighting, refrigeration, heating and cooling, and rail transit. Expanding use of electronics, air-conditioning, and heat pump heating has contributed to electricity's growth. Some foresee further use of electricity for transportation for plug-in vehicles and light-rail transit (see Chapters 10, 13, and 15).

Electricity is versatile not only in its applications but also in its energy sources. It is the only practical way we can currently use coal, nuclear, hydro, wind energy, and solar photovoltaics on a large scale, and we can actually use any other form of energy to produce it, including oil, natural gas, biomass, solar thermal, and geothermal, among others. Although electricity is still our most expensive form of energy, electricity prices have remained relatively stable during the past 30 years when fossil fuels prices have been extremely volatile.

Figure 1.10 shows that total energy for electricity more than doubled from 20 quads in 1975 to 42 quads in 2005. More than 70% of the energy for electricity comes from fossil fuels, and 75% of that is from coal. Energy for electricity comes from coal (52%) and nuclear (20%), but natural gas has grown to 16% of source energy and may overcome nuclear in the

figure 1.11   U.S. Energy Sources for Electricity, 1950–2006

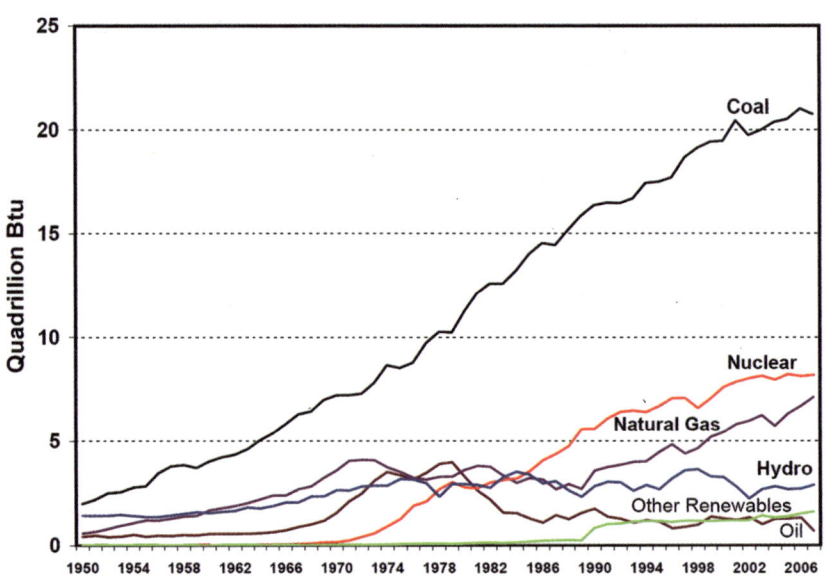

Source: data from U.S. EIA, 2007a, 2007c

next few years (Figure 1.11). Natural gas–fired electricity generating capacity increased by 2 1/2 times between 2000 and 2005 and amounted to 89% of the growth of U.S. generating capacity during that time. Renewable energy contributes only 12% of electricity; this comes from hydro (7%), wood/waste (2.5%), geothermal (1%), and wind/solar (1%).

Excepting hydro, wind, and solar photovoltaic production, 92% of U.S. electricity comes from steam power generation. We will discuss the thermodynamics of power generation later, but suffice it to say here that upgrading low-quality thermal energy of fuel combustion or nuclear reaction to produce steam to spin a turbine and generator to produce high-quality electricity is a losing proposition. As shown in the U.S. electricity energy flow chart for 2005 in Figure 1.12, only 13.01 of 41.6 quads, or 31% of source energy was converted to end-use electricity. The remainder (69%) was lost to thermal conversion (65%) and transmission losses and plant uses (4%).

Because of these large losses, for electricity we make a distinction between end-use energy and primary energy. **End-use energy** is the energy used at the point of use, for example in a building or in a vehicle. **Primary energy** is the original energy needed to produce that end-use energy. For all energy types, there is a difference between primary and end-use energy because it takes some energy to extract, process, and transport energy to the end use. But the difference is by far the greatest for steam-generated electricity, and we must account for the primary energy. In Figure 1.12, end-use electricity in 2005 was 13.01 quads whereas primary energy for electricity was 41.6 quads.

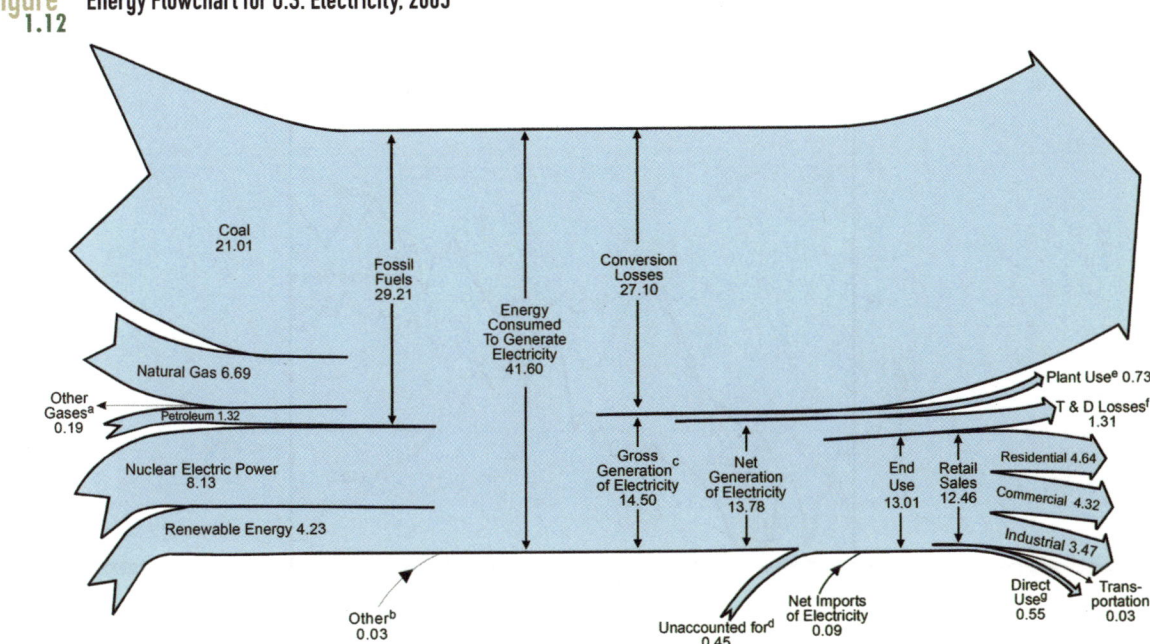

figure 1.12 Energy Flowchart for U.S. Electricity, 2005

Note end-use electricity is only 31% of primary energy.

SOURCE: U.S. EIA, 2006

About three-quarters of our electricity is used in buildings with the remainder used for industrial processes. In buildings, electricity for appliances and equipment, lighting, air-conditioning, and space and water heating amounts to about 40% of end-use energy but 66% of primary energy. Energy losses from electricity generation and transmission constitute the largest energy requirement for commercial and residential sectors and the third largest for industry. There are ways to capture some of these thermal losses through combined heat and power (CHP), but this is limited at large central power stations.

One important consideration of this conversion loss issue is that for every unit of steam-generated end-use electricity saved through greater efficiency, three units of primary energy are saved. Because fossil-fueled steam power, especially coal, is a major source of carbon emissions, improving electricity use efficiency also has a three-fold multiplying effect on carbon emission reductions.

1.4.4 U.S. Energy Production Shortfall and Oil Imports

Energy is one of the most important security problems facing the United States. Figure 1.5(a) showed the growing gap between energy consumption and domestic production that must

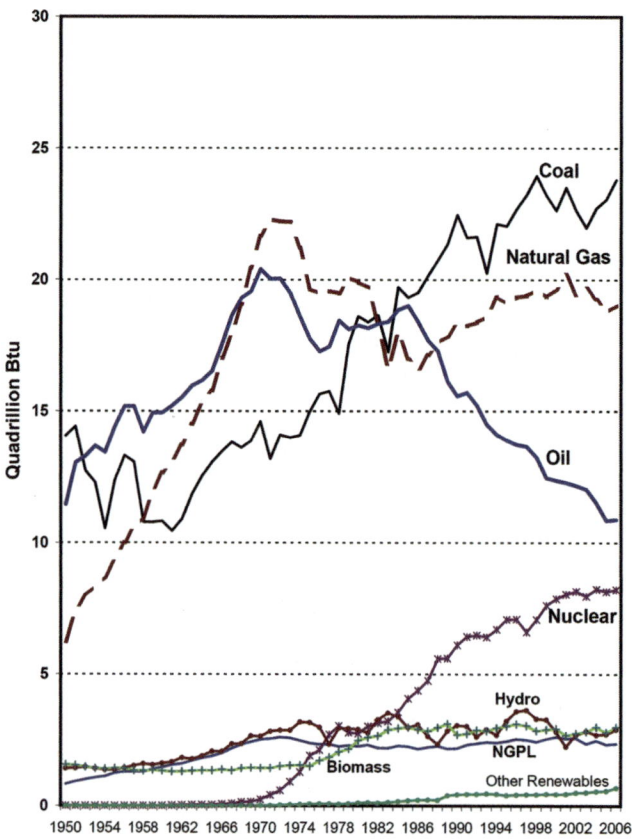

figure 1.13 U.S. Production of Energy by Source, 1950–2006

Despite increases in coal and nuclear, total production has been stagnant at 69–72 quads since 1989. Crude oil peaked in 1970 and has steadily declined after a few years of Alaskan production. Natural gas peaked in 1971 but recovered half of its decline from 1985 to 2000.

SOURCE: data from U.S. EIA, 2007a, 2007c

be filled with imported oil. Because our economy and way of life depend on energy, some believe we need to secure access to energy supplies at all costs, even if these supplies may be in other countries. Others believe that we need to close the gap between our consumption and domestic production through both efficiency improvements to temper growth of consumption and new domestic energy supply.

However, domestic production of all energy in the United States has been flat since the late 1980s at 69–73 quads (it was 69 quads in 2005, slightly less than in 1989). Figure 1.13 shows production by source from 1950 to 2006. While coal and nuclear power from uranium have increased since 1970, petroleum and natural gas production both peaked in the early 1970s. Natural gas production has recovered somewhat since 1985, but like oil, it has not kept up with consumption; net imports of natural gas have increased from 4% of consumption in 1986 to 16% in 2005.

Crude oil production continues its steady decline to 5.1 million barrels per day (mbd) (10.8 quads) in 2006, down from 9.6 mbd in 1970. Simply put, the United States is running out of oil.

This declining oil production, combined with the rising petroleum consumption for transportation shown in Figure 1.14, presents the United States with perhaps its most pressing energy dilemma. Oil consumption has declined in all other sectors since the 1970s, but transportation gasoline and diesel fuels have driven significant growth in overall oil use since 1983. Petroleum continues to be the largest source of energy in the United States at 40% of total consumption. Industry is the second-highest user of oil with 25%, and two-thirds of industrial petroleum is used for material feedstocks. But two-thirds of petroleum is used for transportation, and transportation petroleum use has increased at 2% per year from 1993 to 2006.

The United States must now import two-thirds of its petroleum needs, and that proportion is increasing as consumption rises and production declines. Figure 1.15 tracks the proportions of U.S. oil consumption met by domestic sources and by imports. In 1973, the country supplied 63% of its needs and imported less than 37%; now the United States supplies only 34% of its needs and imports 66%.

The United States imports oil from several exporting countries, headed by Canada (16% in 2005), Mexico (13%), Saudi Arabia (11%), and Venezuela (11%). Total imports in 2006 were 13.6 million barrels per day, of which 40% came from OPEC (Organization of Petroleum Exporting Countries) and 60% from non-OPEC countries. Of OPEC imports, 40% (or 16% of total imports) came from the Persian Gulf, mostly Saudi Arabia and Iraq.

An increasing proportion of future imports will come from the Persian Gulf. This area is home to the vast majority of the world's remaining oil reserves, and continues to be one of the most politically unstable parts of the world.

1.5 Summary

Human-developed science and technology have enabled the conversion of energy sources for productive uses. Through history, the resulting energy use has spurred advancement of human society and civilization. Energy has freed people from slave and animal labor, from agrarian society, and from the constraints of space. It has triggered the development of industry and communications. Only since the mid-nineteenth century and the advance of fossil fuels has energy use enabled unprecedented fourfold growth of human population and a fortyfold increase in the global economy.

But our patterns of energy production and use are not sustainable. World energy use continues to grow rapidly, and 86% comes from carbon-emitting fossil fuels. Petroleum is our largest source and its reserves are concentrated in the politically volatile Persian Gulf. With 20% of the world's population consuming 75% of the world energy and controlling

figure
1.14   **U.S. Transportation Energy, 1950–2006**

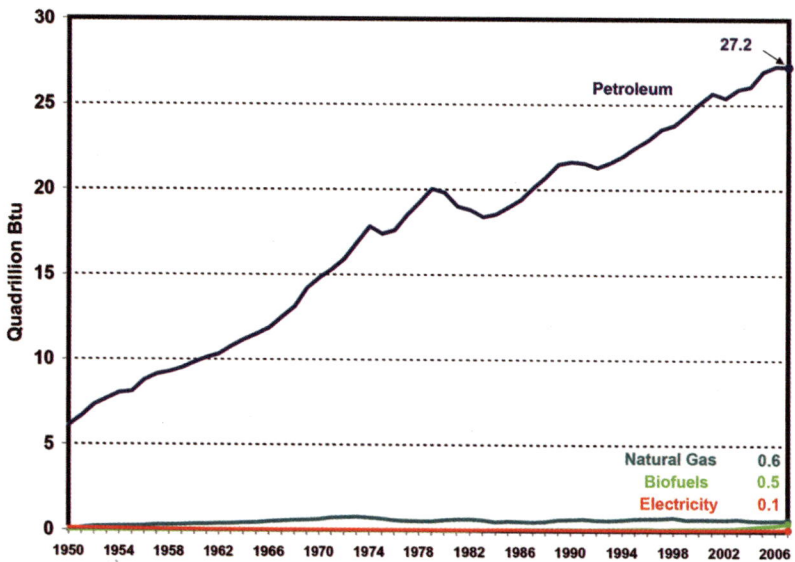

Although petroleum use in other sectors has declined, growth is dramatic in the transportation sector. Thus, oil is still our number-one energy source, contributing 40% of total energy consumption.

Source: data from U.S. EIA, 2007a, 2007c

figure
1.15   **U.S. Petroleum Supply from Domestic Production and Imports, 1973–2006**

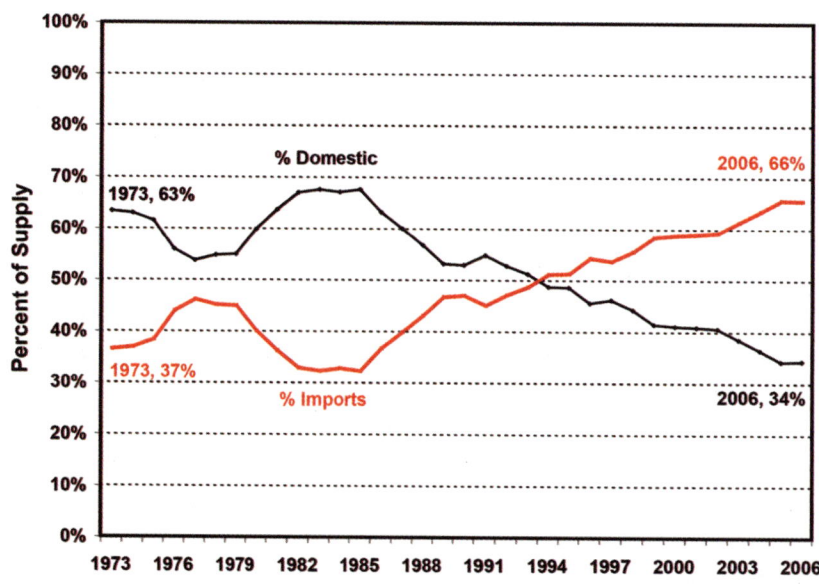

Source: U.S. EIA, 2007a, 2007c

90% of the global economy, while the poorest 80% struggle toward development, we are a long way from an equitable energy and economic system.

The U.S. energy patterns are similar: 86% of energy consumed is fossil fuels. Growth has averaged about 1.2% from 1969 to 2005. During that time, however, the good news is that the United States has increased the economy of energy use with declining energy intensity (energy/$GDP) and stable energy per capita. However, electricity use has increased, and 92% of it is generated by steam power that is inherently inefficient. The United States has increased dependence on imported oil, due to rising consumption and declining domestic production. In 2006, the United States imported 66% of its oil consumption needs, up from 37% in 1973. That trend continues upward as domestic oil production continues to decline and petroleum use for transportation continues to increase at more than 2% per year.

1.5.1 Sustainable Energy: Improve Efficiency, Replace Oil, Reduce Carbon

We simplified our energy problem as three primary issues: oil, carbon, and growing demand. We can also characterize the solutions to our energy problem in three primary objectives or ends and three means to those ends. We need to

1. *Improve efficiency* of energy use to reduce demand growth. We have made progress in improving the efficiency and economic effectiveness of our energy use, but significant opportunities remain.
2. *Replace oil* with other sources to avoid economic and security consequences of oil dependence. We believe that our best immediate opportunities are biofuels and electricity.
3. *Increase carbon-free energy sources,* reduce fossil fuel use, and sequester carbon emissions. We believe that renewable energy sources, including solar, wind, and biomass, offer our best opportunity for carbon-free energy. There is also strong interest in reviving the nuclear industry and in clean coal technology with carbon sequestration, but they face economic, technical, security, and environmental uncertainties.

We can achieve these objectives through three diverse means, all of which are needed for rapid energy market transformation to improve efficiency, replace oil, and increase carbon-free sources:

- Advanced *sustainable energy technologies,* including efficient production and use, renewable energy systems, and selected clean and safe fossil fuel and nuclear technologies.
- *Consumer and community choice* for investment in efficiency and sustainable technologies, and conservation through modifying practices and behavior. Consumer and community choice for sustainable energy is driven by economic, environmental, social, health, security, and other factors, and can take the form of a social movement.
- *Public policies* to develop and deploy technologies and enhance consumer and community choice through investments, incentives, and regulations. Policies can originate in

international agreements and federal, state, and local government market transformation programs.

The remainder of this book explores the constraints and opportunities we face in achieving sustainable energy. It focuses on the three objectives and three means listed above. The book is organized around three dominating energy consuming sectors in Sections III, IV, and V:

- *Buildings* consume nearly half of our energy use, including operating heating, cooling, and electrical appliances and the embodied energy of materials and construction. They contribute 40% of carbon dioxide (CO_2) emissions, the main cause of global climate change. We have made improvements in building energy efficiency, but significant opportunities remain.
- *Electricity* used in buildings and industry requires 40% of our energy consumption and it is growing. More than half of electricity generation comes from coal, one-fifth from nuclear, one-sixth from natural gas, and one-eighth from renewable energy. Electricity generation causes 39% of U.S. carbon dioxide emissions. Wind and solar photovoltaic power have the fastest percentage rates of growth of all sources of electricity.
- *Transportation* uses two-thirds of our oil consumption, and it is 96% dependent on oil. Transportation energy depends on vehicle efficiency, vehicle miles traveled, modal (e.g., car, transit, walking) availability and choice, land use patterns, and the price of fuel. Sustainable transportation must address all of these factors as well as alternative fuels, such as biofuels and electricity.

Before addressing these sectors and energy policy in Section VI of the book, this Section I continues with an introduction to energy sources and constraints in Chapter 2, and different visions of our energy future in Chapter 3. Section II introduces fundamentals of energy science and life-cycle analysis.

CHAPTER 2

Energy Sources and Sustainability

This chapter discusses some of the important life-cycle concerns of our current patterns of energy production and use. Our continued dependence on oil as the largest energy source for our economy poses special geologic, geographic, and political problems. The supply is limited by geologic conditions. And what geologic supply remains is concentrated in the politically unstable Middle East. We have already experienced not only the price shocks associated with a cartel-influenced production market, but also the political and military implications of access to a precious resource.

Just how much of this resource remains is the subject of continuing debate, the so-called "peak oil" debate, centering on the ultimately available quantity of conventional crude oil and when global production will peak. If the world economy remains highly dependent on growing oil demand when that peak occurs, there will be severe economic and political repercussions.

After examining the state of our oil resource and other fossil fuel supplies, this chapter reviews environmental implications of fossil fuel energy sources. It focuses on global climate change, which is triggered in large part by carbon-based fossil fuel combustion and the resulting carbon dioxide emissions. We study the scientific consensus on the topic, and also review other environmental impacts of fossil energy, such as urban air pollution.

Finally, we evaluate our progress in developing non–carbon-based energy sources, including nuclear power and renewable energy, as well as improvements in energy efficiency. Once thought of as the major future source of energy, nuclear power stagnated in the past two decades and amounts to only about 8% of U.S. and 6% of global energy. Recent calls for a renaissance of nuclear power must still confront barriers of public life-cycle concerns over safety and security, long-term waste management, and nuclear weapons proliferation. The implication is that we must continue to look for alternatives.

Efficiency and renewable sources have been viewed with great hope, and indeed best fit the criteria for sustainable energy. However, significant opportunities for efficiency improvements remain untapped and renewable sources still contribute only a small proportion of commercial energy. Still, efficiency is the most cost-effective and environmentally beneficial of energy options; renewable wind and solar electric and biofuels are growing at the fastest rate of all energy sources today; they provide the greatest promise for sustainable energy. These options are introduced in this chapter and are emphasized in the remainder of the book.

Before we dive into the topics of peak oil, climate change, and non-carbon energy in the context of sustainability, it is important to understand that context and introduce some criteria for sustainable energy.

2.1 Criteria for Sustainable Energy

Before reviewing the implications of energy options, it is necessary to discuss further what we mean by sustainable energy. In Chapter 1 we defined sustainability and sustainable energy:

> **Sustainability:** patterns of economic, environmental, and social progress that meet the needs of the present day without reducing the capacity to meet future needs.
> **Sustainable energy:** patterns of energy production and use that can support society's present and future needs with the least life-cycle economic, environmental, and social costs and consequences.

Both definitions emphasize two important criteria:

1. *A broad range of considerations:* Sustainable energy goes beyond short-term economic effects to consider environmental, social, security, and long-term economic implications of energy choices.
2. *The future:* Sustainable energy by definition aims to sustain the availability of energy to meet the needs of future generations. To be sustainable, our actions and choices should neither preclude options nor place undue economic and environmental burdens on those who follow us.

Human history tells us that our predecessors did not think too much about the future, but simply muddled through, doing the best they could and believing that the future took care of itself. Despite calamities, resource shortages, famine, and war, civilization advanced, and here we are.

Many people today think like our predecessors: the future will take care of itself. These "present-thinkers" believe someone will find more oil or discover alternatives. Through technology, they think someone will figure out how to get better at reducing impacts of energy use; at converting coal to energy cleanly; at developing renewable energy and safe nuclear power; at using energy more efficiently. They say, "I'll worry about me and mine, and the greater economic and social system will take care of the rest. That's the way it has always been."

Others think differently. They look at the world around them and see challenges and opportunities. The challenges are inequities and injustices; tensions between security and liberty; local, regional, and global environmental impacts of human activity; and a very uncertain future, among others. The opportunities come from their realization that, unlike our predecessors, we have the power to determine our destiny, to take care of the future ourselves. That power comes from increasing economic wealth, growing democratization, global communication networks, and people's expanding awareness of the world and the future, and their capacity to shape them.

These "future-thinkers" do not see future energy taking care of itself. Today is the future from the perspective of the 1973 oil crisis, and despite great attention by the "greater economic and social system," little has changed in the world's patterns of energy use. Future-thinkers realize that these patterns are not sustainable and that we do not have the luxury of time to wait until that system takes care of the future.

They argue that we need to act quickly, decisively, and collectively to develop more sustainable patterns of energy production and use. The first step to recognizing the need for change is to recognize the problem. This chapter investigates the major constraints to sustainability posed by our current energy patterns.

But how do we make personal, community, and societal choices for more sustainable energy? Given the criteria above, there are several factors to consider:

- Renewability or abundance of the energy resource for long-term reliability
- Life-cycle economic benefits and costs, including cost-effectiveness and national and local economy effects
- Life-cycle environmental benefits and costs, including local, regional, and global effects
- Life-cycle social benefits and costs, including effects on human health, communities, equity, and the disadvantaged
- Life-cycle security benefits and costs, including energy, environmental, and national effects
- Uncertainties of life-cycle benefits and costs

Life-cycle analysis is fundamental to sustainability because it aims to capture full costs and consequences over a long time horizon. We will see in Chapter 5 that life-cycle analysis involves specific techniques, such as net energy analysis and economic and environmental assessment, and a general capacity to think broadly and long-term. For example,

- Life-cycle analysis involves not just considering the carbon emissions from a coal-burning power plant but the full range of economic, environmental, and social costs and benefits of coal mining, processing, and transport; power-plant operations; and waste ash disposal.
- It involves not just the cost effectiveness of a solar photovoltaic array, but the costs and benefits of materials acquisition, production processes, and waste disposal in its production.
- It involves not only the production cost of ethanol from corn or cellulose, but also the energy, fertilizer, irrigation water, and runoff pollution required to produce it; the

carbon and other air emissions from its production and use; the effect on corn and food prices; and other inputs and outputs.

- It includes not only the construction and operating costs and electricity sales revenues from a nuclear power plant, but the full nuclear fuel cycle, from mining to processing to plant operations to long-term waste storage; plant security and safety considerations; ultimate plant decommissioning; and nuclear weapon proliferation concerns.

2.2 The Geologic Limits of Fossil Fuels

Petroleum is a nonrenewable, finite resource subject to depletion. Yet the United States and world economies are dependent on petroleum, which supplies 40% and 37% of their energy, respectively. The big question is: What is the ultimate quantity of this finite resource? A serious scientific debate is raging about when oil production will peak and begin to decline. If it peaks before our economy is able to wean itself from oil, the repercussions will be devastating. Even if that peak is delayed, future supplies of oil will need to come increasingly from the politically volatile Persian Gulf region where the majority of the remaining petroleum resides.

2.2.1 The Peak Oil Debate

For decades, economists and geologists have debated an obscure theory of a former Shell Oil and U.S. Geological Survey geophysicist M. King Hubbert, who, in the mid-1950s, accurately predicted that U.S. domestic oil production would peak in 1970 and decline thereafter, never to rise to that peak again (Figure 2.1). Hubbert also predicted world oil would peak about the year 2000 (Figure 2.2). He died in 1989, but his legacy lives on in a contemporary debate over his theory, its validity, and its implications.

Hubbert's basic theory is that because oil is nonrenewable, under consistent geologic, economic, and market conditions, its production will rise to a peak then fall predictably in a bell curve (Figure 2.2). The peak in production will occur sometime after a peak in the "reserves" of the resource. As shown in Figure 2.1 this lag time was eleven years for the United States. The area under the production bell curve is the **ultimate recoverable quantity** of the resource, or what Hubbert called Q_∞. This turns out to be a critical factor in understanding our energy supply situation, so we need to look at it more closely.

Reserves are the quantity of known deposits that are economically recoverable at today's prices. They are often divided into *proven* and *probable* reserves. Reserves are not static but are depleted by production and added to by new discoveries and by new technologies and higher prices that make deposits once too expensive to extract, profitable to recover (see Figure 2.3). Two prominent industry journals, *Oil & Gas Journal* (*O & GJ*) and *World Oil*, survey companies and governments annually and provide self-reported estimates of oil and natural gas reserves by country, region, and the world as a whole. As shown in Figure 2.3, oil reserves are estimated at 1082 billion barrels (Bbbls) by *World Oil*. The *O & GJ* 2007 estimate of

figure 2.1   **Oil Production Peak for the Continental United States**

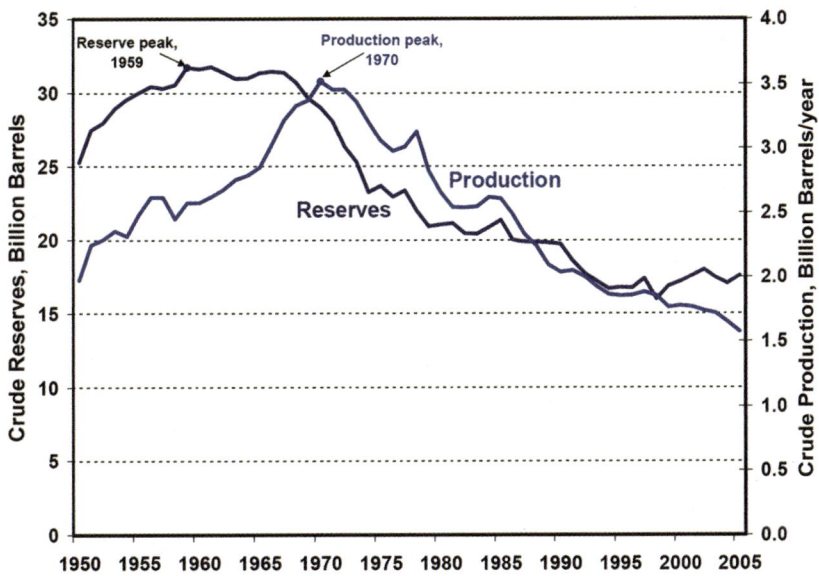

The peak in 1970 followed the peak in reserves by eleven years. In 1960, Hubbert accurately predicted the 1970 peak in production based on the reserve peak.

SOURCE: data from U.S. EIA, 2006a

figure 2.2   **Theoretical Production Curve for World Oil**

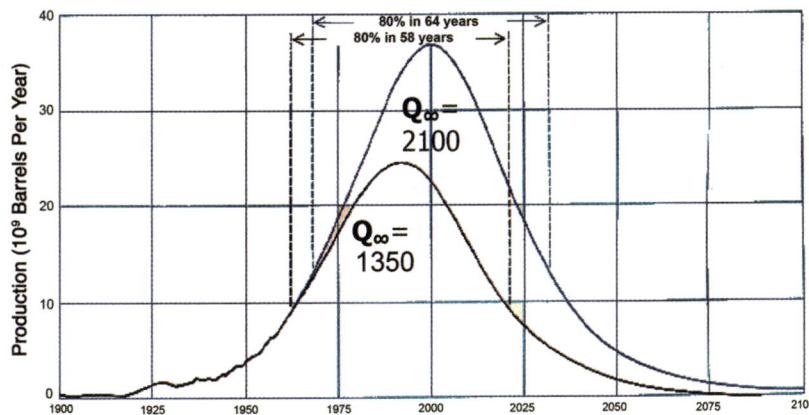

As early as 1949, Hubbert predicted the world oil production peak in 2000 (blue curve) based on 2100 billion barrels ultimate recoverable quantity of the resource (Q_∞) and symmetrical rise and fall of production.

SOURCE: adapted from Hubbert, 1971

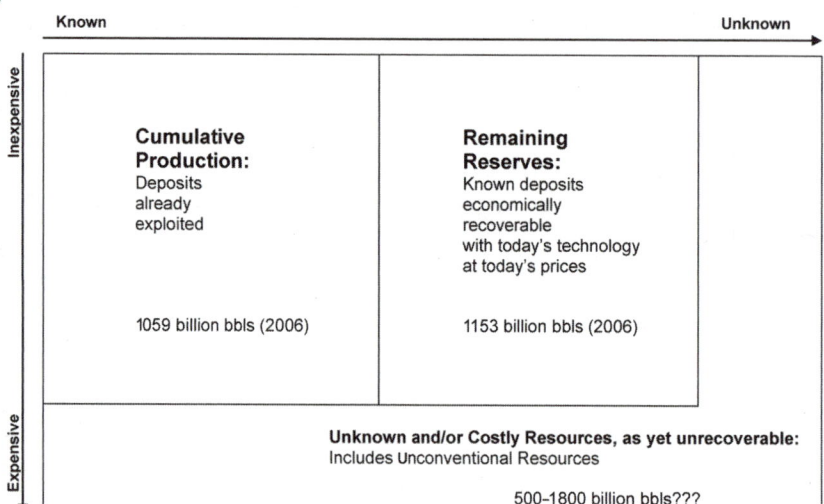

figure 2.3 Ultimate Recoverable Quantity (Q) of a Nonrenewable Resource

Q_∞ is made up of cumulative production to-date, reserves (known and economically recoverable at today's prices), and additional resources that ultimately will be found or made recoverable by new technology or higher prices.

1317 Bbbls includes 174 Bbbls of unconventional oil from oil sands in Canada, a quantity many believe is too difficult to put into production to be considered a reserve.

Reserves are an important measure of today's available deposits, but they do not indicate the ultimate available resource. Figure 2.3 shows that the ultimate available, or the Q_∞ area under Hubbert's curve is made up of (a) current proven reserves (1153 Bbbls in 2007); plus (b) cumulative production to-date (1059 Bbbls through 2006, growing at 32 Bbbls per year in 2007); plus (c) unknown, probable, or not-yet-recoverable deposits that will become future reserves. This latter amount is the most speculative and the primary subject of the peak-oil debate, because it determines how much oil will be recovered and thus when the peak will occur for a given demand. Hubbert's graph given as Figure 2.2 shows the world peak for two different values of Q_∞. The value of 2100 Bbbls gives a peak at the year 2000.

So what is our current best estimate of Q_∞? It depends on whom you ask. In the most definitive government study of oil and natural gas resources to-date, the U.S. Geological Survey (USGS, 2000) provided three estimates: a low estimate of 2248 Bbbls (with 95% confidence), a high estimate of 3896 Bbbls (with 5% confidence), and a mean estimate of 3003 Bbbls. Other recent estimates are in the 1700 to 2400 Bbbl range or close to the USGS low estimate.

Critics of the USGS higher estimates point out that its higher estimates require a rate of discovery or addition to reserves that far exceeds the trends of the past forty years. Figure 2.4 illustrates this point. Most current production is tapping old discoveries and the decline of new discoveries is expected to continue.

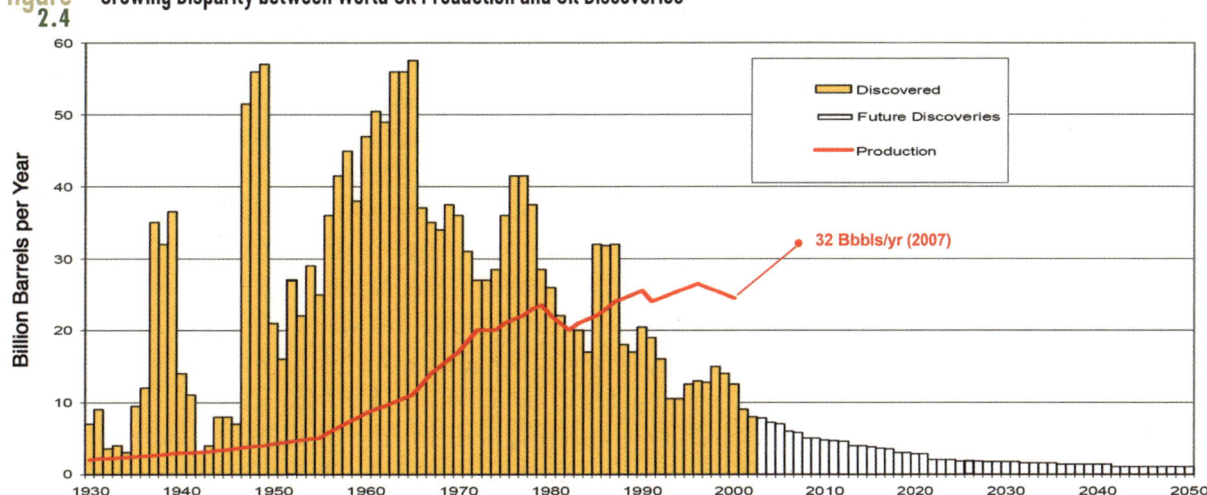

figure 2.4 **Growing Disparity between World Oil Production and Oil Discoveries**

In 2004, the trade publication *Oil & Gas Journal* lamented the declining oil discovery rate in the face of rising production. The gold bars give actual data; the white bars are projections. The 2007 record demand rate of 32 Bbbls/yr is added.

SOURCE: based on *Oil & Gas Journal* data

The two sides of this debate have several legitimate arguments. On one side are the peak oil proponents or "depletionists" who argue that the peak is imminent. Their critics call them doomdayers. The peak-oil skeptics argue there is no reason to worry. Proponent Colin Campbell groups these critics into two camps:

- The "economists" argue that the self-correcting economic system will solve shortage problems. When the supply of a commodity decreases, its price increases, demand decreases, and there is an economic incentive for finding replacements.
- The "pretenders" understand the situation but pretend otherwise for short-term political or economic objectives. They may include government and industry representatives who stand to gain from the status quo.

These critics of peak oil make the following points:

- World oil reserve additions continue to outpace production. *O & GJ* world reserve estimates increased 2% from January 1, 2006, to January 1, 2007.
- Hubbert's theory may have worked for U.S. oil and other cases (such as Pennsylvania anthracite coal) but it has failed in other applications (e.g., U.S. natural gas). Several depletionists like Campbell have had to revise predictions because their predicted global peak date has passed and production continues to increase.

- Although oil resources are finite, no one knows just how finite. Estimates of Q_∞ generally include conventional but not unconventional sources of oil. Unconventional oil includes potential deep-sea deposits (like Chevron's 2006 find in the Gulf of Mexico estimated at 3 to 15 Bbbls), oil shale (estimated 2000 Bbbls in the United States), heavy oil (in Venezuela), and oil sands (in Canada). *O&GJ*'s inclusion of 175 Bbbls of Canadian oil sands in reserves in 2002 indicates that they are at least close to being profitable. Oil sands production in 2007 is 1.3 Mbbls/day or about 0.45 Bbbls per year about 1.3% of world oil production, with expectations to double that by 2015.

- Depletionists argue that Hubbert offers a simple and elegant theory, but the real world is not so simple. Experience has shown that world oil production does not fit the "consistent geologic, economic, and market conditions" that Hubbert's bell shape curve assumes. Indeed, political motivations and market capture by OPEC countries have manipulated the market patterns of oil production and consumption. Instead of rising uniformly after the 1973 oil crisis, production has gone up and down (see Figure 2.4).

- The depletionists are alarmists following a long line of prophets of doom, who have been proven wrong time and time again.

However, the peak-oil proponents argue that Hubbert's theory is not only simple and intuitive, but it has proven itself in several regional studies, such as in the cases of the continental United States (given in Figure 2.1), Alaska, the North Sea, and Russia, among others.

The proponents say that current estimates of reserves do seem to indicate a fairly strong resource base, but closer inspection reveals they are at best uncertain and at worst, wrong. They are self-reported by companies and countries with self-interest for over-reporting. Oil companies' stock values are closely tied to their assets (reserves). In 2004, Shell Oil reevaluated its oil reserves and reduced the estimates by 20%. In internal company memos, the chief of the exploration division said he was "sick and tired of lying about the extent of our reserves." Shell's CEO resigned in disgrace, and its stock dropped 12%.

Similarly, for countries in OPEC, production quotas are linked to their reserves—the more they estimate, the greater their production quota, and the more income they receive. When these rules were established in 1987, the combined reserves of six OPEC countries mysteriously and suspiciously jumped by 300 Bbbls (35% of global reserves at the time) without any major discovery of new fields.

Peak-oil proponents argue that even if Q_∞ is higher than expected (say 3500 Bbbls instead of 2500 Bbbls with the addition of unconventional deposits), the peak will be extended by only a few years as shown in Figure 2.2. Unconventional sources for oil will provide some additions to reserves, but for thirty years they have continued to be out of reach for economic and environmental reasons. When *O & GJ* added Canada's oil sands to reserves in 2002, analysts thought the biggest barriers to development were low oil prices (oil was then $25/bbl) and government environmental regulations. However, by 2007, even with oil approaching $100/bbl, production has just exceeded 1 million barrels per day despite capital investment approaching $50 billion since 2000. Referred to by MIT's *Technology Review* as "dirty oil," extracting bitumen from the sands is an energy-intensive process with far more carbon dioxide emissions and environmental impacts than conventional oil (Bourzac, 2005). Other nonconventional oil sources are likely to encounter similar constraints.

Finally, these proponents argue that even if the peak is uncertain, taking action now will have benefits. We have done very little to arrest our economy's dependence on oil in the years since our first wake-up call in the oil crisis of the 1970s. Reducing our dependence on oil today can postpone the peak, give us more time to transition to other sources, and begin to relieve the environmental and security impacts of our oil dependency.

Both critics and proponents of peak oil agree that this is a serious issue that should be the subject of additional study, that industry and government action should be based on the best available information, and that aggregate reserve data are flawed and a better system of data gathering and verification is needed.

In addition, both sides agree that demand in developing countries will likely push up global demand. World oil production was stagnant from 1977 to 1993, but since then demand for oil has risen by 1.4% per year to 2003 and 4% in 2004 and 2005. Despite a smaller increase in 2006 due to higher prices, demand will likely continue to push oil markets and production capabilities as consumption expands in the less developed countries led by China and India. Still, they disagree about the implications of that growth because they contend different values of Q_∞.

The USGS 2000 study offers relatively optimistic estimates of Q_∞ ranging from 2248 to 3896 Bbbls. U.S. EIA analysis shows that even with these optimistic estimates a production peak may not be far away. The EIA graph in Figure 2.5 assumes the USGS mean estimate (3003 Bbbls), a constant 2% per year production increase to a sharp peak in the year 2016, followed by a 2% per year decline. EIA also ran this analysis with all three USGS estimates of Q_∞ and different production growth rates to peak and a sharp decline at a rate of "R/P" equal to 10. The date of peak varied considerably with different growth rates from 2016 to 2050, but for each growth rate, using the much higher estimate of 3896 Bbbl extended the peak only about 10 years compared to the mean estimate of 3003 Bbbl. We will see in the next chapter that EIA has not considered this analysis in making its energy projections to 2030.

The term **R/P** is Reserves (bbls) over annual Production (bbls/yr) and is called the **static reserve index.** It is an important factor, as it tells us the number of years the current reserves would last if they were produced at current production rates. For example, the U.S. oil reserves in 2006 would last just 11 years if they were produced at the 2006 rate of production. But of course, both current reserves and current production are not static but change each year, so R/P does not give the lifetime of the reserves (as some people assume). However, it is still a good measure of the relative strength of the reserve base. Usually an R/P index of 15 or less indicates a weak reserve base and declining production. Table 2.1 gives 2006 reserves, production, and R/P for selected countries and the world.

Actual world production has not shown a peak, but some countries have, even though they have not followed a nice bell curve. Figure 2.6 shows actual world oil production from 1973 to 2006 and breaks out non-OPEC, OPEC, and Persian Gulf production. Figure 2.7 shows production from selected countries. After readjusting to its new economy, Russia's

figure 2.5   Future World Oil Production

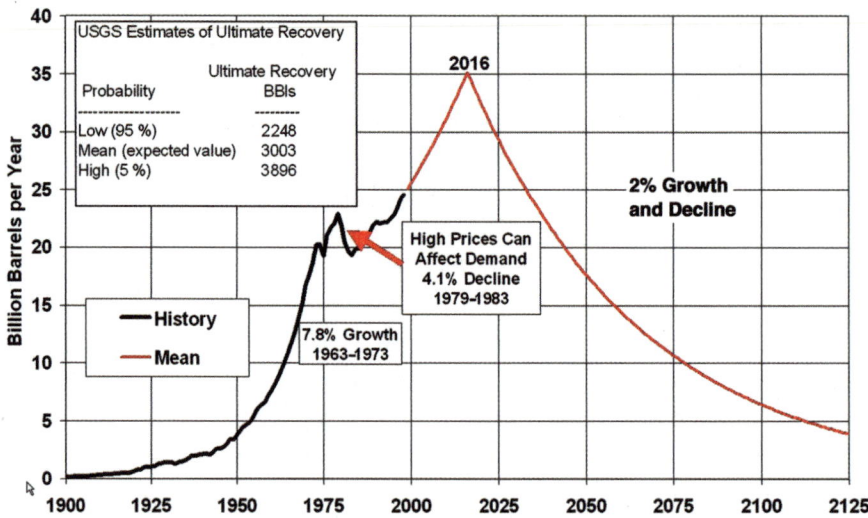

U.S. EIA study reports 2016 peak based on USGS mean Q_∞ (3003 Bbbls) and 2% growth to sharp peak and subsequent 2% decline.

SOURCE: U.S. EIA, 2003

table 2.1   Oil Reserves (2007), Production (2006), and R/P for Selected Countries and the World

Country	2007 Reserves (R) Bbbl/yr	2006 Production (P) Bbbl/yr	Static Reserve Index R/P Years
Saudi Arabia**	260	3.28	79
Iran**	136	1.41	96
Iraq**	115	0.70	164
Kuwait**	99	0.80	112
UAE**	96	0.93	103
Venezuela*	80	0.94	85
Russia	60	3.46	17
Nigeria*	36	0.81	39
China	24	1.35	18
United States	22	1.87	11
Mexico	12	1.19	10
Norway	8	0.90	9
Canada	5 (+175 o.s.[†])	0.91	5 (198)
United Kingdom	4	0.54	7
World Total	1142 (+175 o.s.)	26.5	43 (50)

* Members of OPEC not in Persian Gulf. Other members include Libya, Algeria, and Indonesia.
** Members of OPEC in Persian Gulf.
[†] o.s. = oil sands
SOURCE: data from U.S. EIA, 2007d, and *Oil & Gas Journal,* 2006

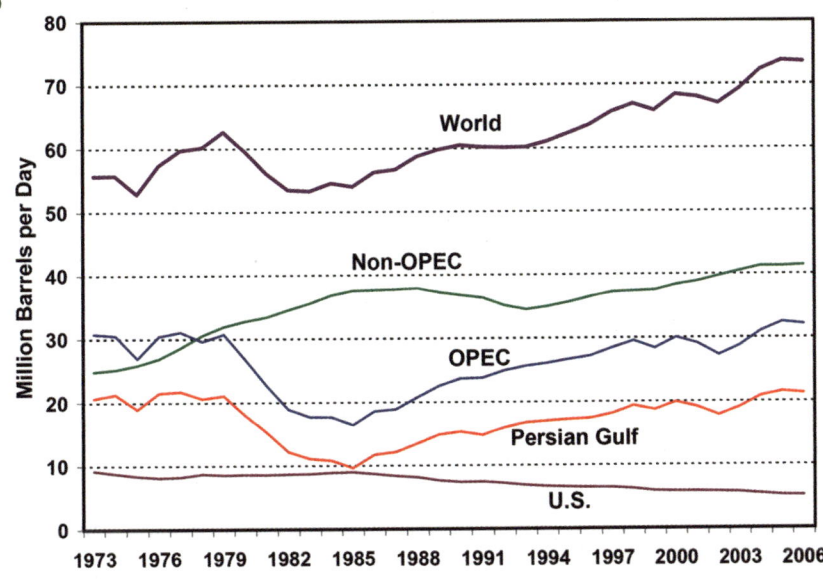

figure 2.6 World Oil Production, Various Regions, 1973–2006

SOURCE: data from U.S. EIA, 2007a

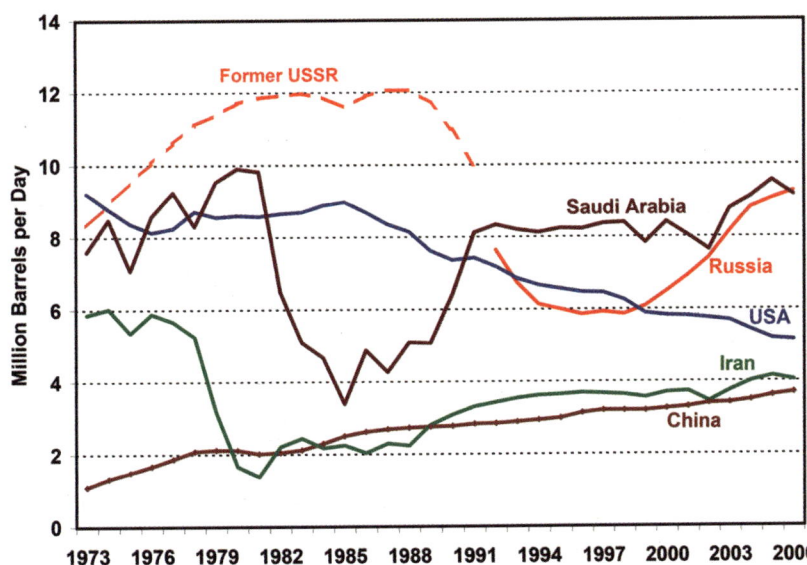

figure 2.7 Oil Production in Selected Countries, 1973–2006

Russia overcame Saudi Arabia as largest producer in 2006.

SOURCE: data from U.S. EIA, 2007a

production has grown and exceeded Saudi Arabia's in 2006. Despite Russia's production increase, non-OPEC production growth is slowing as production in the United States, Mexico, and Norway's and U.K.'s North Sea fields, continued to decline in 2006. Future production increases will come from OPEC and the Persian Gulf where reserves are concentrated. Despite prices pushing $100/bbl, world oil production continued to expand in 2007, growing at 1.7% per year to 88 Mbbl/da or an equivalent annual production of 32 Bbbl/yr.

2.2.2 U.S. Oil Depletion and Dependency

Although there are uncertainties about when world oil production will peak, we know for certain that U.S. oil production peaked in 1970 (Figures 1.13, 2.1, and 2.7), and as a result we are increasingly dependent on foreign sources. We now import two-thirds of the oil we consume (Figure 1.15) and that percentage continues to grow as consumption increases and production declines.

In 2006, 40% of U.S. oil imports came from OPEC members and only 16% of that came from Persian Gulf OPEC members (Saudi Arabia, Iraq, Iran, Kuwait, and UAE), but that is likely to change. As shown in Table 2.1 and Figure 2.8, non-OPEC exporters, with the exception of Russia, have very limited reserves, and three-fourths of global reserves are in OPEC countries. Two-thirds of this OPEC oil is located in Persian Gulf countries, causing the rest of the world to become increasingly dependent on the Persian Gulf. By late 2004,

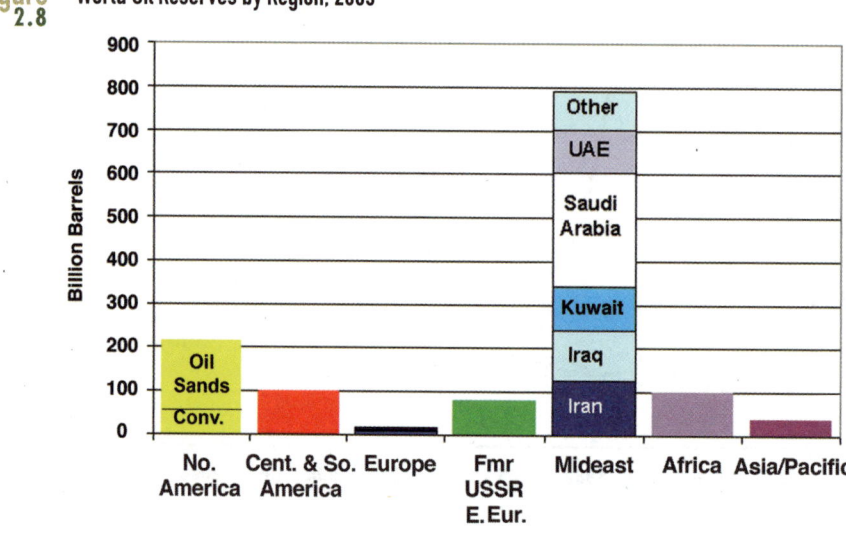

figure 2.8   World Oil Reserves by Region, 2005

The Mideast dominates reserves. Canadian oil sands account for 70% of North American reserves.

SOURCE: adapted from U.S. EIA, 2005; based on *Oil & Gas Journal* data

oil imports accounted for one-third of the U.S. trade deficit, a continuing trend influencing the declining value of the dollar. World oil prices hit a record $96/bbl at this writing in November 2007.

It is ironic that the countries with the greatest oil wealth are among the world's most politically unstable; fast wealth seems to breed corruption, inequity, and repression. Amory Lovins, et al. (2004, p. 18), write "only 9% of the world oil reserves are held by countries considered 'free' by Freedom House, and oil riches correlate well with Transparency International's corruption ratings."

Because access to oil is vital for the future U.S. and global economy, it is no surprise that the Persian Gulf region has attracted so much political attention. Lovins, et al. (2004), argue that "reliance on unstable oil sources incurs costs for both buying it and defending it," and estimate that the continuing cost of military security in the Middle East to protect access to oil is the equivalent of $25/bbl of oil we import. Regarding the Gulf War and the Iraqi War, they speculate:

> Historians will long debate whether the United States would have sent a half-million troops to liberate Kuwait in 1991 if Kuwait just grew broccoli and the United States didn't need it. Decades hence, historians may be better able to say whether an odious tyrant would have been overthrown with such alacrity in 2003 if he didn't control the world's second-largest oil reserves. (Lovins, et al., 2004, pp. 17, 19)

2.2.3 Natural Gas and Coal

As nonrenewable resources, natural gas and coal are ultimately subject to the same supply constraints as petroleum. However, due to lower U.S. and global demand for these fuels, and greater abundance of coal, the potential supply effects are neither as severe nor as immediate. Table 2.2 gives the natural gas reserves and production rates of the same selected countries shown in Table 2.1. Russia has the greatest reserves and the United States is fourth with 204 trillion cubic feet (Tcf), although the U.S. R/P is only 11.

Ultimately, natural gas (NG) resources are estimated to be as large as oil on an energy equivalent basis. The USGS (2000) estimates the exploitable NG resources (Q_∞) at 15,400 Tcf (mean value), of which about 2800 Tcf have been used and 6200 Tcf are listed as reserves. Thus, USGS estimates that about one-sixth of Q_∞ NG has been used, compared to one-third for Q_∞ oil. World NG production has increased by 2.1% per year since 1992, compared to 1.2% for oil. Peak-oil proponent Jean Laherrere estimates ultimate NG resources at 10,000 Tcf; 12,000 if unconventional gas is included. Natural gas thus has more room to grow, but will be subject to the same peak production as oil; however, the NG peak is likely to be a few decades later than oil's peak.

The current U.S. net imports of NG are about 16% of consumption (compared to 4% in 1986). They are mostly by pipeline from Canada. However, like the United States, Canada's reserve base is limited and the United States will need to rely on other sources of imported natural gas to fuel its expected growing demand. According to U.S. EIA projections, future

table 2.2 Natural Gas Reserves (2007), Production (2004), and R/P for Selected Countries and the World

Country	2007 Reserves (R) Tcf/yr	2004 Production (P) Tcf/yr	Static Reserve Index R/P Years
Russia	1680	21.0	75
Iran**	974	2.7	328
Saudi Arabia**	240	2.0	103
United States	204	19.0	11
UAE**	203	1.5	124
Nigeria*	182	0.5	236
Venezuela*	152	1.1	158
Iraq**	112	0.1	1812
Norway	82	2.4	28
China	80	1.2	56
Canada	58	6.6	9
Kuwait**	55	0.3	160
United Kingdom	17	3.6	5
Mexico	15	1.3	10
World Total	**6183**	**98.6**	**63**

* Members of OPEC not in Persian Gulf. Other members include Libya, Algeria, and Indonesia.
** Members of OPEC in Persian Gulf.
SOURCE: data from U.S. EIA, 2006b, and *Oil & Gas Journal,* 2006

imports are expected to increase from liquefied natural gas (LNG) imports to 21% or 5 Tcf by 2025. To put this in perspective, Japan is currently the world's largest LNG importer with 9 Tcf.

Whereas Saudi Arabia has 23% of the world's oil reserves and Russia has 27% of the natural gas reserves, the United States has 27%, or 267 billion short tons (Bt), of the world's coal reserves (997 Bt). The United States produces 18% of the world's coal annually (1.1 billion tons [Bt] giving an R/P = 236 years), about half of China's production that has doubled in the past fifteen years. Although coal is far more plentiful than oil and natural gas (the world coal R/P is 180 years), its solid form complicates its extraction, transport, and use, which limits its applications. More importantly, it has greater carbon content and more impurities than oil and gas, and thus produces more carbon dioxide and air pollution when burned. We have technologies to mitigate some of these effects and more are under development, but as discussed in the next section, these environmental constraints on coal are its greatest limiting factor.

2.3 The Environmental Limits of Fossil Fuels

Energy fuels our economy and quality of life, but it is costly both in monetary terms and in impacts to the natural and human environment. These impacts are part of the "cost of doing business" but to a large extent they are not included in the costs of energy. They are termed externalities. **Externalities** are social costs borne by users and non-users alike, but not internally by the producer and thus are not reflected in the price of goods or services produced. To achieve sustainable energy, we must consider these costs over the fuel or system's life cycle.

These environmental impacts include air pollution from the combustion of fossil fuels, radioactive materials involved in the nuclear fuel cycle, impacts on lands and waters of fuel extraction, and transport and construction of conversion systems. Before addressing these impacts, the section below discusses what appears to be the major environmental constraint facing fossil energy—global climate change triggered by greenhouse gas emissions, primarily carbon dioxide from fossil fuel combustion.

2.3.1 Climate Change

For decades, scientists studying the Earth's energy balance have understood that incoming solar radiation and the Earth's outgoing back-radiation to space are regulated by the atmosphere. A number of atmospheric gases, principally carbon dioxide (CO_2) and water vapor, transmit most of the short solar wavelengths but absorb most of the longer wavelengths of the Earth's back-radiation, holding in energy and warming the Earth's atmosphere and surface—much like the glass in a greenhouse (see Section 4.6.3, Figure 4.8).

Thirty years ago this was a theory. But in the past decade, global warming is now a household term, deemed one of the most difficult problems facing society in the new century. Increasingly sophisticated monitoring has bolstered the theory and revealed disturbing trends:

a. Rising global emissions of CO_2 and other so-called greenhouse gases (GHG), including methane, chlorofluorocarbons (CFCs), and nitrogen oxides
b. Rising global concentrations of CO_2
c. Rising global mean temperature
d. Retreating polar ice caps due to higher temperatures

The most obvious trend is the increase in average temperature. Figure 2.9 shows annual mean temperature from 1880 to 2006. The temperature scale is relative to the base period 1951–1980. The figure shows that global temperature has warmed considerably since 1975. The year 2005 was the warmest year on record, 0.62°C above the 1951–1980 mean. Including 2006, the thirteen warmest years since records began in 1880 have occurred in the past seventeen years since 1990. Data through September indicates that 2007 will be added to this list, rivaling 1998 and 2005 as the hottest.

Figure 2.10 gives trends in the concentration of atmospheric carbon dioxide as measured at the Mauna Loa, Hawaii, site between 1958 and 2004. Preindustrial atmospheric CO_2 concentration (1850) is estimated to have been about 280 parts per million (ppm); ice cores from Greenland have shown that this concentration was fairly constant for the previous

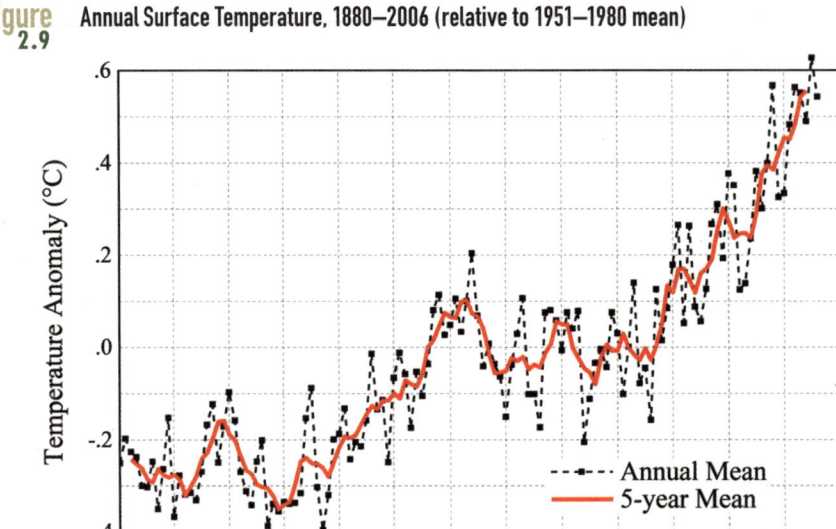

figure 2.9 Annual Surface Temperature, 1880–2006 (relative to 1951–1980 mean)

Note that 2005 was the warmest on record and that the thirteen warmest years since 1880 have occurred in the past seventeen years since 1990.

Source: NASA, 2007

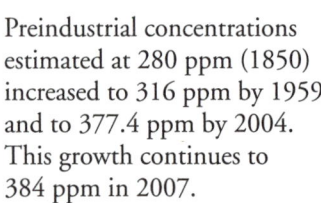

figure 2.10 Atmospheric CO$_2$ Concentrations, Mauna Loa Observatory, Hawaii, 1958–2004

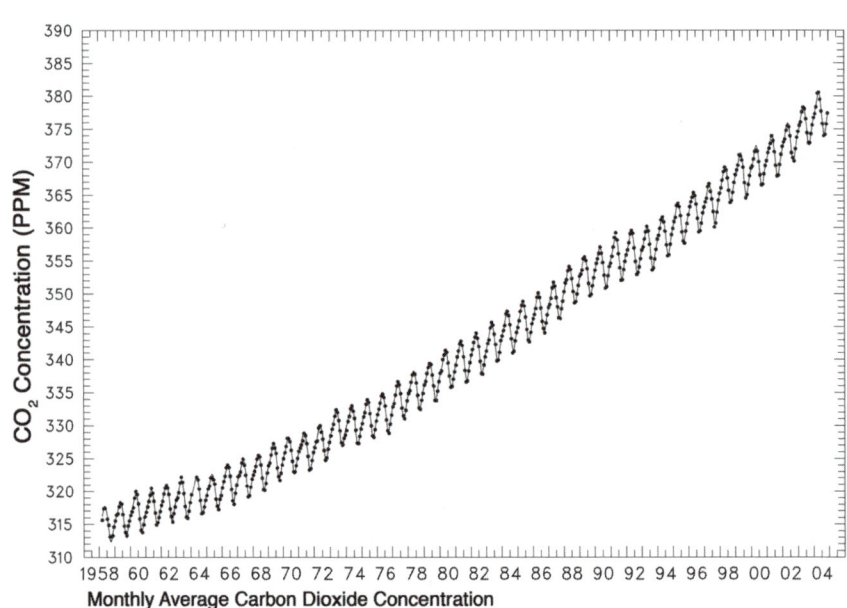

Preindustrial concentrations estimated at 280 ppm (1850) increased to 316 ppm by 1959 and to 377.4 ppm by 2004. This growth continues to 384 ppm in 2007.

Source: Keeling and Whorf, 2005

14,000 years. Mauna Loa readings showed a concentration of 316 ppm in 1959 and a 21.5% increase to 377.4 ppm in 2004. In September 2007 Mauna Loa data showed 384 ppm. From 2001 to 2007, atmospheric CO_2 concentration has risen an average 2.1% per year, a doubling rate of thirty-three years.

Figure 2.11 gives global emissions of carbon from combustion of fossil fuels from 1980 to 2004. Global CO_2 emissions increased by one-third between 1983 and 2004 to 27 billion metric tons (Bmt). Oil in transportation is the largest source followed closely by coal power plants; by the year 2010, coal is expected to exceed oil. Emissions have come mostly from the developed countries, but the significant future increases shown in Figure 2.11 are expected mostly from China and India and other parts of the developing world. Figure 2.12 shows U.S. CO_2 emissions, which hit 6 Bmt in 2005, 22% of the world total. Transportation is shown as the largest source, but buildings (combining residential and commercial) are actually larger.

As a result of these actual trends, there has been a strong response from the scientific and political communities to address some fundamental questions.

- To what extent is the current trend in global warming due to human emissions of GHG?
- What are the prospects for future emissions and effects on CO_2 concentrations and global temperatures?

figure 2.11   Global CO$_2$ Emissions by Fuel Type, 1980–2004

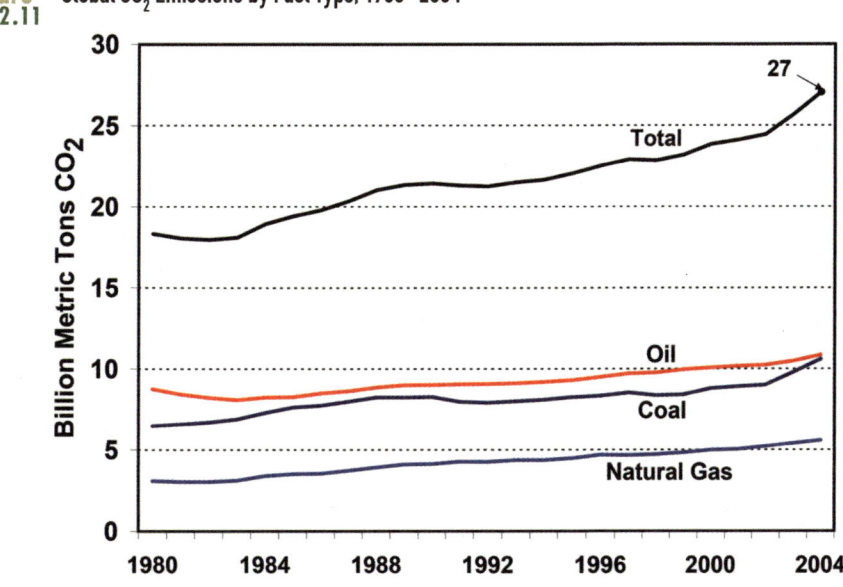

Source: data from U.S. EIA, 2006b

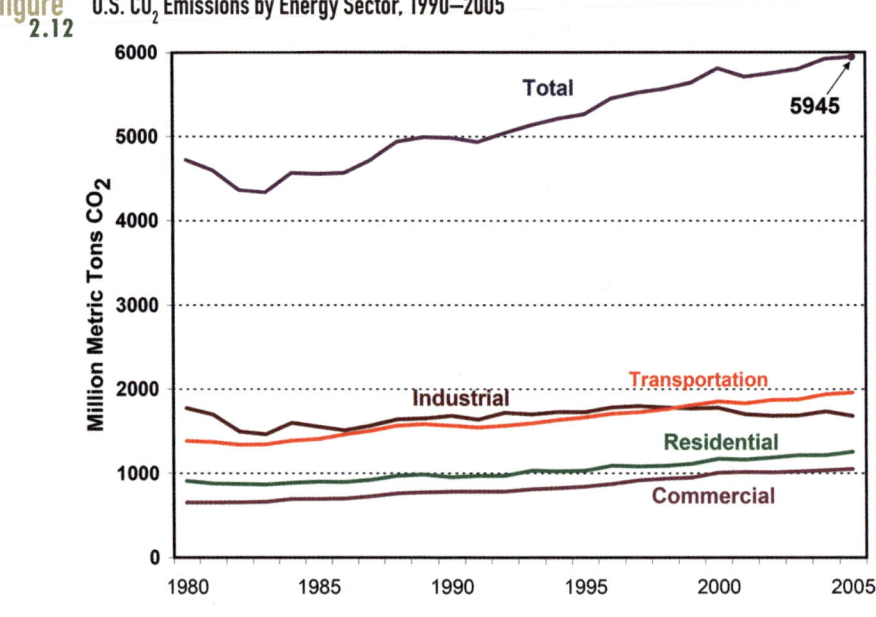

figure 2.12   **U.S. CO₂ Emissions by Energy Sector, 1990–2005**

SOURCE: data from U.S. EIA, 2006a

- What are the potential effects of global warming on weather patterns, food production, ecological systems, sea-level rise, and human settlements?
- What actions are warranted by governments, industries, communities, and citizens to respond to potential impacts, control or reduce emissions, or change patterns of energy use?
- How do we compare the uncertainty and risks of future impacts to the cost of reducing those risks?

Not surprisingly, these questions have fueled considerable controversy due to the high stakes and the uncertainty involved.

2.3.1.1 IPCC: Scientific Consensus on Global Warming

The most authoritative scientific body addressing many of these questions is the Intergovernmental Panel on Climate Change (IPCC), established by the United Nations and the World Meteorological Organization in 1988 "to assess on a comprehensive, objective, open, and transparent basis the scientific, technical and socio-economic information relevant to understanding the scientific basis of risk of human-induced climate change, its potential impacts and options for adaptation and mitigation." For its efforts bringing the science of global climate change to the public and political arena, the IPCC shared the 2007 Nobel Peace Prize with Al Gore.

The IPCC is not intended to conduct research but to engage the best science in assessing and interpreting peer-reviewed scientific and technical studies on the subject. IPCC includes four groups: Work Group (WG) I assesses scientific aspects of the climate system and climate change; WG II assesses the consequences of climate change and options for adapting to them; WG III assesses options for limiting GHG emissions and mitigating climate change; the Task Force on National Greenhouse Gas Inventories runs the GHG inventory program.

The IPCC process is a continuous one. Each work group develops a major report over several years. The WG reports undergo extensive review by scientists and governments and serve as the basis for the IPCC assessment reports developed and approved in a plenary conference. Using this process, IPCC has produced four reports, the First Assessment Report (FAR, 1990), the Second Assessment Report (SAR, 1996), the Third Assessment Report (TAR, 2001), and the Fourth Assessment Report (AR4, 2007). The TAR and AR4 included a "Summary for Policymakers." The six-year AR4 effort involved 450 lead authors, 800 contributing authors, and 2500 reviewers from 130 countries.

Succeeding IPCC reports have become more certain about the occurrence of global warming and human influences:

FAR (1990) "The size of the warming is broadly consistent with predictions of climate models, but the unequivocal detection of the enhanced greenhouse effect from observations is not likely for a decade or more."

SAR (1996) "The balance of evidence suggests a discernible human influence on climate."

TAR (2001) "There is new and stronger evidence that most of the warming observed over the last fifty years is attributable to human activities."

AR4 (2007) "Evidence for warming of the climate system is unequivocal. . . The role of greenhouse gases is well understood and their increases are clearly identified... The net effect of human activities is now quantified and known to cause a warming at the Earth's surface."

Global warming is caused by human activity and its effects are occurring. What sets the 2007 AR4 apart from the previous assessments is (1) the extent of actual observations of climate change and effects; (2) the rising level of certainty that warming is caused by human activity; and (3) the confidence level of predicted impacts. Figure 2.13 from the WG I report shows the extent of change in observations of temperature and physical and biological conditions. The physical and biological observations were taken from 29,000 datasets from 577 studies; 95% of those datasets were from Europe. Nearly all of the observed changes were consistent with the impacts of warming.

The observed effects include the following:

- Increase in glacial melting, the size and number of glacial lakes, and ground instability in permafrost areas and changes in artic/Antarctic ecosystems

figure 2.13 Surface Temperature Changes, 1970–2004, and Significant Changes in Observations of Physical and Biological Systems

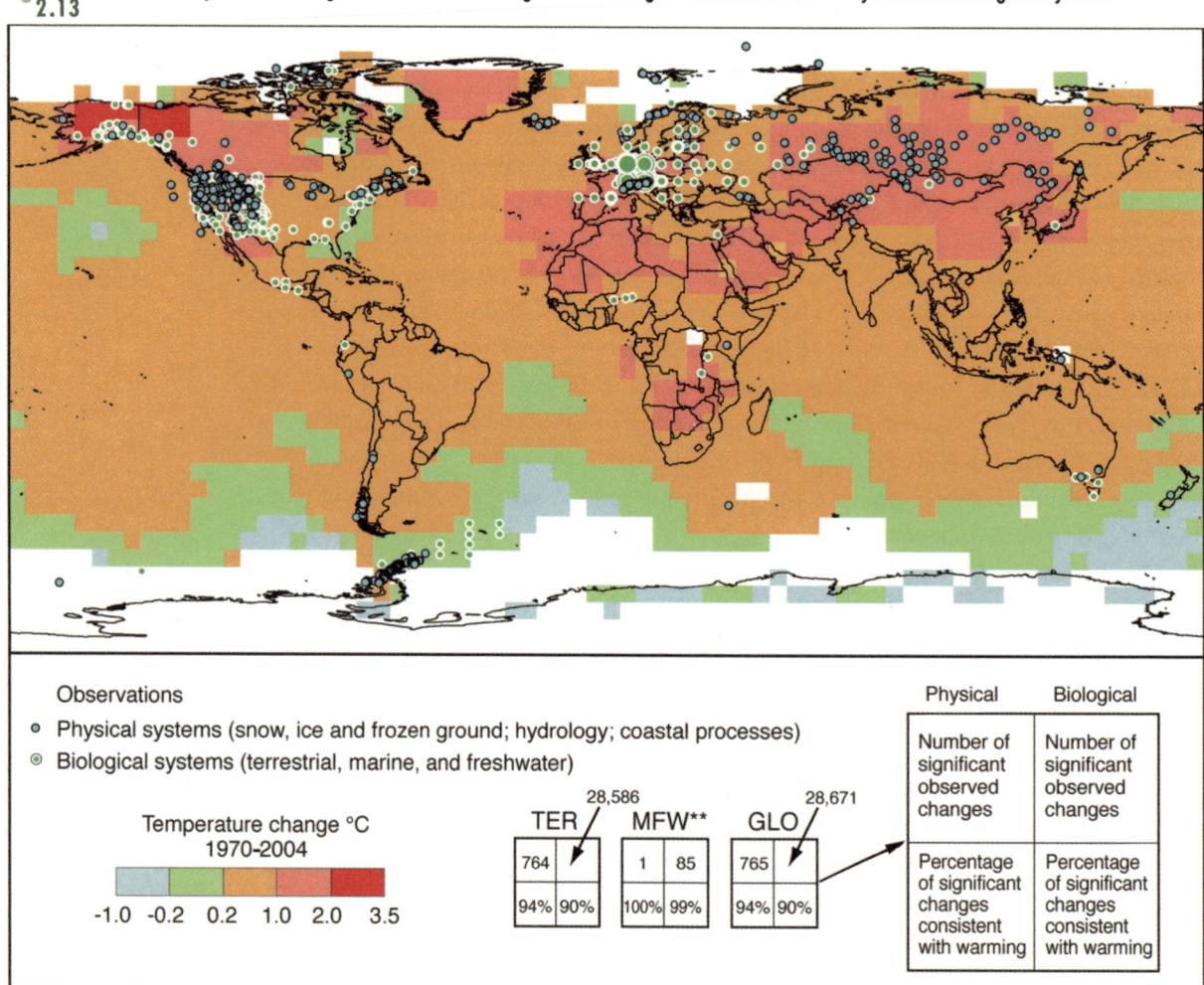

Observations based on 29,000 datasets from 577 studies, of which 28,000 are from European studies. The two-by-two boxes show the number of datasets with significant changes (top row) and the percent of those consistent with warming (bottom row) for terrestrial (TER), marine and freshwater (MFW), and total global (GLO).

Source: IPCC, 2007

- Increased spring runoff and peak discharge in snow-fed rivers, warming of lakes and rivers
- Earlier timing of spring events, such as leaf-unfolding, bird migration, egg-laying
- Poleward and upward shifts in ranges of plant and animal species

Other observed effects are more difficult to pin directly on global warming because of complicating non-climate factors and adaptation. Those observed effects believed caused by global warming with a medium level of confidence (> 50%) include agricultural and forestry changes due to growing season, pests, and fire; and human health impacts from extreme heat and infectious disease vectors.

Suspected effects with lower confidence levels from observation studies include increased flooding in mountainous regions, increased desertification, sea-level rise and effects on coastal wetlands and flooding, and extreme weather events. But there is a higher level of confidence that these effects will occur in the future.

Regarding human influence on these effects, the AR4 report concluded that "most of the observed increase in the globally averaged temperature since the mid-twentieth century is very likely (> 90%) due to the observed increase in greenhouse gas emissions."

Future impacts are significant and more certain. Advanced scientific study and observed evidence of global warming and its effects have helped scientists gain confidence about estimates of future impacts. Table 2.3 from the AR4 WG II report highlights the phenomena associated with global warming, their likelihood, and their impacts on agriculture, forestry, ecosystems, water resources, human health, and society. Some of the effects may be positive (e.g., increased agricultural yields in colder climates), but nearly all pose significant problems for adaptation.

Most of the impacts, such as disruption of agriculture and water resources, human health effects, and dislocation of populations in areas vulnerable to coastal storms and sea-level rise, are likely to affect poorer countries and populations much more than wealthy countries that also have the resources to adapt. Two of the major drivers of impact that are very likely to occur are extreme weather events and sea-level rise.

The Earth's poles will see the most dramatic temperature increases and effects, including receding Arctic sea ice, a reduction in permafrost areas on Arctic lands, and sea-level rise. Figure 2.14 from the Arctic Climate Impact Assessment (2004) shows that the Arctic ice cap already had receded 15% by 2002, and projected melting is more dramatic. Melting of polar ice, especially from land masses such as Greenland and Antarctica, would cause significant sea-level rise throughout the world.

Even if we act soon to reduce CO_2 emissions, future effects are likely to be far-reaching due to CO_2's slow removal times once it accumulates in the atmosphere. Figure 2.15 shows that even if efforts are made to eliminate CO_2 emissions within a century, the damage would already be done: future delayed effects on temperature and sea-level rise and other impacts would likely continue.

2.3.1.2 Responding to Climate Change

Despite the skeptics (see Sidebar 2.1), the increasing certainty about global warming is leading to a growing response by climate scientists; governmental bodies and nongovernmental organizations at global, national, state, and local levels; technology researchers; planners and policy makers; and people on the scale of a social movement. For example, in August 2006, a

table **Major Projected Impacts of Climate Change**
2.3

Phenomena and Direction of Trend	Likelihood	Examples of Major Projected Impacts by Sector			
		Agriculture, Forestry, Ecosystems	Water Resources	Human Health	Industry/ Settlement/Society
Warmer and fewer cold days and nights; warmer or more frequent hot days and nights over most land areas	Virtually certain (> 99% probability)	Increased yields in colder environments; decreased yields in warmer environments; increased insect outbreaks	Effects on water resources relying on snow melt; increased evapotranspiration rates	Reduced human mortality from decreased cold exposure	Reduced energy demand for heating; increased demand for cooling; declining air quality in cities; reduced disruption to transport due to snow, ice; effects on winter tourism
Warm spells/heat waves: frequency increases over most land areas	Very likely (> 90%)	Reduced yields in warmer regions due to heat stress; wild fire danger increase	Increased water demand; water quality problems, e.g., algal blooms	Increased risk of heat-related mortality, especially for the elderly, chronically sick, very young, and socially isolated	Reduction in quality of life for people in warm areas without appropriate housing; impacts on elderly, very young, and poor
Heavy precipitation events: frequency increases over most areas	Very likely (> 90%)	Damage to crops; soil erosion, inability to cultivate land due to water logging of soils	Adverse effects on quality of surface and groundwater; contamination of water supply; water scarcity may be relieved	Increased risk of deaths; injuries; infectious, respiratory and skin diseases; post-traumatic stress disorders	Disruption of settlements, commerce, transport and societies due to flooding; pressures on urban and rural infrastructures
Area affected by drought: increases	Likely (> 66%)	Land degradation, lower yields/crop damage and failure; increased livestock deaths; increased risk of wildfire	More widespread water stress	Increased risk of food and water shortage; increased risk of malnutrition; increased risk of water- and food-borne diseases	Water shortages for settlements, industry and societies; reduced hydropower generation potentials; potential for population migration
Intense tropical cyclone activity increases	Likely (> 66%)	Damage to crops; windthrow (uprooting) of trees; damage to coral reefs	Power outages cause disruption of public water supply	Increased risk of deaths, injuries, water- and food-borne diseases; post-traumatic stress disorders	Disruption by flood and high winds; withdrawal of risk coverage in vulnerable areas by private insurers, potential for population migrations
Increased incidence of extreme high sea level (excludes tsunamis)	Likely (> 66%)	Salinisation of irrigation water, estuaries and freshwater systems	Decreased freshwater availability due to saltwater intrusion	Increased risk of deaths and injuries by drowning in floods; migration-related health effects	Costs of coastal protection vs. costs of land-use relocation; potential for movement of populations and infrastructure; see tropical cyclones above

SOURCE: IPCC, 2007

**figure
2.14**   **Expected Retreat of Polar Sea Ice in This Century**

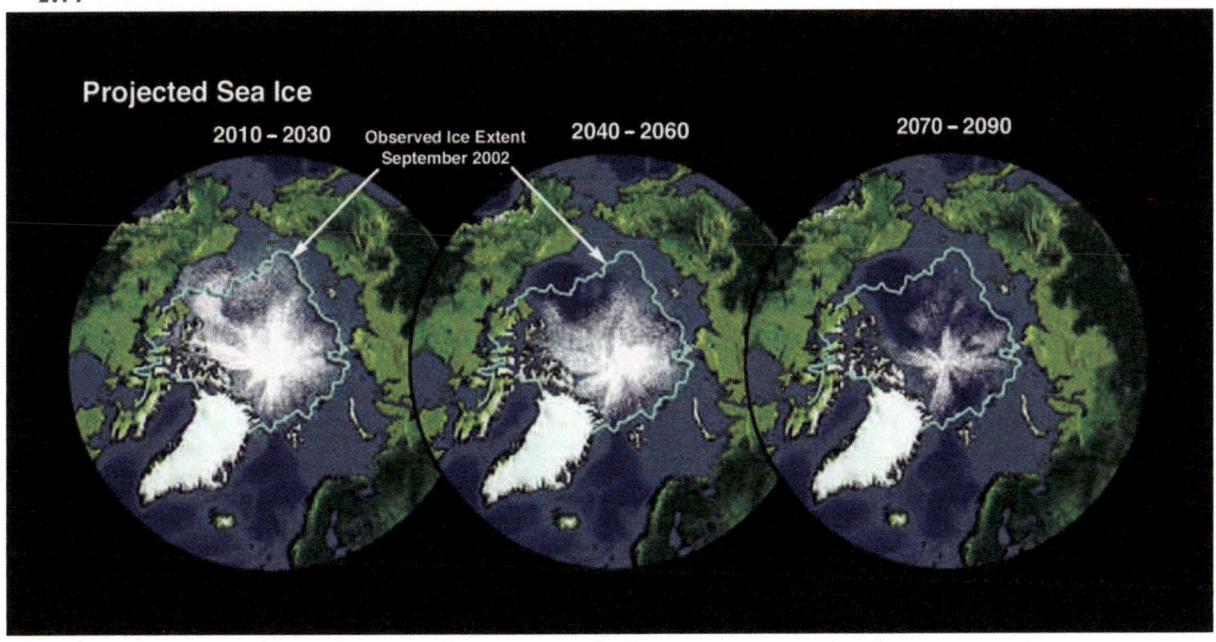

Arctic ice cap had already receded 15% by 2002 and is expected to continue its retreat.

SOURCE: Arctic Climate Impact Assessment, 2004

group of sixty-four prominent California economists, including three Nobel laureates, wrote a letter to the California governor and legislature imploring action on climate change, stating "The Most Expensive Thing We Can Do Is Nothing!"

The emerging diverse response to the increasing effects of climate change can be characterized in two basic approaches:

1. *Mitigating* climate change by reducing GHG emissions through technology, planning, and policy. Examples include developing non-carbon energy sources and establishing targets and mandates for GHG emissions.
2. *Adapting* to climate change by
   a. lessening the impacts using technology and planning, such as building seawalls to counter sea-level rise, expanding irrigation to counter drought conditions, and building more dams and reservoirs to store runoff to make up for reduced snowpack water supply storage; and
   b. anticipating effects and modifying practices and patterns of development and agriculture now so that we can live with those effects in the future, such as relocating populations subject to severe effects of sea-level rise or extreme weather events and formulating new development designs that respond to new regional climatic conditions.

figure 2.15   **Time to Equilibrium: CO₂ and Its Impacts**

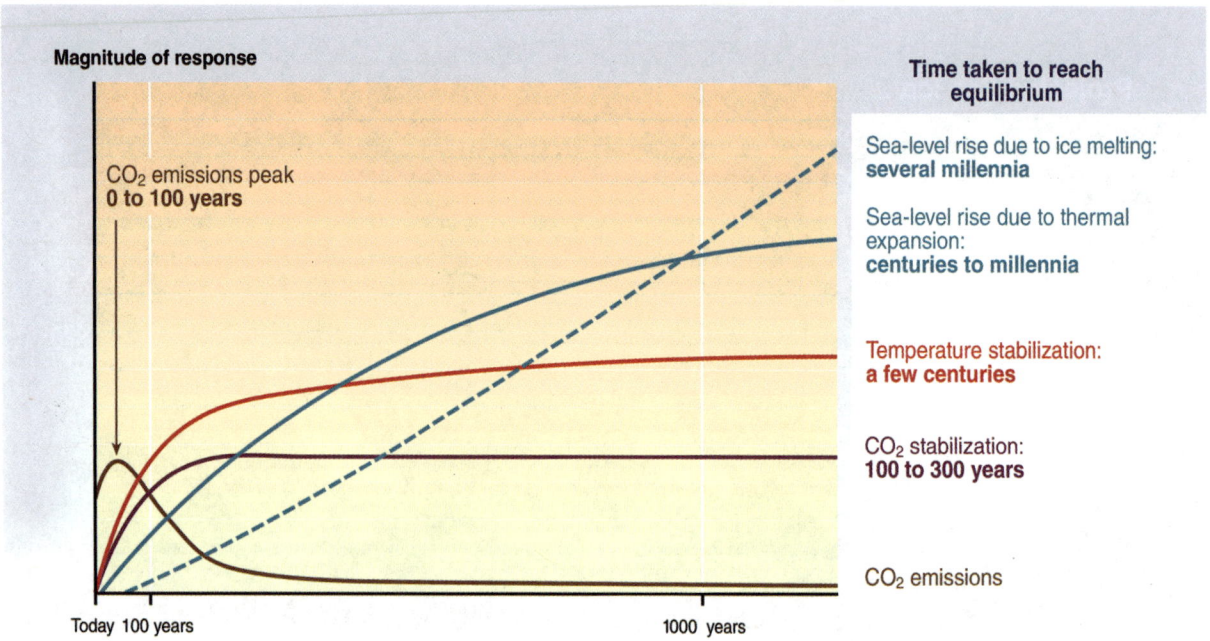

Carbon dioxide is retained in the atmosphere. Even if CO₂ emissions peak then decline, effects on CO₂ concentration, temperature, and sea-level rise may continue.

Source: IPCC, 2002

Mitigating climate change. There are two complementary approaches to mitigating climate change, or reducing CO_2 and other GHG emissions to decrease future effects of global warming. The first approach is through technologies to reduce emissions by using less energy, by replacing high-carbon with low- or zero-carbon energy, or by sequestering carbon. We discuss non-carbon energy sources later in this chapter, and focus on renewable energy and efficiency technologies throughout this book.

The second approach includes a range of energy and carbon policies that aim to accelerate the use of low- or zero-carbon energy technologies by regulation or financial incentives. These policies can be established at the international, national, state, or local level. Several existing and emerging policies of combating climate change are discussed in Chapters 17 and 18. Perhaps the most well-known policy initiative for reducing GHG emissions is the 1997 Kyoto Protocol, the most complex and controversial international agreement yet developed. The Protocol is discussed in Chapter 17, but we introduce it here because of its connection to IPCC.

IPCC's FAR in 1990 prompted the formation of the United Nations Framework Convention on Climate Change (UNFCCC), which was adopted in 1992 and signed by

SIDEBAR 2.1

The Global Warming Skeptics

Although the scientific community has a strong consensus about the evidence of current and future climate change, there are some who disagree. These skeptics are active on the Internet, on talk radio, on the lecture circuit, and in the popular press. They contend one or more of the following:

- That global warming is not occurring
- That if it is occurring, it is the result of normal climatic fluctuations rather than man-induced forcing
- That if it is occurring and possibly caused by man-induced forcing, it is not a serious problem; it is something we can adapt to
- That prospective impacts of GHG mitigation such as restrictions on fossil fuel use would have a far more damaging impact on our economy and society than the impacts of higher temperatures

The handful of scientific skeptics is connected through a number of institutes funded largely by energy industries with a large financial stake in the policy decisions about global warming. Led by S. Fred Singer, Frederick Seitz, Patrick Michaels, Richard Lindzen, and others, with affiliations with the George Marshall Institute, Cato Institute, Tech Central Station Science Foundation, and American Enterprise Institute, this group continues to raise uncertainties about the motives, methods, and results of the scientific process. But increasingly, these skeptics have been shown to receive significant funding from energy companies, including, for example, ExxonMobil as part of the company's 1998 strategy to delay government action on global warming. Seitz and Singer were shown to be wrong in their critiques of climate scientists in recent letters to the *Wall Street Journal* and *Science*. This is not to say that there is no uncertainty or that we need not continue to question our understanding of this complex system and how we monitor it. Countering the skeptics is a group of climate scientists who initiated an Internet blog in the search for truth. See www.RealClimate.org.

154 states and the European Union at the Rio Earth Summit that year; now, 189 countries are party to the Convention. The Convention addresses six GHG including carbon dioxide (82% of total GHG), methane (10%), nitrous oxide (6%), perflourinated hydrocarbons, hydrofluorocarbons, and sulfur hexafluoride.

The 189 countries of the Convention are classified according to their levels of development and their commitments for GHG emission reductions and reporting. They include the following:

- Annex I Parties: Forty developed countries plus EU's fifteen states that aim to reduce emissions to 1990 levels
- Annex II Parties: An Annex I subset of the most developed countries who also commit to help support efforts of developing countries

- Countries with economies in transition (EITs): An Annex I subset that does not have Annex II obligations, mostly made up of eastern and central Europe, and the former Soviet Union
- Non–Annex I Parties: All other, mostly developing countries, which have fewer obligations and should rely on external support to manage emissions

Each year the UNFCCC holds a Conference of Parties (COP). The third COP held in Japan in 1997 produced the Kyoto Protocol, which stated that by the first commitment period (2008–2012), developed countries would have to reduce combined emissions of GHG to at least 5% below 1990 levels. The Protocol would come into force when it was ratified by at least 55 countries, provided they constitute 55% of the CO_2 emissions of Annex I countries. This threshold was reached in November 2004, when Russia ratified the protocol, so it became legally binding to the 128 ratifying parties 90 days hence, on February 16, 2005.

The Protocol sets emissions reduction targets from 1990 for Annex II countries by the first commitment period (2008–2012). These targets range from –8% for many European countries and –7% for the U.S. to +8% for Australia and +10% for Iceland. Those countries with a positive target were allowed an increase in emissions.

Although Europe and other countries have aggressively implemented the Kyoto Protocol, the United States under the administration of George W. Bush decided not to ratify the protocol, arguing that it would seriously impact the U.S. economy and the Protocol would be ineffective without controls on emissions by developing countries. Because the United States is responsible for 21% of the world's carbon emissions, U.S. nonparticipation threatened the viability of the Protocol and the world's ability to meaningfully reduce global carbon emissions. But by 2007, scientific evidence and political pressure were mounting, and U.S. state and local action set the stage for a federal attention to carbon emission reduction. Chapters 17 and 18 discuss the Kyoto Protocol in greater detail, as well as related policies and programs by the European Union, other countries, and U.S. states and localities.

In May 2007, an IPCC international panel agreed in Bangkok to set future limits on emissions beyond Kyoto to try to achieve atmospheric CO_2 concentrations of 445 parts per million (ppm) because of evidence that levels above 450 ppm could trigger severe impacts such as melting of Greenland's ice mass. The target range of 445 to 650 ppm may become the basis for future international agreements for reducing GHG emissions.

Adapting to the effects of climate change. In addition to mitigating global warming through emission reduction, the probable effects in spite of those efforts require us to figure out how to live with climate change. Adaptation measures include lessening the impacts through engineering means without major changes in patterns or locations of development. But attempts to mitigate future impacts in coastal areas, for example, caused by more extreme weather events and sea-level rise, may exceed technological or financial capabilities.

So adaptation must also include anticipating the impacts of climate change and planning for them. This may include emergency preparedness, future land use planning and controls, relocation of existing developments and communities, and alternative water supplies. These measures will be costly and plagued with uncertainty, and will be especially difficult for developing countries that have limited budgets and expertise. Unfortunately, it is these same countries that are likely to experience the most severe impacts.

2.3.2 Local and Regional Air Pollution

Fossil fuel combustion is the major source not only of carbon dioxide emissions, but also of air pollutants affecting human health and ecosystems. All major air pollutants, including fine particulate matter (PM), oxides of sulfur (SO_x) and nitrogen (NO_x), carbon monoxide (CO), ozone (O_3), and some heavy metals such as mercury, are linked to fossil fuel combustion. As shown in Table 2.4, 90% of the major air pollutants in the United States in 2006 were from fuel combustion. Not only do these pollutants affect human health in cities throughout the world, but some are also subject to long-range transport, creating acid rain and other deposition that degrades waters and ecological systems.

The good news is that we have made considerable progress in reducing emissions of air pollutants in the United States, largely through technological controls. Figure 2.16 and Table 2.4 show that aggregate emissions of the criteria air pollutants have been cut in half between 1970 and 2006. Perhaps, surprisingly, most of those gains have been made since 1990: 39% drop in CO emissions between 1990 and 2003, 35% drop in volatile organic compounds (VOC), 40% drop in SO_x, 28% drop in NO_x, and 30% drop in smaller (more hazardous 2.5 micrometer diameter) particulates. This reduction has occurred while energy use, population, vehicle miles traveled, and the economy have all increased (Figure 2.16).

As shown in Table 2.4, stationary energy users such as power plants and industry are major sources of SO_x, PM, and NO_x, as well as several toxic pollutants such as mercury. Power plants, mostly older coal-fired plants, emit 67% of total SO_x, 22% of NO_x, 41% of mercury, and 39% of CO_2. Mobile sources such as automobiles are major contributors of CO, VOC, and NO_x, the latter two of which are the precursors of urban smog and ozone. Technology controls required by government regulations have been incorporated in industrial and power plants and motor vehicles, and progress has been made in reducing air pollutant emissions and resulting episodes of excessive air pollution, especially in industrialized countries.

The bad news is that even with these emissions reductions, nearly half the people in the United States still live in areas not fully attaining clean air standards. Figure 2.17(a) shows the improving overall trend of ozone concentrations in U.S. cities relative to the eight-hour standard. Figure 2.17(b) shows the 124 ozone non-attainment areas. Air quality is far worse in the cities of poor countries, where people continue to experience serious public health hazards from energy-related air pollution. Figure 2.18 gives average annual air pollution levels in twenty Asian cities along with World Health Organization (WHO) standards. WHO estimates that more than 500,000 premature deaths per year in Asia are caused by air pollution.

figure 2.16   U.S. Air Pollutant Emissions Trends, 1970–2006

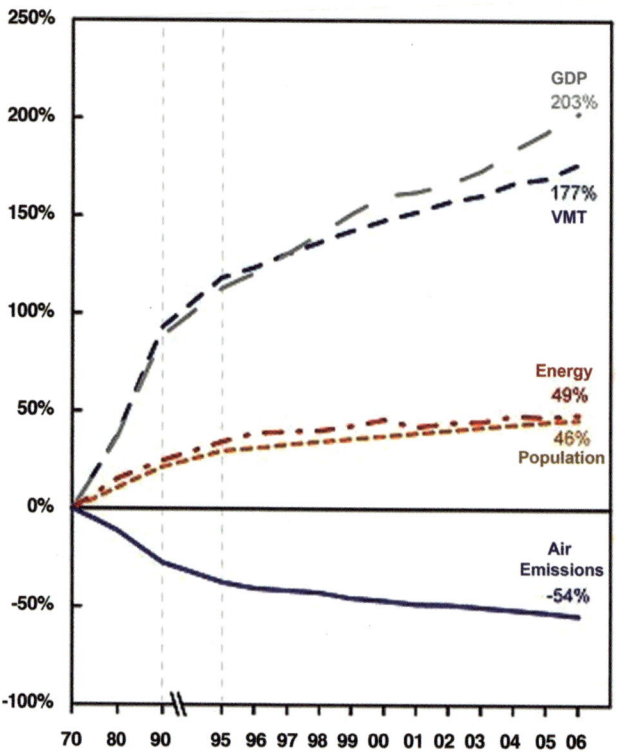

Overall emissions have been reduced by 54% while GDP has more than tripled, vehicle miles traveled have soared, and energy and population have both increased by nearly 50%.

Source: U.S. EPA, 2007

table 2.4   Aggregate U.S. Air Pollutant Emissions (1970, 1990, 2006) and Sources

	1970	1990	2006	% energy	% stationary	% mobile
Sulfur dioxide (SO_x)	31	23	14	87%	83%	4%
Carbon monoxide (CO)	197	144	88	94%	5%	89%
Particulate matter PM 10	12.2	3.2	2.6	60%	39%	21%
Particulate matter PM 2.5	NA	2.3	1.6	63%	40%	23%
Nitrogen oxides (NO_x)	27	25	18	95%	39%	56%
Volatile organic compounds (VOC)	34	23	14	53%	7%	46%
Lead	0.22	0.005	0.002	NA	NA	NA
Total	**302**	**218**	**137**	**90%**	**20%**	**70%**

Source: U.S. EPA, 2007

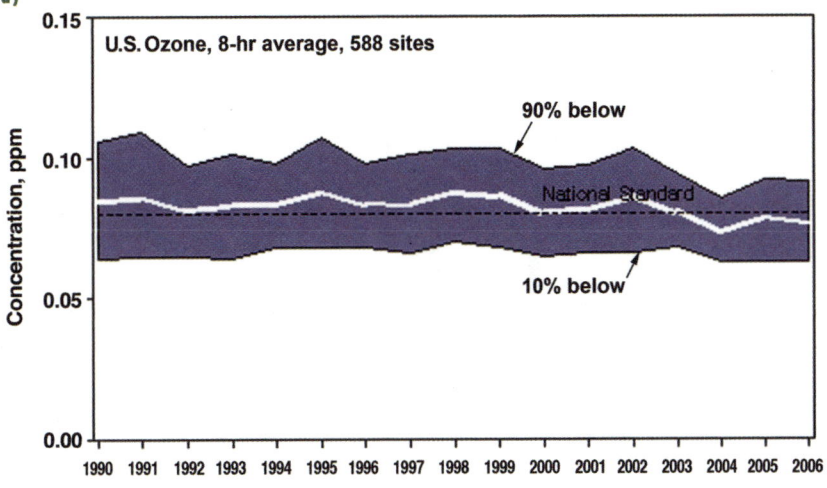

figure 2.17(a) Some Progress in U.S. Urban Ozone Concentrations

SOURCE: U.S. EPA, 2006

figure 2.17(b) U.S. Non-attainment Areas for Ozone

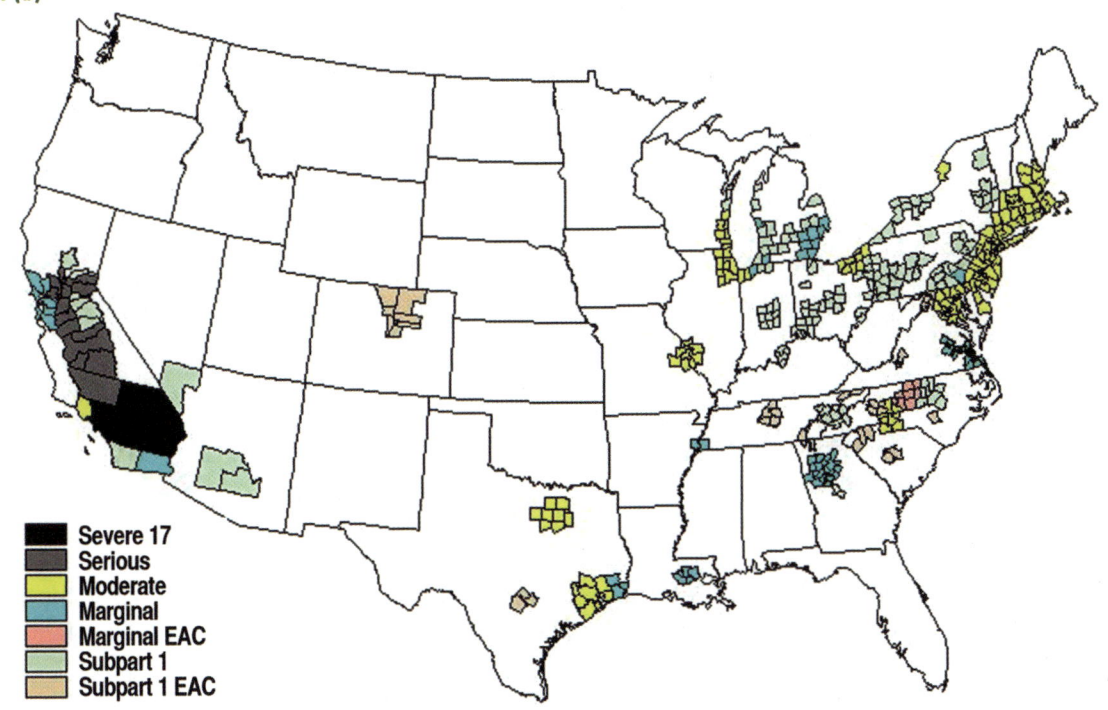

Many U.S. cities are still not attaining health standards for ozone.

SOURCE: U.S. FHWA, 2006

figure 2.18 **Air Pollution Concentrations in Selected Asian Cities, 2000–2004**

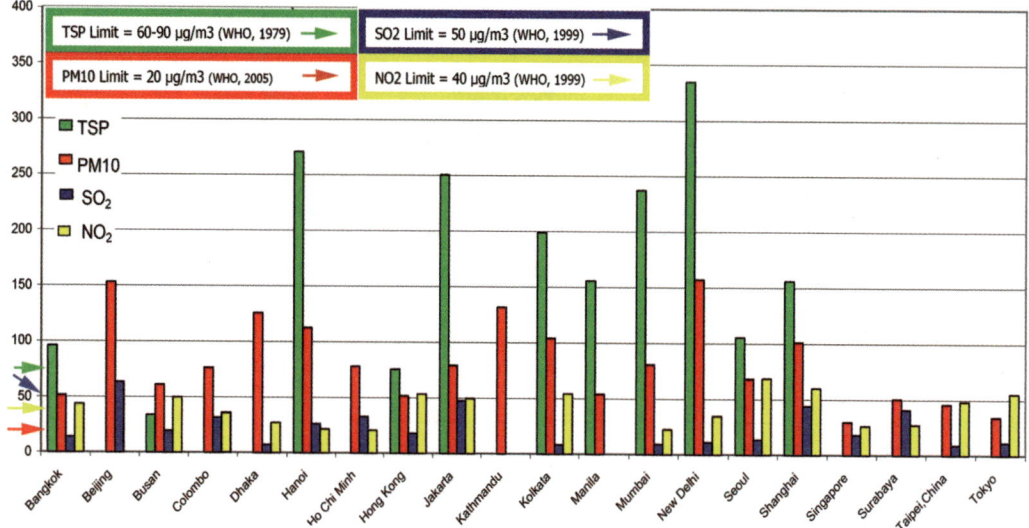

Relative to WHO standards, most cities far exceed particulate matter standards.

Source: CAI-Asia, 2006

According to the *Financial Times,* a 2007 World Bank draft report estimated 750,000 premature air pollution–related deaths in China alone, but this figure was removed from the final report because the Chinese government believed it would cause social unrest.

Improvement of emissions and air quality in the United States has been largely the result of the Clean Air Act, originally passed in 1970 and reauthorized in 1977 and 1990. Congress continues to debate changes in the law, focusing on the pace of additional emission reductions at coal-burning power plants (see Chapter 17).

2.3.3 Other Effects of Fossil Fuels

Our energy use has impacts beyond climate change and air pollution. Table 2.5 highlights many of these environmental impacts by energy source, including fossil fuel, nuclear, and renewable energy. The extraction and transport of energy also impacts the environment and should be considered public costs of our energy system. For example, coal mining has long impacted coal regions: deep mining hazards, strip mining, mountaintop removal and valley fill methods, acid mine drainage, and mineland reclamation have prompted public and political response. Oil transport requires risk of tanker spill. Transport risks for liquefied natural gas (LNG) are different because of the volatile nature of LNG. Some worry about terrorism risks of LNG tankers and facilities and nuclear power plants.

table 2.5 Environmental Impacts of Energy Sources Other Than GHG Emissions

Energy Source	Environmental Impacts	Significance	Mitigation
Coal—electric	Mining and processing impacts on lands/waters	• Severe	• Mineland reclamation
	Combustion: carbon emissions	• Severe	• Carbon sequestration
	Air pollution (part, SO_x, NO_x, mercury); acid/particulate deposition	• Severe	• Technology controls
	Thermal pollution	• Moderate	• Cooling controls
	Ash disposal	• Moderate	• Storage/land application
Coal—synthetic fuels	Mining impacts on lands/waters	• Severe	• Mineland reclamation
	Processing impacts on waters	• Severe	• Technology controls
	Residue disposal on lands	• Severe	• Storage/land application
	Combustion: carbon emissions	• Severe	• Carbon sequestration
	Air pollution (part, SO_x, NO_x)	• Severe	• Technology controls
Petroleum—transportation fuel	Tanker/pipeline spills	• Risk	• Management controls
	Refinery impacts	• Moderate	• Technology controls
	Combustion: carbon emissions	• Severe	• Carbon sequestration
	Air pollution (NO_x, HC, CO)	• Severe	• Technology controls
Oil shale, oil sands	Mining impacts on lands/waters	• Severe	• Mineland reclamation
	Processing impacts on waters	• Severe	• Technology controls
	Residue disposal on lands	• Severe	• Storage/land application
	Combustion: carbon emissions; air pollution (part, SO_x, NO_x)	• Severe	• Technology controls
Natural gas	Pipeline leakage	• Moderate	• Management controls
	Liquefied natural gas transport risks	• Risk	• Management controls
	Combustion: carbon emissions, NO_x	• Moderate	• Technology controls
Nuclear power	Radioactive materials, fuel cycle	• Risk	• Management controls
	Plant safety	• Risk	• Management controls
	Waste storage and disposal,	• Risk	• Technology controls
	Nuclear materials proliferation	• Risk	• Management controls
Hydro—large	Hydro system, fish migration, riparian ecology	• Severe	• Fish passage
	Reservoir flooding impacts	• Severe	• Relocation/compensation
Hydro—small	Hydro system	• Minor	
Solar thermal—on site	Manufacturing impacts	• Minor	
Solar PV—on site	Manufacturing impacts	• Moderate	• Technology controls
Solar PV—farms	Manufacturing impacts	• Moderate	• Technology controls
	Land consumption	• Moderate	• Mixed use
Wind electric—small	Manufacturing impacts	• Minor	
Wind electric—farms	Manufacturing impacts	• Moderate	• Technology controls
	Land consumption, bird mortality, aesthetics	• Moderate	• Mixed use
Biofuels—liquid	Farmland consumption	• Moderate	
	Farming/harvesting impacts on waters	• Moderate	• Technology controls
	Processing impacts	• Moderate	• Technology controls
	Combustion: carbon, NO_x	• Moderate	• Technology controls
Biomass—solid	Forestland consumption	• Moderate	
	Farming/harvesting impacts on waters	• Moderate	• Technology controls
	Processing impacts	• Minor	• Technology controls
	Combustion: carbon, particulates, NO_x	• Moderate	• Technology controls

2.4 Opportunities and Limits for Non-fossil Energy

As discussed above, mitigating climate change by reducing carbon emissions requires technological advances in non-carbon sources and carbon sequestration. There are six principal means to reduce carbon emissions:

1. Reduce fuel combustion and carbon emissions by improving the energy efficiency of buildings, vehicles, appliances, and power generators.
2. Use renewable energy, with wind/solar/geothermal replacing coal electricity and with biofuels replacing petroleum.
3. Shift from high- to low-carbon fuel, in which, for example, natural gas electricity replaces coal electricity.
4. Capture and store CO_2 in deep oceans and deep mine cavities. Carbon sequestration is a principal part of the U.S. policy to reduce CO_2 emissions without major disruptions of the current fossil fuel energy-industrial complex, and possibly to move to a coal-based hydrogen economy.
5. Use nuclear power to replace coal electricity.
6. Sequester carbon via reforestation and agricultural soil conservation. Natural forests and soils are major sinks for global carbon and just as deforestation and conventional agricultural cultivation release carbon to the atmosphere, reforestation and soil conservation practices capture carbon.

Figure 2.19 shows U.S. DOE's projected U.S. GHG emissions and relative means for stabilizing emissions at 2001 levels by 2050. Efficiency and renewables provide the greatest reduction. Chapter 3 discusses Princeton professors Scott Pacala and John Socolow's assessment of CO_2 reductions by these technical methods as well as IPCC future socio-political scenarios for carbon reduction. This section addresses some of the limits and opportunities of non-carbon energy: nuclear, renewables, and efficiency.

2.4.1 Nuclear Power

When commercial use of nuclear power was developed in the late 1950s, it was thought to be the great savior of human civilization. It would become our source of clean, limitless electricity "too cheap to meter" replacing dirty coal and ultimately depletable oil as we moved to the twenty-first century.

But, nuclear power has yet to come close to achieving that promise. Concerns about rising costs, safety, waste disposal, and proliferation of nuclear materials have reduced the favor of nuclear power in the eyes of the public, utilities, investors, and policy makers, and growth of nuclear power has been stagnant for nearly two decades. Some people, including some prominent environmentalists, have called for a renaissance of nuclear power in response

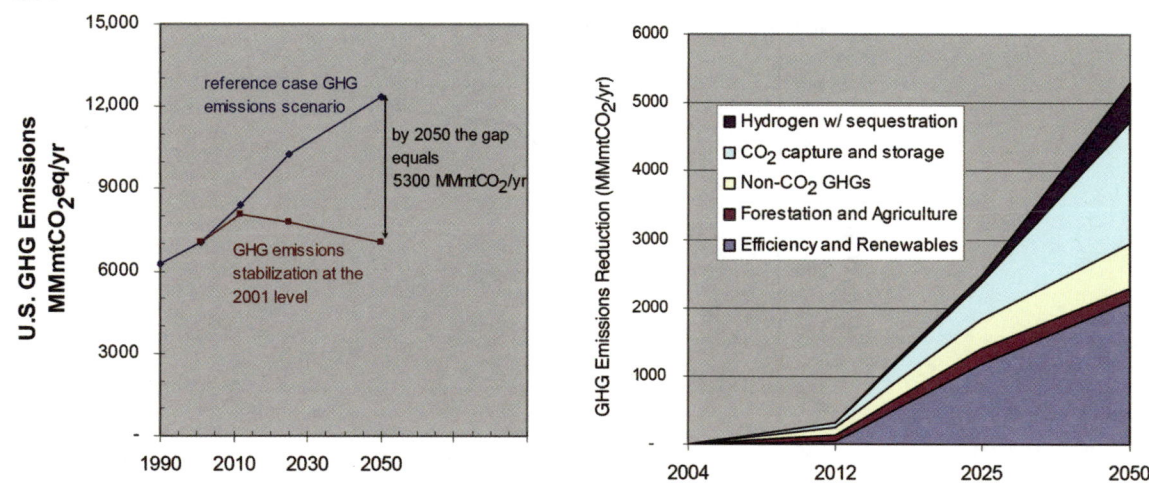

figure 2.19 U.S. DOE Scenario for GHG Emission Stabilization by 2050

Stabilization requires reduction of 5300 MMmt CO_2 below reference-case. The best means of achieving reduction are efficiency, renewables, and carbon sequestration.

SOURCE: U.S. DOE, 2004

to oil and carbon problems, but life-cycle issues of long-term waste management and nuclear weapons proliferation in an unstable world remain significant barriers.

2.4.1.1 From Great Hope to Stagnant Growth

From 1970 to 1990, nuclear power steadily increased its contribution to 19% of U.S. and 17% of global electricity. However, since 1990 the addition of installed nuclear capacity has been stagnant. Figure 2.20 shows the growth of world capacity from 1980 to 2004. Table 2.6 shows that since 1998, capacity growth has been only 0.5% per year when total world energy grew by 2% per year. Thirty countries have nuclear power, but the top ten countries provide 86% of the world nuclear energy. The top three (United States, France, and Japan) generate 57% of the world total. While generation has increased due to improved capacity factor of operation, the nuclear-to-total-electricity ratio of contribution has not changed.

The United States is still the world leader in nuclear power, but no new nuclear power plants have come on line since 1996, and U.S. nuclear capacity of 100 GW is the same as in 1990. Figure 2.21 gives the U.S. nuclear capacity and generation from 1980 to 2006. Although nuclear capacity has not changed since 1990, its increased capacity factor has helped increase generation. **Capacity factor** is the percent of time the plant operates at full capacity (see Chapter 9). Current plants were designed for a lifetime of thirty years and most operating licenses will expire in the next twenty years. The Nuclear Regulatory Commission issued twenty-year

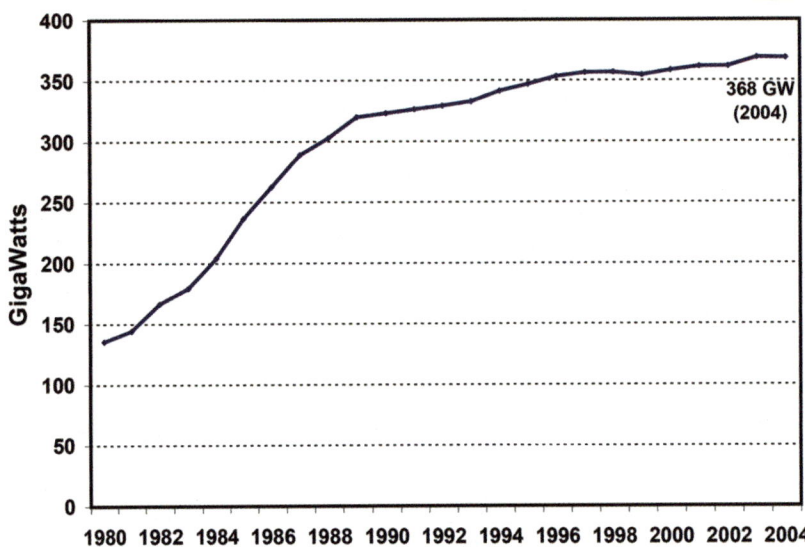

figure 2.20   Global Growth of Nuclear Power, 1980–2004

368 GW (2004)

Growth has been flat since 1990, especially since 1998.

SOURCE: data from U.S. EIA, 2006b

table 2.6   World Nuclear Capacity and Generation

Country	% World Capacity, 2004	% World Generation, 2004	Annual % Change Capacity, 1998–2004
United States	27.0%	30.1%	0%
France	17.2%	16.3%	+0.8%
Japan	12.4%	10.4%	+1.0%
Russia	6.0%	5.2%	0%
Germany	5.8%	6.1%	+0.1%
South Korea	4.3%	4.7%	+7.3%
United Kingdom	3.3%	2.8%	−0.9%
Ukraine	3.2%	3.2%	−2.3%
Canada	2.9%	3.3%	−6.2%
Sweden	2.6%	2.8%	−4.6%
World	**368 GW**	**2219 TWh**	**+0.5%**

license extensions for sixteen plants by 2006, and utilities are expected to apply for nearly all of the operating plants. Although utilities have had safety violations, they have maintained and operated these plants without major incident and hope to extend their lives. However,

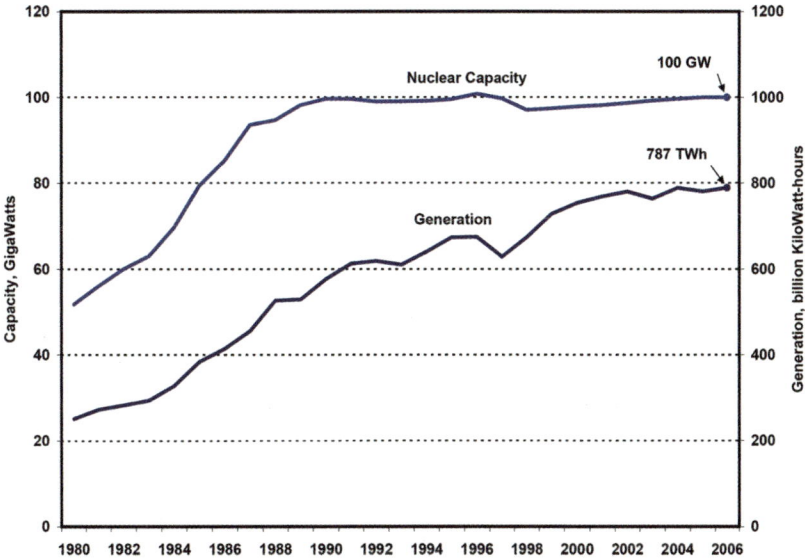

figure 2.21

U.S. Nuclear Capacity and Generation, 1980–2006

U.S. nuclear capacity has been flat since 1990, but improved capacity factor has helped increase generation.

SOURCE: data from U.S. EIA, 2007a

many fear that much like any older machine that is more prone to malfunctioning, an aging nuclear plant increases the likelihood of accidents if it is allowed to operate well beyond its design life.

The future of nuclear power depends on new development, not simply license extensions. With the advent of global warming, many prominent environmentalists, including James Lovelock, Bruce Babbitt, and others, are calling for a reevaluation of nuclear, believing it the only viable energy solution to global warming. Although very few plants have been built in recent years, research on new nuclear plant designs has continued. In the United States, there is a renewed interest in nuclear power; considerable funding, as well as streamlined licensing, was included in the 2005 Energy Policy Act.

2.4.1.2 Barriers to a Renaissance in Nuclear Power

Nuclear power has the benefit of no carbon emissions, but requires the handling and use of radioactive materials that must be kept isolated from humans and other living systems. Exposure to radioactive materials is linked to genetic mutation and human cancers. To prevent release of radioactive materials, the nuclear power industry requires near-perfect management of technologies and human systems involved in the entire nuclear materials cycle from mining, processing, and transport of fuel to power plant operation to waste storage, transport, and disposal.

The biggest obstacles facing a resurgence of the nuclear industry are continuing uncertainty over safety and waste management, uncertainties over costs, increasing concerns over proliferation of nuclear materials in a world plagued by terrorism, and public and investor reluctance. Needed safeguards for safety, waste management, proliferation control, plant decommissioning, and other unknown requirements translate not only into technical, public, and political questions, but also into higher development and operating costs. A major 2003 study by MIT addressed these obstacles and concluded that, while daunting, we need to overcome them so that nuclear can reemerge as a viable option (along with efficiency, renewables, and carbon sequestration) for our energy future to combat global warming (see Sidebar 9.2). Even if many of the challenges facing nuclear power can be overcome and the optimistic scenario of tripling nuclear capacity by 2050 can be realized, it would amount to no more than 20% of global electricity, and cut the expected 100% increase in emissions of carbon by only 12% to 27%. Nor could nuclear provide a direct or substantive answer to our dependence on oil.

2.4.2 Energy Efficiency

Our best bet for both short-term investment to reduce fossil fuel dependency and for long-term sustainability is to increase the efficiency of energy production and use and to develop sustainable renewable energy sources. But energy efficiency comes first because it provides the best short-term opportunity to reduce demand for oil and carbon emissions, is the most cost effective of our energy options, has lasting value as benefits continue to accrue, and resulting lower demand makes supply options easier.

Great progress has been made in the energy efficiency of buildings, appliances, and vehicles in the last three decades, and the global and U.S. economies are far more energy efficient as a result. Figure 1.8 highlighted the improvements in the United States and the estimated $700 billion in energy costs saved as a result of efficiency improvements since the mid-1970s. But artificially low energy prices have constrained investment in efficiency and there is great potential for further improvements. This section introduces some key definitions in efficiency and conservation, and subsequent chapters explore the wide-ranging opportunities for efficiency improvements in buildings, electricity, and transportation.

We have already introduced an important measure of the effectiveness of our energy, that is the **energy intensity** of our economy. Energy intensity of the U.S. and world economies has improved steadily since the mid-1970s (Figures 1.3 and 1.7; Table 1.2). Intensity is measured by energy per dollar GDP, and it has improved by 25% in the world since 1980 and 45% in the U.S. since 1973. Per capita energy increased only 2% during this period for the world and actually decreased by 5% for the United States, all during periods of improved standards of living and economic growth. Recall from Equation 1.1:

Eq. 2.1
$$\text{Energy intensity} = \frac{\text{Energy used}}{\$\text{GDP}}$$

Section 1.4.2 explained that the improvement in energy intensity in the United States during the past three decades resulted from the following:

- Investments in efficiency driven by higher energy prices and government standards mostly until 1985
- Structural changes in the economy driven by advances in the less energy-intensive service and information sectors, mostly since 1995

Energy efficiency is different from energy intensity. **Energy conversion efficiency** is the effectiveness of converting from one form of input energy to another more useful form, such as converting input coal chemical energy to thermal energy in a power plant boiler to mechanical energy in the turbine to useful electrical energy in the power plant generator. If we can convert more useful energy out of a unit of input energy we are converting energy more efficiently.

Eq. 2.2 $$\text{Energy conversion efficiency} = \frac{\text{Useful output energy}}{\text{Input energy required}}$$

Energy functional efficiency is the useful performance we can get out of the energy we consume. We do not really want energy; we want the functions that it provides. We want thermal comfort; lighting; transportation of people and materials; entertainment; food production, preservation, and preparation; industrial processes; mechanical tools; and other life and labor improvements that energy provides us. If we can provide these functions with less energy, we are using energy more efficiently.

Eq. 2.3 $$\text{Energy functional efficiency} = \frac{\text{Functions provided}}{\text{Useful energy consumed}}$$

By its nature, energy conversion and functional efficiency improvements do not require any change in the end result, that is, the functions provided, people's behavior, and standard of living.

Energy conservation is defined here as behavioral changes made by individuals or communities to save energy by cutting back on the functions energy provides. So, for example, in your house:

- Improvement in energy *conversion efficiency* can be realized by replacing an old gas furnace with a new super-efficient one (same useful output, less input).
- Improvement in energy *functional efficiency* can be realized by adding insulation to the house (same function, less useful energy consumed).
- Energy *conservation* can be realized by lowering the thermostat at night during the heating season (less function, less energy).

Since 1975, local and state building codes have improved energy efficiency in new buildings, federal Corporate Average Fuel Efficiency (CAFE) standards have improved auto efficiency, and federal standards have improved electrical appliance efficiency. For example, state and federal efficiency standards for refrigerator efficiency have driven down the energy use per new unit by 75% since 1974, or 5% per year (see Chapter 8).

CAFE standards require automakers to meet an average efficiency in miles per gallon (mpg) for the fleet of cars and light trucks they sell. The standards increased steadily until 1985, but car standards have not increased since. In 2003, light truck standards were increased from 20.2 mpg to 21.0 for model year (MY) 2005, 21.6 for MY2006, and 22.2 mpg for MY2007. But, because of market shift to more light trucks (i.e., SUVs, vans, and pickups), overall new vehicle efficiency has actually decreased since 1985 (see Chapter 13).

The good news is that energy intensity has steadily improved in the United States and throughout the world since the mid-1970s; the bad news is that investments in energy efficiency have slowed in the United States. Reasons for this are as follows:

- Relatively cheap energy prices, especially when compared to rising incomes
- Transaction costs and investment barriers such as uncertainty over future energy costs, new products, and knowledge gaps

However, there is a huge potential for cost-effective energy efficiency improvements in buildings, vehicles, appliances, lighting, and industrial processes. These opportunities are addressed in subsequent chapters.

2.4.3 Renewable Energy

Renewable energy systems avoid many of the problems of conventional energy, including resource depletion, carbon emissions, air pollution, radioactive materials, fuel transport from source to use, and so on. Renewable energy sources are diverse and well suited for a variety of energy applications. They include direct solar thermal energy; solar electricity through photovoltaics; wind electrical generation; hydroelectric generation; biomass energy in gaseous, liquid, and solid forms; geothermal heating and electricity; and tidal and wave energy.

However, renewable energy still contributes very little energy to both the U.S. and world economies. Renewables contribute only about 6%–7% of commercial energy of both the United States and the world. This does not include non-marketed renewable energy, such as wood and other biofuels (estimated to be about 10% of total world energy use) and onsite solar heating and photovoltaic systems. Hydro and wood contribute over 80% of U.S. commercial renewable energy. The primary use of wood is the wood products industry use of residual materials for internal heat and power generation.

Newer sources, such as solar photovoltaics, wind electricity, and liquid and gaseous biofuels, have been plagued for decades by higher capital costs in the face of cheap fossil fuels.

But today these three sources have the fastest growth rate of any energy source, growing at 20% to 40% per year.

Worldwide wind electric generating capacity grew to more than 74 gigaWatts (GW) by the end of 2006, about ten times the capacity in 1997 (Figure 2.22). The annual growth rate of 30% per year during that period makes wind the energy source with the fastest growth rate. Germany leads with 20.6 GW, followed by Spain and the United States both with 11.6 GW.

Photovoltaic (PV) electric system installed capacity grew by 1500 MW in 2005, a 34% annual growth rate (Figure 2.23). World capacity exceeds 6.5 GW at the end of 2006.

Biofuels for transportation liquid fuels have also seen considerable growth. Ethanol production more than tripled in six years to 13.5 billion gallons in 2006 (Figure 2.24). Representing only about 11% of fuel ethanol production, biodiesel production is growing even faster. The United States surpassed Brazil as the world's largest ethanol producer in 2005. Growth continues but U.S. production from corn must be replaced by cellulose-based ethanol if this growth rate is to continue.

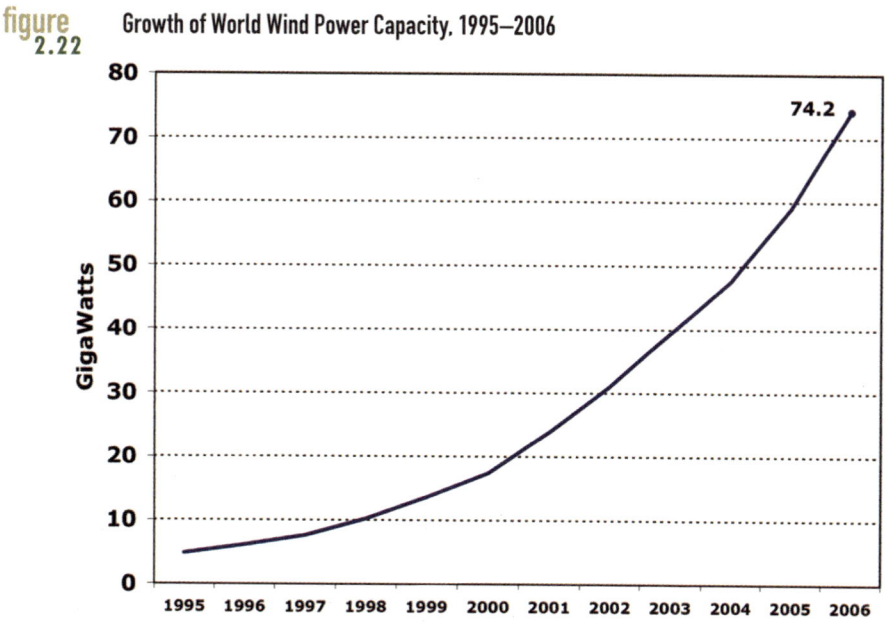

figure 2.22 **Growth of World Wind Power Capacity, 1995–2006**

SOURCE: data from AWEA, 2007

figure
2.23   **Growth of World Photovoltaic Power Capacity, 1994–2006**

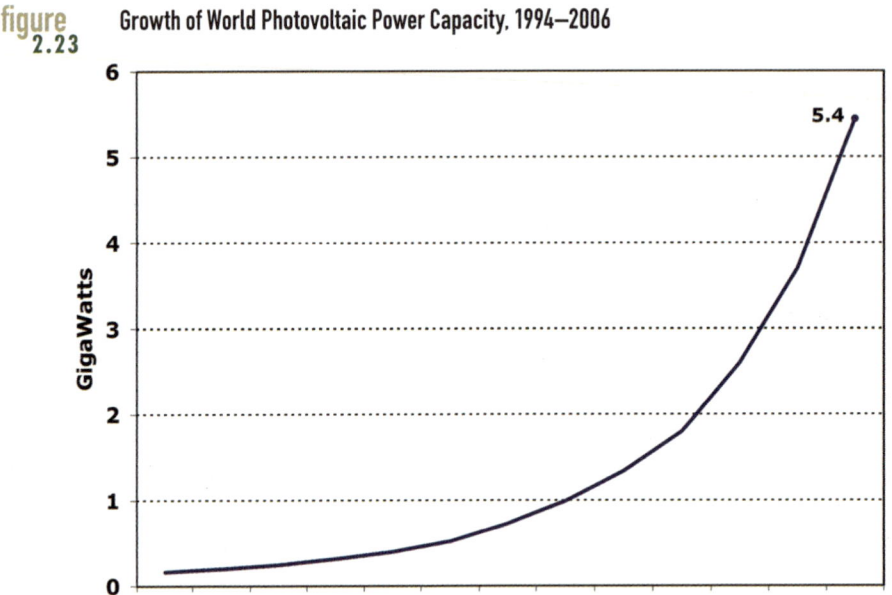

SOURCE: data from SolarBuzz™, 2007

figure
2.24   **Growth of World Biofuels, 1995–2006**

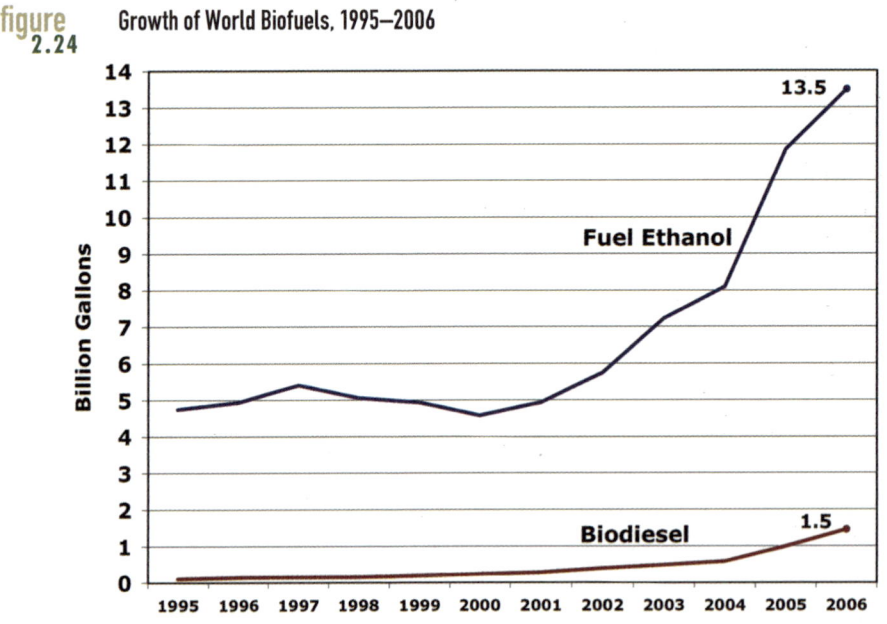

SOURCE: data from RFA, 2007, and NBB, 2007

SOLUTION BOX 2.1

World wind, solar photovoltaic, and biofuel energy currently provide only about 1% of the 463 quads of global marketed energy. But they are growing fast. If worldwide growth of wind, solar photovoltaic, and biofuel energy continues at an annual rate of 30% and global energy use grows at 2% per year, what percentage would these fuels provide in fifteen years in 2020?

Solution: Estimated 2005 wind, solar, biofuels energy is 1% of 463 or 4.63 quads.

Estimated 2020 wind, solar, biofuels energy $= F_{wsb} = P(1 + r)^n$

$$= F_{wsb} = 4.63(1.3)^{15} = 237 \text{ quads}$$

Estimated 2020 total global energy $= F_{ge} = 463(1.02)^{15} = 623 \text{ quads}$

Wind, solar, biofuels percentage in 2020 with current growth $= \dfrac{237}{6323} = 38\%$

Can we continue the current rate of growth of wind, solar, and biofuels? Probably not. But these sources are growing fast and will increasingly affect our future energy mix.

2.5 Summary

The implications of our current patterns of energy use present a troubling picture. The world and the United States remain highly dependent on fossil fuels (86%) and oil (40%). Despite warning signs of the consequences of this dependence on the world economy and environment during the past thirty years, little has been done to alter these patterns of use except for improvement in the energy intensity of our economy. The following points are important to keep in mind as we delve more deeply into our energy system:

- Oil is a nonrenewable resource and production continues to increase to meet growing demand, especially in the less developed countries as their economies develop. While some pessimists surmise that we have already reached peak production or will in the next few years, even some optimists project a peak within twenty years. This has severe implications for the advance of the global economy.
- There is a scientific consensus that global climate change forced by human-induced greenhouse gas emissions is occurring. The vast majority of those emissions are carbon dioxide from the combustion of carbon-based fossil fuels, coal, oil, and natural gas. Scientists believe that the future impacts for climate change will be not only severe but also long-lasting with significant lag effects, so the impacts of emissions today will be felt long into the future.

- Among the United States reservations about the Kyoto Protocol is that they believe there are few viable short-term options to reduce CO_2 emissions by switching to non-carbon energy sources. To some extent they are correct:

  - Nuclear power is stagnant in the United States (and worldwide for that matter). Prospects are improving, but growth will likely be slow before 2030 because of concerns over costs, safety, waste management, and nuclear weapons proliferation.
  - Renewable energy contributes only a small share of U.S. and global commercial energy and its main sources, hydro and wood residue in the forest product industry, have limited growth potential. Still, wind, solar PV, and biofuels are growing quickly.

  - Energy efficiency improvements have helped reduce energy intensity of the economy but significant opportunites for cost-effective improvements have not been achieved.

Add to these points the lessons from Chapter 1:

- Global consumption of energy continues to increase at 2% per year, with prospects for much additional growth as the world's emerging countries, led by China and India, develop modern economies.
- The geopolitical realities of oil have caused increased dependence on the politically unstable Middle East and access to oil has had a significant impact on trade balance and foreign and military policy.
- Fossil fuels with their CO_2 emissions and other environmental effects still provide more than 85% of our energy.

With this troubling array of issues as a backdrop, the following chapter explores a variety of future energy scenarios.

CHAPTER 3

Energy Futures

Chapters 1 and 2 addressed the critical issues of oil, carbon, and demand that confront us as we try to develop more sustainable patterns of energy use. An important aspect of that quest is to think about the future, and prepare a vision that can articulate the possibilities and inform action in planning, policy, consumer choice, and technology development.

In a 2004 presentation on the outlook for energy, Yale economist William Nordhaus reflected on the lessons of the three decades since the first oil crisis:

- On the supply side, there has been little progress, although energy efficiency has increased steadily.
- There are at present no large-scale, environmentally benign, and economical energy resources. No major new energy technologies have emerged in the last three decades.
- Other large-scale options—particularly coal and nuclear power—look distinctly more unattractive today than they did three decades ago.
- The state of international cooperation is at its lowest point in the last century, and effective global agreements to ensure the safety of the nuclear fuel cycle or to slow global warming are moving at an imperceptible pace.

This is not exactly a rosy assessment. The view from early 2008 is slightly more hopeful, but Nordhaus' lesson is clear: Despite diverse economic, financial, security, and environmental signals over the past three decades, energy markets, governments, society, and people have not responded effectively to develop more sustainable patterns of energy production and use.

This is not to say that institutions and people are not thinking about our energy future. Many are studying the issues, analyzing the options, taking out their computer models and crystal balls, and fashioning scenarios and recommendations for future energy sources, technologies, and policies to move toward sustainable energy. But sustainable energy will require more than projections and scenarios—it will require action. Still, the place to start is providing meaningful visions of the future that articulate problems, open our eyes to new opportunities, and galvanize markets, governments, and people to take action.

This chapter begins by introducing methods of thinking about the future, including four basic approaches: projecting and forecasting, developing roadmaps, assessing needs and developing solution "wedges," and developing and using scenarios. The chapter then presents a number of future energy visions developed by government agencies, analysts, and organizations. These examples illustrate the visioning methods and also provide a diverse range of perspectives on our energy future. They are organized into four groups:

1. Official government forecasts, which we call business-as-usual visions of the future
2. Several other visions of U.S. energy or energy sectors
3. Recent visions emphasizing possibilities for renewable energy and efficiency
4. Global visions for reducing carbon emissions

The chapter concludes with the authors' perspective on future energy and a synthesis scenario for global energy, maximizing efficiency and use of renewable energy.

3.1 Planning and Visioning the Future

In thinking about future energy, we need to understand where current trends are taking us, what opportunities and challenges lie ahead, and what actions we can take proactively to determine our destiny rather than having it dealt to us. Four approaches are introduced: (1) Projecting and forecasting emphasize trend analysis; (2) "Roadmapping" seeks to maximize development of a specific technology or objective; (3) Developing "solution wedges" begins with needs assessment then investigates the various means of meeting a portion or "wedge" of those needs; and (4) Developing scenarios embraces the uncertainties of any future visioning, identifies driving forces, and uses storylines to articulate different possible futures.

3.1.1 Projections and Forecasts

A common place to start and the basis for most future visions is projecting and forecasting. There is a difference between the two:

Projections are the extension of past trends into the future. They identify trend lines and simply extend them to the future regardless of constraints or changes. Projections help us to see what extending existing trends into the future may mean and to recognize the need for change. However, they tend to assume the future will replicate the past.

Forecasts are the prediction of future conditions based on projections modified by expected assumptions of driving forces, constraints, and opportunities. Forecasts may depict one possible future or a few possible futures if varying assumptions are desired. Forecasts are usually based on models of several factors, such as energy prices, economic growth, and population.

Predicting future energy is a risky business, and past attempts to do so by government agencies have been largely unsuccessful. More importantly, though, this method of visioning is not very creative. It assumes that driving economic and demographic forces are the major determining factors, and that these can be predicted and modeled. The models seldom include forces for major change, so they tend to predict a future characterized by present trends. They rarely incorporate normative prescriptions or "desired future conditions" that other methods are designed to do.

The U.S. Department of Energy's (DOE) Energy Information Administration (EIA) and the International Energy Agency (IEA) use forecasting models to prepare annual energy outlooks for domestic and global energy, respectively. EIA's Annual Energy Outlook (AEO) is produced each January and looks forward about 25 years. It is based on the National Energy Modeling System (NEMS). That model is largely dependent on prediction of energy prices, and the volatility of actual prices has limited the effectiveness of its results. EIA prepares an annual evaluation of its previous reports to learn how to better forecast energy. Those evaluations as well as several independent assessments of EIA forecasts (see O'Neill and & Desai, 2005; Laitner, et al., 2003; Craig, et al., 2002) indicate room for improvement in forecasting energy prices, production, and use.

Figure 3.1 shows that early energy forecasts (pre-1980) made by EIA and its predecessor agencies overestimated future energy consumption and that recent EIA forecasts also tend to reflect projections of past trends. The figure compares these "official" government forecasts

figure 3.1   **Energy Forecasts versus Reality**

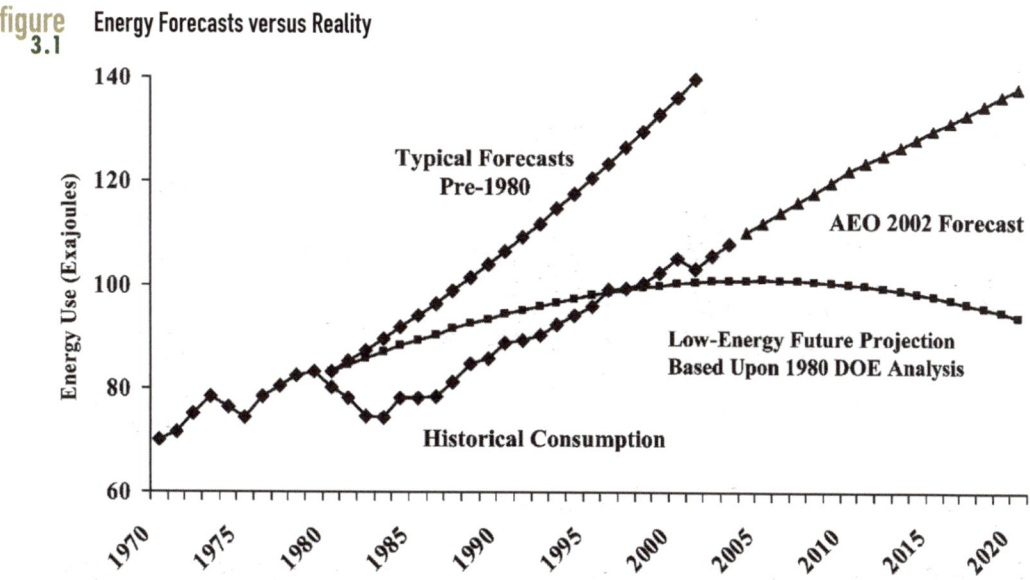

Historical consumption trends prompted overblown typical pre-1980 forecasts. Annual Energy Outlook (AEO) 2002 shows continued use of projections of past trends in more recent forecasts. Low-energy curve illustrates a scenario based on potential for a "desired future condition."

SOURCE: Laitner, et al., 2003. Used with permission from Elsevier Limited.

to a low-energy future scenario developed by DOE in 1980. This latter curve illustrates how future projections and forecasts can be modified by additional analysis driven by a desired future condition. This is typical of the approaches described below.

Although projections and forecasts have their limitations because of inaccuracies caused by future uncertainties and due to their lack of creativity in reflecting trend as destiny, they do provide a useful baseline against which to compare scenarios based on different assumptions or desired future conditions. We use these official energy outlooks as business-as-usual forecasts in the next section.

3.1.2 Technology Roadmaps

Roadmaps or blueprints do not try to predict what *will be* but describe what *could be* by tracing a plausible path to a future maximizing a selected technology or objective. Roadmaps are a popular means for energy advocates to develop and communicate the possibilities for a technology of interest. As the name implies, a roadmap study not only identifies the potential energy production or savings from a specific technology, but the policy and investment pathway to achieve them.

Table 3.1 lists several roadmap studies prepared between 2001 and 2006. The studies vary, but each focuses on a specific energy technology or option, assesses future potential,

figure 3.2   Timeline Given in Nuclear Power Generation IV Roadmap

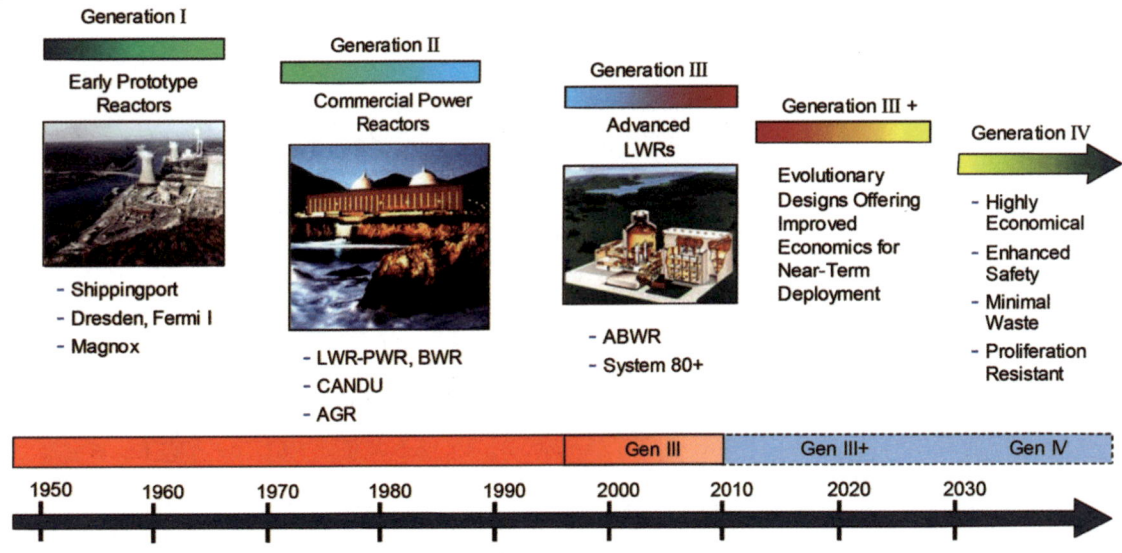

Generation III advanced reactor designs now being developed would ultimately be replaced by more economic Generation IV after 2030.

Source: U.S. DOE, 2002

and outlines the policy, investment, and technology paths to achieve that potential. Some of these studies are described later. Figure 3.2 illustrates the timeline given in the Generation IV Nuclear Roadmap.

3.1.3 Needs Assessment and Solution Wedges

Pacala and Socolow (2004) describe a third approach for developing energy visions using "solution wedges" (see Section 3.5.1); their approach has been applied to other studies. The approach has several steps that are illustrated by an example of eliminating oil imports.

1. Determine a **desired future condition (DFC).** This serves as the primary objective of the study and must be put in quantitative terms. Let's say our DFC is the elimination of oil imports to the United States by 2030. Figure 3.3 shows U.S. petroleum production, consumption, and imports to 2006. As we know, the import gap has grown. If we project production and consumption from recent trends to 2030, consumption of 10,000 million barrels (Mbbls) and production of 1600 Mbbls creates an import gap of 8400 Mbbls. Our DFC is to fill that gap.

table 3.1   Selected Energy Technology Roadmap Studies

Title	Author	Date
25% Renewable Energy for the United States by 2025	University of Tennessee	2006
Energy Efficiency Technology Roadmap	Bonneville Power Administration	2006
A Responsible Energy Plan for America	Natural Resources Defense Council	2005
The U.S. Photovoltaic Industry Roadmap through 2030 and Beyond	Photovoltaic Industry, U.S. DOE-Energy Effciency & Renewable Energy (EERE)	2004
Renewable Energy Future for the Developing World	International Solar Energy Society (Dieter Holm)	2004
America's Oil Shale: A Roadmap for Federal Decision Making	U.S. DOE	2004
Transitioning to a Renewable Energy Future	International Solar Energy Society (Don Aiken)	2003
The U.S. Photovoltaic Industry Roadmap	Photovoltaic Industry, National Center for Photovoltaics	2003
Technology Roadmap: Energy Efficiency in Existing Homes	U.S. HUD-Partnership for Advancing Technology in Housing (PATH); National Association of Home Builders Research Center	2002
Small Wind Energy Roadmap	American Wind Energy Association	2002
National Hydrogen Energy Roadmap	U.S. DOE	2002
Technology Roadmap for Generation IV Nuclear Energy Systems	U.S. DOE	2002
Clean Coal Technology Roadmap	U.S. DOE, Electric Power Research Institute, Clean Coal Utilization Research Council	2001
Clean Energy Blueprint	Union of Concerned Scientists, American Council for an Energy Efficient Economy	2001

Bold = energy efficiency and renewable energy roadmap studies

figure 3.3   Solution Wedges to Eliminate Oil Imports by 2030

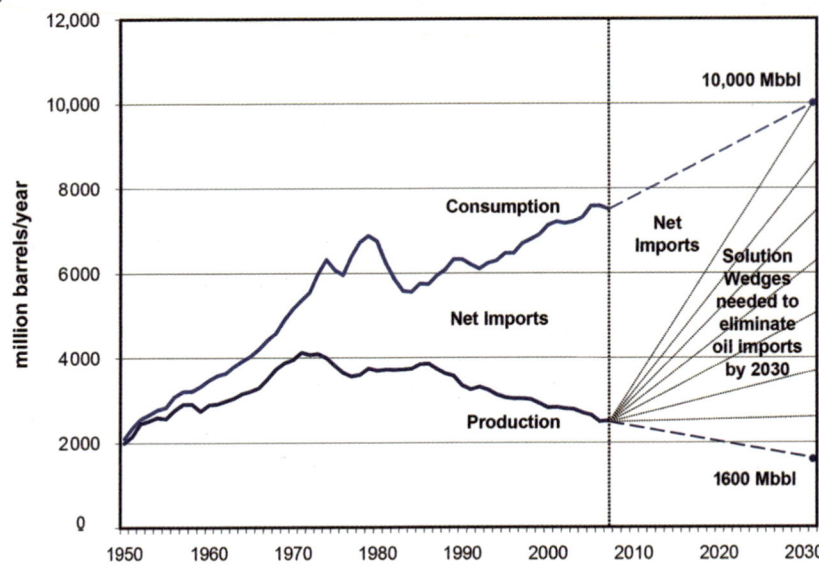

Illustration shows U.S. oil consumption, production, and net imports, 1949–2006, with straight line trend projections to 2030. Seven solution wedges are shown to eliminate oil imports by 2030.

2. Divide into increments or wedges the necessary action needed to achieve the DFC. Finding one solution to filling the gap of 8400 Mbbls by 2030 is daunting, but there are many possible contributing solutions. If we think of multiple solutions each contributing a small share of the requirement, the task is less daunting. In Figure 3.3, the 8400 Mbbls needed is divided into seven wedges each growing to 1200 Mbbls by 2030.

3. Explore multiple options of reducing oil imports and quantify what each could achieve in terms of the solution wedges. In other words, what would it take for each option to achieve one solution wedge? In addition, results from a technology roadmap study might determine how many solution wedges the technology could achieve. For example, the following options would be investigated to reduce imports by 2030:

- Improve efficiency of cars, trucks, and planes.
- Expand domestic oil production by developing offshore oil in currently banned areas (east coast and California).
- Expand domestic oil production by developing unconventional deepwater deposits in the Gulf of Mexico.
- Develop oil production from the Alaska National Wildlife Refuge.
- Develop synthetic oil from oil shale.
- Replace oil by developing coal-to-liquids (CTL).

- Replace oil by developing natural gas-to-liquids (GTL).
- Replace oil by expanding biofuel (ethanol, biodiesel) production.
- Replace oil by electrification of vehicle fleet through plug-in hybrid and electric vehicle options.

4. The analysis of each option would determine what would be needed for each to produce one or more solution wedges. For example, analysis might show that a 50% improvement in vehicle efficiency could contribute two wedges. The cumulative results of all of the options may exceed the total needed, which allows choice among the options to meet the DFC.

Solution Box 3.1 assesses three options for these oil import elimination wedges.

3.1.4 Developing Scenarios

In all types of planning, developing scenarios is a practical tool for articulating future possibilities. It embraces the uncertainties that cannot easily be modeled in forecast models, and expands the discussion from just quantitative analysis to include qualitative considerations. Scenarios can capture the imagination and thus can be a tool to generate discussion and creativity about the future.

Future scenarios are simply different visions of the future depending on different assumptions. In recent years, the approach has become more systematic and is being applied increasingly to a wide range of business, industry, government policy, and community issues. Because of expanding interest, Global Business Network (GBN), a firm that specializes in this approach, has made scenario development services a lucrative business.

The systematic approach uses a participatory process that has several steps:

1. Pose a key focus question about the future.
2. Identify drivers or factors that will affect the answer to that question.
3. Prioritize, cluster, and ultimately combine the drivers into two critical uncertainties that serve as the axes of a two-by-two scenario matrix
4. Develop scenario storylines describing the future associated with each of the four pairs of drivers in the four quadrants of the matrix. To the extent possible, the storylines should reflect accurate technical information.
5. Label each quadrant scenario.

Once the scenarios and storylines are developed they can be used to generate discussion and identify challenges and opportunities that need to be addressed in achieving a desirable scenario or to prevent an undesirable one. Scenario planning can help organize perceptions about the future, remove biases in visioning, focus debates about technology needs, challenge the view that little will change, enable development of different technology portfolios, and foster a probabilistic rather than a deterministic view of the future (EPRI, 2005).

SOLUTION BOX 3.1

Oil Import Wedges from Higher CAFE Standards, Biofuel Production, and ANWR Oil Production

The 2007 Energy Policy Act calls for CAFE standards for light vehicles to increase by 10 mpg to 35 mpg by 2020. The Act also sets a biofuels standard of 36 billion gallons per year by 2022. There have been proposals to exploit the oil potential of the Alaska National Wildlife Refuge (ANWR). What would each of these solution wedges amount to relating to our need to reduce oil imports?

Solution:

a. CAFE standards: The National Commission on Energy Policy estimates that a 10 mpg increase in the CAFE standards would save 4.5 quadrillion Btu per year of energy. How big a wedge is that? Well, a barrel of oil is 5.8 million Btu, so, the oil savings would be:

$$\frac{4.5 \times 10^{15} \text{ Btu}}{5.8 \times 10^6 \text{ Btu/bbl}} = 776 \times 10^6 \text{ bbl/yr}$$

b. The Act's biofuel standard would require an estimated 36 billion gallons of biofuels by 2017. What is this equivalent to in barrels of oil?

$$\frac{36 \times 10^9 \text{ biogal}}{\text{year}} \times \frac{0.7 \text{ gasgal}}{\text{biogal}} \times \frac{\text{bbl}}{42 \text{ gal}} = 600 \times 10^6 \text{ bbl/yr}$$

c. ANWR oil production has been estimated by U.S. EIA to peak at about 876,000 bbl/day by 2025, based on mean estimates of ultimate reserves of 10.4 billion barrels. What is this in bbl/year?

$$876,000 \text{ bbl/day} \times 365 \text{ da/yr} = 320 \times 10^6 \text{ bbl/yr}$$

So the improvement in efficiency standards of vehicles would provide about 2/3 of our 1200 bbl wedge, the biofuel goal would provide about 1/2 of a wedge, and the ANWR production would provide only 1/4 of a wedge, but this latter wedge would be temporary since the ANWR field would be quickly delpleted.

Sidebar 3.1 illustrates a scenario study conducted by GBN for businesses interested in investing in China. The study embraced and articulated the uncertainties about China's future. The focus question was, what will be China's role in the future global economy? The key drivers

SIDEBAR 3.1

Scenario Example: *Four Futures for China, Inc.*

Global Business Network (GBN) developed scenarios about the future of China's economy and role in the world for businesses interested in investing there. For complete storylines see http://www.gbn.com/Article DisplayServlet.srv?aid=38062.

The scenarios:

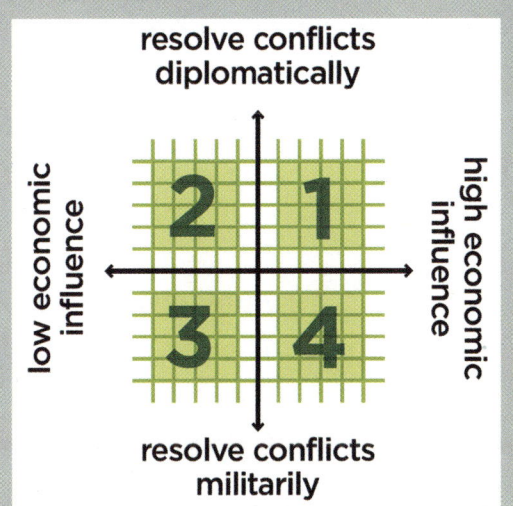

1. *Emperor of Business*—China grows peacefully and plays by the rules.
2. *Emperor's New Clothes*—China's growth rate is short-lived; it becomes a bigger Brazil.
3. *Emperor of Asia*—China grows but only as fast as its neighbors.
4. *Emperor of the World*—China's speedy growth tips all the scales in its favor.

were the country's level of economic influence (from low to high) and the method by which it resolved conflict (from diplomatically to militarily). These drivers define four scenarios in the two-by-two matrix. Storylines and labels are then developed for each scenario.

A community planning example of scenario development using this model is a participatory process used for the Great Valley of California. With the focused question of "what will be the future of the Great Valley?," scenarios were developed for three subregions. In the North Valley workshop, participants brainstormed a long list of key factors and environmental forces. They then prioritized the list to identify a few clusters of closely related issues. Finally, several of those clusters were combined and two critical uncertainties were selected to serve as the axes of a two-by-two scenario matrix: resources (from improving to decreasing) and external forces (from positive to negative). (See inset matrix on next page.) The best part is the storylines for the scenarios, which creatively articulate plausible futures, some good, some bad. You can read the storylines at http://www.greatvalley.org/valley_futures/stories/north/index.aspx.

The scenarios have galvanized considerable discussion in the region, and the program provides tools for conveners and teachers who wish to use the scenarios for their own purposes. They beg for action so that people can determine their destiny, with the motto:

Forewarned = Forearmed

Three energy-related scenario development studies are described in this chapter: the AMIGA model scenarios (3.3.1), EPRI's electric power industry scenarios (3.3.3), and IPCC's scenarios for global development and carbon emissions (3.5.2).

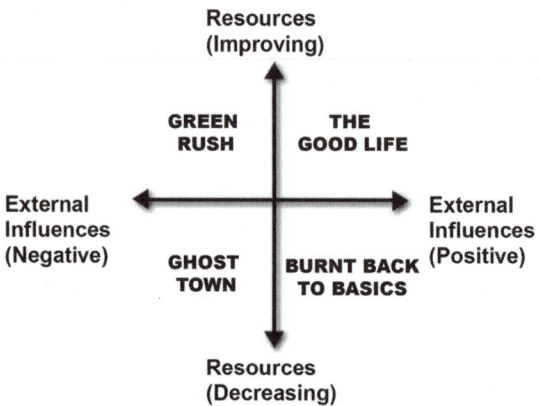

3.2 Business-as-Usual: Energy Outlook of the U.S. Energy Information Administration and the International Energy Agency

As introduced in 3.1.1, each year, the U.S. Energy Information Administration (EIA) and the International Energy Agency (IEA), a member agency of the Organisation for Economic Co-operation and Development (OECD) countries, make forecasts about the next 20 to 25 years based on current trends and model analysis (e.g., U.S. EIA, 2006, 2007; IEA, 2006). While the forecasts are influenced by economic factors, especially forecasts of energy price, they do not incorporate potential constraints of energy, such as peak oil or global warming. However, they do provide useful business-as-usual scenarios and baselines for improvement.

What do they tell us? Figure 3.4 (a–d) gives EIA's International Energy Outlook forecasts to 2030, and Figure 3.5 gives an example of IEA forecasts. EIA forecasts world

figure 3.4(a)   **U.S. EIA 2006 International Energy Outlook Projections of World Consumption to 2030**

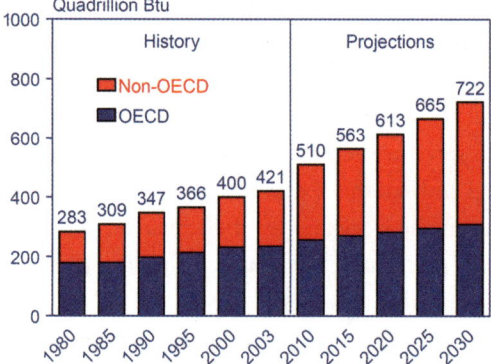

Growth is dominated by non-OECD developing countries.

SOURCE: U.S. EIA, 2007b

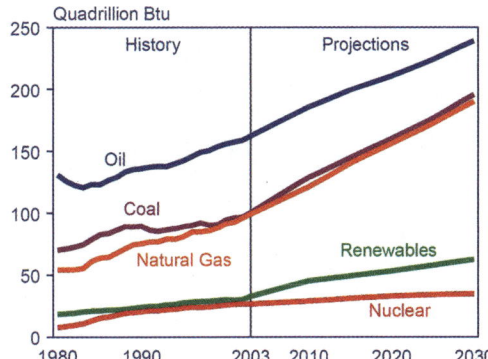

figure 3.4(b) U.S. EIA 2006 Projections of World Energy Sources to 2030

Fossil fuels dominate EIA projections.

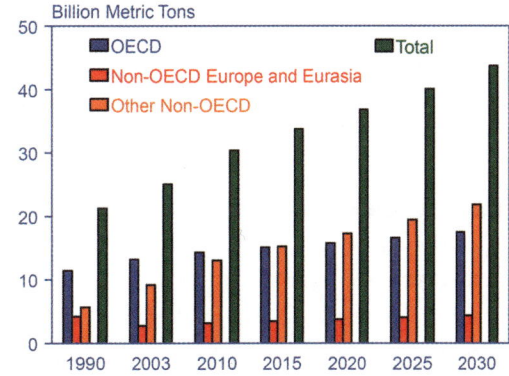

figure 3.4(c) U.S. EIA 2006 Projections of World CO_2 Emissions to 2030

Growth is dominated by non-OECD developing countries.

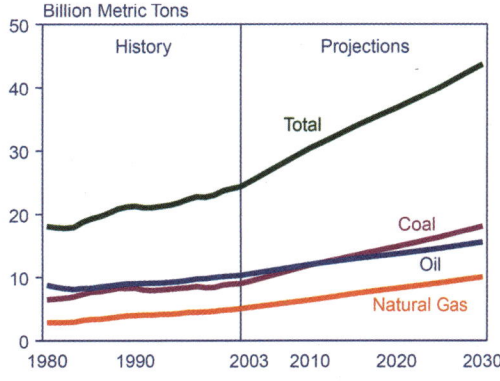

figure 3.4(d) U.S. EIA 2006 Projections of World CO_2 Emissions by Source

Growing coal use exceeds oil as source.

SOURCE: U.S. EIA, 2007b

energy consumption will grow to 722 quads in 2030 and most of that growth would be in non-OECD developing countries. China's energy use is expected to more than double by 2030. According to both forecasts, energy sources are still dominated by fossil fuels with near doubling of renewable energy and essentially no growth in nuclear. This would result in a doubling of CO_2 emissions by 2025. These forecasts appear to assume that neither carbon emission directives nor peak oil symptoms will affect energy use by 2030.

figure 3.5   IEA 2006 World Energy Outlook Projections to 2030

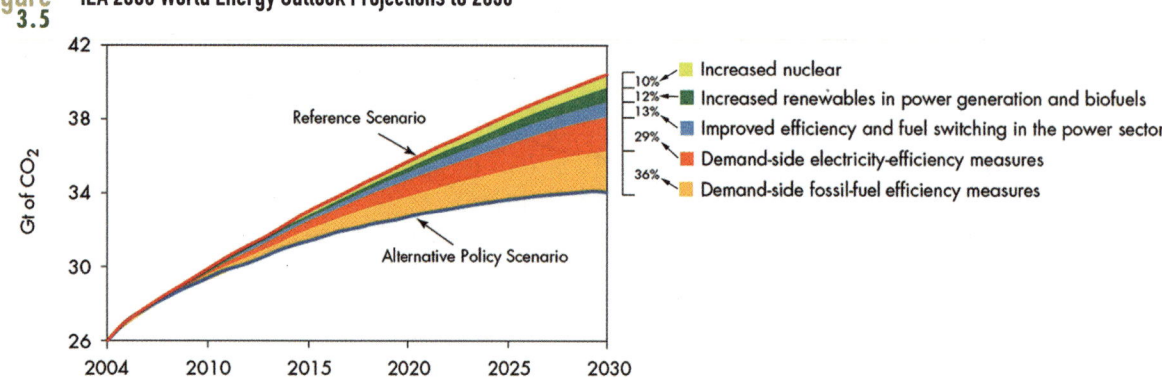

Projections include a reference case and an alternative policy scenario with a 17% reduction of CO_2 emissions compared to the reference.

SOURCE: IEA, 2006

IEA's 2006 World Energy Outlook provided not only a reference case but also an alternative policy scenario that assumes carbon emission reduction actions, including energy efficiency, renewables, and nuclear, that reduce emissions by 17% compared to the 2030 reference case. Similar to the forecasts of EIA, the IEA forecasts see a rise in oil production from OPEC countries unaffected by symptoms of peak oil. In fact, EIA predicts stable world oil prices through 2030 at about $50/bbl in constant 2005$.

The domestic picture isn't much different. Figure 3.6 (a–e) gives EIA's 2007 Annual Energy Outlook forecasts for U.S. energy to 2030, showing business-as-usual growth of energy, with total consumption rising to 131 quads (a), expanding net energy imports (a), continuing reliance on fossil fuels for 86% of energy use (b), rising coal-fired electricity (c), and a 34% increase in CO_2 emissions over 2004 (d). EIA projects an increase in energy use per capita but a continuing decline in energy intensity of the economy (energy/ $GDP) (e).

3.3 Some Visions of U.S. Energy Future

These "official" government projections clearly provide a future glimpse of the energy path we are on, but they do not consider the implications of this path nor its desirability. Other analysts have tried to identify needed changes in our energy production and consumption patterns, and to develop alternative scenarios and the means to achieve them. These visions are based less on past trends and more on future possibilities and constraints. They therefore represent more creativity in thinking, if not always more sophisticated computer modeling, than official government projections.

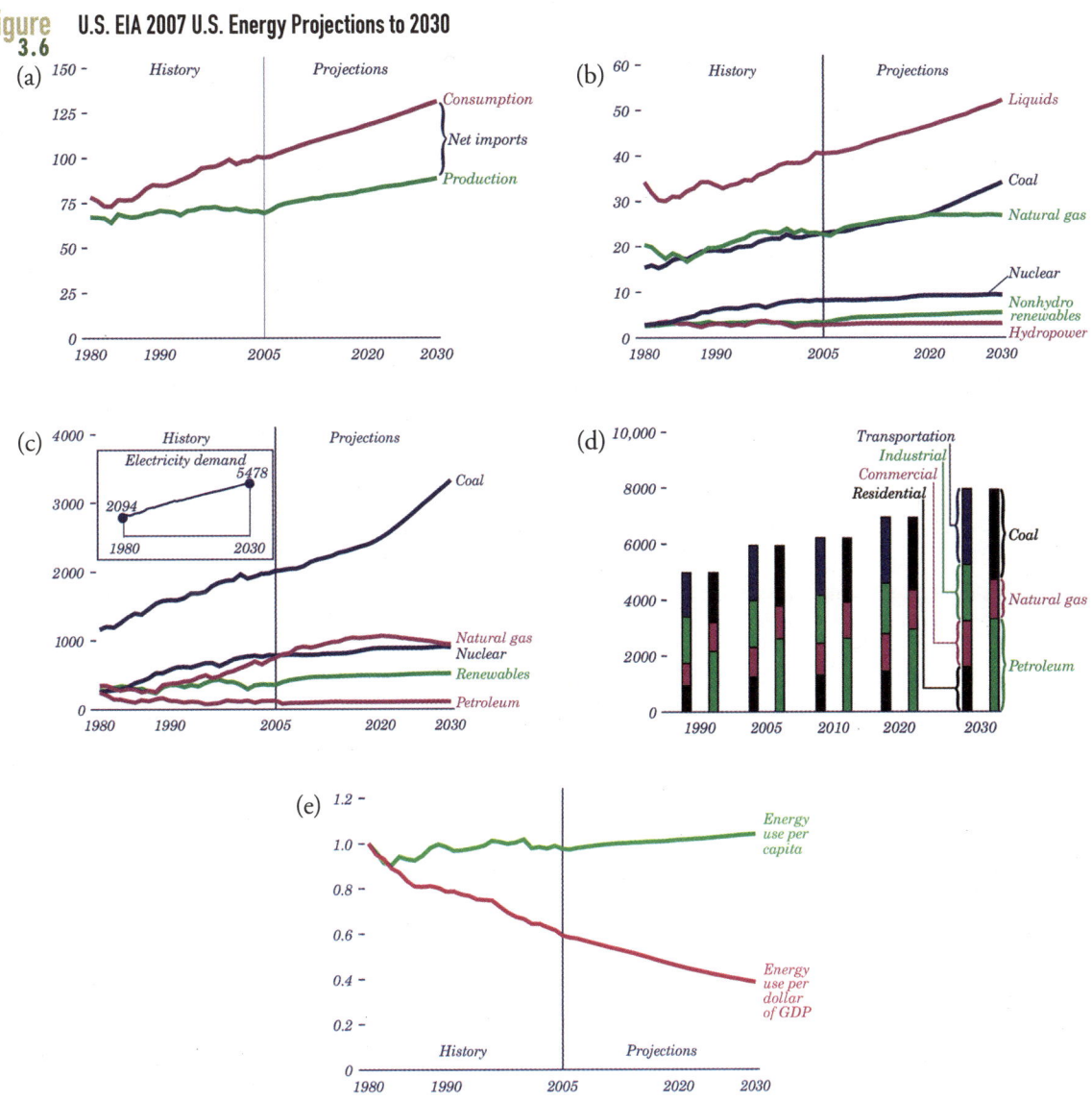

figure 3.6   U.S. EIA 2007 U.S. Energy Projections to 2030

(a) Production, consumption, net imports (in quads); (b) Energy sources (in quads); (c) Electricity generation by fuel (in billion kWh); (d) CO_2 emissions by sector and fuel (in million metric tons); (e) Energy per capita and energy intensity.

Source: U.S. EIA, 2007a

3.3.1 Driving Forces and the AMIGA Model Scenarios to 2050

This assessment combines creative scenario development with a sophisticated model, the comprehensive All Modular Industry Growth Assessment (AMIGA) model developed at

Argonne National Lab. It is an economic model, integrated with energy markets, that is able to simulate a wide range of technologies and policies. Analysts from Argonne, EPA, and DOE used the model to explore driving forces and critical uncertainties affecting U.S. energy markets to 2050 (Laitner and Hanson, 2004; Hanson, et al., 2004). The model results are useful to clarify the future and the range of choices available to industry, consumers, and policy makers.

The Argonne study provides a good example of scenario development using a sophisticated model. The study developed four diverse scenarios with the following themes or storylines:

1. *The Official Future*

   - The reference scenario reflects conventional wisdom about the future patterns of U.S. energy supply and demand and assumes existing U.S. policies, trends in market structure, and the market shares of various technologies extrapolated beyond 2020 to 2050. It is close to the business-as-usual projections of the EIA (Figure 3.6).
   - This scenario's driving forces include rising demand for oil to fuel a growing fleet of U.S. and global automobiles; expanding use of natural gas for electricity production as well as heating; and increasing electrification.
   - The Official Future is an optimistic, surprise-free scenario, a world of "more of the same," with no major discontinuities or disruptive technologies. There are no significant resource shortfalls and no noticeable dislocations to derail the progress of economic growth. Prices remain steady, and adequate supplies of fuels are always available. Foreign countries seek innovative ways to cooperate with the United States in managing global markets, opening their domestic markets to U.S. companies, and helping the United States expand its influence in the world.

2. *Cheap Energy Reigns Supreme*

   - This scenario assumes abundant and inexpensive supplies of oil and gas continue to fuel the engines of economic growth in the United States.
   - U.S. foreign policy is designed to provide continued access to low-cost supplies of oil and gas, placing great emphasis on stability in oil-producing regions.
   - American consumers sustain their historical dependence on cheap fuels and disregard the occasional breakdown of energy supply and delivery systems.
   - Environmental impacts of energy supply and use are considered to be the unavoidable consequences of economic growth.
   - This surprise-free scenario of inexpensive and seemingly limitless supplies of oil and gas exposes the United States to no major discontinuities or disruptive technologies. Oil prices initially fall, then gradually rise and remain stable for most of the scenario period, as OPEC manages the world oil market to ensure cheap fuels for its American patrons. Living comfortably under *Pax Americana*, foreign governments cooperate with the United States in managing most global markets, eagerly opening

their domestic markets to U.S. companies, and seeking favorable trade treatment from the world's one remaining superpower.

3. *Big Problems Ahead*

- This is a chaotic, event-driven scenario, in which domestic policy is disjointed and episodic, buffeted by forces beyond U.S. shores. Foreign governments and terrorist groups see U.S. policies as designed to promote the imperial ambitions of the United States. They take steps to limit U.S. access to resources and to disrupt international trade in energy resources.
- Chronic instability among the Gulf regimes leads to a roller-coaster ride of rapid oil price surges, stressing the U.S. energy sector. Intermittent cutoffs of oil supply from the Gulf cause discontinuities in the path of economic development for both industrialized and developing countries.
- A sense of frustration and malaise characterizes this scenario. Anxiety about the future discourages investment in new technology. Most startup companies fail and most new "miracle" technologies deliver far less than was originally claimed by their proponents. Private R&D investments prove insufficient to overcome the difficulties inherent in efforts to commercialize new technologies. New technologies falter due to unexpected engineering challenges.
- Environmental impacts of the new systems generate significant public resistance to their widespread use. Institutional failures are rampant in a chaotic future beset with shocks, stresses, and discontinuities.
- Economic growth is slowed worldwide. U.S. energy policy is disjointed. Concerns about energy security keep everyone on edge. Rising U.S. oil imports increase U.S. dependence on unstable world regions. And U.S. responses to these challenges make it appear that the United States has become an arrogant and imperial player on the world stage, reducing the inclination toward international cooperation in many countries.

4. *Technology Drives the Market*

- In this scenario, forces converge to reshape the market architecture of the U.S. energy sector.
- Implementation of institutional and regulatory reform sets the new and improved technologies on a level playing field alongside mature technologies in U.S. energy markets, allowing incumbent companies in these markets to embrace the new technologies.
- Engineering advances in the design and development of efficient, low-emissions technologies capture the imagination of business leaders, state officials, and individual consumers.
- Private investment by U.S. companies combines with rapid technical progress and value shifts by U.S. consumers drive the new technologies to rapid market acceptance and widespread commercialization.

- Early in this scenario, state leaders establish an integrated set of tariff policies for energy efficiency systems, renewable energy technologies, and distributed electricity generation schemes. These local micro-grids lower the stress on aging transmission systems and increase the reliability of utility-generating networks.
- Strict environmental permitting standards are applied to both new and traditional technologies, limiting the energy sector's impact on the regional and global environments.
- Several technologies achieve commercial success: building-integrated photovoltaic power systems; medium to large wind machines (i.e., machines with rated capacity of 5–5000 kW); small methane-reforming appliances located at local fueling stations produce hydrogen for fuel cells from natural gas; fuel cells for mobile and stationary applications; and biomass energy systems for heat and electricity.
- "Smart building" technology improves the efficiency of energy end-use in the residential and commercial sector. Changing consumer values create a shift in the trends of housing patterns. Instead of moving farther and farther out from urban core centers and commuting longer distances each day to work, middle-class and working-class households look for "in-fill" housing, limiting urban sprawl and co-locating homes with worksites or telecommuting in their communities.
- As consumer purchasing preferences shift to small and efficient vehicles, oil demand in the U.S. transportation sector plummets while personal mobility is maintained. The new hybrid efficient vehicles use much less gasoline (or diesel) for the same amount of driving, while the new fuel-cell vehicles derive their power from domestic natural gas. This has significant positive implications for energy security as the demand for imported fuel begins to decline steadily.

After running these scenarios, analysts decided to rerun them with a large surprise challenge to see how the outcomes would change. The surprise they incorporated was a clear realization by 2010 that human-forced climate change was active and a global response was desperately needed. In each of their challenge-and-response scenarios, the analysts had U.S. policy makers implement energy policies designed to

- Promote diversity in energy supply
- Decrease U.S. dependence on foreign oil
- Improve U.S. energy security
- Increase efficiency in all energy-intensive sectors of the economy through the introduction of conservation measures and advanced technologies
- Accelerate capital stock turnover particularly in the electricity and transportation sectors
- Sustain economic growth
- Decrease CO_2 emissions resulting from energy supply and use

The resulting base case scenarios and the challenge-and-response scenarios are given in Figures 3.7 and 3.8 and Table 3.2. The base case *cheap energy* scenario, with all of its optimistic assumptions, would lead by 2050 to 65% increase in energy use and 2.8 times the vehicle miles traveled (VMT), largely a result of oil prices assumed to be lower in 2050 than in 2000. The *big problems* assumptions of turmoil in the Middle East predicts much higher oil prices, lower energy use, and constrained economic growth to go along with the economic, political, social, and environmental malaise that accompanies this scenario. The *technology drives* scenario provides huge savings in energy over the "cheap energy" case, while providing the same economic growth and lower prices for oil and natural gas by 2050.

The challenge-and-response case pushes a significant decline of energy use in all three scenarios in order to reduce CO_2 emissions. All three challenge-and-response scenarios would

figure 3.7
Results of AMIGA Model for Future Energy Scenarios: Energy Consumption, 1990–2050

Most resilient of the options is "technology drives the market" with challenge-and-response policy, including investment in energy efficiency and renewable sources.

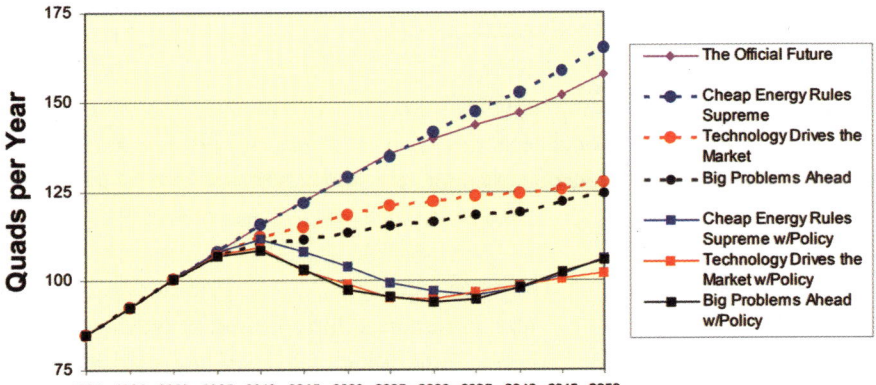

SOURCE: Laitner and Hanson, 2004

figure 3.8
AMIGA Model for Future Energy Scenarios: (a) U.S. GDP in Trillion 2000$; (b) U.S. Carbon Emissions in Million Tons

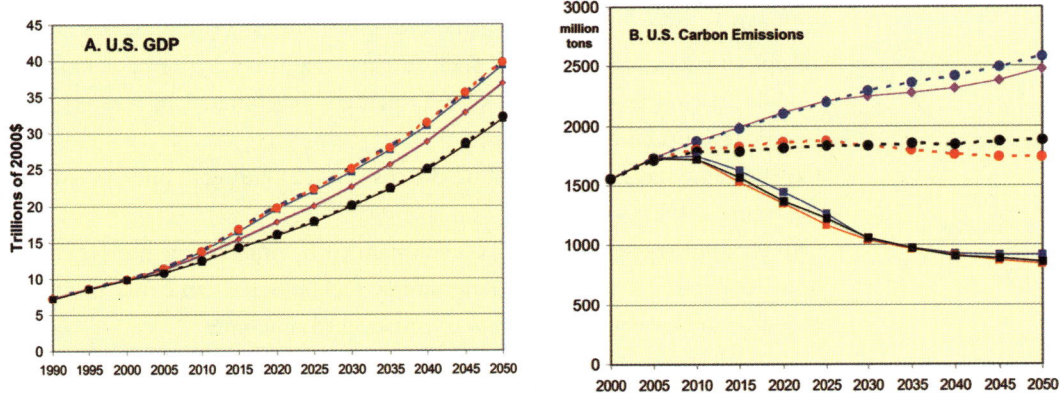

"Technology drives the market" scenario has highest economic growth and lowest carbon emissions.

SOURCE: Laitner and Hanson, 2004

table
3.2   **Summary of AMIGA Model Results for U.S. Energy and Economy Scenarios**

	2000	2050 Base Case			2050 Challenge-and-Response Case		
	Actual	Cheap Energy	Big Problems	Technology Drives	Cheap Energy	Big Problems	Technology Drives
Energy (Q)	100	165	124	128	106	106	102
GDP (T2000$)	$9.9	$39.8	$32.3	$39.8	$39.3	$32.0	$39.7
Oil price ($/bbl)	$27.72	$22.90	$40.50	$21.30	$15.13	$37.80	$18.40
Natural gas price ($/Mcf)	$2.76	$6.13	$6.25	$4.82	$2.42	$4.87	$3.19
Vehicle miles traveled (Bm/y)	2400	6890	3740	3990	3880	3410	3750

Source: Laitner, et al., 2004

have energy consumption in 2050 about the same as 2000. Because the response measures would be an extension of "technology drives" policies, the disruptions and change to its base scenario are minor.

Among the study's conclusions are the following:

- The range of feasible U.S. energy futures is broad, but energy use is expected to grow under all scenarios. Introduction of policies to encourage capital stock turnover and accelerate use of high-efficiency, low-emissions technologies can significantly reduce future primary energy demand.
- Low energy prices can lead to high economic growth. But so can a smart investment path emphasizing energy efficiency improvements and advanced technologies.
- Policies introduced to improve energy efficiency and accelerate the introduction of new technologies do not reduce the prospects for economic growth.
- Public and private choices, along with external events, affect the cost of responding to future surprises. One such "game-changing" surprise is represented by the risk of abrupt climate change. Another such surprise might result from a complete cutoff of Middle East oil exports to the OECD perhaps precipitated by a series of successful Islamic revolutions in the region.
- Expenditures made early can reduce the costs of responding to unexpected problems in the future.
- This project suggests that a smart investment path, emphasizing both energy efficiency improvements and advanced energy supply technologies, can better position the U.S. economy to more quickly respond to unexpected outcomes.

The Argonne study suggests that this "smart" investment path characterized by the "technology drives the market" scenario provides the best opportunity for economic growth and resilience in the face of potentially disrupting external events, such as a massive impact of global warming, a cutoff of Middle East oil, or, one event the study did not consider, a peaking and decline of global oil production before 2050.

3.3.2 Mitigating the Peak-oil Impacts with Carbon-based Alternatives

One study that does address the peak-oil scenario directly is a 2005 report conducted for the National Energy Technology Laboratory (NETL) entitled "Peaking of World Oil Production: Impacts, Mitigation, and Risk Management" (Hirsch, et al., 2005). This study acknowledged the risk of the peaking of world oil production and investigated strategies to mitigate the impacts. Because NETL is the government's primary fossil energy research lab, it is not surprising that the report concludes that the way to solve our looming liquid fossil fuel problem is . . . with more fossil fuels.

The Hirsch study outlines several mitigation strategies and uses the "solution wedge" approach. The authors developed three mitigation strategies: Scenario I assumes that action is not initiated until peaking occurs; Scenario II assumes that action starts ten years before peaking occurs; and Scenario III assumes action starts twenty years before peaking occurs (see Figure 3.10).

The study limited itself to means of creating wedge options that meet the following criteria: It must (1) produce liquid fuel that could displace oil on a massive scale (millions of barrels per day [MMbpd]); (2) use commercial technology; (3) be at least 50% efficient; and (4) be environmentally clean by 2004 standards. They excluded shale oil because of limited technology, biomass fuels such as ethanol because they said they were not commercially competitive, and electric power options such as nuclear, wind, or solar because they could not easily displace liquid fuels. The selected wedge options are given in Table 3.3, along with the delay until first benefits are achieved and benefits ten years later.

Figure 3.9 shows the sum of the wedges after crash programs are developed and Figure 3.10 shows the three scenarios based on different start dates for the programs. The lesson is that if mitigation is started soon enough, it would offset the decline from peak—if only for ten to twenty years.

Critique of study. This study suggests investing trillions of dollars simply to extend our fix on carbon-based liquid fuels without a long-term solution. Most of these options would

table 3.3 Mitigation Options and Impact of Oil Demand and Impact Delay after Initiation

Option	Description	Delay Until Impact, Years	Impact after 10 Years, MMb/d
Fuel-efficient vehicles	CAFE standards: 27.5 mpg to 35.8 mpg (3 years) to 41 mpg (8 years)	3	3
Enhanced oil recovery	3% of peak production 10 years after peak	5	3
Heavy oil/tar sands	Venezuelan heavy oil to 5.5 MMbpd 13 years after start; Canadian tar sands to 2.5 MMbpd 13 years after start	3	8
Coal liquefaction	Five new 100,000 b/d plants per year	4	5
Natural gas-to-liquids	Qatar-type projects to 2 MMbpd 13 yrs after start	3	2

figure 3.9   Peak-oil Solution Wedges for Five Options Applied to Three Scenarios

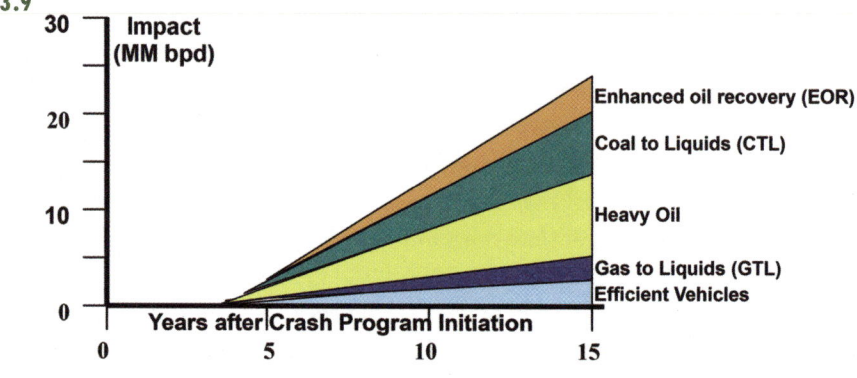

SOURCE: Hirsch, et al., 2005

figure 3.10   Three Scenarios of Timing of Crash Mitigation Program Relative to Time of Peak Oil

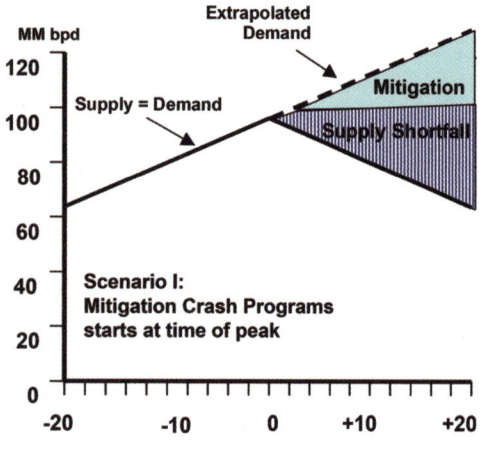

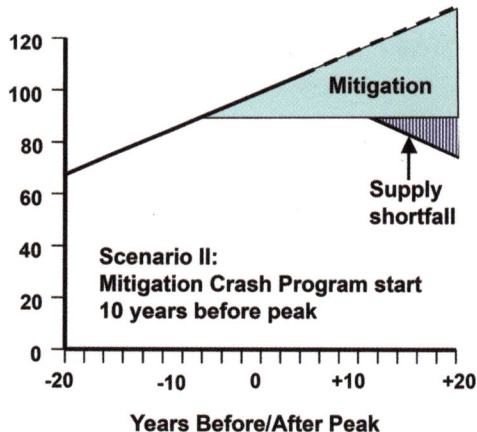

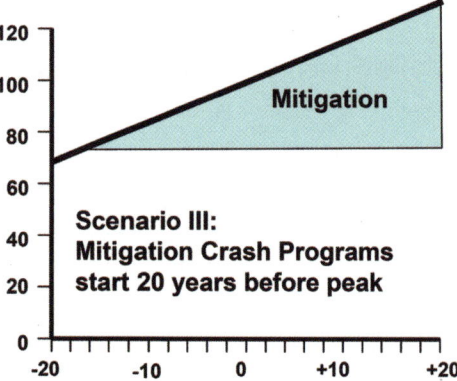

Mitigation assumes solution wedges given in Figure 3.9. Lesson: Early action is better.

SOURCE: Hirsch, et al., 2005

still have to be imported. It shortchanges efficiency to about half the potential stated by the National Commission on Energy Policy, and it discounts biofuels out of hand even though these renewable fuels are produced in worldwide commercial quantities today that exceed any of the selected production options. By 2007, even the Bush administration was calling for 2 MMbpd of biofuels by 2017. The study expounds on the need for risk management, but discounts one of the biggest risks we face—global warming from the carbon emissions, likely to be exacerbated by the study's options.

This study acknowledges the immediate threat of peak oil, but it fails to provide a sustainable response plan. Would it not be better to think a bit more out of the box and invest those trillions in more sustainable solutions that have long-term benefits rather than simply buy us a few years before we are really in trouble?

3.3.3 EPRI's Electricity Future Scenarios

The Electric Power Research Institute (EPRI) began producing electricity technology roadmaps in 1997. In 2005 it decided to use a scenario approach as an input to its latest roadmap because it specifically incorporates uncertainties apparent in power planning. It followed the basic scenario development approach:

- Focus question: How will demand for U.S. energy services and the potential externalities that may result shape electricity technologies over the next twenty years?
- Identify key drivers of change:
  - Evolution of primary fuel markets (natural gas)
  - Changes in societal values in energy industry externalities, especially CO_2
- Create two-by-two matrix with the drivers, develop storylines for the resulting four scenarios, and label the scenarios.

Figure 3.11 shows the four scenarios, which are then characterized by several factors given in Table 3.4:

1. *Digging in Our Heels*—World lives with higher prices and potential effects of CO_2 in a momentum strategy that is not a perfect world but perceived better than higher-cost alternatives.
2. *Supply to the Rescue*—Low cost and abundant natural gas spurs economic growth and development.
3. *Double Whammy*—High gas prices and high societal concerns about environment impact the economy requiring technological efficiency advances to meet world challenges.
4. *Biting the Bullet*—Make painful changes in the near-term to forestall more painful consequences from climate change and other crises. Non-carbon alternatives ultimately drive down cost of carbon fuels.

figure 3.11 Scenario Matrix in EPRI's Industry Technology Scenario Study

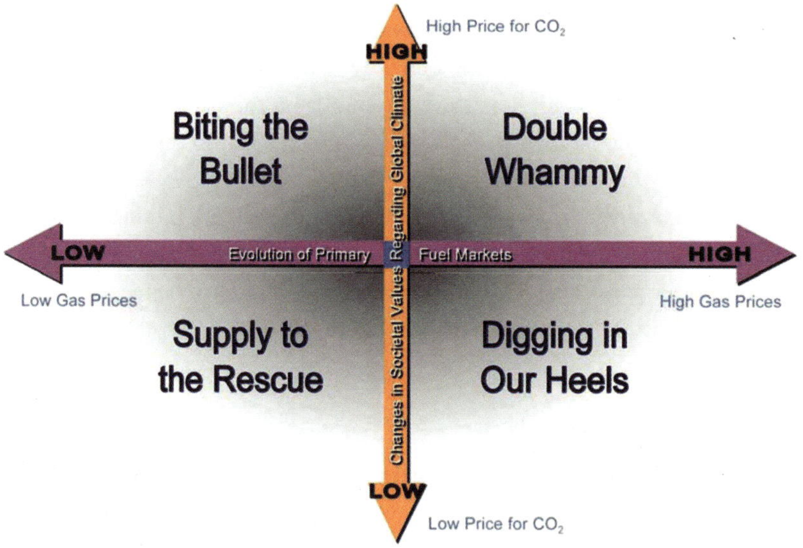

SOURCE: EPRI, 2005. Used with permission.

EPRI's assessment of the scenario analysis indicated the assumption that in 2005 with high prices and low cost for externalities, most of the United States associated with the southeast quadrant (Digging in Our Heels). The authors could imagine, however, movement to the northeast (Double Whammy) quadrant of the matrix if environmental costs rise and prices continue to increase. They also imagine a move to the southwest (Supply to the Rescue) quadrant if fuel prices decline and environmental costs remain low, or to the northwest (Biting the Bullet) quadrant over time if rising environmental costs ultimately cause falling demand for traditional fuels. Movements could be different for different parts of the world and for different states in the United States.

In the study, EPRI's technology recommendations stressed conventional and fossil fuel options including carbon capture, advanced coal and nuclear, enhanced transmission and reliability, and sustaining existing coal, nuclear and natural gas plants. It did suggest better emissions control, but was silent on efficient technologies and renewables, which seem to be logical ingredients of a Double Whammy or Biting the Bullet future.

3.3.4 The Hydrogen Economy Roadmap (and Bumps in the Road)

Many energy prognosticators look to hydrogen as the key to a carbon-free energy future. Hefner (2003) from GHK Company argues that hydrogen will be the product of a natural progression from solid to liquid to gaseous fuels as the basis of our energy economy

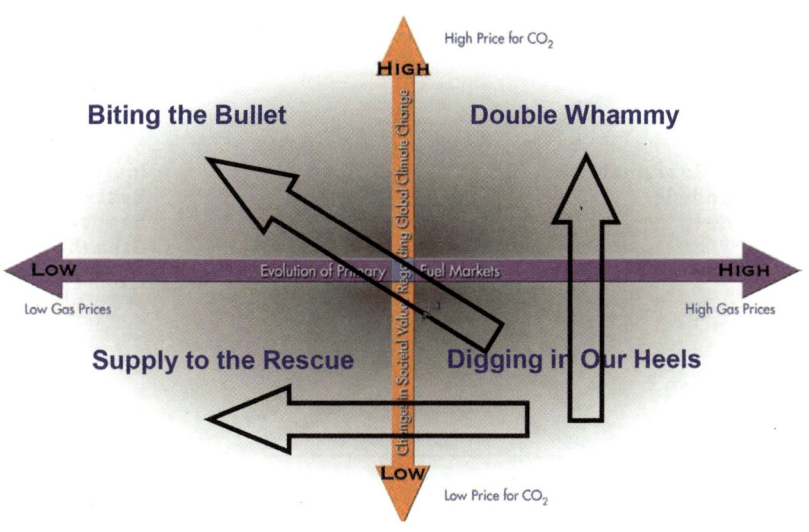

table
3.4

Characterizing EPRI Scenarios by Effects on Selected Factors

	Scenarios			
Effects	Digging in Our Heels	Supply to the Rescue	Double Whammy	Biting the Bullet
Gas prices	Up	Down	Up	Up and down
Reliability	No major events	Increasing	Techno fix	Techno fix
World economy	United States, EU, India, China lead	United States, EU, India, China lead	Rush to wealth grows	India, China grow
U.S. economy	Moderate growth	Strong growth	In transition	Slow growth
Environment	Not number-one priority	Not number-one priority	International priority	Concerns gradually become high priority
Regulation	Keep prices low	Reasonable cost recovery	Return to regulatory compact	DSM and CO_2 reduction mandated
New technologies	No new incentives	LNG, DG, gas technologies	Efficiency	CO_2 sequestration, renewables, nuclear
Consumer demand	Supply, not choice	Increasing	High price of electricity	Constrained by policy
Lifestyles	Price and availability of energy no concern	More stuff	Green	Constrained "forced march"
Liquefied natural gas	"Not in My Backyard" (NIMBY)	Yes	NIMBY and high prices	Yes
Infrastructure	Issue: who pays, who gains	Capital available	To support choice	Select investments
Energy Policy	Slow consensus	Gas supply oriented	Green incentives	By mandates

figure 3.12   **Transition to the Age of Energy Gases**

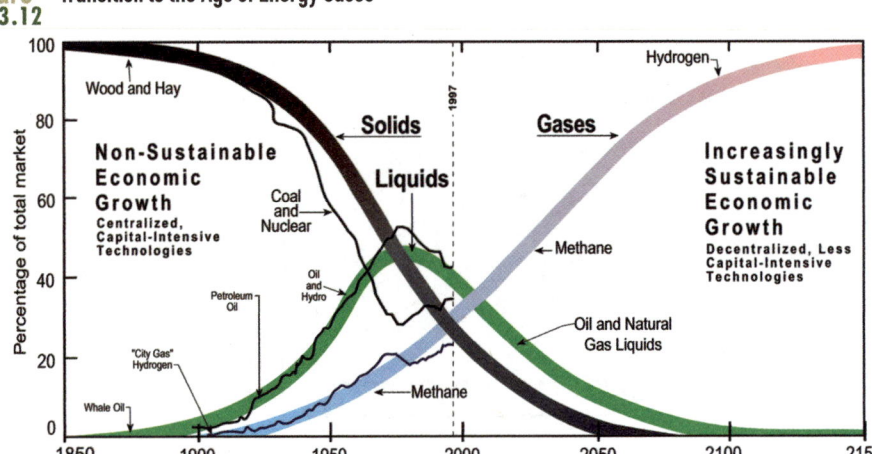

Some see the future of energy based on a delivery network of hydrogen produced from (initially) fossil, (possibly) nuclear, and (ultimately) renewable energy.

Source: Hefner, 2002. Used with permission.

(Figure 3.12). Make no mistake, Hefner works for a natural gas company, and his claim that natural gas is the most abundant of our fossil fuels is not shared by all. His argument that natural gas should serve as the primary transitional fuel to hydrogen is plausible, but also self-serving. Still, the transition to a hydrogen economy for a carbon-emission-free world has wide appeal, with support from environmentalists, industry specialists, and the Bush administration. Others have been more skeptical given the barriers to be overcome.

One important point is that hydrogen is not a *source* of energy, but a storage medium that could be very versatile in terms of both production and use. Molecular hydrogen (H_2) is the perfect fuel, and we have existing and emerging technologies to produce hydrogen from a variety of fossil, renewable, and nuclear energy sources. We may be able to transport hydrogen in our natural gas pipeline infrastructure and, once we have a delivery infrastructure, we can use it in a variety of applications in heating, transportation, industry, and power generation through combustion, gas turbines, and fuel cells. But getting there will not be easy (see Sidebar 3.2).

In 2002, U.S. Department of Energy published the "National Hydrogen Energy Roadmap," the results of a national workshop on the prospects for hydrogen as a key component of future energy in the United States. The roadmap highlights potential means of production, transport and delivery, storage, and application. Production is the key because hydrogen is a storage medium and depends on fossil, nuclear, and/or renewable energy sources.

- *Hydrogen production:* Hydrogen is the most abundant element in the universe and a key element of water. The challenge is splitting it from water or other substances—this requires energy. Current production is by steam methane reforming (SMR) and partial oxidation, both of which extract H_2 from fossil fuels. Electrolysis of water is even more straightforward, but it requires more energy and precious electricity at that. Emerging

SIDEBAR 3.2

Science: "Toward the Hydrogen Economy"

As attractive as a hydrogen economy may be, it will be neither easy nor cheap to achieve. In a special August 2004 issue, "Toward the Hydrogen Economy," the journal *Science* addressed several opportunities and challenges. Here is what *Science* editor Donald Kennedy and issue editors Robert Koontz and Brooks Hanson had to say:

If ever a phrase tripped lightly over the tongue, "the hydrogen economy" does. It appeals to the futurist in all of us, and it sounds so simple: We currently have a carbon economy that produces carbon dioxide, the most prominent of the greenhouse gases that are warming up the world. Fortunately, however, we will eventually be able to power our cars and industries with climate-neutral hydrogen, which produces only water.

Well, can we? There are problems, and they're serious. To convert the U.S. economy in this way will require a lot of hydrogen: about 150 million tons of it in each year. That hydrogen will have to be made by extracting it from water or biomass, and that takes energy. . . . At normal human-scale temperatures, it is an invisible gas: light, jittery, and slippery; hard to store, transport, liquefy, and handle safely; and capable of releasing only as much energy as human beings first pump into it. All of which indicates that using hydrogen as a common currency for an energy economy will be far from simple . . .

[What is needed is a] mix of social and economic changes that might actually reduce current emissions, but current U.S. policy offers few incentives for that. Instead, it is concentrating on research programs [like hydrogen fuel-cell vehicles and hydrogen from fossil fuels] designed to bring us a hydrogen economy that will not be carbon-free and will not be with us any time soon.

The trouble with the plan to focus on research and the future, of course, is that the exploding trajectory of greenhouse gas emissions won't take time off while we are all waiting for the hydrogen economy. Meanwhile, our attention is deflected from the hard, even painful, measures that would be needed to slow our business-as-usual carbon trajectory. Postponing action on emissions reduction is like refusing medication for a developing infection: It guarantees that greater costs will have to be paid later. (*Science*, 13 August 2004, v. 305, pp. 917, 957)

technologies include thermochemical water splitting, solar photolytic processes, and fossil fuel H_2 production with carbon sequestration. It is critical that the production process be carbon-free, energy efficient, and modular to enhance distributed production.

- *Hydrogen transport and delivery* is best provided via the natural gas pipeline system. Household or business gas distribution pipelines and metering could be used for personal fueling of mobile uses like automobiles, but a new delivery infrastructure (refueling stations) would also be needed. Alternatively, existing natural gas delivery infrastructure could feed on-site reformers to produce hydrogen at the point of use.

- *Hydrogen storage* presents technical challenges especially for mobile applications. Hydrogen can be compressed and liquefied, but emerging technologies for metal and chemical hydride storage may offer the best options for stability and volume.

- *Hydrogen conversion* includes traditional combustion in boilers, engines, and turbines, and also conversion to electricity in fuel cells. Advances in fuel-cell technologies, which can convert H_2 to electricity at efficiencies of 60% or more, are critical for hydrogen transportation and electricity generation (see Section 10.9).
- *Hydrogen applications* are diverse, ranging from on-site distributed electricity generation to vehicle transportation, both using efficient fuel cells.

The national roadmap sets a scenario for a 40 million ton (mT) H_2 market, enough to fuel 100 million cars or provide electricity to 25 million homes. Such a market would require a mosaic of H_2 production sources that could include: 100,000 neighborhood electrolyzers (4 mT); 15,000 small reformers in refueling stations (8 mT); 30 coal/biomass gasification plants (8 mT); 10 nuclear water splitting plants (4 mT); and 7 large oil and gas SMR/gasification refineries (16 mT).

3.4 Visions for Renewable Energy and Efficiency

This section reviews a number of visions emphasizing renewable energy and efficiency, including some historic assessments and a recent call for producing 25% of U.S. energy from renewable sources by 2025, so-called **25 × '25.**

3.4.1 Lovins I: "Soft Energy Path"

Any discussion of visions of alternative energy must begin with the soft energy path. In 1976, a young physicist Amory Lovins published an influential article in the prestigious but unlikely journal, *Foreign Affairs*. The article, "Energy Strategy: The Road Not Taken," argued that the domestic energy patterns and policy of the day, what he called the "hard path," with its vision of oil and nuclear, was neither wise nor sustainable. He presented an alternative direction stressing efficiency, matching energy sources to uses, and relying on emerging "soft" technologies. Government projections at the time called for growth in energy to 150 quads by 2000 and 225 quads by 2025.

His alternative path called for just 100 quads by 2000 and a subsequent decline thereafter. Figure 3.13 shows his soft path's total consumption scenario compared to the hard path. It also shows the sources of his soft path energy, the oil and gas wedge shrinks to 2000 and beyond, coal grows slightly as a transitional fuel, then declines after 2000, and soft renewable technologies grow to dominate by 2025. A critical condition for this situation is the reduction of overall consumption to 60–70 quads by 2025.

An important principle of Lovins's vision was increasing not only **conversion efficiency** but also **functional efficiency.** Recall from Chapter 2 that functional efficiency refers to the useful functions (e.g., light, heat, mobility) each unit of energy provides. As shown in Figure 3.14, conversion efficiency brings down energy use, but functional

figure 3.13 Amory Lovins's 1976 Scenarios for Hard and Soft Energy Paths for the United States

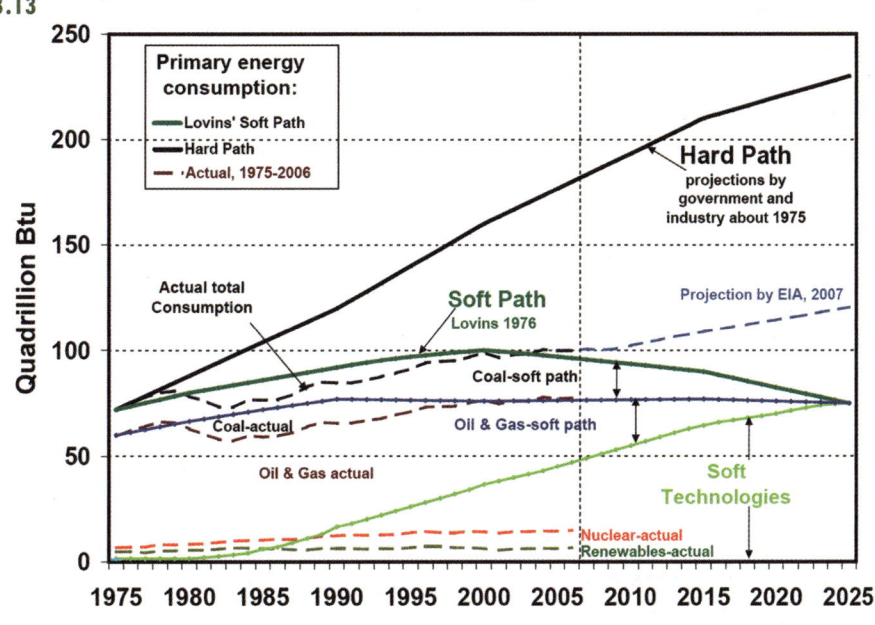

The soft path (green) scenario foresaw fossil fuels dominating until 2000, with efficiency keeping demand down to 100 quads. After 2000, soft renewable technologies take over by 2025, which is when efficiency drives overall demand down to 75 quads. The figure superimposes with dashed line actual data to 2000 for total consumption (which matched the soft path scenario in 2000), renewables (which were much lower than the scenario), nuclear, and coal, oil, and gas. It also shows the 2007 U.S. EIA total consumption projection to 2025.

figure 3.14 Idealized Functions Performed by Energy under Lovins's Hard (a) and Soft (b) Energy Paths

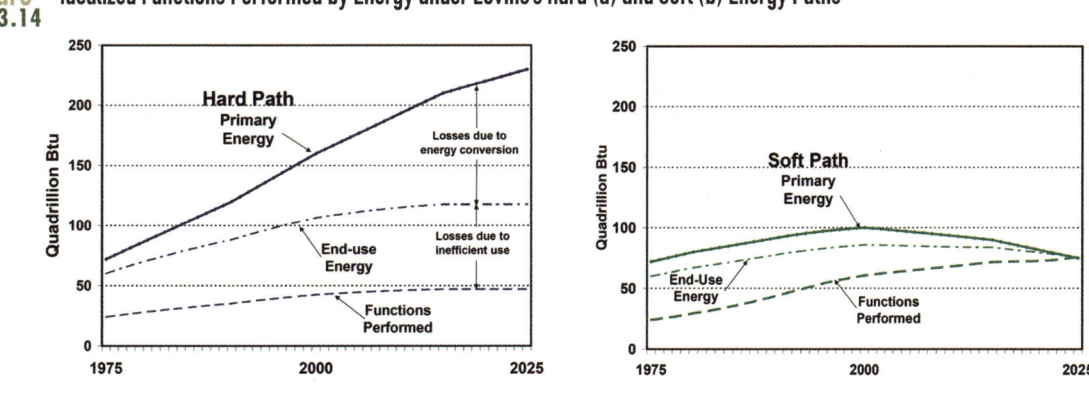

efficiency improvements enhance the utility of energy used, potentially above the utility of the hard path.

Lovins's article prompted a strong response from critics and advocates. Critics called the prospect of only 100 quads by 2000 heresy. But as the black dashed line in Figure 3.13

shows, actual total consumption in 2000 was 98 quads, less than in Lovins's scenario. However, the commitment to energy alternatives that Lovins called for did not materialize, the oil and gas wedges have not begun to close, and renewable energy technologies have not yet grown to the level of his soft energy path.

3.4.2 Lovins II: "Winning the Oil Endgame" with Efficiency and Biofuels

Lovins has remained an influential figure in the energy field. (In 2007, he was awarded the prestigious "Blue Planet" prize from Japan's Asahi Glass Foundation.) In September 2004, Lovins released his latest entry to the foray of energy futures, *Winning the Oil Endgame: American Innovation for Profits, Jobs, and Security.* This treatise offers a scenario for future energy use in the United States, applicable worldwide, that would eliminate oil imports through leadership and investment by business, supported by government incentives not mandates, and cost-effective consumer choices. This optimistic vision (see Sidebar 3.3) generated considerable attention among other energy analysts and in the popular press (see http://www.oilendgame.org/).

The scenario's value is its ability to paint a picture of the steps needed to halve the amount of oil used in the United States through efficiency improvements, and how we can substitute alternatives for the other half. By taking critical steps now, the United States could save as much oil as it gets from the Persian Gulf by 2015; could set the stage by 2025 for the option of transitioning to a hydrogen economy; and could have a flourishing economy without oil by 2050.

The four integrated steps include the following:

1. Double the efficiency of using oil, primarily through ultralight vehicle design.
2. Apply creative business models and public policies for adoption of superefficient vehicles.
3. Provide another one-fourth of U.S. oil needs by a major domestic biofuels industry.
4. Substitute natural gas for oil using gas saved by profitable efficiency techniques that could save half of the projected 2025 use of natural gas.

Although such a major transition would require significant energy disruptions and changes to current business models, future-oriented industrial leaders can take advantage of "disruptive innovations." Already we see transitions in the oil industry led by Shell and BP. Lovins suggests such a transition would require investing $180 billion over the next decade to revitalize strategic industries and eliminate oil dependence with a huge potential payoff of $130 billion gross or $70 billion net, *per year* by 2025.

Lovins outlines several public policies as stepping-stones to an oil-free America:

- *Feebates* (revenue- and vehicle-size neutral fees on inefficient vehicles and rebates for efficient ones) can help shift consumer choice to superefficient vehicles.

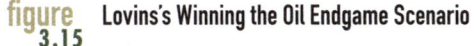

figure 3.15 Lovins's Winning the Oil Endgame Scenario

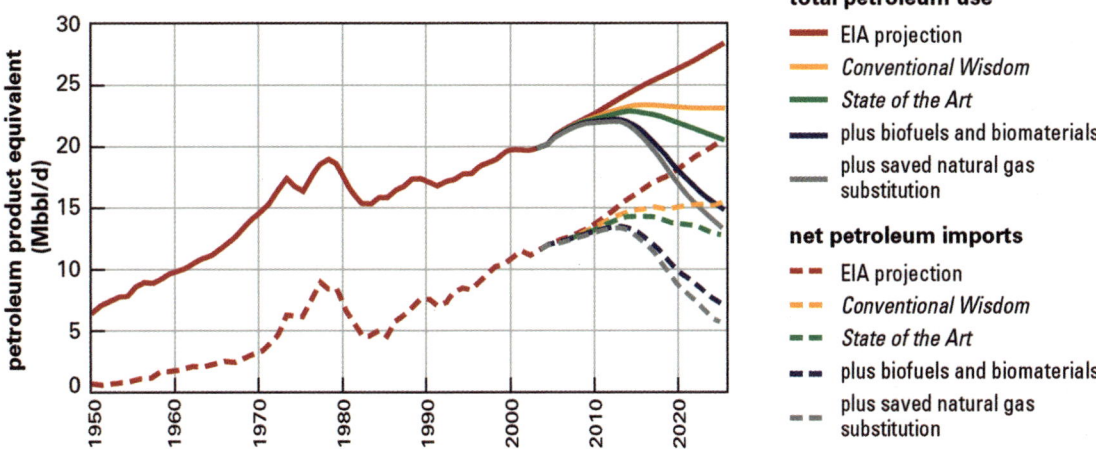

The scenario cuts U.S. EIA projected 2025 oil consumption in half and cuts imports by three-fourths through efficiency improvements and replacement by biofuels and saved natural gas, on the way to an oil-free economy by 2050.

SOURCE: Lovins, et al., 2004

- *"Scrap and replace"* program can lease or sell superefficient vehicles to low-income drivers while scrapping their clunkers.
- Smart government procurement and targeted acquisition can *accelerate manufacturers' conversion.*
- Manufacturers of advanced vehicles and airplanes receive *federal loan guarantees.*
- Policies *hasten competitive evolution* of next-generation biofuel industries.

The vision to 2025 is given in Figure 3.15. Using the U.S. EIA projections as a baseline, Lovins, et al., quantify the reductions in oil consumption and imports provided by conventional wisdom and state-of-the-art efficiency improvements (especially in the transportation sector), a revitalized biofuels industry, and substitution of natural gas freed by efficiency improvement in buildings, industry, and power generation. More than half of the state-of-the-art efficiency savings would come from car and light-truck efficiency.

Baka, Lynn, and Kammen (2006) similarly assessed the potential of reducing dependence on Mideast oil by 2025 by an aggressive combination of efficiency and renewables in transportation. They developed the following four scenarios and compared them to the business-as-usual case as defined by EIA's Annual Energy Outlook, a 1.9% annual increase in oil use:

Scenario 1   Moderate Corporate Average Fleet Efficiency (CAFE) improvements, moderate growth of hybrids, moderate ethanol biofuels

Scenario 2   Moderate CAFE, aggressive hybrids, aggressive ethanol

SIDEBAR 3.3

Winning the Oil Endgame: THE PRIZE

Forever the optimist, Amory Lovins offers this upbeat vision in Winning the Oil Endgame (Lovins, et al., 2004, p. 127):

Imagine a revitalized and globally competitive U.S. motor-vehicle industry delivering a new generation of highly efficient, safe, incredibly durable, fun-to-drive vehicles that consumers want. Imagine equally rugged and efficient heavy trucks that boost truckers' gross profits by $7.5 billion per year.

Consider the pervasive economic benefits of an advanced-materials industrial cluster that makes strong, lightweight materials cheaply for products from Stryker armored vehicles to bicycles to featherweight washing machines you can carry up the stairs by yourself.

Envision a secure national fuels infrastructure based largely or wholly on U.S. energy resources and on vibrant rural communities farming biofuel, plastics, wind, and carbon.

Think of over one million new, high-wage jobs, and the broad wealth creation from infusing the economy with $133 billion per year of new disposable income from lower crude-oil costs.

Recognize with pride that with this new economy, the United States is nearly achieving international greenhouse gas targets as a free byproduct.

Picture increased energy and national security as oil use heads toward zero, and as the United States regains the leverage of using petrodollars to buy what our society really needs rather than handing those dollars to oil suppliers to feed an addiction.

Imagine a U.S. military focused on its core mission of defense, free from the distraction of getting and guarding oil for ourselves and the rest of the world. Imagine that the United States, able once again to practice its admired ideals, has regained the moral high ground in foreign policy.

Finally, **envision** one of the largest and broadest-based tax cuts in U.S. history from eliminating the implicit tax that oil dependence imposes on our country by bleeding purchasing power, inflating military and subsidy costs, and suppressing homegrown energy solutions.

Sound utopian? It is not.

| Scenario 3 | Aggressive CAFE, aggressive hybrids, aggressive ethanol |
| Scenario 4 | Aggressive CAFE, plug-in hybrids, aggressive ethanol |

Their aggressive scenario 4 could achieve 22% reduction (6.3 million barrels per day) of daily oil use of the business-as-usual case, effectively eliminating the need for Persion Gulf imports, reducing CO_2 emissions from light-duty vehicles by 50%, and saving $1.1 trillion (Baka, et al., 2006).

3.4.3 The 25 × '25 Vision for Renewable Energy

In 2004 a group of agricultural and forestry leaders together with the Energy Future Coalition established a goal of achieving 25% of the nation's energy by renewable sources by 2025. The so-called "25 × '25" gathered considerable momentum and by 2007, it was endorsed

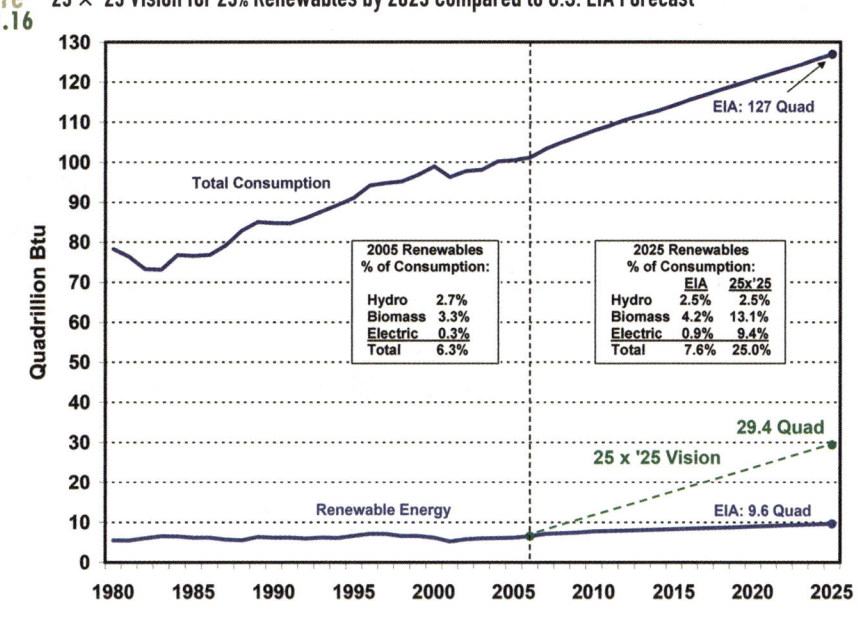

figure 3.16 25 × '25 Vision for 25% Renewables by 2025 Compared to U.S. EIA Forecast

table 3.5 Renewable Energy Required to Meet 25 × '25 Goals

	All Energy (AE)	Electricity Power and Transportation (EPT)
EIA 2006 AEO forecast	127.0 Q	81.6 Q
Adjusted for coal inefficiency	117.7 Q	81.6 Q
25%	29.4 Q	20.4 Q
Non-biomass power 2025	14.0 Q	14.0 Q
Biomass energy	15.4 Q	6.6 Q

by more than 400 organizations, 22 state governors, and resolutions in the U.S. Senate and House of Representatives. Figure 3.16 and Table 3.5 show that the 25 × '25 goal requires 29.4 quads renewable energy or more than three times the forecast by U.S. EIA.

A University of Tennessee study completed in November 2006 analyzed the feasibility of U.S. farms, forests, and ranches to meet this goal while continuing to produce safe, abundant, and affordable food, feed, and fiber. While distributed renewable energy in metropolitan areas can supply some of the goal, most must come from biomass energy and wind and solar farms in agricultural communities (English, et al., 2006).

The Tennessee study assessed two scenarios:

1. **AE goal:** 25% of "all energy" by 2025, or 29.4 quads
2. **EPT goal:** 25% of "electricity power and transportation" by 2025, or 20.4 quads.

figure 3.17   **Biomass Energy Feedstocks Needed to Achieve 25 × '25 AE (a) and EPT (b) Goals**

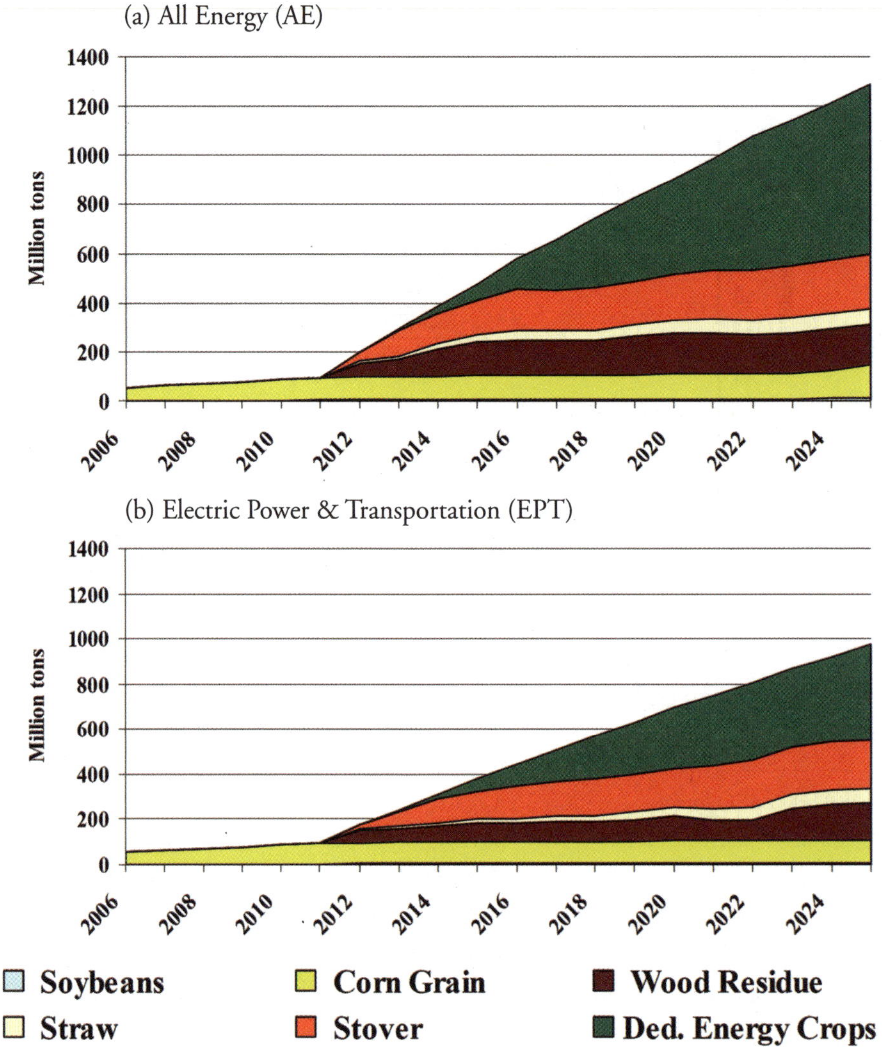

(a) All Energy (AE)

(b) Electric Power & Transportation (EPT)

Nearly all of the increase from 2006 comes from residues and dedicated energy crops such as switchgrass.

☐ **Soybeans**   ☐ **Corn Grain**   ■ **Wood Residue**
☐ **Straw**   ■ **Stover**   ■ **Ded. Energy Crops**

SOURCE: English, et al., 2006

The 15.4 quads comprise 86 billion gallons ethanol, 1.1 billion gallons of biodiesel, and 932 billion kWh electricity. The biomass feedstock required to meet the goals are shown in Figure 3.17.

The Tennessee study found that the 25 × '25 AE goal is achievable. In fact, it can be met while still providing agricultural food, feed, and fiber at reasonable prices, and reaching the goal would have a favorable impact on rural America by generating net farm income of $180 billion.

Some argue that a 25 × '25 future would be too expensive, adding considerably to already increasing energy costs. A November 2006 RAND Corporation study assessed the impact of achieving a 25 × '25 goal on energy expenditures in 2025. The RAND study ran simulations of 1500 scenarios, and the most extreme case for renewables in which renewable prices rose 30% and fossil fuel price fell 50%, produced only a 6% increase in energy expenditures. If renewable costs continue historic trends, there would be essentially no expenditure change by a 25% renewable scenario in 2025 (Bernstein, et al., 2006). However, in December 2006, RAND withdrew the report for revision citing some inadvertent modeling errors, but it is unclear to what extent the results were affected.

In late 2006, the American Council on Renewable Energy reported that the renewable power industry tallied their estimated deliverable capacity of renewable power by 2025 at 550–700 GW, easily enough to meet 25%. The 25 × '25 vision has stressed supply only. Incorporating efficiency improvements could dramatically reduce EIA 2025 demand forecasts and make it much easier for renewables to meet 25%. Solution Box 3.2 calculates how a 25% reduction in projected demand by improved efficiency would affect estimates of 25% renewables by 2025, a "25 – 25 × '25" scenario.

SOLUTION BOX 3.2

What about "25 – 25 × '25"? Adding efficiency to the "25 × '25" initiative.

The 25 × '25 goal of 20.4 quads (EPT) to 29.4 quads (AE) is based on U.S. EIA energy consumption projections of 127 quads with an additional credit of 10 quads because renewable electricity does not have the losses of fossil fuel steam generation. But we know that EIA projections do not include the wide array of opportunities for increased efficiency in vehicles, buildings, and electricity. What if we were able to reduce consumption below EIA estimates by 25%? What would be the resulting requirement to achieve 25% renewable energy?

Solution:
Consumption in 2025 would then be as follows:

$$127 \text{ quads} \times 0.75 = 95 \text{ quads} - 10 \text{ quad credit} = 85 \text{ quads}$$
$$25\% \text{ of } 85 \text{ quads} = 21 \text{ quads for all energy (AE) or } 72\% \text{ of the } 29.4 \text{ quad goal}$$

If EPT energy is same proportion of AE (69%), 14 quads of renewables would be needed to achieve 25% of EPT.

3.5 Carbon Futures and Global Climate Change

3.5.1 Pacala and Socolow's Carbon Stabilization Wedges

In a study reported in *Science* in 2004, Princeton professors Scott Pacala and Robert Socolow suggested that there is a portfolio of technologies available to meet the world's energy needs over the next fifty years that avoids doubling atmospheric CO_2 concentration from pre-industrial times and stabilizes the concentration at 500 ppm by 2125.

They argue that there is no one technology that can achieve this result but that a portfolio of options can meet, and possible exceed, the emissions reductions required so that society may have a choice of which options to pursue. The important point is that society must soon select desirable options and take action.

They illustrate their case with a solution wedges approach. As shown in Figure 3.18, the study starts with a business-as-usual projection of carbon emissions based on current

figure 3.18   Pacala and Socolow's Carbon Stabilization Wedges

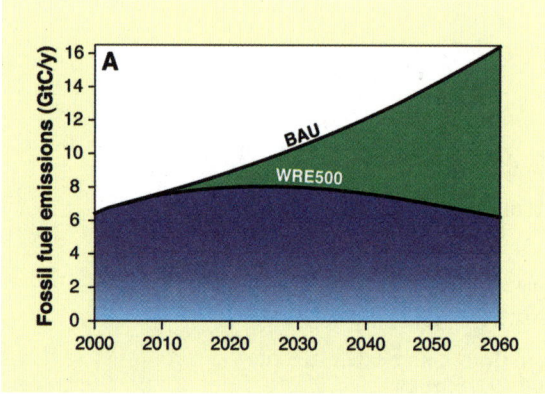
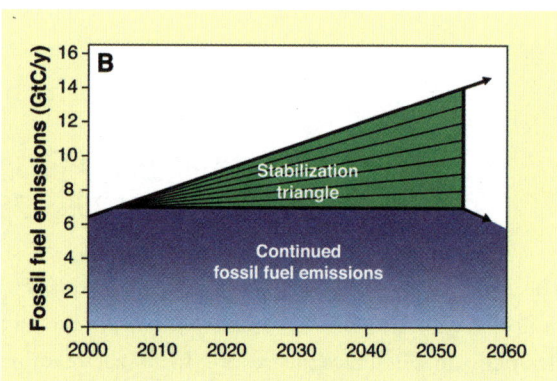

(A) Top curve is a representative of business-as-usual emissions path for global carbon emissions: 1.5% per year growth starting from 7.0 giga metric tons (GtC)/year in 2004. Bottom curve is a CO_2 emissions path consistent with atmospheric CO_2 stabilization at 500 ppm by 2125 akin to the Wigley, Richels, and Edmonds family of stabilization curves. The bottom curve assumes an ocean uptake calculated with the High-Latitude Exchange Interior Diffusion Advection (HILDA) ocean model and a constant net land uptake of 0.5 GtC/year. The area between the two curves represents the avoided carbon emissions required for stabilization.
(B) Idealization of (A): A stabilization triangle of avoided emissions (green) and allowed emissions (blue). The allowed emissions are estimated at 7 GtC/year beginning in 2004. The stabilization triangle is divided into seven wedges, each of which reaches 1 GtC/year in 2054. With linear growth, the total avoided emissions per wedge is 25 GtC, and the total area of the stabilization triangle is 175 GtC. The arrow at the bottom-right of the stabilization triangle points downward to emphasize that fossil fuel emissions must decline substantially below 7 GtC/year after 2054 to achieve stabilization at 500 ppm.

Source: Pacala and Socolow, SCIENCE 305:968–972 (2004). Reprinted with permission from AAAS.

trends to 14 Gt/yr in 2054. Drawing from work by Wigley, Richels, and Edmonds (2000), the study estimates the reduction in carbon emissions by 2054 necessary to achieve stabilization by 2125. This equates to maintaining CO_2 emissions at 2004 levels or 7 gigatons (Gt) per year. They calculate the overall reduction needed in 2054 at 7 Gt/yr (= 14 − 7), and this reduction could be realized by seven emission-reduction wedges, each equivalent to 1 Gt/yr reduction by 2054.

Pacala and Socolow go on to identify fifteen options for carbon emission reduction, and calculate for each option the magnitude needed to achieve one of the seven required 1-GtC wedges by 2054. The fifteen options, their magnitude for 1-GtC reduction, and some related comments are given in Table 3.6. By category, efficiency improvements (options 1–4) could achieve four wedges; renewables (10–13) four wedges; carbon capture and storage (6–8) three wedges; forestry and soil conservation (14–15) two wedges; and fuel switch (5) and nuclear (9) one wedge each.

The Princeton Environmental Institute, which Pacala and Socolow co-direct, produced the "Stabilization Wedges Game," an interactive game that allows users to calculate their own solution to carbon stabilization. Check it out at http://www.princeton.edu/~cmi/resources/stabwedge.htm.

3.5.2 IPCC Climate Change Scenarios

The Intergovernmental Panel on Climate Change (2001) provided visions of the future for several scenarios of development to illustrate some of the options facing global society and implications for future CO_2 emissions, atmospheric temperature change, and sea-level rise.

The scenarios reflect choices not only in energy supply but also in the nature of economic, social, and global development. The driving forces shown in Figure 3.19 include the economic-environmental continuum and the global-regional development continuum. Four scenario "families" are defined within the two-by-two matrix by the extent to which economic or environmental factors determine development and the extent to which global or regional factors drive development. They range from rapid economic growth plus peaked and then declining population (A1) that could be fossil fueled (A1FI) or non–fossil fueled (A1T) to slower-growth regional economies with higher population growth that are less influenced by globalization (A2). B1 and B2 are variations of A1 and A2 with greater emphasis on sustainability.

Figure 3.20 shows the CO_2 and global warming impacts of these development scenarios. The impacts include CO_2 emissions, atmospheric CO_2 concentration, atmospheric temperature change, and sea-level rise. A1FI and A2 show the most growth of all the parameters. A1T and B1 have the quickest reduction of CO_2 emissions and slowest growth of other parameters. High emission scenarios (A1FI and A2) would have significant lag effects on temperature and sea level beyond 2100 (see Figure 2.11). These scenarios and their storylines are discussed further in the next section.

table 3.6 **Potential Wedges: Strategies Available to Reduce the Carbon Emission Rate in 2054 by 1 GtC/year or to Reduce Carbon Emissions from 2004 to 2054 by 25 GtC**

Option	Effort by 2054 for One 1-GtC Wedge (relative to 14 GtC/year business-as-usual)	Comments, Issues
Energy Efficiency and Conservation		
Economy-wide carbon-intensity reduction (emissions/$GDP)	Increase reduction by additional 0.15% per year (e.g., increase U.S. goal of 1.96% reduction per year to 2.11% per year)	Can be tuned by carbon policy
1. Efficient vehicles	Increase fuel economy for 2 billion cars from 30 to 60 mpg	Car size, power
2. Reduced use of vehicles	Decrease car travel for 2 billion 30-mpg cars from 10,000 to 5000 miles per year	Urban design, mass transit, telecommuting
3. Efficient buildings	Cut carbon emissions by one-fourth in buildings and appliances projected for 2054	Weak incentives
4. Efficient baseload coal plants	Produce twice today's coal power output at 60% instead of 40% efficiency (compared with 32% today)	Advanced high-temperature materials
Fuel Shift		
5. Gas baseload power for coal baseload power	Replace 1400 GW 50%-efficient coal plants with gas plants (four times the current production of gas-based power)	Competing demands for natural gas
CO_2 Capture and Storage (CCS)		
6. Capture CO_2 at baseload power plant	Introduce CCS at 800 GW coal or 1600 GW natural gas (compared with 1060 GW coal in 1999)	Technology already in use for H_2 production
7. Capture CO_2 at H_2 plant	CCS at plants producing 250 Mt H_2/year from coal or 500 Mt H_2/year from natural gas (compared with 40 Mt H_2/year today)	H_2 safety, infrastructure
8. Capture CO_2 at coal-to-synfuels plant	Introduce CCS at synfuels plants producing 30 million barrels a day from coal (200 times Sasol)	If half of feedstock carbon is available for capture
Geological storage	Create equivalent of 3500 Sleipner Vest projects (offshore Norway sequestration project)	Durable storage, permitting
Nuclear Fission		
9. Nuclear power for coal power	Add 700 GW (twice the current capacity)	Nuclear issues
Renewable Electricity and Fuels		
10. Wind power for coal power	Add 2 million 1-MW-peak windmills (50 times the current capacity) "occupying" 30×10^6 ha, on land or offshore	Multiple uses of land because windmills are widely spaced
11. PV power for coal power	Add 2000 GW-peak PV (700 times the current capacity): 2×10^6 ha	PV production cost
12. Wind H_2 in fuel cell for gas in hybrid	Add 4 million 1-MW-peak windmills (100 times current capacity)	H_2 safety, infrastructure
13. Biomass fuel for fossil fuel	Add 100 times the current Brazil or U.S. ethanol production, with the use of 250×10^6 ha (one-sixth of world cropland)	Biodiversity, competing land use
Forests and Agricultural Soils		
14. Reduced deforestation, plus reforest, afforestation, and new plantations	Decrease tropical deforestation to zero (from 0.5 GtC/year), and establish 300 Mha of new tree plantations (2 times the current rate)	Agriculture land demands, benefits to biodiversity from forestation
15. Conservation tillage	Apply to all cropland (10 times the current usage)	Reversibility, verification

SOURCE: Pacala and Socolow, 2004

figure 3.19 Four IPCC Development Scenarios from Special Report on Emissions Scenarios (SRES)

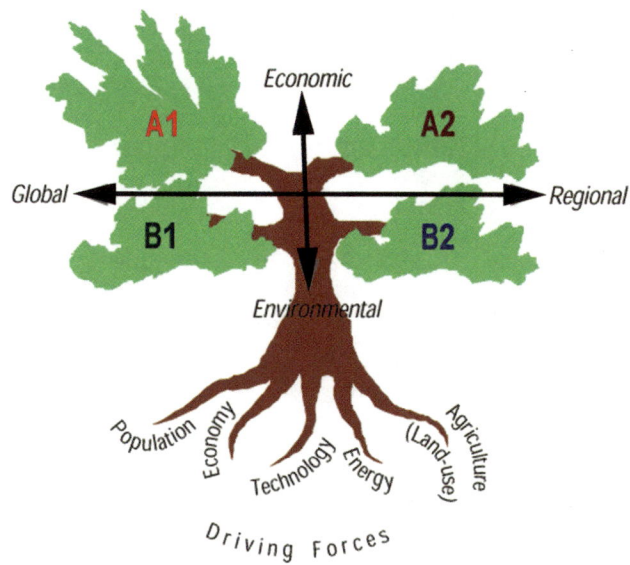

Four scenarios of global development are characterized by the economic-environmental continuum and the global-regional development continuum:

A1: Rapid economic growth, population peaks mid-century then declines, rapid introduction of new technologies: A1FI-fossil-intensive; A1T-nonfossil; A1B-balanced

A2: Heterogeneous world, self-reliance, local identities; increasing population; economic growth and technological change slower and more fragmented

B1: Same population as A1, but convergent world rapidly changing to service and information economy, less materially intensive, clean and efficient technologies; sustainability emphasis

B2: Population increase but lower than A2; local solutions to sustainability; less rapid and more diverse economic and technological change than A1 and B1

3.6 The Global Population-Economy-Energy Conundrum

One of the principal virtues of integrated assessments, such as the IPCC scenarios, is that they explicitly recognize the values and activities of humans and society as a critical variable. The way we choose to use energy in the United States and the world will have a dramatic effect on the global economy, environment, and equitable development. Chapter 1 highlighted the critical role energy played in the advance of civilization and especially in more recent growth of global population and the economy. During just the past one hundred years, energy technologies and availability of fossil fuels facilitated the greatest growth of population and economy in the history of the world. Since 1900, population has more than tripled and the economy has increased by thirty-six times!

Although the momentum of this growth continues, most analysts conjecture that we are in the midst of a transition: a demographic transition to a stabilized population, an economic

figure 3.20 IPCC CO$_2$ Scenarios and Impacts under Different Development Scenarios

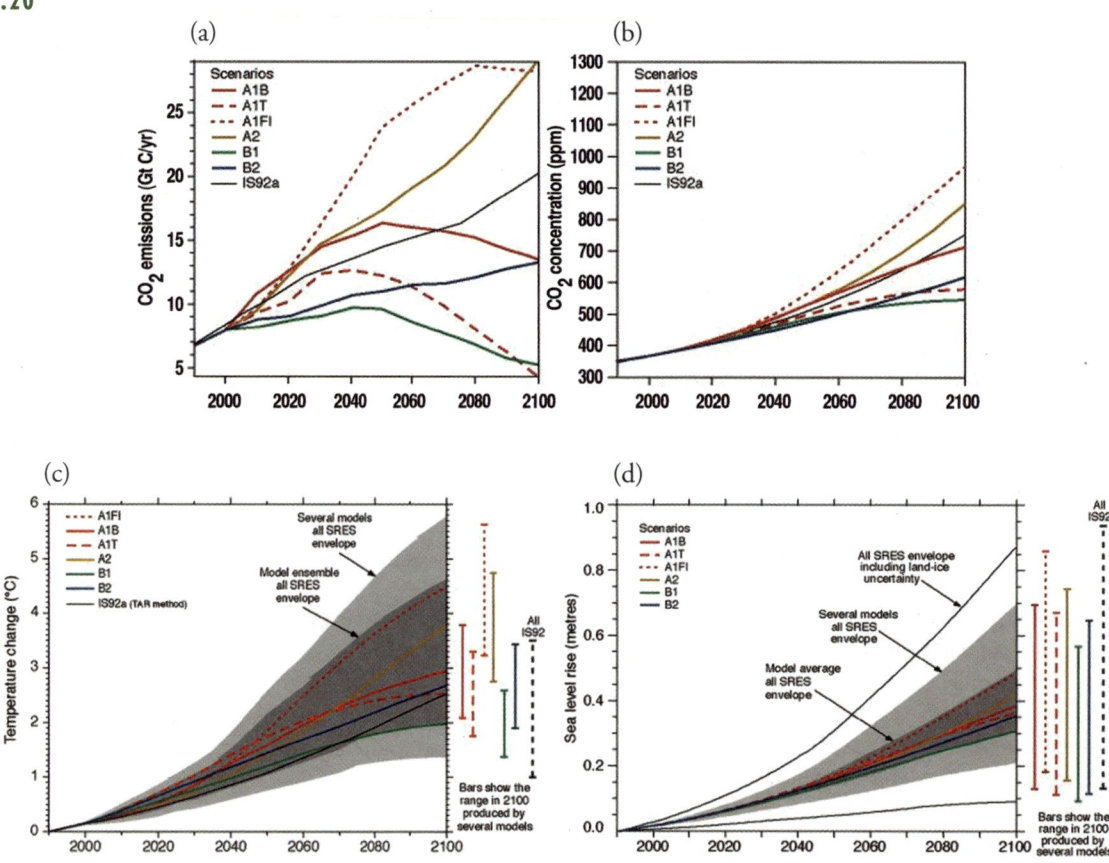

(a) CO$_2$ emissions, (b) CO$_2$ concentrations, (c) global temperature, and sea-level rise (d). Vertical lines in (c) and (d) show uncertainty bands. IS92a is the reference scenario.

Source: IPCC, 2002

transition to a post-industrial capitalist economy, and a democratic transition to greater public self-determination. There are strong indications of this transition. Population growth has slowed, stabilized, and even declined in developed countries, where overall growth is only 0.1% per year. World population has reached the inflection point of its growth curve: while it continues to grow, it has begun to grow at a slower rate. The emerging information and service economy has become less material- and energy-intensive. Free-market capitalism with public oversight has won over centralized economic systems, and democratic political systems are advancing worldwide.

This potential transition is no better displayed than in Figure 3.21, which shows the full history and possible future of human population growth. In this time frame, we are on the steep slope of change, hopefully headed toward a stable population of 11 billion by 2100. Most analysts accept this vision as probable, but others are even more hopeful for a faster demographic transition by 2050 and a subsequent stabilization of or decline in population.

 **History and Future of World Population**

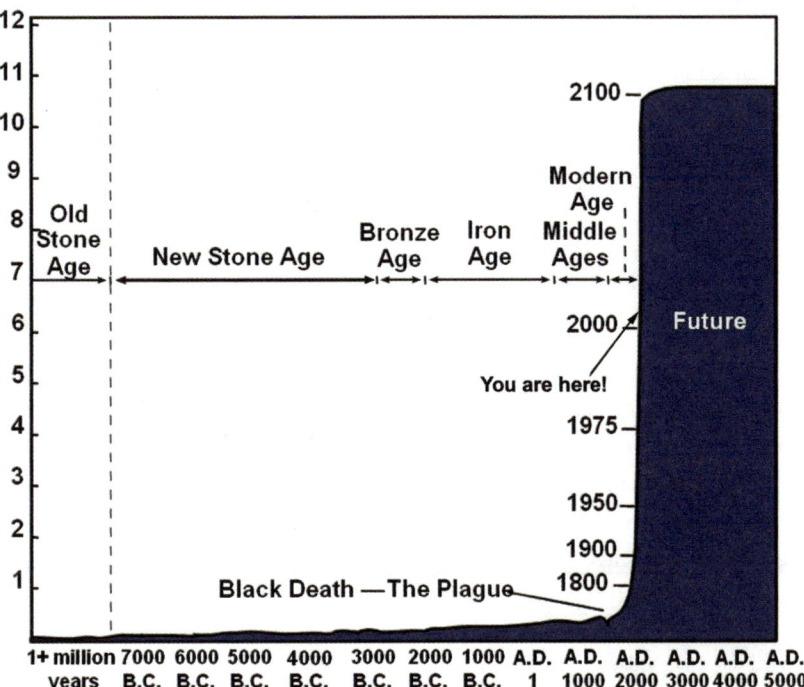

We are at the inflection point of the incredible growth period likely to lead to stabilization at 11 billion by 2100.

SOURCE: Population Reference Bureau, 2004

Much will depend on social and economic development in developing nations and religious and cultural influences on population growth.

Population stabilization is a requisite condition for a sustainable society. While we have turned the inflection corner on the growth curve, we have a long way to go to provide basic needs for the world's poor, a requirement before they too can transition to low birthrates and a stable population. Less developed countries, with 5.2 billion (81%) are still growing at 1.5% per year, and those 3.9 billion poor outside of China are growing at 1.8%. Providing their basic needs will require significant economic development and energy.

In one dark scenario, continuing current patterns of energy, with its climate, supply, and security impacts, would likely inhibit economic development and slow the demographic transition. This results in ever-increasing population, which requires still greater development and energy before population stabilization could be realized. Not only would sustainability be delayed, but the probable economic, environmental, social, political, and military disruptions would likely exceed global capacity to manage them.

A more hopeful scenario sees a faster transition to a stabilized population whose basic needs are met and who can participate in a democratic society and to a less material- and energy-intensive and more equitable global economy. The sooner that transition can be achieved, the lower will be the resulting stable population and fewer and less intense will be the economic and social disruptions. Can one envision a world of a stabilized population of 10 billion or less, an energy- and materials-efficient global economy that engages all regions, and political stability brought about by increased democratization and less competition for access to resources?

There are some hopeful signs: We are seeing slowing population growth, declining energy intensity of the economy, and growing democratization. But we have yet to see a significant shift in our patterns of energy use. We remain dependent on unsustainable fossil fuels for more than 85% of global energy. Nuclear energy is stagnant, and renewable energy, while growing, is still too small a source to have an impact. We are on a path to the dark scenario.

The conundrum we face is the interdependence of energy, economic development, and population: Population stabilization needs economic development, which needs sustainable energy, which needs population stabilization, and so on. This may now seem like a vicious cycle, but it could be turned into a virtuous one if we can solve our energy dilemma. As discussed in the last section, the Intergovernmental Panel on Climate Change (IPCC) tried to articulate this challenge by presenting future scenario storylines on population, energy technologies, and economic growth and assessing their impact on climate change. The three scenarios with the most positive impact on carbon emissions and climate change over more business-as-usual projections were A1T, B1, and B2.

- The **A1T** storyline (dashed red lines in Figure 3.20) describes a future world of very rapid economic growth, global population that peaks in mid-century and declines thereafter, the rapid introduction of new and more efficient technologies with a major emphasis on non-fossil energy, and a high level of globalization. Major underlying themes are convergence among regions, capacity-building, and increased cultural and social interactions, with a substantial reduction in regional differences in per capita income. After an initial increase, the carbon emissions under this scenario drop quickly to pre-1990 levels by 2100.
- The **B1** storyline (green lines in Figure 3.20) describes a convergent world with the same global population that peaks at mid-century and declines thereafter and a high level of globalization, as in the A1T storyline, but with rapid change in economic structures toward a service and information economy, with reductions in material intensity and the introduction of clean and resource-efficient technologies. The emphasis is on global solutions to economic, social, and environmental sustainability, including improved equity. This scenario has the lowest overall carbon emissions through 2100.
- The **B2** storyline (blue line in Figure 3.20) describes a world in which the emphasis is on local solutions to economic, social, and environmental sustainability and a lower level of globalization than A1T and B1. It is a world with continuously increasing global population, intermediate levels of economic development, and less rapid and more diverse technological change than in the B1 and A1T scenarios. While the

scenario is also oriented toward environmental protection and social equity, it focuses on local and regional levels. Because of increased population and more use of fossil energy, this scenario has increased carbon emissions to near-double 1990 levels by 2100.

While any of these three scenarios would move us positively from our current business-as-usual trends, A1T and B1 are the only options toward sustainability, because B2 continues increasing population. B2's more decentralized, local approach to development and change is preferable to many futurists over the more global approaches in A1T and B1 because it could better preserve local culture and identity and foster different, creative, and regionally based responses. However, most believe our need to arrest our current trends of population and energy growth is too immediate and we must mobilize a global effort for change as we race against the climate and oil clocks.

Figure 3.22 extends to the future the historical graph given in Figure 1.1, reflecting positive although uncertain futures for economy, population, and energy consumption. Energy rises to 860 quads by 2100. Population stabilizes at 10 billion and an alternative post-2050 decline to 8.5 billion is also given. Figure 3.23 gives a scenario about how that 860 quads of energy could be met with emphasis on non-carbon renewable energy. Starting from the bottom wedge, petroleum declines after 2020, natural gas increases until 2050 then declines, and coal increases slightly to 2010 then declines slightly to 2100. Hydro and nuclear are relatively constant, although new nuclear begins to gain in 2040. The largest growing

**figure
3.22**   Population, Energy, and Economy, 1800–2000, and One Possible Future Scenario, 2000–2100

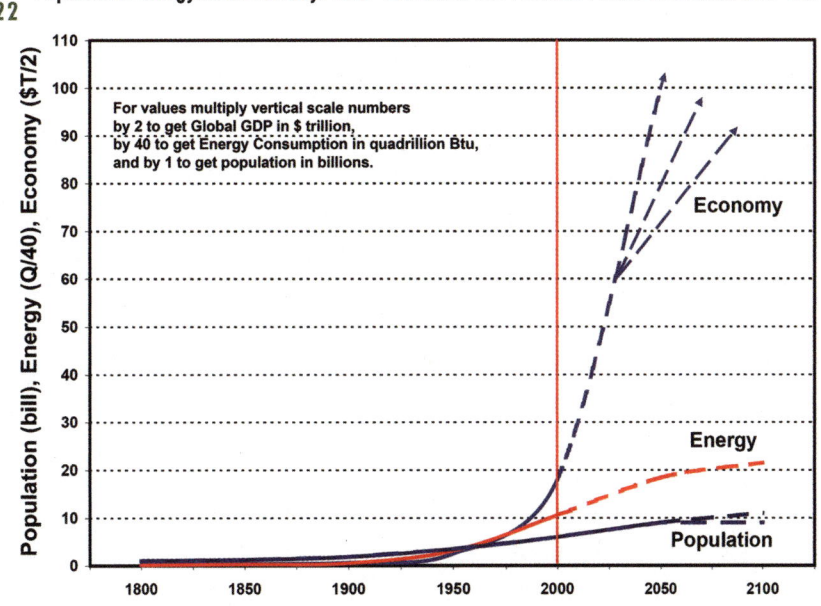

figure
3.23   **Global Energy by Source, 1980–2000, and One Possible Future Scenario, 2000–2100**

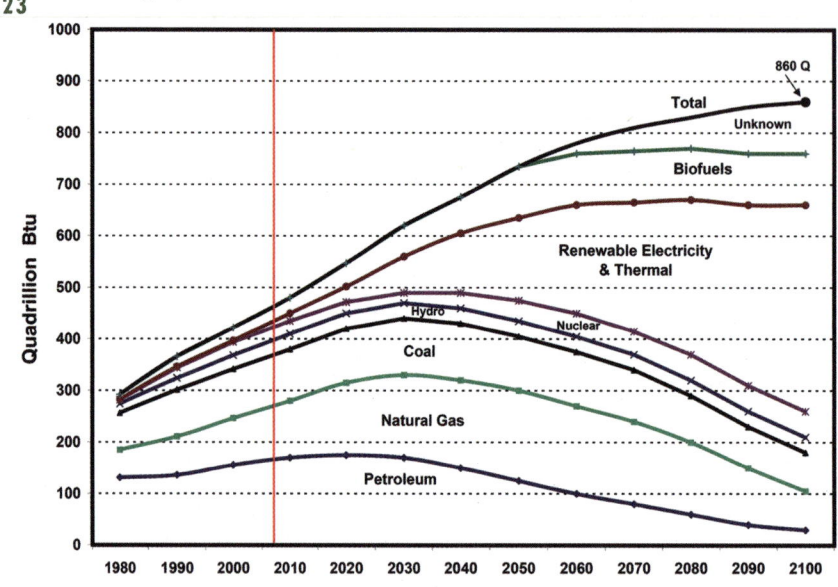

wedge is renewable sources, especially renewable electricity and biofuels. Unknown sources, given in the top wedge, are included in this scenario after 2060.

How do we arrest current trends and move toward greater sustainability of economy, population, and energy? Lester Brown (2003) argues for a mobilization comparable to World War II, which was won by a massive industrial adaptation to new priorities. His and others' prescriptions include the following:

- **Population:** Stabilize population by investment in basic needs and eradication of poverty and disease. Break the vicious poverty-disease-population cycle (poverty begets disease begets delay of demographic transition begets population growth begets poverty begets disease) through a global effort.
- **Energy:** Move toward sustainable energy through improvements in efficiency, reduction of fossil fuels, sequestration of carbon, developing renewable energy, and other carbon-free options. Promote development of decentralized renewable energy as a strategy for less developed countries.
- **Environment:** Sustainable energy will help protect the atmospheric environment, but additional measures are needed to protect the coming resource constraints of water, land degradation, and biodiversity reduction.
- **Economy:** Promote further transition to an information and service economy that is less energy- and material-intensive. Correct market imperfections to better incorporate indirect costs, better value nature's services, and better respect sustainable limits. Foster change through innovation for rapid development of new technology. Use economic mechanisms to create positive change.

- **Policy:** Use public and international policy to correct market imperfections through fees, regulations, and taxes, and therefore prices, to enable economic forces to create positive change. Tax environmental and social impact more than income. Use incentives to reduce risk of developing new technologies. Use global cooperation and agreements to force change that cannot be achieved by domestic and bilateral policies.

- **Political:** Resolve regional and global conflicts and terrorism by addressing root causes rather than by military might and fear alone. Promote democracy and development at an appropriate pace to encourage self-determination, basic-needs provision, and demographic transition.

- **Social:** Enhance education of our common challenges and prospective means of resolving them. Use information technology to advance global communication to advance education and spread and preserve diverse cultural traditions. Promote global perspectives over parochial nationalistic and protectionist perspectives. Enhanced social awareness fosters less-consumptive behavior, "green" markets, and collective efforts for the common good.

On a bad day, we see so many obstacles to these necessary changes to move to a more sustainable global society: the plaguing and constraining problems of worldwide poverty and disease; global terrorism and resulting protectionist nationalism; the paralysis for change brought about by economic and political forces favoring the status quo; and the ticking clock.

But on a good day, we can imagine ourselves on the verge of social, economic, and political change that can address the daunting problems we face: growing global consciousness brought about by information technology; growing consensus about problems we face; expanding public and private economic resources that can be channeled toward positive change; and emerging technologies for energy efficiency and production that can wean our economy from petroleum and transition to a carbon-free economy.

3.7 Summary

If anything, we hope that these first three chapters have made clear the critical nature of energy in our society; the huge stakes energy poses to U.S. and global economic, social, and political development and to the global environment; and the critical uncertainties we face in just the next decade. In this chapter, we have shown the ways future visioning and scenario development can illuminate the challenges we face as well as potential solutions.

It is time for action to modify our patterns of energy production and consumption. In fact many are crying that such action is long overdue. But we seem paralyzed to act as the economic, political, and cultural forces favoring the status quo seem destined to impede the process of change.

If there are signs of hope, they come from the increased attention to the energy challenges we face, the growing consensus on a number of issues from global climate to looming geopolitical risks of oil, and the emerging growth of renewable sources of energy.

Energy Fundamentals

CHAPTER 4

Fundamentals of Energy Science

4.1 Introduction

Before we can explore the array of technologies that will help us make the transition to more sustainable energy systems, we need a clear understanding of energy itself. Energy, primarily from the sun, flows through ecosystems and provides all living things, including our own bodies, the capacity to live, grow, repair tissue, reproduce, and do work. As it does so, its form may change from electromagnetic radiation flowing through space, to chemical energy stored in plants, to heat that keeps us warm, to potential energy as we climb a mountain, and to kinetic energy as we ski back down again. We use solar energy that plants captured and stored eons ago to heat our homes, generate our electricity, and drive our cars. As energy works its way through nature and human societies, it is constantly being transformed from one form to another. Although none is lost as it is stored, converted, and used, its quality is constantly being degraded to less and less useful forms, eventually ending up as relatively useless, low-temperature heat.

Understanding energy is one of the keys to understanding the universe and how physical and living systems work. In fact, a simple definition of *energy* is that it is the capacity to do *work*. Our own sensory perceptions and personal experiences have given us an inherent understanding of energy transformations and flows. We know about heat transfer (if only to put on a jacket or take it off), we know about the chemical energy released during combustion (as we accelerate to the next stoplight), we know about the energy of moving objects (if only to get out of the way of that oncoming car), and we know about radiant energy as we warm ourselves in front of that nice campfire. We even know something about the fairly esoteric concept of entropy, as our workshop gets more and more cluttered with sawdust and scraps as we build that nice piece of furniture for our home. Certainly, we have an intuitive understanding of many important energy systems, such as refrigerators, lights, cars, and furnaces, if only to operate them even though we may have only a vague notion of how they actually work.

Although a thorough explanation of the physics and chemistry of energy is well beyond the scope of this textbook, we can fairly easily develop an intuitive and somewhat quantitative feel for these energy transformations and flows. The vocabulary and basic principles presented here will provide the necessary foundation needed to understand the energy systems to be described in subsequent chapters.

The chapter begins with an introduction to the concept of energy itself, along with some units and conversions. It then explores some basic forms of energy, including mechanical, thermal, chemical, electrical, nuclear, and electromagnetic energy. What society cares about, of course, is not joules or British thermal units, but how we can transform various forms of energy into useful work to cool our beer, heat our homes, and take us where we want to go. The "rules of the road" that dictate what we can theoretically accomplish with a British thermal unit or a joule, as well as what we cannot do (such as create a perpetual motion machine) are introduced using the first and second laws of thermodynamics. With those fundamentals under our belts, we will be prepared for the following chapters, which explore some of the most important energy conversion systems.

4.2 Basics of Energy Science

Just what is energy? A precise answer to that deceptively simple question is surprisingly difficult. A common definition is that energy is "the capacity for doing work." Well, you and I are capable of doing work, does that mean we are energy? Although that may sound funny, Einstein's famous relationship between matter and energy, $E = mc^2$ says yes, we have mass, and mass and energy are inextricably linked to each other. But then, what is work? Work can be defined as the product of the force needed to move an object times the distance that it moves. But isn't thinking sometimes hard work? Moreover, work is not the only form of energy. For example, heat is another form of energy, but then what is heat? Well, heat is energy transferred from one object to another by virtue of their temperature difference. So, what is temperature?

Energy is a complicated concept. But, we can go far just by relying on our intuitive sense that energy is the ability to cause physical things to change. Energy allows us to make things get hotter, move faster, go uphill, and so forth.

4.2.1 Introduction to the First and Second Laws of Thermodynamics

Energy may change forms in any given process, as when chemical energy in wood is converted to heat and light in a campfire, or when the potential energy of water behind a dam is converted to mechanical energy as it spins a turbine, and then into electricity in the generator of a hydroelectric plant. The first law of thermodynamics says we should be able to account for every bit of energy in such processes, so that in the end we have just as much as we had in the beginning. With proper accounting, even nuclear reactions involving conversion of mass to energy can be accounted for.

To apply the first law, it is first necessary to define the system being studied. The system can be anything that we want to draw an imaginary boundary around—it can be a tree, or a nuclear power plant, or a volume of gas emitted from a smokestack. In the context of global climate change, the system could very well be the Earth itself. Quite often, what we really want to know is how efficiently a system converts energy in one form into useful energy in another form. For example, we might like to know how efficient a power plant is when it converts chemical energy in coal into electrical energy delivered to a transmission line. We can write the first law to express this as follows:

Eq. 4.1   **Energy into the system = Useful energy delivered + Wasted energy**

We want useful energy, which leads to the following definition of system efficiency

Eq. 4.2 $$\text{Energy efficiency } (\eta) = \frac{\text{Useful energy out}}{\text{Energy input}}$$

In practice, most of the systems we are interested in involve multiple energy transformations, each with its own efficiency. To find the overall efficiency from start to finish, we simply multiply the individual efficiencies. For example, consider a 35% efficient power plant putting electricity onto 92% efficient transmission lines that deliver electricity to a 5% efficient incandescent lightbulb. As Figure 4.1 indicates, that results in an overall efficiency of only 1.6%.

Figure 4.2 shows the benefits of switching those incandescent lights to energy-efficient, compact fluorescent lamps (CFLs). CFLs require only about one-fourth the electric power of an incandescent, while providing the same illumination. They also last as much as ten times as long. As shown, our 1.6 units of CFL light save a whopping 75 units of fuel into the power plant.

To summarize, then, the first law of thermodynamics tells us that energy is neither created nor destroyed as it makes its way through the universe. In other words, the first law gives us a bookkeeping system that allows us to keep track of *quantities* of energy.

The second law of thermodynamics, on the other hand, tells us that even though no energy is "lost" during transformations, there will invariably be a loss in the *quality* of that energy. The quality of energy has to do with its capability to do useful things for us. For example, electricity is a very high-quality form of energy because it can do everything from powering your TV to heating your house. Low-temperature heat, on the other hand, is a very

figure 4.1

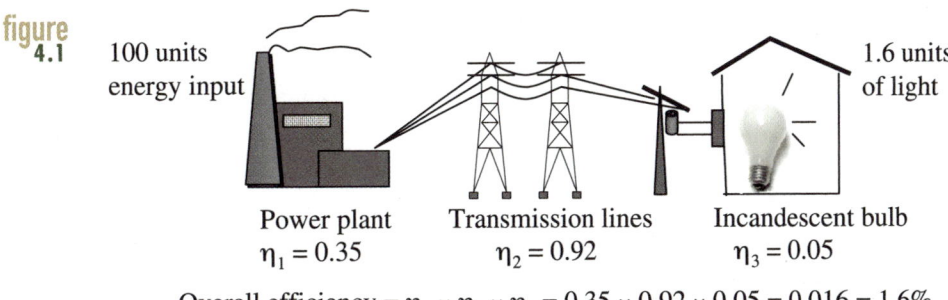

Overall efficiency = $\eta_1 \times \eta_2 \times \eta_3 = 0.35 \times 0.92 \times 0.05 = 0.016 = 1.6\%$

The overall efficiency from fuel to visible light from an incandescent bulb is a mere 1.6%.

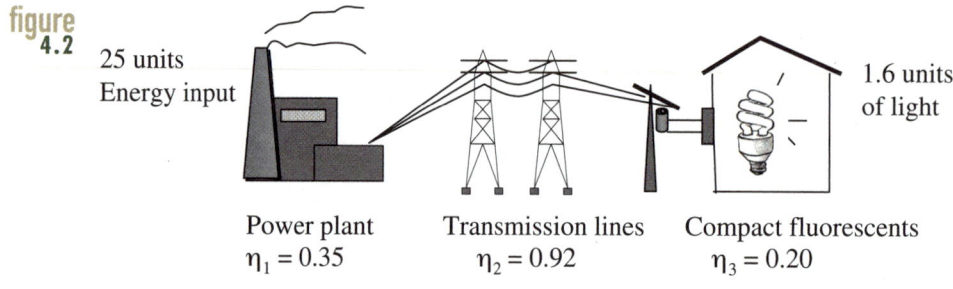

figure 4.2

25 units
Energy input

1.6 units
of light

Power plant
$\eta_1 = 0.35$

Transmission lines
$\eta_2 = 0.92$

Compact fluorescents
$\eta_3 = 0.20$

Overall efficiency with CFLs = $0.35 \times 0.92 \times 0.20 = 0.064 = 6.4\%$

Switching from the incandescents in Figure 4.1 to energy-efficient compact fluorescent lamps (CFLs) saves three-fourths fuel needed at the power plant while delivering the same illumination.

---

low-quality form of energy. You may be able to warm your hands on that cup of coffee, but you certainly can't plug your computer into it. The second law tells us that no matter how hard we may try, every time we do something with energy there is always some loss in energy quality, which usually means some of it ends up as fairly useless waste heat.

We can imagine any number of processes that would satisfy the first law, but which we know realistically cannot occur. I can run some electricity through a heating element to heat my coffee, but I can't heat the element and expect to be able to get an equivalent amount of electricity back again. The second law takes care of such concerns by informing us about the direction in which processes can go. Electricity to heat is easy; heat to electricity is not.

The implications of the second law are profound. It "outlaws" a whole bunch of things, ranging from perpetual motion machines to the possibility of a warm cup of coffee heating itself up by stealing heat from the cool air in your kitchen. It dictates the maximum possible efficiency of your current automobile's engine as well as the fuel cell that may someday replace it. It tells us strange things about continuously increasing disorder in the universe. While you are cleaning up your room, making it more orderly, the power plant making electricity for your vacuum cleaner is creating even more disorder elsewhere as it converts an organized chunk of coal into disorderly gaseous and particulate emissions from its stack. We will explore the second law more carefully in Chapter 9, which discusses heat engines and power plants.

4.2.2 A Word about Units

William Thomson (Lord Kelvin) once stated that

When you can measure what you are speaking about, and express it in numbers, you know something about it; but when you cannot measure it, when you cannot express it in numbers, your knowledge is of a meager and unsatisfactory kind; it may be the beginning of knowledge, but you have scarcely, in your thoughts, advanced to the stage of science.

In the United States, energy units are often reported in both the English system (or, as it is sometimes called, the U.S. Customary System), and the International System (SI), so it is important to be familiar with both. In the SI system, the units of mass, length, and time are the kilogram (kg), meter (m), and second (s). In the English system, they are the pound-mass (lbm), foot (ft), and second (s).

Notice, we have already introduced some potential for confusion. What is this pound-mass business? Isn't a pound a pound? In common American usage, a *pound* is a unit of force (lbf), not a unit of mass. When we say something weighs 6 pounds, we are referring to the force it exerts on a scale, not its mass. If we took that 6-pound object to the moon, where the gravitational force is much lower, it would weigh only about 1 pound. Its mass would, however, be the same on Earth or the moon. As long as we stay at sea level on the Earth's surface, however, a pound is a pound. That is, a mass of 1 lbm does weigh 1 lbf.

The SI system avoids this confusion by always referring to the mass of an object, not its weight. Thus a 1 kg object on Earth has the same mass as a 1 kg object on the moon (see Figure 4.3).

Newton's second law connects mass (*m*) and weight (*W*) with the local gravitational acceleration, which is 9.807 m/s^2 or 32.174 ft/s^2 at the Earth's surface. For computational purposes, we'll call the acceleration g and just round it to 9.8 m/s^2 and 32.2 ft/s^2. Newton's equation is

$$\text{Weight} = mg$$

Eq. 4.3   In SI units: Weight (Newton) = mass (kg) × g (m/s^2)

In English units: Weight (lbf) = mass (slugs) × g (ft/s^2)

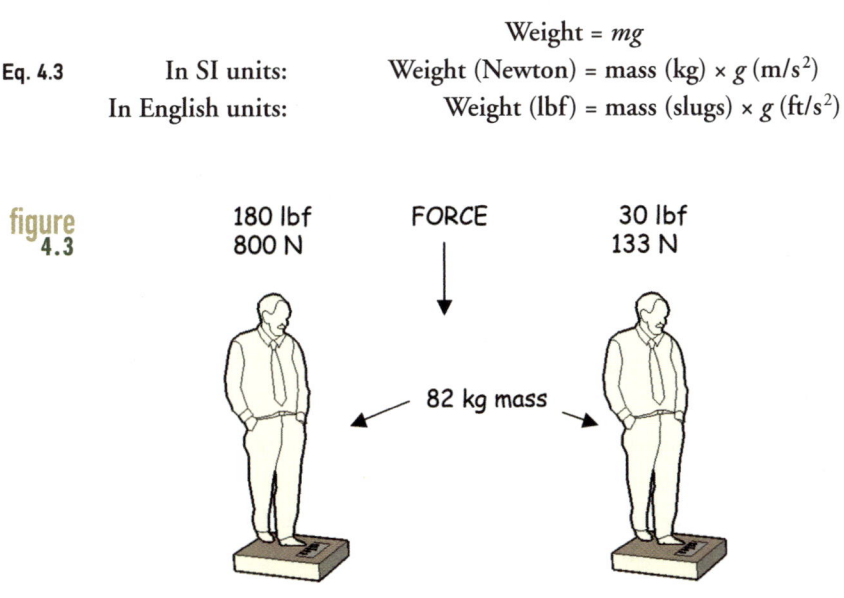

figure 4.3

180 lbf
800 N

FORCE

30 lbf
133 N

← 82 kg mass →

On Earth   On the Moon

A man weighing 180 pounds on Earth will weigh only about 30 pounds on the moon. His mass does not change, however.

If you were to exert a force of 1 Newton on a scale on the Earth's surface, it would register 0.2248 lb, which is about the weight of one small apple. Curious coincidence, isn't it? Newton also gave us his amazingly important relationship between force, mass, and acceleration:

Eq. 4.4   **Force = mass × acceleration or F = ma**

A force of 1 Newton will cause a mass of 1 kg to accelerate at the rate of 1 m/s².

Work, which is a form of energy, can be defined as force times distance. In SI units, force is in Newtons (N) and distance is in meters (m). The product, Newton-meters, is defined as joules:

$$1 \text{ joule} = 1 \text{ Newton-meter}$$

In English and American units, energy is quite likely to be measured in British thermal units (Btu), where a Btu is the energy required to raise 1 pound of water by one degree Fahrenheit. An older heat unit is the calorie, which is defined as the quantity of heat required to raise 1 gram of water by 1 degree Celsius. A calorie is about four times the size of a joule. Notice the interesting difference between the SI interpretation of energy as force times distance, whereas the English system based on the calorie suggests heat as a measure of energy.

4.2.3 The Distinction between Energy and Power

We all have pet peeves, and the constant misuse of energy and power units in popular media can be especially annoying. When we read about a new power plant described as delivering thus and such kilowatts per year, it is a sure tip-off that the writer is not very energy literate. Because you don't want to be considered illiterate, let's get the vocabulary right.

Power is energy per unit of time. It is a rate. For example, in SI units, power is often given in joules per second (J/s). One joule per second is designated as one watt in honor of the Scottish engineer James Watt, who developed the reciprocating steam engine.

$$1 \text{ watt} = 1 \text{ J/s} = 3.412 \text{ Btu/hr}$$

Thus, to describe something in kilowatts per year is like some sort of energy acceleration unit, joules per second squared. It makes no sense at all. An electric heater that uses 10 kilowatts of power for two hours uses 20 kilowatt-hours (20 kWh) of energy. A water heater that uses natural gas at the rate of 16,000 Btu per hour (power) for one-half hour burns 8000 Btu of gas (energy).

Table 4.1 presents some unit conversion factors for both energy and for power. And because numbers can range from extremely small quantities (e.g., nanometers) to extremely large ones (e.g., exajoules), it is handy to have a system of prefixes to accompany the units. Some of the most important prefixes are presented in Table 4.2.

table **Energy and Power Units**
4.1

ENERGY	1 British thermal unit	= 778 ft-lb
		= 252 calories
		= 1055 joules
		= 0.2930 watt-hours
	1 quadrillion Btu	= 10^{15} Btu
		= 1055×10^{15} J
		= 2.93×10^{11} kWh
		= 172×10^6 barrels (42 gal) of oil equivalent
		= 36.0×10^6 metric tons of coal equivalent
		= 0.93×10^{12} ft^3 of natural gas equivalent
	1 joule	= 1 Newton-meter (N-m)
		= 9.48×10^{-4} Btu
		= 0.73756 ft-lb
	1 kilowatt-hour	= 3600 kJ
		= 3412 Btu
		= 860 kcal
	1 kilocalorie	= 4.185 kJ
POWER	1 kilowatt	= 1000 J/s
		= 3412 Btu/hr
		= 1.340 hp
	1 horsepower	= 746 W
		= 550 ft-lb/s
	1 quadrillion Btu per year	= 0.471 million barrels of oil per day
		= 0.03345 terawatt (TW)

4.3 Mechanical Energy

Energy exists in many forms and a major task for engineers is to design systems that transform it from one form to another. We may want to use sunlight (electromagnetic energy) to run our TV (electrical energy), convert gasoline (chemical energy) into motion (kinetic energy), or cause uranium to fission (nuclear energy) to create steam (thermal energy) to run a turbine (rotational energy). Before we can explore such systems we need a brief introduction to these various forms of energy.

The energy systems that are most familiar to us from high school physics classes are often those based on moving weights around—pick them up to gain potential energy, drop them to demonstrate kinetic energy, rotate a wheel to demonstrate gyroscopic forces, and so forth. Those examples of mechanical energy are not only intuitively understandable, but they are also

table 4.2   **Common Unit Prefixes**

Quantity	Prefix	Symbol
10^{-12}	pico	p
10^{-9}	nano	n
10^{-6}	micro	μ
10^{-3}	milli	m
10^{-2}	centi	c
10^{-1}	deci	d
10^{3}	kilo	k
10^{6}	mega	M
10^{9}	giga	G
10^{12}	tera	T
10^{15}	peta*	P
10^{18}	exa	E

* In the United States, "quad" (short for quadrillion) is often used.

easy to analyze and provide an excellent introduction to the astonishing contributions that Sir Isaac Newton made to our understanding of physical science more than 300 years ago.

4.3.1 Potential Energy

It takes energy to lift a weight from one elevation to another, and in the process the object acquires potential energy; that is, it has the potential to do some work if we drop it. That energy required to raise an object, against the force of gravity, is described by the force times distance relationship. The force needed is the object's weight, which is the same as its mass (m) times the local gravitational acceleration (g). When raised to a height (h), its potential energy becomes

Eq. 4.5   Potential energy (P.E.) = **force × distance = weight × height** = $mgh = Wh$

Solution Box 4.1 shows how the units work in both SI and the English system.

4.3.2 Kinetic Energy

If we were to drop a mass (m) from a height (h) in a vacuum (so there is no air friction to slow it down), just before it hit the ground it would be moving at a speed of

Eq. 4.6 $v = \sqrt{2gh}$

SOLUTION BOX 4.1

Potential Energy Gained at the Top of the Mountain

How much potential energy is acquired when a 70 kg (154 lb) man reaches the top of a 2000 m (6562 ft) mountain? How many donuts of energy is that if 1 donut = 150 Calories? Note that food "Calories" (with a capital C) are actually kilocalories of energy.

Solution:

Using Equation 4.5 with SI units, we have

In SI: P.E. = 70 kg × 9.8 m/s^2 × 2000 m = 1.37 × 10^6 J

In English: P.E. = 154 lb × 6562 ft = 1.01 × 10^6 ft-lb

Using Table 4.1, we find that 1 kc = 4.185 kJ, so the donut equivalent is

$$\frac{1.37 \times 10^6 \text{ J}}{150 \text{ kc/donut} \times 4.185 \text{ J/kc}} = 2.2 \text{ donuts}$$

At that point, all of the object's initial potential energy (P.E.) would be converted to kinetic energy (K.E.), so by substituting the value of h given in Equation 4.6 into Equation 4.5 we get the familiar equation for kinetic energy:

Eq. 4.7 $$P.E. = mgh = mg\left(\frac{v^2}{2g}\right) = \frac{1}{2}mv^2 = K.E.$$

4.3.3 Pressure Energy

The defining characteristic of energy is that it enables work to be done. A gas under pressure can force a piston to move and the pressure of water behind a dam can cause a turbine to spin, so pressure is, in essence, another form of potential energy.

The hydroelectric system shown in Figure 4.4 illustrates all three forms of mechanical energy just introduced. The energy in a hydroelectric system starts out as potential energy by virtue of its height above the powerhouse. Water under pressure in the penstock (connecting pipe) is able to do work when released, so there is energy associated with that pressure as well. Finally, as water flows, there is the kinetic energy of moving mass. The hydroelectric system transforms energy through all three forms, from potential, to pressure, to kinetic energy. A turbine and generator in the powerhouse convert that energy to electrical power.

SOLUTION BOX 4.2

Tapping into That Little Spring

Suppose you have a small spring up the hill from your cabin. You estimate it to be 50 feet higher than your house and you think you might be able to deliver about 40 gallons per minute to the powerhouse down near the cabin. You estimate the efficiencies of penstock, turbine, gearing, and generator to be 75%, 80%, 90%, and 90% respectively. How much power would your generator deliver and how much energy would it provide per month?

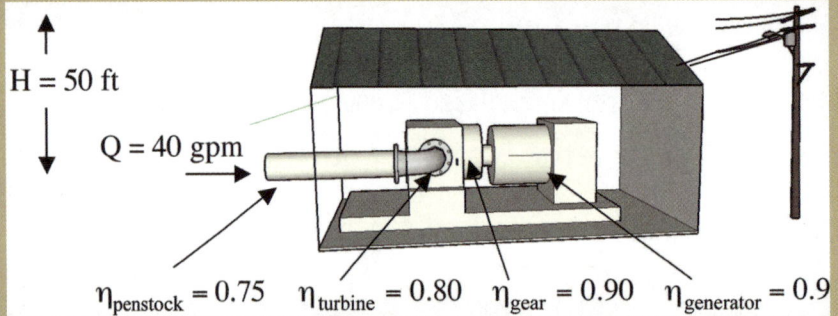

Solution:

From Equation 4.9, the overall efficiency of the hydropower system is

$$\eta = \eta_{penstock} \cdot \eta_{turbine} \cdot \eta_{gears} \cdot \eta_{generator} = 0.75 \times 0.80 \times 0.90 \times 0.90 = 0.49 = 49\%$$

and from Equation 4.8:

$$P(W) = \frac{\eta \, Q(gpm) \, H(ft)}{5.3} = \frac{0.49 \cdot 40 \cdot 50}{5.3} = 185 \text{ W} = 0.185 \text{ kW}$$

In a month with 30 days, the electric energy delivered would be

$$\text{Energy} = 0.185 \text{ kW} \times 24 \text{ hr/day} \times 30 \text{ day/mo} = 133 \text{ kWh/mo}$$

That's only about one-fourth as much electricity as a small house in the city might use, but this is just a cabin in the woods and you would probably have more than enough to keep your beer cold, play some music, and read well into the night.

Notice the hidden complication in this analysis. Suppose the refrigerator needs 1 kW to start its electric motor, but the generator is supplying only 0.185 kW? There isn't enough oomph to start the motor, and it will likely burn out trying. Your system may have sufficient energy to meet your average demand, but not enough power to meet the peak demand. We could solve this problem, however, by adding some batteries to the system, which could easily supply the peak demand.

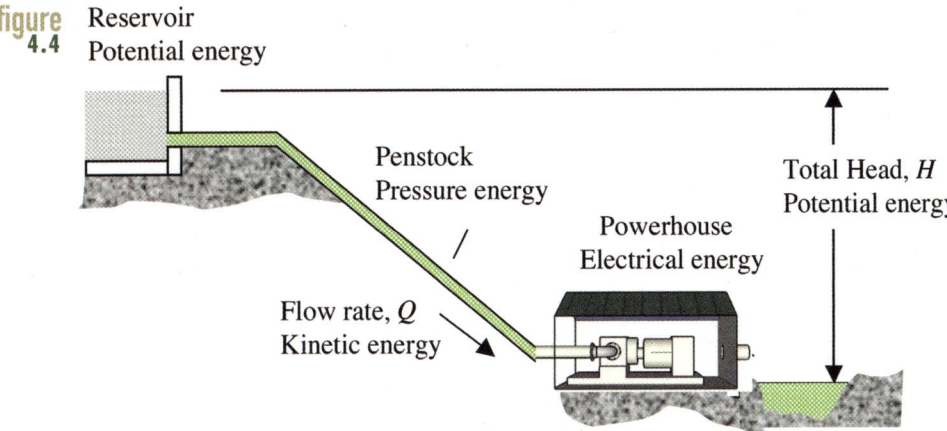

figure 4.4 Reservoir
Potential energy

A hydropower system converts energy from potential, to pressure, to kinetic, and then electrical energy.

The hydropower P potentially available from a site is proportional to the difference in elevation between the source and the turbine, called the *head* (H), times the rate at which water flows from one to the other, Q. In Chapter 10, we will work with the following handy relationship between these three quantities:

Eq. 4.8
$$P(W) = 9810\, \eta\, Q(m^3/s)\, H(m) = \frac{\eta\, Q(gpm)\, H(ft)}{5.3}$$

where η = overall efficiency of conversion from potential energy to electrical energy

That efficiency includes the friction losses in the piping, the turbine efficiency, the efficiency of gears or belts to match the rotational speed of the turbine to the needs of the generator, and the efficiency of the generator in converting rotational energy of its shaft to electrical power delivered to the load. Thus, the overall efficiency is

Eq. 4.9
$$\eta = \eta_{penstock} \cdot \eta_{turbine} \cdot \eta_{gears} \cdot \eta_{generator}$$

Solution Box 4.2 illustrates the use of these equations.

4.3.4 Rotational Energy

Although rotational energy is actually a form of kinetic energy, it is worth considering separately because it is so different from the usual concept of a mass moving along at some linear speed. Old internal combustion engines used to have big, heavy, rotating flywheels to keep the crankshaft turning while they waited for the next burst of exploding fuel. James Watt even had one on his original steam engines. We don't see a lot of these slow-moving flywheels these days, but there is, however, increasing interest in storing energy in high-speed, lightweight flywheels. These new flywheels might replace batteries in such applications as uninterruptible power supplies, backup storage for wind turbines or photovoltaics, or even electric vehicles.

Analogous to Equation 4.7, rotational energy can be written as

Eq. 4.10
$$K.E. \text{ (rotational)} = \frac{1}{2} I \omega^2$$

where I = the object's moment of inertia
ω = the object's rotational speed

For a wheel that is rotating, the relationship between its moment of inertia and its mass (m), radius (r), and and shape (k) results in the following, more useful K.E. expression:

Eq. 4.11
$$K.E. \text{ (rotational)} = \frac{1}{2} kmr^2\omega^2$$

A bicycle wheel, with essentially all of its mass at the outer perimeter, has $k = 1$, whereas a solid disk of uniform thickness would have $k = 0.5$.

Notice that old flywheels had a lot of mass, but very low rotational speeds. With new composite materials, it is possible to store a lot of energy in a relatively lightweight structure by taking advantage of the fact that energy stored increases as the square of the rotational speed. Thus, for example, a carbon composite flywheel spinning at 20,000 rpm can store 4 million times as much energy in the same mass as an old steel flywheel rotating at only 10 rpm.

Flywheels can be used as the heart of an uninterruptible power supply system. Utility power is used to spin up the flywheel, and utility power normally supplies the load (e.g., your building) directly. In the case of a power outage, energy is extracted from the flywheel, through a generator, to supply power to the load during the outage.

4.4 Thermal Energy

When we were talking about mechanical energy, we were focused on doing work. That is, actually moving things around—force times distance and that sort of thing. Now we want to talk about another form of energy, which in everyday terminology we call *heat*. To be a bit more careful, we should call it *thermal energy*, but for our purposes simply calling it heat will do.

4.4.1 Temperature

We all have a pretty good idea of what temperature is, but defining it is a bit trickier. For starters, realize the adjectives we use to describe temperature are often pretty vague. We may say it is "hot" when it is 100°F outside, but if that is the temperature of our coffee we would probably say it was "warm." And if 100°F were the temperature of a charcoal briquette after a barbeque, we would describe it as being "cool." The context leads us describe temperature with words, but just what does the numerical value mean? And just what is it that our measuring device, a thermometer for example, is actually measuring?

Somewhere along the way, you have learned that the temperature scales measured with an ordinary thermometer are based on the temperatures of freezing water, or more precisely, an ice-water mixture, and the boiling point of water (at one atmosphere of pressure). Using the *Celsius scale* (before 1948 it was called the centigrade scale), the freezing temperature is 0° and the boiling temperature is 100°. The Fahrenheit scale assigns the value of 32° to the freeze point and 212° to the boiling point. The relationship between the two is given by

Eq. 4.12 $$T(°F) = 1.8\ T(°C) + 32 \qquad \text{or} \qquad T(°C) = \frac{5}{9}[T(°F) - 32]$$

The Fahrenheit and Celsius scales are based on extrapolating temperatures to any values no matter how high or how low. But, in thermodynamics, there is merit in defining a temperature scale that isn't dependent on the properties of water and that does have an absolute minimum value of zero, below which it is impossible to go. In classical physics, *absolute zero* corresponds to the point at which all molecular motion stops. However, quantum mechanically, molecules cannot cease all motion because that would violate the Heisenberg uncertainty principle, which asserts that you cannot ever know for sure what a particle is doing. So at absolute zero there will still be a certain small, but nonzero, energy known as the zero-point energy.

There are two measurement scales that use absolute zero as their reference temperature. In the SI system it is the Kelvin scale, named after Lord Kelvin (1824–1907). The temperature unit in this scale is the *kelvin* (not degrees Kelvin or °K; the degree designation was officially dropped in 1967). On this scale, absolute zero corresponds to –273.15°C (we'll just round that to –273°C). Note that the temperature interval on the Kelvin scale is the same as on the Celsius scale. That is, a temperature difference of 10°C is the same as a temperature difference of 10 K.

In the English system the temperature scale is named after William Rankine (1820–1872), and the units are designated as R (rankine). Absolute zero corresponds to –459.67°F (which we will round to –460°F) and each degree change is the same on both the Rankine and Fahrenheit scales. That is, a change of 10 R is the same as a change of 10°F.

The following are handy conversions when using an absolute temperature scale:

Eq. 4.13 $$T(K) = T(°C) + 273 \qquad \text{and} \qquad T(R) = T(°F) + 460$$

4.4.2 Internal Energy, Thermal Capacitance

If we turn on the burner under a pot of water we know that energy will be added to the water, raising its temperature. But there will be no work done. That is, no force times distance.

That suggests that there are at least two ways to change the energy of a system: we can move something, doing work, or we can transfer heat thereby changing the internal energy of the system.

Heat can be defined as the energy that is transferred between two systems (the stove and the pot of water) by virtue of a temperature difference. If we want to be picky, we shouldn't say an object has some heat in it, nor should we say we are adding some heat or taking some heat out of something. Objects don't contain heat, even after they have been heated up. What they do contain is referred to as *thermal energy*. Confusing? You bet. The technicality has to do with the "transfer" part of the definition; that is, heat only exists due to the temperature difference between two objects, in which case it is the energy that moves from one to the other. Fortunately, we don't want to be picky in this book, so you can relax and use your intuitive understanding of heat.

Potential and kinetic energy are observable, macroscopic forms of energy that are easily visualized and readily understood. More difficult to envision are microscopic forms related to the atomic and molecular structure of the system being studied. These microscopic forms of energy include the kinetic energies of molecules (that we measure with a thermometer) and the energies associated with the forces acting between molecules, between atoms within molecules, and within atoms. The sum of those microscopic forms of energy is called the system's *internal energy* and is represented by the symbol U. The total energy (E) that a substance possesses can be described then as the sum of its potential energy (P.E.), kinetic energy (K.E.), and its internal energy (U).

Eq. 4.14
$$E = U + \text{K.E.} + \text{P.E.}$$

The first law of thermodynamics can now be rephrased to say that for a *closed system*, that is one in which we don't have to worry about matter passing through the boundary, if we add heat to the system (Q), and the system does work (W), that the net result will be equal to the changes in the internal energy (ΔU), kinetic energy (ΔK.E.) and potential energy (ΔP.E.).

Eq. 4.15
$$Q - W = \Delta U + \Delta \text{K.E.} + \Delta \text{P.E.}$$

Quite often, it is changes in a substance's internal energy caused by changing its temperature that are of interest. For example, we may want to heat some water and store it in a tank so we can take a nice, hot shower, or we may want to design a house with lots of thermal mass to absorb incoming solar energy in the daytime, store it, and then give it back later to help keep the house warm overnight.

For liquids and solids at atmospheric pressure, the change in energy stored (ΔE) when its mass (m) undergoes a temperature change (ΔT) is given by

Eq. 4.16
$$\text{Change in energy stored} = \Delta \text{E} = m \, c \, \Delta T$$

where c = specific heat of the substance

table
4.3
Specific Heat and Volumetric Heat Capacity of Selected Substances

| Substance | Specific Heat | | Density | Heat Capacity |
	(kJ/kg °C)	(Btu/lb °F)	(lb/ft^3)	(Btu/ft^3 °F)
Water	4.18	1.00	62.4	62.4
Air 20°C	1.01	0.24	0.081	0.019
Aluminum	0.90	0.22	168	37
Concrete*	0.88	0.21	144	30
Copper	0.39	0.09	555	50
Dry soil*	0.84	0.20	82	16
Gasoline	2.22	0.53	42	22
Steel*	0.46	0.11	487	54

* Representative values

The specific heat is the energy required to raise a unit of mass by one degree. For example, the specific heat of water is

Eq. 4.17   **Specific heat of water c = 1 Btu/lb°F = 4.18 kJ/kg°C**

Table 4.3 provides some examples of specific heat for several selected substances. It is worth noting that water has by far the highest specific heat of the substances listed; in fact, it is higher than almost all other common substances. This is one of water's very unusual properties and is in large part responsible for the major effect the oceans have on moderating temperature variations of coastal areas.

Also included in Table 4.3 are representative values of density along with the product of density and specific heat, which is known as the *volumetric heat capacity*. Volumetric heat capacity is an important concept in that it tells us how much thermal energy can be stored in a given volume of material as we raise its temperature. Quite often the design challenge is to find a way to pack away as much heat as possible in as small a space as you can. Notice again how water stands out as the substance that stores the most heat in the least volume—more than twice as much as concrete, which is the other most commonly used substance to store heat in passive solar houses.

For an example of how to use these concepts, see Solution Box 4.3.

4.5 Chemical Energy

As you recall from basic science, matter is made up of molecules; molecules are made up of atoms; and atoms are made up of protons, neutrons, and electrons. Although physicists are trying to understand even smaller particles, for our purposes we can stop there and just

SOLUTION BOX 4.3

Storing Heat in Concrete or Water

On a clear, winter day 200 ft² of south-facing window allows 1000 Btu/ft² of solar energy to hit and be absorbed by some thermal mass in a passive solar house. You want to store that energy during the day and release it at night to help keep your house warm. You could use tubes of water behind the window to store the energy or you could let it be absorbed in a concrete floor slab. What volume of each would be required if the sun heats those masses by 20°F? (Ignore any daytime heat losses from the mass.)

Solution:

The total amount of energy to be stored is 200 ft² × 1000 Btu/ft² = 200,000 Btu. Using values from Table 4.3, we find

$$\text{Volume of water} = \frac{200,000 \text{ Btu}}{62.4 \, \frac{\text{Btu}}{\text{ft}^3 \cdot \text{°F}} \times 20\text{°F}} = 160 \text{ ft}^3$$

$$\text{Volume of concrete} = \frac{200,000 \text{ Btu}}{30 \, \frac{\text{Btu}}{\text{ft}^3 \cdot \text{°F}} \times 20\text{°F}} = 333 \text{ ft}^3$$

So, less than half as much water would be required.

If that concrete were 6 inches thick, it would need to cover 667 ft² of floor area. If those water columns were each 8 feet high and 1 foot in diameter, it would take 25 of them.

consider how the energy associated with chemical reactions among atoms and molecules can be captured and utilized.

4.5.1 Atoms and Molecules

You also remember that protons and neutrons form the nucleus of an atom, with the number of protons, called the *atomic number*, identifying which chemical element this is. Protons and neutrons have nearly the same mass (neutrons are very slightly heavier), but the proton carries a positive charge whereas the neutron is electrically neutral. The sum of the number of protons and neutrons is called the *mass number*. Lumped together, the protons and neutrons in a nucleus are referred to as *nucleons*.

Surrounding the nucleus is a swarm (well, just one for hydrogen) of very light, negatively charged electrons, equal in number to the number of positively charged protons, which normally results in an electrically neutral atom. Most of the volume of an atom is empty space. For example, if the orbit of the outer ring of electrons of a typical atom were enlarged to the size of the entire Earth, its nucleus would be only a few hundred feet across.

All atoms with the same chemical name (such as helium) have the same number of protons, but not all such atoms have the same number of neutrons. Elements with the same atomic number but differing mass numbers are called *isotopes*. For example, helium always has 2 protons (by definition), but it may have anywhere from 1 to 8 neutrons. Almost always it has only 1 neutron (Helium-3), roughly one in a million helium atoms has two neutrons (Helium-4), while isotopes with more than 2 neutrons are unstable and occur only for short periods of time during nuclear reactions.

The most common way to describe a given isotope is by giving its chemical symbol with the mass number written at the upper left and the atomic number at the lower left. For example, the two most important isotopes of uranium (which has 92 protons) are

$$^{235}_{92}\text{U} \qquad \text{and} \qquad ^{238}_{92}\text{U}$$
$$\text{Uranium-235} \qquad\qquad \text{Uranium-238}$$

When referring to a particular element, it is common to drop the atomic number subscript because it adds little information to the chemical symbol. Thus "U-238" and "Helium-3" are common ways to describe those isotopes.

When atoms are close enough together that their outer electrons can interact with each other, forces can be set up between those atoms that are strong enough to hold them together. Nearby atoms share outer electrons forming *covalent bonds*, with each of a pair of atoms providing an electron that the other shares. *Molecules*, then, are made up of atoms held together by these covalent bonds.

4.5.2 Solids, Liquids, and Gases

The forces of attraction between atoms within a molecule are very strong. Molecules, in turn, exert small, but appreciable attractions on one another, which are called *van der Waals attractions* (after nineteenth-century Dutch physicist Johannes van der Waals). That is, molecules are slightly "sticky." At low temperatures, molecules are rather docile; that is, they don't have much kinetic energy, and those sticky forces are sufficient to hold them pretty much in place in an orderly array. These molecules are said to be in *solid, crystalline state*.

When those molecules are heated, they begin to vibrate more and at some point, called the *melting point* temperature, they have acquired sufficient energy to break loose from the crystal and can now begin to slide around past each other. These warmed molecules are still touching, but they are now mobile enough to fill out the shape of the container in which they are held. That is, the substance has made a transition from the solid state to the liquid state. The energy needed to melt a substance is called the *latent heat of fusion*. For example, the latent heat of fusion required to melt ice is 333 kJ/kg (144 Btu/lb). It is also the amount of heat that has to be removed from the substance to cause it to go from the liquid to the solid state; that is, to cause it to freeze (see Solution Box 4.4).

If still more heat is added, molecules can acquire so much kinetic energy that those puny van der Waals attractions can no longer compete, at which point the molecules can fly off in any direction they want to. This is the gas phase. A gas adapts to the shape of its container, just as liquids do, but gases differ in that they are easily expanded or compressed. As we shall see later when we talk about air-conditioning systems, the expansion and compression of gases is of crucial importance. The temperature at which the transition from the liquid to gaseous state occurs is called the *boiling point* temperature, which for water is 100°C (212°F). The energy required to do so is called the *latent heat of vaporization*. The latent heat of vaporization needed to cause water at 100°C to make the transition to steam at 100°C is 2257 kJ/kg (972 Btu/lb).

4.5.3 Stoichiometry: Mass Balance in Chemical Reactions

The principle of conservation of mass can be applied to a chemical reaction to tell us how much of each compound is involved to produce the results shown. The balancing of equations so that the same number of each kind of atom appears on each side of the equation, and the subsequent calculations to determine amounts of each compound involved, is known as *stoichiometry*.

Consider the following simple reaction in which methane (CH_4) is oxidized (burned) to produce carbon dioxide and water. Because a significant fraction of our energy comes from natural gas, and because natural gas is mostly methane, this reaction is extremely important.

Eq. 4.18
$$CH_4 + 2\,O_2 \rightarrow CO_2 + 2\,H_2O$$

Let's use Equation 4.18 to help us review a little bit of chemistry. First, notice the reaction as written is balanced; that is, there are as many atoms of carbon, hydrogen, and oxygen on the left side of the reaction as there are on the right.

Now, we can interpret the equation to say that 1 molecule of methane reacts with 2 molecules of oxygen to produce 1 molecule of carbon dioxide plus 2 molecules of water. It is of more use, however, to be able to describe this reaction in terms of the mass of each substance (for example, How many grams of CO_2 are put into the atmosphere when so many

---

Cooling Your Room with a Ton of Ice

Suppose you want to cool your room by dragging in a ton of ice (2000 lb) and letting it melt. How much heat would the ice need to take from the room to cause it all to melt? If it took 24 hours for that to happen, what average rate of cooling would you have achieved?

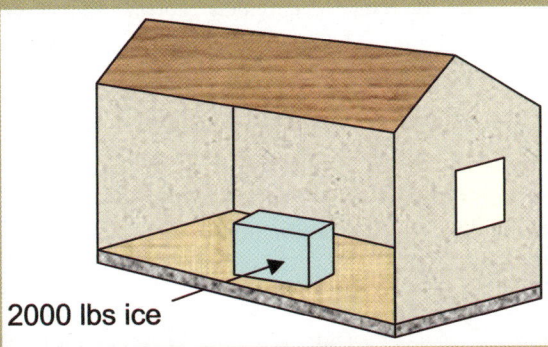

2000 lbs ice

Solution:

We have learned that the latent heat of fusion for water is 144 Btu/lb, so the amount of heat needed to melt 1 ton of ice would be

$$2000 \text{ lb} \times 144 \text{ Btu/lb} = 288,800 \text{ Btu}$$

If it takes 24 hours to melt that ton of ice, the average rate of heat removal from the room, and hence cooling of the room, would be

$$\frac{288,800 \text{ Btu}}{24 \text{ hr}} = 12,000 \text{ Btu/hr}$$

As it turns out, the standard way to rate the cooling capacity of an air conditioner in the United States is by the tons of cooling it can provide, where 1 ton = 12,000 Btu/hr. Thus, for example, a 3-ton home air conditioner provides 36,000 Btu/hr of cooling, equivalent to melting 3 tons of ice per day.

---

grams of CH_4 are burned?, and so forth). To do so, we need to define the atomic weight and molecular weight of atoms and molecules, and we need to lump large numbers of molecules into chunks called *moles*.

The *atomic weight* of an element is expressed in *atomic mass units* (amu), where one amu is defined to be exactly one-twelfth the mass of carbon-12. (C-12 has 6 protons and 6 neutrons.) Why bother measuring mass in amu's rather than just expressing it with the mass number (total protons plus neutrons)? To answer that, recall that chemical elements often have various isotopes in nature—carbon, for example, comes with 6 neutrons in most atoms (C-12), but some have 8 neutrons (C-14). That suggests the naturally occurring mix of carbon should have an atomic weight a bit higher than 12. Because the mass number and the atomic weights of an element are so similar, we will henceforth ignore that distinction and follow the somewhat standard engineering practice of rounding off the atomic weights (e.g., C will be 12 amu rather than 12.011).

The molecular weight of a molecule is simply the sum of the atomic weights of the constituent atoms. Thus, the (slightly rounded) molecular weight of CH_4 is $12 + 4 \times 1 = 16$ amu. If we divide the mass of a substance by its molecular weight the result is the mass expressed in moles (mol). For example, 32 grams of CH_4, divided by 16 g/mol, is 2 moles of methane.

The special advantage of expressing chemical reactions using moles is that one mole of any substance contains exactly the same number of molecules (1 g-mole = 6.02×10^{23} molecules), which gives us another way to interpret a chemical reaction such as that given in Equation 4.18:

$$CH_4 + \quad 2\,O_2 \rightarrow \quad CO_2 \quad + \quad 2\,H_2O$$
$$1 \text{ mol } CH_4 + 2 \text{ mol } O_2 \rightarrow 1 \text{ mol } CO_2 + \quad 2 \text{ mol } H_2O$$

We can now describe the reaction by saying that one mole of CH_4 reacts with two moles of O_2 to produce one mole of CO_2 and 2 moles of H_2O. We can translate that into grams by first converting each constituent into moles using their (rounded) atomic weights:

$$CH_4 = 12 + 4 \times 1 = 16 \text{ g/mol}$$
$$O_2 = 2 \times 16 = 32 \text{ g/mol}$$
$$CO_2 = 12 + 2 \times 16 = 44 \text{ g/mol}$$
$$H_2O = 2 \times 1 + 16 = 18 \text{ g/mol}$$

We now have a third way of expressing the oxidation of methane:

$$CH_4 + \quad 2\,O_2 \quad \rightarrow CO_2 \quad\quad + \quad 2\,H_2O$$
$$16 \text{ g } CH_4 + 64 \text{ g } O_2 \rightarrow 44 \text{ g } CO_2 \quad + \quad 36 \text{ g } H_2O$$

Notice that mass is conserved in this last expression; that is, there are 80 grams on the left and 80 grams on the right.

4.5.4 Enthalpy: The Energy Side of Chemical Reactions

Most of the energy we use to power our industrial societies is obtained by burning fossil fuels; mainly, coal, oil, and natural gas. Intuitively, we know that burning a fuel converts energy stored in chemical bonds into heat that we can use to do work. Chemical reactions in which heat is liberated, such as those that occur when fuel is burned, are called *exothermic* reactions. Reactions that want to go in the other direction; that is, in which heat is required to make them happen, are called *endothermic* reactions.

Just as we use stoichiometry to do mass balances on chemical reactions we can use something called *enthalpy* to help us do energy balances. As is often the case with thermodynamic properties of substances, the precise definition of enthalpy is rather subtle and does not lend itself to simple interpretation. One way to think about it, however, is that it is a measure of the energy that it takes to form a substance out of its constituent elements, in which case it is called the *enthalpy of formation*.

Table 4.4 lists some enthalpies of formation for selected substances. Notice that gaseous oxygen (O_2) and hydrogen (H_2) are listed as having zero enthalpy. Enthalpy needs a reference point, just as other forms of energy do (think potential energy), and the zero point on the enthalpy scale is applied to the stable form of a chemical element under standard temperature and pressure (STP) conditions (25°C and 1 atmosphere of pressure). Because, for example, the stable form of oxygen under STP conditions is molecular O_2, it is given an enthalpy value of zero, but atomic oxygen O, which is not stable, does not have zero enthalpy.

Also notice that enthalpies depend on the state of the substance; that is, the enthalpy of liquid water is different from that of gaseous water vapor. In the case of H_2O, that difference is very important, as we shall see later.

In chemical reactions, the difference between the enthalpies of the products (right side) and the reactants (left side) tells us how much energy is released or absorbed in the reaction. When there is less enthalpy in the final products than in the reactants, heat is liberated; that is, the reaction is exothermic.

By combining stoichiometric and enthalpy analyses of a chemical reaction we can, with modest effort, determine such important characteristics of a fuel as the energy and carbon released when the fuel is burned. This is demonstrated in Solution Box 4.5.

table 4.4   Enthalpy of Formation for Selected Substances (STP)

Substance	Formula	State	Enthalpy, H (kJ/mol)
Hydrogen	H_2	Gas	0
Oxygen	O	Gas	247.5
Oxygen	O_2	Gas	0
Water	H_2O	Liquid	−285.8
Water vapor	H_2O	Gas	−241.8
Methane	CH_4	Gas	−74.9
Carbon dioxide	CO_2	Gas	−393.5
Methanol	CH_3OH	Liquid	−238.7
Ethanol	C_2H_5OH	Liquid	−277.7
Propane	C_3H_8	Gas	−103.9
Octane	C_8H_{18}	Liquid	−250.0
Glucose	$C_6H_{12}O_6$	Solid	−1260

SOLUTION BOX 4.5

Heat and Carbon Emissions from Combustion of Methane

To demonstrate the use of enthalpy, let us return to the combustion of methane given in Equation 4.18 but now let us write the associated enthalpies under each substance as given in Table 4.4. Notice we have to decide whether the water released will be in the vapor form or liquid. In this case, we assume it is vapor (later we'll see that this means our result will be what is called the lower heating value, LHV, of methane).

$$CH_4 + 2\,O_2 \rightarrow CO_2 + 2\,H_2O(g)$$
$$(-74.9) \quad 2 \times (0) \quad (-393.5) \quad 2 \times (-241.8)$$

The change in enthalpy associated with this reaction is then

$$\Delta H = (\Sigma\,H \text{ of products}) - (\Sigma\,H \text{ of reactants})$$
$$\Delta H = [-393.5 + 2 \times (-241.8)] - [(-74.9) + 2 \times (0)] = -802.2 \text{ kJ/mol of } CH_4$$

Because the enthalpy change is negative, that means the reaction is exothermic. It releases 802.2 kJ of heat for every mole (16 g) of methane burned; that is, 50.14 kJ/g.

To determine carbon emissions, we note that one mole of CH_4 (16 g) produces one mole of CO_2 (44 g). Combining that with the 50.14 kJ of energy released per gram of CH_4 gives

$$\frac{44 \text{ g } CO_2}{12 \text{ g } CH_4} \times \frac{\text{g } CH_4}{50.14 \text{ kJ}} = 0.05485 \text{ g } CO_2/\text{kJ} = 54.85 \text{ kg } CO_2/\text{MJ}$$

Quite often, emissions are expressed in kg of C (not CO_2) per MJ of energy. That's easy to fix. Because CO_2 has 12 g C per 44 g CO_2, we can adjust the emission rate to be

$$\frac{54.85 \text{ kg } CO_2}{MJ} \times \frac{12 \text{ kg C}}{44 \text{ kg } CO_2} = 15.0 \text{ kg C/MJ}$$

4.5.5 Heat of Combustion: HHV and LHV

When a fuel is burned, the magnitude of the enthalpy difference $\Delta(H)$ between the reactants and the products is called the *heat of combustion* (or, sometimes, *heat content*) of the fuel. Table 4.5 presents some values for various commonly used fuels, expressed in typical U.S. units.

Notice that Table 4.5 qualifies the heat content of those fuels by saying it is the HHV value. Also given is an HHV to LHV ratio. What does that mean?

table 4.5   **Heat of Combustion for Selected Fuels (Average HHV Values)**

Fuel	Heat of Combustion (HHV)	HHV/LHV
Coal (bituminous)	14,000 Btu/lb	1.05
Fuel ethanol	84,262 Btu/gallon	1.11
Fuel oil (#2)	140,000 Btu/gallon	1.06
Gasoline	125,000 Btu/gallon	1.07
Hydrogen	61,400 Btu/lb	1.18
Kerosene	135,000 Btu/gallon	1.06
Methanol	64,600 Btu/gallon	1.14
Natural gas	1025 Btu/cu. ft.	1.11
Pellets (for pellet stove; premium)	8250 Btu/lb	1.11
Propane	91,330 Btu/gallon	1.09
Wood (20% moisture)	7000 Btu/lb	1.14

* For comparison, the ratios of HHV to LHV are given as well.

Recall from section 4.5.2 that it takes quite a lot of energy, called the latent heat of vaporization, to cause water to change state from a liquid to a gas. When a fossil fuel is burned, some of its energy ends up as latent heat in the water vapor produced. Whether the latent heat is, or is not, included leads to two different values of the heat of combustion. The higher heating value HHV, also known as the *gross* heat of combustion, includes the latent heat of the water vapor. That is, HHV includes all of the heat that could possibly be captured if you could condense the water vapor and utilize it for your process. The lower heating value (LHV), also known as the *net* heat of combustion, does not include latent heat. For example, the 802.2 kJ/mol calculated for the combustion of methane in Solution Box 4.5 didn't include condensing the water vapor, so that is the LHV for methane. If we were to repeat that example using the enthalpy of liquid water, an HHV of 890.2 kJ/g for methane would have been found.

Solution Box 4.6 illustrates the enthalpy calculations that lead to both the HHV and LHV of hydrogen. Notice the difference between the LHV and HHV values is about 18%, which is pretty sizeable. For comparison, the difference between LHV and HHV for natural gas is about only 11%, whereas for coal it is less than 6%. In general, the simpler the fuel molecule—that is, how little carbon it contains—the more important it becomes to be careful about distinguishing between LHV and HHV.

When calculating efficiencies of various processes, it is important to specify whether the answers obtained are based on LHV or HHV—especially when there are significant differences between the two systems, as illustrated in Solution Box 4.7.

Unfortunately, the distinction between LHV and HHV often leads to considerable confusion. If you know what LHV and HHV mean, and it is not mentioned in someone's description of an energy transformation, you could be somewhat uncertain about how to interpret that person's results. On the other hand, if it is mentioned and you don't know anything about

SOLUTION BOX 4.6

HHV and LHV for Hydrogen

The oxidation of hydrogen, either by burning, or perhaps through a chemical reaction in a fuel cell, can be described as follows:

$$\text{LHV} \qquad H_2 + \frac{1}{2}O_2 \rightarrow H_2O \text{ (gas)} \qquad \Delta H = -241.8 \text{ kJ/mol}$$

$$\text{HHV} \qquad H_2 + \frac{1}{2}O_2 \rightarrow H_2O \text{ (liquid)} \qquad \Delta H = -285.8 \text{ kJ/mol}$$

Notice how easy it was to write those reactions. Because the enthalpy of formation of H_2 and of O_2 is zero, the overall energy released is just the numerical value of the enthalpy of formation of water vapor (for LHV) or liquid water (for HHV).

Because there are 2 g per mole of hydrogen, its LHV is 241.8/2 = 120.9 kJ/g and its HHV is 285.8/2 = 142.4 kJ/g.

LHV and HHV, you again could be left scratching your head. To compound the problem, all energy data given the Energy Information Administration's *Annual Energy Reviews* (the principal source in the United States) are HHV values, whereas most of the rest of the world uses LHV as its standard. Moreover, even in the United States, for small power plants—fuel cells, microturbines, and the like—LHV is often used. So, which makes more sense?

The argument many make for LHV values is that the latent heat in the water vapor produced during combustion is almost always lost out the stack, along with the other combustion gases, so, it shouldn't be counted. On the other hand, there are, for example, very efficient home furnaces these days that capture latent heat by purposely cooling exhaust gases enough to cause water vapor to condense. On an HHV basis, these condensing furnaces can have efficiencies of well over 90%. If we were to use the LHV basis for the fuel, such furnaces can have an efficiency of over 100%, which sounds a little strange even though it is correct.

The bottom line is that for casual computations, in most circumstances the difference between LHV and HHV is only a few percent and can usually be ignored. When more precision is required the most important thing is to be consistent and base everything on either LHV or HHV, and be careful to specify which system you are using.

4.6 Solar Energy

The source of energy that keeps our planet at just the right temperature, powers our hydrologic cycle, creates our wind and weather, and provides our food and fiber is a relatively insignificant yellow dwarf star some 93 million miles away. Powering our sun are thermonu-

---

SOLUTION BOX 4.7

LHV and HHV Efficiency of a Fuel Cell

Suppose a fuel cell converts 80 g of hydrogen into 1 kWh (3600 kJ) of electricity. Using the lower heating value (LHV) for H_2, what is the energy efficiency of the fuel cell? What is its efficiency on an HHV basis?

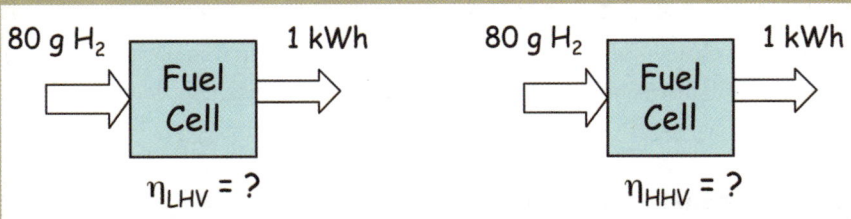

Solution: Using the results found in Solution Box 4.6, 80 g of hydrogen can provide 120.9 kJ/g × 80 g = 9672 kJ on an LHV basis (no condensation). The energy delivered by the fuel cell is 1 kWh, which is 3600 kJ (Table 4.1). So, on an LHV basis the fuel-cell efficiency would be

$$\text{LHV efficiency} = 3600 \text{ kJ}/9672 \text{ kJ} = 0.372 = 37.2\%$$

On an HHV basis, 80 g of H_2 can provide 142.4 kJ/g × 80 g = 11,392 kJ. So the HHV efficiency is

$$\text{HHV efficiency} = 3600 \text{ kJ}/11,392 \text{ kJ} = 0.316 = 31.6\%$$

Thus, the same fuel cell, producing the same output from the same amount of hydrogen, could be described as having an efficiency of either 37.2% or 31.6% depending on whether efficiency is determined using an LHV or HHV basis. Most often, fuel-cell efficiencies are, perhaps not surprisingly, stated using the LHV basis.

---

clear reactions in which hydrogen atoms fuse together to form helium. In the process, about 4 billion kilograms of mass per second are converted into energy following Einstein's famous relationship $E = mc^2$. This fusion has been continuing reliably for the past 4 or 5 billion years and is expected to continue for another 4 or 5 billion.

At the center of the sun, temperatures are estimated to be on the order of 15 million K, but those incredibly high temperatures are degraded as energy works its way some 400,000 miles from the core to the surface. The surface of the sun radiates about 3.8×10^{20} MW of electromagnetic energy outward toward space with wavelengths that correspond closely to what would be expected from an object around 5800 K.

Just outside of our atmosphere, the Earth receives an average of about 1.35 kW/m² of radiant energy. Some of that energy is reflected back into space and some is absorbed in the atmosphere, and of course half of the time at every spot on Earth there is no sunlight at all. The result is that an average of about 160 watts per square meter of solar energy actually reaches the Earth's surface. To put that into perspective, the sunlight falling onto roads in just the United States alone is equivalent to the entire global rate of consumption of fossil fuels. And, some claim that no country on Earth uses as much energy for all purposes as is contained in the sunlight striking its buildings.

Subsequent chapters will explore the conversion of sunlight into useful forms of energy, but for now we will just examine the characteristics of electromagnetic energy itself.

4.6.1 Electromagnetic Radiation

Electromagnetic radiation can be described in terms of discrete zero-mass particles of energy, called photons, or in terms of electromagnetic waves of various wavelengths and frequencies. Both are correct, and the descriptor to use is mostly a matter of which one is more convenient for the particular phenomenon being described. The origin of electromagnetic radiation is the motion of electrically charged particles. It occurs when atoms collide with each other or when absorption of incoming photons causes atoms to temporarily transition to higher-energy, unstable states. As excited atoms relax, photons are released. Radiation can also be the result of nuclear decay reactions or other nuclear and subnuclear processes. Electromagnetic radiation can travel through empty space, or through air or other substances.

When described as a wave phenomenon, such as is shown in Figure 4.5, the wavelength and frequency of vibration of electromagnetic radiation are related as follows:

Eq. 4.19
$$\lambda = \frac{c}{v}$$

where λ = wavelength (m)
         c = speed of light (3×10^8 m/s)
         v = frequency (hertz, i.e., cycles per second)

When radiant energy is described in terms of photons, the relationship between frequency and energy is given by

Eq. 4.20
$$E = h v$$

where E = energy of a photon (J)
         h = Planck's constant (6.6×10^{-34} J-s)

Equation 4.20 indicates that photons with higher frequency (shorter wavelengths) have higher energy.

figure
4.5

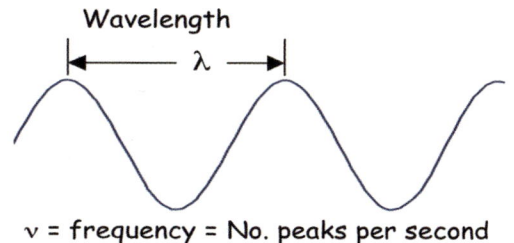

Electromagnetic radiation as a wave phenomenon is described in terms of frequency and wavelength.

Radio waves, microwaves, visible light, and x-rays are all examples of electromagnetic radiation. Each is characterized by wavelengths that range over some portion of the electromagnetic spectrum. As shown in Figure 4.6, the range extends from long-wave radio waves, with wavelengths thousands of meters long, to gamma rays, with wavelengths down in the vicinity of 10^{-12} m. The solar portion covers only a small fraction of these wavelengths, 200 to 2500 nanometers (1 nm = 10^{-9} m), with the visible portion—the part that enables us to see things—extending from about 400 to 700 nm.

4.6.2 The Solar Spectrum

Every object emits thermal radiation, the characteristics of which depend on the object's temperature. The usual way to describe how much radiant energy the object emits, as well as the characteristic wavelengths of that electromagnetic energy, is to compare it to a theoretical abstraction called a *blackbody*. A blackbody is defined to be a perfect emitter and a perfect absorber. As a perfect emitter, it radiates more energy per unit of surface area than any real object at the same temperature. As a perfect absorber, it absorbs all radiant energy that strikes its surface; that is, none is reflected and none is transmitted through it.

figure
4.6   **The Electromagnetic Spectrum**

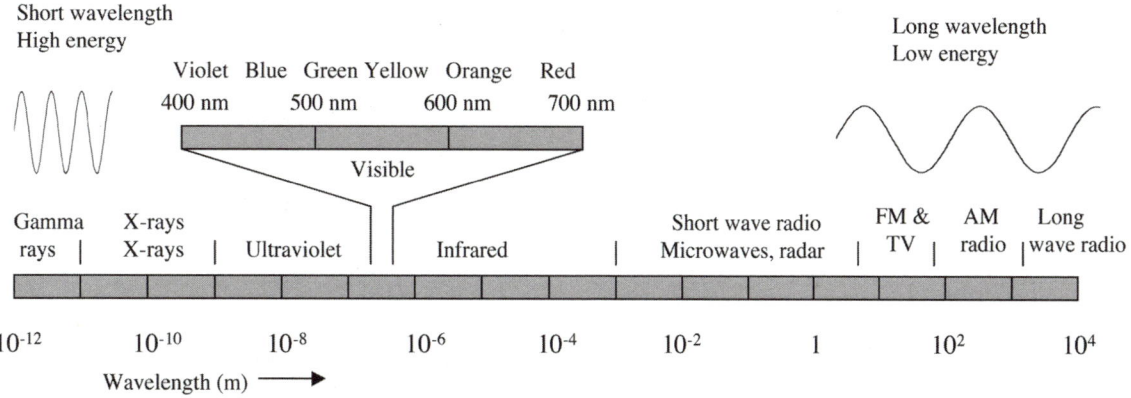

The total rate at which energy is emitted from a blackbody with surface area A and absolute temperature T is given by the *Stefan-Boltzmann law of radiation*:

Eq. 4.21 $$E = \sigma A T^4$$

where E = total blackbody emission rate (W)
σ = the Stefan-Boltzmann constant = 5.67×10^{-8} W/m^2 – K^4
T = absolute temperature (kelvins K, where K = 273 + °C)
A = surface area of the object (m^2)

Actual objects do not emit as much radiation as our hypothetical blackbody, but most are quite close to this theoretical limit. The ratio of the amount of radiation an actual object emits, to that of a blackbody, is called the object's *emissivity*. The emissivity of desert sand, dry ground, and most woodlands is estimated to be 0.90, whereas water, wet sand, and ice have an emissivity of roughly 0.95. A human body, no matter what pigmentation, has an emissivity of around 0.96.

Although Equation 4.21 gives us the total emissive power of a blackbody, it doesn't describe the range of wavelengths associated with that radiation. A more complex equation, known as *Planck's law*, provides that spectral distribution of energy, an example of which is shown in Figure 4.7. In that figure, the spectrum of a 5800 K blackbody is compared with the actual spectrum of solar energy arriving just outside of Earth's atmosphere. The closeness

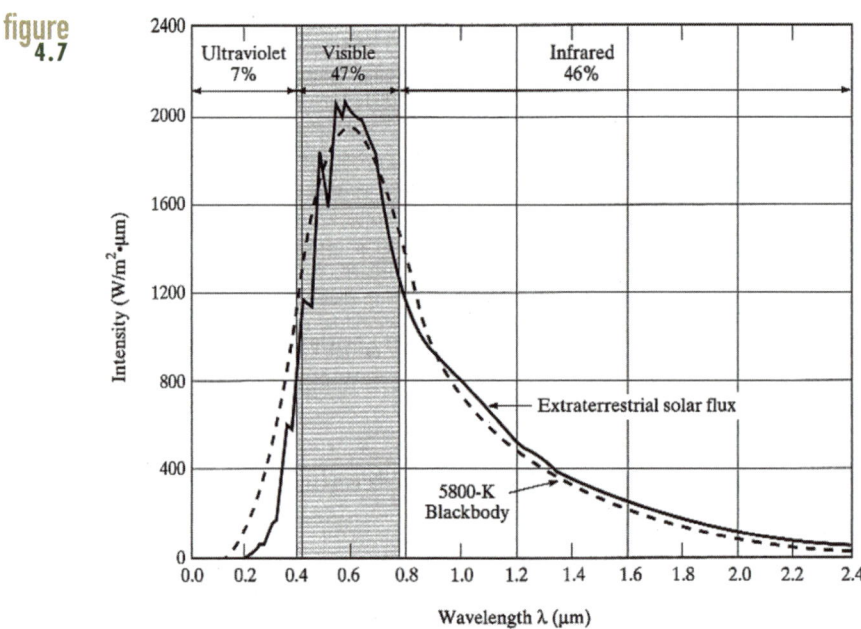

figure 4.7

The extraterrestrial solar spectrum (solid line) compared with the spectrum of a 5800 K blackbody (dashed). Also shown is the fraction of solar energy that falls with the ultraviolet, visible, and infrared portions of the solar spectrum.

of the match demonstrates quite clearly that the sun can quite reasonably be mathematically modeled as if it were a 5800 K blackbody.

The area under a spectral distribution curve between any two wavelengths is the total radiant power within that region. In Figure 4.7, the solar spectrum is divided into three regions: ultraviolet, visible, and infrared. Roughly half of the energy in the extraterrestrial solar spectrum (47%) is contained within the visible wavelengths (0.38 to 0.78 μm). The ultraviolet (UV) is only 7%, but because those are such short wavelength photons they pack a more energetic punch and can be especially damaging to living things. Fortunately, stratospheric ozone filters most of those dangerous wavelengths out of the spectrum before they reach the Earth's surface (unless you happen to live under the seasonal ozone hole). The infrared portion makes up 46% of the total. These photons help keep us warm, but they don't contribute to our ability to see anything. Later, when we describe energy-efficient building technologies, we will see that window coatings for office buildings can be designed to transmit just the visible wavelengths for natural daylighting, while blocking out the UV to help keep fabrics from fading, and reflecting the infrared to help reduce the air-conditioning load.

4.6.3 The Greenhouse Effect

The wavelength at which the spectrum peaks is sometimes a handy way to characterize blackbody radiation. Objects with higher temperatures have their peak at a shorter wavelength as described by *Wien's displacement rule*:

Eq. 4.22
$$\lambda_{max}(\mu m) = \frac{2898}{T(K)}$$

This equation predicts that the sun, at 5800 K, has a peak at about 0.5 μm, which agrees with Figure 4.7. The Earth, with its surface at about 15°C (298 K), should show its spectral peak at about 10 μm, which it does (Figure 4.8).

That large difference between incoming solar wavelengths (short) and outgoing wavelengths radiated to space from the Earth's surface (long) is critical to understanding the greenhouse effect. As it turns out, our atmosphere is relatively transparent to the short wavelengths coming from the hot sun, but it is relatively opaque to longer infrared wavelengths trying to work their way from the Earth's surface, through the atmosphere, and back to outer space. By preferentially absorbing outgoing radiation, greenhouse gases, including CO_2, CH_4, N_2O and water vapor, act like an insulating blanket around the Earth causing it to be about 19°C warmer than it would be without those gases. As Solution Box 4.8 demonstrates, without the naturally occurring greenhouse effect, the Earth would likely be a virtually uninhabitable frozen planet. Our current worry about global warming is of course based on human interactions that are increasing the greenhouse effect, with uncertain, but potentially devastating, consequences.

The calculation in Solution Box 4.8 indicates that without the greenhouse effect the average temperature of the Earth would be about 34°C below its actual 15°C temperature. That is, the Earth would be well below zero and all water would be frozen solid.

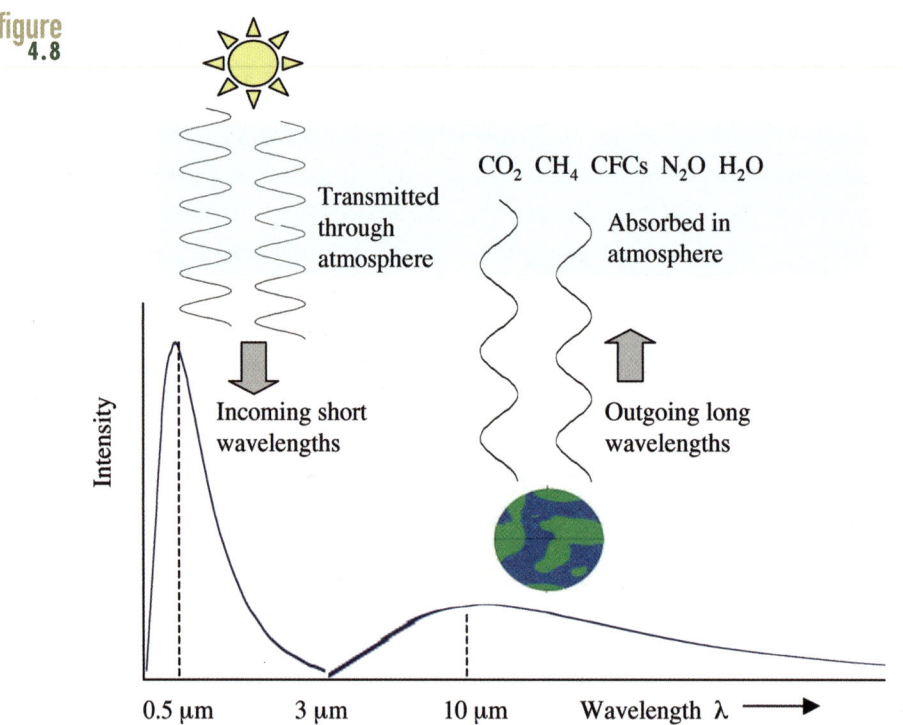

Incoming solar wavelengths more easily pass through the atmosphere than outgoing long-wave radiation from the Earth's surface.

4.6.4 Solar Energy for Living Things

Although the emphasis in this book is on capturing, converting, and using energy for material things in life (hot showers and cold beer), without the energy services that nature freely provides for us life as we know it today would not exist. Sunlight provides the foundation for the biosphere. Green plants bottle up sunlight in energy-rich bonds using just water and carbon dioxide as raw materials, from which they build sugars that are the energy basis for essentially all life on Earth. And, as a bonus, they pump fresh oxygen into the atmosphere. The same chlorophyll photon-trapping processes that enable life to thrive on Earth today are responsible for the ancient solar energy, bottled up hundreds of millions of years ago, that we now exploit in the form of fossil fuels.

The process of photosynthesis utilizes chlorophyll, the green pigment in the chloroplasts of plants and some algae. Chlorophyll absorbs sunlight from the red portion of the solar spectrum (around 0.7 micron) whereas other pigments absorb shorter-wavelength blue light. Note that green light is not absorbed, and is instead reflected, which is why the leaves of plants appear green to us because those green wavelengths are reflected to our eyes.

A simple representation of photosynthesis is the following in which photons from the sun provide the energy needed to convert simple water and CO_2 molecules into sugars, in this case glucose, $C_6H_{12}O_6$, with the leftover oxygen being released into the atmosphere.

SOLUTION BOX 4.8

Estimating the Earth's Temperature without the Greenhouse Effect

Let's do a calculation for the expected temperature of the Earth if it were a blackbody without an atmosphere, and hence, without any greenhouse effect. We will find the equilibrium temperature that would cause the energy radiated to space from the Earth's surface to be equal to the solar energy absorbed by the surface.

Solution:

Imagine a transparent hoop located between the sun and the Earth. The Earth and the hoop have the same radius R and the hoop is set up so that any solar energy (S, W/m²) that passes

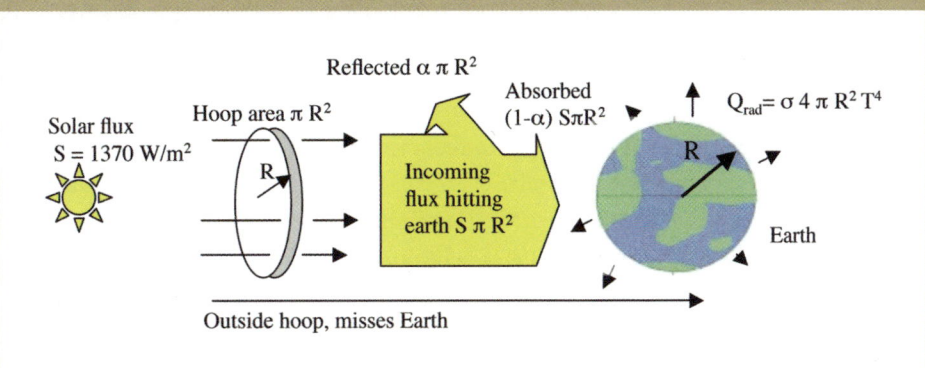

through the hoop hits the Earth and any that misses the hoop also misses the Earth. Some fraction of incoming solar radiation, called the *albedo* (α), is reflected directly back into space.

Incoming Solar Energy:

Solar flux that passes through the hoop and hits the Earth = $S \pi R^2$

Flux hitting the Earth that is reflected = $\alpha S \pi R^2$

Flux hitting the Earth that is absorbed by the Earth = $(1 - \alpha) S \pi R^2$

Outgoing Energy Radiated from Earth:

Assume the Earth is a blackbody with uniform surface temperature = T

Area of Earth's surface = $4 \pi R^2$

Radiation from Earth toward space = $\sigma A T^4 = \sigma 4 \pi R^2 T^4$

(continued on next page)

Energy Balance:

Incoming solar energy = Outgoing energy radiated from Earth

$$(1 - \alpha)\, S\, \pi\, R^2 = \sigma\, 4\, \pi\, R^2\, T^4$$

The Earth's α is about 0.31 (31%) and the incoming solar flux just outside of the atmosphere, called the solar constant S, is about 1370 W/m². Solving the above equation for the equilibrium temperature for Earth gives:

$$T = \left[\frac{S(1-\alpha)}{4\sigma}\right]^{1/4} = \left[\frac{1370 \text{ W/m}^2 \,(1-0.31)}{4 \times 5.67 \times 10^{-8} \text{ W/m}^2\text{K}^4}\right]^{1/4} = 254 \text{ K} = -19°C = -2°F$$

Eq. 4.23 $6\,CO_2 + 6\,H_2O + \text{light} \rightarrow C_6H_{12}O_6 + 6\,O_2$ ΔH = 2820 kJ/mol glucose

As indicated, the enthalpy change is positive, meaning that 2820 kJ of energy per mole of glucose must be provided to make this reaction take place. At 180 g/mol that works out to be 15.7 kJ/g stored. The annual capture of solar energy by plants is enormous, something like 3000 exajoules (3000 × 10¹⁸ J/yr), which is almost ten times the rate of human energy use. In Chapter 14, the use of some of this energy in the form of biomass fuels will be explored.

The overall efficiency at which solar energy is converted to ecosystem biomass varies widely. Some estimates suggest the theoretical maximum to be about 5%, but actual ecosystems are usually far below that value. Whereas a healthy cornfield may capture 1%–2% of the summer's sunlight, a prairie may net only a few tenths of one percent. Globally, the fraction of sunshine that ends up as stored energy in plants is just a few hundredths of a percent.

The reverse of Equation 4.23 summarizes the respiratory process by which living things use stored chemical energy to provide for their own energy needs.

Eq. 4.24 $C_6H_{12}O_6 + 6\,O_2 \rightarrow 6\,CO_2 + 6\,H_2O$ ΔH = –2820 kJ/mol

Respiration uses up some of the energy that plants capture during photosynthesis, resulting in less energy available for animals that eat those plants. As we work our way up the food chain, the fraction of consumed energy that is needed to meet the respiratory needs of the next trophic level increases. A common rule-of-thumb estimate is that every time we move one notch up the food chain, from plant to herbivore, from herbivore to first-level carnivore, and so on, only about 10% of one level's energy is transferred to the next. This argument can be restated to suggest that it takes about ten times the land area to feed carnivores than herbivores, which is an often-cited reason why some people choose to be vegetarians.

The magic of both photosynthesis and respiration is provided by a number of enzymes and nucleotides, such as adenosine diphosphate (ADP) and adenosine triphosphate (ATP), that transport and accept electrons, permitting the chemical reactions for biosynthesis of

table 4.6 **Examples of Food Calories (kcal)**

Food	Calories	Food	Calories
Pizza (3.2 oz)	260	Granola bar	120
Big Mac hamburger	560	Yogurt, nonfat (6 oz)	100
Milkshake (10 oz)	350	Milk, nonfat (8 oz)	85
French fries (large)	400	Cheerios (1¼ cup)	110
Oat bran muffin	330	Carrot, raw	30
Cake donut	270	Lettuce, iceburg (1 head)	70
Twinkies, each	160	Bagel, plain	150
Pecan pie (1/8 of 9-in.)	430	Bread, whole wheat (1 slice)	70
Peanut butter cups, each	140	Apple, medium	70
Ice cream, vanilla soft	380	Halibut, broiled (4 oz)	195

DNA and proteins, assembly of cell structures, transport of solute molecules, neurological information transfer by the nerves and senses, muscle contraction and motion, and all the other physical miracles of life.

4.6.5 Food Calories

Sunshine becomes the calories we eat, so let's take a brief look at food. The nutritional community traditionally uses *Calories* (capital C) to indicate the metabolizable energy content of our food, where 1 Calorie = 1 kilocalorie = 4.185 kJ. Often they don't stick with the convention of using a capital C, so just realize when someone talks about food calories they really mean kilocalories. Table 4.6 provides some examples of the Calories in typical foods we eat.

Table 4.7 provides some estimates of the rate at which an adult will burn Calories while performing various activities. The baseline for a body at rest is called the basal metabolism rate, which corresponds to the energy needed just to provide basic functions such as breathing and blood circulation. Also shown is the equivalent amount of heat given off by a 70 kg (154 lb) individual, expressed in watts. For example, a 154 lb adult watching TV is equivalent to an 86-W heater sitting in the room. Later, when we look at the heating and cooling requirements for a building we will see that we need to account for the heating load provided by people and appliances.

Combining Tables 4.6 and 4.7 allows us to do simple calculations that are rather illuminating. For example, we might wonder how long it would take for a 180 lb man to jog off that Big Mac and milkshake that he had for lunch.

$$\text{Calories} = 560 + 350 = 910 \text{ kcal}$$
$$\text{Jogging} = 3.61 \text{ kcal/hr/lb} \times 180 \text{ lb} = 650 \text{ kcal/hr}$$
$$\text{Time} = 910 \text{ kcal}/650 \text{ (kcal/hr)} = 1.4 \text{ hr}$$

table 4.7	Approximate Calories Used by an Adult, per Unit of Body Weight		
Activity	kcal/hr per kg	kcal/hr per lb	Watts per 70 kg (154 lb)
Basal metabolism	1.06	0.48	86
Watching TV	1.06	0.48	86
Driving a car	2.65	1.20	215
Swimming (slow)	4.24	1.93	345
Walking (4 mph)	6.35	2.89	517
Jogging (5 mph)	7.94	3.61	646
Fast dancing	8.82	4.01	718
Bicycling (13 mph)	9.40	4.27	765
Swimming (fast)	12.65	5.75	1030
Running (8 mph)	13.76	6.26	1121

4.7 Nuclear Energy

We are all very comfortable with the concept of gravity. Using some calculus and the simple notion that the gravitational attraction between two objects is proportional to the product of their masses and inversely proportional to the square of their separation, Newton and Kepler were able to predict the behavior of the solar system. Similarly, in the electrical world, we know opposite charges attract with a force proportional to their charges and inversely proportional to the distance between them. Both gravitational and electrical forces exist within a nucleus, so we might wonder how these forces play out against each other. The gravitational attraction between nucleons pulls them together while the electrical forces between like-charged particles try to shove protons apart. The gravitational attraction is far weaker than the electrical repulsion, which suggests the nucleus should fly apart. But, of course, it doesn't.

There are, it turns out, extremely powerful forces that hold a nucleus together. The energy that would be required to break apart that nucleus, separating it into individual protons and neutrons, is called the nuclear *binding energy*. And this is where Einstein's famous relationship comes in.

Eq. 4.25
$$E = mc^2$$

where E = energy (kJ)
          m = mass (kg)
         c = the speed of light = 2.998×10^8 m/s

If we imagine building an atom out of individual electrons, protons, and neutrons we would find out that the resulting atom has less mass than the sum of the masses of its individual constituents. That difference in mass, converted to energy units using Equation 4.25 is the binding energy of the atom.

4.7.1 The Nature of Radioactivity

For many people, the mere mention of radioactivity conjures up images of cancer, birth defects, and mushroom clouds. But radioactivity also benefits humankind. Radioactive elements (called *radioisotopes* or *radionuclides*) are used as labels or tags to help unravel the complexities of chemical reactions; indeed, such use was crucial to developing our present understanding of the utterly important process of photosynthesis. Radioisotopes also provide an accurate way to date past events, both historical and geological. And, radioisotopes offer an important way to focus radiation directly onto tumors to help cure those afflicted with cancer.

The term *radioactivity* refers to an atomic nucleus that is unstable. In trying to reach a more stable configuration of protons and neutrons in its nucleus, a radioactive atom emits various forms of radiation, transforming itself from one chemical element into another. The first experiments demonstrating the spontaneous decay of a naturally occurring element were reported by Henri Becquerel in 1896 using a uranium-containing ore called pitchblende. The radiation that Becquerel observed was the emission of what is now called an *alpha* (α) particle. An alpha particle is a two-proton, two-neutron helium nucleus, and the reaction he observed can be described as follows:

Eq. 4.26 $^{238}_{92}U \rightarrow {}^{234}_{90}Th + {}^{4}_{2}He$ (α particle) $T_{1/2} = 4.51 \times 10^9$ yrs

Recall that the number to the lower-left of a chemical symbol is the number of protons in the nucleus whereas that to the upper-left is the sum of protons plus neutrons. When a U-238 atom (with 92 protons) emits an alpha particle, it loses two protons so it becomes a new element—thorium, which has 90 protons. And because the nucleus has lost four nucleons, its mass number drops from 238 to 234.

Also shown in Equation 4.26 is the *half-life* of U-238, which is about 4.5 billion years. That is, if we had 1 kg of U-238 today, in 4.5 billion years half of that will be transformed into thorium and we would have left only 0.5 kg of uranium. After 9 billion years, we would have half of that left, or 0.25 kg. Figure 4.9 illustrates the concept of half-life.

As an alpha particle passes through an object, its energy is gradually dissipated as it interacts with other atoms. Its positive charge attracts electrons in its path, raising their energy levels and possibly removing them completely from their nuclei (*ionization*). Alpha particles are relatively massive and easy to stop. Our skin is sufficient protection for sources that are external to the body, but taken internally, such as by inhalation, alpha particles can be extremely dangerous.

A radionuclide can decay in other ways besides the emission of an alpha particle. Many radionuclides have too many neutrons relative to their number of protons and decay by converting one of those neutrons into a proton plus an electron, with the electron being emitted from the nucleus. These negatively charged electrons are referred to as *negative beta* (β^-) particles. Emission of a β^- results in an increase in the atomic number by one, whereas

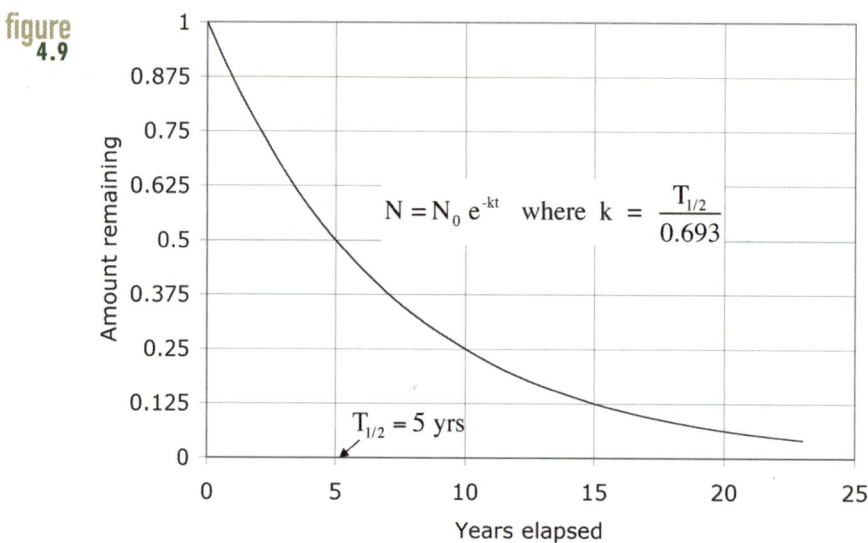

figure
4.9

$$N = N_0\, e^{-kt} \quad \text{where} \quad k = \frac{T_{1/2}}{0.693}$$

$T_{1/2} = 5 \text{ yrs}$

Amount remaining

Years elapsed

The half-life of a radioactive isotope ($T_{1/2}$) is the time required for half of the material to be converted to another element. As shown, a material with a half-life of 5 years would have half remaining after 5 years, one-fourth after 10 years, one-eighth after 15 years, and so forth.

the mass number remains unchanged. The following reaction shows the decay of strontium-90 into yttrium-90:

Eq. 4.27 $^{90}_{38}\text{Sr} \rightarrow {}^{90}_{39}\text{Y} + \beta^-$ $T_{1/2} = 29 \text{ yrs}$

It is also possible for a nucleus to have too many protons, in which case a proton may change into a neutron while ejecting a positively charged electron, or positron, designated as β^+. An example of this reaction is the conversion of nitrogen-13 into carbon-13:

Eq. 4.28 $^{13}_{7}\text{N} \rightarrow {}^{13}_{6}\text{C} + \beta^+$ $T_{1/2} = 9.96 \text{ min}$

As β particles pass through materials, they are also capable of ionizing atoms in our tissues, and they may do so at much greater depths. While alpha particles may travel less than 100 μm into tissue, β radiation may travel several centimeters. They can be stopped with a modest amount of shielding, however. For example, a centimeter or so of aluminum is sufficient.

Equations 4.26–4.28 illustrate the spontaneous emission of actual particles having mass; usually there will also be electromagnetic *gamma* (γ) radiation released. Gamma rays have very short wavelengths in the range of 10^{-11} to 10^{-13} m. Having such short wavelengths means that individual photons are highly energetic and easily cause biologically damaging ionizations. These rays are difficult to contain and may require several centimeters of lead to provide adequate shielding.

All of these forms of ionizing radiation are dangerous to living things. The electron excitations and ionizations that are caused by such radiation cause molecules to become unstable, resulting in the breakage of chemical bonds and other molecular damage. The chain of chemical reactions that follows creates new molecules that did not exist before the irradiation. Exposure to ionizing radiation can result in everything from cancer, leukemia, sterility, cataracts, and reduced lifespan to mutations in chromosomes and genes that will be transmitted to future generations.

4.7.2 Nuclear Fission

All elements having more than 83 protons are naturally radioactive, spitting out combinations of alpha, beta, and gamma radiation. In addition, other nuclear reactions offer us the tantalizing potential to tap into the energy within the nucleus itself. Just before World War II, nuclear scientists discovered that one particular isotope of uranium, U-235, can *fission*, or break apart, when bombarded with neutrons. As suggested in Figure 4.10, when U-235 absorbs a neutron it becomes unstable U-236, which almost instantly splits apart, discharging two *fission fragments* along with two or three neutrons and an intense burst of gamma (γ) rays. Most of the energy released is in the form of kinetic energy in the fission fragments. In a nuclear reactor, that kinetic energy is used to boil water to make steam to spin a turbine and generator.

The fission fragments produced are always radioactive, and concerns for their safe disposal have created much of the controversy surrounding nuclear reactors. Typical fission fragments include cesium-137, which concentrates in muscles and has a half-life of 30 years, and strontium-90, which concentrates in bone and has a half-life of 8.1 days. The half-lives of fission fragments tend to be no longer than a few decades, so after a period of several hundred years their radioactivity will decline to relatively insignificant levels.

An example of one such fission is the following reaction:

Eq. 4.29 $$\,_{92}^{235}\text{U} + \text{n} \rightarrow \,_{55}^{143}\text{Cs} + \,_{37}^{90}\text{Rb} + 3\text{n} + 184 \text{ MeV}$$

Notice that there is a balance of nucleons on each side of the reaction and there is an amount of energy released, here expressed in millions of electron-volts (MeV). Although Equation 4.20 expresses the conversion of mass to energy using units of kJ, that unit is hardly ever used in calculations associated with nuclear reactions because it is just too big to be convenient. The more common measure is *electron-volts* (eV), or million electron-volts (MeV), where $1 \text{ eV} = 1.60 \times 10^{-19}$ J. The basis for the electron-volt measurement will be described in section 4.8, "Electrical Energy" and an example of its use is demonstrated in Solution Box 4.9.

Immediately apparent from Figure 4.10 is that it took a single neutron to cause one atom of U-235 to fission, and in doing so two or three new neutrons are emitted. Those neutrons can go on to cause other fissions, leading to the possibility of a controlled, self-

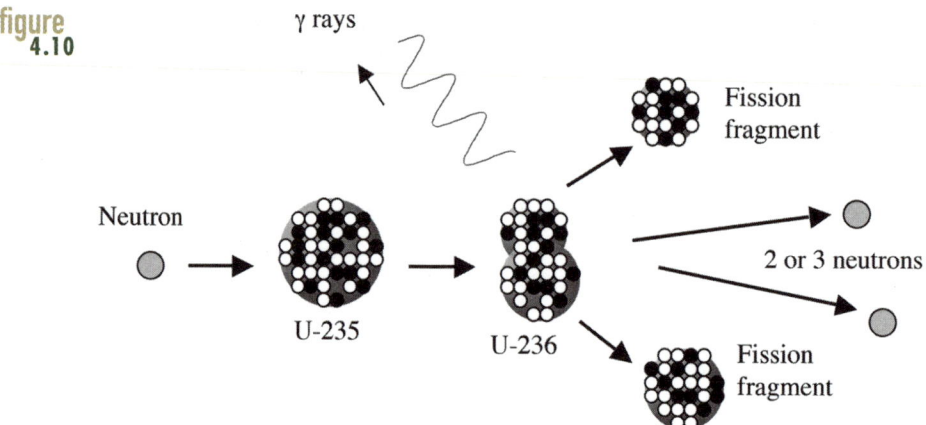

figure 4.10

γ rays

Neutron

U-235

U-236

Fission fragment

2 or 3 neutrons

Fission fragment

The fissioning of U-235 creates two radioactive fission fragments plus two or three neutrons and gamma rays.

sustaining chain reaction for a nuclear power plant, or an uncontrolled, explosive chain reaction as occurs in an atomic bomb. To create a uranium bomb, however, requires a much more concentrated supply of U-235 than is present in a conventional nuclear power plant.

Uranium-235 is the only fissile material occurring in nature, but it makes up only about 0.7% of naturally occurring uranium. The other 99.3% is mostly U-238, which does not fission. To build a nuclear reactor requires a U-235 concentration of about 3%, but to build a bomb it needs to be enriched to more than 90%. That enrichment process is extremely difficult and is a major deterrent to creating bomb-grade material from U-235. There is, however, another approach that is based on creating fissile plutonium in a conventional reactor and then separating it out from the reactor wastes. Indeed, this is how the bomb that destroyed Nagasaki was built, and it is the process used by every country that now possesses nuclear weapons.

Most of the uranium in a reactor is the isotope U-238, but it is not capable of fissioning. However, when it absorbs a neutron, it can be transformed through the following reactions to plutonium, and plutonium is a fissile material (Figure 4.11).

Eq. 4.30

$$^{238}_{92}\text{U} + \text{n} \rightarrow \,^{239}_{92}\text{U} \xrightarrow{\beta} \,^{239}_{93}\text{Np} \xrightarrow{\beta} \,^{239}_{94}\text{Pu}$$

Plutonium is radioactive, emits alpha particles when it decays, and has a half-life of 24,360 years. It does not occur in nature and the plutonium we produce will be around for many tens of thousands of years. It is a worrisome material not only because it can be used for nuclear weapons, but also because it is considered by many to be the most toxic substance known to humankind. Its presence in nuclear wastes, along with other long-lived actinides, contributes to the need to isolate those wastes, in essence, forever.

By converting exceedingly small amounts of mass into exceptionally large amounts of energy, nuclear energy gives us the potential to wean ourselves from our fossil fuel

---

Nuclear Energy versus Chemical Energy

A typical fission reaction releases about 200 MeV per atom of U-235. How does this compare with the energy released by burning methane?

Solution:

For the U-235, first convert that 200 MeV per atom to joules:

U-235: $200 \text{ MeV} \times 1.60 \times 10^{-22} \text{ kJ/eV} \times 10^6 \text{ eV/MeV} = 3.2 \times 10^{-14} \text{ kJ/atom}$

Those uranium atoms are very heavy, so converting to a per unit of mass basis

$$\frac{3.2 \times 10^{-14} \text{ kJ/atom} \times 6.023 \times 10^{23} \text{ atoms/mol}}{235 \text{ g/mol}} = 82.02 \times 10^6 \text{ kJ/g U-235}$$

Recall from Solution Box 4.5 that burning a mole of CH_4 liberates 802 kJ, which translates to

$$CH_4: \frac{802 \text{ kJ/mol}}{(12 + 4 \times 1) \text{ g/mol}} = 50.125 \text{ kJ/g } CH_4$$

Wow! The nuclear reaction with U-235 releases more than 1.6 million times as much energy per gram as the chemical reaction with CH_4.

A similar assessment for coal, accounting for the concentration of U-235 in nuclear fuel, and a representative type of coal, works out to about 1 ton of nuclear fuel providing the same energy as roughly 100,000 tons of coal.

---

dependence with all of its environmental and national security disadvantages. As usual, however, we can't get something for nothing and nuclear power presents its own challenges having to do with disposal of radioactive wastes, human and financial risks should there be a reactor accident, and the increasing worry about terrorists gaining access to plutonium to make their own atomic bombs.

4.7.3 Nuclear Fusion

Nuclear fission relies on the ability to break atoms apart; the opposite occurs in nuclear fusion reactions. With fusion, it is the loss of mass associated with merging nuclei that creates the energy

figure
4.11

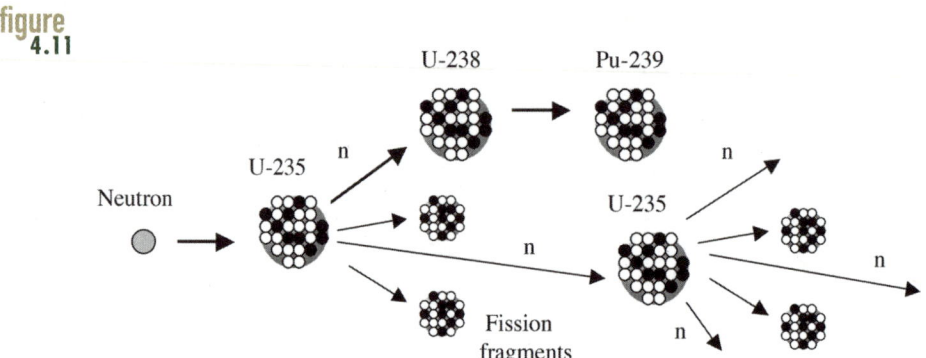

Plutonium created in a nuclear reactor is a source of fissile material that can be used either as a reactor fuel or for nuclear weapons.

we are after. The sun is the perfect example of a safe, working nuclear fusion device; a not-so-perfect example is a thermonuclear bomb. The reactions occurring within the sun involve a number of steps, but can be summarized as four hydrogen atoms combining to form helium-4:

Eq. 4.31
$$4\,{}^{1}_{1}H \rightarrow {}^{4}_{2}He + 2\beta^{+} + \text{energy}$$

The most promising fusion reactions for us down here on Earth involve various isotopes of hydrogen, which combine to form helium. Hydrogen has only one proton, and in nature it almost always has no neutrons. A very small fraction, however, about 0.015% of naturally occurring hydrogen, has a single neutron in the nucleus along with the proton, and this isotope is given a special name, *deuterium*. Another important isotope has 2 neutrons, and it is called *tritium*. A sample deuterium-deuterium (D-D) reaction is the following:

Eq. 4.32
$$\,{}^{2}_{1}H + {}^{2}_{1}H \rightarrow {}^{3}_{2}He + n + 3.3\ \text{MeV}$$

Because roughly one out of every 5000 hydrogen atoms in water is deuterium, there is enough deuterium in the oceans to supply all of the world's energy needs for millions of years into the future. This deuterium-deuterium (D-D) reaction presents an extreme challenge because the reactions need to take place under conditions that are even hotter than the interior of the sun.

A much more promising first-generation fusion reactor is based on fusing deuterium and tritium in the following D-T reaction:

Eq. 4.33
$$\,{}^{2}_{1}H + {}^{3}_{1}H \rightarrow {}^{4}_{2}He + n + 17.6\ \text{MeV}$$

The trick in this case is finding enough tritium. Tritium has a half-life of only 12 years and does not exist in any great quantity in nature, so if this reaction is to provide much energy for the future we will have to find some way to manufacture it. One way to produce tritium is to expose lithium-6 to neutron bombardment in a fission reactor, as follows:

Eq. 4.34
$$\text{n} + {}^{6}_{3}\text{Li} \rightarrow {}^{3}_{1}\text{H} + {}^{4}_{2}\text{He} + 4 \text{ MeV}$$

Unfortunately, there isn't that much lithium in the world either. The United States, for example, might be able to produce enough lithium from its own resources to power the country for only a few hundred years. That's pretty good, of course, but not as exciting as the dazzling potential of D-D fusion.

Significant progress has been made on D-T fusion, to the point that in mid-2005, it was announced that a consortium made up of the European Union, the United States, Russia, Japan, South Korea, and China had agreed to build a demonstration reactor called the International Thermonuclear Experimental Reactor (ITER), to be constructed in Cadarache, France. It will be something like building our own small star, with reaction temperatures hotter than our own sun, around 100 million kelvins. If all goes well, it will become operational in 2015 at a cost of around $15 billion, and could eventually lead to commercial reactors that would have the potential to supply enormous amounts of energy without greenhouse gases and far less radioactivity than current fission reactors.

4.8 Electrical Energy

At last count, there were only four fundamental forces in the universe. The first is gravitational force. Two others are forces within atoms, called the strong nuclear force and the weak nuclear force. And the fourth is the electrical force that one charged object exerts on another. That fourth force was first explained by the French physicist Charles-Augustin de Coulomb (1736–1806) and the result is known as *Coulomb's law:*

Eq. 4.35
$$F = k\frac{q_1 q_2}{d^2}$$

where, in SI units, F is in newtons (N), q_1 and q_2 are the electrical charges of each object, and are given in coulombs (C). The distance d between the two objects is in meters, and the coefficient k is 9×10^9 (N-m^2/C^2).

To visualize this phenomenon, it helps to imagine an invisible *electric field* surrounding every charge. When a test charge is brought into an electric field we can measure the force exerted by the field on the charge, so we know it is there even if we can't see it. This simple notion that electric fields exist and they can exert forces on charges has implications that extend throughout the field of electrical engineering.

4.8.1 Electric Current

One coulomb of charge is equal to the charge on 6.242×10^{18} electrons—that's a lot of electrons—but, when it comes to having electrons do work for us, some are much more helpful than others. Most electrons are too tightly bound to their nuclei to do us much good, but others are far enough away that the attraction of any particular nucleus can easily be overcome. These *free electrons* easily wander from one atom to another, and if subject to even a slight electric field they can be made to bounce along in a particular direction. That flow of charge constitutes an electric current.

In general, charges can be negative or positive. In a copper wire, the only charge carriers are electrons, with negative charge. In a neon light, however, under the influence of an electric field positive ions move in one direction and negative electrons move in the other. Each contributes to current and the total current is their sum. By convention, the direction of current flow is taken to be the direction that positive charges would flow, whether or not positive charges happen to be in the picture. Thus, in a wire, electrons moving in one direction constitute a current flowing in the opposite direction, as shown in Figure 4.12.

If we imagine a wire, when one coulomb's worth of charge passes a given spot in one second the current is defined to be one *ampere* (abbreviated A), named after the nineteenth-century physicist Andre Marie Ampere. In equation form, current i is the flow of charge q past a point, or through an area, per unit of time t.

Eq. 4.36
$$i = \frac{q}{t}$$

When charge flows at a steady rate in one direction only, it is said to be *direct current*, or *dc*. A battery, for example, supplies direct current. When charge flows back and forth sinusoidally, it is called *alternating current*, or *ac*. In the United States the ac electricity delivered by the power company has a frequency of 60 cycles per second, or 60 hertz (abbreviated Hz). Examples of dc and ac are shown in Figure 4.13.

figure
4.12

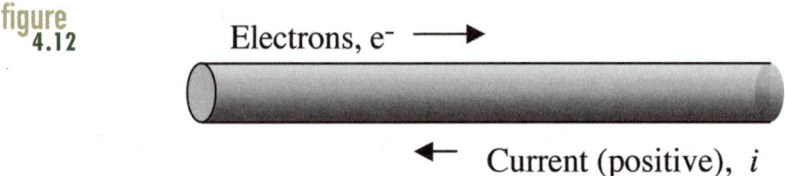

Electrons, e⁻ ⟶

⟵ Current (positive), i

By convention, in a wire the direction of positive current flow is the opposite of the actual flow of electrons.

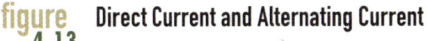 **Direct Current and Alternating Current**

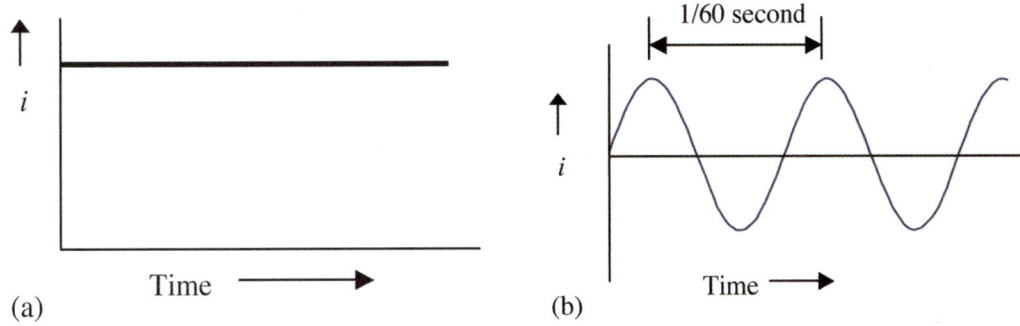

(a) Steady, direct current (dc). (b) 60 Hz alternating current (ac).

4.8.2 Voltage

Electrons won't flow through a circuit unless they are given some energy to help them along their way. That "push" is measured in volts, where voltage is defined to be the amount of energy (w, joules) given to a unit of charge, q.

Eq. 4.37
$$v = \frac{w}{q}$$

For example, a 9-volt battery provides 9 joules of energy to each coulomb of charge that it stores. Voltage describes the potential for charge to do work. Just as mechanical forms of potential energy are always measured with respect to some reference, so too is voltage. Thus, the positive terminal of a 9-volt battery is 9 volts higher than the voltage of the negative terminal. The negative terminal may be grounded by connecting a wire from it to a copper rod pounded into the Earth, in which case we would say the positive terminal is 9 volts with respect to ground.

4.8.3 The Concept of an Electrical Circuit

A simple circuit consists of a *source* of energy, say a battery; a *load* such as a lightbulb, toaster, or something else that you want to power with electricity; and some connecting wire to carry current from the source to the load and back again. Notice that a circuit has to have a return path for the electrons. They can't just go one way to the lightbulb and then drop off onto the floor.

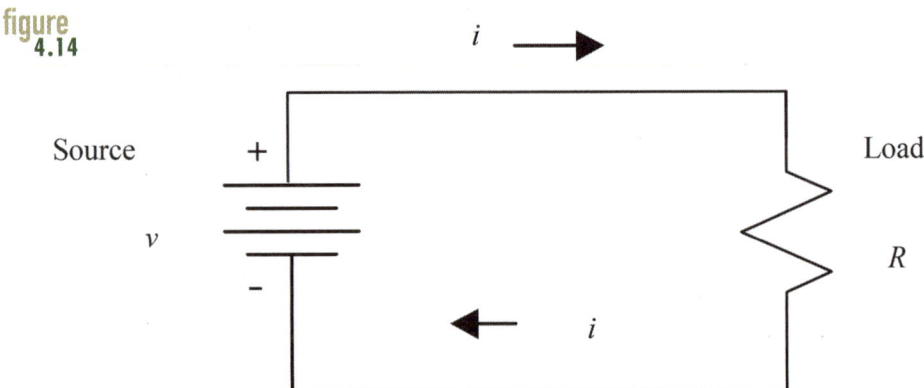

figure
4.14

A simple dc circuit consisting of a voltage source, a resistive load, and a connecting wire through which current flows.

In a simple dc circuit, such as the one shown in Figure 4.14, the load can be characterized by its *resistance* to the flow of electrons through it. The wires, on the other hand, are often assumed to be perfect conductors that offer no resistance at all to electron flow. The relationship between the voltage v applied to this circuit, the current that flows through the circuit i, and the resistance offered by the load to that current flow R (ohms, Ω) is given by

Eq. 4.38 $$v \text{ (volts)} = R \text{ (ohms)} \times i \text{ (amps)}$$

This deceptively simple relationship is known as Ohm's law in honor of the German physicist, Georg Ohm, whose original experiments led to this incredibly important relationship.

4.8.4 Electrical Power and Energy

Electric circuits can have many different goals, but they can be distinguished by whether their purpose is to transmit information or power. Most of the circuitry in your laptop is there to process information by detecting the presence or absence of a voltage (0s and 1s). In that context, power consumption is a bad thing because it makes your battery drain faster. In the context of this book, however, our goal is to generate and transmit large amounts of electrical energy and power to do real work, work that can not only charge those laptop batteries, but also power whole factories, cities, and regions of the country.

If we go back to the simple circuit of Figure 4.14, we can derive a few important relationships between voltage, resistance, and current and then relate those to power and energy. Recall that voltage is the energy given to a unit of charge ($v = w/q$) and current is the rate at which charge moves around a circuit (q/t). Combining these two with the definition of power, which is the rate of delivering energy ($p = w/t$), we get:

Eq. 4.39
$$p = \frac{w}{t} = \frac{w}{q} \cdot \frac{q}{t} = vi$$

So, the power delivered to a load is the product of the voltage across the load times the current through the load. Combining Equation 4.39 with Ohm's law (4.38) gives us three ways to express power to a resistive load:

Eq. 4.40
$$p = vi = \frac{v^2}{R} = i^2 R$$

When v is in volts, i is in amps, and R is in ohms, the above expressions give power in watts (1 W = 1 J/s).

Multiplying power watts (W) or kilowatts (kW) by the length of time that the power is consumed (hours) we get energy in watt-hours (Wh) or kilowatt-hours (kWh). The example in Solution Box 4.10 illustrates the above electrical relationships.

In Chapter 11, we'll be looking at how much electrical power and energy various household appliances use and then figuring out how much that will cost. The process is conceptually simple. Just multiply the watts used by an appliance by the hours of operation and you have watt-hours. Household utility bills are based on how many kilowatt-hours you use, so divide that by 1000 to get kWh.

As the example in Solution Box 4.11 illustrates, some appliances use power whether they are turned on or not. Look around your home and see how many devices have glowing red or green lights turned on all the time. Your TV and other remote controlled appliances consume standby power while their electronics wait for you to push the "on" button from your remote. In fact, upwards of two-thirds of the energy used in many home electronic devices is used for standby power.

4.9 Summary

In this chapter on fundamentals, we have tried to lay the groundwork you need to understand the wide range of energy technologies that will follow. You should now understand energy units, conversions between them, and ways to describe very small values (e.g., nanometers) and very large quantities (e.g., exajoules). We also made an important distinction between energy and power. Power is a rate; that is, it is energy per unit of time. Some power units sound like rates (e.g., Btu per hour) whereas others do not (e.g., kilowatts). You can lose credibility if you use them wrongly (e.g., kilowatts per hour).

We have explored the ways that nature manages to store energy in the potential and kinetic forms associated with work, in heat energy that is transferred from warm objects

SOLUTION BOX 4.10

Power to an Incandescent Lamp

Suppose a heavy-duty flashlight uses a 6-volt, lead-acid battery to deliver power to the filament of an 18-watt bulb. Find the resistance of the filament and the current that flows.

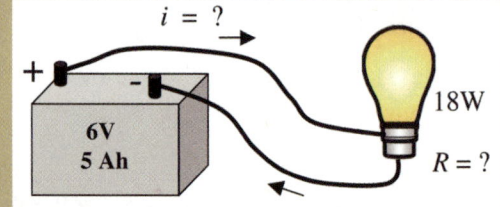

Solution:

Because power is volts times amps, the current drawn by the bulb will be

$$i = \frac{18 \text{ W}}{6 \text{ V}} = 3 \text{ A}$$

From Ohm's law, $v = R\,i$, the filament resistance must be

$$R = \frac{v}{i} = \frac{6 \text{ V}}{3 \text{ A}} = 2 \ \Omega$$

Batteries are usually rated by their amp-hour (Ah) capacity, which roughly speaking translates to the product of the amps they deliver multiplied by the hours they can provide those amps. If the battery in this flashlight is rated at 5 Ah, how long could it keep the flashlight going? How much energy would have been delivered?

$$\text{Time} = \frac{5 \text{ Ah}}{3 \text{ A}} = 1.67 \text{ hr}$$

$$\text{Energy} = 18 \text{ W} \times 1.67 \text{ hr} = 30 \text{ Wh}$$

to cold ones, in the chemical form stored in the bonds between atoms, in the electromagnetic form from the sun that powers essentially all living things, in the form released when fission breaks apart atoms or fusion joins them together, and finally in the electricity that transformed the twentieth century, enabling all of the technology marvels we so enjoy today.

In nature, and in physical systems that humans construct, energy is constantly being shuttled about from one form to another. The first law of thermodynamics tells us that energy is conserved; that is, we can draw up balance sheets that account for every bit of it, including the conversion of mass to energy. To say we have consumed, or "used up," some energy is therefore a bit of a misnomer. We merely transform it, but while we do so we are constantly degrading the quality of energy, which is where the second law of thermodynamics comes in.

That Satellite TV System

A typical satellite TV system consumes 16 watts while turned off and 17 watts when it is in use. Similarly, a typical 20-inch TV consumes 5 watts while off and 68 watts when on.

Suppose you watch TV 4 hours per day, 30 days per month. What fraction of your electricity is used while the TV is turned off?

16 W off, 17 W on   5 W off, 68 W on

Solution:

TV on: (68 W + 17 W) × 4 h/d × 30 d/mo = 10,200 Wh/mo = 10.2 kWh/mo

TV off: (5 W + 16 W) × 20 h/d × 30 d/mo = 12,600 Wh/mo = 12.6 kWh/mo

Total = 22.8 kWh/mo

So, 12.6/22.8 = 0.553 = 55.3% of your electricity is used while the TV is off.

If the utility charges 10¢ per kWh, how much does your TV cost you in a 30-day month? What is that cost per hour of actual TV use?

At $0.10/kWh × 22.8 kWh/mo = $2.28/mo

Per hour of use that works out to 228¢/120 hr = 1.9¢/hour

The first law deals with energy quantities and the second with energy quality. As we manipulate energy, the quality of that energy is constantly degraded; that is, more and more of it ends up as relatively useless waste heat. The second law also provides information on the direction in which energy flows may take place. On your kitchen table, it has no problem with hot coffee cooling off and cold beer warming up, but it frowns on allowing the opposite to spontaneously occur. A more subtle interpretation of the second law describes the universe as constantly moving toward greater and greater disorder. The orderliness of hot separated from cold, becomes the random sameness of warm. The orderly arrangement of molecules in ice becomes a collection of disordered, randomly arranged molecules when ice melts.

In subsequent chapters, the energy flow concepts introduced here will be applied to real systems. Questions such as how we size systems, estimate their performance, evaluate their environmental implications, and determine their economic value will be addressed.

CHAPTER 5

Energy Analysis and Life-Cycle Assessment

As individuals, as communities, and as a society, we want to make smart energy choices and investments that are cost-effective and that can help make our energy economy sustainable. To do that we must address some questions that are basic but not always easy to answer.

For example, as an individual, should I buy a hybrid vehicle or a high-efficiency refrigerator or furnace? Should I put more insulation in my attic? Should I put solar panels on my roof? To answer these questions effectively we need to know the personal financial effects of these decisions, which requires information on energy savings and investment costs. We may also wish to assess the global environmental effects of the energy options we face, and then compare them with the financial effects. This is not always easy.

As a community, should our municipal utility invest in wind farms or an energy effciency program? Should we strengthen our building energy codes or provide incentives for efficiency improvements in existing buildings? Again this requires energy and economic analysis to evaluate energy savings and costs. But we may also justify community investments if they have additional long-lasting effects, such as local economic development or reduced impacts on the local or global environment. Again, these are not easy assessments.

As a society, should we commit to greater efficiency through vehicle, appliance, or building efficiency standards? Should we accelerate use of renewable energy through a renewable portfolio standard for electricity or a requirement for greater use of ethanol fuels? How should we balance subsidies and tax incentives and disincentives for fossil fuels, nuclear, renewable energy, and efficiency?

These questions can be answered effectively only with good information on energy, economic cost-effectiveness, and environmental costs and benefits. This chapter introduces four basic analytical methods to provide the rational information on which to base energy decisions:

1. Life-cycle assessment
2. Energy analysis
3. Economic cost-effectiveness
4. Environmental assessment

Life-cycle assessment is fundamental to "sustainability analysis" and gives us a broad framework for energy analysis in terms of both time and criteria. It forces us to look at the full range of energy, economic, and environmental impacts from "cradle to grave."

Energy analysis is the first step to determine and compare energy consumption and production of different options. This can involve complex life-cycle net energy analysis, but often the most useful energy analysis is done by calculating simple energy consumption or conversion efficiency on the back of an envelope. These calculations require some boilerplate or monitored energy data, some knowledge of energy conversion (like that described in Chapter 4), some algebra, and dimensional analysis to get the units right. We can enhance these simple calculations with more elaborate methods, even computer models that incorporate more detailed data and operating assumptions, but the simple approach will be the mainstay of this book.

Economic cost-effectiveness defines energy analysis in terms of economic and financial costs and benefits. Choices among energy options require investment, and we want to use our limited financial resources wisely. Economic cost-effectiveness methods require energy analysis to know how much energy is required; the economic value of energy supplied, produced, and/or consumed; the capital and operating costs of the system option; and the time-value of money.

Environmental assessment looks beyond economic effects and determines the impacts of energy options on the natural and human environment. It can use a range of impact indicators, such as greenhouse gas and air pollutant emissions, toxic effluents, land and water requirements, human health effects, risk and uncertainty, aesthetic impacts, and ecological effects, to name a few. Some of these measures can be put in economic terms (and incorporated into economic assessment), but others cannot.

Before exploring methods of energy, economic, and environmental analysis, we first introduce some basic principles of life-cycle assessment.

5.1 Some Principles of Life-Cycle Thinking and Sustainability Analysis

Too often we make decisions based on our perceptions of costs and benefits today without thinking of costs and benefits over the long term. For example, we may think we're smart buying a 50¢ incandescent lightbulb, rather than a $2 compact fluorescent lamp, because it is cheaper. But life-cycle thinking tells us the opposite is true: the life-cycle cost of the electricity to operate the bulbs makes the compact fluorescent lamp far cheaper despite its higher initial price.

Considering both initial "capital" cost and operating cost in making decisions is the first step to life-cycle thinking. But we are interested not only in the effects of buying and using a product but also in the full costs and benefits of acquiring materials, manufacturing, transporting, installing, operating, and ultimately disposing of a product, the so-called "cradle-to-grave" costs and benefits.

We assume that the price we pay for energy and other products includes these full costs, but most often it does not. Coal-fired power is the cheapest electricity in the United States today, but the price we pay for it does not include the full costs of mining on communities; of mercury, NO_x, SO_x, and particulate emissions on human health; of CO_2 emissions on global climate change; and of ash disposal on land and water. We have tried to enact environmental regulations that integrate those costs into the costs of business, but they fall short of considering the full life-cycle costs and benefits.

Life-cycle **cradle-to-grave** thinking expands our thinking both backward and forward along the full process of product development and disposal. William McDonough and others extend this thinking further, imagining final waste products as opportunity resources to regenerate into other uses, or what they call **"cradle-to-cradle"** (see Chapter 8).

We will later apply life-cycle assessment to the following:

- "Embodied" energy or the energy it takes to develop, process, manufacture, and transport the materials used in a building or other product (Chapter 8)
- "Well-to-wheels" assessment to compare full energy, economic, and environmental costs for different transportation options from the fuel wellhead to the vehicle or passenger miles traveled (Chapter 13)
- "Gate-to-gate" assessment to compare energy and environmental costs for different materials manufacturing or processing from entry of raw materials to manufacturing or processing plant to exit of the product

Although most energy and environmental regulations do not promote full life-cycle thinking, the recent development of voluntary certification systems and labeling systems has incorporated life-cycle costs and impacts. These include the International Organization for Standardization (ISO) 14000 family of standards for environmental management systems (EMS) for industry, the United States Green Building Council's LEED green building certifications, and a large number of green labeling systems that are popular in Europe. ISO 14001 certificates increased by 37% in 2004 to more than 90,000 in 127 countries. The protocol's prescribed standards call for extensive use of life-cycle assessment.

The process of life-cycle assessment. Life-cycle analysis assesses the performance of an activity or product over its life cycle. Measures of performance include energy use, economic cost, social effects, and environmental impact. ISO 14000 standards specify a four-step process:

1. Define goals and scope, including system boundary and impact indicators.
2. Inventory impact activities.
3. Assess impacts.
4. Evaluate and interpret results.

The heart of the process is inventory and assessment. Life-cycle inventory involves detailed tracking of all the flows in and out at various stages of the system from cradle to grave. Figure 5.1 shows the system of wood products for house construction from cradle to grave including inputs and outputs. "Gate-to-gate" addresses flow in and out for product manufacturing only.

figure 5.1   Life-Cycle Inventory and Analysis

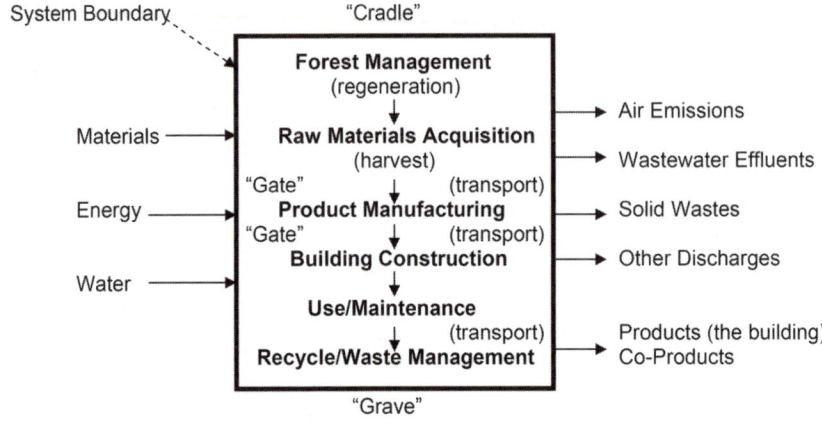

This life-cycle inventory and analysis uses an example of wood products for house construction.

Source: adapted from CORRIM, 2005

Because of the number and complexity of the processes involved and the lack of detailed information, this is not an easy task. It is simplified by focusing on a few impact indicators, such as energy used, carbon emissions, and pollutant emissions. These indicators are given as the product of inventory data (quantity of material) and impact coefficients or characterization factors (impact per quantity):

Eq. 5.1   Inventory data (e.g., lb steel) × Impact coefficient (e.g., lb CO_2/lb steel)
= Impact indicator (e.g., lb CO_2)

This is simple enough, but the challenge is finding accurate inventory data and reliable coefficients. We discuss some of these in Section 5.5.

5.2 Energy Analysis

Energy analysis begins by applying the principles of energy engineering to measure, estimate, or predict energy consumption and energy efficiency. For example, in Chapter 6 we will learn how to predict the heat loss for a building during winter using the laws of heat transfer, information on the size and material envelope of the building, and how cold the winter is at our location. We can use that information to calculate heating fuel requirements and compare those requirements for different assumptions of building insulation and materials to inform our choice of building design or whether we want to add insulation or a high-efficiency furnace.

In addition, we can use "boilerplate" specifications on the efficiency of lightbulbs, automobiles, air conditioners, refrigerators, and other consumer products, to calculate the operating energy requirements and useful energy outputs. We can use that energy information to calculate cost-effectiveness and environmental impacts so we can make smart choices about the products we buy and use.

We can also use energy analysis to calculate the energy it takes to produce energy. This is the energy we need, for example, to grow corn and process it into biofuel ethanol; to extract, transport, and refine Persian Gulf petroleum into gasoline at the pump; or to manufacture and install photovoltaic modules to produce electricity. By comparing input energy including indirect inputs to output energy, we can understand the viability of energy options. If it takes more energy to produce energy than the energy we get out, we should take pause, unless perhaps in the process we can replace depletable oil or carbon-emitting fossil fuels with some type of renewable energy.

Most of this book is on applications of energy analysis, and it is useful here to give an overview of the various measures used. Ultimately, we want useful energy (and the functions it provides) from our energy sources and systems, and we want to know what it takes in energy, dollars, and environmental costs to get it. In general, energy analysis compares useful energy outputs to necessary energy inputs, either by fraction (division) or difference (subtraction). **Net energy analysis** goes further by assessing indirect energy inputs. Two things to be wary of in energy analysis: (a) you must pay careful attention to units of energy and time by using dimensional analysis, and (b) you must pay careful attention to time period because some energy outputs and inputs are continuous and some are one-time only. Often a period of one year (annual output and input) is used as the time unit of analysis.

Before discussing specific metrics used in energy analysis, we need to define some terms. Figure 5.2 helps define the terms and introduces the metrics used in energy analysis of a conversion device or system. The figure is a useful reference as we discuss each metric.

figure 5.2　**Terms and Metrics Used in Energy**

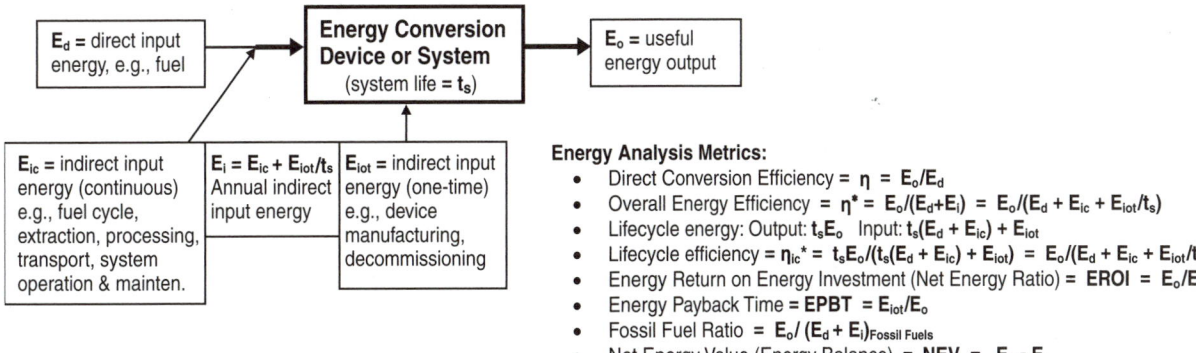

E_d = direct input energy, e.g., fuel

Energy Conversion Device or System (system life = t_s)

E_o = useful energy output

E_{ic} = indirect input energy (continuous) e.g., fuel cycle, extraction, processing, transport, system operation & mainten.

$E_i = E_{ic} + E_{iot}/t_s$ Annual indirect input energy

E_{iot} = indirect input energy (one-time) e.g., device manufacturing, decommissioning

Energy Analysis Metrics:
- Direct Conversion Efficiency = η = E_o/E_d
- Overall Energy Efficiency = η^* = $E_o/(E_d+E_i)$ = $E_o/(E_d + E_{ic} + E_{iot}/t_s)$
- Lifecycle energy: Output: t_sE_o　Input: $t_s(E_d + E_{ic}) + E_{iot}$
- Lifecycle efficiency = η_{lc}^* = $t_sE_o/(t_s(E_d + E_{ic}) + E_{iot})$ = $E_o/(E_d + E_{ic} + E_{iot}/t_s)$
- Energy Return on Energy Investment (Net Energy Ratio) = **EROI** = E_o/E_i
- Energy Payback Time = **EPBT** = E_{iot}/E_o
- Fossil Fuel Ratio = $E_o/(E_d + E_i)_{Fossil Fuels}$
- Net Energy Value (Energy Balance) = **NEV** = $E_o - E_i$

E_o = **useful energy (or power) output**, usually energy/time (e.g., energy/year)

E_d = **direct input of energy (or power) during operation** (e.g., fuel and/or electricity direct inputs at point of use, usually energy/time, e.g., energy/year)

E_i = **indirect input of energy (or power)** used to produce and transport to point of use (a) the conversion device (one-time cost) and (b) the direct input energy or power (continuous cost)

E_{ic} = **continuous energy costs** is usually given as energy/year.

E_{iot} = **one-time energy costs** is given by one-time energy, but can be converted to annual energy by dividing by lifetime of system in years to get energy/year. In other words,

$$E_i = \frac{E_{ic} + E_{iot}}{t_s}$$

where E_{ic} = continuous indirect energy inputs (energy/year)

E_{iot} = one-time indirect energy inputs (energy)

t_s = lifetime of system (years)

In previous chapters, we discussed mostly E_o, the useful energy output, and E_d, direct input energy or the fuel or energy input to an energy conversion system. However, it takes energy to make the direct input energy and the conversion systems we use, and "indirect input energy" tries to take that into account. For example:

- For a diesel generator, the direct input energy (E_d) is the diesel fuel. The indirect input energy (E_i) includes the energy required to extract, refine, and transport petroleum and its diesel product to the site of the generator as well as the energy required to manufacture the generator, transport it to sale, and transport it to the site of use. The indirect input energy of the fuel is a continuous cost (E_{ic}), whereas the indirect input energy to make the generator is a one-time cost (E_{iot}).
- For a photovoltaic battery system, the direct input energy is sunlight, essentially free, so $E_d = 0$. The indirect energy input is the energy required in the manufacture of the photovoltaic modules, batteries, and other components and their transport and installation on site, a one-time cost (E_{iot}).

The following sections describe each of the energy analysis metrics given in Figure 5.2, beginning with our old favorite, direct energy conversion efficiency.

5.2.1 Direct Conversion Efficiency (η)

Direct conversion efficiency is the most useful metric of efficiency and performance for energy system assessment. It describes the efficiency of a system to convert direct input energy to output energy.

Eq. 5.2   Direct conversion efficiency $= \eta = \dfrac{E_o}{E_d}$ $0 \leq \eta \leq 1$

This metric is especially useful to compare options when we start and end at common points in the energy conversion process. For example, we can compare the cumulative efficiencies from "power plant to wheels" of a fuel-cell car with electric grid electrolysis for hydrogen versus an all-electric car with electric grid charging of lithium-ion batteries. We can find cumulative direct conversion efficiency by multiplying each component efficiency from the same start point (power plant) to the same end point (energy to the wheels). This is similar to how we determined the efficiency of a small hydro plant in Solution Box 4.2.

The third column of Table 5.1 gives component efficiencies for the two systems. Cumulative efficiencies are simply the product of component efficiencies. What is the cumulative efficiency of each option?

$$\eta = \eta_{cumulative} = \textbf{Product of component } \eta$$

Fuel-cell car: $\eta_{cumulative} = \eta_{pp} \times \eta_{trans} \times \eta_{electrol} \times \eta_{H_2comp} \times \eta_{fuelcell} \times \eta_{motor/wheels}$

$\eta_{cumulative} = 0.33 \times 0.96 \times 0.81 \times 0.90 \times 0.40 \times 0.92 = 0.085 = 8.5\%$

All-electric car: $\eta_{cumulative} = \eta_{pp} \times \eta_{trans} \times \eta_{charger} \times \eta_{battery} \times \eta_{motor/wheels}$

$\eta_{cumulative} = 0.33 \times 0.96 \times 0.93 \times 0.93 \times 0.92 = 0.252 = 25.2\%$

Table 5.1 tracks the cumulative efficiency through the different components. Given these assumptions, the all-electric car is three times more energy efficient than the fuel-cell car.

Direct conversion efficiency has limitations when we try to compare systems and sources that do not have the same starting and ending points. For example, it is straightforward to calculate the direct conversion efficiency for a diesel generator and a photovoltaic battery system:

- What is the direct conversion efficiency for a 10,000-watt diesel generator that consumes 1 gallon of diesel fuel per hour (138,000 Btu/gal)?

$$\eta = \frac{E_o}{E_d} = \frac{10,000 \text{ W}}{1 \text{ gal/hr} \times 138,000 \text{ Btu/gal} \times \dfrac{\text{Wh}}{3.412 \text{ Btu}}} = 0.25 = 25\%$$

- What is the direct conversion efficiency for a 100 m^2 silicon cell photovoltaic battery that produces 10,000 watts of electricity in full sun (1000 w/m^2)?

$$\eta = \frac{E_o}{E_d} = \frac{10,000 \text{ W}}{1000 \text{ W/m}^2 \times 100 \text{ m}^2} = 0.10 = 10\%$$

So, the diesel generator has 25% efficiency and the PV system has 10% efficiency. Does that tell the whole story? The two systems end at the same point but do not start the same. The diesel system needs a steady supply of nonrenewable fuel that takes energy to produce

table 5.1 **Comparing Direct Conversion Efficiency for Fuel–Cell Vehicle and All–Electric Vehicle**

System	Conversion	Component Efficiency	Fuel-Cell Car Cumulative Direct Conversion Efficiency	All-Electric Car Cumulative Direct Conversion Efficiency
Power plant	Fuel to kWh	33%	33.0%	33.0%
Transmission	kWh to kWh	96%	31.7%	31.7%
Electrolysis	kWh to H_2	81%	25.7%	—
H_2 compressor	H_2 to H_2	90%	23.1%	—
Fuel cell	H_2 to kWh	40%	9.2%	—
Charger	kWh to A-hr	93%	—	29.5%
Battery	A-hr to A-hr	93%	—	27.4%
Motor to wheels	kWh to ft-lb	92%	8.5%	25.2%

and get to the site. The PV system has no direct or continuous indirect energy requirements, but needs indirect one-time energy initially to manufacture the cells, modules, and batteries and transport and install them on site.

Other metrics consider indirect energy and differences in energy sources and systems.

5.2.2 Overall Energy Efficiency (η^*)

Overall energy efficiency takes into account both direct and indirect input energy.

Eq. 5.3 $$\text{Overall energy efficiency} = \eta^* = \frac{E_o}{E_d + E_i} \qquad 0 \leq \eta^* \leq 1$$

If the indirect input energy E_i includes one-time costs included in the manufacture and transport of conversion systems, those energy costs need to be spread over the life of the system.

Eq. 5.4 $$\eta^* = \frac{E_o}{E_d + E_{ic} + \dfrac{E_{iot}}{t_s}}$$

For example, the overall efficiency of our diesel generator would include the energy it takes to extract crude oil, refine it, and transport it to the generator. Let's say that indirect energy is 10% of the energy content of the resulting fuel. The resulting overall efficiency would be

$$\eta^* = \frac{E_o}{E_d + E_i} = \frac{10{,}000 \text{ W}}{1.1 \times 1 \text{ gal/hr} \times 138{,}000 \text{ Btu/gal} \times \dfrac{\text{Wh}}{3.412 \text{ Btu}}} = 0.227 = 22.7\%$$

Because η^* has the same numerator as η but a larger denominator, the fraction is smaller than η.

5.2.3 Life-Cycle Energy Efficiency ($\eta_{lc}*$)

Life-cycle analysis is useful because it consciously takes a long-term view of inputs and outputs. Life-cycle efficiency is essentially the same as $\eta*$ but considers the output energy for the life of the system and all indirect input energy for the life of the system. Nuclear power systems, for example, have energy requirements for the fuel cycle (including waste management) and plant decommissioning that may be extensive and long-lasting, so their life-cycle costs include the following:

- The fuel cycle (E_{ic})
- Construction and materials (E_{iot})
- Operation/maintenance (E_{ic})
- Decommissioning (E_{iot})

Eq. 5.5 $\qquad$ **Life-cycle energy output (lco) = $t_s E_o$**

where $\quad E_o$ = constant annual output, energy/year
$\qquad$ t_s = system life, years

Eq. 5.6 $\qquad$ **Life-cycle energy input (lci) = $t_s(E_d + E_{ic}) + E_{iot}$**

where $\quad E_d$ = constant annual direct inputs, energy/yr
$\qquad$ E_{ic} = constant annual indirect inputs, energy/yr
$\qquad$ E_{iot} = one-time indirect inputs, energy

Eq. 5.7 $\qquad$ **Life-cycle efficiency = $\eta_{lc}* = \dfrac{t_s E_o}{t_s(E_d + E_{ic}) + E_{iot}} = \dfrac{E_o}{E_d + E_{ic} + \dfrac{E_{iot}}{t_s}}$**

5.2.4 Energy Return on Energy Investment (EROI)

This measure ignores direct energy input and compares useful output energy to indirect input energy or the energy it takes to get energy. It indicates how much energy other than direct fuel energy input must be invested to get a unit of useful energy.

Eq. 5.8 $\qquad$ **Energy return on energy investment = EROI = $\dfrac{E_o}{E_i}$**

Ideally, EROI is greater than one. If it is less than one, the system takes more indirect input energy to produce useful output energy and it may not be worth it, unless we can replace depletable oil and other carbon-emitting fossil energy with renewable energy in the process.

The analysis time period or product unit must be consistent in the numerator and denominator. If E_i includes only continuous inputs, E_o and E_i can be measured as energy per year or energy per unit product (such as gallon of equivalent fuel or kWh of electricity). If E_i is primarily a one-time input (such as for wind or PV systems) then life-cycle E_o and E_i should be used.

Eq. 5.9
$$EROI = \frac{t_s E_o}{E_{iot}}$$

when $E_i = E_{iot}$, and E_o in energy/year

This metric can illustrate the diminishing returns of our energy investment in energy production as we move toward depletion of conventional energy sources. It is taking more energy to make energy. In the early days of oil production, for example, reservoirs were easy to tap and indirect energy investments were low. As we have depleted these reservoirs, we have had to go deeper and farther to get oil and it has taken more energy to do that. The same has been true for coal and natural gas. According to a study by Cleveland et al. (2005), the EROI of U.S. oil production in the 1930s was 100 to 1. In 2000, it was 20 to 1. For new discoveries, it is 8 to 1. Coal production had an EROI of 100 in 1950; in 2000, it was about 80.

As we move toward unconventional fossil fuels, this becomes even more of an issue. Deep offshore oil deposits, heavy crude deposits, oil sands, and oil shale require more energy to extract and process, reducing EROI. One estimate of Canadian oil sands gives an EROI of 3. This is a double whammy for carbon emissions because more fossil fuel combustion is needed just to produce useful fuels so they can be burned.

EROI can also be used to assess the viability of new energy options on an energy return basis, such as biofuel ethanol, coal gasification, hydrogen fuel-cell systems, and others. Table 5.2 gives the results of several studies of EROI. The top five entries show the diminishing returns of the U.S. oil, gas, and coal production industries. Undocumented estimates of oil shale and oil sands EROI are 3 to 3.5. The next nine entries give EROI for several electricity generating options. Hydro has a high EROI of 205, and some renewable sources of electricity (wind [80], sawmill wastes [27], and photovoltaic [9]) fare better than coal with scrubbers (5) and natural gas combined cycle with 2000 km pipeline (5). Nuclear EROI is estimated at 16; EROI of solar photovoltaics is estimated at 7 to 14 depending on the technology.

Of course, all the results depend on the assumptions of indirect energy inputs that vary from one study to the next, so it is difficult to compare one result to another. This is dramatically clear in the current debate over biofuels. In a battle of scientific studies, academic and government researchers have calculated EROI for corn-based ethanol ranging from 0.78 to 1.29, and for switchgrass-based ethanol from 0.79 to 10.3. See Section 5.2.8 on energy analysis of ethanol.

5.2.5 Energy Payback Time (EPBT)

Energy Payback Time (EPBT) gives the time it takes an energy system to recover its one-time input energy with output energy. It divides one-time input energy by annual energy output. It is a particularly useful measure for renewable energy systems such as wind and photovoltaics, the costs of which are dominated by one-time development. For such systems, it is equivalent to 1/EROI.

table 5.2 Energy Return on Energy Investment (EROI) for Various Energy Sources/Systems

Source/System	EROI	Literature Source
U.S. oil and gas production, 1930	100	Cleveland (2005)
U.S. oil and gas production, 2000	20	Cleveland (2005)
U.S. gasoline production	7	Cleveland (2005)
U.S. coal production, 1950	100	Cleveland (2005)
U.S. coal production, 2000	80	Cleveland (2005)
Electricity from hydro with reservoir	205	Gagnon, et al. (2002)
Electricity from wind	80	Gagnon, et al. (2002)
Electricity from sawmill wastes	27	Gagnon, et al. (2002)
Electricity from nuclear	16	Gagnon, et al. (2002)
Electricity from PV modules	9	Gagnon, et al. (2002)
Electricity from coal (with SO_2 scrubbers)	5	Gagnon, et al. (2002)
Electricity from natural gas CC (2000 km del)	5	Gagnon, et al. (2002)
Electricity from biomass plantation	5	Gagnon, et al. (2002)
Electricity from fuel cell, H_2 from NG reform	2	Gagnon, et al. (2002)
PV modules (thin-film CIS)	14	Knapp and Jester (2001)
PV modules (crystalline silicon)	7	Knapp and Jester (2001)
U.S. ethanol fuel from corn	1.67	Shapouri, et al. (2004)
U.S. ethanol fuel from corn	0.78	Pimentel and Patzek (2005)
U.S. ethanol fuel from corn	1.29	Farrell, et al. (2006)
U.S. ethanol fuel from switchgrass	0.79	Pimentel and Patzek (2005)
U.S. ethanol fuel from switchgrass	10.3	Wang (2005)
U.S. ethanol fuel from switchgrass	8.3	Farrell, et al. (2006)
U.S. biodiesel from soybeans	3.20	U.S. DOE, USDA (1998)
U.S. biodiesel from soybeans	0.67	Pimentel and Patzek (2005)

Eq. 5.10
$$\text{Energy payback time} = \text{EPBT} = \frac{E_{iot}}{E_o}$$

Eq. 5.11
$$\text{For } E_i = E_{iot}, \text{ EPBT} = \frac{1}{\text{EROI}}$$

A good example of EPBT is Knapp and Jester's (2001) study of photovoltaic modules. They estimated total indirect input energy for manufacturing and installing two types of PV modules:

- Crystalline silicon PV modules: $E_i = E_{iot} = 5600 \text{ kWh/kW}_p$
  (kilowatt-hour electric input per kilowatt peak power output of module)
- Thin-film copper indium diselenide (CIS) modules: $E_i = E_{iot} = 3100 \text{ kWh/kW}_p$

The kW_p of the module occurs at peak sun (1 kW/m^2). A typical average value for solar energy falling on a site in the United States is $1700 \text{ kWh/m}^2/\text{year}$. That means that each kW_p will produce 1700 kWh per year on average for the United States. System losses due to wires, inverters, operating temperatures, and so on, were accounted for by a performance ratio (PR) of 0.8. The study calculated EPBT for the two types of PV modules.

Given these values, the EPBT calculations are easy. Without the performance ratio considered,

$$\text{EPBT (silicon)} = \frac{E_{iot}}{E_o} = \frac{5600 \text{ kWh/kW}_P}{1700 \text{ kWh/kW}_P} = 3.3 \text{ years}$$

$$\text{EPBT (thin-film CIS)} = \frac{E_{iot}}{E_o} = \frac{3100 \text{ kWh/kW}_P}{1700 \text{ kWh/kW}_P} = 1.8 \text{ years}$$

To consider the performance ratio (PR), E_o is multiplied by PR or EPBT is divided by PR:

$$\text{EPBT w/PR (silicon)} = \frac{\text{EPBT}}{\text{PR}} = \frac{3.3 \text{ years}}{0.8} = 4.1 \text{ years}$$

$$\text{EPBT w/PR (thin-film CIS)} = \frac{\text{EPBT}}{\text{PR}} = \frac{1.8 \text{ years}}{0.8} = 2.2 \text{ years}$$

If the expected life of these modules is 30 years, both PV modules have an energy payback period well within their lifetime.

We can then calculate the EROI for these modules, assuming the PR of 0.8:

$$\text{EROI} = \frac{E_o}{E_i} = \frac{t_s E_o}{E_{iot}} = \frac{t_s}{\text{EPBT}}$$

$$\text{EROI (silicon)} = \frac{30 \text{ years}}{4.1 \text{ years}} = 7$$

$$\text{EROI (thin-film CIS)} = \frac{30 \text{ years}}{2.2 \text{ years}} = 14$$

These are the EROI values that appear in Table 5.2.

5.2.6 Net Energy Value (NEV) or Energy Balance

Net energy value (NEV) is similar to EROI, but it uses the difference rather than the ratio to compare useful output energy to input energy.

Eq. 5.12   Net energy value = NEV = $E_o - E_i$

Because this is an absolute value rather than a dimensionless ratio, it is computed as energy per unit of energy or fuel, such as Btu/gal. Farrell, et al. (2006), argue that the NEV metric is more robust than EROI, especially when there is question about how to treat co-product energy, or the energy content of feed and fuel outputs other than the primary product. For example, production of ethanol not only produces ethanol liquid fuel, but also co-product feed and solid fuel that have energy value. The EROI value depends significantly on whether the co-product energy is treated as a negative input (in the denominator of the ratio) or a positive output (in the numerator). With NEV it doesn't matter.

NEV and EROI calculations vary depending on differing assumptions of indirect input and output energy and EROI depends on treatment of co-product energy. These assumptions can be quite contentious, as we see in the continuing debate about the net energy of corn-based ethanol (see Section 5.2.8).

5.2.7 Fossil Fuel Ratio (FFR) and Petroleum Input Ratio (PIR)

Like EROI, fossil fuel ratio (FFR) is the ratio of energy output to energy input, but it includes both direct and indirect fossil fuel input energy. Petroleum input ratio (PIR) is similar but it focuses on direct and indirect petroleum input and inverts the ratio to energy input to output. These are useful measures because they reflect relative dependency of petroleum or fossil fuels (see Figures 5.3 and 5.4).

Eq. 5.13 $$\text{Fossil fuel ratio} = \text{FFR} = \frac{E_o}{(E_d + E_i)_{FF}}$$

where $(E_d + E_i)_{FF}$ = direct + indirect fossil fuel energy input

Eq. 5.14 $$\text{Petroleum input ratio} = \text{PIR} = \frac{(E_d + E_i)_{Petro}}{E_o}$$

where $(E_d + E_i)_{Petro}$ = direct + indirect petroleum energy input

Michael Wang from Argonne National Lab argues that FFR and PIR are better measures than EROI in addressing our national objectives to reduce oil use and the greenhouse gas emissions from combustion of fossil fuels. Although gasoline has a better EROI (8) than corn ethanol (1.2 to 1.7), gasoline's FFR includes the petroleum fuel energy and the FFR ratio is 0.81, while ethanol's FFR is 1.79. Figure 5.3 illustrates Wang's analysis of the fossil energy to produce 1.0 million Btu of fuel: it takes 0.56–0.74 million Btu of fossil fuels for corn ethanol (depending on energy credit for co-products) and twice that, or 1.23 million Btu of fossil fuels, for gasoline.

5.2.8 Applying Energy Analysis to Biofuel Ethanol

The 2005 and 2007 Energy Policy Act, provide significant incentives for the production of biofuel ethanol. Compared to 2004, oil companies are mandated to double their use of ethanol in transportation fuels by 2012 and increase it by ten times by 2022. Is this a good idea?

Well, it depends on whom you talk to. And that makes ethanol a good example of the importance of assumptions made in energy analysis studies. David Pimentel from Cornell has long argued that ethanol fuel from corn is an energy loser (i.e., it takes more energy to produce it that it would replace). On the other hand, Hosein Shapouri from USDA and others have come up with different results showing ethanol to have an EROI greater than one and

figure 5.3   Fossil Fuels Needed to Produce One Million Btu of Fuel Ethanol and Gasoline

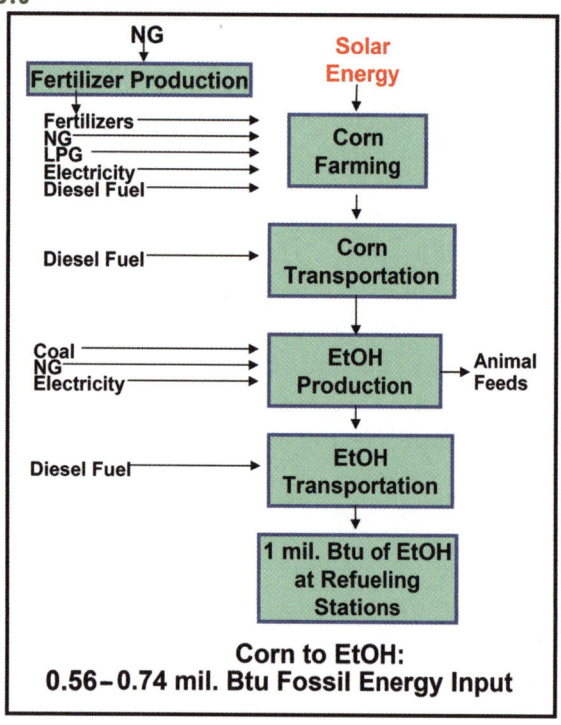
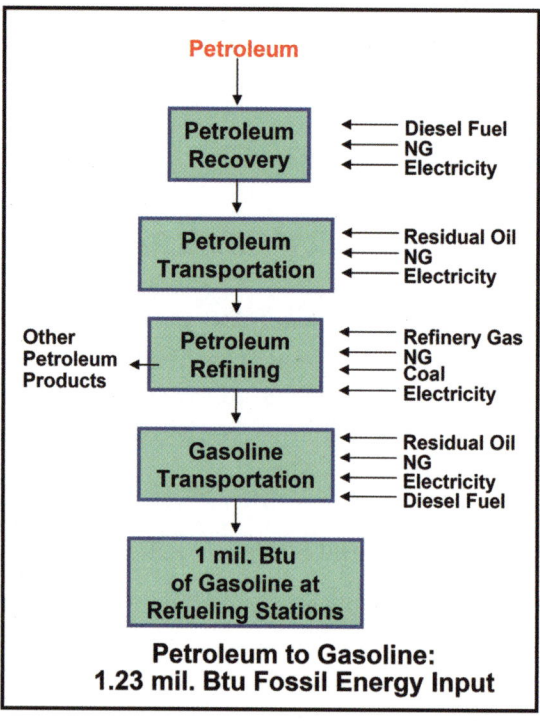

Corn to ethanol (EtOH) takes 0.56–0.74 million Btu (range depends on energy credits for co-products): FFR = 1/0.56 = 1.79. Petroleum to gasoline takes 1.23 million Btu: FFR = 1/1.23 = 0.81.

NG = natural gas; LPG = liquid petroleum gas
Source: Wang, 2005

a positive NEV. Why the difference? Ethanol advocates argue Pimentel uses outdated corn production and ethanol processing data, discounts co-product energy, and wrongly takes the analysis back too far (e.g., to indirect energy required in manufacturing the farm equipment needed to grow the corn). Pimentel argues that Shapouri and others omit some of the energy inputs in the ethanol production process. Who is right?

Farrell, et al. (2006), at the Energy and Resources Group (ERG), University of Calfornia-Berkeley, tried to bring some order to this debate by adjusting various studies with common assumptions. They also conducted their own analysis using their ERG Biofuel Analysis Meta-Model (EBAMM). The comparison of net energy and petroleum input ratio (Figure 5.4) shows positive net energy with the exception of Pimentel and Patzek. PIR values also vary, but all are 80% to 95% better than gasoline. The Berkeley group's assessment of cellulosic ethanol has significantly higher net energy than all of the studies of corn-based ethanol.

Calculating net energy and EROI for ethanol. Some calculations for corn- and cellulose-based ethanol can illustrate the process of net energy analysis and the importance of calculated or assumed values of energy inputs and outputs to the results. In addition, we can

figure 5.4 Net Energy and Petroleum Input for Fuel Ethanol and Gasoline

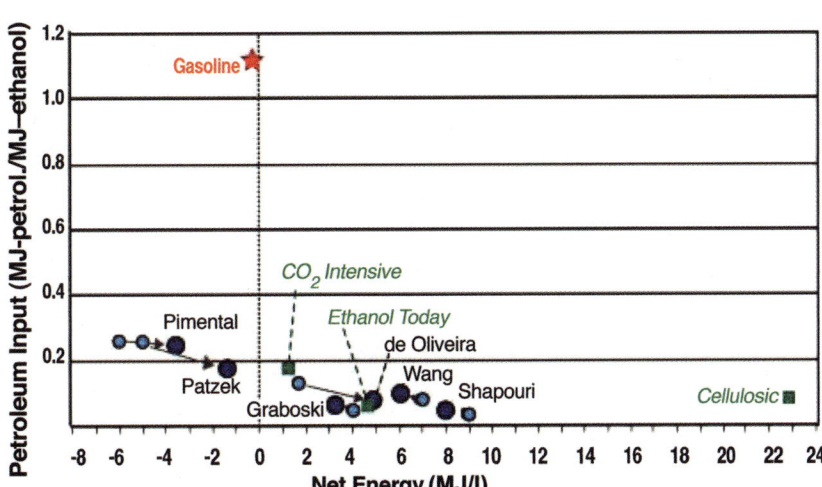

Farrell et al. (2006) plot PIR and NEV results from various studies of corn feedstock ethanol, adjusted to common assumptions (arrows to large dots), as well as gasoline and their own EBAMM cases—cellulosic ethanol, ethanol today, and CO_2-intensive ethanol. All ethanol cases show considerable petroleum savings over gasoline. Cellulosic ethanol has much higher net energy than corn ethanol.

SOURCE: Farrell, et al., SCIENCE 311:506–508 (2006). Reprinted with permission from AAAS.

also see that EROI ratio depends on whether co-product energy is treated as negative input or positive output, especially when it is large compared to input energy as it is in the Farrell et al. cellulose case. Table 5.3 gives the values of the input, output, and co-product energy estimated in three well-cited studies. We then step through calculations of net energy (NEV) and EROI. We use Btu/gallon of ethanol, but then convert to megajoules per liter (MJ/l) (1 Btu/gal = 2.79×10^{-4} MJ/l) to be consistent with the values in Figure 5.4.

We employ our Equations 5.12 and 5.8:

$$NEV = E_o - E_i$$
$$EROI = \frac{E_o}{E_i}$$

Pimentel and Patzek (2005) estimate higher input energy and no credit for co-product energy:

$$NEV = 77,053 \text{ Btu/gal} - 99,096 \text{ Btu/gal} + 0 \text{ co-product} = -22,043 \text{ Btu/gal} = -6.1 \text{ MJ/l}$$
$$EROI = \frac{77,053 \text{ Btu/gal}}{99,096 \text{ Btu/gal}} = 0.78$$

Shapouri, et al. (2004), estimate lower input energy and a significant credit for co-product energy:

table 5.3	Contradictory Studies on NEV and EROI for Fuel Ethanol from Corn				
Study	E_o, Output Ethanol Energy, Btu/gal	E_i, Input Energy, Btu/gal	Co-Product Energy, Btu/gal	$E_o - E_i$, Net Energy Value (NEV), Btu/gal	$\frac{E_o}{E_i}$ = EROI
Shapouri, et al. (2004)	76,330	71,800	26,000	30,528	1.67/1.43
Pimentel and Patzek (2005)	77,053	99,096	0	−22,043	0.78
Farrell, et al. (2006), corn	76,060	74,265	14,700	10,505	1.29/1.22
Farrell, et al. (2006), cellulose	76,060	11,120	17,200	82,140	8.3

$$NEV = 76{,}330 \text{ Btu/gal} - 71{,}800 \text{ Btu/gal} + 26{,}000 \text{ co-product} = 30{,}528 \text{ Btu/gal} = 8.5 \text{ MJ/l}$$

$$EROI = \frac{76{,}330}{71{,}700 - 26{,}000} = 1.67 \text{ (if co-product treated as negative input)}$$

$$EROI = \frac{76{,}330 + 26{,}000}{71{,}700} = 1.43 \text{ (if co-product treated as positive output)}$$

Farrell, et al. (2006), estimate slightly higher input and lower co-product than Shapouri:

$$NEV = 76{,}060 - 74{,}265 + 14{,}700 = 16{,}495 \text{ Btu/gal} = 4.6 \text{ MJ/l}$$

$$EROI = \frac{76{,}060}{74{,}265 - 14{,}700} = 1.29 \text{ (if co-product treated as negative input)}$$

$$EROI = \frac{76{,}060 + 14{,}700}{74{,}265} = 1.22 \text{ (if co-product treated as positive output)}$$

Farrell, et al., also estimate inputs and co-products for cellulosic ethanol:

$$NEV = 76{,}060 - 11{,}120 + 17{,}200 = 82{,}140 \text{ Btu/gal} = 22.9 \text{ MJ/l}$$

$$EROI = \frac{76{,}060}{11{,}120 - 17{,}200} = -12.5 \text{ (meaningless) (if co-product treated as negative input)}$$

$$EROI = \frac{76{,}060 + 17{,}200}{11{,}120} = 8.3 \text{ (if co-product treated as positive output)}$$

5.2.9 Accounting for Energy Quality in Net Energy Analysis

Net energy analysis and other metrics estimate direct and indirect input energy needed to produce useful energy. This input energy can vary from raw fuel to electricity. Does the type of input energy matter? Are all joules created equal? The second law of thermodynamics says that they are not. Higher quality, lower entropy energy is more precious than lower quality thermal energy. If, for example, an energy source or system requires input energy of high-quality electrical joules, whereas another requires an equal amount of input energy joules as low-quality heat, the first will be harder to come by. So we might try to account for that in our energy analysis.

This is easier said than done. Analysts have developed approaches to aggregate input energy by quality using *emergy* and *exergy* as measures of quality. Emergy analysis was first developed by Howard Odum. The standard units of emergy are solar emjoules (SEJ) or the solar energy needed to produce another type of energy. For example, each heat unit of electricity (1 joule) is 1.59×10^5 SEJ, which is derived from the sunlight required to produce 1 joule of standing wood (3.23×10^4 SEJ) as well as losses incurred in the harvesting, transport, and steam cycle to convert the wood to electricity.

Exergy embraces the second law and is defined as the potential for mechanical work that can be extracted from a type or flow of energy. For example, the exergy of a fuel or heat source is calculated by multiplying its heat equivalent by the Carnot factor ($1 - T_a/T_o$), where T_a is ambient temperature and T_o is output temperature, both in kelvin (see Chapter 10 for discussion of Carnot efficiency). High-quality mechanical and electrical energy are not constrained by this factor because they can be transformed directly to useful work. Using exergy analysis, input energy can be differentiated by quality. Similar to emergy, however, exergy has conceptual appeal, but limited practical value.

A third approach used to incorporate energy quality in energy analysis uses energy prices as a proxy for energy quality. Energy prices reflect value and value reflects quality. On a per-energy-unit basis, electricity has a higher price than natural gas, natural gas has a higher price than petroleum products, and petroleum products have a higher price than coal. Using price as an indicator for quality, the EROI equation can be rewritten as a "quality corrected EROI" or the ratio of the sum of individual energy outputs and inputs each multiplied by a quality factor based on energy price.

These heroic efforts to be true to the second law and incorporate energy quality in energy analysis are valiant, but we should first try to find common ground on basic assumptions on energy inputs in simple net energy assessments before complicating the analysis with energy quality considerations.

5.3 Energy Monitoring and Energy Audits

We wish to use the results of energy analysis to make smart energy choices, to design and manage energy systems, and to use energy more efficiently. Energy analysis requires good information and the best data come from physical monitoring of systems, of energy consumption, and of functions performed.

Energy monitoring can be as simple as reviewing your monthly electric utility bill or jotting down your car's odometer reading when filling up with gas so you can compute miles per gallon. It can be as complicated as installing a multifunction computer datalogger that retrieves data on energy use and ambient conditions and sends the results to a distant receiving location via wireless technology. Whatever means are used, the point to remember is this: the better the data, the better are the analysis and results, and the better informed are the decisions that follow.

Energy monitoring is an important component of an **energy audit,** an analytical approach to assess energy consumption and identify potential efficiency improvements.

Before reviewing some basic energy monitoring methods, we need to introduce energy audits.

5.3.1 Energy Audits

Energy audits apply energy analysis methods to evaluate patterns and trends of energy consumption and efficiency opportunities in households, government agencies, and private commercial and industrial firms. Auditing is applied mostly to buildings (see Chapter 6) but also to transportation fleets and industrial processes. It is an important first step in energy management services. Although we will discuss methods used in energy auditing in later chapters, a brief introduction is useful here.

According to the American Society of Heating, Refrigerating, and Air-Conditioning Engineers (ASHRAE), which sets standards for building energy systems and audits, there are three levels of energy audits:

Level 1: Walk-through or visual assessment—Rapid assessment looks for energy problems and solutions that are easily identified and helps scope out needed monitoring and analysis.

Level 2: Energy survey and analysis—This standard audit includes monitoring of historical utility billing data, submeter data, and use of monitoring and diagnostic equipment where necessary. The standard audit will identify potential energy conservation measures (ECMs) and often calculate their cost-effectiveness.

Level 3: Detailed analysis of capital intensive modifications—This extensive audit goes beyond basic analysis and may employ computer simulations, more detailed monitoring, and more sophisticated economic assessment of major modifications to the building or industrial process.

A basic procedure for Level 1 and 2 audits includes the following steps:

1. Perform preliminary walk-through to identify audit goals and objectives.
2. Analyze billing data from energy suppliers to determine energy consumption trends.
3. Use submetered data as available.
4. Review specifications, mechanical drawings, and other information on energy systems and equipment, building envelope, lighting, and so on, to assess opportunities for efficiency improvements.
5. Perform facility walk-through and diagnostics, including interviews with users and use of diagnostic devices such as lighting monitors, blower door (for air leakage), and so on.
6. Monitor energy systems and equipment using submetering devices and data loggers.
7. Synthesize results and findings.
8. Identify potential ECMs, conduct economic analysis of options, and prepare final report with recommendations.

We discuss step 2 using billing data and step 6 on monitoring below, step 8 on economic analysis in the next section, and other auditing methods for buildings in Chapters 6–8.

5.3.2 Monitoring with Energy Billing Information

Energy utilities and companies that provide fuel and electricity monitor energy sales for billing purposes. Electric and natural gas utilities have cumulative kWh and gas meters on our houses that they read monthly to determine our consumption and to bill us accordingly; fuel oil distributors fill our tanks, using a flow meter to measure their sale in gallons; gas stations also use flow meters to measure the gallons we buy at the pump. We can use this monitored energy sale information to calculate consumption and efficiency of use. Solution Box 5.1 gives my first assessment of fuel economy of my Toyota Prius.

In our homes, our utility bills are our best source of energy monitoring data, and as discussed in the last section, one of the first data sources in energy audits. These bills give us a monthly record of consumption of electricity (kWh) and natural gas (therms = 100,000 Btu, or about 100 cubic feet [ccf]). We can plot the results to see variation from month to month, from heating to cooling season, from year to year. We can see changes resulting from new appliances or from energy conservation measures.

Some more sophisticated methods have been developed to track and analyze utility bill data. For example, PRISM, an analytical computer software developed at Princeton, uses data from weather records and utility bills to assess historical heating and cooling energy use. This method has been useful in evaluating efficiency intervention, such as weatherization retrofit, when other monitoring methods are not used. Because the billing data on energy consumption are stored with the utility, they can be accessed at any time. Samples of energy data before and after the intervention can be corrected for weather variation and compared to see what savings were achieved by the weatherization.

Several energy service vendors market utility bill data tracking software and online services to help companies and institutions manage their energy use. Examples include UtiliVision's Energy Watchdog (http//www.energywatchdog.com) and Abraxas Energy Consulting's Metrix Utility Accounting System (http://www.abraxasenergy.com/metrix.php).

5.3.3 Energy Data Logging

Data logging involves the use of meters and loggers to measure energy use and functions performed for energy analysis and evaluation studies. The most commonly used meters are the same ones used by utilities for billing purposes, such as the kWh (Figure 5.5) and gas meters on a building or housing unit. We may want more detailed or site-specific data than these meters provide, so we submeter smaller units or individual equipment or appliances to get this detail. Some meters measure electricity consumption, but others measure the time that electricity is flowing. These latter **"run-time" meters** simply measure the cumulative

SOLUTION

SOLUTION BOX 5.1

Using Gasoline Receipts to Monitor Vehicle Fuel Economy

I bought a Toyota Prius in 2005 and wanted to know its fuel economy compared to the EPA ratings.

Solution:

I kept my gasoline receipts for the first four tankfuls in my Prius and recorded on each the odometer mileage when I filled it up. I bought the car with only 5 miles on the odometer, and the dealer filled it up at that time. I logged the date, gallons, and mileage in the following table, then calculated the miles per gallon (mpg) efficiency for each tankful and cumulative.

$$mpg = \frac{miles\ per\ tank}{gallons\ per\ tank\ fill\text{-}up}$$

The miles per tank is the odometer reading minus the odometer reading at last fill-up.

For the first tankful: Tankful #1: $mpg = \frac{(410 - 5)}{8.5} = 47.6$ mpg

The cumulative mpg = total miles/total gallons = $\frac{(1698 - 5)}{36.4} = 46.5$ mpg

Date	Miles	Fill-Up Gallons	Miles/Tankful	Miles per Gallon
4/30	5	??	NA	NA
5/20	410	8.5	405	47.6
6/15	833	8.9	423	47.5
7/7	1255	9.2	422	45.9
8/1	1698	9.8	443	45.2
Total		36.4	1693	46.5

How do I interpret these results? The mpg is less than the EPA estimate for the Prius (60 city, 51 highway), but I've heard this is typical for most cars. The mpg decreased since my first tankful. I began to monitor the type of driving that dominated each tankful, such as highway, city, short trips, long trips, as well as tire pressure and the way I drive. By using my monitoring results, I began to fine-tune my driving to improve efficiency. Two years later I am getting 50+ mpg on an average tankful.

time the equipment is on. They are useful for thermostatically controlled devices such as oil and gas furnaces and refrigerators. For oil and gas furnaces, if we know the run-time and the furnace firing rate (fuel volume per minute), we can calculate fuel use. Sidebar 5.1 describes several submeters on the market.

SIDEBAR 5.1

Examples of Submeters for Energy Monitoring

A. **Honeywell Programmable Digital Thermostat** not only allows you to program temperature settings by time but also logs run-time of the furnace or heat pump and thus can be used for monitoring energy.

B. **Kill-a-Watt submeter** is simply plugged into the outlet, and it measures power draw (W), energy used (Wh), and run-time of device plugged into it.

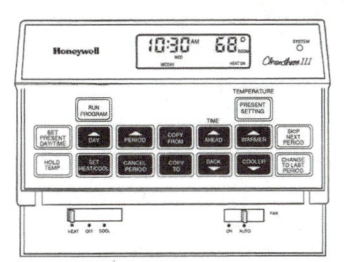

C. **Electric Usage Monitor** is a submeter with current clips that can be clipped to any circuit and measure power (kW), energy (kWh), and run-time. It can also be programmed with electrical rates to read off dollars instead of energy and can be used to project energy use over a long period of time.

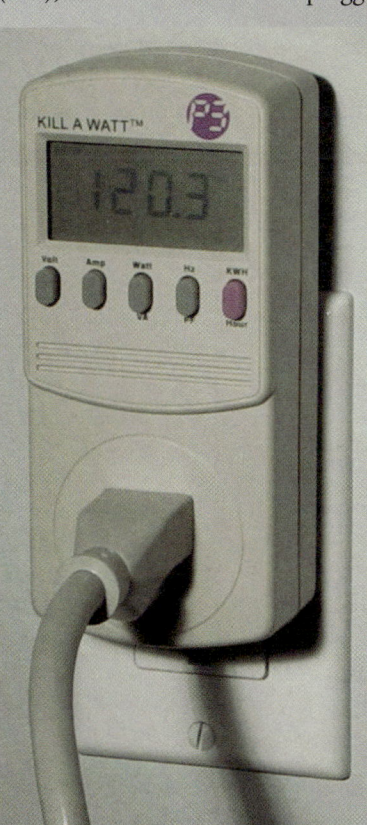

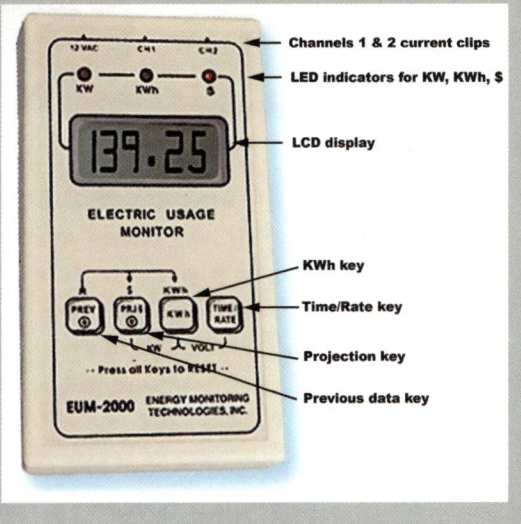

Dataloggers are electronic devices that store digital data retrieved from different sensors. These sensors can measure temperature, pressure, light, run-time, weather conditions, and energy parameters. The data can be easily downloaded to a computer and converted to spreadsheet form. Some loggers are equipped with modems or telemetry systems where phone

SIDEBAR 5.2

Examples of Dataloggers for Energy Monitoring

A. The **HOBO series** of dataloggers by Onset are the size of a matchbox and they can be programmed to collect various data at desired time intervals. Built-in or remote plug-in sensors can be used. After monitoring, the HOBO can be plugged into a computer USB port to download data to spreadsheet software.

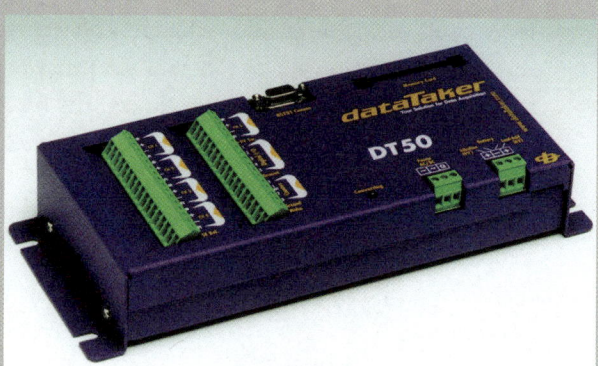

less technology and downloaded automatically to a receiver or receiving computer. The battery for the datalogger and transceiver is trickle charged by a small PV array.

B. **DataTaker datalogger** is a conventional datalogger with multiple ports for various data sensors. Retrieved data can be downloaded to a computer.

C. Onset's **Solar Stream wireless data transceiver** can monitor energy use, weather, and other data at remote locations. The data are transmitted via wire-

lines are not available so that data can be retrieved remotely. Sidebar 5.2 illustrates some of the dataloggers used for energy analysis. The simplest is the HOBO datalogger, which is the size of matchbox. The HOBO can be programmed to collect various data, placed at a location to collect the data, and then plugged into a computer USB port to download the data.

5.4 Economic Analysis of Energy Systems

We have been talking about various means of measuring energy consumption and efficiency. For many reasons we want to minimize energy use and maximize efficiency and the functions

figure
5.5   **Kilowatt-hour Meter for Utility Billing**

This meter is also a useful monitor for energy analysis. Although the dials read cumulative kWh used, the meter can be used to measure power draw. The wheel rotates at a watt-hour per revolution equal to the Kh factor given on the face of the meter. If you turn on only those devices you wish to measure and count the revolutions per minute (rpm), you can calculate the power drawn by the device(s) power draw by multiplying the rpm by the Kh factor and 60 sec/min to get the kilowatts: **Power (kW) = Rev/min × Kh Wh/Rev × 60 min/h = kW.**

energy provides for us. We want to accelerate our use of clean, renewable energy sources, and reduce the pollution and other impacts of conventional energy. Some people will turn to sustainable energy not because it will save them a lot of money but because they think it is good for the future of the planet or is just fun to do.

However, if we want everyone to turn to sustainable energy, we better make sure it is worth it financially. We as individuals, communities, and society have limited economic resources, and we need to invest them wisely or we will have little left for other needs of life. Although economic analysis does not capture all of the values we have as individuals and society, it is a necessary first step to see if certain options are worth doing.

In this section we take our energy assessment methods a step further to include economic analysis. Remember, though, it all starts with energy analysis. We need to quantify energy use and efficiency first. We can then put economic value on those energy numbers based on energy rates and prices, and then proceed to calculate cost-effectiveness and economic feasibility.

Before discussing those measures of cost-effectiveness, it is important to introduce monetary value of energy as well as life-cycle costing and the time-value of money.

5.4.1 Economic Value of Energy

Before going too far into economic analysis, we need to talk about putting energy into monetary terms. This is pretty easy because energy is bought and sold in the marketplace. Energy prices are set by the market as influenced by government policy. For example, fuel oil is primarily market-based, gasoline is market-based with federal and state taxes (more of a fee because they are used largely for road building and maintenance), and retail natural gas and electricity are regulated markets (because customers are largely captured by their providing utilities). Government subsidies, taxes, and environmental policies have a range of effects on the prices of different energy sources. We will talk more about the effect of policy on energy prices later.

Energy prices have been volatile and increasing. Table 5.4 gives nominal prices for selected fuels and electricity for 1998 to 2007 in price per unit and commensurate price per million Btu. In those nine years, gasoline prices increased an average 10% per year, natural gas 6.5% per year, and electricity 2.6% per year. Gasoline prices have more than doubled since 2002. Higher-quality electricity costs much more per unit of energy than natural gas or fuel oil but that gap has closed a bit in the last few years. Higher natural gas prices are affecting the cost of natural gas–fired electricity.

Figure 5.6 shows the national average retail and wholesale price of gasoline and crude oil for August 2006 to August 2007. Prices turned up in late 2007 when oil approached $100 per barrel or $2.38 per gallon. Other trends of energy prices are available from U.S. EIA's energy price reports (http://www.eia.doe.gov); click on energy type and then on retail prices to find EIA's full database.

We use these energy prices to translate energy consumption into economic costs. Solution Box 5.2 gives an example showing that translating heating energy to economic cost depends on the fuel type.

Of course, these costs may be more next year if prices keep going up. If we want to plan ahead or predict future costs, we have to consider how prices will change in the future, and for this and other reasons time becomes an important factor in energy and economic analysis.

table 5.4   **Energy Prices for Selected Sources, 1998–2007**

Year	Gasoline all grades $ per gal	Natural Gas residential $/ccf (100 ft³)	Electricity residential ¢ per kWh	Gasoline all grades $/10⁶ Btu	Natural Gas residential $/10⁶ Btu	Electricity residential $/10⁶ Btu
1998	1.07	6.82	8.26	8.56	6.61	24.21
1999	1.18	6.69	8.16	9.44	6.49	23.92
2000	1.52	7.76	8.24	12.16	7.53	24.15
2001	1.46	9.63	8.58	11.68	9.34	25.15
2002	1.37	7.89	8.44	10.96	7.65	24.74
2003	1.60	9.63	8.72	12.80	9.34	25.56
2004	1.90	10.75	8.95	15.20	10.43	26.23
2005	2.31	12.84	9.45	18.48	12.45	27.70
2006	2.62	13.75	10.40	20.96	13.35	30.48
2007	2.80*	14.10**	10.62***	22.40	13.68	31.13

* through October: **through August ***through July

Source: U.S. EIA, 2007

figure 5.6   Retail and Wholesale Gasoline and Crude Oil Prices, August 2006–August 2007

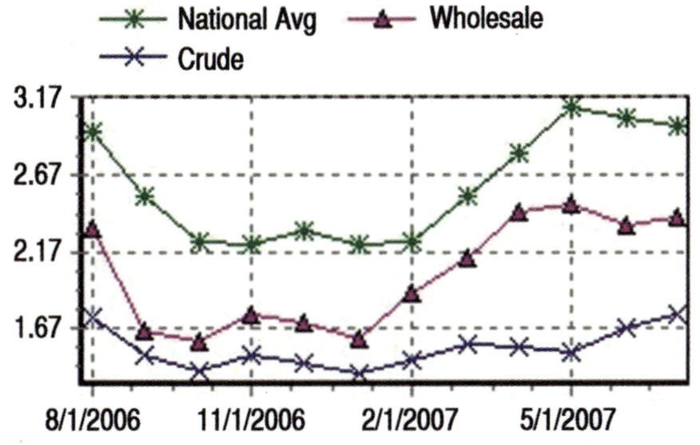

SOURCE: AAA Fuel Gauge Report, 2007, www.aaa.org

5.4.2 Life-Cycle Costing and Time-Value of Money

5.4.2.1 Life Life-Cycle Costing

We introduced life-cycle thinking in Section 5.1 and life-cycle energy requirements in Section 5.2.4. If we are to understand the full energy requirements of a system, we need to look down the road and consider longer-term commitments. For example, for a large commitment to nuclear power, we have to consider the future energy needs and monetary costs of not only plant construction and operation and fuel mining and enrichment but also plant decommissioning and disposal of wastes that may last longer than our human history to-date. In August 2005, in response to a 2004 court case, the EPA proposed radiation exposure standards for the Yucca Mountain nuclear waste site that would protect people in the area for one million years! Once approved, the standards would allow DOE to file an application to construct and operate the site, the next step forward for the $58 billion facility. Planning for one million years has presented new challenges to the agency's energy and economic analysts.

The same life-cycle cost issue is even more germane to mundane personal decisions such as buying lightbulbs. We are all tempted to choose energy-consuming items that are initially cheaper, when in fact they will be far more expensive over the life cycle of the product. Solution Box 5.3 illustrates the simple calculations for life-cycle cost of incandescent lightbulbs and compact fluorescent lamps. This exercise is not complete without considering the full costs of manufacturing and disposal of the bulbs, but we assume that they are the same for the two cases.

5.4.2.2 Time-Value of Money

Economic analysis recognizes that money has a time dimension. If we borrow money, we will have to pay back more than we borrow due to interest charges. If we invest our precious cash, we should consider the risk-free return we could get from simply putting it in a savings account. As a result, a dollar tomorrow is considered less valuable than a dollar today. We

SOLUTION BOX 5.2

Translating Energy Consumption to Economic Cost

I estimate that it takes the same amount of energy (40 million Btu per year) to heat my house and my neighbor's house (we'll discuss how to calculate this in Chapter 6). She has baseboard electric heat and I heat with natural gas. Her baseboard electric operates at 100% efficiency (all the electricity ends up as heat in the house), whereas my natural gas furnace and forced-air system operates at about 80% efficiency (20% is lost in exhaust gases and losses in the duct system). If I pay $1.30 per therm (1 therm = 100,000 Btu) for natural gas and she pays 10.5¢/kWh for electricity, how much does each of us pay per year for heating?

Solution:

My House:

$$\eta = \text{useful heating energy/natural gas (NG) energy input}$$

or

$$\text{NG energy input} = \frac{\text{useful heating energy}}{\eta}$$

$$= \frac{40 \times 10^6 \text{ Btu/yr}}{0.80}$$

$$= 50 \times 10^6 \text{ Btu/yr}$$

$$\text{NG cost} = \frac{\text{NG input Btu} \times \text{cost}}{\text{NG Btu}}$$

$$= 50 \times 10^6 \text{ Btu/yr} \times \frac{\$1.30}{10^5 \text{ Btu}} = \$650$$

My Neighbor's House:

$$\text{Electricity cost} = 40 \times 10^6 \text{ Btu/yr} \times \frac{\$0.105}{\text{kWh}} \times \frac{\text{kWh}}{3412 \text{ Btu}} = \$1231$$

"discount" future dollars to "present value" by a "discount rate" using the classic present value equation:

Eq. 5.15

$$P = \frac{F}{(1 + d)^n}$$

SOLUTION BOX 5.3

Life-Cycle Economic Cost of Lightbulbs

When I need lightbulbs, I face a choice between bulbs that produce the same amount of light or lumens, one for $0.30 each, the other for $2.50 each. Which do I choose?

Solution:

My "buy-it-cheap" self tells me to grab the $0.30 bulbs, but my smarter "buy-it-least-cost" self makes me pause and after some thought (and a quick run of the numbers), the opposite answer emerges. The $0.30 incandescent bulb will ultimately cost much more than the $2.50 compact fluorescent lamp (CFL) over the life cycle, considering electricity and replacement costs.

What are the life-cycle costs of these two lightbulb alternatives: a $0.30, 60 W incandescent bulb with 1000-hour life or a $2.50, 11 W CFL with 10,000-hour life? Both produce the same lumens of light. We can assume a low electricity rate of $0.07/kWh and set a common time period for comparable analysis at 10,000 hours.

Incandescent Lightbulb:

1. Bulb cost: **$0.30/bulb × 1 bulb/1000 hr × 10,000 hr = $3.00**
   (10 bulbs needed for 10,000 hr)
2. Energy cost: **60 W × 10,000 hr × kWh/1000 Wh × $0.07/kWh = $42**
3. Life-cycle cost = $45 (ignoring labor cost to change the bulb 9 times)

Compact Fluorescent Lamp (CFL):

1. Lamp cost: **$2.50/lamp × 1 bulb/10,000 hr × 10,000 hr = $2.50**
   (1 lamp needed for 10,000 hr)
2. Energy cost: **11 W × 10,000 hr × kWh/1000 Wh × $0.07/kWh = $7.70**
3. Life-cycle cost = $10.20

Life-cycle cost savings of CFL over incandescent bulb: $45.00 − $10.20 = $34.80

The "cheap" incandescent bulb cost $4\frac{1}{2}$ times or $35 more than the "expensive" compact fluorescent lamp.

where　P = present value dollars

　　　　F = future value dollars at year n

　　　　d = discount rate

This is the inverse of the compound interest equation that calculates the growth of dollars if invested at an annual interest rate:

Eq. 5.16
$$F = P(1 + i)^n$$

where P = present dollars
F = future dollars at year n
i = interest rate

If I have $100 and put it in a certificate of deposit (CD) with a 4% annual interest rate of return, what will the CD be worth in 10 years? Interest compounded annually increases each year's balance by 4% by the end of the year. The table steps this growth forward for each year:

Today	End of Year 1	Year 2	Year 3	Year 4	Year 5	Year 6	Year 7	Year 8	Year 9	Year 10
$100	$104	$108.16	$112.48	$116.99	$121.67	$126.53	$131.59	$136.86	$142.33	$148.02

Alternatively, the compound growth equation can be used to calculate future value:

$$F = P(1 + i)^n = \$100(1 + 0.04)^{10} = \$148.02$$

If I expect to get a payment of $100 in 10 years, what is the present value assuming a discount rate of 4%?

$$P = \frac{F}{(1 + d)^n} = \frac{\$100}{(1 + 0.04)^{10}} = \$67.56$$

The time-value of money is important to consider for energy analysis that involves long time periods and high discount rates. For short time periods and for low discount rates, ignoring the time-value of money is usually not a problem. The following examples illustrate this guideline.

What is the present value of future $100 ten years (long) from now if the discount rate is 0.5% (low)?

$$P = \frac{F}{(1 + d)^n} = \frac{\$100}{(1 + 0.01)^{10}} = \$95.13$$

What is the present value of future $100 six months (short) from now if the discount rate is 10% (high)?

$$P = \frac{F}{(1 + d)^n} = \frac{\$100}{(1 + 0.10)^{0.5}} = \$95.31$$

What is the present value of future $100 ten years (long) from now if the discount rate is 10% (high)?

$$P = \frac{F}{(1+d)^n} = \frac{\$100}{(1+0.10)^{10}} = \$38.55$$

In each case, ignoring the discount rate and setting $P = F$ for the low d and/or short n would give only a small (5%) error, but ignoring it for high d and long **n** would give a large (62%) error.

How do we choose a discount rate? Many factors contribute to an appropriate discount rate: prevailing interest rates (and those in the future), rates of overall inflation, inflation or deflation of fuel and electricity prices, expected returns from alternative investments, and so on. All these factors are uncertain so a bit of guesswork is included. Given the uncertainties, a simple approach is appropriate. The major factors are interest rate and fuel price inflation.

Eq. 5.17 $d = i - r$

where d = discount rate
       i = interest rate
       r = inflation rate for energy prices

Interest rate (i) may depend on the situation. If money is borrowed for an energy-saving improvement through a simple-interest loan, i is the loan interest rate. If cash savings are used, i should be based on expected return from an alternative investment. You can make this more complicated than it is worth. For example, if money is borrowed through a home equity loan for which a tax deduction is made on interest payments, the accurate i should be the equity loan interest rate times (1 minus % tax bracket). But this level of detail is usually unnecessary given the other uncertainties involved, such as the fuel price inflation rate. Because of price volatility (see Table 5.4), it is hard to estimate fuel inflation.

The literature has debated the appropriate discount rate to use for evaluating energy systems and programs. Generally a value of 6%–8% is used for program evaluation (see Chapter 16). The time-value of money based on the discount rate is used in most of the methods of economic analysis described below.

5.4.3 Economic Measures of Cost-Effectiveness

Social and environmental factors drive some people to make sustainable energy choices, but economic factors are most important in driving widespread investment in more efficient and renewable energy systems and energy conserving behavior. There are several metrics used to assess economic cost-effectiveness and compare investments. These measures require energy analysis to evaluate energy savings of one option over another and the dollar value of energy saved. (For energy production systems, "energy savings" is the conventional energy avoided.) Most of these measures discount future dollar savings to present value.

Simple Payback Period of Low-Flow Showerheads

When my family of five moved into our 50-year-old house, it had old high-flow shower-heads. I thought of replacing them with low-flow showerheads, but I wanted to calculate the payback on the investment just to be sure it was a winner. Each new showerhead cost $10. What is the simple payback period?

Solution:

To do the energy analysis, I measured the flow rates of the old and new showerheads: it took the old head 1 minute to fill a 5-gallon bucket, and it took the low-flow head 2.5 minutes to fill it. I stuck a thermometer in a comfortable shower and it was 100°F; when I ran just cold water it was 60°F. With my three teenage sons the average shower time was 10 minutes for the 25 showers per week the family took. I assumed my gas water heater operated at 75% efficiency, and I paid $1 per therm for natural gas (NG) (1 therm = 100,000 Btu). Here are the energy analysis results:

Energy for Old Showerheads:

Flow rate = 5 gal/min

Flow time = 25 showers/wk × 10 min/shower × 52 wk/yr = 13,000 min/yr

Hot water flow = 5 gal/min × 13,000 min/yr = 65,000 gal/yr

Energy for hot water: Specific heat Equation 4.16

$E = mc\Delta T$ = 65,000 gal/yr × 8.34 lb/gal × 1 Btu/lb-°F × (100°F − 60°F) = 21.7 × 10⁶ Btu/yr

NG energy for hot water:

$$\eta = \frac{\text{useful hot water energy}}{\text{NG energy input}}$$

5.4.3.1 Simple Payback Period

The simple payback period (SPP) gives the number of years an energy efficiency improvement or production system will take to pay for its initial capital cost based on its energy and economic savings. It holds true for short time periods and/or low discount rates because it ignores the time-value of money and for minor operation and maintenance costs because it usually ignores them as well. Despite these limitations, SPP is one of the most intuitive and useful measures of cost-effectiveness.

$$\text{NG energy input} = \frac{\text{useful heating energy}}{\eta} = \frac{21.7 \times 10^6 \text{ Btu/yr}}{0.75} = 28.9 \times 10^6 \text{ Btu/yr}$$

Energy for New Showerheads:

Flow rate = 5 gal/2.5 min = 2 gal/min

Hot water flow = 2 gal/min × 13,000 min/yr = 26,000 gal/yr

Energy for hot water: Specific heat Equation 4.16

$E = mc\Delta T$ = 26,000 gal/yr × 8.34 lb/gal × 1 Btu/lb-°F × (100°F − 60°F) = 8.68 × 10^6 Btu/yr

NG energy for hot water:

$$\text{NG energy input} = \frac{8.68 \times 10^6 \text{ Btu/yr}}{0.75} = 11.56 \times 10^6 \text{ Btu/yr}$$

Energy Savings:

$\text{NG energy}_{old} - \text{NG energy}_{low\text{-}flow}$ = 28.9 − 11.56 = 17.34 × 10^6 Btu/yr = AES

So now I can calculate the SPP:

$$\text{SPP} = \frac{\text{IC}}{\text{AES} \times \text{Pr}} = \frac{\$10/\text{head} \times 3 \text{ heads}}{17.34 \times 10^6 \text{ Btu/yr} \times \$1/10^5 \text{ Btu}} = 0.173 \text{ yr} = 63 \text{ days}$$

This investment would be recovered by natural gas monetary savings in only 2 months, after which the savings would continue to accrue. If I had an electric water heater, the savings would be even more because electricity is more expensive per unit energy than natural gas (Table 5.4). Additional savings would come from reduced water bills.

Eq. 5.18 $\text{Simple payback period (SPP, in years)} = \text{SPP} = \dfrac{\text{IC}}{\text{AES} \times \text{Pr}}$

where   IC = Initial capital cost, $ (or cost difference between two options)
        AES = Annual energy savings (e.g., kWh/yr, Btu/yr)
        Pr = Energy price (e.g., $/kWh, $/Btu)

Solution Box 5.4 gives an example of the SPP of low-flow showerheads; we will carry this example through the various measures of economic cost-effectiveness.

5.4.3.2 Return on Investment

Return on investment (ROI) is a popular economic measure that is equal to the inverse of SPP. It tells what percentage of an investment will be returned in the first year. Generally an ROI greater than 10% is a good investment. We calculate it as follows:

Eq. 5.19   Return on Investment (%/year) = ROI $= \dfrac{100(\text{AES} \times \text{Pr})}{\text{IC}} = \dfrac{100}{\text{SPP}}$

where   AES = annual energy savings
P = price of saved energy
IC = initial capital cost

Returning to our example above, we can figure out the ROI of the investment in low-flow showerheads:

$$\text{ROI} = \frac{100}{\text{SPP}} = \frac{100}{0.173\text{yr}} = 580\% \text{ per year}$$

This of course is the kind of ROI most CEOs would be quite happy to report to their shareholders.

5.4.3.3 Cost of Conserved/Produced Energy

If we want to compare the cost of an energy investment to present or future energy prices, we want to know the cost of conserved energy (CCE), which measures the cost per unit of energy saved or produced by an efficiency or production investment over its lifetime. Annual operation and maintenance costs, if any, can be included. This measure considers the time-value of money through the capital recovery factor (CRF) using a discount rate. The CRF is the classic mortgage rate factor that spreads out a one-time dollar expense (like the price of a house or in our case the initial capital cost of an energy investment). The CRF and other useful economic analysis factors are defined in Sidebar 5.3. We can compute CCE as follows:

Eq. 5.20   Cost of conserved energy ($/energy unit) = CCE $= \dfrac{\text{IC} \times \text{CRF} + \text{O\&M}}{\text{AES}}$

where   IC = initial capital cost
CRF = capital recovery factor
O&M = annual operation and maintenance cost, $
AES = annual energy savings, energy unit/year

If O&M = 0, CCE $= \dfrac{\text{IC} \times \text{CRF}}{\text{AES}}$

The cost of conserved energy is an extremely useful economic measure because it calculates dollar/unit-energy that can be compared to existing or expecting energy rates or prices. It incorporates the time-value of money and annual operation and maintenance costs, if any. Solution Box 5.5 illustrates CCE calculation using our low-flow showerhead example.

SIDEBAR 5.3

Economic Analysis Factors

(Usually given as function buttons on a business calculator)

Compound Growth Factor (CGF) $= (1 + d)^n$	Multiplied by dollar amount, CGF will grow that amount at annual compound rate d to year n. Inverse of PVF.
Present Value Factor (PVF) $= \dfrac{1}{(1 + d)^n}$	Multiplied by dollar amount, PVF will discount that amount back **n** years at discount rate d. Inverse of CGF.
Capital Recovery Factor (CRF) $= \dfrac{d(1 + d)^n}{(1 - d)^n - 1}$	Multiplied by dollar amount, it spreads that one-time cost over **n** years with equal annual payments; used to calculate annual mortgage payments. Inverse of UPVF.
Uniform Present Value Factor (UPVF) $= \dfrac{(1 - d)^n - 1}{d(1 + d)^n}$	Takes an annual payment or monetary savings over **n** years and converts it with discounting to a lump present value. Inverse of CRF.

SOLUTION BOX 5.5

Cost of Conserved Energy of Low-Flow Showerheads

What is the cost of conserved energy for the low-flow showerhead investment given in Solution Box 5.4?

Solution:

Let's assume the life of the shower heads is 20 years, the discount rate is 3%, and O&M = 0.

$$\text{CRF} = \frac{d(1 + d)^n}{(1 - d)^n - 1} = \frac{0.03(1.03)^{20}}{(1.03)^{20} - 1} = \frac{0.054}{0.806} = 0.067$$

$$\text{CCE} = \frac{\text{IC} \times \text{CRF}}{\text{AES}} = \frac{\$30 \times 0.067}{\left(17.34 \times 10^6 \text{ Btu/yr} \times \dfrac{\text{therm}}{10^5 \text{ Btu}}\right)} = \$0.01/\text{therm}$$

So for 20 years, this investment will conserve natural gas at an equivalent price of about 1¢ per therm compared to current price of $1 per therm. Anything lower than the current price is a good investment, but 1% of current price is a no-brainer.

5.4.3.4 Present Value Life-Cycle Savings

Present value savings (PVS) calculates the total life-cycle dollar savings of the energy investment put in present-day dollars, based on the assumed discount rate (d). It can include annual operation and maintenance cost if appropriate. The future net annual dollar savings (assumed to be uniform each year) are discounted to present value by the uniform present value factor (UPVF) (see Sidebar 5.3) PVS can be compared to the total cost of the investment. Here's how it is calculated:

Eq. 5.21　　Present value savings $= \text{PVS} = (\text{AES} \times \text{Pr} - \text{O\&M}) \times \text{UPVF}$　　($)

> where　AES = annual energy savings, energy unit/yr
> 　　　　Pr = current energy price, $/energy unit
> 　　　O&M = annual operation and maintenance cost, $/yr
> 　　　UPVF = uniform present value factor, based on d

What is the PVS of the low-flow showerhead investment given above? Assume the life of the showerheads to be 20 years, a discount rate of 3%, and O&M = 0.

$$\text{UPVF} = \frac{(1-d)^n - 1}{d(1+d)^n} = \frac{(1.03)^{20} - 1}{0.03(1.03)^{20}} = \frac{0.806}{0.054} = \frac{1}{\text{CRF}} = 14.9$$

$$\text{PVS} = (\text{AES} \times \text{Pr} - \text{O\&M}) \times \text{UPVF} = (17.34 \times 10^6 \text{ Btu/yr} \times \$1/10^5 \text{ Btu} - 0) \times 14.9 = \$2584$$

Compared to the initial cost of $30, this PVS of $2584 looks mighty good!

5.4.3.5 Net Present Value over Life-Cycle

Net present value (NPV) is simply the difference between PVS over the life cycle and the initial capital cost of the investment. This gives us a simple measure of profit or earnings from the investment, considering the time-value of money. If the NPV is positive, the system is said to be cost-effective. Obviously, the larger the NPV, the better. It is calculated as follows:

Eq. 5.22　　　　　Net present value ($) $= \text{NPV} = \text{PVS} - \text{IC}$
> where　PVS = present value savings ($)
> 　　　　IC = initial capital cost ($)

What is the PVS of the low-flow showerhead investment given in Solution Box 5.4?

$$\text{NPV} = \text{PVS} - \text{IC} = \$2584 - \$30 = \$2554$$

This still leaves our initial $30 investment looking pretty good.

5.4.3.6 Benefit-Cost Ratio

The benefit-cost (B/C) ratio compares annualized dollar savings and annualized costs to provide a ratio of benefits to costs. If the B/C is greater than one, it indicates a cost-effective investment. Obviously, the larger the ratio, the better.

Eq. 5.23
$$\text{Benefit-cost ratio} = \frac{B}{C} = \frac{PVS}{IC}$$

where   PVS = present value savings
        IC = initial capital cost ($)

What is the B/C ratio for the low-flow showerhead investment given above?

$$\frac{B}{C} = \frac{PVS}{IC} = \frac{\$2584}{\$30} = \frac{86}{1} = 86$$

5.4.4 Performing Economic Analysis with Spreadsheets

These calculations are not too complicated, but they can be tedious given the number of economic factors. It is not surprising that economic analysts long ago produced tables, and more recently calculator functions and Internet multipliers to ease these menial calculations. We would like a means to perform these calculations easily so that we can vary our assumptions and produce different scenarios. Spreadsheets give us that capability, and we will use them in many of our analyses.

Table 5.5 gives a spreadsheet developed for energy and economic analysis. The beauty of a spreadsheet is that once the master is set up, it can be used for the analysis at hand. As the new data entries are made, new calculations are performed automatically. So it is easy to change an assumption, enter that value, and see all of the new results without actually doing a calculation. The spreadsheet is applied in our low-flow showerhead example. As you can see, the AES and Pr values are entered with the appropriate matching units. All values match our results from the previous sections.

5.4.5 Cost-Effectiveness and Market Penetration

So how do we interpret the results of economic analysis? What do we mean by the term *cost-effective*? How does cost-effectiveness drive economic behavior? At what point do people choose to invest in energy efficiency or renewable energy?

Technically, cost-effectiveness is defined as net positive economic value. Using our economic measures, cost-effectiveness is:

- Benefit-cost ratio greater than one, or when benefits exceed costs.
- Net present value greater than zero, indicating a positive bottom line.
- Present value savings greater than cost of investment.
- Cost of conserved energy (CCE) less than current or expected energy price. We can compare the CCE for a wide range of measures as well as their energy savings with the conservation supply curve, introduced in the following section.
- Simple payback period less than the life of the investment, understanding that future savings are neither inflated for energy price escalation nor discounted to present value.

table 5.5	Economic Analysis of Energy Systems and Efficiency Improvements			
	Using a spreadsheet to calculate cost-effectiveness, with the low-flow showerhead example			
	Abbreviation	Value	Units	Formula in Value Column
Annual energy savings or production	AES	17.34	10^6 Btu/year	
Total cost	IC	30	$	
Energy price	Pr	10	$/$10^6$ Btu	
Annual operation and maintenance cost	O&M	0	$	
Energy price escalation rate	r	0.03	%/100	
Interest rate	i	0.06	%/100	
Discount rate*	d	0.03	%/100	$i - r$
Number of years	n	20	years	
Capital recovery factor*	CRF	0.0672		$d(1 + d)^n/[(1 + d)^n - 1]$
Compound growth factor*	CGF	1.8061		$(1 + d)^n$
Present value factor*	PVF	0.5537		$1/(1 + d)^n$
Uniform present value factor*	UPVF	14.8775		$[(1 + d)^n - 1]/d(1 + d)^n$
Today's value	P	100.00	$ or units	
Compound growth of present value*	F	180.61	$ or units	F = P * CGF
Present value of discounted future value*	P	100.00	$	P = F * PWF
Simple payback period*	SPP	0.173	years	SPP = IC/(AES * Pr)
Cost of conserved energy*	CCE	0.116	$/$10^6$ Btu	$CCE = \dfrac{IC * CRF + O\&M}{AES}$
Present value savings*	PVS	2580	$	PVS = (AES * Pr − O&M) * UPWF
Net present value*	NPV	2550	$	NPV = PVS − IC
Benefit-cost ratio*	BCR	86.0		$BCR = \dfrac{(AES * Pr - O\&M)}{(IC * CRF)}$

* Calculated values

We know that people face choices on how they invest their money. If we are to change our energy patterns on a large scale, renewable energy and efficiency must compete effectively against other investment choices. We need to know how cost-effectiveness translates to consumer choice and to market penetration. But we also know that people make choices based on other values. Full life-cycle analysis includes assessment of environmental effects, which we discuss later in this chapter. We will explore the broader issues of market penetration and transformation in Chapter 16.

5.4.6 Conservation Supply Curves for Efficiency and Savings

The conservation supply curve (CSC) was first popularized by Arthur Rosenfeld and Amory Lovins in the 1980s. It shows how different efficiency or production measures contribute to energy savings or supply, ordered by their cost-effectiveness measured by cost of conserved energy (CCE). Recall from Section 5.4.3.3 that CCE is annualized cost divided by annual energy savings produced and is given in cost/energy, such as ¢/kWh.

Figure 5.7 gives a hypothetical CSC, which illustrates the way the graph is developed. As shown in the inset, each point is plotted by a measure's CCE on the y-axis starting with the lowest-CCE, most cost-effective measures. The x-axis gives the cumulative energy savings of the measures, so that the graph becomes a step function. The CSC is elegant in its ability to combine economic and technical potential as well as to see both incremental and aggregate savings and costs.

Figures 5.8 and 5.9 give two examples of conservation supply curves from well-known studies.

Figure 5.8 is from a study of statewide electric efficiency potential in California funded by the Energy Foundation/Hewlett Foundation in 2002, right after the California electricity crisis (Rufo & Coito, 2002). As the figure shows, the study estimated that 10% of the state's projected base energy consumption could be saved at less than 5¢/kWh and 14% at less than 10¢/kWh.

Figure 5.9 gives an aggregate of 200 CSCs developed in 1993 representing U.S. national residential electricity measures and their potential savings. The study estimated that 400 TeraWh could be saved through efficiency measures costing less than 8¢/kWh, the national average residential electricity rate at the time (Stoft, 1995; Rosenfeld, 1993).

figure 5.7 **Hypothetical Conservation Supply Curve**

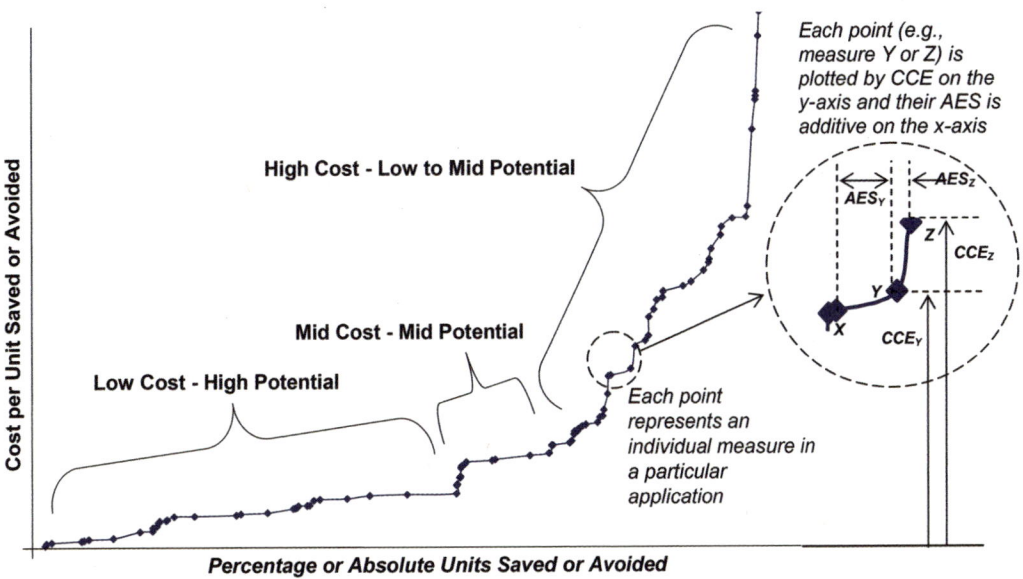

Plots energy measures' cost of conserved energy versus cumulative energy saved or avoided.

SOURCE: Adapted from Rufo, 2003

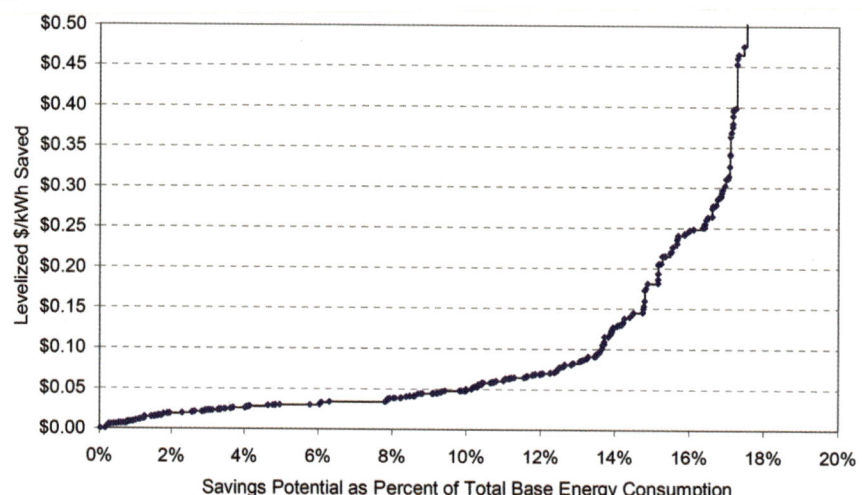

**figure
5.8**   **Conservation Supply Curve for California Electricity**

Graph shows a 10% savings for less than 5¢/kWh, 14% for less than 10¢/kWh.

Source: Rufo and Coito, 2002

**figure
5.9**   **Conservation Supply Curve for U.S.
Residential Electricity**

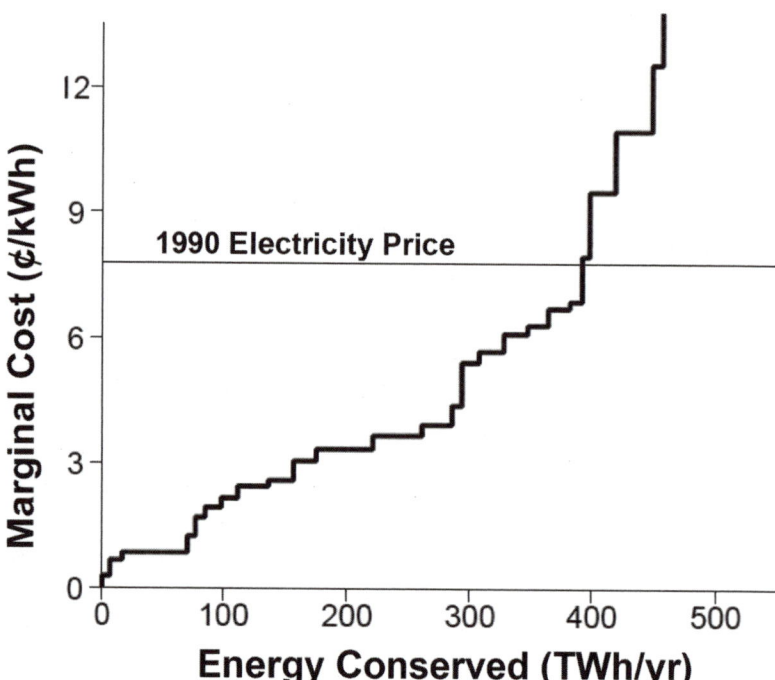

For less than the existing residential
rates (at that time), 400 TeraWh
could be saved.

Source: Stoft, 1995; Rosenfeld, 1993

The conservation supply curve gives an excellent snapshot of the most cost-effective technologies, prioritizes them by both cost and energy benefit, and compares them to the existing price of energy or stated policy goals. However, some analysts have been critical of the wide use of CSC (Stoft, 2002) and CCE (Golove & Eto, 1996), and some of their concerns, listed below, should be considered when developing or using CSCs.

- A measure's energy savings is often overestimated.
- Energy savings from different measures are not always additive as CSC assumes.
- CCE usually underestimates the cost of implementing the savings by ignoring maintenance costs, transaction costs, and other barriers that must be overcome.
- It is difficult to predict accurately the future including future cost of other fuels to which CCE is compared, and the future capital cost or price of new technologies.

5.5 Environmental Analysis of Energy and Materials Systems

In addition to economic assessment, life-cycle analysis aims to evaluate the environmental implications of the energy and materials options. Energy extraction, transport, and use have a wide range of impacts on the environment. As we look to embodied energy in materials, we know that the acquisition, processing, transport, use, and disposal of various materials (e.g., steel, wood products, concrete, chemicals, agricultural products) have input energy and output environmental impact. Some of these impacts are regulated by government, and the cost of these regulations is usually passed on to consumers so it is reflected in the cost of energy. However, many of these effects are not regulated, appear as economic externalities, and are not fully accounted for in economic markets. Therefore, if we wish to allocate energy and material resources sustainably, we need to apply life-cycle analysis that incorporates both economic and environmental analysis and the short- and long-term cradle-to-grave impacts of energy and materials choices.

As discussed in Section 5.1.3 and shown in Figure 5.1, life-cycle analysis requires defining a system boundary and process as well as the inputs (e.g., materials, energy, water) and outputs (e.g., useful products as well as emissions, effluents, and waste) of the system from cradle to grave. The analysis requires inventory data of materials and energy used, and impact coefficients to determine the effects according to Equation 5.1:

$$\text{Inventory data} \times \text{Impact coefficient} = \text{Impact indicator}$$

This section discusses sources and examples of environmental impact coefficients used for the life-cycle and environmental analysis. These coefficients give environmental impact per unit of energy or unit of material for various sources and uses. Just as we needed energy prices (e.g., $ per kWh) of various energy options to calculate economic costs of energy, we need environmental coefficients (e.g., impact per kWh) for different energy sources and materials to calculate environmental costs of energy and materials.

5.5.1 The NREL Life-Cycle Inventory

U.S. DOE's National Renewable Energy Laboratory (NREL) has partnered with ATHENA and Franklin to create a database of life-cycle inventory coefficients to be used in life-cycle analysis. This is a work in progress, but already the project has generated useful data for the long list of products and processes given in Table 5.6.

For each product or process, input and output coefficients are provided in the format given in Table 5.7, which shows petroleum refining. For each 1000 pounds of petroleum products, the table gives the materials and energy inputs, the emissions and waste outputs, and product and co-product outputs. The data are provided in downloadable spreadsheet format for easy access and integration into life-cycle analyses. See http://www.nrel.gov/lci.

How do we use the life-cycle inventory? For example, if I drive 15,000 miles per year and get 20 mpg, I can figure out the crude oil and electricity, emissions of SO_x and NO_x, chemical oxygen demand (COD) effluent, and solid waste associated with refining the gasoline I consume each year.

For 15,000 mpy, 20 mpg, and gasoline at 6.25 lb/gal, I can calculate my gasoline use:

$$\frac{15,000 \text{ mi/yr}}{20 \text{ mi/gal}} = 750 \frac{\text{gal}}{\text{yr}} \times 6.25 \frac{\text{lb}}{\text{gal}} = 4688 \frac{\text{lb}}{\text{yr}} \text{ gasoline (this is my inventory data)}$$

We can find impact coefficients for gasoline from Table 5.7 by multiplying the overall coefficients by 0.421, the proportional output of gasoline for each pound of petroleum products. So my use requires:

$$\text{Crude oil} = \frac{1034 \text{ lb crude/1000 lb prod.}}{421 \text{ lb gas/1000 lb prod.}} \times \frac{4688 \text{ lb gas}}{\text{year}} = \frac{11,514 \text{ lb crude/yr}}{7.5 \text{ lb crude/gal} \times 42 \text{ gal/bbl}}$$
$$= 36.5 \text{ bbl crude}$$

table 5.6 Selected Processes and Products Included in NREL Life-Cycle Inventory Database

Agricultural Products

Corn production
Rapeseed production
Soybean production

Building and Construction Products

Glue laminated beam (Glulam), PNW at mill gate
Oriented strand board (OSB), SE at mill gate
Plywood, PNW at mill gate
Softwood lumber (dry), PNW at mill

Electricity Generation
Materials Used in Manufacturing of Automobiles/Durables

Cold rolled sheet production
Primary aluminum production

Primary Fuel Combustion

Biomass combustion in utility turbines
Bituminous coal combustion in utility boilers
Distillate oil combustion in utility boilers
Natural gas combustion in utility boilers
Wood combustion

Primary Fuel Production

Bituminous coal production
Crude oil extraction
Natural gas extraction and processing
Petroleum refining
Uranium fuel production

Transportation

Cargo plane transportation
Diesel-fueled single unit truck transportation

table 5.7   Sample Data from NREL Life–Cycle Inventory Database: Inputs and Outputs from Petroleum Refining

1000 lb Products	Name Unit	Category	Units	Petroleum Refining
Inputs from technosphere (for each 1000 lb petroleum output)	Crude oil		lb	1034.00
	Electricity		kWh	62.66
	LPG		gal	0.11
	Natural gas		ft^3	146.60
	Residual oil		gal	2.69
	Barge		ton-mi	0.37
	Ocean freighter		ton-mi	1472.00
	Pipeline		ton-mi	136.00
Inputs from nature				
Outputs to nature (for each 1000 lb petroleum output)	Aldehydes	air	lb	0.04
	Ammonia	air	lb	0.02
	Carbon monoxide	air	lb	13.30
	HC (other than CH_4)	air	lb	2.03
	Methane	air	lb	0.07
	NO_x	air	lb	0.33
	Particulate	air	lb	0.24
	SO_x	air	lb	2.35
	COD	water	lb	0.23
	Nitrogen (as NH_3)	water	lb	0.02
	Oil/grease	water	lb	0.01
	TSS	water	lb	0.03
	Solid waste	waste management	lb	5.60
Product/co-product outputs (and fraction of outputs to nature)	Distillate fuel oil (0.219)		lb	219.00
	LPG (0.027)		lb	27.00
	Gasoline (0.421)		lb	421.00
	Residual fuel oil (0.049)		lb	49.00
	Asphalt/road oil (0.037)		lb	37.00
	Kerosene/jet fuel (0.091)		lb	91.00
	Petroleum coke (0.060)		lb	60.00
	Still gas (0.045)		lb	45.00
	Other (0.052)		lb	52.00
	Total petroleum products (1.000)		lb	1000.00

This is the crude oil required to refine the gasoline I use. There are other useful petroleum co-products produced in the refining process, and although this entire amount is needed for my gasoline, the oil associated with just my gasoline is 36.5 bbls/yr × 0.421 = 15.4 bbls/yr.

- **Electricity** = 62.66 kWh/1034 lb crude × 11,514 lb crude = 698 kWh
  (698 × 0.421 = 294 kWh associated with my gasoline)
- **SO_x** = 2.35 lb SO_x/1034 lb crude × 11,514 lb crude = 26 lb SO_x
  (11 lb for my gasoline)
- **NO_x** = 0.33 lb NO_x/1034 lb crude × 11,514 lb crude = 3.7 lb NO_x
  (1.5 lb for my gasoline)
- **COD** = 0.23 lb COD/1034 lb crude × 11,514 lb crude = 2.6 lb COD
  (1.1 lb for my gasoline)
- **Solid waste** = 5.6 lb SW/1034 lb crude × 11,514 lb crude = 62 lb SW
  (26 lb for my gasoline)

As NREL's LCI and other life-cycle databases are developed further, we will be able to enhance assessment of energy options and better recognize their environmental implications.

5.5.2 Air Pollutant and Carbon Emissions from Combustion of Fossil Fuels

Perhaps the most severe of the environmental impacts of current energy use is the air pollution emitted from the combustion of fossil fuels. As shown in Table 2.4, energy use accounts for 90% of air pollution emissions in the United States, about 20% from stationary sources, mostly power plants, and 70% from mobile sources, mostly passenger vehicles. Much progress has been made to reduce emissions through technological controls during the past 30 years. Indeed, total emissions are less than half of what they were in 1970 despite increases in population, vehicle miles traveled, and the economy. Still, many cities have not attained national clean air standards, and additional progress is needed.

Increasing attention has been given to emissions of carbon dioxide and other GHG from fossil fuels, because of the impacts associated with global climate change. These emissions are not yet regulated in the United States, but they have become an important part of environmental accounting for energy use. Indeed, the "carbon footprint," or the carbon emissions associated with a person's or a community's energy use patterns, has become a useful overall indicator of environmental impact of energy use. Table 5.8 gives CO_2 emission rates for various fuels. Of the fossil fuels, natural gas has the lowest rate, about 75% of oil products and 57% of coal. Biomass combustion is considered carbon neutral because it emits "biogenic" CO_2 or CO_2 recently absorbed by vegetation from the atmosphere.

Let's look at some emissions impact coefficients or rates associated with two important energy uses, electricity and buildings. We will discuss transportation emissions rates in Chapter 13. At the end of this section we explore the increasingly popular approach to calculate a household's "carbon footprint" calculations and estimating remediation to offset those footprint emissions through Green Tags and tree planting.

table 5.8 CO_2 **Emission Rates, Various Fuel Combustion**

Fuel	Pounds CO_2 per unit	Pounds CO_2 per 10^6 Btu
Motor gasoline	19.56 per gal	156.4
Distillate and diesel	22.38 per gal	161.4
Natural gas	120.6 per 10^3 ft^3	117.1
Coal (bituminous)	4931 per short ton	205.3
Biomass*	0	0

* Biomass contains "biogenic" carbon. Under international greenhouse gas accounting methods developed by the Intergovernmental Panel on Climate Change, biogenic carbon is part of the natural carbon balance and it will not add to atmospheric concentrations of carbon dioxide.
SOURCE: U.S. EIA: http://www.eia.doe.gov/oiaf/1605/coefficients.html

5.5.2.1 Emissions Rates for Electricity

Table 5.9 gives emissions rates for electricity generation for coal, oil, and natural gas, as well as national average rates and rates for non–fossil fuel generation. Coal has the highest emissions rates for all pollutants. Compared to coal, the natural gas emission rate is 62% for CO_2, 28% for NO_x, and less than 1% for SO_x. Wood and MSW generation produce NO_x and CO_2, but the CO_2 emitted is "contemporary" not "fossil" carbon, so it is considered part of the natural contemporary carbon cycle. In other words, the carbon in wood and paper (the main combustible part of MSW) was recently atmospheric CO_2 photosynthesized in wood products, and it is released back to the atmosphere during combustion. MSW plants give off trace amounts of mercury and, depending on the constituents of the waste stream, may release potentially toxic emissions such as dioxins.

Of course, the national averages may not be too precise for a given location that is usually served by a unique mixture of power sources from the grid. EPA's **eGRID database** gives emissions rates for each state based on the state's mix of power generation. Table 5.10 gives rates from the database for four states. The rates reflect the mix of sources of power in Washington (76% hydro, 9% nuclear, 8% coal), California (49% natural gas, 19% nuclear, 17% hydro, 12% wind/geothermal), Virginia (51% coal, 37% nuclear), and West Virginia

table 5.9 **U.S. Average Emissions Rates for Different Sources of Electricity (lb per MWh)**

	National Average	Coal	Natural Gas*	Oil	MSW	Biomass	Hydro	Nuclear	Solar/Wind	Geothermal
CO_2	1392	2249	1135	1672	1500**	1500**	~0	~0	~0	~0
SO_x	6	13	0.1	12	negl	negl	~0	~0	~0	negligible
NO_x	3	6	1.7	4	2.0	2.0	~0	~0	~0	~0
Mercury (lb/GWh)	0.03	0.06	–	–	trace**	–	–	–	–	–

* Rates for combustion turbine; combined cycle systems have one-third lower emissions rates
** Not considered GHG-CO_2 because it is part of contemporary carbon cycle. Also, potential trace amounts of other toxin emissions such as dioxins.

<table>
<tr><td>table
5.10</td><td colspan="5">Average Emissions Rates for Selected States' Generation of Electricity, lb per MWh</td></tr>
<tr><td></td><td>Washington</td><td>California</td><td>Virginia</td><td>West Virginia</td></tr>
<tr><td>CO_2</td><td>287</td><td>633</td><td>1232</td><td>2027</td></tr>
<tr><td>SO_x</td><td>1.6</td><td>0.17</td><td>5.8</td><td>12.9</td></tr>
<tr><td>NO_x</td><td>0.6</td><td>0.56</td><td>2.6</td><td>5.8</td></tr>
<tr><td>Hg (lb/GWh)</td><td>0.006</td><td>0.002</td><td>0.02</td><td>0.05</td></tr>
</table>

(98% coal). See the eGRID database for emissions rates for other states (http://www.epa.gov/cleanenergy/egrid/index.htm).

The example in Solution Box 5.6 shows that environmental impacts of electricity obviously depend not only on consumption, but also on the source of that power. Check out EPA's "Power Profiler," an interactive online calculator that performs impact calculations for users by simply submitting their zip code and monthly electricity use (see http://www.epa.gov/cleanenergy/powerprofiler.htm).

5.5.2.2 Carbon Emissions Rates for Buildings

Emissions from building operations are more difficult to assess because of the variety of operating conditions. For example, gas and oil heating systems give off NO^+, PM, and CO_2, but their emission rates are very dependent on the system size, type, age, and condition. Electricity use in buildings does not emit air pollutants or CO_2 directly, but that use may require significant emissions back up the transmission wire at the power plant. Emissions from electricity use in buildings can be determined from the emissions rates described in Section 5.5.2.1. The CO_2 emission rates per unit of fuel for various fuels used in buildings are given in the CO_2 calculator shown in Table 5.12. We discuss methods of calculating building heating fuel use in Chapter 6 and building electricity use in Chapter 8.

5.5.2.3 Emissions Rates for Transportation

Transportation mobile sources are the main source of urban air pollution. They contribute about half of the nation's NO_x and volatile organic compounds (VOC), which combine to form photochemical smog. Transportation vehicles also contribute about one-third of the CO_2 emissions in the United States. Chapter 13 discusses emissions rates and standards for vehicles.

5.5.3 Assessing Other Environmental Impacts of Energy Use

In addition to air quality and climate change, there are many other environmental impacts of energy use but they are not as easy to assess as air emissions. Table 2.5 illustrates the wide range of impacts and qualifies them in severity and risk.

SOLUTION BOX 5.6

Calculating Emissions from Electricity Consumption

Let's say the authors' households each consume an average of 500 kWh per month. What are the annual emissions attributed to electricity consumption in each household?

Solution:

Masters spends time in both Washington and California, so we'll calculate emissions for both states. Randolph lives in Virginia, but is served by American Electric Power, which generates most of its power in West Virginia, so we should use the West Virginia emissions rate.

Their annual electricity consumption is 500 kWh/mo × 12 mo = 6000 kWh = 6 MWh. In Washington the CO_2 emissions for that use are 287 lb/MWh × 6 MWh = 1722 lb. In West Virginia, the CO_2 emissions are 2027 lb/MWh × 6 MWh = 12,162 lb **or 8 times that of Washington!** Table 5.11 gives solution results for other emissions and for California.

table 5.11 **Annual Emissions (lb) Attributable to 500 kWh/mo Electricity Consumption**

	Washington	California	West Virginia
CO_2	1722	3798	12,162
SO_x	9.6	1.0	77.8
NO_x	3.6	3.4	34.8
Hg (10^{-3} lb)	0.04	0.01	0.30

5.5.4 Calculating Your Carbon Footprint

With the increased interest in global climate change, more people are interested in determining the effect their energy consumption has on greenhouse gas (GHG) emissions, and then taking measures to reduce or offset those emissions. They can reduce emissions by employing energy efficiency and conservation and on-site renewable energy systems or by buying "green power" from an electricity supplier. Not all consumers have access to green power, but all consumers can buy Green Tags (also called Renewable Energy Certificates), which are a proxy for green power. Consumers buy these Green Tags in 1000 kWh bundles and the revenues are using to develop renewable electricity. Green power and Green Tags are discussed in Chapter 18. Consumers can also offset their CO_2 emissions by planting trees.

To help consumers assess their carbon emissions, several groups have developed carbon calculators, applying the concept of the ecological footprint to carbon emissions. The ecological footprint approach aims to calculate a person or household's impact on the

table **5.12**   Carbon Dioxide Emissions Calculator and Offsets from Green Tags and Tree Planting

CO_2 Emission Coefficients	Pounds CO_2		
Petroleum Products	**per gal**	**per 10^6 Btu**	Other coefficients:
Motor gasoline	19.6	156.4	33.4 aviation passenger mile per gal
Distillate/diesel fuel	22.4	161.4	0.63 lb CO_2 per aviation passenger mile
Jet fuel	21.1	156.3	1400 lb CO_2 offset per 1000 kWh Green Tag
Kerosene	21.5	159.5	667 lb CO_2 offset per tree planted
Liquefied petroleum gases	12.8	139.0	
Residual fuel	26.0	173.9	
Propane	12.7	139.2	
E-85*	3.7	29.7	
B-20 biodiesel*	17.9	129.1	
Gaseous Fuels	**per 1000 ft³**	**per 10^6 Btu**	
Methane	116.4	115.3	
Flare gas	133.8	120.7	
Natural gas	120.6	117.1	
Coal	**per short ton**	**per 10^6 Btu**	
Anthracite	3852.2	227.4	
Bituminous	4931.3	205.3	
Subbituminous	3715.9	212.7	
Lignite	2791.6	215.4	
Electricity	**per MWh**	**per 10^6 Btu**	
National average	1392	408.0	
Coal	2249	659.1	
Natural gas	1135	332.6	
Oil	1672	490.0	
MSW*	1500	439.6	
Biomass*	1500	439.6	
Geothermal energy	0	0	
Wind	0	0	
Photovoltaic and solar thermal	0	0	
Hydropower	0	0	
Nuclear	0	0	

Table 5.12 Continued

Energy Use Carbon Footprint and Remediation Offsets

	Energy Use or Activity	Efficiency	CO_2 Coefficient	CO_2 Emissions & Offsets*
Electricity	11,256 kWh/yr		1.392 lb/kWh	15,668 lb/yr
Natural gas	831 therms/yr		11.7 lb/therm	9728 lb/yr
Fuel oil	0 gal/yr		22.4 lb/gal	0 lb/yr
Propane	0 gal/yr		13.9 lb/gal	0 lb/yr
Vehicle 1—gasoline	13,900 mi/yr	25 mpg	19.6 lb/gal	10,898 lb/yr
Vehicle 2—diesel	0 mi/yr	35 mpg	22.4 lb/gal	0 lb/yr
Air travel	962 mi/yr		0.63 lb/pass. mi	606 lb/yr
Total				36,900 lb/yr
Offset—Green Tags = total CO_2/offset rate			1400 lb/1000 kWh	26 Green Tags
Offset—Tree Planting = total CO_2/offset rate			667 lb/tree	55 Trees planted

* Emissions depend on make-up of waste or biomass. Actual emissions are given, but biomass contain "biogenic" carbon. Under international greenhouse gas accounting methods developed by the Intergovernmental Panel on Climate Change, biogenic carbon is part of the natural carbon balance, and it will not add to atmospheric concentrations of carbon dioxide.

environment in terms of consumption of materials and energy and generation of emissions and wastes. The carbon footprint focuses on the CO_2 emissions from energy consumption. The calculation is useful to see the magnitude of impact and ways to reduce household CO_2 emissions, but it can also determine the mitigation offsets needed through Green Tags, tree planting, or other mitigation to become "carbon neutral." A carbon neutral household is one in which its mitigation measures offset its carbon emissions.

Table 5.12 gives a spreadsheet giving CO_2 emissions coefficients for various energy sources. The portion of the spreadsheet on this page gives an energy use carbon footprint and remediation offsets. Given energy use for household heating and electricity, vehicle use, and air travel, the spreadsheet calculates total emissions per year and the Green Tags and trees planted that would offset the CO_2 emissions.

The default values given for energy use for electricity, natural gas, vehicle gasoline, and air travel are U.S. national averages. The average household produces 36,900 lb CO_2/yr. These emissions could be offset by the purchase of twenty-six Green Tags or the planting of fifty-five trees. The spreadsheet can be used for any energy use for fuels and sources for which emissions coefficients are given. As we know from Section 5.5.2.1, electricity emissions rates vary for different states, so specific state rates from EPA's eGRID database should be used.

5.6 Summary

This chapter described several methods of life-cycle, energy, economic, and environmental analysis that are important to compare energy and materials options and make informed decisions. Energy analysis is important to understand how much energy is used, the efficiency of use, and also how much energy it takes to produce energy. Once energy use requirements are known, economic analysis evaluates relative cost-effectiveness. Several techniques are available, but the most straightforward is simple payback period. Discounting future savings is important for long time periods and high discount rates. Spreadsheets are very useful in performing economic analyses, especially with varying assumptions.

Environmental assessment adds an important sustainability dimension to energy and economic analysis. Assessing air and carbon emissions of energy options using emissions factors or coefficients is more advanced than assessing other environmental impacts, such as water and land pollution.

Life-cycle analysis combines energy, economic, and environmental analysis to assess the broad impacts of energy and materials options from cradle-to-grave or from the first step in resource acquisition to the last step of deconstruction and waste disposal. Life-cycle analysis is not yet fully integrated into common practice, but recent developments indicate improved analytical tools and data for what may come to be called "sustainability analysis."

Market penetration of new energy-saving technologies tends to require very short payback periods because of competing investment opportunities. Improved information, access to capital, and government policies can help overcome transaction costs and other barriers to penetration of efficient and renewable energy technologies. Market penetration and the role of public policy are discussed in Chapter 16.

Buildings and Energy

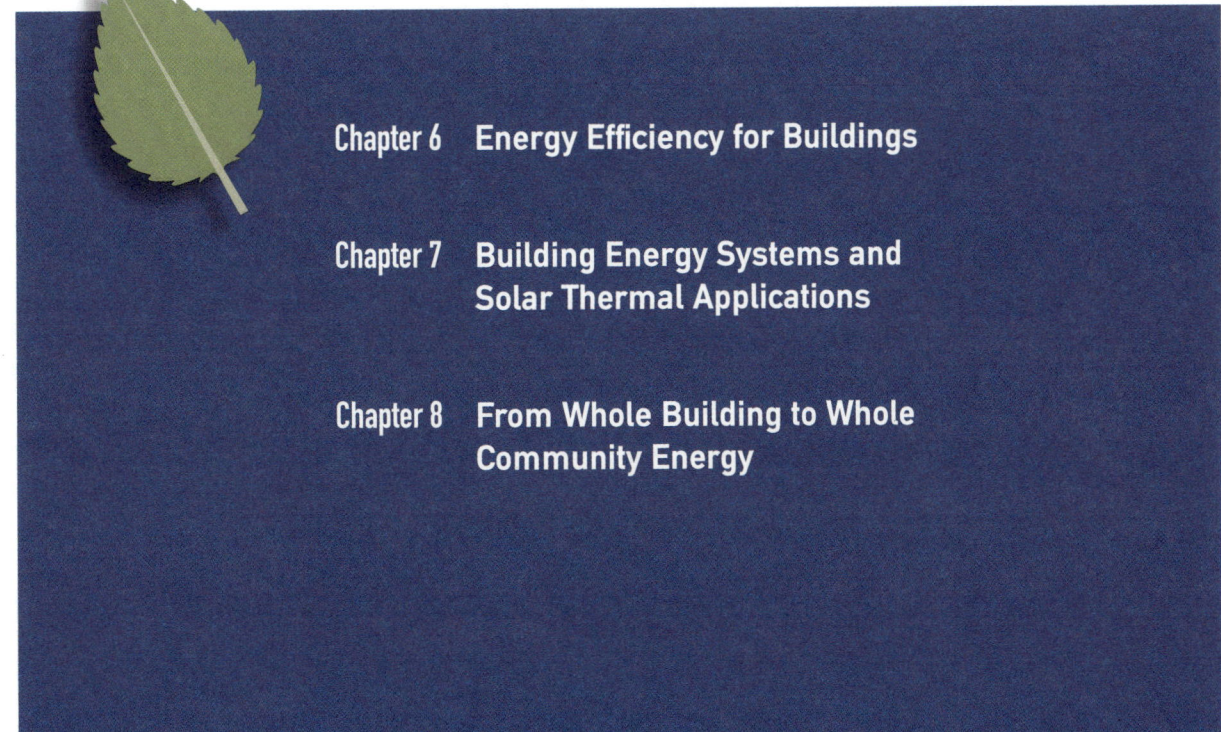

CHAPTER 6

Energy Efficiency for Buildings

Energy for buildings is the most important sector of energy demand in the United States. To provide the heating, cooling, lighting, water heating, as well as all of the other things that we use energy for in our residential and commercial buildings accounts for 39% of U.S. primary energy demand. Transportation accounts for 29%, and the category labeled "industry" uses 32%.

Although the energy required to operate buildings is sizable, there is an additional energy burden in the building sector associated with the energy needed to manufacture building materials, transport them to the site, and actually construct the building. As Figure 6.1 shows, when that *embodied energy,* along with energy for industrial buildings (not for processes within those buildings) is included, the total energy demand for the building sector grows to close to half of all U.S. primary energy. This embodied energy, along with some very

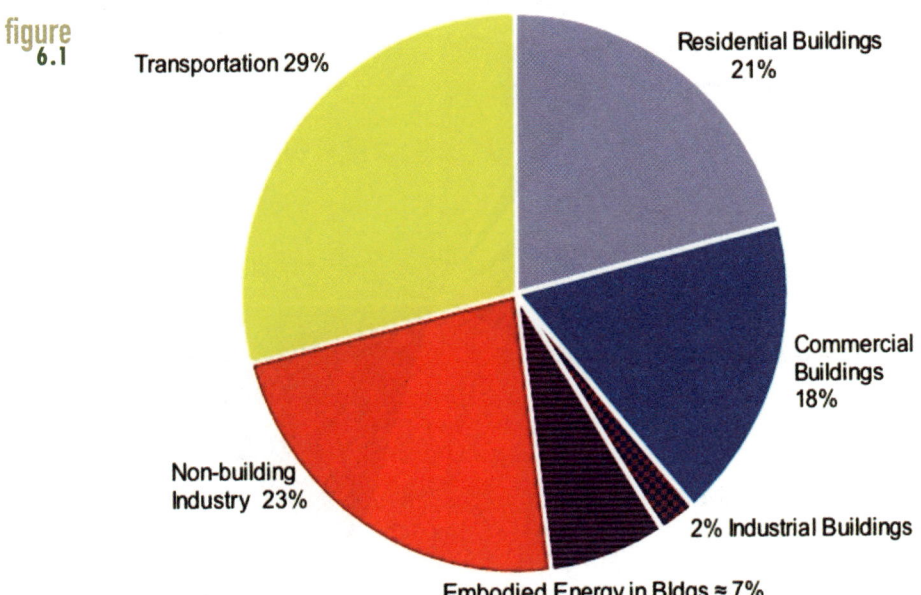

figure 6.1

When embodied energy is included, residential, commercial, and industrial buildings account for close to half of all U.S. primary energy.

important policy and planning approaches to improving building energy efficiency will be introduced in Chapter 8.

Buildings also use almost three-fourths of the electricity generated in the United States, which means that buildings are responsible for a corresponding fraction of the carbon emissions and other pollution associated with power plants. In other words, buildings are a really big deal in terms of energy and pollution in the United States.

In this chapter, we will look carefully at how better windows, more insulation, tighter building envelopes, better ducts, and other features can reduce the energy demand for heating and cooling in buildings. We'll then describe conventional heating and air-conditioning systems used to satisfy the remaining space-conditioning demand. In the next chapter, we will focus on using solar energy to help heat buildings in the winter without overheating them in the summer, provide natural daylight for illumination, and supply energy for water heating.

6.1 Residential and Commercial Buildings

Figure 6.2 presents the principal end uses for energy in buildings. One-fourth of the demand is simple space heating that helps us keep comfortably warm in our homes and commercial buildings. Perhaps surprisingly, the second biggest sector of demand is lighting, which accounts for about one-fifth of the total, followed by space cooling and water heating.

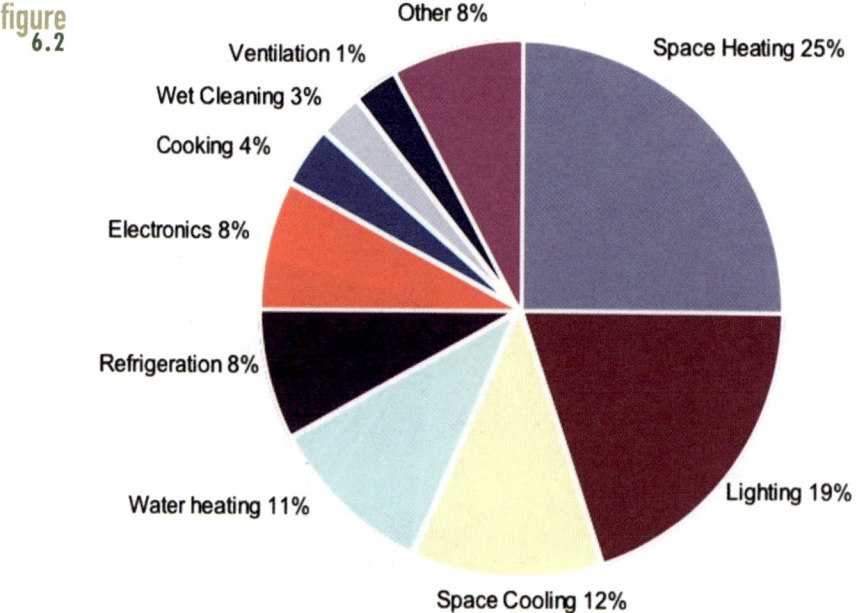

figure 6.2

Primary energy demand for U.S. residential and commercial buildings. The total demand in 2006 was estimated to be just over 100 quads.

Figure 6.1 showed the breakdown of U.S. energy demand between residential and commercial buildings, with the residential sector being the larger of the two. We live in houses and apartments, so we all have considerable intuitive knowledge of the energy demands of those dwellings, but it is important to note that commercial buildings, such as banks, schools, shopping centers, and office buildings, have energy characteristics quite different from the ones we normally focus on in relatively small residential structures. An office building is not just a "big house."

Larger buildings have a smaller surface-to-volume ratio, which reduces the importance of heat transfer through the building envelope per unit of floor space. Moreover within those square feet of floor space there is a lot more going on. Office buildings are illuminated much more brightly and uniformly than homes to accommodate the needs of workers, and most of that energy ends up as waste heat. Moreover, the density of people, copy machines, computers, and other plug loads increases, driving up the internal thermal gains, which means these buildings can potentially pretty much heat themselves until the ambient temperature drops well below the indoor temperature.

One way to see the impact of these differences between big buildings and small ones is to plot out the major energy demands for residential versus commercial buildings, as shown in Figure 6.3. As can be seen there, plain old residential space heating is the dominant single use of energy in buildings (that's why we will spend so much time on it in this chapter). The next most important sector is lighting for commercial buildings. We'll discuss that in Chapters 7 and 8. So, when do we need lights in most commercial buildings? It is in the

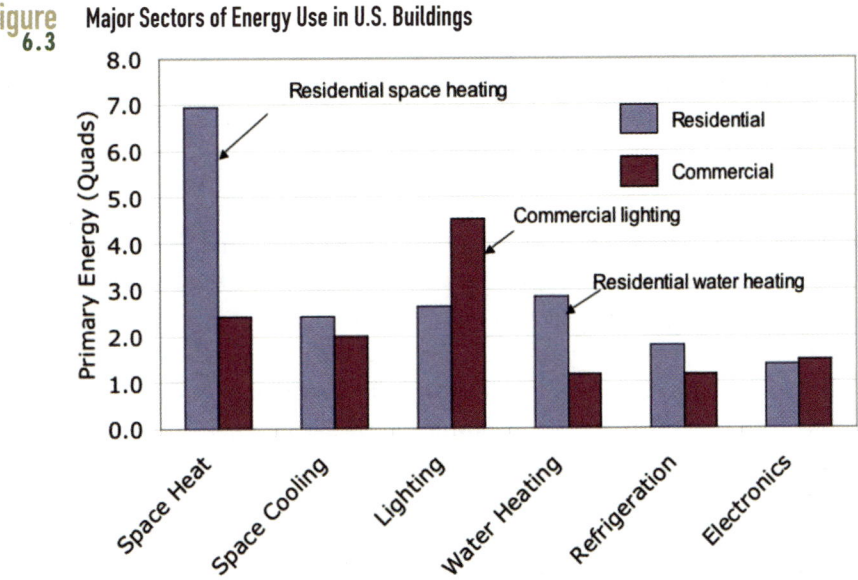

figure 6.3   **Major Sectors of Energy Use in U.S. Buildings**

Residential space heating and water heating and commercial lighting dominate energy use in buildings.

daytime, when there is an abundance of light outside, provided cleanly and freely by the sun. If we could take greater advantage of that natural daylighting, we could make a significant dent in the commercial building energy demand.

Figure 6.3 also shows the importance of residential water heating, but we will spend some time on that load in the next chapter. Interestingly, cooling isn't the dominant energy demand for either type of building. That conclusion is based on national statistics, but of course in certain climate areas it is overwhelmingly important. Another factor that makes cooling more important than Figure 6.3 might make it appear to be is that it drives up the electricity demand in the middle of those hot days when the grid is running at maximum capacity. Reducing cooling loads can help avoid blackouts and also reduce our need to build and operate inefficient peaker power plants.

6.2 Introduction to Heat Loss Calculations

As Figures 6.2 and 6.3 indicate, simple space heating is the biggest single category of energy demand in U.S. buildings, accounting for one-fourth of all building energy. Most of that is needed to heat our homes. With that much at stake, it is important to look somewhat carefully at this sector to see how greater efforts to improve the energy efficiency of the building envelope can dramatically reduce those demands. Then, in the next chapter we will look at how some simple passive solar ideas coupled with the most efficient heating and cooling systems can help us drive our energy bills as low as possible. To get started, we need to learn something about basic heat transfer.

6.2.1 Basic Heat Transfer through the Building Envelope

Figure 6.4 shows the basic problem we want to attack. It is assumed to be winter so we are trying to maintain some nice, warm inside temperature T_i while the ambient is colder at T_a. There will be heat loss through the walls, windows, doors, ceiling, and floors. In addition, there will be heat loss associated with warm room air leaking out of the building, which is replaced by colder outside air coming in that must be heated up to keep us comfortable. This latter contribution to the heat load is called *infiltration*. The difference between infiltration and ventilation is that ventilation is fresh air that we bring in on purpose whereas infiltration is air that leaks in and out of cracks and holes, whether we like it or not.

Our approach is simple. We just figure out the heat loss rate through each component (wall, window, door, and so on) of the building envelope. Using small letters q to represent heat loss *rates* (in Btu/hr), we write down that the total heat loss is the sum of the losses through each path:

Eq. 6.1 $$q_{tot} = q_{walls} + q_{windows} + q_{ceiling} + q_{floor} + q_{doors} + q_{infiltration}$$

figure
6.4   **Heat Loss Paths for Our House in Winter**

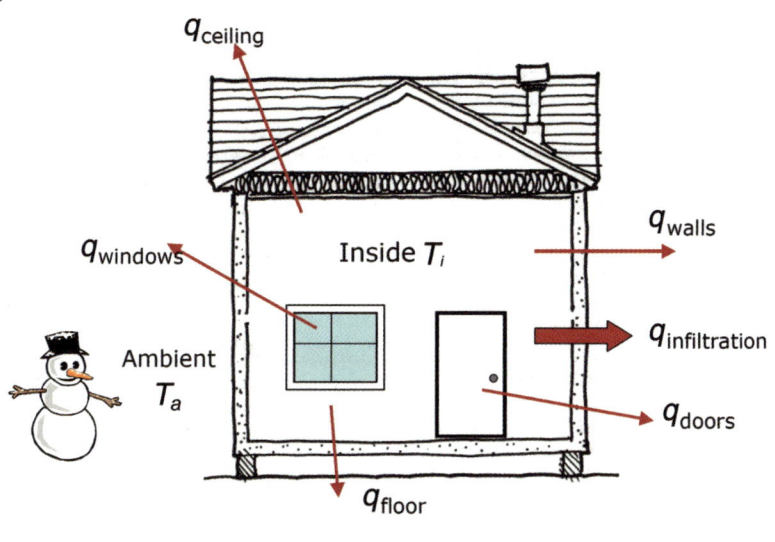

$$q_{tot} = q_{walls} + q_{windows} + q_{ceiling} + q_{floor} + q_{doors} + q_{infiltration}$$

Our task now is to evaluate each of those heat transfer pathways. For all but infiltration, we will use a sort of "Ohm's law" of heat transfer, which suggests that the heat loss rate is proportional to the area, *A*, through which the heat passes, times the temperature difference $(T_i - T_a)$, which drives that heat transfer. The proportionality constant is called the thermal conductance, or U-value. For areas measured in ft² and temperatures in °F, the units of the U-value are Btu/hr of loss per square foot of area, per degree Fahrenheit of temperature between the inside air and the outside air:

Eq. 6.2
$$q(\text{Btu/hr}) = U\left(\frac{\text{Btu/hr}}{\text{ft}^2 \cdot {}^\circ\text{F}}\right) A(\text{ft}^2)(T_i - T_a){}^\circ\text{F}$$

An alternative description of the proportionality constant introduces the concept of the **R**-value of the building component. The **R**-value is just the inverse of the U-value. Thus we have

Eq. 6.3
$$q = UA\Delta T = \frac{1}{R} A\Delta T$$

Equation 6.3 might remind you of an electrical equivalent in which current is voltage divided by the resistance. In fact, it is common to draw the thermal resistance using the same symbol used for electrical resistances, such as is shown in Figure 6.5.

figure
6.5

$$q = \frac{A}{R}\left(T_i - T_a\right) \qquad\qquad i = \frac{1}{R}\left(V_A - V_B\right)$$

The electrical and thermal analogies. Both involve some sort of flow at a rate that is inversely proportional to some resistance.

For example, let's find the heat loss rate through 1000 ft² of "**R-20**" wall when the inside air temperature is 70°F and it is 10°F outside. Using Equation 6.3, we have

$$q = \frac{A}{R}(T_i - T_a) = \frac{1000\ \text{ft}^2}{20(\text{hr-}^\circ\text{F-ft}^2/\text{Btu})}\,(70{-}10)^\circ\text{F} = 3000\ \textbf{Btu/hr}$$

The **R**-value, mentioned above, summarizes a rather complex combination of three heat-transfer mechanisms that include radiation, convection, and conduction. Figure 6.6 introduces these three basic processes in the context of a wall or a window. Notice there are four temperatures to track: inside air temperature, labeled T_i; ambient (outside) air temperature, labeled T_a; surface temperature of the wall or window on the room side, labeled T_1; and surface temperature of the wall or window on the ambient side, T_2. Also shown in the figure is a series combination of thermal resistances with nodes that correspond to these four temperatures.

6.2.2 Heat Transfer by Conduction

The most familiar mode of heat transfer is conduction. Using the window as an example, the inside surface of the glass (T_1) in Figure 6.6 is warmer than the outside surface (T_2). Atoms within the glass near the warmer inside surface are more agitated than their cooler counterparts near the cold outside surface. As warmer atoms bang into adjacent, cooler atoms, energy is transferred from one to the other and the resulting heat flow from the warm side of the glass to the cool side is said to occur by conduction. The rate of heat flow will be proportional to that temperature difference, as well as the surface area of the glass, A, and be inversely proportional to the glass thickness. This is spelled out in Equation 6.4 in which the proportionality constant, k, is called the *thermal conductivity*.

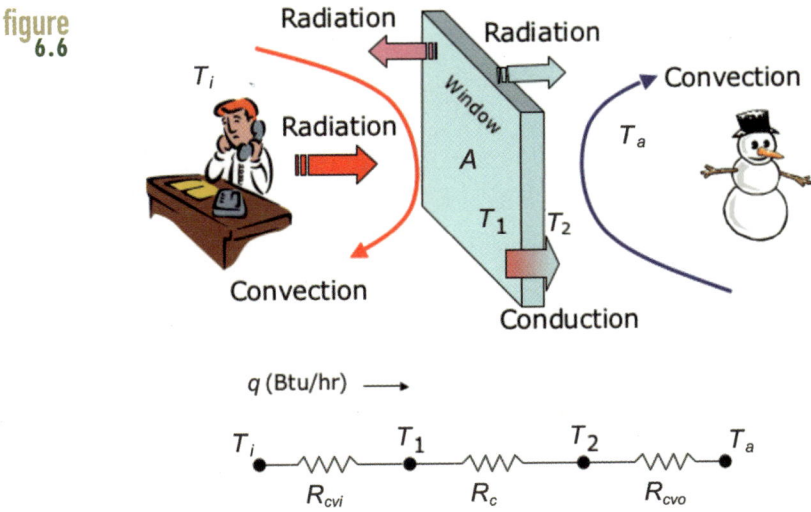

Heat transfer through the window involves radiation, convection, and conduction. Total resistance to heat flow can be modeled as a series combination of thermal resistances.

Eq. 6.4   *Conduction:*
$$q = \frac{kA}{t}(T_1 - T_2)$$

In conventional American units,

A = surface area (ft²)

t = material thickness (in.)

T_1 and T_2 = surface temperatures (°F)

k = thermal conductivity (Btu-in/hr-ft²-°F)

Table 6.1 gives typical values of thermal conductivity for selected materials, ranging from a high of 1400 for aluminum, which means it very readily conducts heat, to a low of 0.14 for a rigid foam material called polyisocyanurate, which is an excellent insulator. Notice the rather odd-looking American units of Btu-in/hr-ft²-°F have the advantage of simplifying calculations for material thicknesses measured in inches.

By comparing Equations 6.3 and 6.4, we can define the conductive thermal resistance of a homogeneous material, such as glass, steel, or wood, to be

Eq. 6.5   *Conductive resistance:*
$$R_C = \frac{t}{k}$$

Solution Box 6.1 illustrates the use of this simple relationship for conductive resistance.

A very convenient measure of thermal resistance is the **R**-value per inch thickness of a material, which from Equation 6.5 is just $1/k$. Table 6.2 provides some **R**-values per inch

table 6.1 Thermal conductivity (*k*) of selected materials

MATERIAL	*k* (Btu-in/hr-ft²-°F)
Air (stationary, no convection)	0.18
Argon (stationary, no convection)	0.14
Aluminum	1400
Brick, common	5
Brick, face	7
Concrete: lightweight aggregate	6
Concrete: sand, gravel, stone aggregate	12
Glass	5.5
Gypsum, solid	3
Plaster, cement	8
Polyisocyanurate board	0.14
Soil, 10% moisture	13
Steel	310
Wood, fir	0.8
Wood, oak	1.1

for a number of common building materials, including many of the most popular forms of insulation, along with some overall **R**-values for typical material thicknesses.

It is particularly interesting to note how well concrete conducts heat. When used as thermal mass, heat from sunlight shining on its surface can be conducted into its depths quite well. On the other hand, with an **R**-value of about 1-per-foot, it is a very poor insulator. In fact, it would take about 12 feet of concrete wall thickness to have the same (steady-state) **R**-value as a simple fiberglass-insulated wall framed with two-by-fours.

6.2.3 Heat Transfer by Convection

Convection is similar to conduction in that excited, hot molecules transfer heat to adjacent cooler ones, but in this case the substance is a liquid or gaseous fluid such as air or water. Warm molecules in a fluid are free to move about and when they encounter something colder they can transfer some of their heat to the colder object. In our example, warm air in the room that comes in contact with the cold window surface transfers some heat to the window, which cools the air and increases its density. That colder, dense air falls toward the floor, encouraging warmer room air above to move toward the window where it too dumps some heat onto the window surface. That cold window creates convective air currents in the room, which can be a source of discomfort. If you were sitting near the floor next to this cold window, you would likely feel an uncomfortable cold stream of air flowing down the window

---

SOLUTION BOX 6.1

R-Value of Wall Studs

What are the **R**-values of 2 × 4 and 2 × 6 wood framing members (studs) in a wall?

Solution:

First, we have to real-
ize that a "two-by-four"
piece of construction
lumber does not have
dimensions of 2 inches
by 4 inches. In fact, a
two-by-four is actually
$1\frac{1}{2}$ inches × $3\frac{1}{2}$ inches.
Similarly, a "two-by-six"
is $1\frac{1}{2}$ inches × $5\frac{1}{2}$ inches.
Using Table 6.1, we
find that an ordinary
softwood, such as fir or
pine, has a thermal con-

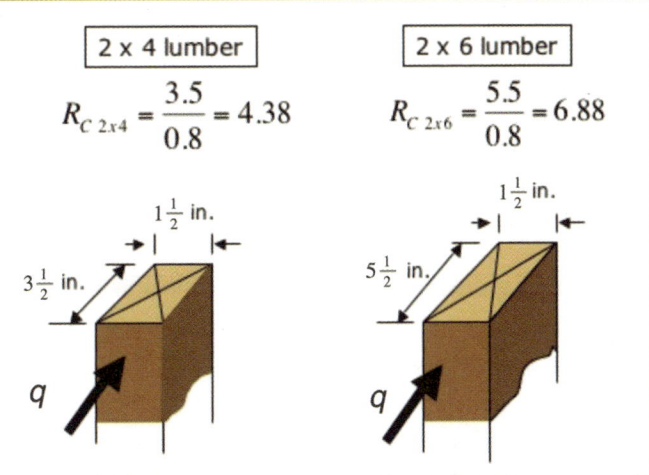

$$R_{C\,2x4} = \frac{3.5}{0.8} = 4.38 \qquad R_{C\,2x6} = \frac{5.5}{0.8} = 6.88$$

2 x 4 lumber — $3\frac{1}{2}$ in., $1\frac{1}{2}$ in., q

2 x 6 lumber — $5\frac{1}{2}$ in., $1\frac{1}{2}$ in., q

ductivity k of about 0.8 Btu-in/hr-ft^2-°F. As studs in a wall, heat flow is through the $3\frac{1}{2}$-inch
or $5\frac{1}{2}$-inch direction, which gives the two values of R shown above. Notice how those funny
k units make R-value calculations so easy.

and across the floor. To help counteract those drafts many heating systems purposely locate
warm-air registers directly beneath windows, which may seem a little odd because it means a
lot of that heat literally goes right through the window.

The convective heat transfer associated with air coming in contact with a building
surface is influenced by several factors, one of which is the speed at which air moves along
the surface. Clearly, faster-moving air blowing against the outside of your house can transfer
heat more easily than slow-moving, indoor "still" air. By convention, heating and cooling
calculations are based on an assumed 15 mph wind speed outside during the winter and
7.5 mph during the summer.

Convective heat transfer also depends on the direction of heat flow (heat transfer is
easier through the ceiling where warm air wants to collect than at the cold floor), and the
surface roughness (bumpy surfaces encourage heat transfer). In the next section, we'll see how
these factors affect the **R**-value associated with convection.

table **R-values-per-inch and typical total R-values for selected building materials**
6.2

R-Values per Inch	(hr-ft²-°F/Btu-in.)
Acoustic ceiling tile	2.9
Brick	0.2
Cellulose insulation	2.7
Concrete: 140 lb/ft³, sand/gravel/stone aggregate	0.08
Expanded polystyrene (EPS)	4.0
Extruded polystyrene (XPS)	5.0
Fiberglass, mineral wool insulation	3.5
Gypsum board	0.9
Hardwood	0.91
Particleboard	1.1
Plywood, softwood	1.25
Polyisocyanurate	7.0
Polyurethane foam sprayed roofing	4.5
Typical Total R-Values	**(hr-ft²-°F/Btu)**
Carpet and fibrous pad	2.1
Concrete block (8-in. lightweight), perlite cores	6
Concrete block (8-in. normal wt.), hollow core	1.1
Hardwood flooring	0.7
Mineral fiber insulation, 3.5-in.	13
Mineral fiber insulation, 5.5-in.	22
Wood shingles	1.0

6.2.4 Radiation Heat Transfer

The third mode of heat transfer is radiation. Whereas conduction and convection require a substance for heat to flow through, such as the room air and window glass in our example, radiant heat transfer can occur even in a vacuum.

As was described in Chapter 4, every object constantly radiates energy toward its surroundings at a rate proportional to the object's surface area and its temperature raised to the fourth-power. When you are sitting in a room you radiate energy, some of which is absorbed by surfaces that surround you, such as the window. The window, in turn, radiates energy back into the room, and you will absorb some of that. The net heat loss by radiation from you to the window will be related to the temperature difference between you and the surface temperature of the window. Whereas heat transfer by radiation is highly nonlinear, for the relatively small temperature differences experienced in a building we will get by with a linear approximation that will allow us to use the "Ohm's law" heat transfer relationship introduced in Equation 6.3.

Radiation emitted by one object can be absorbed by other surfaces in the line of sight. For almost all typical surfaces, the absorptivity (α) for thermal radiation is well above 90%, which means almost all of the radiation that hits a wall or a window will be absorbed, leaving very little to be reflected or transmitted. The principal exceptions are shiny, metallic surfaces, which do a great job of keeping heat in the room by reflecting outgoing radiation back into the room. That's why a lot of insulating materials have a tinfoil-like surface, and why high-quality ("low-e") windows have a thin film of metal on them to help reflect thermal radiation back into the room. The low-e designation, by the way, means "low emissivity," which is a little confusing until we recall that the absorptivity and emissivity for any surface are numerically equal for any given wavelength. A shiny low-e surface is therefore also a low-absorptance surface; and if it is low absorptance it means it has high reflectance.

6.2.5 The Combined Convective-Radiative R-Value

It is conventional practice when doing heat loss calculations to lump the convective and radiative components of heat transfer into a single, simple R-value. The radiative-convective R-value for indoor surfaces is designated R_{cvi}; the value for outside surfaces is R_{cvo}. The standard value for R_{cvo} for winter heating calculations (based on an assumed wind speed of 15 mph) is 0.17 hr-ft^2-°F/Btu; the corresponding 7.5 mph summer value used for cooling calculations is 0.25 hr-ft^2-°F/Btu. The indoor values of R_{cvi} depend on the direction of heat flow as well as whether they are ordinary surfaces (emissivity, $\varepsilon \approx 0.9$), somewhat reflective surfaces such as aluminum-covered paper or bright galvanized steel ($\varepsilon \approx 0.20$), or very shiny surfaces such as bright aluminum foil ($\varepsilon \approx 0.05$). Table 6.3 summarizes these convective/radiative R-values and Figure 6.7 illustrates their application in a building.

We now have the basis for doing heat loss calculations for a building. For a simple, homogeneous material, such as a window, the total R-value is just the sum of the R-values due to conduction, convection, and radiation. As we shall see later, for other building surfaces,

table 6.3 **Convective-Radiative R-Values (hr-ft²-°F/Btu) for Various Building Surfaces**

Indoor	Non-Reflective	Reflective	
Air, R_{cvi}	$\varepsilon = 0.90$	$\varepsilon = 0.20$	$\varepsilon = 0.05$
Ceiling	0.61	1.10	1.32
Wall, window, door	0.68	1.35	1.70
Floor	0.92	2.70	4.55
Outdoor Air, R_{cvo}	**15 mph** (winter)	**7.5 mph** (summer)	
Any surface	0.17	0.25	

SOURCE: based on ASHRAE Handbook, 1993 Fundamentals

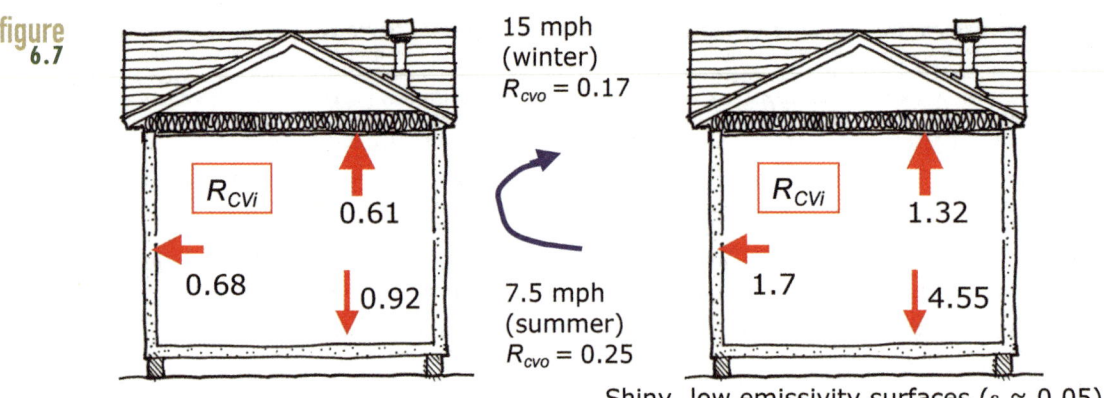

Commonly used convective-radiative heat transfer coefficients, R_{cv}, for ordinary building surfaces and for shiny, low-emissivity surfaces.

such as a wall consisting of a mix of wood framing members, insulation between the studs, wall sheathings, and so forth, there are parallel heat conduction paths that make the analysis a little more complicated.

6.3 Heat Loss through Windows

Windows, of course, serve many purposes in a building. They let us see what's going on outside, which, it turns out, has been identified as a major contributor to our sense of well-being when we are living or working indoors. Windows can also let in sunlight, which can help heat a building in the winter as well as provide natural daylight to offset the artificial lighting load all year long. But, we pay for these benefits with increased heat losses through windows in the winter and increased cooling loads if direct sunlight is allowed to shine through those windows in the summer.

6.3.1 Single-Pane Window Analysis

Windows are one of the biggest sources of heat loss in buildings, and simple, single-glazed windows (one sheet of glass) are the biggest losers of all. Unfortunately, more than half of the window area in the existing stock of buildings in the United States is still single-pane glass. Fortunately, in new construction almost all windows now have at least two sheets of glass (referred to as insulating glass, double-glazing, or double-pane windows).

    Using our basic heat transfer tools, we can easily analyze the performance of various window types (see Solution Box 6.2).

SOLUTION BOX 6.2

R-Value of a Simple Window

Find the R-value for a single-glazed window (one sheet of glass) made with 3/16-inch-thick glass. Then find the rate at which heat would be lost through a 2 × 3 foot window when the indoor temperature is 70°F and it is a chilly 25°F outside.

Solution:

We can model this as a series combination of R-values as shown below. From Table 6.1, the thermal conductivity of the glass is 5.5 Btu-in/hr-ft²-°F, and from Table 6.3, $R_{cvi} = 0.68$ and $R_{cvo} = 0.17$ (winter).

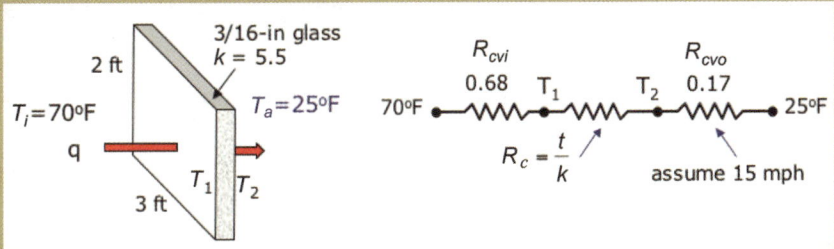

The conductive resistance of the glass itself is therefore

$$R_C = \frac{t}{k} = \frac{3/16 \text{ in.}}{5.5 \text{ Btu-in/hr-ft}^2\text{-}°F} = 0.034 \text{ hr-ft}^2\text{-}°F/\text{Btu}$$

so the total R-value of the window is

$$R_{tot} = R_{cvi} + R_c + R_{cvo} = 0.68 + 0.034 + 0.17 = 0.884 \text{ hr-ft}^2\text{-}°F/\text{Btu}$$

Notice almost none of that R-value is due to the glass itself (just 0.034 out of 0.884). The heat loss rate will be

$$q = \frac{A}{R}(T_i - T_a) = \frac{2 \text{ ft} \times 3 \text{ ft}}{0.884 \text{ hr-ft}^2\text{-}°F/\text{Btu}}(70 - 25°F) = 305 \text{ Btu/hr}$$

6.3.2 Discomfort and Condensation Problems with Cold Windows

Although the heat loss rate through windows is our primary concern, there is another concern that is quite important as well. With a single sheet of glass between you and the cold outdoors, the inside surface temperature of the glass can be very cold. You radiate energy to the window, but the window, being so cold, does not radiate much back. The net effect is that even though you may be sitting in a room with a supposedly comfortable 70°F air temperature, you are likely to be uncomfortable sitting next to that cold window.

Using our newly developed heat transfer skills, it is easy to figure out how cold that glass will be. If we go back to Figure 6.6, we can write that the heat loss rate (Btu/hr) from the room at T_i to the inside surface of the glass at T_1 will be the same as the heat loss rate from the room all the way to the outside ambient temperature T_a. That is,

Eq. 6.6
$$q = \frac{A}{R_{cvi}} (T_i - T_1) = \frac{A}{R_{tot}} (T_i - T_a)$$

from which, we can solve for the inside surface temperature of the glass:

Eq. 6.7
$$T_1 = T_i - \frac{R_{cvi}}{R_{tot}} (T_i - T_a)$$

Notice that we can use this formula for more complicated windows with multiple panes of glass as well as other building components such as walls and water heaters. An application of Equation 6.7 is given in Solution Box 6.3.

Cold window surfaces not only decrease occupant comfort, they also can cause condensation problems on the windows. The temperature below which air can no longer hold the moisture it contains is called the dew point. If the room-side surface of the glass drops below the dew point, water vapor condenses on the glass and drips down onto the windowsill, which not only causes a mess, it also can contribute to wood rot. If it is cold enough, it can even freeze on the glass. A plot of temperature at which condensation will occur for various window types is shown in Figure 6.8. For example, with indoor relative humidity at 50%, condensation will form on a single-pane R-0.88 window when the ambient temperature drops below about 45°F, whereas an R-3.1 double-pane, low-e window won't show condensation until the ambient temperature is well below 0°F.

6.3.3 Better Windows

After double-pane windows became more common, the next improvement was the addition of low-emissivity (low-e) coatings usually deposited onto an inside surface of one of the

---

Glass Temperature for Single-Glazing versus Low-e, Double-Pane Windows

Find the inside surface temperature of the single-pane, R-0.884 window from Solution Box 6.2 and compare it with an R-3.1, low-e, double-pane window. That 3.1 R-value, by the way, includes the usual R_{cvi} = 0.68. Room temperature is 70°F and it is 25°F outside.

Solution:

Using Equation 6.7 for the single-pane window gives us

$$T_1 = 70° - \frac{0.68}{0.884}(70 - 25°) = 35°F$$

Repeating the calculation for the R-3.1 window gives

$$T_1 = 70° - \frac{0.68}{3.1}(70 - 25°) = 60°F$$

The low-e window will be a heck of a lot more comfortable.

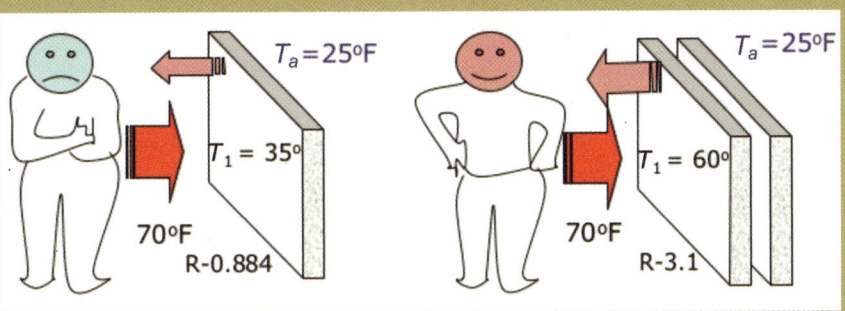

---

panes of glass. The low-e coating helps reduce radiation from the warm pane of glass to the colder one, which means if you are trying to keep heat in the house the low-e surface should be on the inside pane of glass. On the other hand, in very hot climate zones the goal might be to keep heat out of the building, in which case the low-e surface should be on the outside pane. Low-e windows improve the R-value of conventional double-glazing from about R-2 to about R-3.

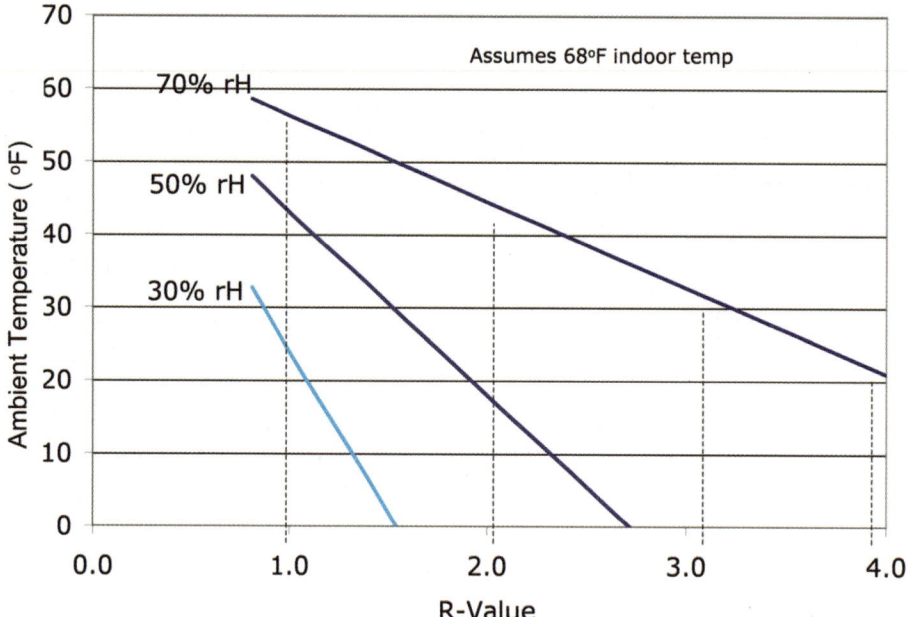

Ambient temperature at which condensation will occur as a function of window R-value, for various values of room relative humidity at 68°F room temperature.

Double-pane, low-e windows can be further improved by filling the air gap with low-conductivity argon or krypton gas instead of air. Argon improves the R-value beyond conventional low-e by about 0.6, and the more expensive krypton can increase it by another 0.6.

One manufacturer produces exceptionally high-performance glazing systems by including layers of low-emissivity coatings on polyester films suspended inside an insulating glass unit. These "Heat Mirror" windows create more air spaces without increasing the weight of the glazing system. Figure 6.9 shows two of these products with astonishing R-values of 9.1 and 12.5 in the center-of-glass area away from the frame. Figure 6.10 illustrates the range of window R-values for a variety of types of glazing systems.

6.3.4 Center-of-Glass R-Values and Edge Effects

Window heat losses are complicated by the edge effects associated with the window frame. Window frames, especially if they are metal, act as thermal short-circuits that cool the edge of the glass near the frame. The advantages of high-efficiency glazing systems can be negated to a large extent by a poor choice of window framing materials.

The R-value (or U-value, which is just the inverse of the R-value) in the center of the window, well away from the edges, can be very different from what it is near the frame. For example, a double-glazed, low-e window has a center-of-window R-value of about 3.1, but if

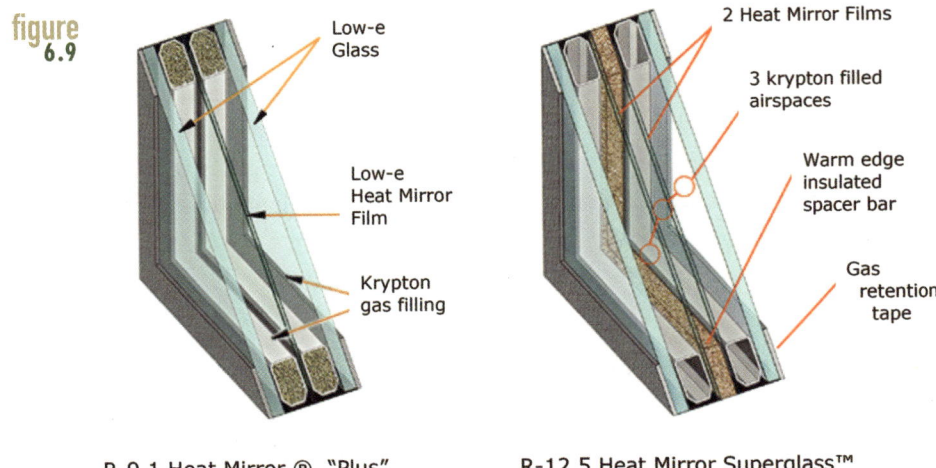

figure 6.9

R-9.1 Heat Mirror ® "Plus"   R-12.5 Heat Mirror Superglass™

Heat Mirror™ "Plus" with krypton gas and low-e coatings on the glass and interior film makes an R-9.1 window. The Superglass™ system with three krypton-filled air spaces can be as high as R-12.5.

SOURCE: Courtesy of Southwall Corporation

figure 6.10 **Example Center-of-Glass R-Values of Various Glazing Types**

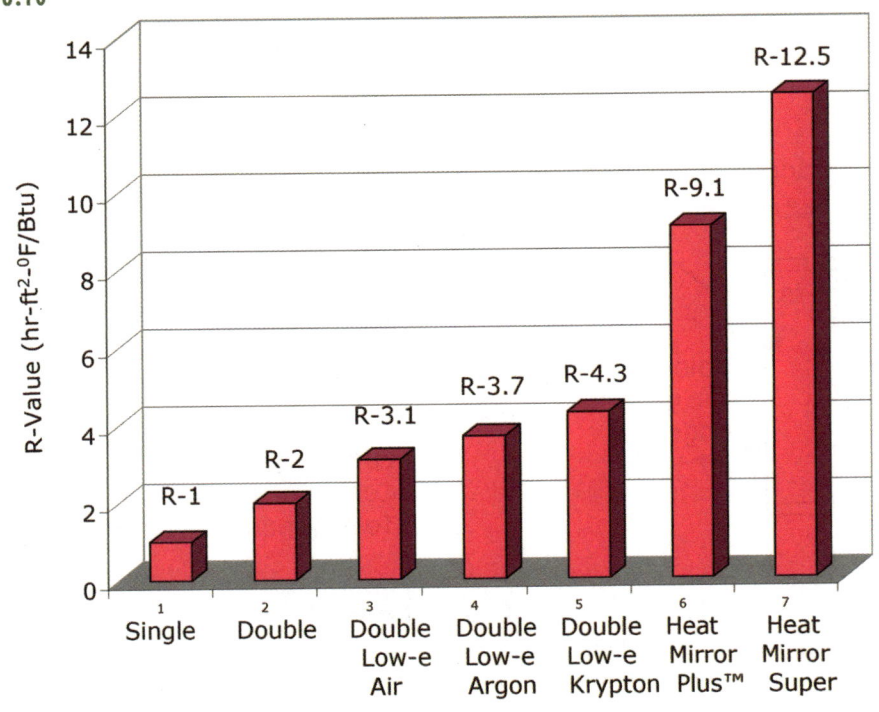

it is built with an aluminum frame with metal spacers the average **R**-value in the vicinity of the frame (defined to be the first 2.5 inches of glass) drops to about 2.1. The aluminum frame itself has an **R**-value of about 0.6. For a 3 × 5 foot, double-hung window, that combination of **R**-values results in an overall **R**-value of only 1.5. If instead, that same window used a high-quality fiberglass or vinyl frame with insulated spacers holding the panes apart, the overall **R**-value would be about 2.9, which is nearly twice as good as the aluminum-framed window. Figure 6.11 illustrates this example.

6.4 Heat Loss through Walls, Ceilings, and Floors

Heat loss calculations for walls, ceilings, and floors are a bit more complicated than those demonstrated for our simple window. For walls, we have to account for parallel conduction paths through framing members and through wall cavities filled (ideally) with insulation. For ceilings, we may want to account for the differences between a ceiling and a roof, with most houses having an attic in between. Floors are especially tricky because they may be concrete slabs poured on grade, or framed floors over crawl spaces, or floors over basements that may or may not be heated. Fortunately, many of these complications can be handled with tables—often provided by the American Society of Heating, Refrigerating and Air-Conditioning Engineers (ASHRAE). Even though these tables provide convenient shortcuts, it is worth doing a few simple calculations on our own to gain insight into energy-efficient construction techniques.

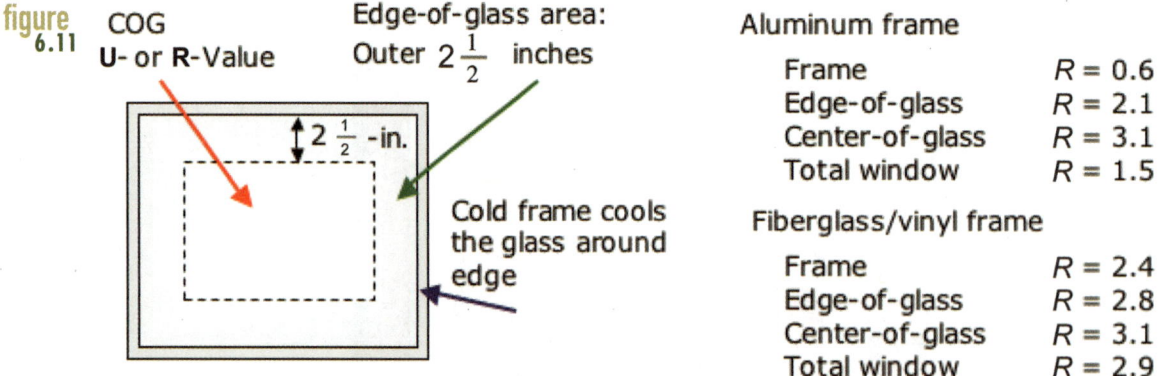

figure 6.11

The window frame can have an enormous influence on the overall R-value of a window. This example illustrates the impact of an aluminum frame versus a fiberglass/vinyl frame for a 3 × 5 foot window. Several smaller windows are more affected by edge effects than a single larger window with the same total area.

6.4.1 Walls

Let's begin with the wall shown in Figure 6.12. Walls consist of framing members usually made of wood (e.g., two-by-fours or two-by-sixes with center-to-center spacings between studs of 16 or 24 inches), a cavity area between framing members that should be filled with insulation, and various levels and types of sheathing on the inside and outside surfaces. If you were looking at the wall from inside of the house, you can imagine an area of wall behind the sheetrock that is framing, A_F, and an area of wall that is cavity, A_C, with the total wall area, A_{TOT}, being the sum of the two. To analyze the heat loss rate for parallel heat conduction pathways, such as occur in the wall, the mathematics is slightly easier if we use U-values instead of R-values in Equation 6.3. Letting U_F be the U-value through framing, U_C through the cavity, and U_{AVG} be the average for the wall, we can write the total heat loss rate q_{WALL} as

Eq. 6.8 $$q_{WALL} = q_C + q_F = U_F A_F (T_i - T_a) + U_C A_C (T_i - T_a) = U_{AVG} A_{TOT} (T_i - T_a)$$

from which, we can write the average U-value for the wall as follows:

Eq. 6.9 $$U_{AVG} = U_F \left(\frac{A_F}{A_{TOT}} \right) + U_C \left(\frac{A_C}{A_{TOT}} \right)$$

Eq. 6.10 $$F_R = \text{Framing factor} = \frac{A_F}{A_{TOT}}$$

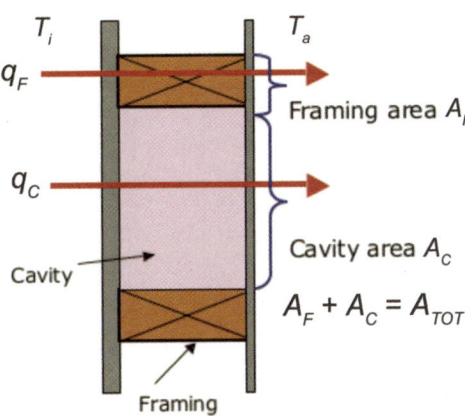

Looking down onto a wall, showing two parallel heat conduction pathways. One is through framing members (q_F) and one is through the cavity between framing (q_C).

The total average U-value for the wall is therefore

Eq. 6.11
$$U_{AVG} = U_F F_R + U_C (1 - F_R) \text{ and } R_{AVG} = \frac{1}{U_{AVG}}$$

Or, if you prefer R-values, the more complicated formula for average R-value is

Eq. 6.12
$$R_{AVG} = \frac{R_F R_C}{R_C F_R + R_F (1 - F_R)}$$

Simple changes in the way walls are framed can reduce the amount of wood materials needed and increase the cavity area available for insulation. Figure 6.13 shows framing for a typical 2 × 4, 16-inch on-centers wall compared with a more carefully thought out 2 × 6, 24-inch on-centers wall-framing system. Even though the 2 × 6 wall has more wood per stud, there are fewer of them, which saves wood. In addition to having more cavity area in the 2 × 6 wall, it has also been designed to carefully align windows with studs to eliminate unnecessary framing.

Figure 6.14 shows the cross section of a wall and Table 6.4 shows a simple spreadsheet analysis for the wall under three assumptions: (1) No cavity or exterior insulation (an unreflective air gap is about R-1.01) with 2 × 4 studs, 16-inch on-centers (o-c) with an estimated framing factor of 15%; (2) a 2 × 4, 16" o-c wall with R-13 fiberglass insulation and no sheathing; (3) a 2 × 6, 24" o-c wall, 12% framing factor, with R-22 cavity insulation and 2 inches of isocyanurate exterior sheathing. The results show enormous R-value advantages to rather simple, improved construction techniques. Whereas the uninsulated, 2 × 4 wall has an R-value of less than 4, the wall with 2 × 6 framing and easily incorporated insulation beefs the R-value to almost 35!

figure 6.13

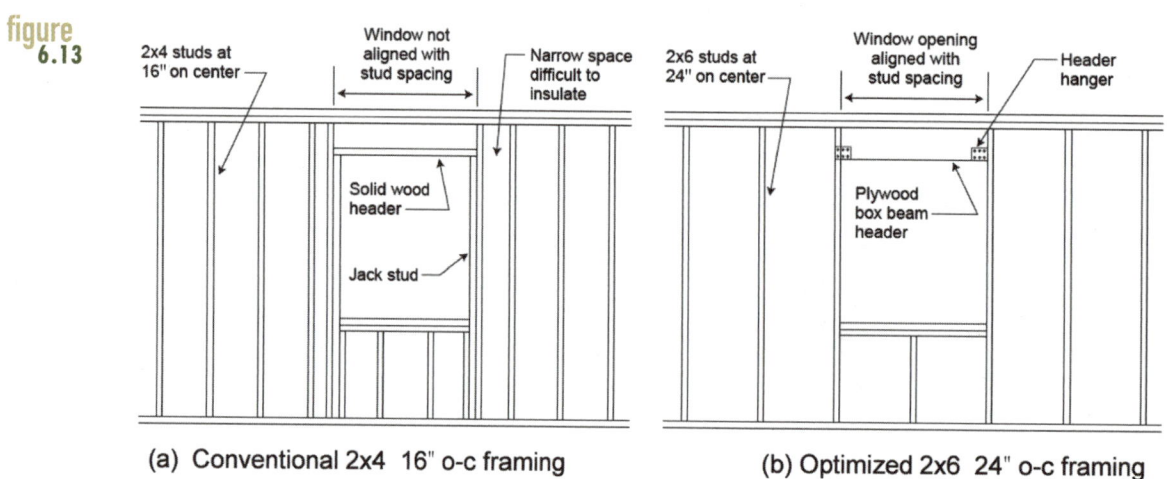

(a) Conventional 2x4 16" o-c framing   (b) Optimized 2x6 24" o-c framing

The framing factor can be reduced by using two-by-sixes on 24-inch centers and more carefully aligning window openings with stud spacing.

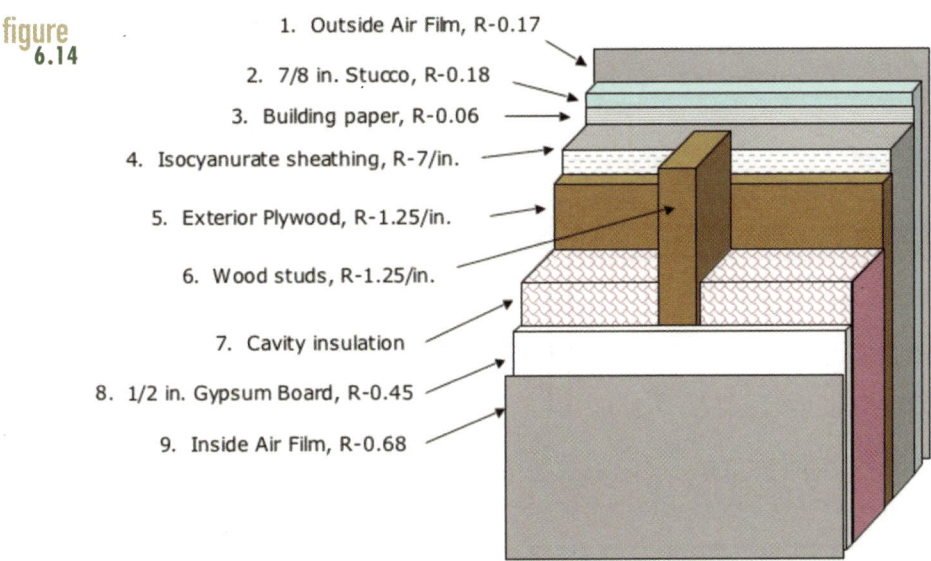

figure 6.14

1. Outside Air Film, R-0.17
2. 7/8 in. Stucco, R-0.18
3. Building paper, R-0.06
4. Isocyanurate sheathing, R-7/in.
5. Exterior Plywood, R-1.25/in.
6. Wood studs, R-1.25/in.
7. Cavity insulation
8. 1/2 in. Gypsum Board, R-0.45
9. Inside Air Film, R-0.68

An example wall, variations of which are analyzed in Table 6.4.

table 6.4   R-value calculations for three simple walls (ft²-°F-hr/Btu):

A 2 × 4 wall, no insulation, 16-in o.c., FR = 15% (W.0.2 × 4.16)
A 2 × 4 wall, R-13 cavity insulation, 16-in o.c., FR = 15% (W.13.2 × 4.16)
A 2 × 6 wall, R-21 cavity insulation, 2-in sheathing, 24-in o.c., FR = 12%
(W.35.2 × 6.24)

Construction Components	W.0.2 × 4.16, FR 15%		W.13.2 × 4.16, FR 15%		W.35.2 × 6.24, FR 12%	
	Cavity	Framing	Cavity	Framing	Cavity	Framing
1. Outside air film (15 mph)	0.17	0.17	0.17	0.17	0.17	0.17
2. 7/8-in, stucco	0.18	0.18	0.18	0.18	0.18	0.18
3. Building paper	0.06	0.06	0.06	0.06	0.06	0.06
4. 2-in, isocyanurate @ R-7/in.	–	–	–	–	14.0	14.0
5. 1/2-in plywood	0.625	0.625	0.625	0.625	0.625	0.625
6. Cavity	1.01	–	13.0	–	21.0	–
7. Framing	–	4.38	–	4.38	–	6.88
8. 1/2-in, gypsum board	0.45	0.45	0.45	0.45	0.45	0.45
9. Inside air film (still air)	0.68	0.68	0.68	0.68	0.68	0.68
Sum of thermal resistances	3.18	6.55	15.17	6.55	37.17	23.05
U-values	0.315	0.153	0.066	0.153	0.027	0.043
Fraction of wall	0.85	0.15	0.85	0.15	0.88	0.12
Avg. U-Value (Btu/ft²-hr-°F)	**0.291**		**0.079**		**0.029**	
Total R-Value (ft²-hr-°F/Btu)	**3.44**		**12.66**		**34.62**	

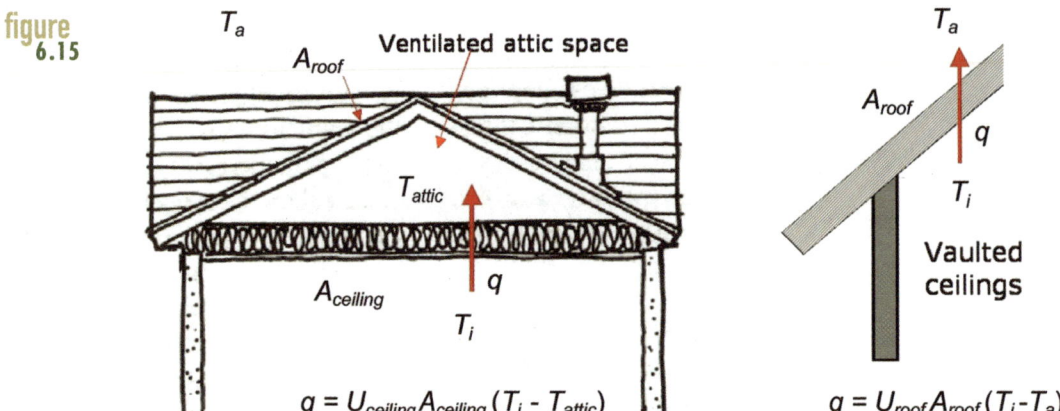

figure 6.15

$$q = U_{ceiling} A_{ceiling} (T_i - T_{attic})$$

$$q = U_{roof} A_{roof} (T_i - T_a)$$

For cathedral ceilings, use the actual roof area; for homes with ventilated attic spaces, assume the attic is at ambient temperature and use the ceiling area for heat loss calculations.

6.4.2 Ceilings and Roofs

Heat loss through a vaulted ceiling is a straightforward extension of the calculations used for walls. Because the ceiling area is the same as the roof area, just find an appropriate **R**-value and use the ceiling (roof) area to find heat loss rate as shown in Figure 6.15.

For homes with ventilated attic spaces, the heat loss calculation is slightly complicated by the fact that heat loss from the room is through the ceiling into the attic and from there to the great outdoors. It is a convenient assumption to treat the attic as if it is at the same temperature as ambient, in which case it is the **R**-value of the ceiling and the ceiling area (not the roof area) that should be used.

6.4.3 Floors

Heat loss calculations for floors are complicated in part because there are so many ways to build floors. Figure 6.16 shows floors built over crawl spaces, those built over basements, and slab-on-grade floors. Even very careful calculations for any of these involve a number of assumptions and approximations. We'll keep it simple here and relax knowing that heat loss through floors is only a modest fraction of the total anyway so if we're off a bit it won't matter that much.

For a floor built over a crawl space, the rate of heat loss between T_i and ambient T_a is determined by the **R**-value of the flooring and insulation plus some equivalent **R**-value of the crawl space itself. One approximation that is sometimes used is simply to add R-6 to account for the impact of the crawl space (see Solution Box 6.4). For unheated

<figure>figure 6.16</figure>   **Heat Losses through a Variety of Floor Types**

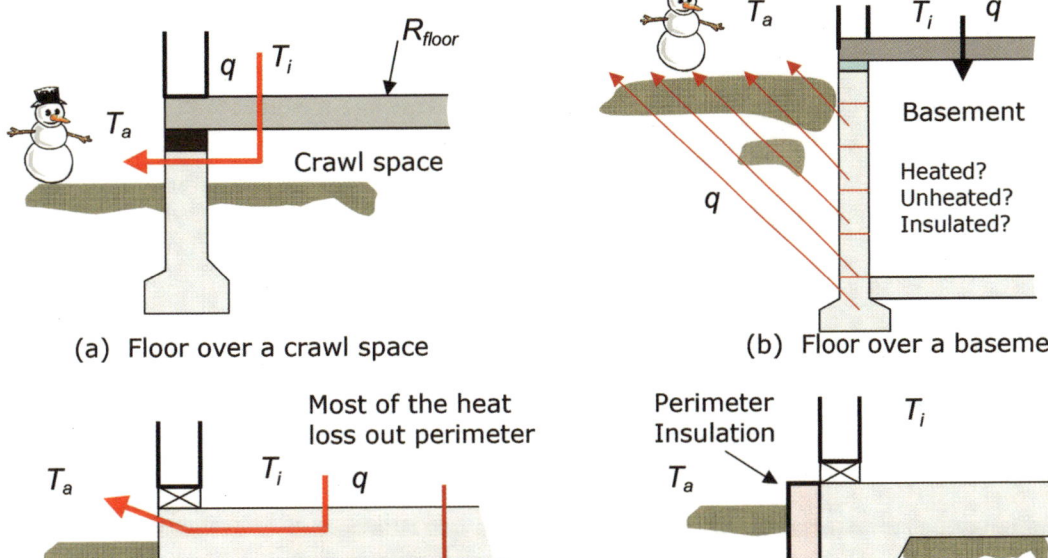

(a) Floor over a crawl space

(b) Floor over a basement

(c) Uninsulated slab-on-grade floor

(d) Insulated slab-on-grade

SOLUTION BOX 6.4

Heat Loss Approximation for a Floor Over a Crawl Space

A house with 2000 ft^2 of floor space built over a crawl space has an R-20 floor. If the indoor temperature is 70°F and ambient is 30°F, estimate the floor heat loss rate.

Solution:

We'll just use the approximation in which we add R-6 to account for the crawl space, giving a heat loss rate of

$$q = \frac{A_f}{R_f + R_{cs}}(T_i - T_a) = \frac{2000 \text{ ft}^2}{(20 + 6) \text{ ft}^2\text{-hr-°F/Btu}}(70 - 30) \,°\text{F} = 3076 \text{ Btu/hr}$$

basements, which are much more complicated, this trick provides a reasonable first approximation as well.

For relatively small slab-on-grade floors (that means homes, not commercial buildings), heat transfer is dominated by losses through the perimeter of the slab to ambient. Recall concrete is a pretty good conductor with an R-value of about 1 per foot so heat entering the slab zips along horizontally more easily than it flows into the dirt below. To reduce those losses, slabs are often insulated along the perimeter as shown in Figure 6.16(d).

The equation used to estimate slab losses is therefore based on the linear feet of perimeter rather than floor area:

Eq. 6.13
$$q = F_2 P (T_i - T_a)$$
where P = perimeter (ft)
  F_2 = heat loss factor (Btu/hr-ft-°F)

The factor F_2 depends on the R-value of the insulation and its vertical dimension (referred to as its "depth"). Table 6.5 provides some examples of the F_2 factor and Solution Box 6.5 illustrates its use.

6.5 Heat Loss Due to Infiltration

A major load for your furnace is heating up cold air leaking into your house while warm indoor air leaks out. These infiltration losses are driven in part by the difference in indoor-to-outdoor temperature (stack-driven infiltration) and in part by pressure differences caused by wind blowing against the side of the house (wind-driven infiltration) as shown in Figure 6.17.

Stack-driven infiltration is caused by warm, buoyant indoor air that rises and creates a higher pressure indoors, near the ceiling, than exists outdoors. As that warm air finds holes and leaks out, it creates a low-pressure suction that draws in cold air near the floor. We can imagine a pressure profile in which there is a high-pressure zone near the ceiling and a low-pressure zone near the floor; somewhere in between is a neutral plane such as is shown in Figure 6.17(a). The most important leaks to plug to control stack-driven infiltration are near the ceiling and floor; holes near the neutral plane don't matter as much. In certain parts

table 6.5 Values of Slab Heat Loss Factor, F_2 (Btu/hr-ft-°F)

Slab Surface	Insulation Depth (in.)	No Insulation	R-7 Insulation
No carpet	0	0.9	0.9
No carpet	18	0.9	0.55
Carpet	18	0.72	0.49

SOLUTION BOX 6.5

An Insulated Slab

Suppose a 40 × 50 foot slab has 18 vertical inches of R-7 edge insulation along its perimeter. With indoor temperature 70°F and ambient at 30°F, estimate the heat loss through the carpeted floor.

Solution:

From Table 6.5, we find F_2 = 0.49 Btu/hr-ft-°F. From Equation 6.13 we find

$$q = F_2 P(T_i - T_a)$$
$$= 0.49 \text{ Btu/hr-ft-°F} \cdot (40 + 40 + 50 + 50) \text{ ft} \cdot (70 - 30)°F = 3528 \text{ Btu/hr}$$

Notice this heat loss rate is just a bit higher than the 3076 Btu/hr rate that we found for the R-20 floor over a crawl space in Solution Box 6.4.

---

of the country there is another reason for controlling stack-driven infiltration. Some soils naturally emit radon gas, which if drawn into a house can expose residents to a quite potent cancer-causing substance if inhaled. Stack-driven infiltration helps draw that radon out of the soil and into the house so plugging floor leaks is the first and most important method of helping to control that exposure.

Wind-driven infiltration is also caused by a pressure difference, but this time it is holes in the walls, not the floor and ceiling, that are most important to plug.

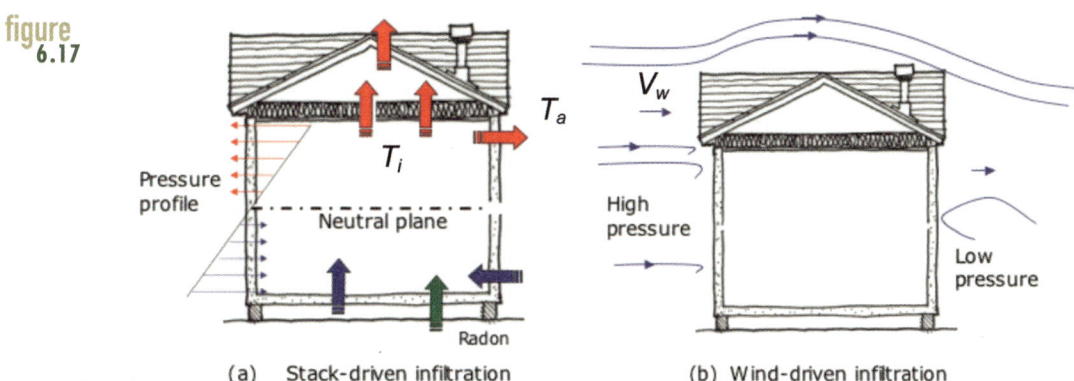

figure 6.17

(a) Stack-driven infiltration   (b) Wind-driven infiltration

The most important leaks to plug to control stack-driven infiltration are near the floor and ceiling. For wind-driven infiltration, the leaks in the walls are most important.

6.5.1 Estimating Infiltration Rate

To estimate heat loss due to infiltration, it is convenient to express the leakage rate in terms of how many complete air changes occur per hour (abbreviated as ach). We can begin with the following:

Eq. 6.14
$$q_{inf} = \rho\, c n\, V\,(T_i - T_a)$$

where q_{inf} = heat loss due to infiltration (Btu/hr)
ρ = density of air (0.075 lb/ft³)
c = specific heat of air (0.24 Btu/lb-°F)
n = number of air changes per hour (ach)
V = volume of air per air change (ft³/ac)

The product of density and specific heat given above is 0.018 Btu/ft³-°F, resulting in

Eq. 6.15
$$q_{inf} = 0.018 n V(T_i - T_a)$$

The trick, of course, is to somehow estimate n (ach). We can begin with some rules of thumb. Years ago, there wasn't much attention paid to weatherizing houses to control leakage and it was quite common to find infiltration rates on the order of 1.5 ach. A reasonable estimate for a good, tight new home these days is perhaps 0.5 ach. Much lower rates are physically possible, but when a home is tightened below about 0.35 ach, indoor air quality becomes an issue and it may be worth considering a heat-recovery ventilator (HRV), which is an air-to-air heat exchanger that warms incoming fresh air with heat from the outgoing, stale room air.

Figure 6.18 illustrates the most common way to actually measure infiltration rates by artificially pressurizing or depressurizing a house using a large fan and nylon door insert. In order to create a given house pressure, a leaky house will require more airflow through the fan than will a tight house. In a standard test, the airflow rate required to create a 50 Pascal pressure difference (about 1 lb/ft²) between the inside and outside of the house is measured. By dividing that flow rate by the volume of the house, an infiltration rate at 50 Pa is determined. Using conventional units the air change rate at 50 Pa (ACH50) is given by

Eq. 6.16
$$\text{ACH50} = \frac{\text{Air flow rate at 50 Pa}}{\text{House volume}} = \frac{\text{CFM50 (ft}^3/\text{min)} \times 60\ \text{min/hr}}{V\,(\text{ft}^3\ \text{per air change})}$$

After literally thousands of blower door tests, the following convenient rule of thumb has emerged to estimate annual average infiltration

Eq. 6.17
$$\text{Annual average air change rate } n\ (\text{ach}) \approx \frac{\text{ACH50}}{20}$$

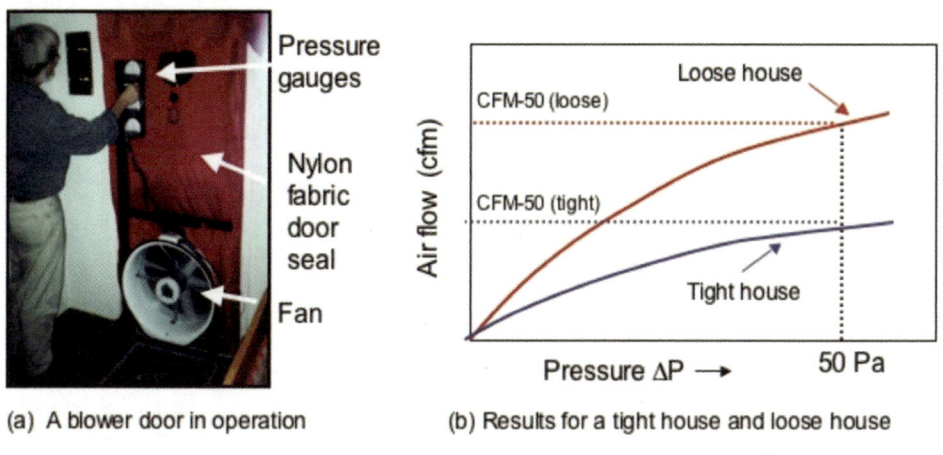

figure
6.18   Infiltration Measurement Using a Blower Door

(a) A blower door in operation   (b) Results for a tight house and loose house

An example of how to use ACH50 to estimate heat loss from infiltration is given in Solution Box 6.6.

6.5.2 Controlling Leaks

How do we tighten a house to reduce infiltration? In new construction, we need to employ good building materials and practices, including vapor barriers and sealed joints. In existing buildings, we can find leaks using the blower door. With the blower door pressurizing or depressurizing the house, leaks are accentuated and can be easily found by touch or with a smoke generator. Figure 6.19 shows typical sources of infiltration in a home. Major leaks in attics and basements or crawl spaces, which contribute to stack-driven infiltration, can be easily plugged using expanding spray foams. Window and door leakages that lead to wind-driven infiltration are quite easily controlled with weatherstripping and caulk. With some effort, densely packed blown cellulose insulation can be forced between studs in walls to control infiltration and increase thermal resistance.

6.6 The Overall Heat Loss Factor

We have described all of the pieces contributing to the heat loss of a house. Let us now put them together into an overall heat loss rate based on just summing up the contributions of each of the individual factors (windows, walls, floors, ceilings, doors, infiltration). To help with this exercise, Table 6.6 summarizes typical R-values for each of these components. Notice they are numbered to help us describe components for our evaluations.

SOLUTION BOX 6.6

Calculating Infiltration

A blower door test performed on a 2000-square-foot house showed 4000 cfm of airflow were required to pressurize the house to 50 Pa. If the average floor-to-ceiling height is 8 feet, estimate its average infiltration rate. Use that estimate to find the heat loss rate due to infiltration when it is 70°F inside and 30°F outside.

Solution:

From Equation 6.16

$$\text{ACH50} = \frac{\text{CFM50}}{V} = \frac{4000 \text{ ft}^3/\text{min} \times 60 \text{ min/hr}}{(2000 \text{ ft}^2 \times 8 \text{ ft})/\text{air change}} = 15 \text{ ach}$$

and from Equation 6.17 the average infiltration would be about

$$n \approx \frac{\text{ACH50}}{20} = \frac{15 \text{ ach}}{20} = 0.75 \text{ ach}$$

This is a pretty leaky house that could be significantly improved. Using Equation 6.15 we can estimate the heat loss rate due to infiltration.

$$q_{\text{inf}} = 0.018 \frac{\text{Btu}}{\text{ft}^3 \cdot {}^\circ\text{F}} \times 0.75 \text{ ach} \times \frac{2000 \text{ ft}^2 \times 8 \text{ ft}}{\text{ac}} \times (70 - 30)^\circ\text{F} = 8640 \text{ Btu/hr}$$

We will take a spreadsheet approach to evaluating the heat loss rate for the entire building. Each row will summarize a major building element, such as a wall, giving its area A, R-value, U-value, the UA product (Btu/hr-°F), and its percentage of the total UA-value. Infiltration isn't strictly speaking the product of a U-value and an area, but we can set it up to have the same Btu/hr-°F units so that it can easily be accounted for in the table. The infiltration loss will be included by converting it into an equivalent UA product using the relationship

Eq. 6.18 $(UA)_{\text{inf}} = 0.018 \text{ nV (Btu/hr-}^\circ\text{F)}$

Finally, for really tight houses, a heat recovery ventilator (HRV) may be included to increase ventilation without losing all that valuable exhaust heat. The UA equivalent for the HRV is given by

figure 6.19   Typical Points of Air Leakage

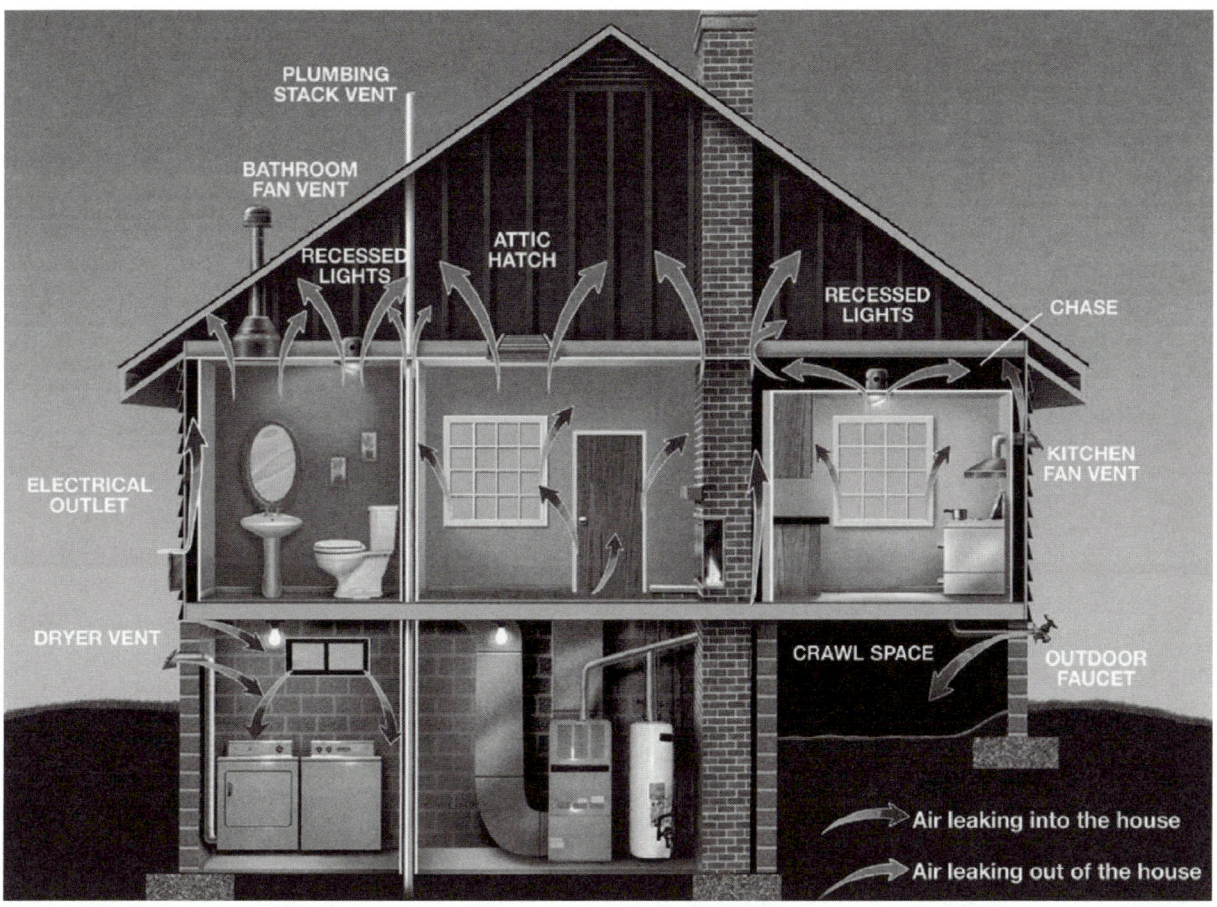

Leaks to the basement and attic can create a chimney effect and exacerbate infiltration energy loss.

SOURCE: U.S. EPA, 2000

Eq. 6.19
$$(UA)_{HRV} = 0.018 \, n_{HRV} \, V(1 - \eta_{HRV})$$

where n_{HRV} = induced ventilation (ach)
η_{HRV} = efficiency of the heat recovery ventilator

Table 6.7 shows an example of a spreadsheet created for a small, well-insulated, 1500 ft² house that would likely meet most building codes in existence today. It features 2 × 6 walls with R-21 insulation (#3 in Table 6.6), double-pane aluminum-frame windows (#21), R-30 insulation in the ceiling (#13), and R-21 floor insulation (#7, to which we will add R-6). The infiltration rate is a reasonable 0.6 ach and there is no HRV. Notice the two dominant sources of heat loss are the windows (37%) and infiltration (27%).

table 6.6 Default R-values of Common Home Construction Techniques

COMPONENT	CONSTRUCTION	Continuous Insulated Sheathing		
		None	1" R-5 XPS	2" R-14.4 ISO
Walls	1. Wood framed, 2 × 4, 16" o-c, R-13	12.7	17.5	27.0
	2. Wood framed, 2 × 4, 24" o-c, R-13	12.8	17.9	27.0
	3. Wood framed, 2 × 6, 16" o-c, R-21	19.2	24.4	33.3
	4. Wood framed, 2 × 6, 24" o-c, R-21	20.0	25.0	34.5
	5. Metal framed, 2 × 4, 16" o-c, R-13	8.0	13.0	22.2
	6. Masonry brick over block	5.9	7.8	
Floors	7. 2 × 6 joists, 16" o-c, R-21	23.3	27.8	37.0
Add R-6 for	8. 2 × 8 joists, 16" o-c, R-25	27.0	32.3	41.7
crawlspaces	9. 2 × 10 joists, 16" o-c, R-30	31.3	37.0	45.5
	10. Concrete over unheated space	2.6	7.6	16.9
Attics	11. Standard wood frame, no insulation	1.7		
	12. Standard wood frame, R-19	19.6		
	13. Standard wood frame, R-30	30.3		
	14. Standard wood frame, R-49	47.6		
Roof w/o attic	15. No insulation between rafters	2.5	7.5	18.9
	16. R-13 between rafters	13.7	18.9	28.6
	17. R-30 between rafters	29.4	34.5	43.5
Doors		**Door**	**+ Storm Door**	
	18. 1-3/4" Hollow-core	2.2	3.4	
	19. 1-3/4" Panel door with 1-1/8" panels	2.6	3.8	
Windows (4'6" × 2'8" dimensions)		**Aluminum**	**Wood/Vinyl/Fiberglass**	
	20. Single pane	0.8	0.9	
	21. Double pane, air gap	1.4	2.0	
	22. Double pane, low-e	1.6	2.6	
	23. Double pane, low-e, argon gas	1.7	3.0	
	24. Triple-pane, low-e, argon gas	–	6.7	

SOURCE: ASHRAE *90.1 Code-Compliance Manual*, 1995; Kolle, 1999

Also included in Table 6.7 is a normalized measure of the efficiency of this house, which we'll call the *thermal index*. The thermal index lets us take into account differences in floor-space area of houses when comparing the efficiency of one house with another. The floor space area, by the way, is the total area on all of the floors of the house. Thus, for example, a two-story house with a 1000-ft² footprint has 2000 ft² of floor space. The thermal index is defined as

Eq. 6.20
$$\text{Thermal index} \left(\frac{\text{Btu}}{\text{ft}^2\text{-}°\text{F-day}} \right) = \frac{24 \text{ hr/day} \times (\text{UA})_{tot} \text{ Btu/hr-}°\text{F}}{\text{Floor space area (ft}^2)}$$

The thermal index for this example house is 7.7 Btu/ft² per *degree-day* (later we'll see how degree-days play into our estimates of total annual heating loads). Older houses often

table
6.7 **Heat Loss Spreadsheet**

House #1: Conventional

Component	Area (ft²)	Insulation	R	$U = 1/R^*$	UA*	% of Total*
Ceiling	1500	R-30 #13	30.3	0.033	49.5	10%
Windows	250	dbl Al #21	1.4	0.714	178.6	37%
Doors	60	No storm #19	2.6	0.385	23.1	5%
Walls	970	R-21 #3	19.2	0.052	50.5	10%
Floors	1500	R-21 #7	29.3	0.034	51.2	11%
	ACH	Volume (ft³)	Efficiency			
Infiltration	0.6	12,000	0		129.6	27%
Ventilation	0	12,000	70%		0	0%

TOTAL UA-value = 482 Btu/hr-°F

Thermal Index = 7.7 Btu/ft²-°F-day

* Columns are calculated Values. All other information is data entry.

have a thermal index of around 15, new code-built houses have values around 8, and super-insulated houses can easily have values down around 4.

6.7 Let's Size a Furnace

Furnaces are sized to be able to deliver enough heat to keep a house at some desired thermostat set point while the ambient drops to the coldest temperature likely for that location. That ambient temperature is usually based on meeting the expected load 99% of the time (called the *99 percent design temperature*). Table 6.8 provides a brief sample of design temperatures for a few U.S. cities along with their Heating Degree Days, which is a measure to be introduced later of how cold their annual climates are.

Furnaces are usually oversized, somewhat, both to assure an adequate heat supply as well as to allow a little extra boost, called the *pick-up factor*, to heat things up quickly enough when you are first bringing the house up to temperature. Furnaces are most efficient if they run rather continuously rather than turning on and off frequently, so for efficiency oversizing is often restricted to about 40% above the design load.

The furnace converts fuel to heat that is delivered to the house through a heat distribution system as illustrated in Figure 6.20. Accounting for furnace losses, distribution system losses that don't provide heat to conditioned spaces, and an adequate pick-up factor, we can write the furnace output needed for a house as

Eq. 6.21
$$q_{furnace} = \frac{(UA)_{tot} \cdot (T_{set} - T_{design}) \cdot pick\text{-}up\ factor}{\eta_{distribution}}$$

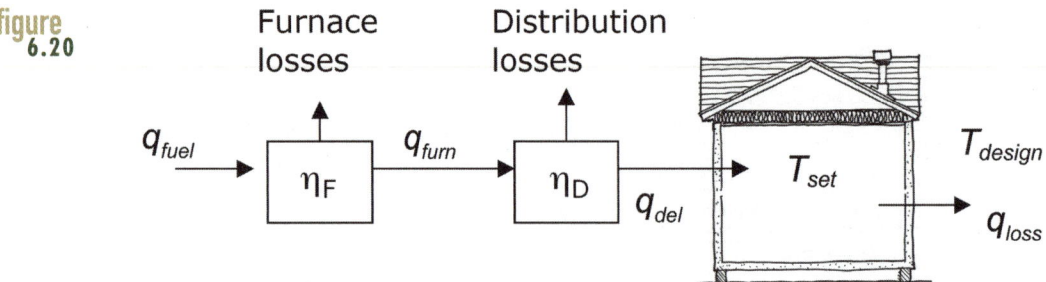

figure 6.20

In sizing a furnace the distribution system losses must be included. Because furnaces are rated by their output (q_{furn} Btu/hr), picking the right furnace size doesn't depend on its efficiency. Solution Box 6.7 illustrates the use of Equation 6.21 to size a furnace.

table 6.8   99% Design Temperatures and Annual Heating Degree Days (Base 65°F) for selected cities.

CITY	T_{des} (°F)	HDD65 (°Fd/yr)	CITY	T_{des} (°F)	HDD65 (°Fd/yr)
Phoenix	31	1552	Reno	12	6022
San Francisco	42	3042	Albuquerque	14	4292
Denver	−2	6016	New York	11	4848
Miami	45	206	Cleveland	2	6154
Atlanta	18	3095	Medford	21	4930
Boise	4	5833	Charleston	26	2146
Chicago	−3	6127	Memphis	17	3227
Topeka	3	5243	Houston	29	1434
New Orleans	32	1465	Salt Lake City	5	5983
Detroit	4	6228	Blacksburg	−5	5052
Minneapolis	−14	8159	Seattle	28	5185

6.8 Annual Cost of Heating

The ultimate goal of our somewhat careful analysis of R-values, furnace efficiencies, and so on, is to be able to estimate the value of improving the efficiency of a house. That means we need to introduce something about the local climate because we want annual heating demand, not just instantaneous heat rates. And we need to incorporate the cost of fuel.

6.8.1 Internal Gains

People and appliances in homes provide enough heat to keep the indoor temperature several degrees above the ambient temperature. One rule of thumb suggests that these *internal gains*

SOLUTION BOX 6.7

Sizing a Furnace for Blacksburg, VA

Suppose the $(UA)_{tot}$ = 482 Btu/hr-°F house shown in Table 6.7 has a 75%-efficient forced-air distribution system. Using a 70°F thermostat set point, size a furnace for Blacksburg, VA, using a 1.4 pick-factor.

Solution:

Table 6.8 show the 99% design temperature for Blacksburg is a very cold –5°F. Using Equation 6.21, we get

$$q_{furance} = \frac{482 \text{ Btu/hr-°F } [70 - (-5)]\text{°F} \cdot 1.4}{0.75} = 67,620 \text{ Btu/hr}$$

We could then look up available furnaces and pick the smallest, most efficient one that has at least that much output. For example, a list of efficient furnaces and efficiencies is available at http://hes.lbl.gov/hes/aceee/furnace.html.

provide something like 3000 Btu/hr in a typical house. As a quick check on that, imagine three people, each delivering 350 Btu/hr in personal heat, who are in the house twelve hours per day. That averages out to about 525 Btu/hr. Add to that the typical indoor electricity use in a home in the winter, about 500 kWh/mo, which, if all of that ends up as heat, works out to another 2400 Btu/hr. That suggests 3000 Btu/hr is not a bad estimate.

The degree to which that 3000 Btu/hr of internal gains heats the house depends on the UA-value of the house. For our 482 Btu/hr example house, those internal gains would raise the temperature above ambient by

Eq. 6.22
$$\Delta T = \frac{\text{Internal gains}}{(UA)_{tot}} = \frac{3000 \text{ Btu/hr}}{482 \text{ Btu/hr-°F}} = 6\text{°F}$$

What that means, for example, is if you set your thermostat to 70°F, your furnace would have to heat the house only to 70° – 6° = 64°F, and the internal gains would raise it the rest of the way. The more efficient the house is, the less fuel you have to buy for the furnace.

The temperature that the furnace has to raise the house to, with internal gains providing the rest, is called the *balance point temperature*. It is found using

Eq. 6.23   Balance point temperature: $T_b = T_{Set} - \dfrac{q_{int}}{UA_{tot}}$

6.8.2 Heating Degree-Days

Heating degree-days (HDD) are a measure of how cold and how long the heating season is for a given location. Historically, heating degree-days were based on an assumed 65°F balance point temperature so every day that the average temperature was above 65°F, the house could probably heat itself with internal gains. Meteorologists, using min/max thermometers, used the average of the minimum and the maximum as their estimate of the average temperature for that day. Each day $65°F - T_{avg}$ is determined, and if that quantity is positive it contributes to the degree-days for that month or year. If it is negative there are no degree-days. For example if January 1 has an average temperature of 50°F, then 15 degree-days are accumulated that day. If January 2 has an average of 70°F it contributes no degree-days. Tallying up these quantities for a year gives the yearly degree-days. Table 6.8 provides a few examples of annual HDD65 values for a short list of cities.

Quite often, especially with more efficient houses, the old choice of 65°F as the base temperature for degree-days is too high. The following empirically derived adjustment of the base temperature seems to work pretty well.

Eq. 6.24
$$HDD(T_b) = HDD65 - (0.021 \cdot HDD65 + 114)(65 - T_b)$$

For example, applying Equation 6.24 to Denver (HDD65 = 6016) to convert to a 60°F base gives

Eq. 6.25
$$HDD60 = 6016 - (0.021 \cdot 6016 + 114)(65 - 60) = 4814°F\text{-day/yr}$$

which is just a few percent off the recognized value of 4733. Figure 6.21 shows a graph of Equation 6.25 and the cities from which it was derived.

6.8.3 Annual Heating Load

The annual heating load that needs to be delivered by the furnace and heat distribution system is given by

Eq. 6.26
$$Q_{del}(\text{Btu/yr}) = 24 \text{ hr/day} \times (UA)_{tot} \text{ Btu/hr-°F} \times HDD(T_b) \text{ °F-d/yr}$$

Notice we are using capital Q to mean total amounts (Btu per year or month) and lowercase q to represent Btu/hr rates. Although Equation 6.26 tells us how much heat the furnace must provide, what we really want is how much fuel we need to purchase. The fuel needed is given by

Eq. 6.27
$$Q_{fuel} = \frac{Q_{del}}{\eta_{furn} \cdot \eta_{distribution}}$$

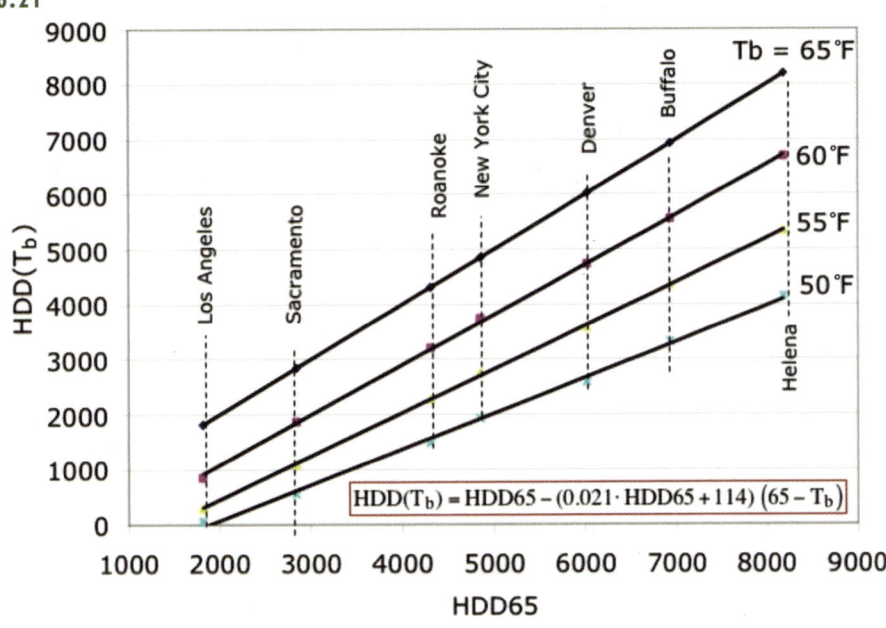

figure 6.21 Equation and Graph for Converting Annual HDD65 to Annual HDD(T_b)

$$HDD(T_b) = HDD65 - (0.021 \cdot HDD65 + 114)(65 - T_b)$$

Once we know the cost of fuel, we can figure out the expected annual fuel bill and then we can determine the value of energy savings as we modify components in the home heating analysis spreadsheet. Table 6.9 provides rough unit costs for the most commonly used home heating fuels.

We can now put all of this information into a spreadsheet to streamline our approach to evaluating home heating costs. The beauty of spreadsheets, of course, is that you can easily change any of the assumptions and immediately determine the implications of that change. The spreadsheet shown in Table 6.10 begins with the 482 Btu/hr-°F example house that we analyzed in the heat-loss spreadsheet shown in Table 6.7. Remember this is a pretty good house that would readily satisfy most current energy efficiency building codes.

table 6.9 Energy Content and Rough Cost of Fuels

Fuel	Units	Rough Costs	$/10⁶ Btu
Natural gas	100,000 Btu/therm	$1.20/therm	$12.00
Propane	92,000 Btu/gal	$2.20/gal	$23.90
Fuel oil	140,000 Btu/gal	$3.00/gal	$21.40
Electricity	3412 Btu/kWh	$0.10/kWh	$29.30
Electric Heat Pump (COP = 2.5)	8530 Btu/kWh	$0.10/kWh	$11.72

table Complete spreadsheet for calculating annual fuel bills and furnace sizing for an example "conventional"
6.10 new house in Blacksburg, VA

House #1: Conventional house in Blacksburg, VA

Component	Area (ft²)	Insulation	R	$U = 1/R^*$	UA*	% of Total*
Ceiling	1500	R-30 #13	30.3	0.033	49.5	10%
Windows	250	dbl Al #21	1.4	0.714	178.6	37%
Doors	60	No storm #19	2.6	0.385	23.1	5%
Walls	970	R-21 #3	19.2	0.052	50.5	10%
Floors	1500	R-21 #7	29.3	0.034	51.2	11%
	ACH	**Volume (ft³)**	**Efficiency**			
Infiltration	0.6	12,000	0		129.6	27%
Ventilation	0	12,000	70%		0	0%

TOTAL UA-value = 482 Btu/hr-°F
Thermal Index = 7.7 Btu/ft²°F-day

* Columns are calculated values. All other information is data entry.

Annual Fuel Consumption and Furance Sizing

Fuel Price	$12	per million Btu (Table 6.9)
Furnace Efficiency	80%	
Distribution Efficiency	75%	
Internal gains, q_{int}	3000	Btu/hr
Indoor set point, T_i	70	°F
HDD65	5052	°F-d/yr for Blacksburg, VA (Table 6.8)
T_{design}	–5	°F (Table 6.8)
Furn. pick-up factor	1.4	

Calculations:

* Balance Point Temp	63.8	°F	$T_{bal} = T_i - q_{int}/(UA)$
* HDD @ T_{bal}	4784	°F-d/yr	$HDDT_b = HDD65 - (0.021 * HDD65 + 114)$ $(65 - T_{bal})$
* Q_{del}	55.4	Million Btu/yr	$Q_{del} = 24(UA)HDD$
* Q_{fuel}	92.3	Million Btu/yr	$Q_{fuel} = Q_{del}/(\text{furn eff} \times \text{dist eff})$
* **Annual fuel bill**	**$ 1,108**	per year	$\$/yr = Q_{fuel} \times \text{fuel price}$
* **Furnace output**	67,546	Btu/hr	$\text{Furnace} = (UA)_{tot} \times (T_i - T_{design}) \times \text{pickup/dist eff}$

* Rows are calculated values. All other information is data entry.

6.9 Impacts of Improving Efficiency

With the spreadsheet in hand, it is easy to explore the potential energy savings associated with modifying the construction. We can easily determine the impact of improving the efficiency of windows, or the furnace, or plugging leaks in the distribution system, or tightening the house and including a heat recovery ventilator. You could go through these improvements one at a time, or do them all at once. Table 6.11 shows the spreadsheet for all of the

table 6.11   The final spreadsheet for the super efficient house

House #2: Super efficient house

Component	Area (ft²)	Insulation	R	U = 1/R*	UA*	% of Total*
Ceiling	1500	R-47.6 #13	47.6	0.021	31.5	11%
Windows	250	dbl low-e Arg #23	3	0.333	83.3	29%
Doors	60	Storm #19	3.8	0.263	15.8	6%
Walls	970	24"-o-c 2" ISO #4	34.5	0.029	28.1	10%
Floors	1500	R-21 + 2" ISO #7	37	0.027	40.5	14%
	ACH	**Volume (ft³)**	**Efficiency**			
Infiltration	0.3	12,000	0		64.8	23%
Ventilation	0.3	12,000	70%		19.4	7%

TOTAL UA-value = 284 Btu/hr-°F

Thermal Index = 4.5 Btu/ft²-°F-day

* Columns are calculated values. All other information is data entry.

Annual Fuel Consumption and Furnace Sizing

Fuel Price	$12	per million Btu (Table 6.9)
Furnace Efficiency	95%	Btu/hr
Distribution Efficiency	90%	°F
Internal gains, q_{int}	3000	°F-d/yr for Blacksburg, VA (Table 6.8)
Indoor set point, T_i	68	°F (Table 6.8)
HDD65	5052	
T_{design}	–5	
Furn. pick-up factor	1.4	

Calculations:

* Balance Point Temp	57.4	°F	$T_{bal} = T_i - q_{int}/UA$
* HDD @ T_{bal}	3384	°F-d/yr	$HDDT_b = HDD65 - (0.021 * HDD65 + 114)(65 - T_{bal})$
* Q_{del}	23.0	Million Btu/yr	$Q_{del} = 24(UA)HDD$
* Q_{fuel}	26.9	Million Btu/yr	$Q_{fuel} = Q_{del}/(\text{furn eff} \times \text{dist eff})$
* **Annual fuel bill**	**$ 323**	per year	$\$/yr = Q_{fuel} \times \text{fuel price}$
* **Furnace output**	**32,197**	Btu/hr	$\text{Furnace} = UA_{tot} \times (T_i - T_{design})/\text{dist eff}$

* Rows are calculated values. All other information is data entry.

improvement measures taken at once. With all of the improvements made, this superefficient house has a fuel bill 70% lower than that of the original house. That's quite an accomplishment given that the original house was already quite efficient.

Figure 6.22 shows the results of the step-by-step approach starting with fixing those leaky ducts that are losing so much heat in the distribution system, followed by replacing the furnace with one that is more efficient. Those two steps alone save 30% of the original energy demand. Moreover, if the original house already exists, these are fairly straightforward retrofit projects. Once those are done, the marginal benefit of each additional efficiency measure is diminished, which points out an interesting difficulty with doing a cost-benefit analysis on a step-by-step basis If we improve the furnace and ducts first, the marginal benefit of fixing the windows, for example, is less than if we fix the windows first and then replace the furnace and ducts. Treating the whole set of measures together, especially if it is new construction, may make more sense.

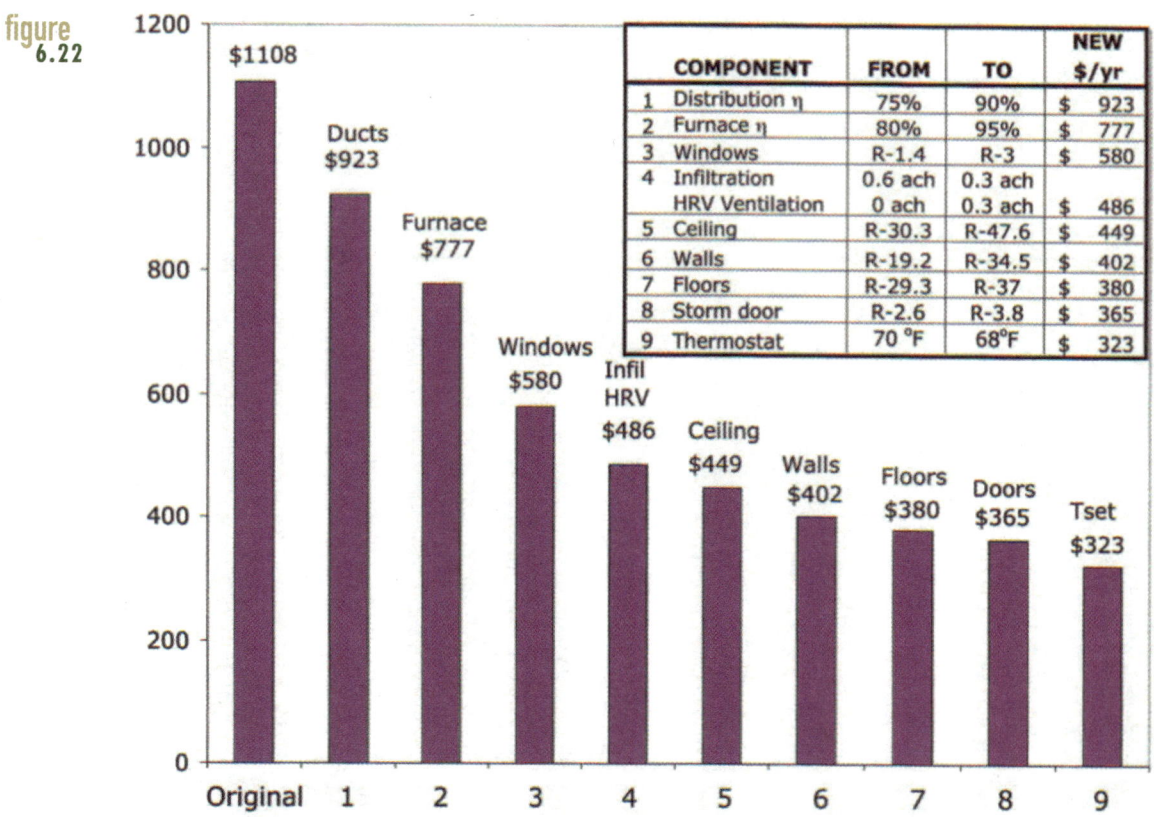

figure
6.22

	COMPONENT	FROM	TO	NEW $/yr
1	Distribution η	75%	90%	$ 923
2	Furnace η	80%	95%	$ 777
3	Windows	R-1.4	R-3	$ 580
4	Infiltration	0.6 ach	0.3 ach	
	HRV Ventilation	0 ach	0.3 ach	$ 486
5	Ceiling	R-30.3	R-47.6	$ 449
6	Walls	R-19.2	R-34.5	$ 402
7	Floors	R-29.3	R-37	$ 380
8	Storm door	R-2.6	R-3.8	$ 365
9	Thermostat	70 °F	68°F	$ 323

Step-by-step efficiency improvements added to the original house in Table 6.10 cut the annual fuel bill more than two-thirds from $1108 to $323.

6.10 Heating, Ventilating, and Air-Conditioning (HVAC) Systems

After doing everything possible to keep heat inside in the winter and outside in the summer with a tight building envelope, and then being clever about using passive solar ideas described in Chapter 7 to provide some winter heating and avoid summer solar gains, there will still be times when your building needs a full-size heating and/or cooling system.

6.10.1 Forced-Air Central Heating Systems

Almost two-thirds of all homes in the United States, and about 90% of new homes, use a central furnace with forced-air distribution such as is shown in Figure 6.23. The furnace may be gas-fired or oil-fired, it may use straight electric-resistance heating elements, or it may be an electric heat pump. A blower, or fan, pushes warm supply air through ducts that distribute the heat, typically through floor registers, throughout the house. Return air is sucked through high-return registers back into the furnace.

figure 6.23   **Central Furnace with Forced-Air Distribution**

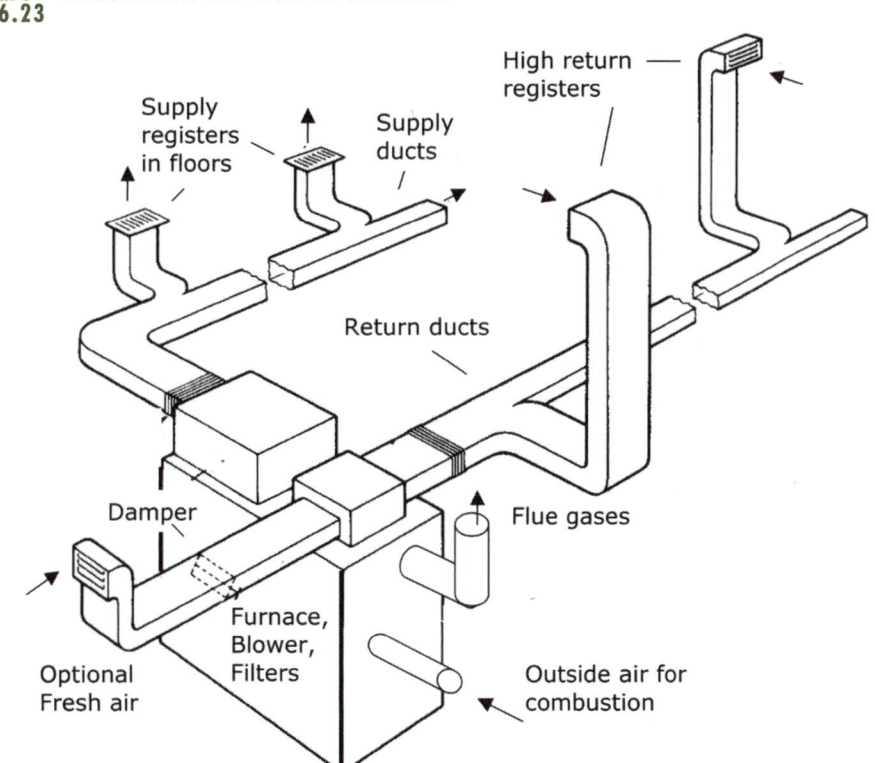

Several advantages of forced-air systems help explain their popularity. They provide rapid delivery of heat throughout the house; adjusting the indoor temperature is quick and easy. They also are versatile because the same duct system can be used for both heating and cooling as well as air humidification and filtration. On the downside, blowers and ducts can be noisy, drafts created by all that moving air can be uncomfortable, and controlling heat delivery to just certain zones in the house is difficult.

Federal minimum-efficiency standards for furnaces took effect in 1992, requiring that new furnaces have an Annual Fuel Utilization Efficiency (AFUE) of at least 78%. In comparison, many old furnaces have estimated AFUE ratings of only 55 to 65%. Energy Star–qualified furnaces must have an AFUE of 90% or higher. Natural gas–fired condensing furnaces, which capture latent heat from exhaust gases, have efficiencies as high as 97%. Oil-fired furnaces tend to have lower efficiencies, typically in the range of 78 to 82%.

From an energy-efficiency perspective, however, the weak link in forced-air systems can be those ducts, which are all too often poorly insulated and plagued by leaks—especially at connections to supply registers, which sometimes aren't even connected at all. It is not uncommon for duct losses to increase heating and cooling bills by 20 to 30%, which in some estimates adds $10 to $15 billion per year to fuel bills in single-family residences. In addition, pushing air down long, winding, narrow supply ducts requires quite a bit of energy for the blower motor.

6.10.2 Hydronic Systems

Hydronic systems use boilers to heat, and pumps to distribute, hot water. Whereas forced air systems require big fan motors to distribute heat through bulky ducts, hydronic systems get by with very small, efficient pumps circulating hot water through relatively tiny piping systems. They are quiet, don't create drafts, and distribution losses can be negligible. They can well be described as the most comfortable heating systems available.

The actual heat delivery component of hydronic systems can be either baseboard radiator/convectors that run along walls, or through tubing embedded typically in concrete floor slabs as shown in Figure 6.24a. Both have disadvantages. Baseboard systems are not very attractive and may interfere with furniture placement. Radiant floor-slab systems are slow to respond due to the thermal momentum of all that concrete mass. Radiant floor heating systems can also be embedded into conventional wood subflooring, but their performance is somewhat diminished when wood flooring and carpeting are put on top.

Boiler efficiencies are typically a bit lower than their gas-fired furnace counterparts, with AFUE ratings usually in the range of 80 to 85%. They make up for this with higher distribution efficiencies as well as lower thermostat settings made possible by the lack of drafts common in forced-air systems.

Hydronic systems also suffer from the difficulty associated with trying to use the same distribution system for both heating and cooling. This is especially problematic in humid areas because those cold surfaces can cause unwanted condensation. In environments where

figure
6.24   **Hydronic Heating and Cooling**

(a) Distribution tubing being installed in a floor slab for heating. (b) A ceiling-mounted hydronic radiator for cooling.

humidity is well controlled, such as an office building, hydronic cooling using ceiling-mounted radiators can be quite effective (Figure 6.24b).

6.10.3 Compressive Air Conditioners

Compressive air-conditioning systems are based on the simple principle that a compressed gas can get very cold when it is allowed to expand. They consist of four major components: a compressor to pressurize the refrigerant, a condenser coil that rejects heat from the hot compressed refrigerant to the environment, an expansion valve that allows the pressurized refrigerant to quickly expand and cool, and a cold evaporator coil that cools warm return air blown across the coil by a blower. *Packaged* systems put all of these components into a single box. A *split system*, such as is shown in Figure 6.25, puts the compressor and condenser outside, and the evaporator and blower are housed indoors in an air-handling unit. The air handler will usually also contain a burner and heat exchange coils so that it can be used as both a heater and an air conditioner.

The units used to describe the cooling capacity of air conditioners date back to the days when they were compared to cooling using melting ice. Air conditioners are sized based on their "tons of cooling" capacity, where 1 ton of cooling is equal to a heat absorption rate equivalent to melting 1 ton (2000 lb) of ice in a 24-hour period. Because it takes 144 Btu to melt 1 pound of ice, a 1-ton air-conditioning unit cools at the rate of

Eq. 6.28 $$1 \text{ ton of cooling capacity} = \frac{2000 \text{ lb}}{24 \text{ hr}} \cdot \frac{144 \text{ Btu}}{\text{lb}} = 12{,}000 \text{ Btu/hr}$$

A typical home air conditioner may be rated in the range of about 3 to 5 tons.

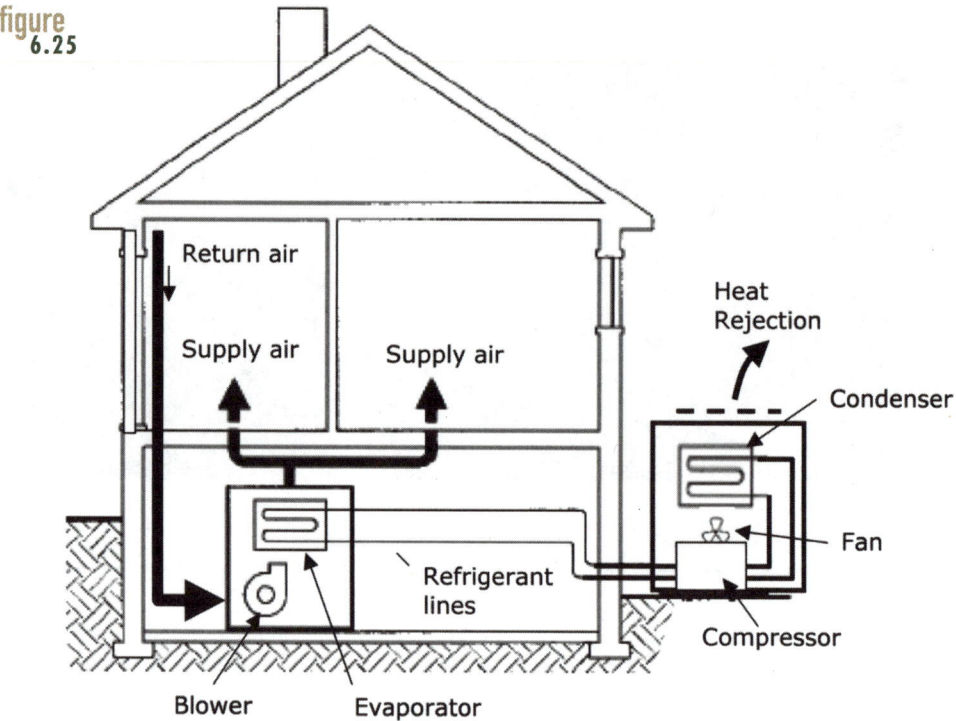

A split system air conditioner has the compressor and condenser outside of the building envelope.

6.10.4 Evaporative Cooling and Absorption Cooling

Most air conditioners use the compressive cycle described above. However, other methods of providing cooling are less electricity intensive. Evaporative coolers have thousands of years of history behind them. Earthen jugs filled with water purposely ooze water out of their sides, which removes latent heat as it evaporates, cooling the jug and any air that blows against it. That principle of evaporative cooling continues to be used in less-humid regions of the world.

Another approach is based on the same compressive cycle used in conventional air conditioners, but heat is used instead of electricity to compress the refrigerant. In these *absorption cycle* air conditioners, the expansion valve, evaporator, and condenser are pretty much identical to those in a conventional air conditioner, but the power-hungry compressor is replaced by an absorption unit, as shown in Figure 6.26. Within the absorber, refrigerant gas is dissolved into a liquid. When that liquid is heated in the generator, pressure builds up, compressing the dissolved gas. After passing through the condenser, the expansion valve relieves the pressure, allowing the compressed, dissolved gas to emerge from the liquid in much the same way that carbon dioxide bubbles out of beer when your warm hand heats the bottle. The expanding gas

figure 6.26   Lithium-Bromide Absorption-Chiller Schematic

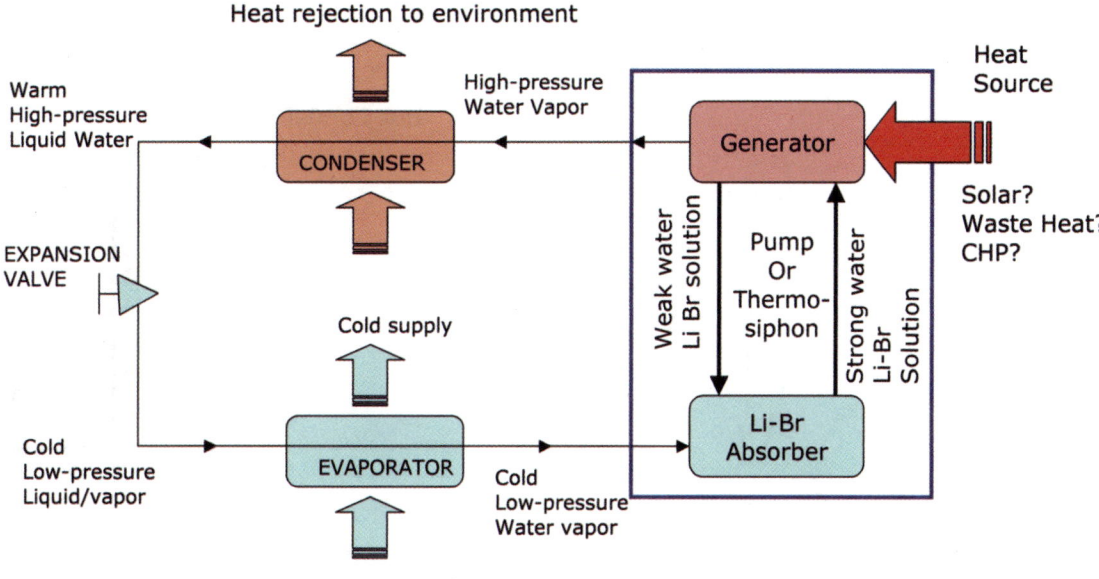

The compressor in a standard compressive refrigeration system has been replaced with a generator and absorber that perform the same function but do so with heat instead of electrical power.

cools and then heads to the cooling evaporator coil, which provides the heat exchange surface for cooling your building.

The most promising attribute of absorption cooling is that it runs on any source of heat, which means it can provide a use for waste heat from on-site cogeneration systems such as the fuel cell and microturbine systems to be described in Chapter 10.

6.10.5 Heat Pumps

An air conditioner takes heat from a relatively cool space (e.g., your home at 70°F) and dumps it into a hot place (e.g., the outdoors at 95°F). If it can do that, why not just reverse it in the winter and have it pump heat from relatively cool outdoor air and send it into your warmer house interior? That is, in fact, what a heat pump does. If this sounds somehow like magic, just imagine taking the door off of your refrigerator and then rolling the refrigerator over to an outside doorway (Figure 6.27). When you want to cool your house, face the condenser coils outside and let the fridge draw heat out of your house just the way it normally sucks heat out of your warm beer. Where's that heat go? Out the condenser coils on the back

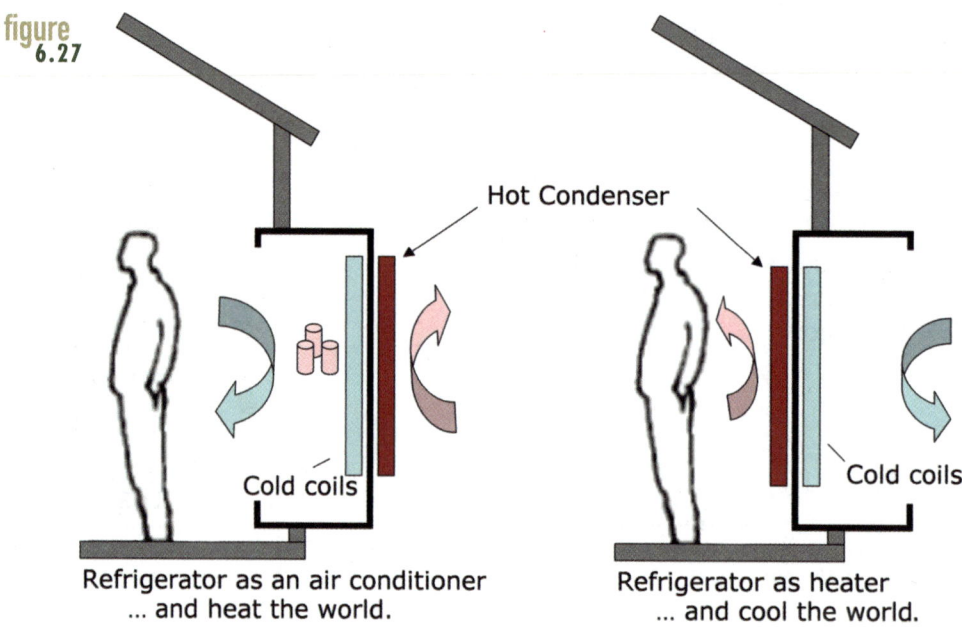

figure 6.27

Hot Condenser

Cold coils

Cold coils

Refrigerator as an air conditioner
... and heat the world.

Refrigerator as heater
... and cool the world.

You could remove your refrigerator door, back the refrigerator up to an outside doorway, and then use it as a heat pump to heat or cool your house.

of your fridge. Then when you want to heat your house, reverse the fridge, face the condenser into the kitchen, and pump heat from the cold outdoors into your house.

A heat pump is therefore just a reversible air conditioner in which the roles of evaporator and condenser switch with each other (without physically moving them around) when the seasons change. Figure 6.28 illustrates the concept, including quantities representing heat flow between warm and cold sinks (Q_H and Q_L) and work done by the compressor W.

Because heat pumps work as both heater and air conditioner, they have two separate efficiency measures. As an air conditioner, the performance is the same Seasonal Energy Efficiency Ratio (SEER), which tells us the number of Btu that can be extracted per watt-hour of input power (Btu/Wh). As a heater, they are rated by their *Heating Season Performance Factor* (HSPF), which is the seasonal heat delivered (Btu) divided by the seasonal electrical input (Wh); thus it also has units of Btu/Wh. In 2006, a new minimum standard for central heat pumps went into effect, requiring them to have a minimum SEER of 13 and a minimum HSPF rating of 7.7.

Whereas SEER and HSPF are defined for an assumed climate, another measure often used for thermodynamic analysis under any conditions is the dimensionless *coefficient of performance* (COP). The COP is the ratio of the quantity we want (e.g., heat delivered in the winter, heat removed during the summer) divided by the input work done by the compressor (see Figure 6.28).

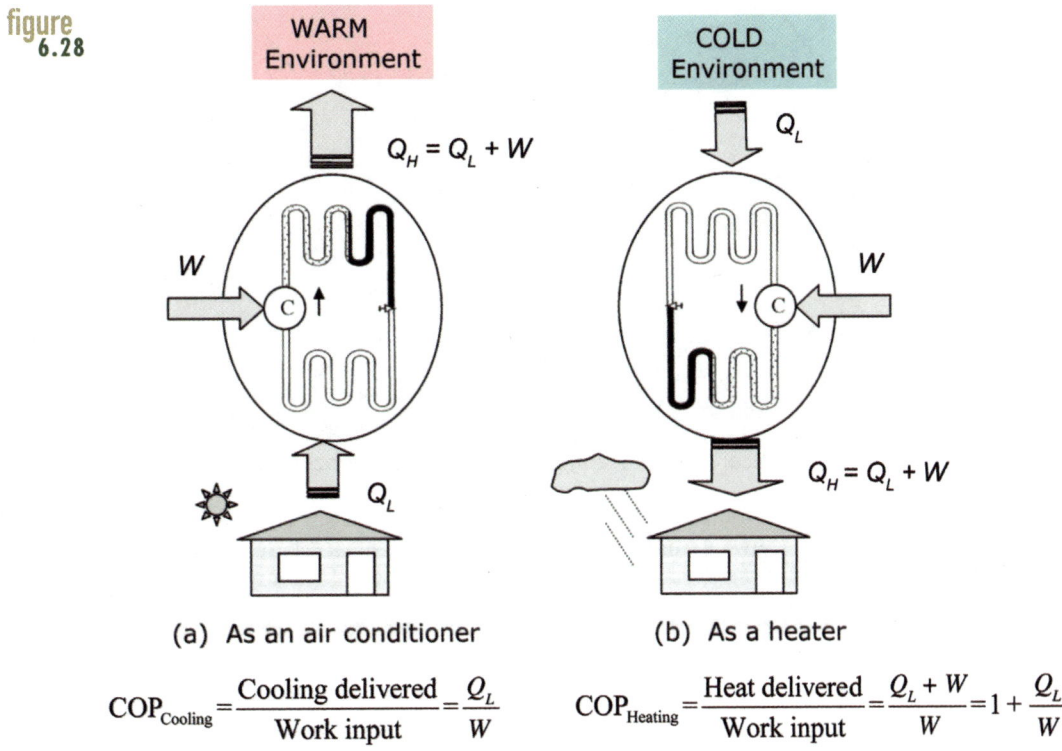

figure
6.28

$$\text{COP}_{\text{Cooling}} = \frac{\text{Cooling delivered}}{\text{Work input}} = \frac{Q_L}{W}$$

$$\text{COP}_{\text{Heating}} = \frac{\text{Heat delivered}}{\text{Work input}} = \frac{Q_L + W}{W} = 1 + \frac{Q_L}{W}$$

Heat pumps have reversing valves to allow the heat exchangers to operate as either evaporators or condensers to heat or cool a building.

6.10.6 Geothermal Heat Pumps

The efficiency (e.g., COP) of conventional air-source heat pumps declines in the winter when it gets really cold; it also declines in the summer when it gets really hot. Geothermal heat pumps (also known as ground-coupled heat pumps, ground-source heat pumps, or geo-exchange heat pumps) avoid those temperature extremes by taking advantage of the relatively constant, mild temperatures underground, which means their efficiencies can be far better than their air-source counterparts. A geothermal heat pump (GHP) is essentially a conventional heat pump with one of its heat exchangers located in a series of deep boreholes or sometimes in a horizontal array not far beneath the surface (Figure 6.29). If you are lucky enough to live next to a pond, it can be used as the source and sink. Energy Star GHPs can be 40% to 60% more efficient than conventional air-source heat pumps. They are also quieter and can provide hot water as well as space conditioning.

On the downside, GHPs cost about twice as much as their air-source counterparts. The extra cost of a GHP can be more than offset by the decrease in energy costs, especially in areas that are characterized by having both hot summers and cold winters.

 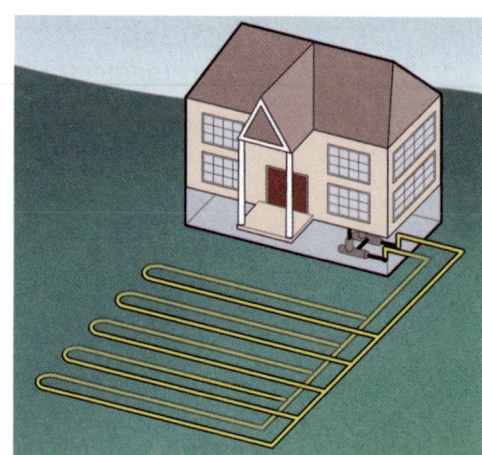

a. Vertical heat exchangers   **b. Horizontal heat exchangers**

A geothermal heat pump has one of its heat exchangers located in the relatively-constant-temperature earth. Arrays can be vertical in deep boreholes, or relatively shallow in horizontal arrays.

The comparison of various heating systems in Figure 6.30 reaffirms the Environmental Protection Agency's conclusion that geothermal heat pumps can be the most efficient of all heating systems.

6.11 Software Packages for Building Energy Analysis

A considerable amount of system analysis can be done with the simple calculations and spreadsheets presented in this chapter. In addition, a number of software packages out there can systematize the calculations and derive more robust results. At last count, there were 324 software tools for building analysis listed in the DOE Web site http://www.eere.energy.gov/buildings/tools_directory/, but some have risen to the top in terms of ease of use and confidence in results. Most of the thermal analysis in these programs is based on hour-by-hour simulations driven by typical meteorological year (TMY) data for the site. DOE-2 and BLAST were early simulation programs that are still at the heart of today's most powerful software packages.

- **Energy-10** is a program for small commercial and residential buildings (less than 10,000 ft^2) that integrates daylighting, passive solar heating, and low-energy cooling strategies with energy efficient envelope design and efficient mechanical equipment. New packages include solar water heating and photovoltaic simulations.
- **EnergyPlus** is a powerful building energy simulation program for modeling large-building heating, cooling, lighting, and ventilating performance as well as photovoltaic system analysis.

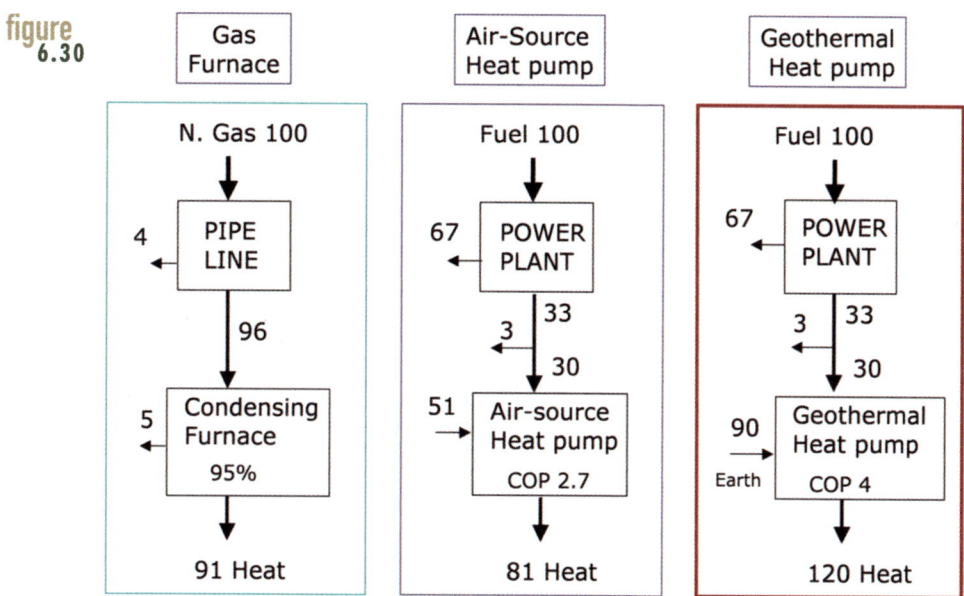

figure 6.30

For the same 100 units of fuel input to each of these three heating systems, the geothermal heat pump delivers far more heat.

- **eQuest** was funded by California utilities to enable designers to perform detailed analysis of state-of-the-art building design technologies. It is versatile, widely used, and free.
- **Google SketchUp** is an easy-to-learn 3D design tool that allows you to quickly draw buildings, view them from any angle, and see sunlight and shadow patterns on and in the building throughout the year. Great for designing overhangs and other fenestration details, it is free, runs on Macs and PCs, and it works with Google Earth so you can place your building in its actual location on Earth.
- **Home Energy Saver** (http://hes.lbl.gov) is a Web-based program that allows you to easily enter R-values, furnace efficiencies, and so on for your home to predict energy usage and utility bills. It is easy to use—just enter your zip code, and it lets you estimate dollar savings associated with upgrading building components.

6.12 Summary

Advances made in building science have helped us understand better how buildings work and what we can do to increase energy efficiency and occupant well-being. This chapter focused on analytical methods used to evaluate the heat loss characteristics of the building envelope. Thermal efficiency depends not only on the envelope but also on the heating and cooling systems themselves. We have seen the impacts of more efficient furnaces and ducts and how they complement building envelope improvements. Simple spreadsheets were developed to help us perform iterative calculations to test different efficiency measures.

In the next chapter, we will explore the potential impacts of solar energy on the three biggest categories of building energy demand shown in Figure 6.3; that is, residential space heating, commercial lighting, and residential water heating. Then, in Chapter 8, we will consider the life-cycle performance of buildings, which leads into the concept of Green Building rating systems and zero-energy buildings.

CHAPTER 7

Solar Energy for Buildings

The single biggest demand for energy in the building sector is residential space heating, the next largest is lighting for commercial buildings, and the third most important is residential water heating. In this chapter we will explore the use of solar energy to satisfy a significant fraction of these energy loads. We will also pay attention to cooling loads by seeing how buildings can be designed to minimize the impact of unwanted solar gains in those hot summer months. Building orientation, use of overhangs to shade windows, and selection of window coatings that admit natural daylight while rejecting unwanted thermal gains, are important techniques used to minimize cooling loads.

7.1 The Solar Resource

With just a rudimentary understanding of where the sun is, season-by-season, at various times of the day, we can both let it help heat our buildings in the winter and keep it from overheating our buildings in the summer. With a little more effort, we can quantitatively evaluate its impact on building energy demand.

7.1.1 Solar Angles to Help Design Overhangs

For starters, let's locate the sun in the sky. It rises in the east, reaches some maximum height in the sky when it is along our own line of longitude (that time is called *solar noon*), and it sets somewhere in the west. Figure 7.1 identifies the two key angles of interest: the sun's altitude angle β and its azimuth angle ϕ.

Of particular interest is the altitude angle of the sun at solar noon; that is, when it is either due south or due north of your location. We know this altitude angle will vary considerably as the seasons change and we want to learn how to take advantage of that variation. Equation 7.1 tells us just what we want to know:

figure 7.1

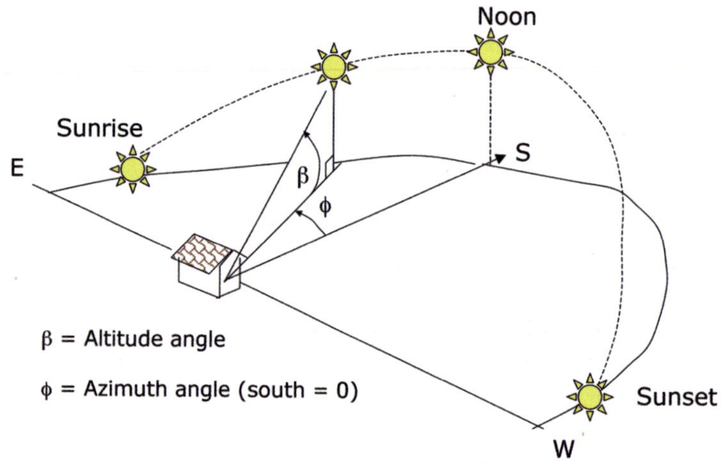

The sun's position can be described in terms of its altitude angle β and its azimuth angle φ.

Eq. 7.1
$$\beta_N = 90 - L + \delta$$

where β_N = the altitude angle of the sun at solar noon
 L = the local latitude
 δ = the solar declination

The solar declination δ is the line of latitude over which the sun revolves on that particular day; that is, if you drew a line from the center of the sun to the center of the Earth it would pass through that line of latitude all day long. The equation for δ is given below. The only variable is n, which is the day number (where $n = 1$ is January 1, and $n = 365$ is December 31).

Eq. 7.2
$$\delta = 23.5 \sin\left[\frac{360°}{365°}(n-81)\right]$$

Notice the solar declination varies sinusoidally between ±23.5 and on day number 81, which corresponds to the spring equinox, March 21, $\delta = 0$ and the sun is directly over the equator. (Notice, by the way, that we simplify the narrative of solar angles by assuming we are in the northern hemisphere.) The sun reaches its highest point at solar noon on the summer solstice, June 21, and its lowest noon angle occurs on the winter solstice, December 21. Figure 7.2 shows this range of angles.

The analysis of overhangs is especially simple for south-facing surfaces, which is convenient because that's where you want to place solar-gain windows that will help heat your house in the winter. As shown in Figure 7.3, an overhang that projects out a distance P casts a shadow that ends at a distance y down the south-facing vertical surface. At solar noon, we can easily write that

Eq. 7.3
$$y = P \tan \beta_N$$

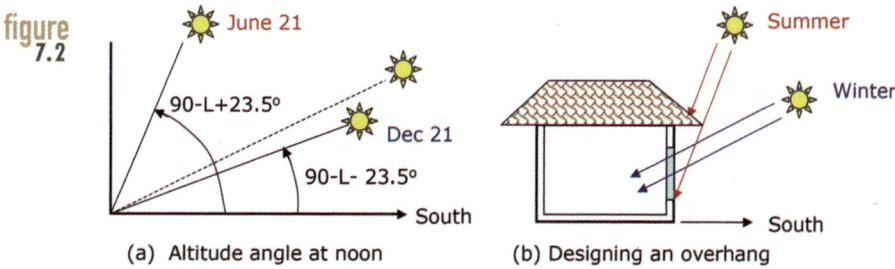

(a) Altitude angle at noon   (b) Designing an overhang

We can use the altitude angle of the sun at noon to design overhangs that allow sunlight to hit south-facing windows in the winter while blocking unwanted sun in the summer.

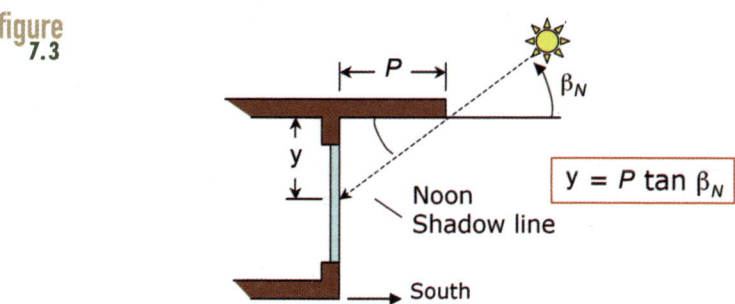

$$y = P \tan \beta_N$$

Locating the shadow line at solar noon for a south-facing window.

The example shown in Solution Box 7.1 illustrates the use of Equation 7.3 to design an overhang.

7.1.2 Site Surveys Using Sun-Path Diagrams

The equations used to locate the sun at any time of day, any day of the year are fairly cumbersome (see for example Masters, 2004, for a more careful analysis), but graphs that show the sun's location are readily available on the Internet. The University of Oregon Web site is especially handy: http://solardat.uoregon.edu/SunChartProgram.html. Just enter your location either by zip code or latitude and longitude (longitude matters only if you want to adjust for local time). Figure 7.4 shows a sample sun-path diagram along with some potential obstructions that shade the site at certain times of the year.

Sun-path diagrams are especially useful for doing a quick site analysis to determine whether obstructions such as trees or buildings may cast shadows onto your proposed location. The altitude and azimuth angles of potential obstructions can be measured using a simple protractor and plumb bob along with a compass (corrected for local magnetic declination) as

SOLUTION BOX 7.1

Design an Overhang

Design an overhang for Palo Alto, California, latitude 37.5°, that will completely shade south-facing, sliding-glass doors at noon on the summer solstice, June 21. Check the shadow line on the winter solstice, December 21.

On the solstices, the declination is ±23.5°, so using Equation 7.1 the altitude angle of the sun at solar noon will be β_N (June 21) = 90 − 37.5 + 23.5 = 76° while on December 21 it is 90 − 37.5 − 23.5 = 29°. It is now easy to figure out the overhang and its impact in the winter:

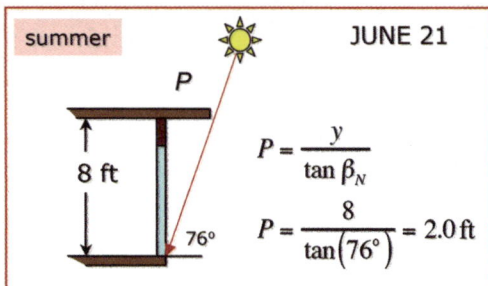

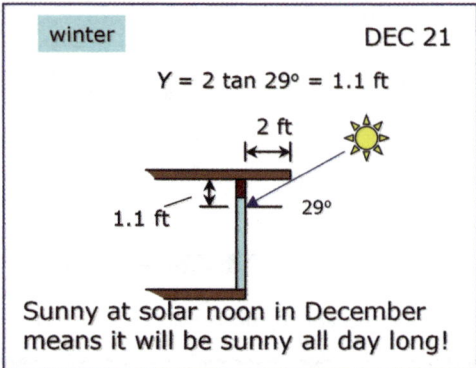

summer	JUNE 21

$$P = \frac{y}{\tan \beta_N}$$

$$P = \frac{8}{\tan(76°)} = 2.0 \text{ ft}$$

Shaded at solar noon in June means it will be shaded all day long!

Sunny at solar noon in December means it will be sunny all day long!

So, a 2-foot overhang completely shades the glass doors at noon in June and completely exposes the windows to the warm rays from the sun at noon in December. Designing overhangs for solar noon provides the fortuitous result that a south-facing window shaded at noon in the summer will be shaded all day long, and a window exposed to the sun in winter at noon will be exposed to the sun all day long.

suggested in Figure 7.5. Sketching the obstructions directly onto a sun-path diagram makes it easy to determine the monthly hour-by-hour shading problems at the site. For example, the site diagramed in Figure 7.4 will have full sun from February through October, but will be shaded from roughly 8:30 to 9:45 a.m., and after about 3:00 p.m., from November through January.

7.1.3 Incident Solar Radiation (Insolation)

Sun angles and site surveys are very useful, but in most circumstances what we really need to know is how much solar energy is available to generate electricity and to heat our buildings

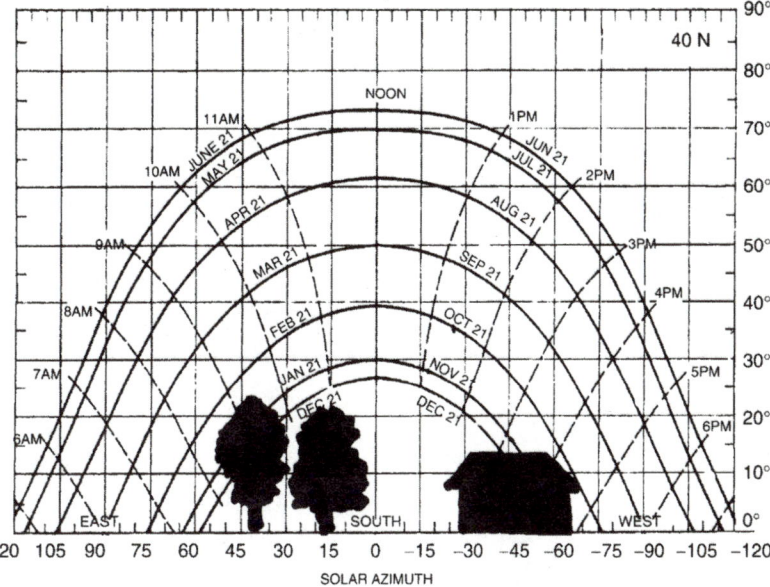

figure 7.4

A sun-path diagram showing obstructions that shade the site in the morning and late afternoon from November through January.

figure 7.5

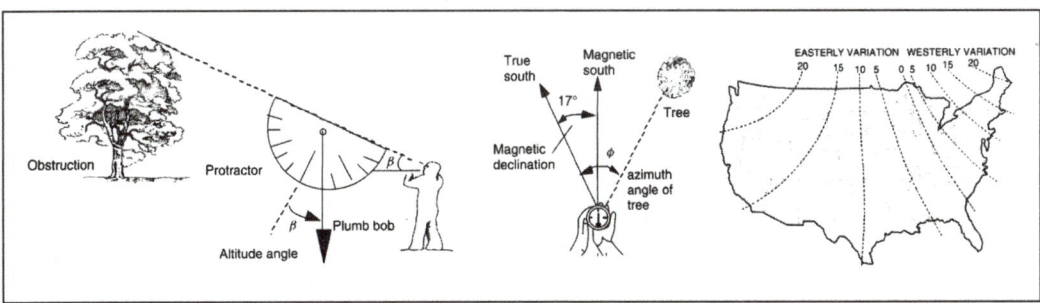

A solar site survey can be made using a simple compass, protractor, and plumb bob. The example shows the compass correction for San Francisco, which has a declination of 17°E.

and hot water. In addition, we need to know how much extra cooling our building will need because too much sun is coming through the windows. The incident solar radiation on a collector, referred to as the *insolation*, is often measured in kW or kWh for photovoltaic calculations and Btu or Btu/hr for solar thermal applications. Table 7.1 provides conversion factors between these and other insolation units. Notice one of the entries is based on the idea that the solar intensity on a surface normal to the incoming rays in mid-afternoon on a clear day is very close to 1 kW/m^2. That is defined as "1-sun" insolation. A site with say 5 kWh/m^2 per day of insolation is said to have 5 hours of 1-sun insolation (this will be very handy in Chapter 11 when we look at generating power with photovoltaics).

table 7.1	Conversion Factors for Various Insolation Units	
1 kW/m^2	=	316.95 Btu/hr-ft^2
	=	1.433 langley/min
	=	"1-sun" of insolation
1 kWh/m^2	=	316.95 Btu/ft^2
	=	85.98 langleys
	=	3.60×10^6 joules/m^2

In some circumstances, especially for summer air-conditioning load calculations, clear sky insolation is all we need to know. These values can be calculated for any spot on Earth at any time of any day using a rather cumbersome set of equations (see, for example, Masters, 2004). Whereas clear-sky insolations are useful for a number of purposes, more often what is needed is actual data for a given location. There are, perhaps surprisingly, only 56 primary data-collection stations around the United States for which long-term solar measurements have been made. Those have been supplemented with estimates made from meteorological models for another 183 sites. All of these data have been compiled by the National Renewable Energy Laboratory and are readily available on the Internet.[1] Table 7.2 presents representative insolation data for several cities, just to give you a sense of what is available.

7.2 Passive Solar Heating

As Figure 6.3 pointed out, simple heating of houses is the biggest single category of energy demand in the entire building sector. So, why not let the sun do some of that job? There are two ways to try to do that. The *passive solar* approach is based on encouraging sunlight to pass through windows and other solar apertures to provide needed heat. Passive solar systems are simple, cheap, and reliable. The second approach, *active solar*, uses special solar-thermal collectors to collect heat, which is then moved to storage and distribution systems using pumps and blowers. Although active systems provide greater control of heat flow, their high cost and uncertain reliability have led to less widespread acceptance.

The basic design guidelines for passive solar are pretty simple.

1. Maximize envelope efficiency (we did that in Chapter 6).
2. Try to orient the building along an east-west axis to control solar gains.

---

[1] *Solar Radiation Data Manual for Flat-Plate and Concentrating Collectors* (http://rredc.nrel.gov/solar/pubs/redbook/) and *The Solar Radiation Data Manual for Buildings* (http://rredc.nrel.gov/solar/old_data/nsrdb/bluebook/).

table 7.2 Average Daily Insolation (Btu/ft^2-day) on South-facing Surfaces for Various Tilt Angles (Latitude −15°, Latitude, Latitude +15°, and 90°)

LOS ANGELES, CA: LATITUDE, 33.93 N

TILT	JAN	FEB	MAR	APR	MAY	JUN	JLY	AUG	SEP	OCT	NOV	DEC	YEAR
LAT −15°	1204	1426	1743	2028	2028	2028	2250	2155	1870	1585	1331	1141	1743
LAT	1395	1585	1807	1997	1933	1902	2092	2092	1902	1712	1490	1331	1775
LAT +15°	1490	1616	1775	1870	1712	1648	1838	1902	1807	1743	1585	1426	1712
90	1299	1299	1204	1046	792	697	761	951	1141	1331	1363	1299	1109

BOULDER, CO: LATITUDE, 40.02 N

TILT	JAN	FEB	MAR	APR	MAY	JUN	JLY	AUG	SEP	OCT	NOV	DEC	YEAR
LAT −15°	1204	1458	1712	1933	1965	2092	2092	1997	1870	1616	1268	1109	1712
LAT	1395	1616	1775	1902	1870	1933	1933	1933	1902	1775	1458	1331	1743
LAT +15°	1521	1680	1775	1775	1648	1648	1680	1743	1838	1807	1521	1426	1680
90	1426	1458	1363	1141	887	824	856	1014	1268	1458	1395	1363	1204

BOSTON, MA: LATITUDE, 42.37 N

TILT	JAN	FEB	MAR	APR	MAY	JUN	JLY	AUG	SEP	OCT	NOV	DEC	YEAR
LAT −15°	951	1204	1458	1648	1807	1902	1902	1807	1585	1299	887	792	1426
LAT	1078	1331	1490	1585	1680	1743	1775	1743	1616	1363	983	919	1458
LAT +15°	1141	1363	1458	1490	1490	1521	1553	1585	1553	1395	1046	983	1395
90	1078	1236	1173	983	887	824	887	983	1109	1141	951	919	1014

ATLANTA, GA: LATITUDE, 33.65 N

TILT	JAN	FEB	MAR	APR	MAY	JUN	JLY	AUG	SEP	OCT	NOV	DEC	YEAR
LAT −15°	1078	1331	1616	1902	1965	1997	1933	1870	1680	1553	1204	1014	1585
LAT	1204	1458	1680	1838	1838	1838	1807	1807	1712	1648	1331	1173	1616
LAT +15°	1299	1490	1616	1712	1648	1616	1585	1648	1616	1680	1553	1236	1553
90	1109	1173	1109	951	761	697	697	856	1014	1268	1204	1109	983

SOURCE: Masters, 2004, based on NREL data

3. Provide south-facing glazing systems to admit solar energy.
4. Design proper overhangs to protect south-facing windows in the summer.
5. Provide sufficient thermal mass to absorb solar energy in excess of daytime needs.

We will now explore these important concepts.

7.2.1 The Importance of Building Orientation

The orientation of a building will greatly affect our dual goals of letting the sun get into a building in the winter, while minimizing excessive solar gain during the summer. We have

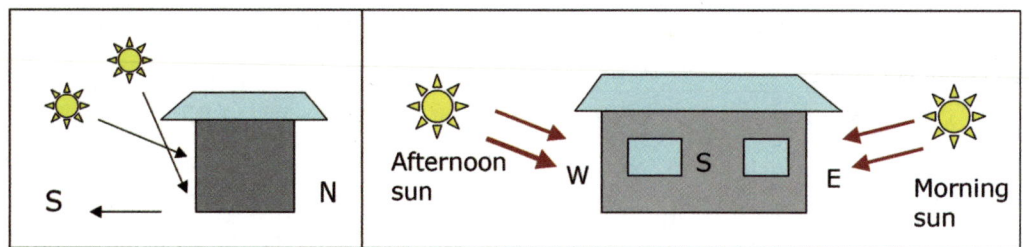

figure
7.6

Solar gains through south-facing windows are easy to control using overhangs, but east- and west-facing windows can cause overheating in the summer.

already seen how easy it is for overhangs to be designed for south-facing windows (north-facing for those of you in the southern hemisphere) to do just that. But what about east- and west-facing windows? As shown in Figure 7.6 those windows are exposed to nearly horizontal morning and afternoon solar radiation, so overhangs don't do much good to protect the building from summer overheating.

To help introduce the importance of orientation, consider Figure 7.7, which compares monthly clear-sky insolation striking south-facing windows with amounts hitting east and west windows as well as horizontal surfaces (at 40° latitude). Solar energy in the winter helps heat your building, but in the summer it drives up air-conditioning costs. In that sense,

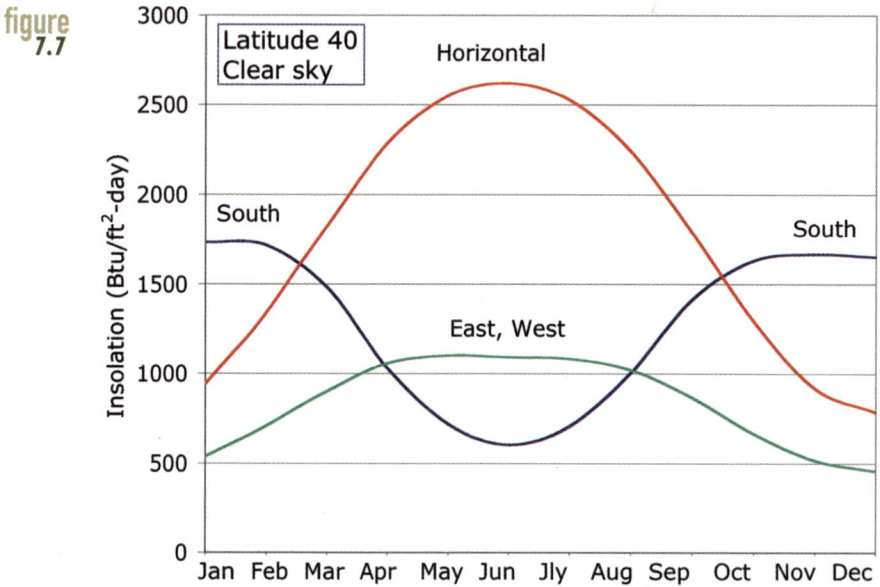

figure
7.7

South-facing windows provide solar gains in the winter when you want it; east and west windows provide it when you don't. Horizontal skylights are the worst for air-conditioning loads.

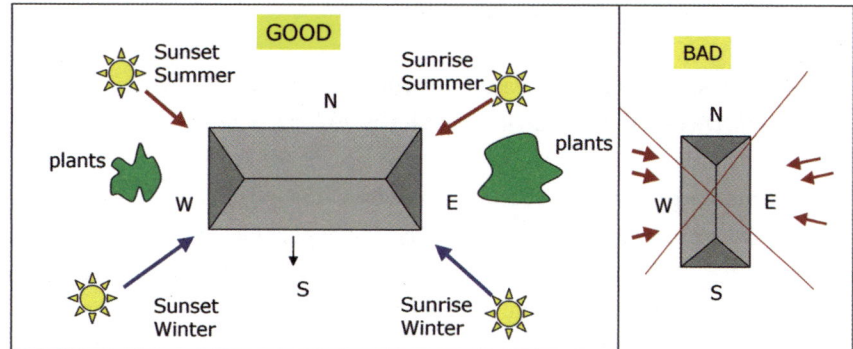

figure
7.8

A building with its long axis along the east-west direction helps maximize solar gains in the winter while minimizing morning and afternoon solar exposure in the summer. Vegetation along the east and west sides can help control summer overheating without affecting winter solar gains.

south-facing windows are pretty ideal. East and west windows, on the other hand, can cause overheating in the summer and they're not very helpful in the winter. Also note how problematic horizontal skylights are in the summer.

   The best starting point to satisfy the simultaneous needs of providing winter heat and minimizing summer cooling loads is to orient the building with its main axis along the east-west direction as shown in Figure 7.8. That maximizes the amount of wall area available for easily controlled solar gains and minimizes the area of wall and windows exposed to morning and afternoon summer heating. Vegetation along those east and west walls can beautifully and effectively control those summer gains. One way to encourage proper orientation of houses is to plan neighborhood streets to run in the east-west direction rather than north-south, as suggested in Figure 7.9.

figure
7.9

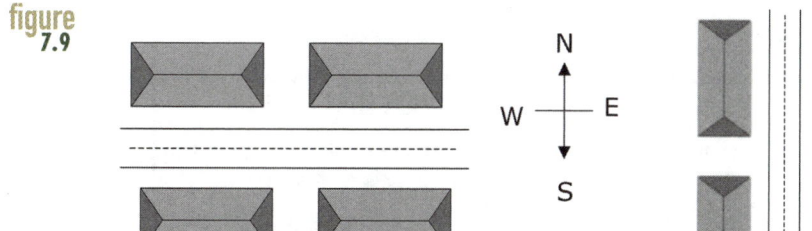

a. East-West street orientation

b. North-South

East-west street orientation increases the potential for passive solar and decreases cooling loads.

7.2.2 South-Facing Windows for Solar Gains

It is true that south-facing windows can bring in a lot of solar energy during a sunny day, but they also lose energy all day and all night all winter long. The question of whether they can be net energy providers is important, and easily answered.

We know how to calculate thermal losses using the U-factor of the window. To find solar gains we need a measure of the fraction of solar energy that hits the window that is transmitted into the interior of the building, called the *solar heat gain coefficient* (SHGC). A high SHGC is desirable for passive solar heating, whereas a low value is especially important in commercial buildings with significant cooling loads. The National Fenestration Rating Council (NFRC) labels provide both the U-factor and SHGC as well as a measure of the transmittance of visible wavelengths and the air leakage of the window, an example of which is shown in Figure 7.10. Solution Box 7.2 illustrates the usefulness of several of these important factors.

Virtually anywhere in the country, solar gains for double-glazed, south-facing windows exceed thermal losses all winter long. That suggests putting in a lot of them. There is some risk, however, if too much south-window area is provided. For one, recall from Chapter 6 how cold those windows may feel at night when you are sitting next to them. The other reason is that you may actually overheat the house in the daytime with so much solar power coming in. That can be avoided by adding extra thermal mass, such as a concrete floor, to absorb some of that heat in the daytime so it can be given back at night. Providing enough thermal mass is one of the principal challenges in passive solar design.

figure 7.10   Sample of a National Fenestration Rating Council (NFRC) Window Label

SOLUTION BOX 7.2

Net Gain for a Solar Window in Boulder, Colorado

Compare the average January solar gain for a clear, double-glazed, south-facing window with its average thermal losses. The window has a U-factor of 0.50 and an SHGC of 0.71. It is located in Boulder, Colorado, where the month-long average January temperature is 30°F. Assume the interior of the house is kept at 70°F.

Solution:

From Table 7.2 we find the insolation hitting the window in January averages 1426 Btu/ft²-day. The solar gains and thermal losses are easy to find:

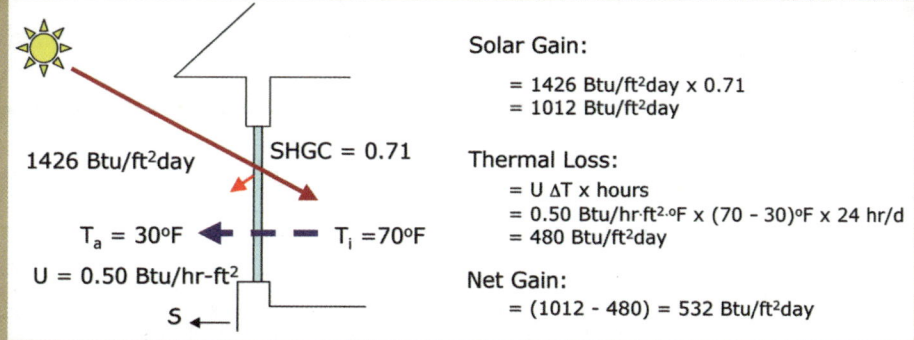

Solar Gain:

$$= 1426 \ Btu/ft^2day \times 0.71$$
$$= 1012 \ Btu/ft^2day$$

Thermal Loss:

$$= U \ \Delta T \times hours$$
$$= 0.50 \ Btu/hr{\cdot}ft^{2}{\cdot}°F \times (70 - 30)°F \times 24 \ hr/d$$
$$= 480 \ Btu/ft^2day$$

Net Gain:

$$= (1012 - 480) = 532 \ Btu/ft^2day$$

Figure labels: 1426 Btu/ft²day; SHGC = 0.71; $T_a = 30°F$; $T_i = 70°F$; U = 0.50 Btu/hr-ft²; S

That means every square foot of this window acts like a little furnace providing a net positive 532 Btu/ft² per day—better than the best possible insulated wall, which will always lose some energy.

7.2.3 A "Sun-Tempered" House

A natural question is, How much south glass can we put into an ordinary lightweight framed house that has no extra thermal mass before we risk causing overheating on sunny days? One rule of thumb suggests that as long as the solar gain area is less than about 7% of the floor space area, the inherent mass of the house itself is probably sufficient. A simple, conservative way to analyze such *sun-tempered* houses is to ignore the south window area when doing heat-loss calculations. That is, find the UA-value of the house as if the south-facing windows have infinite R-value. We'll call that $(UA)_{ST}$. Table 7.3 demonstrates this procedure using the "House #1" spreadsheet developed in Table 6.10. For this particular 1500 ft² house, the 100 ft² of south-facing solar-gain windows provide enough heat to drop the annual fuel bill about 20% from $1108 per year down to $897. That's a sizable chunk for such a simple thing to do.

7.2.4 The Importance of Thermal Mass

Because we can imagine south-facing windows as net sources of heat, it is tempting to add more and more of them in a passive solar house design. However, as the area of south-facing glass increases, there will come a point when the inherent house mass will be insufficient to avoid excessive indoor temperature swings. For example, a computer simulation of an

table 7.3   Building Efficiency: Sun-Tempered UA-Value House #1 with 100 ft² of south window out of total 250 ft²

Component	Area (ft²)	Insulation	R	$U = 1/R$*	UA*	% of Total*
Ceiling	1500	R-30 #13	30.3	0.033	49.5	12%
South Windows	100	dbl Al #21	Infinite	0.000	0.0	0%
Non-S Windows	150	dbl Al #21	1.4	0.714	107.1	26%
Doors	60	No storm #19	2.6	0.385	23.1	6%
Walls	970	R-21 #3	19.2	0.052	50.5	12%
Floors	1500	R-21 #7	29.3	0.034	51.2	12%

	ACH	Volume (ft³)	Efficiency			
Infiltration	0.6	12,000	0		129.6	32%
Ventilation	0	12,000	70%		0	0%

TOTAL $(UA)_{ST}$ = 411 Btu/hr°F

The fuel bill for this "sun-tempered" house drops about 20% when the solar gains of its 100 ft² of south-facing windows are accounted for. Compare this with the original house in Table 6.10.
* Columns are calculated values. All other information is data entry.

ANNUAL FUEL CONSUMPTION

Fuel Price	$12	per million Btu (Table 6.9)
Furnace Efficiency	80%	
Distribution Efficiency	75%	
Internal gains, q_{int}	3000	Btu/hr
Indoor set point, T_i	70	°F
HDD65	5052	°F-d/yr for Blacksburg, VA (Table 6.8)

Calculations:

* Balance Point Temp	62.7	°F	$T_{bal} = T_i - q_{int}/UA_{ST}$
* HDD @ T_{bal}	4546	°F-d/yr	HDDTb = HDD65 - (0.021 * HDD65 + 114)(65 - T_{bal})
* Q_{del}	44.8	Million Btu/yr	Q_{del} = 24(UA)HDD
* Q_{fuel}	74.7	Million Btu/yr	$Q_{fuel} = Q_{del}$/(furn eff × dist eff)
* Annual fuel bill	$ 897	per year	$/yr = Q_{fuel} × fuel price
*** House without solar**	**1,108**	per year	using all 250 ft² of windows

* Rows are calculated values. All other information is data entry.

ordinary lightweight house without extra thermal mass is shown in Figure 7.11. With a south window area equal to 7% of the floor area, the sun warms the house to a reasonable 70°F in the afternoon. If we doubled the window area to 14%, the temperature would rise above 90°F, which is way too hot, which means we need additional thermal mass. With 14% glazing and enough additional thermal mass the house remains comfortable day and night without an auxiliary heat source.

In most passive solar houses, heat storage in extra thermal mass is achieved by letting the sun warm up concrete, tile, or some other dense masonry material, or occasionally, water. A useful measure of the ability of a material to store heat is called *volumetric capacitance*, which is the product of the material's density ρ (lb/ft^3) times its specific heat c (Btu/lb-°F). Let's compare concrete and water for volumetric capacitance:

Concrete: 140 lb/ft^3 × 0.2 Btu/lb-°F = 28 Btu/°F-ft^3

Water: 62.4 lb/ft^3 × 1.0 Btu/lb-°F = 62.4 Btu/°F-ft^3

That means you can store more than twice as much heat by raising a cubic foot of water by 1°F than by raising a cubic foot of concrete by that same amount. In spite of that advantage, very few passive solar houses attempt to use water for thermal mass. Solution Box 7.3 demonstrates a useful guideline for sizing thermal mass.

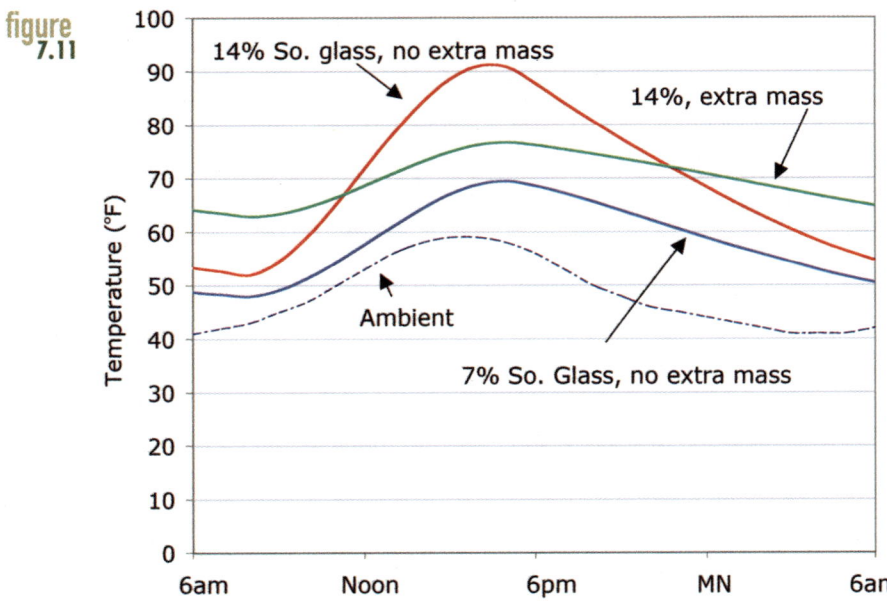

figure 7.11

A computer simulation of a small, lightweight house showing the impact of having too much south-facing solar gain area and not enough mass. With enough mass, 14% of the floor area in south-facing glass remains comfortable with no external heat source.

SOLUTION BOX 7.3

Sizing Thermal Mass

How much thermal mass do we need for a passive solar house?

Solution:

One recommendation for the amount of thermal mass needed in a passive solar house is to provide at least 30 Btu/°F of storage capacity per square foot of solar gain area. For a house with 200 ft^2 of solar aperture, that would suggest needing a thermal capacitance of

$$C = 30 \text{ Btu/°F-ft}^2 \times 200 \text{ ft}^2 = 6000 \text{ Btu/°F}$$

If we use concrete with a volumetric capacitance of 28 Btu/ft^3-°F, that says we need about

$$V = \frac{6000 \text{ Btu/°F}}{28 \text{ Btu/ft}^3\text{-°F}} = 214 \text{ ft}^3 \text{ of concrete}$$

If we use a concrete slab for mass, and estimate it is just the top 4 inches that is effective, that suggests a floor area of about

$$\text{Area of concrete} = \frac{214 \text{ ft}^3}{4/12 \text{ ft}} = 640 \text{ ft}^2$$

Another way to state this result is that you should have about 3 ft^2 of surface area of concrete thermal mass in the vicinity of the windows per square foot of solar glazing.

There is another approach to thermal mass that is quite promising. Rather than storing heat by changing the temperature of a substance, you instead change its state from solid to liquid when it absorbs heat, and transform it back to liquid to give back that heat. One such *phase change material* is made with hydrated calcium chloride, which absorbs 82 Btu/lb when it melts at 81°F. With a density of 97 lb/ft^3, that means it can store about 8000 Btu/ft^3 when it changes phase. That's about 14 times better than a 20°F temperature swing in a cubic foot of concrete.

7.2.5 Types of Passive Solar Heating Systems

There are three "generic" types of passive solar heating systems: *direct gain*, *mass wall*, and *sunspace*. These are illustrated in Figure 7.12. The essence of a direct gain system is just

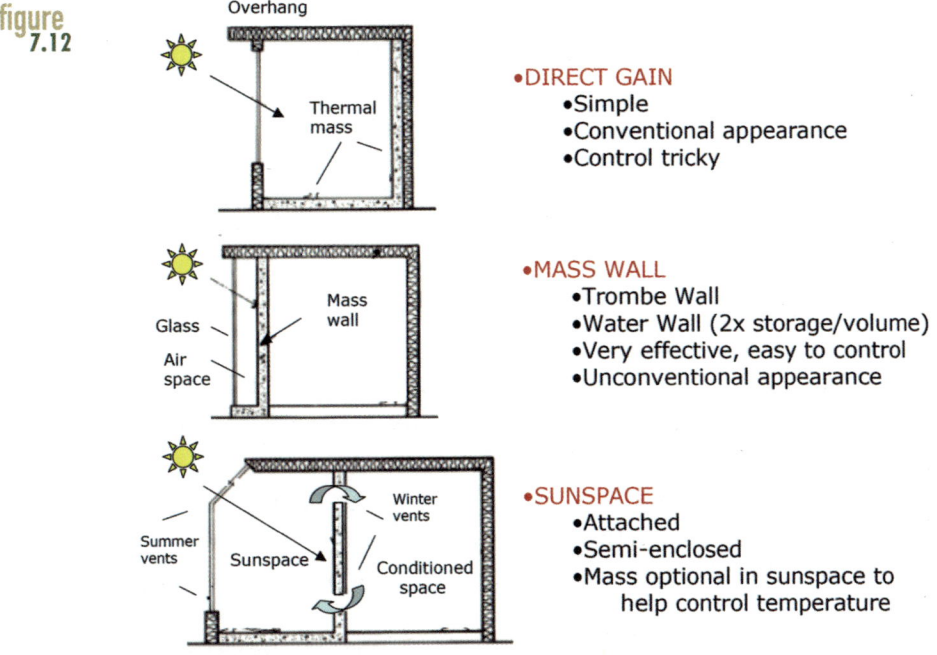

figure
7.12

•DIRECT GAIN
- •Simple
- •Conventional appearance
- •Control tricky

•MASS WALL
- •Trombe Wall
- •Water Wall (2x storage/volume)
- •Very effective, easy to control
- •Unconventional appearance

•SUNSPACE
- •Attached
- •Semi-enclosed
- •Mass optional in sunspace to help control temperature

Three "generic" types of passive solar systems.

south-facing glass, an overhang to shade the glass in the summer, and enough thermal mass to adequately back up the solar gains. These systems are simple and result in a conventional appearance to the house. The trick is to provide enough mass to keep the temperature swings under control.

A second approach is based on putting the thermal mass directly behind the glazing, with an air gap to help keep the heat from dissipating to the ambient at night. The mass is usually a dark-colored concrete wall, in which case it is often called a *Trombe wall* named after a French engineer, Felix Trombe, who popularized the idea in the 1960s. The concept is simple and very effective. The mass not only absorbs solar energy, but it also provides a helpful delay between the time the sun hits the wall during the day and the time that the heat works its way through the mass so that it can radiate into the interior space in the evening. Their main disadvantage is aesthetic. We like to look out those windows. To mitigate that, many installations have a combination of direct gain and mass wall to let daylight in and provide views to the outside.

The third category of passive heating is based on an attached sunspace, or greenhouse. While there are many variations on sunspaces, the common denominator is that the range of temperatures allowed within the sunspace is much greater than would be tolerated in the rest of the house. Sunspaces are meant to get pretty warm during the daytime to provide heat that can be directed into the adjacent conditioned space, and then they are usually allowed to cool off quite a bit at night. The range of temperatures through which the sunspace swings can be controlled by the amount of mass used. The flip side is that more mass means more of the heat stays in the sunspace and therefore doesn't provide as much heat to the conditioned space inside the house.

7.2.6 Estimating Solar Performance

A careful analysis of passive solar performance is complicated and challenging, but we can make reasonable estimates based on work done in the 1970s by a team at the Los Alamos National Laboratory in New Mexico.[2] We'll use their simplest procedure, called the *Load-Collector Ratio* method, which results in a very rough estimate of the annual heat load for a south-facing, passive solar house. We begin by identifying the solar aperture area, A_p, which for direct-gain and mass-wall systems is just the south-facing glazing area. For a sunspace, A_p is the projected area of the glass onto the south wall as shown in Figure 7.13.

In this procedure, the solar aperture area is treated as if it is thermally neutral (in essence infinite R-value). That makes the UA-value the same as the one we used for a sun-tempered house, $(UA)_{ST}$. The key parameter that describes a passive solar house is called the load-collector ratio (LCR). The numerator of LCR is a heat loss term whereas the denominator is a solar gain term. Therefore, a low LCR is indicative of better solar performance.

Eq. 7.4
$$LCR = \frac{24(UA)_{ST}}{A_p}$$

figure 7.13

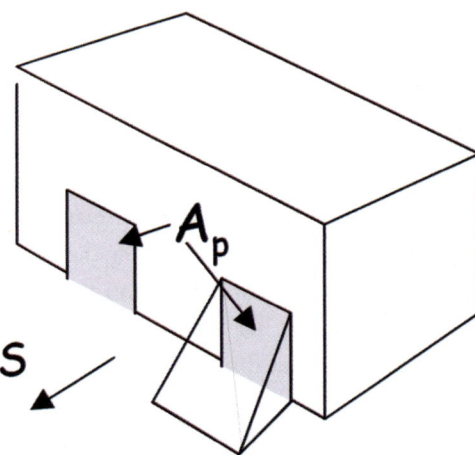

Identifying the solar aperture area, A_p.

[2] Balcomb, D., R. Jones, C. Kosiewicz, G. Lazarus, R. McFarland, and W. Wray, *Passive Solar Design Handbook, Vol. 3.* (New York: American Solar Energy Society, 1983).

The full LCR procedure is based on finding the closest match between your house design and a number of "standard" passive solar designs defined in the passive solar design handbook. You then pore over huge tables of numbers that use your LCR to find the *solar savings fraction* (SSF) for your particular house in your particular location (see, for example, Stein and Reynolds, 1992, for an extensive set of LCR tables). The energy that needs to be delivered by your heating system is then

Eq. 7.5

$$Q_{del} \text{ (Btu/yr)} = 24(UA)_{ST} \text{ HDD65 } (1 - SSF)$$

where   HDD65 = degree-days at a 65°F base temperature

For illustrative purposes, we have simplified their procedure and their tables considerably. Table 7.4 allows you to pick a generic passive solar design (direct gain, sunspace, Trombe wall) for a particular city. For that design, find values for LCR2 and LCR5, then plug them into the following empirically derived relationship to find SSF:

Eq. 7.6

$$SSF = 0.18 + 0.3 \, \frac{\log_{10} \text{(LCR2/LCR)}}{\log_{10} \text{(LCR2/LCR5)}}$$

Our simplified LCR method gives a quick estimate of the annual heating demand. The analogous month-by-month analysis is called the *solar load ratio* (SLR) method. There are several software packages built around the SLR method, including *BGW20004* from Solaequis and *BuilderGuide*.

Do these passive solar ideas really work? One very early study that carefully monitored forty actual houses produced the following persuasive answer to that question. In Figure 7.14 the normalized heating requirements (like our thermal index in Equation 6.20) broken down into the amount supplied by solar gains, internal gains, and back-up furnace are shown. For most of these houses, the furnace is required to supply less than 2 Btu/ft^2 per degree-day; most new houses today require something like 6–8 Btu/ft^2-°F day.

7.3 Cooling Loads

For a number of reasons, cooling calculations are more difficult to make than those for heating. Not only are there UA heat gains through the building envelope, there are also unwanted solar gains through windows, roofing that absorbs sunlight, infiltration and ventilation air that needs to be dehumidified as well as cooled, and finally, those internal gains that help heat a building in the winter but work against us when we need cooling (Figure 7.15).

7.3.1 Avoiding Cooling Loads

The best way to reduce air-conditioning load, of course, is to focus on avoiding the need for cooling in the first place. A good starting point is building orientation. We have already seen

table 7.4   Simplified Parameters for the LCR Method

		Trombe Wall (TWB1)		Direct Gain (DGB1)		Sunspace (SSB1)		HDD65	T_{avg}
		LCR2	LCR5	LCR2	LCR5	LCR2	LCR5	°F-d/yr	°F
Birmingham	AL	101	20	80	20	125	28	2844	62
Phoenix	AZ	308	68	274	92	353	87	1552	70
Tucson	AZ	281	63	248	85	331	84	1752	68
Los Angeles	CA	351	76	314	102	457	110	1819	62
Mount Shasta	CA	74	13	54	3	99	19	5890	50
Oakland	CA	213	46	192	58	289	66	2909	57
Red Bluff	CA	141	27	116	30	165	34	2688	63
Sacramento	CA	145	28	122	31	179	37	2843	60
San Diego	CA	391	86	350	117	500	123	1507	63
Santa Maria	CA	212	49	191	64	298	77	3053	57
Denver	CO	86	18	68	18	106	25	6016	50
Pueblo	CO	92	19	73	20	111	26	5394	53
Washington	DC	42	7	22	1	57	11	5010	54
Jacksonville	FL	248	54	213	71	294	73	1327	68
Atlanta	GA	92	18	72	17	116	26	3095	61
Boise	ID	69	10	50	1	85	14	5833	51
Chicago	IL	27	1	15	1	39	4	6127	51
Des Moines	IA	34	4	20	1	45	7	6710	49
Wichita	KS	58	10	38	1	73	14	4687	57
New Orleans	LA	212	44	182	57	257	61	1465	68
Boston	MA	32	4	20	1	44	7	5621	51
Kansas City	MO	50	8	30	1	64	12	5357	54
Omaha	NB	40	6	17	1	51	9	6601	49
Las Vegas	NV	207	45	180	58	233	56	2601	66
Reno	NV	104	22	84	23	133	30	6022	49
Albuquerque	NM	117	26	97	30	141	35	4292	57
Los Alamos	NM	79	17	60	16	105	25	6359	48
Charlotte	NC	97	19	77	19	119	27	3218	61
Raleigh-Durham	NC	84	17	65	15	104	23	3517	59
Tulsa	OK	80	16	60	13	97	22	3680	60
Medford	OR	61	7	40	1	81	12	4930	53
North Bend	OR	98	21	83	21	135	31	4688	52
Portland	OR	55	1	31	1	73	8	4792	53
Charleston	SC	142	29	120	34	172	40	2146	65
Nashville	TN	57	9	36	1	74	15	3696	59
Austin	TX	180	38	153	47	219	52	1737	68
Dallas	TX	136	29	113	34	164	39	2290	66
Houston	TX	183	38	153	47	221	53	1434	69
Salt Lake City	UT	73	13	55	1	92	18	5983	51
Roanoke	VA	65	12	47	1	83	18	4307	56
Seattle	WA	51	1	25	1	69	1	5185	51
Cheyenne	WY	69	14	51	11	87	20	7255	46

The full table is available in the *Passive Solar Design Handbook, Vol. 3*. Also given are base 65°F degree days and annual average temperature. Notes: LCR2 = LCR @SSF = 0.20; LCR5 = LCR @SSF = 0.50. DGB1 is double glazed, 3:1 mass-to-glazing ratio, 6" mass, no night insulation; TWB1 is 6", vented, double glazed; SSB1 is attached, 90/30, masonry common wall, opaque end walls.

SOLUTION BOX 7.4

Applying the Simplified LCR Method to a Direct-Gain House

A house in Denver with a total UA-value of 400 Btu/hr-°F has 200 ft² of U-0.50 direct gain windows (with suitable mass). If the furnace and ducts are each 90% efficient and the fuel is natural gas at $1.20/therm, estimate the fuel bill. Denver has HDD65 = 6016 °F-day/yr.

Solution:

The sun-tempered UA-value subtracts those solar windows

$$(UA)_{ST} = 400 \text{ Btu/hr-°F} - 200 \text{ ft}^2 \times 0.50 \text{ Btu/hr-ft}^2\text{-°F} = 300 \text{ Btu/hr-°F}$$

The load collector ratio is $LCR = \dfrac{24(UA)_{ST}}{A_p} = \dfrac{24 \times 300}{200} = 36$

From Table 7.4, LCR2 = 68 and LCR5 = 18, so using Equation 7.6 we get

$$SSF = 0.18 + 0.3 \frac{\log_{10}(68/36)}{\log_{10}(68/18)} = 0.324$$

Using Equation 7.5 gives

$$Q_{del} = 24 \text{ hr/day} \times 300 \text{ Btu/hr-°F} \times 6016 \text{ °F-day/yr} \times (1 - 0.324) = 29 \times 10^6 \text{ Btu/yr}$$

Accounting for furnace and duct losses, the final fuel bill will be about

$$\text{Fuel} = \frac{29 \times 10^6 \text{ Btu/yr}}{0.90 \times 0.90} \times \frac{\$1.20}{10^5 \text{ Btu}} = \$430/\text{yr}$$

---

the importance of east-west orientations to reduce wall and window areas exposed to the sun. Buildings can also be oriented to take advantage of prevailing winds that can help provide natural ventilation. As suggested in Figure 7.16, if breezes approach at a bit of an angle, rather than head on, ventilation efficiency is improved. Locating rooms that produce heat and humidity, such as kitchens and laundry rooms, on the leeward side of the house helps keep their heat and humidity from affecting other areas of the house. If there is an attached garage, it too should be placed on the leeward side of the house to keep it from blocking needed airflow in the rest of the building.

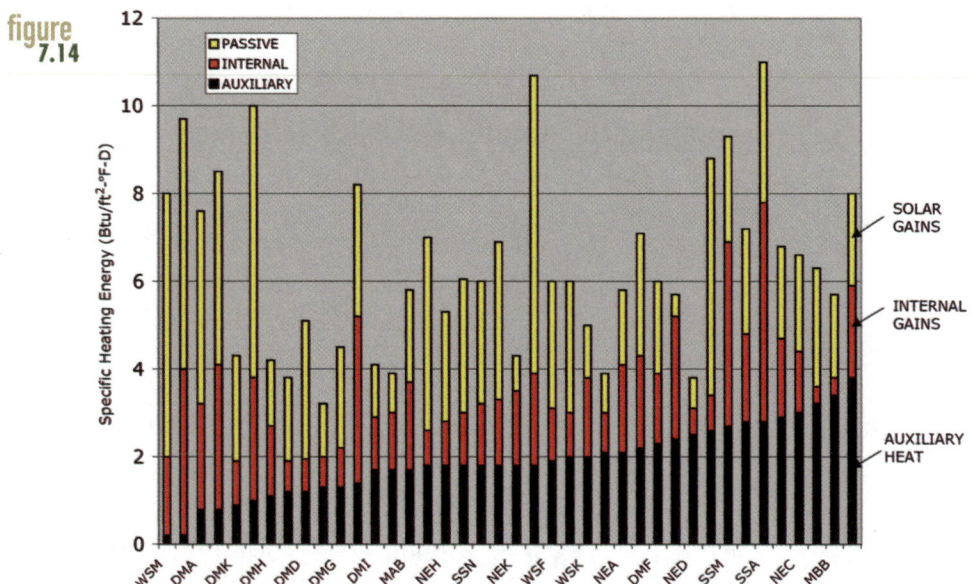

Measured specific heating energy for forty passive solar homes showing most requiring less than 2 Btu/ft² per degree-day of auxiliary heating.

Source: Swisher (1985)

Cooling loads are trickier to analyze.

A major cause of overheating in a building is the sun beating down all day long onto rooftops. As Figure 7.17 indicates, light-colored roofs may be as much as 70°F cooler than dark ones. Dark roofs not only increase the cooling load of a building, they also help increase

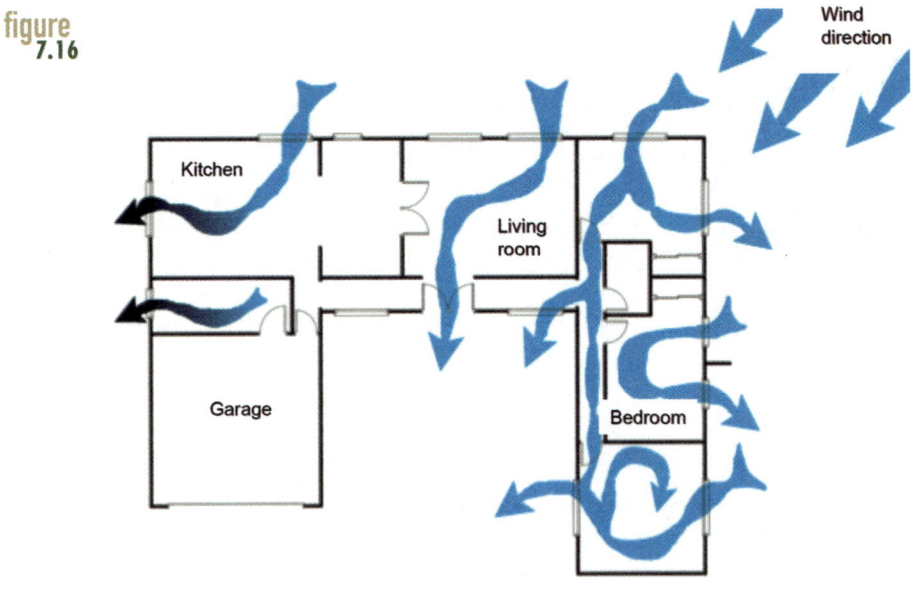

figure 7.16

Designing for good cross ventilation.

SOURCE: DBEDT, 2001

the ambient temperature of the whole neighborhood, a phenomenon called the *heat island effect*. Researchers at Lawrence Berkeley National Labs and at the Florida Solar Energy Center have monitored buildings with lightly colored, more reflective roofs and concluded that those buildings used up to 40% less energy for cooling than buildings with darker roofs. Extrapolating their results across the United States suggests savings on the order of $750 million per year could be realized if cool roofing materials were to be used.

Newly developed roofing materials that enable color retention while increasing the overall solar reflectivity suggest we don't really have to have white roofs to capture these savings. Recall that roughly half of incoming solar energy is in the infrared region and about half in visible wavelengths. These new materials attempt to reflect as much of that IR as possible, and then reflect just enough in the visible range to provide the desired color. Figure 7.18 shows this concept along with a comparison of the reflectivity and colors for cool roofing tiles versus standard tiles.

In addition to more reflective roofing materials, attic temperatures can be dramatically reduced by using continuous ridge and soffit vents augmented with shiny, foil radiant barriers under the roof rafters or on top of ceiling joists (Figure 7.19).

In commercial buildings, proper east-west axis orientation, choosing glass that reduces solar gains while letting in natural daylight, careful design of fins and overhangs, and thoughtful placement of vegetation to shade windows and roofs all have a significant effect. Reducing internal gains by using energy-efficient lighting, office machines, and appliances, is especially important in commercial buildings where such gains are a major driver of air-conditioning loads. In areas with hot days and cool nights, additional thermal mass to absorb "coolth" at

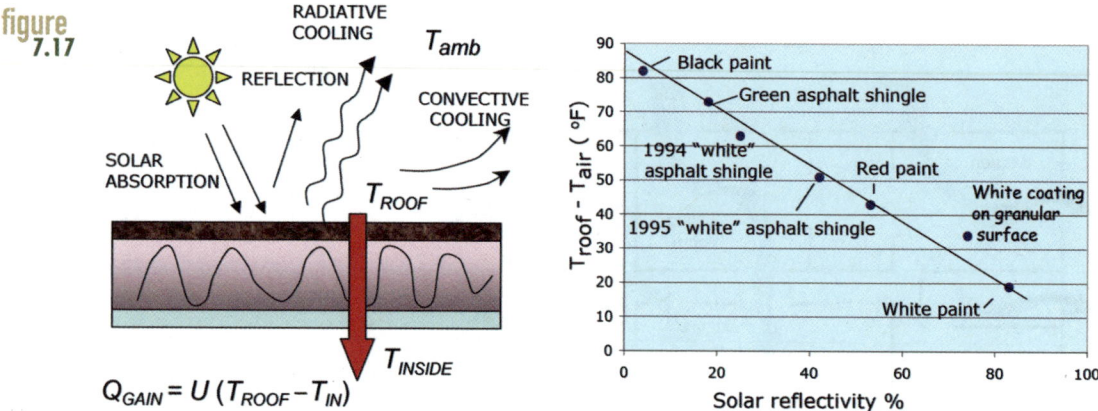

figure
7.17

Heat gain can be greatly increased due to the elevated temperature of roofing materials.

SOURCE: based on LBNL http://eetd.lbl.gov/HeatIsland/CoolRoofs/

night can help carry the building through the next day using a sort of thermal flywheel effect. Another approach to cool roofing takes advantage of the insulating and shading ability of plants growing on the roof itself. These *green roofs* are becoming especially popular in areas where stormwater runoff is a problem. When designed properly, green roofs help avoid storm surges by absorbing and temporarily storing rainwater.

7.3.2 The Building Envelope and Cooling Degree-Days

Having briefly explored ways to reduce cooling loads, we will now spend a little time trying to quantify some of the ways cooling loads can be calculated. For the building envelope, we

figure
7.18

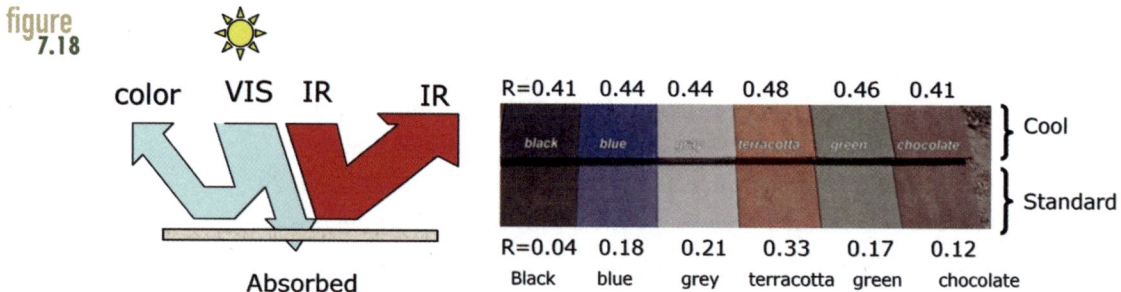

Cool roofing reflects as much infrared as possible and enough visible to provide color. The cool roofing tile samples shown increase reflectance by about 0.3.

SOURCE: American Rooftile Coatings

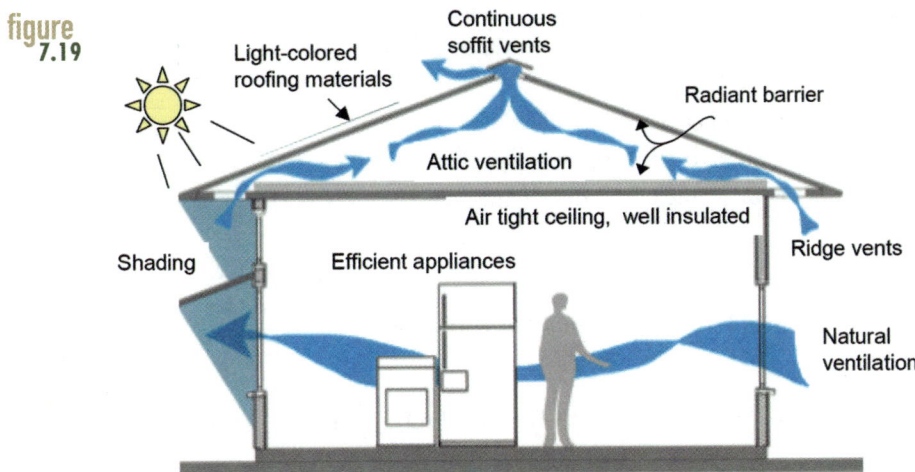

figure 7.19

A combination of construction techniques can dramatically reduce the need for cooling.

SOURCE: DBEDT, 2001

can use the same UA-value analysis developed in Chapter 6 to find heat gains due to ambient-to-indoor temperature differences. Notice, however, that the temperature differences driving heat gains (maybe 20°F–30°F) are often considerably smaller than those for heat loss calculations (often 40°F–60°F).

We can also use the same sort of degree-day analysis for annual cooling loads as we did for heating loads. The procedure for determining cooling degree-days is analogous to that used for heating degree-days. For every day that the average temperature is above the base temperature (T_b), cooling degree-days (CDD) are accrued; for every day they are below the base temperature, heating degree-days (HDD) are accrued.

The relationship between annual HDD and CDD is as follows:

Eq. 7.7
$$\mathrm{CDD_{Tb}} = \mathrm{HDD_{Tb}} - 365(T_b - T_{AVG})$$

where T_{AVG} = annual average temperature (e.g., Table 7.4)

We can use CDD for annual loads in just the same way that we used HDD:

Eq. 7.8 $$Q_{Envelope}\ (\mathrm{Btu/yr}) = 24\ \mathrm{hr/day} \times (UA)\ (\mathrm{Btu/hr\text{-}°F}) \times \mathrm{CDD}\ (\mathrm{°F\text{-}day/yr})$$

To figure out the annual cost of cooling, we need to account for the efficiency of the air conditioner (AC) and the inefficiencies of the ducts. The efficiency of air conditioners is given by a quantity called the seasonal energy efficiency ratio (SEER), which is the cooling output (Btu/yr) divided by the electrical energy input (Wh/yr) for a hypothetical average U.S. climate.

Eq. 7.9
$$\text{SEER (Btu/Wh)} = \frac{\text{Annual cooling (Btu/yr)}}{\text{Electrical input (Wh/yr)}}$$

The first federal efficiency standard for central ACs in 1992 required a minimum SEER of 10. New standards raised the SEER requirement to 13 in 2006. Many older ACs have a SEER of only 6 or 7. Including duct efficiency, the electric energy input for an AC is given by

Eq. 7.10
$$\text{Annual Envelope Cooling Electricity (Wh/yr)} = \frac{24(\text{UA})\text{CDD}}{\text{SEER} \cdot \eta_{\text{Ducts}}}$$

Solution Box 7.5 demonstrates the use of CDD and SEER.

7.3.3 Controlling Solar Gains with Better Windows

We already know the importance of orienting a building to minimize east- and west-facing windows, and we know something about using overhangs to shade south-facing windows. Even after exhausting the advantages of these simple design guidelines, there will almost always be windows that will be hit with solar radiation in those hot months of the year. Especially in large buildings, some of that radiation is useful in that it brings in daylight that helps reduce the need for artificial lighting. Table 7.5 presents some month-by-month clear-sky numbers for insolation striking windows with various orientations.

Recall from Figure 6.3 that the critical loads for commercial buildings are lighting and air-conditioning (AC). As Figure 7.7 suggests, to avoid solar gains, window orientation is of paramount importance. Compared to south-facing windows (easy to shade), east- and west-facing windows are typically exposed to twice the solar radiation in the summer, and that factor increases to 4 or 5 for horizontal skylight glazing. East- and west-facing glass can drive up air-conditioning loads; horizontal glazing is even worse.

The traditional approach to controlling solar gains for commercial buildings has been to choose windows that are either tinted (e.g., bronze) or reflective. The downside to tinted windows is that they block solar radiation by absorbing it in the glazing itself. The resulting hot glass can be extremely uncomfortable to sit next to as it radiates its heat into the interior space. Shiny, reflective windows avoid the heated-glass problem, but by indiscriminately blocking all wavelengths they provide very little natural daylight, which increases the need for artificial lights.

The more modern approach to controlling solar gains is with *spectrally selective* glass. Recall the solar spectrum introduced in Chapter 4 and redrawn here as Figure 7.20. About 2% of the insolation is in the ultraviolet (UV) portion of the spectrum (gives us skin cancer and fades materials), about 47% is in the visible range (helps us see things), and 49% is in the near infrared (provides heat but no light). The far infrared region shown corresponds to wavelengths emitted by objects at ordinary room temperatures.

SOLUTION BOX 7.5

Cooling Load in Houston Due to Envelope Gains

What would the cost of cooling be (envelope portion only) for a UA = 500 Btu/hr-°F house in Houston with 70% efficient ducts, an AC with SEER = 10, and electricity that costs $0.10/kWh? From Table 7.4, Houston has HDD65 = 1434 °F-day/yr and annual average temperature of 69°F.

Solution:

First find CDD65 from Equation 7.7.

$$CDD65 = HDD65 - 365\left(T_b - T_{AVG}\right) = 1434 - 365(65 - 69) = 2894 \text{ °F-day/yr}$$

From Equation 7.10 the electricity needed by the air conditioner will be

$$\frac{24(UA)CDD}{SEER \cdot \eta_{Ducts}} = \frac{24 \text{ hr/day} \times 500 \text{ Btu/hr °F} \cdot 2894 \text{ °day/yr}}{10 \text{ Btu/Wh} \cdot 0.70} \times \frac{1 \text{ kW}}{1000 \text{ W}} = 4961 \text{ kWh/yr}$$

At $0.10/kWh, the cooling load due to the building envelope would be about $496/yr.

---

table 7.5   Daily Clear-sky Insolation (Btu/ft²) on Vertical Surfaces (e.g., Windows)

	LATITUDE 30 N			LATITUDE 35 N			LATITUDE 40 N			LATITUDE 45 N		
	S	SE/SW	E/W	S	SE/SW	E/W	S	SE/SW	E/W	S	SE/SW	E/W
JAN	1786	1336	654	1784	1315	599	1733	1268	536	1604	1164	445
FEB	1624	1339	834	1687	1349	777	1715	1333	707	1704	1293	631
MAR	1225	1200	946	1363	1257	925	1481	1312	900	1575	1351	868
APR	706	1001	1036	872	1097	1049	1031	1184	1058	1186	1262	1061
MAY	404	840	1064	554	928	1085	717	1027	1101	880	1120	1112
JUN	318	769	1051	457	857	1075	603	948	1092	763	1044	1104
JLY	402	825	1044	549	911	1066	706	1007	1082	864	1098	1093
AUG	698	971	1002	855	1061	1013	1006	1144	1021	1152	1217	1023
SEP	1179	1153	913	1307	1203	892	1414	1252	865	1498	1285	832
OCT	1570	1285	792	1619	1285	732	1632	1262	663	1611	1219	592
NOV	1739	1300	635	1730	1277	581	1667	1221	515	1526	1108	421
DEC	1808	1329	601	1769	1291	541	1652	1197	456	1445	1038	344
Annual	408,418	405,541	321,581	441,456	420,321	314,477	466,229	430,233	304,182	479,978	431,592	290,024

Conventional low-e windows focus on ordinary building and ambient temperature radiation. These low-e windows are designed to keep far infrared (thermal) radiation inside the building in the winter and outside in the summer. On the other hand, spectrally selective glazings have window coatings that are designed to block as much of the solar near-IR sunlight as possible, while allowing a controllable amount of visible solar wavelengths into the building. These spectrally selective windows, combined with automatic dimming systems to modulate the artificial lighting to account for the availability of natural daylight, can dramatically reduce lighting loads in commercial buildings.[3] Figure 7.21 presents the idea.

As shown in Figure 7.10, window labels now contain information on the fraction of visible energy transmitted as well as U-value, and solar heat gain coefficient (SHGC). Table 7.6 presents a number of window types with representative values of these important characteristics. All the double-pane, low-e windows have comparable R-values, but a good choice for passive solar heating might be the low-e 178 with #3 surface coating (SHGC = 0.63). For daylighting and reduced cooling loads you might choose the low-e 128, which has a low SHGC of 0.24. Figure 7.22

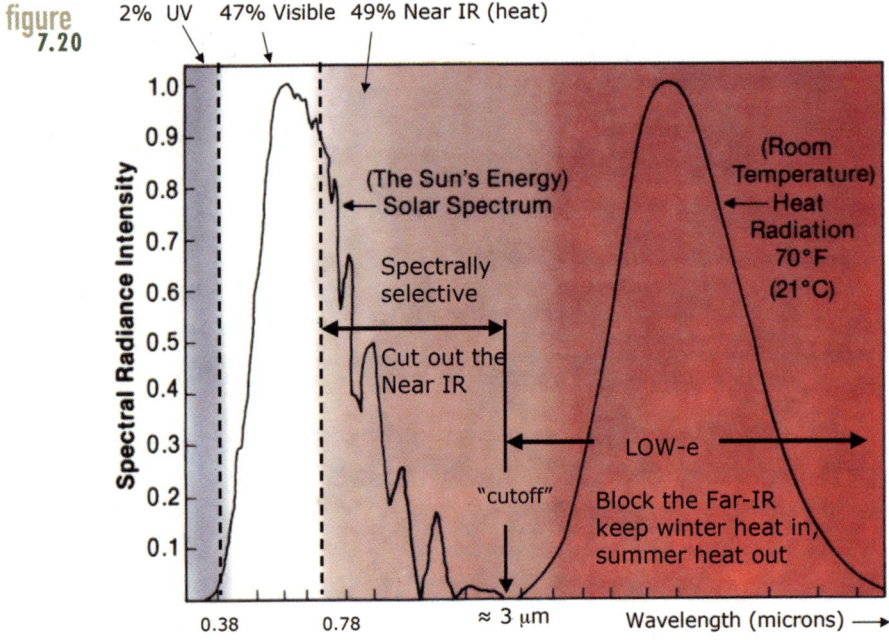

figure 7.20

The solar spectrum and the spectrum for objects near room temperature (the near-IR). Low-e windows reflect the far infrared, whereas spectrally selective, low-e coatings try to block both the near and far infrared.

---

[3] Two excellent sources of information on daylighting systems are the International Energy Agency's 2000 report *Daylight in Buildings: A Source Book on Daylighting Systems and Components,* and LBNL's *Tips for Daylighting with Windows.*

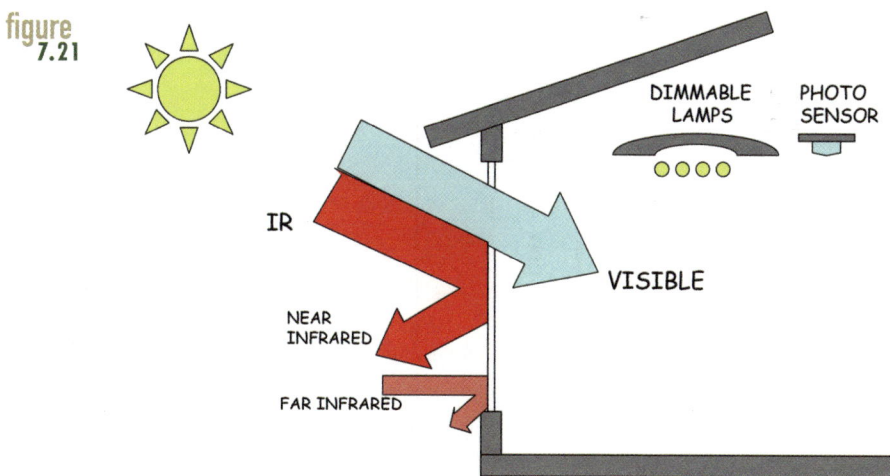

figure
7.21

Spectrally selective, low-e windows coupled with automatically dimming interior lighting systems take advantage of available natural daylight to reduce both lighting and cooling loads.

table
7.6

Examples of Center-of-Glass Visible Transmittance, Solar-Heat-Gain Coefficients, U- and R-Values

Type	Glass Coating	Visible Light Transmittance	SHGC	U-factor Air	R-Value Air	U-factor Argon	R-Value Argon
1-pane	Clear	0.90	0.86	1.04	0.96		
1-pane	Bronze Tinted	0.68	0.73	1.04	0.96		
2-pane	Clear	0.81	0.76	0.48	2.08		
2-pane	Bronze Tinted	0.61	0.63	0.48	2.08		
2-pane	Lo e 178 #3 Surface	0.78	0.63	0.31	3.23	0.27	3.70
2-pane	Lo e 172 #2 Surface	0.72	0.41	0.30	3.33	0.25	4.00
2-pane	Lo e 170 #2 Surface	0.70	0.37	0.30	3.33	0.25	4.00
2-pane	Lo e 140 #2 Surface	0.40	0.25	0.30	3.33	0.26	3.85
2-pane	Lo e 128 #2 Surface	0.38	0.24	0.30	3.33	0.25	4.00
2-pane	Lo E 172 #2 Surface	0.66	0.38	0.22	4.55	0.19	5.26
3-pane	Lo e 172 #2 & #3 Surfaces	0.58	0.35	0.16	6.25	0.13	7.69

shows the surfaces on which low-e coatings should be placed depending on whether controlling the cooling load or encouraging heating is the more important goal.

7.3.4 Dehumidification

A major part of the cooling load in many parts of the country is the energy required to dehumidify infiltration or ventilation air. It takes about 1060 Btu to evaporate one pound of

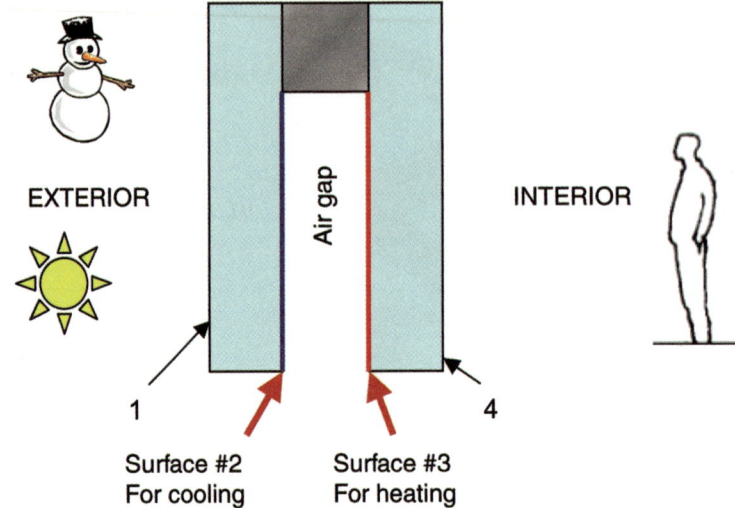

figure 7.22   **The Glazing Number System**

Windows designed to emphasize cooling reduction have the low-e coating on surface #2, whereas those primarily controlling winter heat loss coat surface #3.

water, converting it from liquid water to water vapor. That means to dehumidify moist air, we have to remove 1060 Btu of *latent heat* for every pound of water removed.

To understand this dehumidification process, we need to become familiar with the simplified psychrometric chart shown in Figure 7.23. A "psych chart" is a plot of the moisture content of air (called the absolute humidity) versus dry bulb temperature (what you measure with a thermometer). The parameter is relative humidity (rH), which is the fraction of the maximum moisture that air can hold at a given dry bulb temperature.

The example in Figure 7.23 shows the process taken to cool air from an outside temperature of 80°F with 40% rH (point A) to a desired temperature of 65°F at 40% rH (point D). Imagine the air conditioner beginning to lower the temperature along the A-to-B path on the psych chart. The line is horizontal because the amount of moisture in the air hasn't changed yet (the absolute humidity is constant). When the air reaches 53°F (point B), it is fully saturated and any further decrease in temperature will cause moisture to condense out of the air. Dropping the temperature to 40°F moves the operating point from B to C and in the process enough moisture condenses to lower the absolute humidity from about 0.009 to 0.0055 lb H_2O/lb air. As the 40°F air at point C absorbs heat, it moves horizontally on the chart to the desired temperature and humidity (point D). Solution Box 7.7 demonstrates this process.

The examples in Solution Boxes 7.5 through 7.7 illustrate three primary sources of cooling load for buildings: heat transfer through the building envelope, solar gains through windows exposed to the sun, and dehumidification. Although they were illustrated for residential buildings, they apply equally well to larger, commercial buildings.

SOLUTION BOX 7.6

Cooling Load in Houston from West-Facing Windows

Suppose a house in Houston (latitude 29° N) has 100 ft² of unprotected, clear double-pane, west-facing windows. Estimate the cost of summertime cooling due to those windows if the AC has an SEER of 10, the ducts are 70% efficient, and electricity costs $0.10/kWh.

Solution:

From Table 7.5 the daily insolation on east- or west-facing windows during the summer is about 1000 Btu/ft²-day. Let's imagine 100 such days. From Table 7.6, clear double-pane windows have an SHGC of 0.76. A reasonable estimate for summer solar gains would be

$$\text{Gain} = 100 \text{ ft}^2 \times 1000 \text{ Btu/ft}^2\text{-day} \times 100 \text{ days} \times 0.76 = 7.6 \times 10^6 \text{ Btu}$$

That heat has to be removed by the SEER 10 AC through 70% efficient ducts.

$$\text{Cost} = \frac{7.6 \times 10^6 \text{ Btu}}{10 \text{ Btu/Wh}} \times \frac{1}{0.70} \times \frac{1 \text{ kW}}{1000 \text{ W}} \times \$0.10/\text{kWh} = \$108$$

figure 7.23

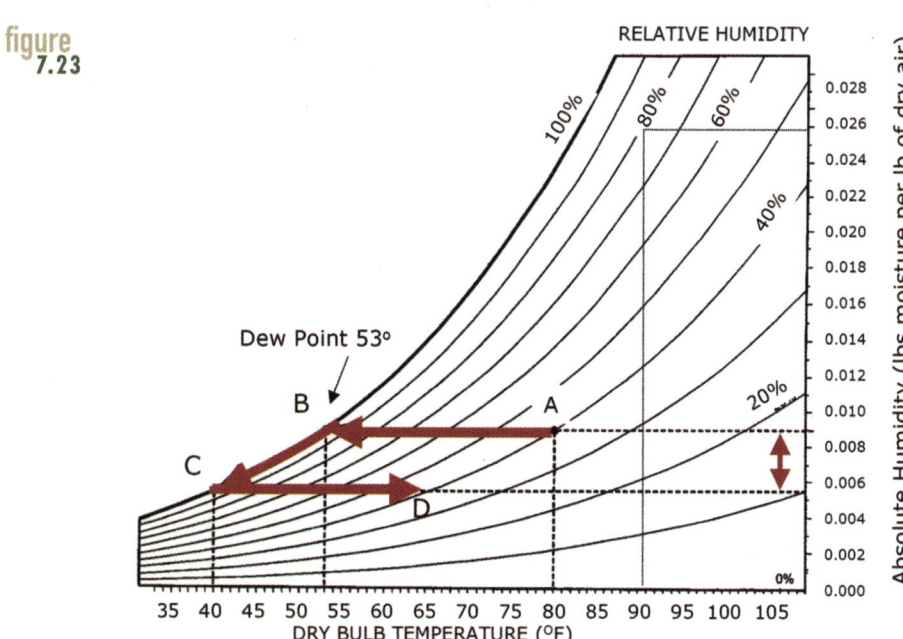

A simplified psych chart showing the path taken to cool air at 80°F, 40% rH down to 65°F and 40% rH. Each pound of water condensed liberates about 1060 Btu.

SOLUTION BOX 7.7

Cooling Load Due to Latent Heat

Suppose a 2000 ft² house with 8-foot ceilings has an infiltration rate of 0.5 ach. Find the cooling load associated with dehumidifying infiltration air from ambient 90°F, 80% rH to a nice AC supply at 65°F, 40% rH. Assume air density is 0.075 lb/ft³.

Solution:

The rate at which infiltration air entering the building must be cooled and dehumidified is

$$0.5 \text{ air changes/hr} \times (2000 \times 8) \text{ ft}^3/\text{air change} \times 0.075 \text{ lb/ft}^3 = 600 \text{ lb/hr}$$

From Figure 7.22, to drop the absolute humidity of outside air at 90°F, 80% rH to the desired 65°F, 40% rH requires the removal of about

$$(0.026 - 0.005) \text{ lb H}_2\text{O/lb air} \times 600 \text{ lb air/hr} \times 1060 \text{ Btu/lb} = 13,038 \text{ Btu/hr}$$

If we imagine an SEER = 10, 70% ducts, $0.10/kWh, AC handling the dehumidification of infiltration for 100 days of summer, it would cost

$$\text{Cost} = \frac{13,038 \text{ Btu/hr} \cdot 24 \text{ hr/d} \cdot 100 \text{ d/yr}}{10 \text{ Btu/Wh} \cdot 0.70} \times \frac{1 \text{ kW}}{1000 \text{ W}} \times \frac{\$0.10}{\text{kWh}} = \$447/\text{yr}$$

7.4 Domestic Water Heating

Figure 6.3 showed the importance of residential water heating. As a single category, only residential space heating and commercial lighting account for more primary energy. We are all familiar with the conventional storage-tank versions of water heaters, but there are a number of other approaches that can provide hot water with greater energy efficiency, but perhaps at a higher first cost. Options include demand (also known as tankless or instantaneous) water heaters, solar water heaters, and heat-pump water heaters (Figure 7.24).

Conventional household storage water heaters typically range in size from about 20 gallons to 80 gallons and are fueled with natural gas, electricity, propane, or fuel oil. Their efficiency is indicated by an energy factor (EF), which includes the energy required to heat the water in the first place under an assumed 64-gallon/day use rate, plus standby and nearby piping losses. The most efficient conventional natural gas–fired storage water heaters have an EF of 0.60 to 0.65, which corresponds to an annual gas demand of around 230 therms. Condensing water heaters, which capture some of the latent heat in the exhaust gases, have energy factors as high as 0.86. Solution Box 7.8 shows how to use these EFs to help estimate the cost of water heating.

Electric water heaters have much higher energy factors, about 0.93 to 0.95, which is a reflection of the fact that the conversion of electricity to heat in resistive elements immersed in the water is nearly 100%. At that EF, and 64 gallons/day, their (expensive!) electricity demand is around 4500 kWh/yr. Heat pump water heaters using less than half as much electricity as conventional electric resistance water heaters are finally becoming commercially available. They are expensive, but their extra first cost pays for itself by way of reduced electricity bills.

Demand water heaters avoid the standby losses associated with conventional storage water heaters. In these units, water is heated instantaneously as hot water is drawn from the tap. For gas-fired units, pilotless demand heaters have longer lifetimes and higher EF ratings than conventional storage water heaters, which gives them a sizable life-cycle cost advantage.

Table 7.7 presents levelized annual cost estimates for various types of water heaters. As would be expected, natural gas–fired storage water heaters have much lower life-cycle costs than those fired by expensive electricity. However, when electricity is used in a heat pump water heater or as a backup for a solar water heater, annualized costs are comparable.

figure 7.24

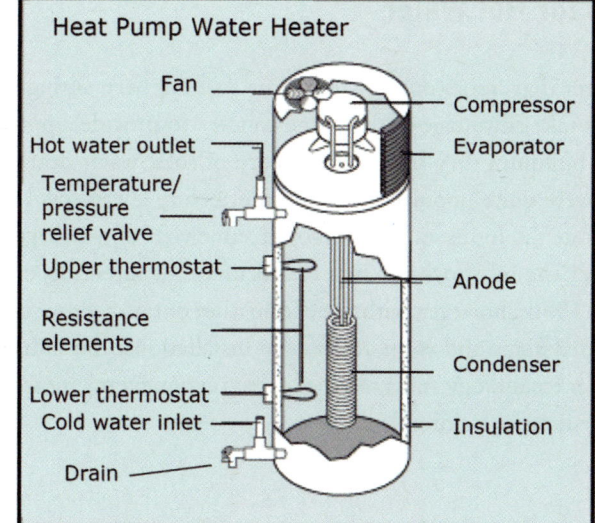 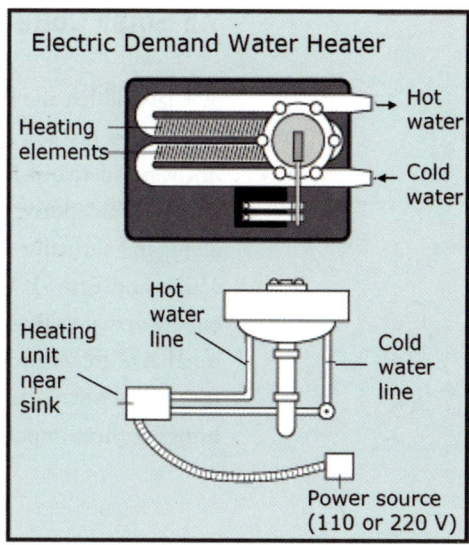

Heat pump water heaters and demand water heaters are two variations on conventional hot water systems.

SOURCE: www.eere.energy.gov/consumer/your_home/water_heating, 2006

SOLUTION

SOLUTION BOX 7.8

Cost of Water Heating

Estimate the annual fuel cost to heat 64 gallons per day from 55°F to 130°F in an EF 0.57 storage water heater fueled by $1/therm natural gas.

Solution:

Begin by figuring out the energy that would be required if the system were to be 100% efficient:

$$Q = 64 \text{ gal/day} \times 365 \text{ day/yr} \times (130 - 55)°F \times 1 \text{ Btu/lb-}°F \times 8.34 \text{ lb/gal}$$
$$= 14.6 \times 10^6 \text{ Btu/yr}$$

Dividing by the system efficiency and converting to therms gives an annual cost of

$$\text{Cost} = \frac{14.6 \times 10^6 \text{ Btu/yr}}{0.57} \times \frac{1 \text{ therm}}{100,000 \text{ Btu}} \times \frac{\$1.00}{\text{therm}} = \$256/\text{yr}$$

7.5 Solar Collectors for Hot Water

We have all felt the hot water that comes out of a dark hose that has been sitting in the sun. Solar water heating systems take advantage of that same concept to provide a portion of the hot water demand for the buildings they serve. The history of solar water heating systems is somewhat spotty. They were quite popular 100 years ago, before cheap fossil fuels came along and virtually wiped out the industry. They enjoyed somewhat of a resurgence in the 1970s and early 1980s after the oil shocks of that era focused our attention on reducing energy demand. By the late 1980s, however, with the elimination of tax credits, coupled with unreliable freeze protection systems and some really ugly installations, the industry pretty much died once again. Better technology and a renewed focus on our energy future are finally bringing these important systems back into play.

7.5.1 Flat-Plate Solar Collectors

At the heart of most solar water heating systems is the simple, flat-plate collector, consisting of a black absorber plate in an insulated box with a glass top to let in the sun (Figure 7.25).

table
7.7   Levelized Costs for Various Types of Water Heaters

Water Heater Type	Energy Factor	Installed Cost	Annual Energy	1st yr Energy Cost	Life (years)	Levelized Cost $/yr
GAS FIRED			**Therms**			
Conventional storage	0.57	$380	256	$256	13	$351
High-efficiency storage	0.65	$525	225	$225	13	$329
Instantaneous demand	0.70	$650	209	$209	20	$243
High-efficiency pilotless demand	0.84	$1,200	174	$174	20	$255
Solar (SF = 0.7) with gas back-up	0.57	$2,500	77	$77	20	$271
ELECTRICITY			**kWh**			
Conventional storage	0.90	$350	4758	$428	13	$557
High-efficiency storage	0.95	$440	4508	$406	13	$539
Instantaneous demand	0.95	$600	4508	$406	13	$556
Electric heat pump	2.20	$1,200	1947	$175	13	$340
Solar (SF = 0.7) with electric back-up	0.90	$2,500	1427	$128	20	$318

Assumptions: 64 gpd demand, 75°F delta T, natural gas @ $1/therm, electricity @ 9¢/kWh, 3% fuel escalation, 5% discount rate, solar savings fraction 70%.

A pump, or sometimes just natural buoyancy, causes water to circulate from a storage tank to the collector. If it is pumped, the system needs a controller with sensors on the tank outlet and the collector outlet. If the collector is warmer than the tank, the controller turns on the pump. Each pass through the collector raises the water temperature by about 5°F or 10°F so that by the end of a sunny day you have a tank of hot water.

figure
7.25

Typical hydronic flat-plate collector and system diagram for a solar water heater.

A fairly straightforward thermal analysis of flat-plate collectors begins with an energy balance as shown in Figure 7.26. We have insolation I striking the collector area A, some fraction of that, $(\tau\alpha)$, is transmitted through the glazing system and absorbed by the plate. Some of the absorbed heat is then lost, mostly through the glazing, and some is taken away in the form of heat delivered to the storage tank. Using lowercase q to represent a heat rate, we can write

Eq. 7.11
$$q_{del} = q_{ABS} - q_{loss}$$

Eq. 7.12
$$q_{del} = IA(\tau\alpha) - U_L A\left(T_P - T_{amb}\right)$$

where the quantity U_L = an overall heat loss factor Btu/hr-°F per ft² of collector area

If we divide the delivered energy by the incident radiation, we get the following equation for the efficiency of the collector:

Eq. 7.13
$$\eta = \frac{q_{del}}{q_{incident}} = \frac{IA(\tau\alpha) - U_L A\left(T_P - T_{amb}\right)}{IA} = (\tau\alpha) - U_L\left(\frac{T_P - T_{amb}}{I}\right)$$

This awkward looking equation turns out to be remarkably straightforward when we plot efficiency versus $(T_P - T_{amb})/I$, as has been done in Figure 7.27a. It is just a straight line with y-axis intercept = $(\tau\alpha)$ and slope U_L. The x-axis is a normalized measure of how hot the collector is, how cold the environment is, and how much sunlight is hitting the panel. With just those three quantities we can find the collector efficiency.

Figure 7.27b shows a plot of collector efficiency that is almost exactly the same as 7.27a, but notice the x-axis parameter uses the inlet temperature of the collector T_{in} instead of the average plate temperature T_p. The collector inlet temperature is much easier to measure than the average plate temperature, so this little adjustment makes for far easier collector analysis. To legitimatize this switch, a little fudge factor, labeled F_R is shown in Figure 7.27b. Manufacturer descriptions of collector efficiency are specified by simply providing the "y-axis intercept" $F_R(\tau\alpha)$ and "slope factor" $F_R U_L$.

figure 7.26   Energy Analysis of a Flat–Plate Collector

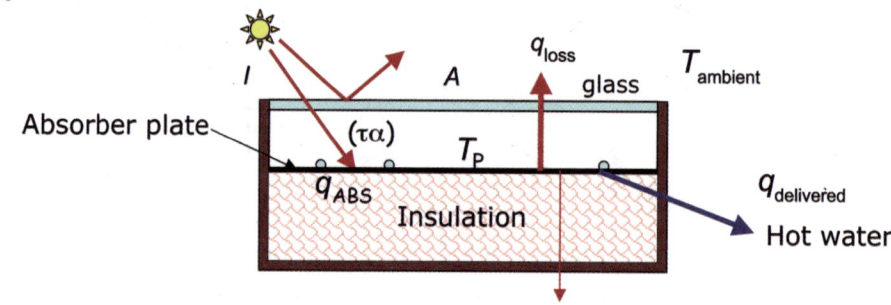

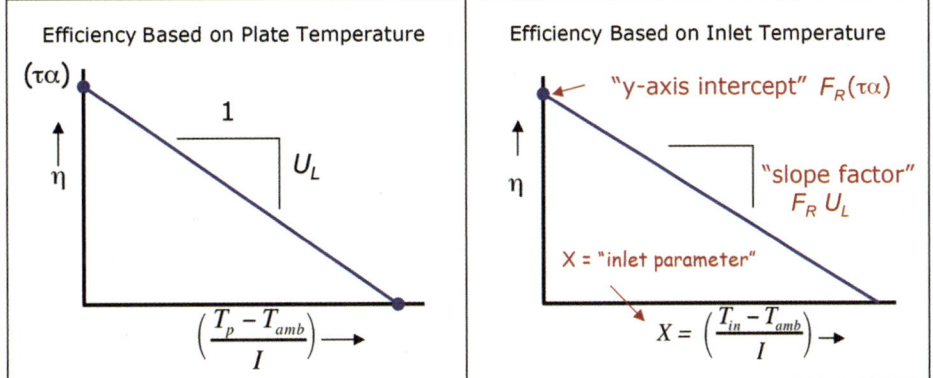

Collector efficiency is specified by the "y-axis intercept" and "slope factor" with the x-axis assumed to use T_{in} rather than T_p.

Flat-plate collectors come with many options. They may have no glazing, or single-glazing, or double; they may have a *selective surface* with high absorptivity to solar wavelengths and low emissivity for the longer wavelengths radiated by the hot absorber. To gain some intuition into these options consider Figure 7.28, which shows efficiency curves for three versions of glazing under representative midday insolation and ambient temperature. With added glazing the y-axis intercept drops as less and less insolation reaches the absorber; on the other hand, with more glazing the heat loss represented by the slope decreases. The graph suggests that for low temperature applications, such as swimming pool heating, no glazing at all is probably most efficient; for average temperatures, such as water heating, single-glazing is probably best; and for solar space heating, double-glazing is usually the best choice.

7.5.2 Solar Water Heater System Sizing

When collectors are tested, insolation is normal to the glazing. In real life, of course, the sun moves around all day long and the incidence angle changes with time and seasons. The more the sun's rays hit the collector at an off-axis angle, the greater the reflection off the glass, which reduces $(\tau\alpha)$. To account for that reduction, an incidence angle modifier, called $K_{\tau\alpha}$, is used, which shifts the average y-axis intercept down a bit in the standard efficiency curve. A reasonable default value for single-glazed collectors is a $K_{\tau\alpha}$ of 0.93.

In sizing solar water heater systems, we need to include such factors as location, energy demand, insolation, collector tilt, collector parameters, ambient temperature, and system losses. Solution Box 7.9 demonstrates how these factors can be integrated into a collector sizing exercise.

The spreadsheet in Table 7.8 continues this example to show how an overall annual solar savings fraction (SF) for a solar water heater could be found. Although the above procedure is

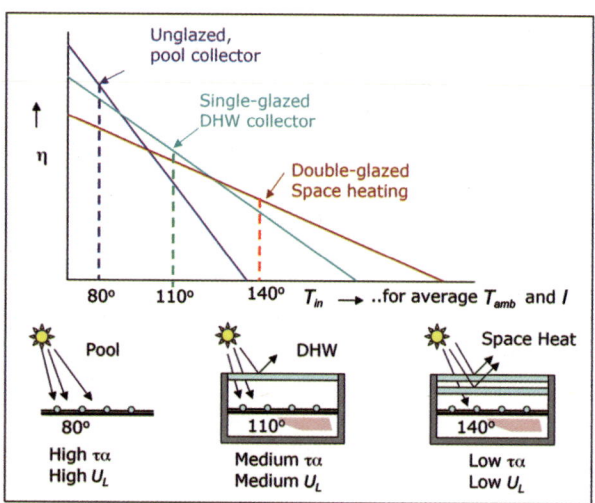

figure 7.28

Unglazed collectors in mild weather may be best for pools; single-glazing makes sense for domestic hot water (DHW); double-glazing is needed for space heating.

adequate for hand or spreadsheet calculations, there are computer programs available on the Internet that may be worth purchasing if you want greater confidence in the results or if you are going to do this sort of sizing over and over again. One such program is called f-chart and it is downloadable from http://www.fchart.com.

7.5.3 Evacuated-Tube Solar Collectors

A relatively new, and quite exciting, product has been recently introduced that puts the absorber plate in a vacuum inside long cylindrical tubes. The vacuum eliminates convective losses from plate to glass, greatly improving efficiency especially at high plate temperatures and/or low ambient temperatures.

Besides the basic vacuum technology, these collectors use a clever technique to transfer heat from the absorber, within the evacuated tube, to the header above. They use very efficient heat pipes to do this. As shown in Figure 7.29, heat from the absorber plate vaporizes a working fluid in the heat pipe. That working fluid travels by convection to a heat-exchange bulb at the top of the tube, where it condenses and transfers its heat to the water circulating through the header. The evacuated tubes with their heat-exchange bulbs are installed by inserting them, one by one, into the copper header that runs along the top of the collector (Figure 7.30).

As can be seen from Figure 7.30, there is a considerable gap between tubes, which means on a per-unit-of-overall area, they have relatively low efficiency. But the tubes themselves have efficiencies that remain high and relatively constant over a wide range of temperatures, which means these can be used not only for mid-temperature water heating but also for higher-temperature space heating, and even higher temperature absorption cooling systems.

SOLUTION BOX 7.9

A Simple Solar Water Heater Sizing Estimate for Atlanta

Size a solar collector to heat 64 gallons/day of water from 60°F to 140°F in Atlanta in June using a flat-plate collector with x-axis intercept $F_R(\tau\alpha) = 0.75$, a slope factor $F_R U_L = 1.18$ Btu/hr-°F-ft², and an incidence angle modifier $K_{\tau\alpha} = 0.93$.

Solution:

First we'll work on estimating the average collector efficiency using

$$\bar{\eta} = F_R(\tau\alpha)K_{\tau\alpha} - F_R U_L \left(\frac{\bar{T}_{in} - \bar{T}_{amb}}{\bar{I}} \right)$$

Let's use a LAT −15° tilt angle, which for latitude-34° Atlanta is 19°, fairly close to that of a typical roof surface so it should look pretty nice. From Table 7.2 (or from NREL's Web site http://rredc.nrel.gov/solar/pubs/redbook/), average June insolation is 1997 Btu/ft², which we will imagine is spread out over roughly a 12-hour period giving us an average hourly rate of about 1997/12 = 166 Btu/ft²-hr. That same NREL source tells us the average daily maximum temperature in June is 86°F. So let's guess the average daytime temperature is about 80°F. The inlet temperature climbs from 60° in the morning to 140° by the end of the day, so we will use an average inlet temperature of 100°F. The average efficiency of the collectors will therefore be about

$$\bar{\eta} = 0.75 \cdot 0.93 - 1.18 \text{ Btu/hr-°F-ft}^2 \left(\frac{100° - 80°\text{F}}{166 \text{ Btu/hr-ft}^2} \right) = 0.55 = 55\%$$

To estimate the daily energy collected, we should probably account for dirt accumulation on the glazing plus piping losses and so forth. Let us say losses are 15%.

The collector area should therefore be

$$A = \frac{64 \text{ gal/day} \times 8.34 \text{ lb/gal} \times 1 \text{ Btu/lb°F} \times (140 - 60)°\text{F}}{1997 \text{ Btu/ft}^2\text{-day} \times 55\% \times (1 - 0.15)} = 46 \text{ ft}^2$$

That is, a collector area of around 50 square feet would be in the right ballpark.

table 7.8   Solar Water Heating Spreadsheet for the Atlanta Example in Solution Box 7.9

Location	Atlanta		Solar Water Heater Analysis				
$F_{R(\tau\alpha)}$	0.75						
$F_R U_L$	1.18	Btu/ft²-hr-°F					
$K_{\tau\alpha}$	0.93						
T_h	140	°F					
T_c	60	°F					
Load	64	gal/day					
Losses	15%						
Load*	42,701	Btu/day					
Area	50	ft²					

Monthly Solar Fraction

Month	Insolation Btu/ft²day	T_{amb} °F	Day Length hr/day	Efficiency* %	Delivered* Btu/day	Usable* Btu/day	% Solar*
Jan	1,078	44	9	15%	6,837	6,837	16%
Feb	1,331	49	9	29%	16,467	16,467	39%
Mar	1,617	58	10	39%	26,972	26,972	63%
Apr	1,902	67	10	49%	39,672	39,672	93%
May	1,965	74	11	52%	43,654	42,701	100%
Jun	1,997	80	12	55%	47,057	42,701	100%
Jly	1,934	82	11	58%	47,381	42,701	100%
Aug	1,870	81	11	57%	45,005	42,701	100%
Sep	1,680	76	10	53%	37,698	37,698	88%
Oct	1,553	67	10	44%	29,336	29,336	69%
Nov	1,205	57	9	32%	16,445	16,445	39%
Dec	1,014	48	9	15%	6,582	6,582	15%

* items are calculated; all else is data entry.
Annual Solar Fraction = 68%

7.5.4 Integral Collector-Storage Panels

In relatively mild climates, very simple collectors that combine collector and storage into a single unit have some significant advantages over conventional pumped systems. These integral-collector-storage (ICS) systems (also known as "batch" systems) are, in essence, just big, fat black hoses sitting up on the roof. The only time water moves through them is when someone turns on a hot-water tap in the house, at which time cold city water goes up to the ICS and pushes solar heated water down into the regular cold-water inlet to your conventional water heater where it is topped up in temperature if necessary (Figure 7.31). They have no moving parts, no pumps, no controllers, and no sensors to break down, which means they are inexpensive and extremely reliable. On the down side, water that heats up in the daytime unfortunately cools down at night because it is still sitting up on the roof exposed to the cold

figure
7.29

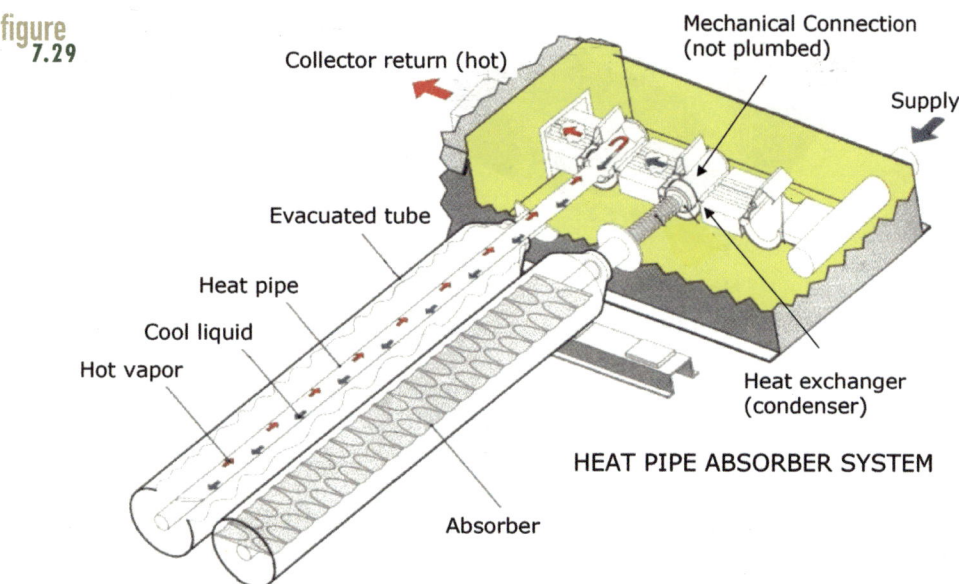

An evacuated-tube solar collector showing how heat pipes deliver heat to the header.

SOURCE: Courtesy of Sunda

figure
7.30

Inserting tubes into the header into the of an evacuated tube collector.

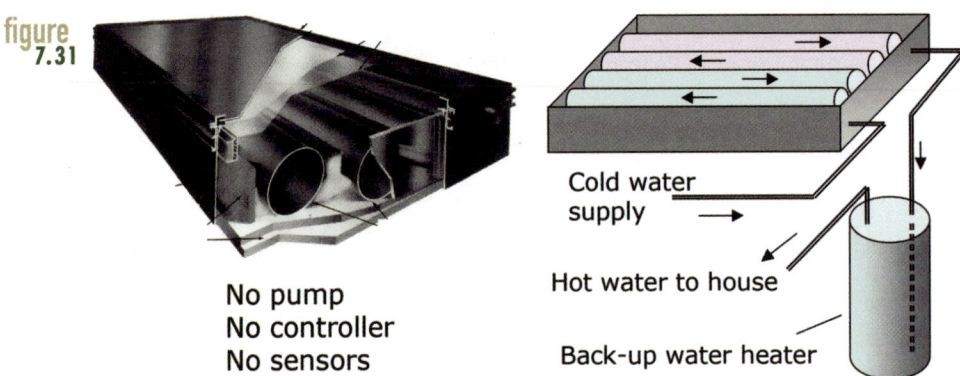

figure
7.31

No pump
No controller
No sensors

Cold water
supply

Hot water to house

Back-up water heater

An integral-collector-storage system needs no pumps, controllers, or sensors. Whenever someone turns on a hot-water tap, cold water pushes heated ICS water into the regular backup water heater.

night sky (though you could imagine putting some night insulation over the glass cover to help keep heat in). With these systems it is better to take your shower at night than in the morning.

A computer simulation of an ICS system showing the impact of evening hot water use, before the collector has had a chance to cool down, versus morning use, is shown in Figure 7.32. For this example a 40-gallon demand taken from a 40-gallon ICS at 8:00 a.m. is compared with the same demand taken at 8:00 p.m. The ICS provides more than enough heat to satisfy the entire demand for the evening load but only 40% if the demand is postponed until the morning.

7.5.5 Solar System Variations

A number of system variations for solar water heaters are available. Many of the differences have to do with how the collectors are protected from freezing on cold nights.

- **Thermosiphoning systems** avoid the use of pumps and controllers by placing the storage tank above the collectors (Figure 7.33). When hot water in the collectors becomes more buoyant than the colder water at the bottom of the tank, the warm water rises and the colder water sinks creating a convective circulation loop.
- **Closed system with antifreeze:** Even if temperatures only occasionally dip below freezing, solar collectors need to be protected from the possibility of freezing. When water freezes it expands and can easily break open a flat plate collector with potentially disastrous consequences. Many systems have a closed loop containing a nontoxic antifreeze, such as propylene glycol, with a heat exchanger to transfer heat to the potable hot water system (Figure 7.34).

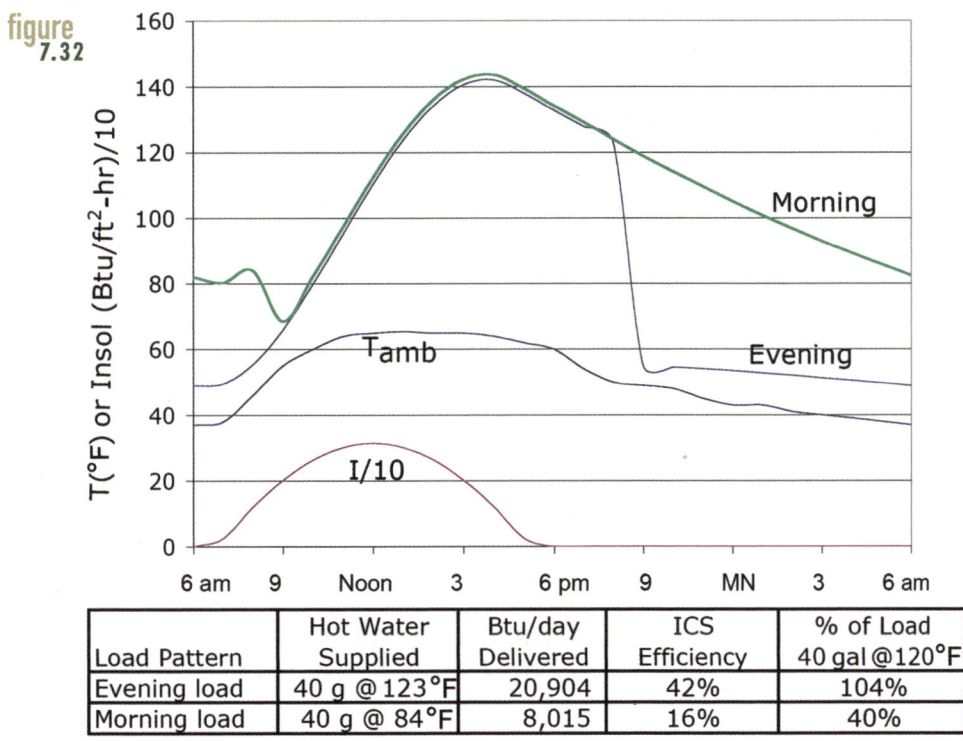

Load Pattern	Hot Water Supplied	Btu/day Delivered	ICS Efficiency	% of Load 40 gal @120°F
Evening load	40 g @ 123°F	20,904	42%	104%
Morning load	40 g @ 84°F	8,015	16%	40%

Computer simulation of a 40-gallon ICS serving a 40-gallon, 120°F load taken at 8:00 a.m. versus 8:00 p.m.

figure 7.33 **Thermosiphon Water Heater System**

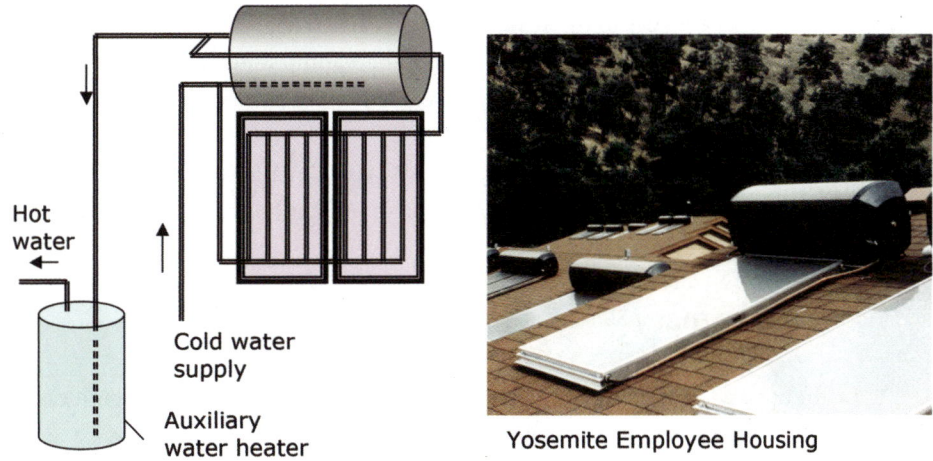

Yosemite Employee Housing

With the tank placed above the collector, thermosiphoning systems don't need pumps or controls.

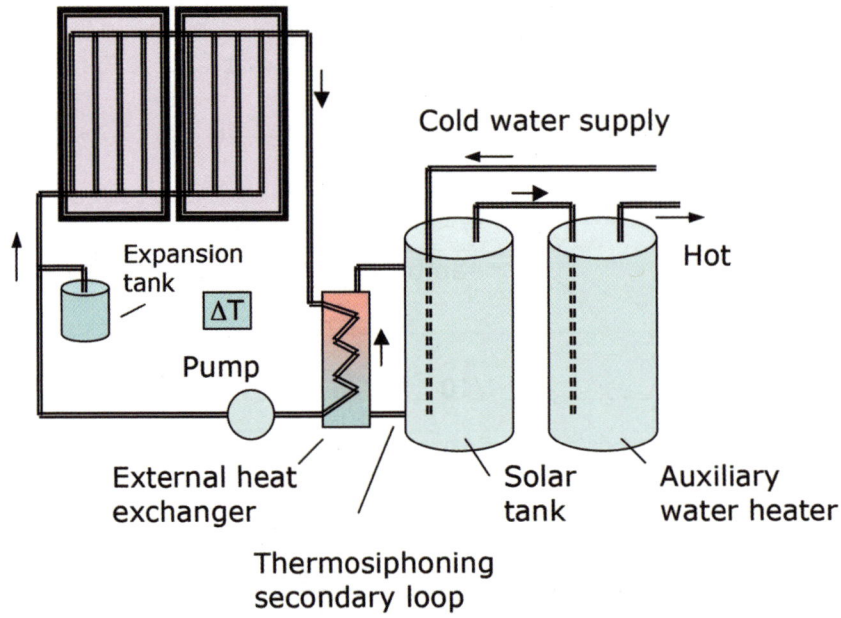

figure
7.34
Antifreeze System with External Heat Exchanger

A closed-loop, antifreeze-protected solar water heater system.

- **Drainback systems** have an unpressurized tank containing water or an antifreeze solution that is pumped around the collector loop (Figure 7.35). When the pump stops, water in the collector simply drains back into the tank. When a user draws hot water from a tap, it first passes through a heat exchanger in the open loop to preheat the water before it goes to the auxiliary water heater. These are relatively simple, reliable systems, but the pump must be strong enough to pump water against gravity up to the collector, which reduces the net energy provided by the system.

There are many choices of system type and method of freeze protection for solar water heating systems. Choosing the right one for a given household can be a challenge.

7.6 Summary

In this and the previous chapter, we have explored an array of design ideas that can significantly reduce the thermal energy demand in the buildings sector. With an aggressive effort to provide more insulation, better windows, tighter ducts, and more efficient heating and cooling systems, the heating demand can be reduced well below current building codes. And, in many locations, passive solar heating can nearly drive that heating demand toward zero.

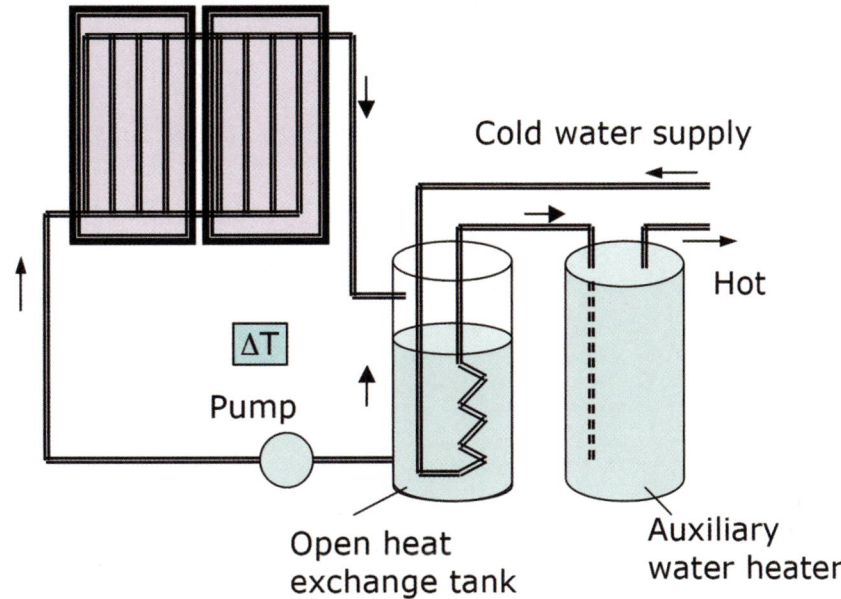

figure
7.35 **Drainback System**

Drainback systems are freeze-protected by having an open loop that allows the collectors to drain whenever the pump is turned off.

With care in building orientation, use of overhangs, natural ventilation, spectrally selective windows, and new cool roofing materials, cooling loads can be kept under control even in quite challenging climate zones. Finally, solar energy can effectively cut the demand in the very important water heating sector as well.

In the next chapter, we will shift our attention to the electricity demands in buildings for lighting and appliances. At that point, we will take on the concept of a Whole Building Life-Cycle assessment, including the embodied energy required to construct the building in the first place. That will lead us toward Green Building rating systems and the ultimate goal of zero-energy, zero-carbon buildings for the future.

CHAPTER 8

From Whole Building to Whole Community Energy

8.1 The Evolution toward Green Buildings and Communities

Energy in buildings has long focused on the thermal envelope and HVAC, mainly heating systems, because these have been the major energy uses. However, with the growth of appliances and equipment, including air-conditioning, greater attention has been recently given to electricity, because of its increasing proportion of *primary* energy among end uses in buildings. Figure 6.2 showed that building space and water heating amount to one-third of building primary energy; electricity use for cooling, refrigeration, electronics, and other appliances constitutes two-thirds. This fact has led to a more comprehensive approach to building energy, what Don Aitken calls the **Whole Building** approach.

More recently, related occupant health and environmental impacts of building operations have emerged as important considerations in design and construction. And it has become evident that the extraction, processing, manufacture, construction, and demolition of buildings and their component materials also consume considerable energy and impact the environment. These considerations for health, embodied energy and materials, and waste management, have been integrated into the Green Building movement, what we call **Whole Building Life-Cycle.**

Table 8.1 characterizes these evolving considerations in building energy from the **Building Envelope** emphasis of the 1960s and 1970s, when codes and guidelines focused on insulation levels. As air-conditioning became a dominant practice in new buildings in the 1970s and infiltration and ventilation were recognized as important operating factors, codes and guidelines began to reflect them. More recently, with the recognition of the importance of electrical appliances and lighting to building operating energy, some codes and guidelines have begun to take a Whole Building approach.

In some guidelines, such as Green Building rating systems, although not yet in codes, Whole Building Life-Cycle considerations are emerging, with the added consideration of occupant health, embodied energy, and life-cycle environment impacts.

table 8.1 **Evolving Considerations in Building Energy**

Period	Emphasis	Considerations
1960s–1970s	Building envelope	Heating operating energy
1980s	Building envelope, HVAC, infiltration	Heating + air-conditioning (AC) operating energy
1990s–2000s	Whole Building Energy	Heating + AC + appliances + lighting operating energy
2000s+	Whole Building Life-Cycle	Heating + AC + appliances + lighting operating energy + environmental/health impacts + life-cycle embodied energy
2010s+	Whole Community Energy	All of the above + on-site generation + site/neighborhood design + regional connectivity

However, we do not believe this evolution will be complete until codes and guidelines consider the broader role buildings play in community energy, including on-site distributed energy, site and neighborhood design, and regional connectivity. This extends the energy considerations of the building itself to the role it plays and can play in distributed generation of electricity, transportation energy, and land use. We call this broader perspective **Whole Community Energy,** which uses buildings as the centerpiece of sustainable community energy, enhancing the efficiency of building design and operations, materials, and embodied energy, on-site generation, and efficient land use and transportation. We touch on Whole Community Energy in this chapter, introducing "zero net energy" buildings using on-site generation, and provide greater detail on distributed energy in Chapters 10 and 11. Chapter 15 addresses the transportation, connectivity, and land use considerations of Whole Community Energy.

These evolving considerations are part of an emerging approach to building design, materials, construction, and operation, the so-called **Green Building** movement. This movement has been driven by several factors:

- **New building technologies** and practices, including windows; HVAC systems; "cool roofs"; and value-engineered framing, lighting, and appliances have made efficiency improvements more available and affordable.
- **Improved methodologies** for energy, economic, and life-cycle analysis have helped clarify the economic and environmental value of improved efficiency in buildings.
- **Government information and incentive programs** have helped educate builders and consumers and reduced initial cost of efficiency investments.
- **Third-party rating and certification programs** have translated an overwhelming amount of information on energy options into clear and accountable guidelines for builders and consumers.
- **Voluntary action by innovative architects and builders** in response to better technology and information has improved building design and practice.

- **A growing consumer market** for energy efficient, healthy, and environment-friendly buildings, informed by better information and rating systems, has pushed designers and builders to produce more efficient buildings.
- **Improving building energy codes and equipment standards** continue to adopt the innovations in technology, design, practice, and rating systems for all new buildings and products.

Sidebar 8.1 traces important developments in the Green Building movement, from early building codes to more recent rating and certification programs. Along the way, the driving forces of new technology, information and ratings programs, standards and codes, and the market have reinforced one another to increase diffusion and market penetration of building energy efficiency. New technologies are first incorporated into information and rating systems, then they are adopted by innovative builders and consumers, and ultimately they are reflected in building codes for all new construction.

This interaction has led to an iterative process of improved efficiency in new buildings, a so-called virtuous cycle, in which new technologies are reflected in ratings and the market and ultimately in regulatory codes. This is illustrated hypothetically in Figure 8.1, which shows improvement in energy efficiency and environmental impact over time. Technological potential has the lowest energy use and impact. Energy rating systems, such as ENERGY

figure 8.1   Improvements in Building Energy Technology, Ratings, Market, and Codes Reinforce One Another in an Iterative Process

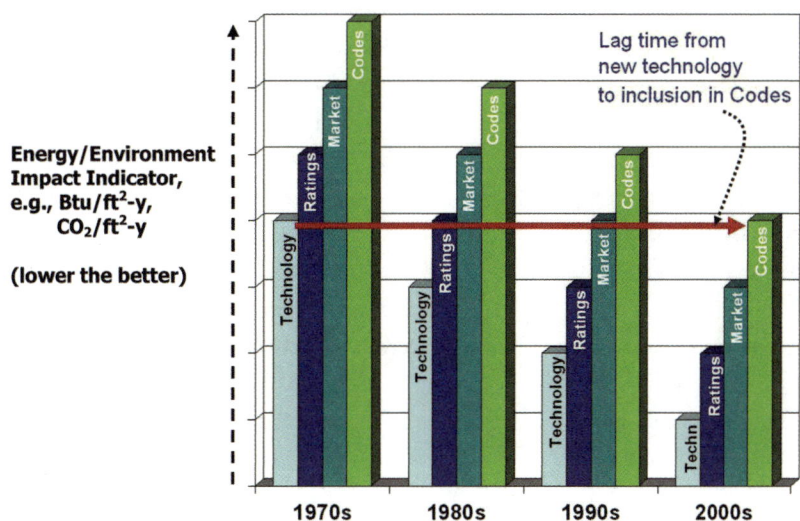

Over time technological innovation improves rating systems, which affects the market and ultimately is incorporated into codes.

SIDEBAR 8.1

Evolution of Energy Efficient and Green Buildings:

From Codes to Consumer Choice and from Building Envelope
to Whole Building to Life-Cycle Analysis

1915 Building Officials and Code Administrators (BOCA) formed in response to need for fire and electrical safety.

1920s First codes introduced for furnaces and ducts. First insulation codes for buildings not developed until the 1970s.

1974 **California Energy Resources Conservation Act** establishes first appliance energy standards.

1975 **ASHRAE** Standard 90-1975; split to 90.1 (commercial) and 90.2 (single-family residential); subsequent versions: 1980, 1989, 1999, 2004.

1975 Federal Energy Policy and Conservation Act established State Energy Conservation Program (**SECP**) providing funding to states for energy offices and plans *if* states adopted building energy code based on ASHRAE 90-1975.

1978 **California Title 24** consolidates all building codes; energy code based in part on Davis code; subsequent versions, most recently 2005.

1983 BOCA and Council of American Building Officials (CABO) develop Model Energy Code (**MEC**-1983); subsequent versions: 1986, 1989, 1992, 1993, 1995.

1985 Austin's (Texas) municipal utility **Austin Energy** establishes Energy Star program to label appliances and buildings meeting minimum efficien-

cy; program is renamed Austin Green Building Program in 1991.

1986 **Federal Appliance Energy Conservation Act** followed California's lead in creating national efficiency standards for appliances and equipment; standards are updated on a continuing basis.

1992 **Federal Energy Policy Act** required state energy building codes meeting at a minimum, MEC-1992 and ASHRAE 90.1-1989.

1992 EPA establishes **ENERGY STAR** program, a certification and labeling program to promote energy efficiency, first for office equipment, then expanding over the next decade to residential appliances, HVAC, and homes. EPA ENERGY STAR partners with DOE in 1996 and HUD in 2002.

1993 **U.S. Green Building Council** (USGBC) established by some building organizations and equipment manufacturers "to promote buildings that are environmentally responsible, profitable, and healthy." USGBC worked initially to develop certification protocol with American Society of Testing and Materials (ASTM) before deciding to develop its own **LEED** (Leadership in Energy and Environmental Design) protocol in 1998.

1994 **International Code Council** (ICC) is formed by BOCA, CABO, and other organizations to better coordinate building code activities.

(continued)

1995 **ENERGY STAR Homes** program is established by EPA for new houses exceeding MEC-1993 by 30%.

1995 The Residential Energy Services Network (**RESNET**) is founded by the National Association of State Energy Officials and Energy Rated Homes of America to develop a national market for home energy rating systems (HERS) and energy efficient mortgages. RESNET's HERS is used for ENERGY STAR homes.

1996 **ISO 14000** standards become effective, establishing certification standards for energy management systems (14001, first version in 1996), environmental auditing (14011, 1996), environmental labels (14021, 1999), and life-cycle assessment (14041, 1998).

1998 ICC develops the International Energy Conservation Code (**IECC**), a model code that replaces MEC. IECC is incorporated into ICC's International Residential Code (IRC). Subsequent versions: 2000, 2001, 2003, 2006. IECC-2006 incorporates ASHRAE 90.1-2004.

1998 USGBC releases **LEED version 1.0** for new construction and pilots the protocol in 1999. LEED-NC 2.0 is released in 2000, 2.1 in 2003, and 2.2 in 2005.

2004 U.S. Circuit Court of Appeals reverses Bush administration attempt to roll back Clinton administration's **SEER-13** efficiency standard for new central air-conditioning to SEER-12. SEER-13 standard becomes effective January 2006.

2004 California Executive Order S-20-04 announces state Green Building Action Plan to promote LEED certification of all state buildings and improve commercial energy use by 20% by 2015.

2005 USGBC develops pilot versions of new protocols for homes (**LEED-H**), core and shell buildings (**LEED-CS**), and neighborhood development (**LEED-ND**). LEED-H incorporates many ENERGY STAR criteria.

2006 Upgrades in state building energy codes, federal SEER-13 standard, and new technologies prompt ENERGY STAR home rating system to adopt "whole house" operating energy criteria and revised HERS.

2007 National Association of Home Builders predicts Green Building market will expand from $7.4 billion in 2006 to $38 billion in 2010. The 2007 survey of professional builders revealed that 70% believe green building is important to their market strategy.

2007 California legislature passes AB 1058 calling for all home builders to build green by 2013, but Governor Schwarzenneger vetoes the bill arguing building codes should not be statutory. Meanwhile, California Energy Commission (CEC) considers revised Title 24 standards for 2008, including more lighting efficiency measures, cool roofs, demand response programmable communication thermostats, photovoltaic (PV) compliance option. Legislature passes SB 1 calling for CEC to examine if PV should be required in future building codes. California Public Utility Commission (CPUC) orders utilities to prepare "next generation" efficiency plans so that all new residential buildings will be zero net energy by 2020, and all new commercial buildings will be zero net energy by 2030.

STAR and LEED, are less efficient than this full potential, but more efficient than the average market and the codes. Because the average market includes some innovators, its energy use is less than the codes, which are the minimum standard or maximum energy use.

Over time, technology, design, and practice improve, and rating systems are modified to reflect those improvements. Confidence grows in ratings and practice with improved energy and life-cycle analysis methods. The market also responds with better average performance. Energy codes and standards follow these trends; they are currently being upgraded about every 2 to 5 years (see Figures 8.3 and 17.1). As codes change, ratings systems, which are often set as a percentage improvement over code, follow suit. The efficiency bar is continually raised (energy use and impact bar continually lowered as in Figure 8.1) as these factors influence one another, and the virtuous cycle continues.

This chapter discusses the evolution of energy considerations in buildings moving from the "envelope, HVAC, infiltration" approach emphasized in Chapters 6 and 7, to Whole Building and Whole Building Life-Cycle approaches now appearing in building rating systems and to a lesser extent in building codes. We first review building codes because these regulations are critical to wide adoption of efficiency technologies in new construction. The following section presents recent developments in electrical lighting and equipment technologies and efficiency standards, an important component of the Whole Building approach.

The chapter then turns to Whole Building Life-Cycle issues of embodied energy and materials, then describes how these considerations are being reflected in Green Building rating systems. The chapter concludes by discussing progress toward "zero net energy" buildings through whole building efficiency and on-site generation, the first step toward Whole Community Energy.

8.2 Building Energy Codes and Standards: Toward Whole Building Energy

As Table 8.1 suggests, building energy codes first focused on the building envelope then added consideration of ventilation, infiltration, and heating and cooling systems. Most codes do not yet address Whole Building Energy including electrical appliances and hot water, but there is some movement in that direction, as we will see in Section 8.3. This section presents background and emerging trends in building energy codes.

8.2.1 Building Energy Code Development and Adoption

Nearly all state and local governments in the United States and developed countries regulate the construction of buildings to protect public health, safety, and welfare. Building codes date back to the founding of Building Officials and Code Administrators (BOCA) in 1915 to coordinate code activities across the country. The first energy codes for insulation and fenestration did not appear until 1974, the same year the American Society of Heating Refrigerating and Air Conditioning Engineers (ASHRAE) developed its Standard 90 for building energy efficiency. ASHRAE has updated this standard on about a five-year cycle, the latest in 2004. BOCA and other members of the Council of American Building Officials (CABO) developed their own Model Energy Code (MEC) in 1983. By 1998, the MEC was converted into the International Energy Conservation Code (IECC). Subsequent versions of the IECC were developed in 2000, 2001, 2003, and 2006. The 1992 federal Energy Policy Act required all states to adopt a building energy code based on the 1992 MEC for residential buildings and the ASHRAE standard 90.1-1989 for commercial buildings.

Figure 8.2 gives the status of state energy codes as of August 2007. There is considerable variability across the states, but most have adopted versions of the model IECC or ASHRAE codes. Some states still do not have a statewide code, and several that do, depend on voluntary adoption by local jurisdictions. Some states, notably California, Washington, Oregon, and Florida, have developed their own codes that exceed these model codes. Most state codes are uniform across the state, but California's Title 24 Code is unique not only because it is the most stringent, but also because it has different prescriptions for the state's sixteen different climatic zones (see Sidebar 8.2).

8.2.2 Typical Building Energy Code Requirements

Energy codes can usually be met by inclusion of specified **prescriptive measures**, such as U-values of building envelope components; by demonstration of building energy **performance**

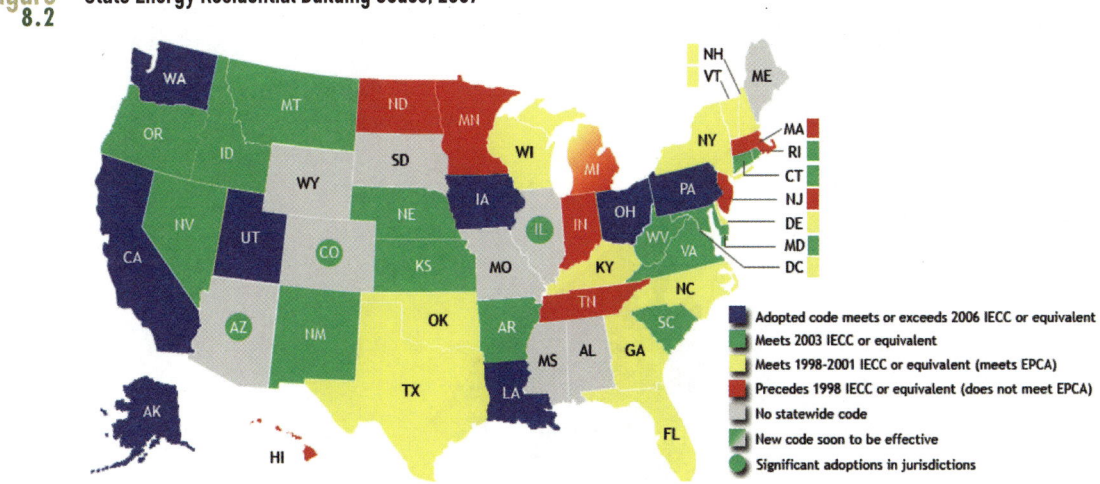

figure 8.2   State Energy Residential Building Codes, 2007

- ■ Adopted code meets or exceeds 2006 IECC or equivalent
- ■ Meets 2003 IECC or equivalent
- ■ Meets 1998-2001 IECC or equivalent (meets EPCA)
- ■ Precedes 1998 IECC or equivalent (does not meet EPCA)
- ■ No statewide code
- ■ New code soon to be effective
- ● Significant adoptions in jurisdictions

Most states' codes are based on the model International Energy Conservation Code (IECC) or its equivalent.

SOURCE: Building Codes Assistance Project (BCAP), 2007; for updates, see www.bcap-energy.org

SIDEBAR

California's 2005 Title 24 Building Energy Code

Mandatory Measures:

- Insulation and fenestration (see Table on next page)
- Infiltration: Seal all penetrations to unconditioned space; exhaust fans must have back draft damper; fireplace must have closeable doors.
- Space Conditioning: AFUE 78% min; SEER 10.0 min; duct insulation R-4.2; no un-ducted plenums/chases; UL181 approved tape and sealants; no cloth-back tape allowed; setback thermostats; California Certified equipment; installation certification

- Water Heating: if EF < 0.58, R-12 blanket; low-flow faucets and showerheads; if recirculation pumps, insulate hot-water loop; insulate first 5 feet hot and cold pipes; California Certified equipment; installation certification
- Lighting: > 50% of kitchen/bath/utility lighting high efficacy; all other lighting high efficacy except those with dimmers, motion sensors

Table shows Package D for natural gas and heat pump.

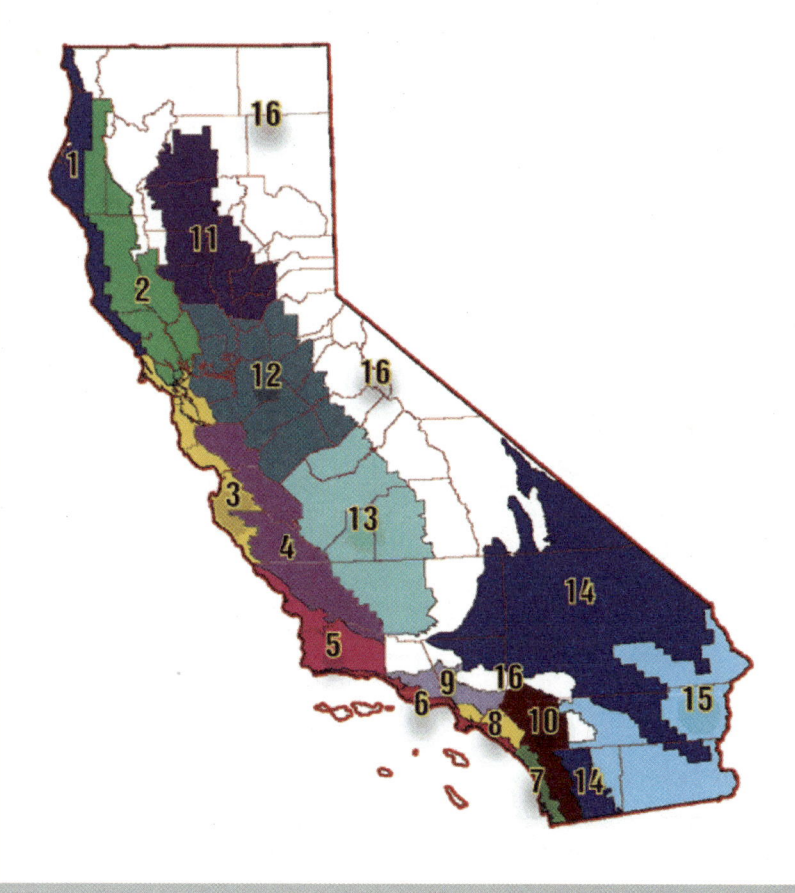

Prescriptive Measures for Sixteen Climatic Zones (see map for 16 zones):

NR = not required; REQ = required; REQ* = HERS rater verification and diagnostic testing required. In lieu of diagnostic testing of ducts and refrigerant charge, better glazing and higher efficiency space conditioning are given in lower table.

Climatic Zone	1	2	3	4	5	6	7	8	9	10	11	12	13	14	15	16
BUILDING ENVELOPE																
Ceiling	R38	R30	R30	R30	R30	R30	R30	R30	R30	R30	R38	R38	R38	R38	R38	R38
Walls	R21	R13	R13	R13	R13	R13	R13	R13	R13	R13	R19	R19	R19	R21	R21	R21
Heavy mass walls	R4.76	R2.44	R2.44	R2.44	R2.44	R2.44	R2.44	R2.44	R2.44	R2.44	R4.76	R4.76	R4.76	R4.76	R4.76	R4.76
Below-grade walls	R0	R0	R0	R0	R0	R0	R0	R0	R0	R0	R0	R0	R0	R0	R0	R13
Slab floor perimeter	NR	NR	NR	NR	NR	NR	NR	NR	NR	NR	NR	NR	NR	NR	NR	R7
Raised floors	R19	R19	R19	R19	R19	R19	R19	R19	R19	R19	R19	R19	R19	R19	R19	R19
Concrete raised floors	R8	R8	R0	R0	R0	R0	R0	R0	R0	R0	R8	R4	R8	R8	R4	R8
Radiant barrier	NR	REQ	NR	REQ	NR	NR	NR	REQ	REQ	REQ	REQ	REQ	REQ	REQ	REQ	NR
GLAZING																
Maximum U-factor	0.65	0.65	0.75	0.75	0.75	0.75	0.75	0.75	0.75	0.65	0.65	0.65	0.65	0.65	0.65	0.60
Maximum total area	16%	16%	20%	20%	16%	20%	20%	20%	20%	20%	16%	16%	16%	16%	16%	16%
SOLAR HEAT GAIN COEFFICIENT																
South-facing glazing	NR	0.40	NR	0.40	NR	NR	0.40	0.40	0.40	0.40	0.40	0.40	0.40	0.40	0.40	NR
SPACE-COOLING SYSTEM																
Refrigerant charge and airflow testing of TXV	NR	REQ*	NR	NR	NR	NR	NR	REQ*	REQ*	REQ*	REQ*	REQ*	REQ*	REQ*	REQ*	NR
SPACE CONDITIONING DUCTS																
Duct sealing	REQ*	REQ*	REQ*	REQ*	REQ*	REQ*	REQ*	REQ*	REQ*	REQ*	REQ*	REQ*	REQ*	REQ*	REQ*	REQ*

In lieu of duct and refrigerant charge testing, use these standards—glazing and space conditioning.

	1	2	3	4	5	6	7	8	9	10	11	12	13	14	15	16
GLAZING																
Maximum U-factor	0.55	0.40	0.55	0.40	0.55	0.55	0.40	0.40	0.40	0.40	0.40	0.40	0.40	0.40	0.40	0.55
SHGC	NR	0.35	NR	0.35	NR	NR	0.35	0.35	0.35	0.35	0.35	0.35	0.35	0.30	0.30	NR
SPACE CONDITIONING																
SEER	Min	Min	Min	Min	Min	Min	Min	Min	11.0	11.0	11.0	11.0	12.0	12.0	13.0	Min
AFUE or	90	Min	Min	Min	Min	Min	Min	Min	Min	Min	Min	Min	Min	Min	Min	90
HSPF	7.6	Min	Min	Min	Min	Min	Min	Min	Min	Min	Min	Min	Min	Min	Min	7.6

by simulation or calculation; and/or **certification** by licensed evaluators. Nearly all current codes take a "building envelope, infiltration, HVAC" approach, but California's Title 24 is one of the few codes to include appliance and lighting requirements (see Section 8.3.3.3), thereby moving toward a Whole Building approach. For its 2008 Title 24 revisions, the California Energy Commission is considering additional lighting efficiency measures, cool roof requirements, demand response programmable thermostats, and a PV compliance option, among others.

California's 2005 Title 24 is illustrated in Sidebar 8.2. It includes the mandatory measures and prescriptive requirements for residential buildings with natural gas heat or heat pumps. Certified diagnostics are required for duct compliance and refrigerants or, in lieu of such certification, stricter glazing and space conditioning efficiency requirements apply. California's and the IECC-based codes offer a performance approach that allows builders to customize designs with different features so long as they can demonstrate by approved calculation methods that energy use is equivalent to the prescriptive requirements.

Figure 8.3 illustrates the changing nature of code requirements and efficiency. It shows the heating and cooling energy loads under 1983 MEC and the 2003 IECC for three cities of varying winter and summer design temperatures: Minneapolis (–16°F winter, 89°F summer), Kansas City (–2°F, 94°F), and San Antonio (24°F, 97°F). The figure also shows the relative impact of windows and other building envelope components on energy loads. Both winter

 **figure 8.3   Building Codes Improve over Time**

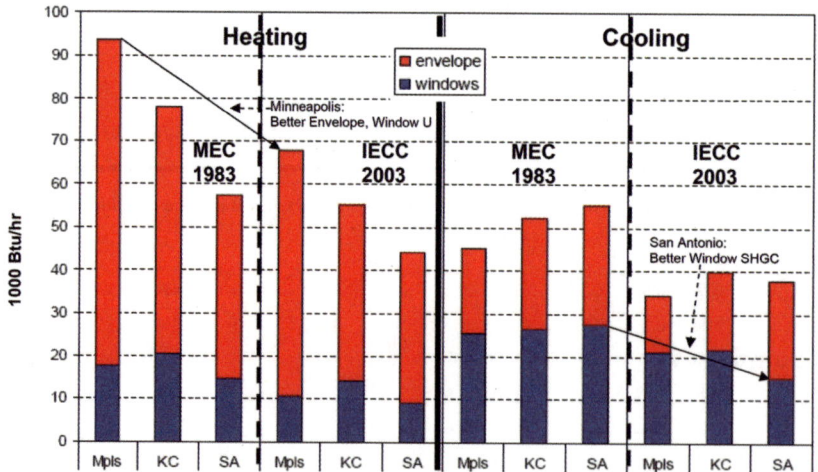

Heating and cooling energy loads for design conditions in Minneapolis (Mpls), Kansas City (KC), and San Antonio (SA) under 1983 MEC and 2003 IECC code requirements. IECC window SHGC requirements apply only to southern cities such as San Antonio; as a result, San Antonio shows greater reduction of cooling load from the 1983 code to the 2003 code.

SOURCE: adapted from Larsen, 2003. Used with permission.

and summer loads under the codes have improved significantly in twenty years as a result of improved envelope insulation and window U-value requirements. IECC has window SHGC (solar heat gain coefficient) requirements only for southern cities including San Antonio. This explains the additional improvement in San Antonio's cooling energy load from 1983 to 2003 relative to the other two cities.

8.2.3 Assessing Code Compliance: RES*check*, COM*check*

DOE's building energy codes program recognizes that codes can maximize efficiency only when they are fully embraced by users and supported through education, implementation, and enforcement. Therefore the program provides some innovative online methods to assist users in assessing compliance. RES*check* and COM*check* are online and downloadable interactive software methods that evaluate building design features against applicable residential building codes and commercial building codes, respectively. These software packages require the following:

- General understanding of Windows-based computer software
- Basic information about the home to be constructed
- House plans including areas of exterior walls, glazing, roof/ceiling, basement walls, and so on
- Insulation R-values, glazing and door U-values, and so on
- Heating and cooling system efficiencies

Figure 8.4 illustrates the RES*check* software windows, data entry, and results. To learn more about the software and to try it out, see http://www.energycodes.gov/.

Building energy codes are critical to improving new building efficiency on a large scale. Codes continue to improve as model codes such as IECC and ASHRAE standards incorporate new technologies and economics, but codes still focus on the building envelope and HVAC systems and not Whole Building Energy including lighting and appliances, the subject of the next section.

8.3 Whole Building Energy: Electrical Appliances and Lighting in Buildings

Whole Building Energy looks beyond thermal energy requirements to consider electricity needs in building operation. Taken together, heating, cooling, ventilation, and water heating account for 70% of residential and 60% of commercial energy used in U.S. buildings. That means that 30% of residential and 40% of commercial buildings' end-use energy is for other purposes, including appliances, electronics, and lighting. Because most of these uses require electricity, their percent of "primary" energy is much higher: 45% of residential and 52%

**figure
8.4**   **U.S. DOE's RES*check***

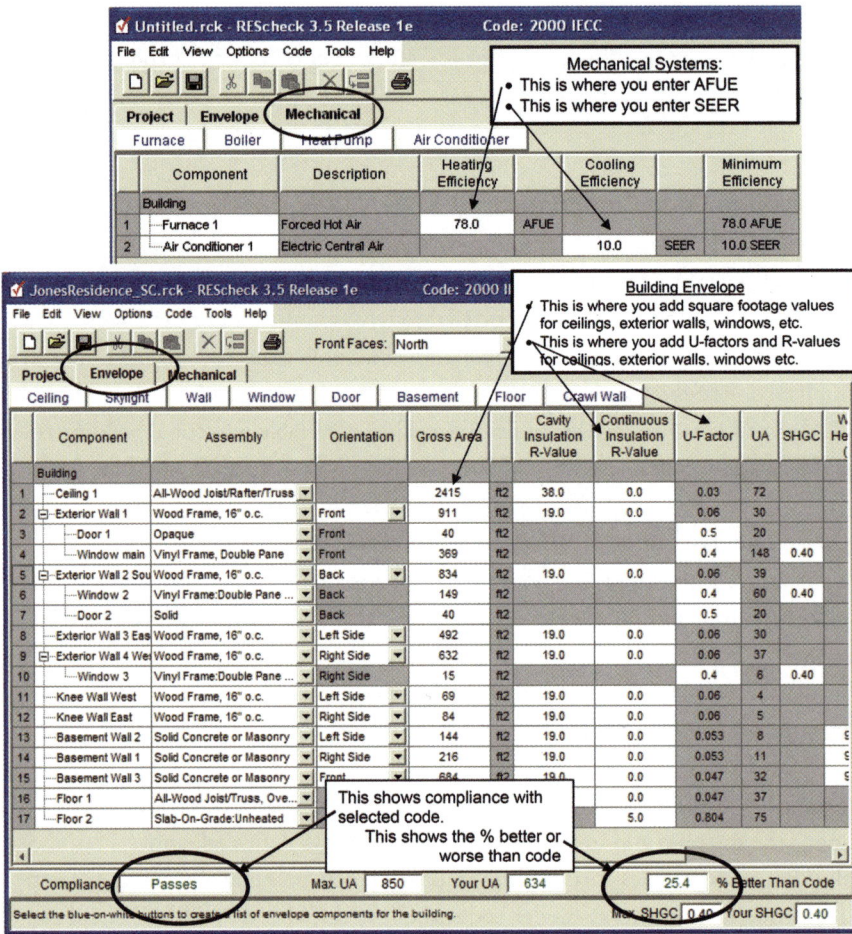

A code-compliance assessment tool that evaluates building design features against code requirements. Users can select from a variety of codes including versions of MEC and IECC and several state codes. This figure gives sample software windows showing data entry for building envelope components and mechanical HVAC systems information, as well as results including compliance and percentage better than code.

SOURCE: RES*check* http://www.energycodes.gov/rescheck/

of commercial primary energy is for purposes other than space conditioning. That is why the Whole Building Energy approach is so important. Efforts to improve building energy efficiency must consider opportunities for lighting, appliances, and equipment. The importance of these non–space conditioning uses is shown in Table 8.2. Refrigerators and lighting are the biggest nonthermal energy uses in residential buildings, whereas lighting and office

table 8.2 Percentage Energy Use in Residential and Commercial Buildings

End Use	Residential		Commercial	
	End Use Energy	Primary Energy	End Use Energy	Primary Energy
Water heating	17.0%	13.6%	6.0%	3.4%
Lighting	3.5%	5.9%	13.5%	18.7%
Refrigeration	5.4%	9.0%	5.0%	6.9%
Office equipment	n/a	n/a	10.5%	14.5%
Other non–space conditioning	21.0%	30.2%	11.6%	11.5%
Total non–space conditioning	46.9%	58.7%	46.6%	55.0%
Space conditioning	53.1%	41.3%	53.4%	45.0%

equipment are the biggest uses in commercial buildings. Table 8.3 details the use of electricity in residential buildings in the United States. Of the more than 1.1 trillion kWh consumed in all U.S. households in 2001, refrigerators (14%), lighting (9%), and other appliances (35%) are major users.

In this section, we look more closely at the progress of technology and standards in electrical appliances and lighting efficiency. First, we need to review some basics of determining building electrical use, so we can assess the effect of these improved technologies.

table 8.3 U.S. Electricity End Use in Residential Sector, 2001

End Use	kWh/yr per Household having the Appliance	Billion kWh	% Household Electricity
Air-conditioning	2800	183	16%
Refrigerators	1462	156	14%
Space heating	3524	116	10%
Water heating	2552	104	9%
Lighting	940	101	9%
Electric cooking	1185	72	6%
Clothes drying	1079	66	6%
Freezer	1150	39	3%
Furnace fan	500	38	3%
Television	137	33	3%
Dishwasher	512	29	2%
Other	—	203	18%
Total	10,656	1140	100%

8.3.1 Determining Building Electricity Needs

In Chapter 4 we introduced some basics of electricity, and we will describe home electrical systems in Chapter 10. For our purposes here, it is important to note that electrical power (or energy per unit time, measured in watts) is delivered by utility distribution lines to the electrical meter on our house. The meter measures how much power and time we draw grid power and shows the cumulative electrical energy used in watt-hours (or power times time). The utility reads the meter monthly to determine our electrical bill (or kilowatt-hours [kWh] times the billing rate [$/kWh]). We can use our utility bills to monitor our own electricity use and our electric meter to measure energy use and power demand (see Sections 5.3.3 and 10.5.1).

We use the electricity by connecting load devices (lights, appliances, etc.) to our internal wiring systems of outlets and connections. These lights and appliances have a power rating (in watts), and their energy use is simply their rating times the number of hours they are on.

For example, if we have five 60-watt lights that are on for 5 hours/day, what is their monthly electrical energy use?

$$\text{5 lights} \times \text{60 W/light} \times \text{5 hr/day} \times \text{30 day/mo} = 45,000 \text{ Wh} = 45 \text{ kWh}$$

What is the maximum electrical power demand for these lights when they are on at the same time?

$$\text{5 lights} \times \text{60 W/light} = 300 \text{ W}$$

Most residential electricity bills are based only on energy (kWh) used, but large users such as commercial buildings are billed on both energy used and peak power demand.

So, if we know the power rating of our lights and appliances and know how long they are on, we can determine our electrical energy consumption. This is easy for some uses, such as lights, but is difficult for appliances that are controlled by a thermostat, such as refrigerators, air conditioners, ovens, and furnace fans. Chapter 5 described submetering devices that can measure the time such an appliance is on (its run-time) or actual energy used (see Section 5.3.3). Alternatively, but more crudely, we can get typical energy use estimates from EnergyGuide labels for our appliances (see Section 8.3.2.2).

Table 8.4 gives a simple spreadsheet for estimating electrical energy use and cost in a home. Power rating and average hours of use are used to calculate kWh for many appliances, and EnergyGuide estimates are used for the refrigerator and the oven. As with our other spreadsheets, we can simply enter new values (e.g., compact fluorescent instead of incandescent lights, high-efficiency refrigerator, etc.) and immediately see the impact on electricity use and cost.

8.3.2 Appliance Efficiency

Appliances and electronic equipment use more than one-fourth of end-use and one-third of primary energy in residential and commercial buildings. Appliance efficiency has increased

table 8.4	Spreadsheet for Calculating Home Electricity Use				
Electricity Rate		0.08	$kWh		
Appliance	Wattage	Hours/da	kWh/mo	kWh/yr	% of use
Lighting 1	60	15	27	324	4
Lighting 2	100	8	24	288	3
Lighting 3	75	5	11	135	2
Refrigerator				900	11
Electric range	5000	0.5	75	900	11
Electric oven	6800	0.5	102	1224	15
Microwave	1580	0.2	9	114	1
Toaster oven	1200	0 1	4	43	1
Coffee maker	1025	0.5	15	185	2
Clothes washer	1000	0.5	15	180	2
Clothes dryer	5000	0.5	75	900	11
Dishwasher	1200	1	36	432	5
TV 1	270	5	41	486	6
TV 2	200	2	12	144	2
Computer	65	4	8	94	1
Stereo	50	6	9	108	1
Clock radio	20	24	14	173	2
Air conditioner 1	500	4	60	720	9
Air conditioner 2	500	4	60	720	9
Furnace fan	250	2	16	180	2
Total				8249	100
Annual $ Cost				646	

significantly in the past twenty-five years as a result of improved electronic technologies, federal and state efficiency standards, and consumer choice assisted by better information, such as U.S. EPA's ENERGY STAR program. We can use this information to evaluate options for improved efficiency.

8.3.2.1 Appliance Efficiency Standards

California enacted the first appliance efficiency standards in 1974 and the federal government followed suit in the Appliance Energy Conservation Act of 1986. Fourteen appliances were addressed by this act, and the list has grown to thirty by amendments in the energy acts of 1992, 2005, and 2007. Overall, the appliance standards are expected to reduce total 2030 U.S. energy by 7 quads (6%), electricity by 577 billion kWh, peak demand by 177 GW (or 600, 300-MW power plants), and CO_2 emissions by 45 MMt, with a net economic benefit through 2030 of more than $250 billion (Nadel et al., 2006; ASAP/ACEEE, 2007).

Refrigerator efficiency dramatically improved due to California's appliance efficiency standard of the mid-1970s. Figure 8.5 shows that the average energy use of new refrigerators was 520 kWh/yr in 2002 compared to 1726 kWh/yr in 1972. The "best available" new full-size refrigerator-freezer efficiency in 2002 was 428 kWh/yr. Average energy for refrigerator-freezers in use was 1462 kWh/yr in 1990, 1319 kWh/yr in 1997, and 1220 kWh/yr in 2001 (U.S. DOE, 2005).

Table 8.5 gives efficiency standards for several appliances. Refrigerator and freezer efficiency standards are based on annual electricity consumption (kWh/yr) and vary for size and model features. Other appliance standards are based on an efficiency factor (EF) or energy efficiency ratio (EER in Btu per Wh). The table relates these standards to typical annual electricity consumption. The revision of the federal standards is a continual process. For example, stricter new standards for clothes washers took effect in January 2007, and the 2005 and 2007 energy acts call on the U.S. DOE to update standards for 24 appliances.

The tables show a marked improvement in minimum efficiency of new appliances, and can be used to compare expected energy and cost reduction from a new appliance versus an old one. These appliance standards have turned out to be one of the most effective federal policies for saving energy and related economic and environmental costs. Ten states have ap-

figure 8.5   **Trends in Refrigerator Energy Consumption, Size, and Cost, 1947–2002**

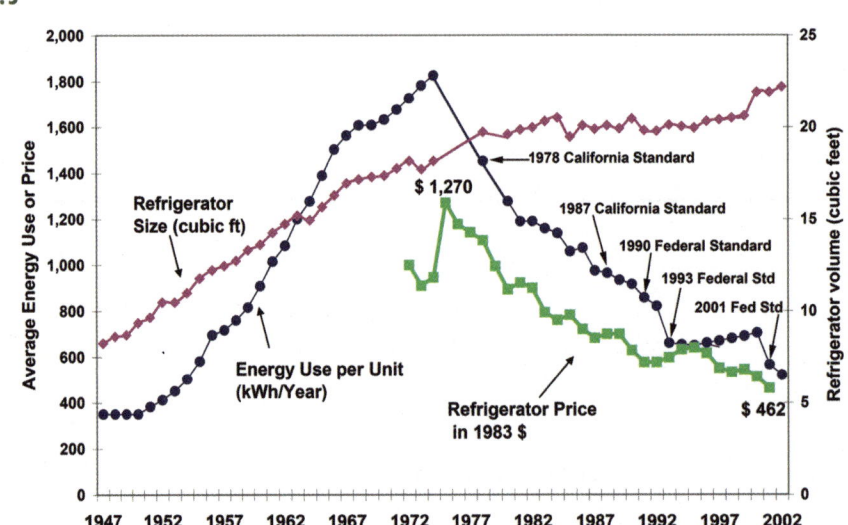

Arthur Rosenfeld's dramatic graph (after Goldstein) showing significant reduction in energy consumption following California and federal appliance standards, despite larger average size and reduced cost.

Source; Rosenfeld, 2006

table 8.5 Appliance Efficiency Standards and Related Annual Energy Consumption

A. Refrigerator–Freezers (Auto Defrost)

	Adjusted Volume (ft³)	Rated Maximum Electricity Use (kWh)		
		1990	1993	2001
Top freezer without through-the-door ice service	20.6	955	685	478
Side freezer without through-the-door ice service	25.1	1183	797	631
Bottom freezer without through-the-door ice service	25.1	1183	781	574
Side freezer with through-the-door ice service	28.5	1428	992	694

B. Freezers

	Adjusted Volume (ft³)	Rated Maximum Electricity Use (kWh)		
		1990	1993	2001
Upright freezer with manual defrost	25.7	702	529	452
Upright freezer with automatic defrost	30.0	1103	838	699
Chest freezers	24.8	590	433	389

C. Room Air Conditioners

	Minimum EER	Typical Maximum Electricity Use (kWh)*
Less than 6000 Btu/h	9.7	464
6000 to 7999 Btu/h	9.7	541
8000 to 13,999 Btu/h	9.8	842
14,000 to 19,999 Btu/h	9.7	1314
20,000 Btu/h or more	8.5	1765

*Based on 750 hours of operation

D. Clothes Dryers

	Minimum EF (lb/kWh)	Typical Maximum Energy Use
Electric, standard	3.01	835 kWh
Gas	2.67	32 therms

E. Clothes Washers

	Minimum EF (ft³/kWh per cycle)	Minimum Modified EF (ft³/kWh per cycle)		Typical Maximum Electricity Use (kWh)
	1994	2004	2007	2007
Top-loading, standard	1.18	1.04	1.26	1265*
Front-loading	n/a	1.04	1.26	731*

* Includes electricity for water heating and clothes drying   *(continued)*

F. Dishwashers

	Minimum EF (cycles/kWh)	Typical Maximum Electricity Use (kWh)
Standard dishwasher	0.46	498

G. Water Heaters

	Minimum EF* Btu (hot water)/Btu (fuel)			Typical Maximum Energy Use per Year		
	1990	1991	2004	1990	1991	2004
Gas-fired	0.54	0.54	0.59	208 therms	208 therms	191 therms
Oil-fired	0.51	0.51	0.51	155 gallons	155 gallons	155 gallons
Electric resistance	0.90	0.88	0.92	3456 kWh	3534 kWh	3380 kWh

* Minimum EF is based on size; list assumes 40-gallon tank

Source: U.S. DOE, 2005

pliance standards that exceed the federal minimums. We discuss efficiency standards more in Chapters 17 and 18.

8.3.2.2 ENERGY STAR Appliances and Labeling

To encourage performance that surpasses the minimum standards, EPA started the ENERGY STAR program in 1992 to certify and label products that exceed the federal standards by 10%–15% or more. This rating and information program helped enhance the market for the most efficient appliances and equipment.

The relationship between federal efficiency standards and ENERGY STAR certification is illustrated in Table 8.6, which shows a few of the more than 1200 brands and models of energy-efficient refrigerators that meet the ENERGY STAR criteria (≥15% better than the federal standard). The standard is based on size, configuration, and features such as automatic defrost and ice maker.

Table 8.7 gives the sales of ENERGY STAR labeled appliances from 1997 to 2006. These appliances are now about one-third of all room air conditioners, refrigerators, and washing machines sold, and more than 90% of dishwashers. It looks like it is time to increase the federal standard for dishwashers. The biggest energy use of dishwashers and clothes washers is the energy to heat hot water, and the most efficient models are designed to use less water. The most efficient clothes washers are front-load models that use much less hot water.

The federal appliance standards and ENERGY STAR program provide better information on appliance performance and energy and cost savings than ever before. This information is easily available from government Web sites, such as the following:

- http://www.eere.energy.gov/buildings/appliance_standards/
- http://www.energystar.gov/

table 8.6 Sample of Energy Star Qualified Refrigerator–Freezers

Brand	Model	Freezer	Ice Service through Door?	Volume (ft³)	Adjusted Volume (ft³)	E.S. (kWh/year)	Federal Standard (kWh/year)	% Better
Kenmore	7519#	Bottom	No	20.7	24.5	465	572	19%
Kenmore	7555#	Bottom	No	25	29.60	499	595	16%
General Electric	GSH25KGR	Side-by-side	Yes	25	30.98	610	719	15%
General Electric Profile	PTS25LBS	Top	No	24.6	29.10	475	561	15%
KitchenAid	K55548QM	Side-by-side	Yes	29.8	36.7	659	777	15%

For complete listing of ENERGY STAR refrigerators, see: http://www.energystar.gov/index.cfm?c=refrig.pr_refrigerators

For top-rated refrigerators, see ACEEE: http://www.aceee.org/consumerguide/topfridge.htm

table 8.7 Sales (in Thousands) of ENERGY STAR Labeled Appliances and Percentage of Total Sales, 1997–2006

Year	Room Air Conditioners		Refrigerators		Clothes Washers		Dishwashers	
	ENERGY STAR	% of Total	ENERGY STAR	% of Total	ENERGY STAR	% of Total	ENERGY STAR	% of Total
1997	474	12%	2008	25%	226	4%	265	6%
1998	589	13%	1705	19%	392	6%	955	19%
1999	835	13%	2218	24%	624	9%	664	12%
2000	1230	19%	2489	27%	697	9%	595	11%
2001	642	12%	1610	17%	758	10%	1119	20%
2002	2195	36%	1956	20%	1262	16%	2262	36%
2003	2369	29%	2570	26%	1879	23%	1290	20%
2004	2632	35%	3628	33%	2405	27%	5437	78%
2005	NA	52%	NA	33%	NA	36%	NA	82%
2006	NA	36%	NA	31%	NA	38%	NA	92%

SOURCE: U.S. EPA, DOE, 2007, http://www.energystar.gov/index.cfm?c=manuf_res.pt_appliances

- http://www.energy.ca.gov/efficiency/appliances/ (California's appliance program)
- http://www.aceee.org/consumerguide/mostenef.htm (American Council for an Energy-Efficient Economy's consumer guide)

These and other sources give good information for the most efficient appliances, but the federal **EnergyGuide** label, which you have probably seen in appliance stores, gives efficiency information for *every* appliance. Figure 8.6 shows labels for two 25 ft³ refrigerators in the same class of features. The one on the left is an ENERGY STAR model. The label shows it will consume 611 kWh/yr and cost $51/yr for electricity at 8.29¢/kWh. The label on the right is for a less efficient model that will consume 715 kWh/yr and cost $59/yr to operate.

figure 8.6   EnergyGuide Label

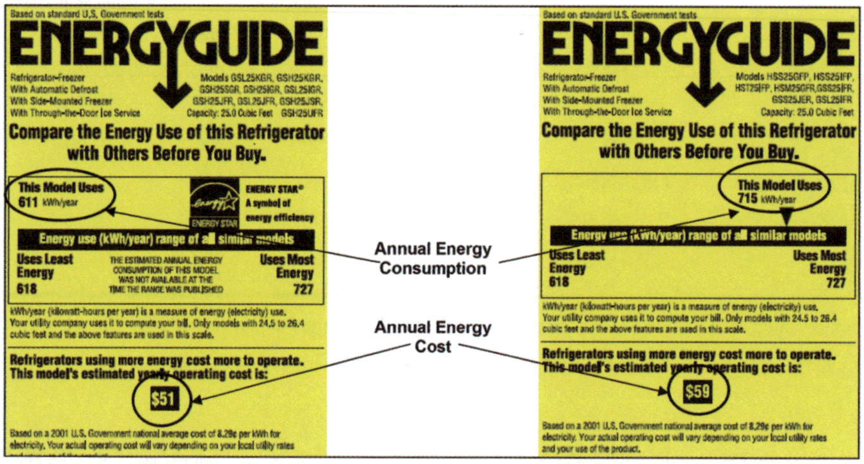

Federal mandatory labels for all appliances give expected annual energy use and cost and the range of energy use for comparable models on the market. Label on the left is for an ENERGY STAR model.

Making choices about appliances. With improvement in appliance efficiency, there are opportunities for making good economic and environmental choices when selecting new appliances. Major appliances last about 10 to 15 years, and estimates of average stock age in 2001 were 8 to 12 years for refrigerators, freezers, and water heaters. An estimated 40 million major appliances were replaced in the United States in 2005 (U.S. DOE, 2005).

What are the reasons for considering a new high-efficiency appliance? We can quantify the economic and environmental costs and benefits using our assessment methods from Chapter 5 by calculating the energy savings of the high-efficiency appliance versus the option of keeping an old low-efficiency appliance or the option of buying a standard efficiency. Once we have energy savings, we can calculate dollar savings and then cost-effectiveness measures, such as simple payback period (SPP), if we know the initial cost of our options. The best way to explain this assessment is with the example in Solution Box 8.1.

8.3.3 Energy for Lighting

Lighting contributes only 3.5% of residential end-use energy, but a significant 13.5% of commercial end-use and nearly 19% of commercial primary energy. It is estimated that lighting requires about 20% of all electricity in the United States. Lighting technology has greatly improved in the past few decades with the advance of compact fluorescent, low-voltage halogen, and light-emitting diode (LED) bulbs. But lighting is more than bulbs, and efficient lighting

---

SOLUTION BOX 8.1

Should I Replace My Old Refrigerator with a New Energy-Efficient Model?

I have an older side-by-side refrigerator-freezer and I'm thinking of replacing it. It has a through-the-door ice dispenser that malfunctions often and I've never liked it anyway. I'm thinking about a new ENERGY STAR rated model. My father-in-law says forget replacing it (and especially the ENERGY STAR business) and just wait until it dies. Fortunately my wife wants to improve our energy and environmental efficiency as well as get desired features in a refrigerator. I would just go ahead and buy an ENERGY STAR model, but I want to show my father-in-law that it's the right thing to do.

To do this I need to do some energy and economic analysis. To start, I need the energy use of my current unit, and the cost and energy use of prospective replacements.

a. **Energy use and cost of existing unit:** I can't just find kWh usage from the power rating (kW) of the unit, because it is thermostatically controlled and I don't know how many hours it runs per day. I can use a Kill-A-Watt submeter (see Sidebar 5.1) to measure power draw, energy use, and run-time by simply plugging it into the wall outlet and plugging the refrigerator into it. I do this for exactly a week (168 hours) of normal use, and the Kill-A-Watt meter reads 30 kWh, 500 W draw, and 60 hours of run-time. I can calculate the annual energy use and cost, assuming national average electricity rate of 9.3¢/kWh:

Energy/year = Energy/wk × 52 wk/yr = 30 kWh/wk × 52 wk/yr = 1560 kWh/yr
Cost/year = kWh/yr × cost/kWh = 1560 kWh/yr × 9.3¢/kWh = $145

b. **What are my choices?** Well, I decide to forgo the through-the-door ice dispenser option and my wife wants a bottom freezer. I can view the ENERGY STAR and ACEEE most efficient models on their Web sites, and I can research the prices of these and other less-efficient models on dealer Web sites. Lowe's posts online EnergyGuide labels for its appliances. In the table below, I compare my options using the assessment of my fridge and product information, and our simple payback period

(continued)

---

systems need to consider lighting design, daylighting, and lighting controls, all of which can enhance lighting effectiveness while saving energy. Commercial daylighting was discussed in Chapter 7. This section introduces basic concepts of lighting energy and discusses opportunities for additional efficiency improvements.

SOLUTION BOX 8.1 *(continued)*

(SPP) formula from Chapter 5. I compare my existing unit to two new ones with about the same size and features, one rated ENERGY STAR and the other not, and to two new ones with bottom freezer and no through-the-door dispenser, one rated ENERGY STAR and one not. The ENERGY STAR refrigerators cost about $50 more but use 13%–15% less electricity.

Refrigerator Choice	1. Cost to Replace	2. kWh/yr to Operate	3. Energy Cost/yr	4. Savings: Replacement $/yr	5. Savings: ENERGY STAR vs. not ENERGY STAR	6. SPP: Replacement	7. SPP: ENERGY STAR vs. not ENERGY STAR	8. 13-year CO_2 Emissions
My existing unit	$0	1560	$145	n/a	n/a	n/a	n/a	46,000 lb
New, not ENERGY STAR	$947	715	$67	$78	n/a	12 yr	n/a	21,000 lb
New, ENERGY STAR	$997	611	$57	$88	$10	11 yr	5 yr	18,000 lb
Bottom freezer, not ENERGY STAR	$847	559	$52	$93	n/a	9 yr	n/a	16,000 lb
Bottom freezer, ENERGY STAR	$897	488	$45	$100	$7	9 yr	7 yr	14,000 lb

1. Retail price of new refrigerator-freezer
2. Annual energy use for existing from energy assessment, for new from EnergyGuide labels
3. Annual cost to operate, $/yr = kWh/yr (col. 2) × 9.3¢ /kWh
4. Annual $ savings by replacing = $145 – $/yr (col. 3)
5. Annual savings of ENERGY STAR compared to non–ENERGY STAR =
   cost to operate non-ENERGY STAR – cost to operate ES
6. Simple Payback Period for replacement = cost to replace (col. 1)/$ savings by replacing (col. 4)
7. Simple Payback Period ENERGY STAR vs. not ENERGY STAR =
   (cost difference ENERGY STAR – non–ENERGY STAR)/$ savings ENERGY STAR (col. 5)
8. CO_2 emissions for 13-year life = kWh/yr (col. 2) × 2.27 lb CO_2/kWh

So, if a new refrigerator will last for the average life of 13 years, it is easy to justify not only replacing it now, but also investing in an ENERGY STAR model. For a new model, the added efficiency would recoup the added cost in 5 to 7 years. Further, reducing electricity consumption will also reduce my household's emissions of carbon dioxide for the lifetime of the refrigerator by 25,000–32,000 lb compared to my existing refrigerator, and 2000–3000 lb for ENERGY STAR models compared to less-efficient new models.

8.3.3.1 Lighting Efficacy and Quality

Four important characteristics of artificial lighting are energy efficiency, light quality, lamp lifetime, and cost. Our prime interest is efficiency, but effective lighting systems must consider all four factors. The inefficient standard incandescent lightbulb is more of a heater than a light source, but it still provides about 85% of household illumination. Better alternatives can provide more light with less energy. The efficiency of lighting is measured by "efficacy" or the amount of light produced per unit of electric power:

Lighting efficacy = lumens/watt (lm/w)
where lumen = measure of luminous flux or brightness
1 lumen/ft^2 = 1 foot-candle

But the brightness of light is not all that matters. Because light is gauged in reference to the sensitivity of the human eye on the one hand and the natural light spectrum on the other, the spectral characteristics or quality of artificial light matters. The quality of light is measured by the color rendering index (CRI), a 0–100 point scale, where CRI = 100 is natural sunlight. There are four categories of light quality:

Color Rendering Index (CRI), 0–100

Low CRI   = 10–69

Medium CRI = 70–79

High CRI   = 80–89

Very High CRI = 90–100

There are several artificial lighting bulb technologies that vary in efficacy, CRI, cost, and lifetime. Table 8.8 gives these parameters for different technologies, including incandescent, fluorescent, high intensity discharge (HID), and emerging light emitting diode (LED) technology (see Figure 8.7).

- Standard and halogen incandescent bulbs have high CRI and low cost, but they have relatively poor efficacy and lifetime.
- Fluorescents have much higher efficacy and lifetime, but have slightly lower CRI.
- HID sodium vapor, metal halide, and mercury vapor lamps have long lifetime, high efficacy at high wattage, and low cost, so they are good high-wattage options for large commercial and industrial building spaces and outdoor street, parking lot, or stadium lighting, where their relatively lower CRI is not a problem.
- Very high efficacy LED technology is advancing and many think it will compete with current technologies over the next decade (see Section 8.2.5).

8.3.3.2 Lighting Applications and Use

How are these lighting technologies used in the United States? Table 8.9 gives electricity consumption for different bulb types in different consuming sectors in 2001. Inefficient

table 8.8   Efficacy, Lifetime, CRI, and Cost for Different Lamp Technologies

	Efficacy (lumens/watt)	Typical Rated Lifetime (1000 hours)	CRI	Lamp Cost ($ per kilolumen (klm))
Incandescent				
Standard	8–15	0.75–2	95+	0.56
Torchiere halogen	2–14	2	95+	3.00
Tungsten halogen	18–33	2–4	95+	9.19–13.11
Fluorescent				
T-8 tube	83	14–18	68–80	0.74–1.45
T-12 tube	68	7.5–24	49–92	0.50–2.89
Compact	55–65	10–20	82–86	5.33–10.59
High-intensity discharge				
Mercury vapor	25–50	24+	22–52	1.53
Metal halide	50–115	6–20	65–92	1.22
High-pressure sodium	80–140	16–24	21–80	0.71
Low-pressure sodium	110–150	12–18	0–18	1.20
Emerging Technology: Solid State Lighting				
Light-emitting diodes (LED)	20–30	50	75	160

SOURCE: U.S. DOE, 2005; ADL, Inc., 2001

incandescent bulbs still dominate the market with 90% of residential lighting and 42% of total lighting electricity use. T-8 and T-12 fluorescents in the commercial and industrial sectors account for 38% of the total. T-8 lamps are 22% more efficient than T-12 lamps with about the same CRI, lifetime, and cost, so many lighting efficiency programs swap out T-8s for T-12s.

Efficient compact fluorescent lamps (CFL) accounted for only 2%. CFLs have become a symbol of improving energy efficiency, and many community and university campus groups have engaged in "make the switch" and other campaigns to replace incandescents with CFLs. There are moves in many states and in proposed federal legislation to mandate use of CFLs.

Solution Box 8.2 calculates the savings if half of the standard incandescent lightbulbs in the United States were replaced with compact fluorescent lamps.

But what about light quality? Do CFLs provide the right CRI for residential and commercial use? Table 8.10 gives the results of a national lighting inventory indicating the CRI demand in different sectors. CFLs are high CRI (82–86), so they could meet 90% of lighting quality demand.

figure 8.7   **Lamp Types**

(a) Standard incandescent

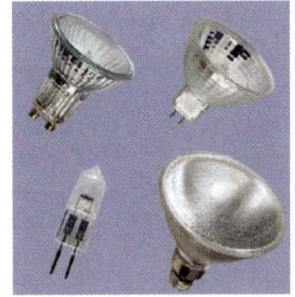

(b) Halogen fluorescent

(c) Compact fluorescent

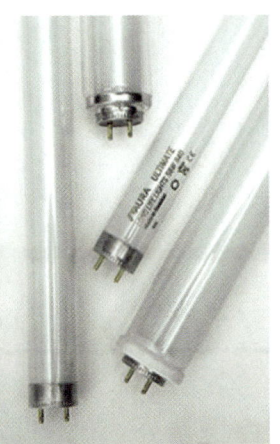

(d) T-8 and T-12 fluorescent

(e) Sodium and mercury vapor

(f) LED cluster

(g) Electroluminescent sheet

8.3.3.3 Improving Lighting Efficiency

Efficient lamp use and replacement. Improving efficiency of lighting requires replacing low-efficiency options with higher-efficiency technologies, but still maintaining light quality. The halogen and tungsten-halogen lamp is a type of incandescent that has a gas filling and heat-reflective coating that make the filament burn brighter and produce more light with

table 8.9   Lighting Electricity Consumption by Sector and Technology, 2001 (10^9 kWh/yr)

	Residential	Commercial	Industrial	Other	Total	% Total
Incandescent						
Standard	176	103	2	5	287	38%
Halogen	6	21	0	1	28	4%
Fluorescent						
T-8 tube	n/a	50	23	0	71	9%
T-12 tube	n/a	157	49	0	206	27%
Compact	1	13	1	n/a	14	2%
Miscellaneous	18	0	0	1	19	3%
High-intensity discharge						
Mercury vapor	1	7	3	12	22	3%
Metal halide	n/a	34	25	4	62	8%
High-pressure sodium	0	6	5	30	41	5%
Low-pressure sodium	na	0	0	3	3	0%
Total	202	391	108	56	756	100%
% Total	27%	52%	14%	7%	100%	

Source: U.S. DOE, 2005

table 8.10   Lumen–Hour Output by Sector and Color Rendering Index, 2000 (Tera–lm–hr/yr)

	Residential	Commercial	Industrial	Other	Total	% Total
Low CRI (10–69)	9	3097	2016	2119	7241	25%
Medium CRI (70–79)	1095	13,508	3833	59	18,495	63%
High CRI (80–89)	51	421	70	66	608	2%
Very high CRI (90–100)	1875	913	27	81	2895	10%
Total:	3030	17,939	5946	2325	29,240	100%

Source: ADL, Inc., 2001

less energy. The very high-quality (CRI 95+) halogen light is best for task lighting. However, they are a poor option in popular halogen torchiere lamps that provide indirect light reflected from ceilings.

We just saw the incredible savings of replacing incandescent bulbs with compact fluorescent lamps in medium- to high-quality lighting applications not requiring dimming controls. Not only do CFLs have much higher efficacy rating than incandescent bulbs, they last considerably longer. In fact the longer lifetime makes CFLs far more cost-effective than incandescents in commercial applications because of reduced labor cost of bulb replacement. Recall from Sidebar 5.11 in Chapter 5 we compared the life-cycle cost of incandescent and CFLs and showed that even though we assumed the CFL cost of $2.50 versus $0.30 for the incandescent, over the life of the bulbs the CFL cost $35 less than the incandescent considering energy and replacement costs.

Lighting efficacy is not yet included in many building codes, but when California's Energy Commission (CEC) revised its state Title 24 Building Code in 2005, it added lighting standards. For residential lighting the new standard defines "high efficacy" lighting for different wattage, because higher wattage lamps generally have higher efficiency of light output. Incandescent bulbs (10 lm/W) including halogens (25 lm/W) do not qualify as high efficacy (see Table 8.9). The Title 24 Code requires, for example, that 50% of the lighting in new residential kitchens be high efficacy.

California 2005 Title 24 Definition of High Efficacy Lighting

< 15 watts: ≥ 40 lm/watt

$15–40$ watts: ≥ 50 lm/watt

> 40 watts: ≥ 60 lm/watt

SOLUTION BOX 8.2

Savings from Replacing Half of Standard Incandescent Bulbs with Compact Fluorescent Lamps

How much electrical energy would be saved if 50% of current standard incandescent lamps in the residential and commercial sectors in 2001 were replaced by CFLs?

Solution:

To estimate the savings, we use the data in Tables 8.8 and 8.9. The efficacy of a CFL is about 60 lumens/Watt (lm/W) and that of an incandescent is about 10 lm/W, so the CFL power demand and energy use is about one-sixth of an incandescent. Table 8.9 shows that residential and commercial standard incandescent lighting consumed 279×10^9 kWh/yr in 2001 (176 + 103). If half of those bulbs were replaced with lamps using one-sixth of the energy, the savings would be five-sixths or

$$\text{Savings} = 0.5 \times 279 \times 10^9 \, \text{kWh/yr} \times \frac{5}{6} = 116 \times 10^9 \, \text{kWh/yr}$$

This is 15% of total U.S. lighting energy and 3% of the total U.S. electricity consumption. At 8.3¢/kWh, this amounts to a savings of $10 billion per year. Wow!

Of course, CFLs are more expensive than incandescents, as much as 5–10 times the price, but they last 5-10 times as long so this initial cost is offset by their longer life, which also reduces relamping labor costs.

In 2007, California enacted Assembly Bill 1109 calling on the CEC to develop regulations to reduce average residential lighting energy by 50% by 2018. Other states have proposed banning incandescent lamps. The federal energy act passed at the end of 2007 also incorporates lighting standards. The act requires minimum efficacy for lumen ranges phased in between 2012 and 2014 and prohibits the sale of 100 W incandescent lamps by 2012 unless their efficacy is at least 60 lm/watt.

Lighting design and controls. Design of lighting systems also affects efficiency. It is important to match the fixture to the lighting function, such as ambient lighting, task lighting, and accent lighting. Ambient lighting can be reduced if task lighting is provided. Very high-quality task lighting can be provided effectively by halogen lights. Table 8.11 gives the design protocol for lighting for ENERGY STAR certified homes. ENERGY STAR qualified fixtures include a number of commercial products that use at most one-third of the energy of standard lighting and distribute light more effectively. The ENERGY STAR Web site gives many examples: http://www.energystar.gov/index.cfm?c=fixtures.pr_light_fixtures.

Finally, lighting controls can improve efficiency by matching the amount of light to the desired function, and turning off lights when not needed. Dimmer switches can produce a desired effect and reduce energy use of incandescent bulbs, but they cannot be used with most CFLs (although dimmable CFLs are now available). Motion-detection, timer, and photo switches can control lights automatically, turning them on only when needed. California's Title 24 Code requires motion-detection switches in new residential bathrooms, utility rooms, and garages.

8.3.3.4 Advanced Lighting Technologies: Solid State Lighting

Compact fluorescent lamps illustrate how new technology can create significant opportunities for energy efficiency in lighting. One emerging technology that holds great promise comes from solid state electronics. Light emitting diodes (LED) have been used in specialized applications during the past decade, especially those requiring long life and red, green, or blue color because the diodes emit principally those colors. For example, they are now common in traffic lights and "exit" signs.

table 8.11   Lighting Requirements for ENERGY STAR Certified Homes

Room Category	Specific Rooms in Category	Minimum Percentage of Required ENERGY STAR Qualified Fixtures per Room Category
High-use rooms	Kitchen, dining room, living room, family room, bathrooms, halls, stairways	50% of total number of fixtures
Medium/low-use rooms	Bedroom, den, office, basement, laundry room, garage, closets, all other	25% of total number of fixtures
Outdoor	Outdoor lighting affixed to home or free-standing except landscaping or solar lighting	50% of total number of fixtures (including all flood lighting)

However, with technical improvements that have resulted in reduced cost, higher efficacy, and especially higher "white light" CRI, many now look to LEDs as the next leap forward in lighting efficiency. Table 8.8 gave key parameters for LEDs now available. Although lifetime (50,000 hr), efficacy (20–30 lm/W), and CRI (75) are all good, cost ($160/klm) is prohibitive. In a study of the prospects for LED market penetration for U.S. DOE, Arthur D. Little, Inc., speculated that with both technical and price breakthroughs, LEDs could compete with incandescent and fluorescent technologies and result in significant energy savings by 2020.

Table 8.12 indicates potential for medium CRI (75) efficacy and price compared to conventional technology. Industry representatives report that 75 lm/W at $20/klm could be achieved by 2007 and 150 lm/W at $5/klm with 100,000-hr life could be achieved as soon as 2012 (Krames, 2003). ADL (2001) considers the current transition from T-12 to T-8 fluorescents as the biggest barrier to LED market penetration because T-8s are considered "high" CRI with a value of 80. ADL does not expect that LEDs will be able to compete in that category.

A further advance in LED comes from "organic" or thin-film LED. These electroluminescent sheets that are used in those cool 0.03 W flat nightlights, could provide walls or panels of light in large markets if they become cost effective at high efficacy and quality.

Electric lighting, appliances, and equipment are important primary energy users in buildings and the Whole Building Energy approach recognizes the huge potential for reducing energy and cost by improving efficiency. Although this potential is not yet addressed in most building codes, appliance efficiency standards and ratings have been very effective in both new and replacement appliance markets, saving consumers billions of dollars in energy costs. Codes and ratings are starting to address efficiency gains of new lighting technologies.

8.4 Whole Building Life-Cycle: Embodied Energy in Buildings

Building codes and appliance standards are beginning to address Whole Building Energy considerations, but have not yet incorporated embodied energy and life-cycle environmental impact factors that are part of the Whole Building Life-Cycle approach to building efficiency introduced in Table 8.1. Recall from Chapter 5 that life-cycle analysis considers cradle-to-grave energy, economic, and environmental costs and benefits to inform decisions. These considerations are becoming part of Green Building rating and certification systems, but before discussing them, it is useful to describe recent progress in life-cycle analysis of buildings.

The energy used to operate buildings amounts to more than 40% of U.S. annual energy consumption (see Figure 6.1). But this underestimates the energy cost of buildings because it does not include the energy required to make them; the embodied energy involved in building materials manufacture, transport to site, and construction; as well as the manufacture, transport, and installation of mechanical and electrical systems and interior furnishings and finishes. Embodied energy in buildings can be as much as 5 to 15 times a building's annual

table 8.12   **Scenarios for LED Light Source Attributes, 2020**

Technology	Scenario	Efficacy (lm/W)	Price ($/klm)
Medium CRI (75) LED	Base case	50	$8.30
	Technical breakthrough	120	$8.00
	Price breakthrough	120	$0.50
Conventional technology	Incandescent	14	$0.55
	Fluorescent	69	$0.70
	High-intensity discharge (HID)	99	$0.61

SOURCE: ADL, 2001

operating energy. This one-time embodied energy use can be a small amount (~5%–10%) of a building's life-cycle energy if the building lasts 100 years, but many buildings, especially commercial ones, do not last that long in the United States.

Analyzing embodied energy of buildings is important as we improve life-cycle analysis and consider the energy and environmental effects of the materials we use. This is critical to the Green Building movement. The analysis of embodied energy is not nearly as well developed as that for operating energy, but it is improving.

8.4.1 Life-Cycle and Embodied Energy in Buildings and Materials

Embodied energy is mostly a one-time initial cost, but there are also recurring embodied energy requirements for building maintenance and renovation. Figure 8.8 traces cumulative embodied and operating energy over a 100-year life cycle in gigajoules (GJ) for a hypothetical building having three different operating costs. The high-operating case is least efficient; the low-operating case is most efficient. The cumulative embodied energy starts at about 800 GJ and increases slightly over the 100 years due to recurring renovation and maintenance.

The cumulative operating energy increases dramatically for all three cases, but especially for the high-operating case. Note where the operating energy line crosses the embodied energy line: this is the number of years it takes the cumulative operating energy to reach the embodied energy. For these hypothetical cases, it takes about 10 years for the high-operating case and 20 years for the low-operating case, which is typical for actual buildings studied in London, Vancouver, Toronto, and Australia.

We discussed life-cycle analysis that is used to assess embodied energy in Chapter 5. Figure 5.1 showed the process of life-cycle analysis of wood products for house construction. Section 5.5.1 described the NREL Life-Cycle Inventory and Table 5.5 listed several processes and products, for which impact coefficients are available, including building and construction products such as varieties of plywood, softwood lumber, and other wood products.

There are some general rules of thumb to consider in reducing embodied energy in building:

figure 8.8 **Embodied Energy versus Operating Energy**

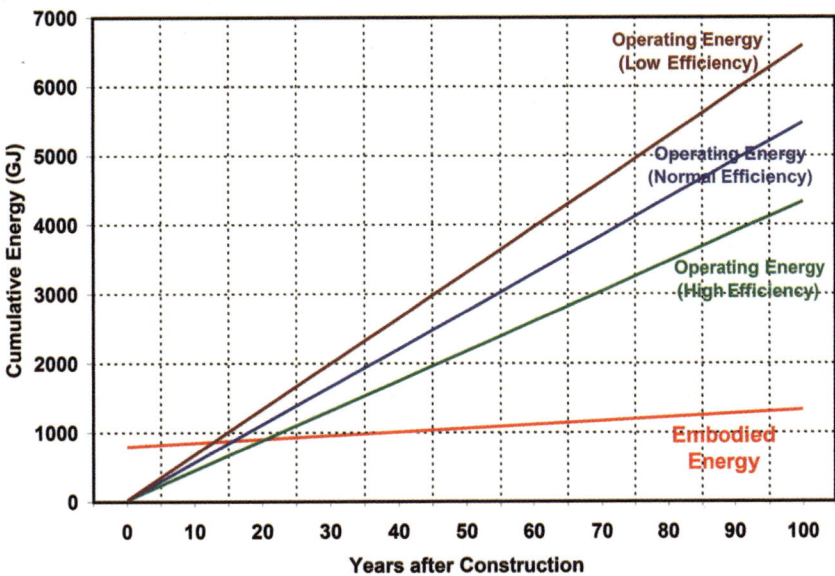

Cumulative embodied energy and operating energy for high, normal, and low operating cases.

- Recycled materials and indigenous materials found locally tend to have much less embodied energy than virgin or distant materials.
- Of virgin materials, wood, especially air-dried timber, generally has the least embodied energy on a per mass basis. Brick has the next lowest, about 4 times (x) that of wood, then concrete (5x), plastic (6x), glass (14x), steel (24x), and aluminum (126x). However, embodied energy per unit mass or per unit volume does not always enable comparing applications, such as flooring, walls, and roofing that are best compared on a per unit area basis.
- Engineered materials, such as plywood, particleboard, or other laminated or composite wood products, have higher embodied energy than lumber, especially rough or air-dried timber, but they can take advantage of residual material that would otherwise be wasted.
- Plastics and composites generally have higher embodied energy than natural materials.

The Consortium for Research on Renewable Industrial Materials (CORRIM), a group of twenty-one U.S. universities, was established to investigate the life-cycle energy and environmental costs of various wood products relative to other building materials such as steel and concrete. The group used a "cradle-to-gate" assessment (see Figure 5.1) to estimate the energy for acquisition, transport, and processing of materials from the forest to shipment from the product manufacturing plant. Figure 8.9 shows results for embodied energy of different wood products. Figure 8.10 compares lumber, concrete-, and steel-framed wall and floor subassemblies for the cold and warm locations. In addition, the third bar from the left (lumber with substitutes) shows potential reductions if wood product substitutes are made for non-wood components, and air drying and biofuels are used rather than fossil fuels in processing.

The CORRIM Phase I research used these data to assess embodied energy and life-cycle costs of hypothetical houses built to code in Minneapolis and Atlanta with wood, steel, and concrete framing units. Table 8.13 compares the total embodied, maintenance, and demolition energy, as well as the operating heating and cooling energy and construction and operating costs for the different locations and configurations.

CORRIM also assessed carbon emissions and sequestration for wood products and their potential substitution for other building products. They showed that using wood products in buildings can control GHG carbon emissions in three ways:

1. Capture atmospheric carbon dioxide in growing trees, harvest the trees, and put the wood into buildings where that sequestered carbon will sit for the life of the building.
2. Reduce carbon emissions by using wood instead of carbon-emission intensive concrete and steel.

figure 8.9   Cradle-to-Gate Energy for Various Wood Products

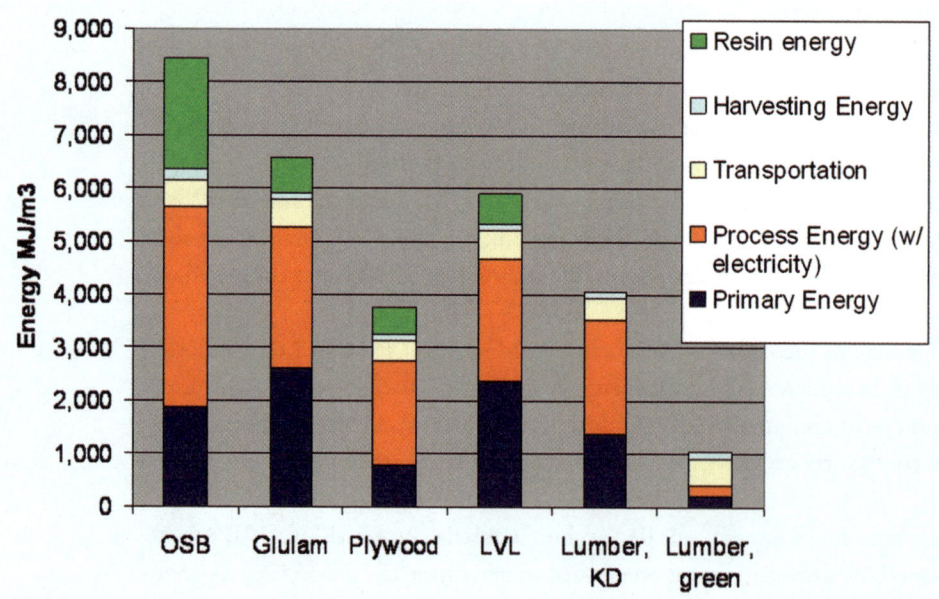

Embodied energy for various wood products: OSB = oriented strand board; Glulam = glue-laminated; LVL = laminated veneer lumber; KD = kiln-dried.

SOURCE: CORRIM, 2005. Used with permission.

table
8.13 **Energy Used in Representative Building Life-Cycle Stages**

	Minneapolis House		Atlanta House	
	Wood	Steel	Wood	Concrete
Energy, structure (GJ)	646	759	395	456
Energy, maintenance (GJ)	73	73	110	110
Energy, demolition (GJ)	7	7	7	9
Energy subtotal (GJ)	727	840	512	573
Energy, heating/cooling, 75 years (GJ)	7800	7800	4575	4575
Energy total, 75 years (GJ)	8527	8640	5087	5148
% Embodied/demolition of life-cycle energy	8.5%	9.7%	10.1%	11.1%
House cost	$168,000	$168,000	$135,000	$135,000
Construction cost	$92,000	$92,000	$74,000	$74,000
Annual cost, heating/cooling	$692	$692	$491	$491
Present Value (PV) heating/cooling cost, 75 yrs, discount rate = 5%	$13,490	$13,490	$9,565	$9,565
PV heating/cooling cost / construction cost	14.7%	14.7%	12.9%	12.9%

figure
8.10 **CORRIM Embodied Energy Estimates for Wood, Concrete, and Steel Wall and Floor Subassemblies in Cold and Warm Climates**

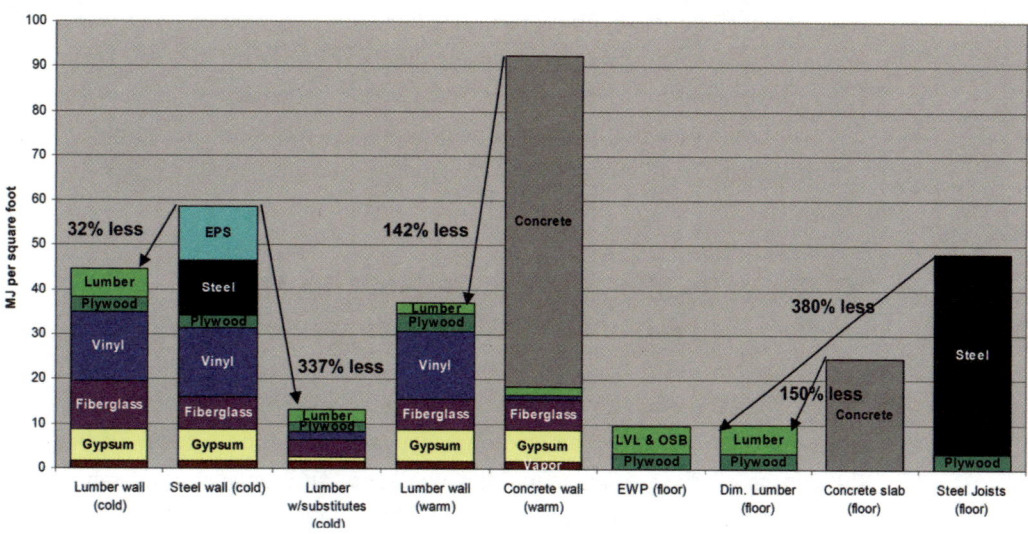

Wood subassemblies have less energy especially with plywood and cellulose substitutes for vinyl, gypsum, and fiberglass, and biofuels for processing.

SOURCE: CORRIM, 2005. Used with permission.

3. Reduce carbon emissions by using biofuels instead of fossil fuels in the manufacturing of wood products.

8.4.2 Tools for Embodied Energy and Life-Cycle Analysis of Buildings

Analyzing embodied energy and life-cycle energy and environmental costs isn't easy, but methods are improving, and with that has come the opportunity to integrate life-cycle energy and environmental criteria into rating systems, codes, and standards. These methods include better models and conceptual approaches, better data, and better techniques and software for analysis. The basic approach to life-cycle and environmental assessment was discussed in Chapter 5, including the NREL Life-Cycle Inventory currently under development (see http://www.nrel.gov/lci).

With regard to buildings, two methods are worth mentioning here, although this is an evolving field and it is best for readers to investigate new developments as time goes on. The Athena Sustainable Materials Institute *Environmental Impact Estimator* software, uses the Institute's large life-cycle database covering more than 90 structural and envelope materials for analysis of building life-cycle energy and environmental impact. CORRIM uses the Athena software in its assessment of embodied energy and life-cycle cost.

The National Institute for Standards and Technology (NIST) *BEES* software uses the life-cycle assessment specified in ISO 14000 standards to measure the environmental performance of building products, including all stages of the life of a product: raw material acquisition, manufacture, transportation, installation, use, and recycling and waste management. Economic performance is measured using the ASTM standard life-cycle cost method, which covers the costs of initial investment, replacement, operation, maintenance and repair, and disposal. Environmental and economic performance measures are combined into an overall performance measure using the ASTM standard for Multi-Attribute Decision Analysis.

8.5 Green Building Ratings: Helping the Market Advance Whole Building Life-Cycle

"Green" buildings aim to use new technologies, materials, and design and construction innovations to maximize the environmental and economic benefits of buildings, reduce the impacts of their construction and operation, and enhance the livability, health, and productivity of their occupants. Energy use has perhaps the greatest influence on these objectives because it affects air pollutant and GHG emissions, embodied materials sustainability, occupant thermal comfort, and indoor air quality. Green Building embraces the following five strategies, which aim to advance building considerations from "building envelope and HVAC" to Whole Building and Whole Building Life-Cycle (Table 8.1):

1. **The Site:** Land use and site planning affect energy and environment and include considerations for stormwater management, landscaping, and ecosystem protection.

2. **Water Conservation:** Efficiency of plumbing fixtures reduces hot water energy and water and wastewater and the energy to collect and treat them. Efficient irrigation and rainwater collection also save water.
3. **Energy Use:** Greater efficiency and use of renewable energy reduces energy consumption and related emissions including air pollutants, greenhouse gases and ozone-depleting gases.
4. **Materials and Resources:** Use of reused, recycled, and indigenous materials and sustainable forest products combined with reduced household and construction waste reduces embodied energy and saves energy and environmental impacts of materials from cradle to grave.
5. **Indoor Environmental Quality:** Occupant health, comfort, and productivity are enhanced by better indoor air quality, thermal comfort and control, daylighting, ventilation, and reduced toxic materials.

Unlike codes and standards, the Green Building movement has been largely voluntary. Similar to the ENERGY STAR program, Green Building programs go beyond minimum standards. Through accountable certification, they provide consumers with confidence about the buildings they occupy and the homes they buy or rent. They provide builders and designers accepted criteria and certification to market their buildings and services. With greater reliability and less uncertainty about the efficiency and quality of buildings, the market can respond much more effectively. In 2007 we are seeing an ever-expanding market for Green Buildings from homes to office complexes. Before discussing the most prominent Green Building certification in the United States, the U.S. Green Building Council's LEED system, we first review the ENERGY STAR rating system for residential homes.

8.5.1 ENERGY STAR Homes and the Home Energy Rating System (HERS)

In addition to appliances and office equipment, EPA's ENERGY STAR program has a rating system for residential housing. Like the appliance program, it requires ENERGY STAR homes to exceed existing standards, in this case building energy codes. The ENERGY STAR rating has been based on a Home Energy Rating System (HERS) score. This Home Energy Rating System is developed by the Residential Energy Services Network (RESNET), which was founded in April 1995 to develop a national market for home energy rating systems and energy-efficient mortgages.

Until 2005, the HERS score had a 0–100 point scale, where 100 was zero energy use, 80 points was the reference house meeting the 1992 MEC for heating and cooling, and each point above 80 representing a 5% reduction in energy use for these purposes. An ENERGY STAR qualified home had a minimum score of 86 or a 30% improvement in efficiency.

In 2005, however, because of progress toward the Whole Building approach, advances in other rating systems such as LEED, new SEER-13 standards for central air-conditioning, and new federal incentives for building efficiency under the 2005 National Energy Policy Act, RESNET and the ENERGY STAR program updated the criteria. The HERS was completely changed, replacing the 0–100 HERS score with a 0–500 HERS index. Whereas the former scoring system gave greater efficiency as higher numbers (score of 100 = zero energy use), the HERS index gives greater efficiency as lower numbers (index of 0 = zero energy use). Table 8.14 compares the score and index and relates them to a graduated "Star" rating.

The 2005 HERS index and the 2006 ENERGY STAR rating incorporate appliance and lighting efficiency, so they consider Whole Building efficiency. This is illustrated in Figure 8.11, which shows the pre-2006 ENERGY STAR criteria characterized by "envelope, infiltration, HVAC" to the 2006 updated ENERGY STAR criteria.

Paths to ENERGY STAR rating. The ENERGY STAR rating system has two paths to certification—the Performance Path and the Builder Option Package (BOP). Described

figure 8.11   Changing Criteria for ENERGY STAR Homes: (a) Before 2006; (b) After 2006

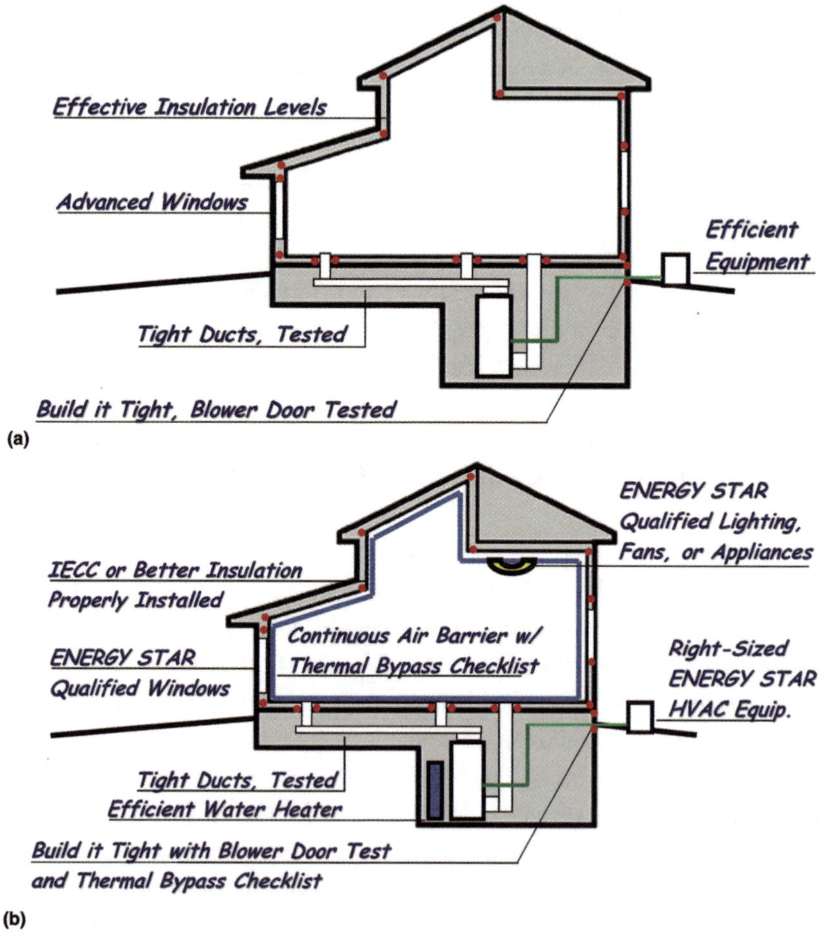

Old criteria represent an "envelope, infiltration, HVAC" approach, whereas new criteria take a Whole Building approach.

SOURCE: U.S. EPA, 2006

table 8.14   Home Energy Rating System "Score" (up to 2005) and "Index" (after 2005)

Stars	HERS Score (< 2006)	Efficiency Change to Reference House	HERS Index (≥2006)	Relative Energy Use to Reference House
*	≥ 0 and < 20	≥ −400% and < −300%	≤ 500 and > 400	≤ 500% and > 400%
*₊	≥ 20 and < 40	≥ −300% and < −200%	≤ 400 and > 300	≤ 400% and > 300%
**	≥ 40 and < 50	≥ −200% and < −150%	≤ 300 and > 250	≤ 300% and > 250%
**₊	≥ 50 and < 60	≥ −150% and < −100%	≤ 250 and > 200	≤ 250% and > 200%
***	≥ 60 and < 70	≥ −100% and < −50%	≤ 200 and > 150	≤ 200% and > 150%
***₊	≥ 70 and < 80	≥ −50% and < 0%	≤ 150 and > 100	≤ 150% and > 0%
****	≥ 80 and < 83	≥ 0% and < 15%	≤ 100 and > 85	≤ 0% and > −15%
****₊	≥ 83 and < 86	≥ 15% and < 30%	≤ 85 and > 70	≤ −15% and > −30%
*****	≥ 86 and < 90	≥ 30% and < 50%	≤ 70 and > 50	≤ −30% and > −50%
*****₊	≥ 90 and < 100	≥ 50%	≤ 50 and ≥0	≤ −50% and ≥ −100%

Source: RESNET, 2005

in Figure 8.13, the Performance Path depends on a maximum HERS index depending on climate zone and a set of minimum standards. The BOP, shown in Figure 8.12, sets HVAC efficiency depending on climate zone and requires specific measures for the building envelope as well as lighting and appliances.

With lighting and appliances included, ENERGY STAR is moving to a Whole Building approach. To qualify as ENERGY STAR, a home must meet the requirements specified, be verified and field-tested in accordance with the HERS standards by a RESNET-accredited provider, *and* meet all applicable codes. For more information on the new ENERGY STAR requirements, including notes for tables in Figures 8.12 and 8.13, see the ENERGY STAR Web site: http://www.energystar.gov.

8.5.2 USGBC's LEED Certification Program

With its 2006 update, EPA's ENERGY STAR Homes program is moving to a Whole Building approach, but falls well short of Whole Building Life-Cycle. However, other Green Building rating programs have moved beyond Whole Building to include embodied energy and life-cycle environmental impacts including indoor air quality.

About eighty Green Building programs operate in the United States, mostly on the state and local level, and many more operate abroad. There are notable programs in Colorado, California, Austin (Texas), and Boulder (Colorado). But the best known and the emerging standard in national Green Building certification is the LEED (Leadership in Energy and Environmental Design) program developed by the U.S. Green Building Council (USGBC).

The USGBC is a national nonprofit membership organization "leading a national consensus for producing a new generation of buildings that deliver high performance inside

figure 8.12 2006 ENERGY STAR Builder Option Package (BOP)

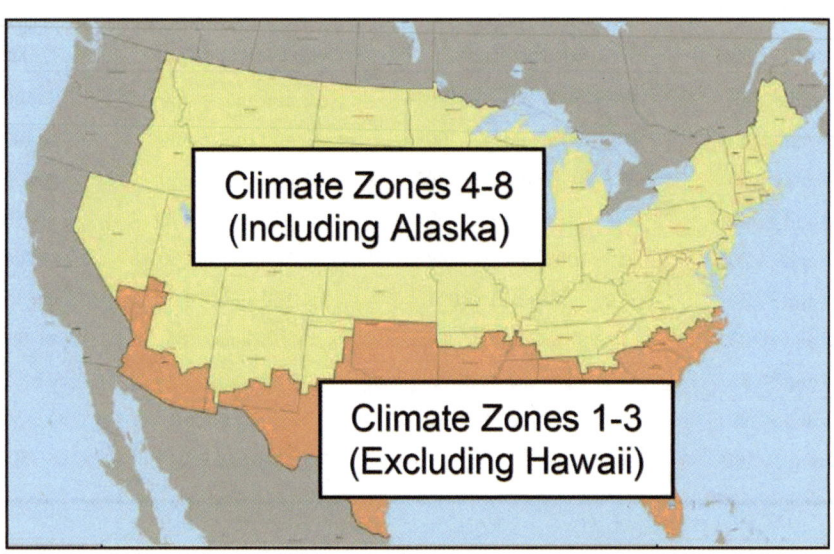

Climate Zones 4-8
(Including Alaska)

Climate Zones 1-3
(Excluding Hawaii)

	Hot Climates (2004 IRC Climate Zones 1,2,3)	Mixed and Cold Climates (2004 IRC Climate Zones 4,5,6,7,8)
Cooling Equipment (Where Provided)	Right-Sized : • ENERGY STAR qualified A/C *(14 SEER / 11.5 EER)*; <u>OR</u> • ENERGY STAR qualified heat pump *(14 SEER / 11.5 EER / 8.2 HSPF)*	Right-Sized : • 13 SEER A/C; <u>OR</u> • ENERGY STAR qualified heat pump *(14 SEER / 11.5 EER / 8.5 HSPF)*
Heating Equipment	• 80 AFUE gas furnace; <u>OR</u> • ENERGY STAR qualified heat pump *(14 SEER / 11.5 EER / 8.2 HSPF)*; <u>OR</u> • 80 AFUE boiler; <u>OR</u> • 80 AFUE oil furnace	• ENERGY STAR qualified gas furnace *(90 AFUE)*; <u>OR</u> • ENERGY STAR qualified heat pump *(See Note 3 for specifications)*; <u>OR</u> • ENERGY STAR qualified boiler *(85 AFUE)*; <u>OR</u> • 85 AFUE oil furnace
Thermostat	ENERGY STAR qualified thermostat (except for zones with radiant heat)	
Ductwork	Leakage : ≤ 4 cfm to outdoors / 100 sq. ft.; <u>AND</u> R-6 min. insulation on ducts in unconditioned spaces	
Envelope	• Infiltration (ACH50): 7 in CZ's 1-2 | 6 in CZ's 3-4 | 5 in CZ's 5-7 | 4 in CZ 8; <u>AND</u> • Insulation levels that meet or exceed the 2004 IRC ; <u>AND</u> • Completed Thermal Bypass Inspection Checklist	
Windows	ENERGY STAR qualified windows or better (additional requirements for CZ2 and CZ4)	
Water Heater	Gas (EF): 40 Gal = 0.61 | 60 Gal = 0.57 | 80 Gal = 0.53 Electric (EF): 40 Gal = 0.93 | 50 Gal = 0.92 | 80 Gal = 0.89 Oil or Gas : Integrated with space heating boiler	
Lighting and Appliances	Five or more ENERGY STAR qualified appliances, light fixtures, ceiling fans equipped with lighting fixtures, and/or ventilation fans	

Includes detailed criteria for hot and mixed/cold climates. To qualify as ENERGY STAR using this BOP, a home must meet the requirements specified, be verified and field-tested in accordance with the HERS standards by a RESNET-accredited provider, *and* meet all applicable codes.

Source: U.S. EPA, 2006

figure 8.13 2006 ENERGY STAR Performance Path Criteria

ENERGY STAR Mandatory Requirements:

Envelope	Completed Thermal Bypass Inspection Checklist
Ductwork	Leakage ≤ 6 cfm to outdoors / 100 sq. ft.
ENERGY STAR Products	Include at least one ENERGY STAR qualified product category: • Heating or cooling equipment ; <u>OR</u> • Windows ; <u>OR</u> • Five or more ENERGY STAR qualified light fixtures , appliances , ceiling fans equipped with lighting fixtures, and/or ventilation fans
ENERGY STAR Scoring Exceptions	• On-site power generation may not be used to decrease the HERS Index to qualify for ENERGY STAR. • A maximum of 20% of all screw-in light bulb sockets in the home may use compact fluorescent lamps (CFLs) to decrease the HERS Index for ENERGY STAR compliance. CFLs used for this purpose must be ENERGY STAR qualified.

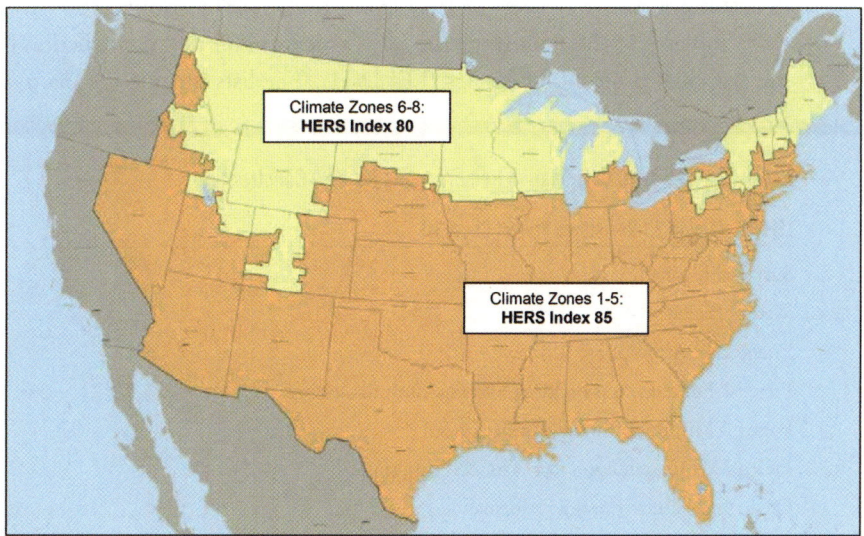

Includes maximum HERS index and mandatory measures. To qualify as ENERGY STAR, a home must meet the minimum requirements specified below, be verified and field-tested in accordance with the RESNET standards by a RESNET-accredited provider, *and* meet all applicable codes.

SOURCE: U.S. EPA, 2006

and out." Through its 4500 members and external review process, USGBC has developed a number of Green Building protocols that define "green" and are used to evaluate, score, and certify buildings. The protocols rate design, materials, and practices using an extensive set of criteria on environment, energy, health, and livability. The building rating systems have been developed for new construction (LEED-NC), existing buildings (LEED-EB), and commercial interiors (LEED-CI), and are under development for homes (LEED-H) and neighborhood development (LEED-ND). LEED-ND is described in Chapter 15.

The LEED protocols are becoming the international standard for Green Building and development. Still, relatively few buildings have been LEED certified. In early 2007, there are only about 230 LEED-certified buildings in North America. But the number registered (the first step in the certification process) has grown from 1900 in 2005 to more than 3000 in early 2007. Because LEED protocols are still focused on commercial buildings, owners of registered buildings include governments (45%), private corporations (26%), nonprofit organizations (19%), and others (10%). One-third of the registered building use is public, 27% multi-use, 20% commercial, 6% industrial, and only 5% residential. Figure 8.14 gives some examples of LEED-certified buildings.

LEED protocol and checklists. The LEED process requires registration of the building, then a certified LEED evaluator checks the building against established checklists of energy and environmental measures. Points are accumulated to achieve a rating of "certified" (40% of maximum points), "silver" (50%), "gold" (60%), or "platinum" (80%).

Tables 8.15–8.16 illustrate the LEED protocol for new construction (LEED-NC), and the proposed protocol for homes (LEED-H). The lists include the main scoring categories

table 8.15   LEED-NC Certification Requirements for New Construction

LEED-NC Version 2.1 Registered Project Checklist	Points
Sustainable Sites	14
Water Efficiency	5
Energy and Atmosphere	17
Prereq 1 Fundamental Building Systems Commissioning	Required
Prereq 2 Minimum Energy Performance	Required
Prereq 3 CFC Reduction in HVAC&R Equipment	Required
Credit 1 Optimize Energy Performance	1 to 10
Credit 2.1 Renewable Energy, 5%	1
Credit 2.2 Renewable Energy, 10%	1
Credit 2.3 Renewable Energy, 20%	1
Credit 3 Additional Commissioning	1
Credit 4 Ozone Depletion	1
Credit 5 Measurement and Verification	1
Credit 6 Green Power	1
Materials and Resources	**13**
Indoor Environmental Quality	**15**
Innovation and Design Process	**5**
Project Maximum	**69**

Certified: 26–32 points; **Silver:** 33–38 points; **Gold:** 39–51 points; **Platinum:** 52–69 points

Source: U.S. Green Building Council, 2006

figure 8.14 Examples of LEED–Certified Buildings

(a) Alberici Constructors

(f) CalEPA

(b) Atlanta Community Food Bank

(c) Blakely Hall

(d) Yellowstone Park

(e) California Health Services

(a) Alberici Constructors Headquarters, St. Louis, Missouri: LEED-NC Platinum. (b) Atlanta Community Food Bank Atlanta, Georgia: LEED-NC Silver. (c) Blakely Hall, Issaquah Highlands Urban Village, Issaquah, Washington: LEED-NC Silver. (d) Yellowstone National Park, Employee Housing: LEED-NC. (e) California Health Services Buildings, Sacramento, California: LEED-NC. (f) California CalEPA Office Building, Sacramento, California: LEED-EB.

SOURCE: U.S. GBC, 2007. For additional LEED project profiles, see http://www.usgbc.org/DisplayPage.aspx?CMSPageID=1721.

table 8.16   Proposed LEED–H Certification Requirements for Homes

LEED for HOMES	Points
Location and Linkages	**10**
Sustainable Sites	**14**
Water Efficiency	**12**
Indoor Air Quality	**14**
Materials and Resources	**24**
1.1 **Home Size** Home that is Smaller than National Average	10
2.1 **Material Efficient Framing** No Extra Uses of Lumber for Aesthetic Purposes	Required
2.2 Advanced Framing Techniques	2
3 **Local Sources Materials** Extracted / Manufactured / Produced within 500 Miles	3
4.1 **Durability Plan** Detailed Durability Plan; (Pre-Construction)	Required
4.2 Third-Party Verification of Implementation of Durability Plan	3
5.1 **Environmentally Preferable** Tropical Hardwoods, if used, must be FSC	Required
5.2 Products Select Environmentally Preferable Products from List	4
6.1 **Waste**: Max of 2.5 Lbs Per Sq. Ft. of Construction Waste Sent to Landfill	Required
6.2 0.5 Pts for Each Additional 0.5 Lbs Per Square Foot Reduction	2
Energy and Atmosphere	**29**
1.1 **ENERGY STAR Home** Meets ENERGY STAR Homes w/Third-Party Testing	Required
1.2 Exceeds ENERGY STAR for Homes, 2 Pts per HERS Point > HERS 86	16
2.1 **Insulation** Third-Party Inspection of Insulation Installation, At Least HERS Grade II	Required
2.2 Third-Party Inspection of Insulation Installation, At Least HERS Grade I	1
2.3 OR Above Code Insulation; At Least 5% > Local Code Per REScheck	1
3.1 **Air Infiltration** Third-Party Envelope Air Leakage Tested 0.35 ACH	Required
3.2 Third-Party Envelope Air Leakage Tested # 0.25 ACH	1
3.3 OR Third-Party Envelope Air Leakage Tested # 0.15 ACH	2
4.1 **Windows** Windows Meet ENERGY STAR for Windows (See Table)	Required
4.2 Windows Exceed ENERGY STAR for Windows by ≥ 10% (See Table)	1
4.3 OR Windows Exceed ENERGY STAR for Windows by ≥ 20% (See Table)	2
5.1 **Duct Tightness** Third-Party Duct Leakage Test # 5.0 CFM25/100 SF to Outside	Required
5.2 Third-Party Duct Leakage Tested # 3.0 CFM25 / 100 SF to Outside	1
5.3 OR Third-Party Duct Leakage Tested # 1.0 CFM25 / 100 SF to Outside	2
6.1 **Space Heating and Cooling** Meets ENERGY STAR for HVAC	Required
6.2 Exceeds ENERGY STAR for HVAC by ≥ 10%	1
6.3 OR Exceeds ENERGY STAR for HVAC by ≥ 20%	3
7.1 **Water Heating** Improved Hot Water Distribution System	3
7.2 Improved Water Heating Equipment	3

(continued)

table 8.16	Proposed LEED-H Certification Requirements for Homes (*continued*)	
LEED for HOMES		**Points**
8.1 **Lighting** Energy Efficient Fixtures and Controls		1
8.2 OR ENERGY STAR Advanced Lighting Package		3
9.1 **Appliances** Select Appliances from List		2
9.2 Very Efficient Clothes Washer (MEF > 1.8, AND WF < 5.5)		1
10 **Renewable Electric Generation System** (1 Point /10% Ann. Load Red.)		6
11 **Non-HCFC Refrigerant** Use Only Non-HCFC Refrigerants for HVAC Appliances		1
Homeowner Awareness		1
Innovation and Design Process		4
Project Maximum Points:		108

Certified: 30–49 points; **Silver:** 50–69 points; **Gold:** 70–89 points; **Platinum:** 90–108 points

SOURCE: U.S. Green Building Council, 2006

and the points assigned to each, but the detailed criteria are given only for the "Energy and Atmosphere" category because it is most relevant here. The "Materials and Resources" category is also included in the LEED-H because it addresses embodied energy. Because of the availability of more established guidance for residential homes, the LEED-H has more specific energy criteria and references several guidelines discussed earlier in this book on ENERGY STAR lighting, windows, appliances, insulation, and infiltration ACH.

The LEED protocol reflects a Whole Building Life-Cycle approach by addressing (in addition to the building envelope) HVAC, infiltration, appliances, lighting, embodied energy and life-cycle costs of materials, indoor air quality, and other environmental impacts.

8.5.3 Cradle-to-Cradle?

The Whole Building Life-Cycle approach aims to consider operating energy as well as embodied energy, indoor air quality, and environmental impacts of materials in design, construction, operation, renovation, and demolition of buildings. This is a tall order considering codes and norms that guide current practices.

But this is not sufficient for sustainability, and architect William McDonough and others are attempting to extend this thinking even further by considering the regenerative and productive reuse of products and materials in what they call a "cradle-to-cradle" approach. This new paradigm puts the "cycle" back in life-cycle, applying to human industrial processes and products and the natural principles of material cycles in nature. They apply this not only to buildings, but also to industrial activities. They have gotten the attention of large industries, such as Ford Motor Company, Nike, and BASF. We have a long way to go to achieve the cradle-to-cradle (C2C) dream of "waste equals food" but the concept has been an inspiration for designers of sustainable products, processes, and buildings.

The application of the cradle-to-cradle approach in buildings incorporates much of our discussion so far: reduced operating energy and embodied energy, durability and low maintenance, renewable energy and materials, passive solar heating and cooling, daylighting, minimized life-cycle waste and emissions, healthy environments, water management, and regenerative sites and landscapes.

A 2005 C2C building design competition in Roanoke, Virginia, challenged participants to incorporate these concepts into residential house designs on four donated sites in the city, then worked to construct winning designs on the sites in 2006.

8.5.4 Really Green Buildings: Green Roofs and Other Natural Building Materials

In the spirit of C2C, there is a growing interest in incorporating natural organic materials into building design. We discussed the carbon sequestering benefits of building wood products in Section 8.4.1. Some designers and builders have demonstrated the thermal-mass benefits of rammed earth and adobe structures. Straw-bale-walled buildings also have had growing interest for their thermal insulation and affordability. But both approaches have only niche custom markets.

Green or vegetated roofs have seen a growing market, especially in urban multifamily residential, commercial, and industrial buildings where roofs are flat, green space is limited, and green roof benefits are maximized. Green roof installation includes several layers: waterproof membrane and root barrier, drainage layer, filter fabric, and growth membrane. We introduced the cooling benefits of green roofs in Chapter 7. The full list of benefits are as follows:

- Reduced heating and cooling costs through roof insulation and evapotranspiration
- Reduced total runoff volume through rainwater storage and evapotranspiration
- Reduced peak runoff discharge rates
- Extended roof life
- Improved building appearance
- Reduced "heat island" effect and improved local air quality
- Enhanced recreational benefits

Chicago, Portland (Oregon), Madison (Wisconsin), Seattle, and other cities have embraced green roofs because of the community benefits. The Greenroof Projects database managed by greenroofs.com lists 661 projects in the United States totaling 445 million ft^2.

8.5.5 Green and Affordable?

Some advocates of affordable housing have criticized the Green Building movement, alleging higher costs and a focus on upscale markets. However, energy efficiency improvements can reduce operating costs, and the concept of "affordable comfort" has gained greater saliency in both new construction and weatherization retrofit markets. As discussed in Section 8.4.1, the Residential Energy Services Network (RESNET) developed the HERS rating system to facilitate energy-efficient home mortgages that recognize that owners of efficient houses will pay less in monthly operating costs and should then qualify for higher mortgage payments.

HUD's PATH program and the NextGen program of DOE's Building America program both aim to apply new efficiency and green technologies to the affordable housing market. The affordable Habitat for Humanity house in Denver shown in Figure 8.16 is a good example. It illustrates the movement toward Whole Community Energy by integrating on-site generation into buildings, the subject of the section below.

8.6 Zero-Energy Buildings: Toward Whole Community Energy

It would be nice if we could design buildings that used zero energy, but it probably makes little practical or economic sense to do so. But it does make sense to design buildings that minimize their energy demand, make productive use of solar heating and natural cooling, and generate on-site energy through solar photovoltaics on rooftops or other technologies that can be hooked into the local electrical grid. We will discuss these "distributed energy" technologies in some detail in the next section of this book, but it is useful to introduce the concept of on-site generation here in the context of zero-energy buildings.

On-site generation is part of the next step in our evolution of energy in buildings, moving from the Whole Building Life-Cycle to Whole Community Energy. This next step is a recurring theme in subsequent chapters as we discuss distributed electricity in the community, land use and transportation, and community transit systems. Whole Community Energy aims to integrate energy considerations for buildings, electricity, and transportation. Buildings are but one part of Whole Community Energy, but they are a central part. Buildings account for nearly half of our energy use, and we have spent the last three chapters exploring building efficiency in our quest for sustainable energy. But in addition, building location and land use largely determine our transportation needs and vehicle miles traveled, and efficient land use and building location are key parts of improving transportation energy. We discuss this role further in Chapter 15. And finally, buildings can play an important part in community energy production, not only through thermal uses of the sun in passive and active solar systems, but in on-site generation of electricity. We discuss this distributed generation in detail in Chapters 10 and 11.

Here, it is useful to consider on-site generation as part of our strategies for building energy. In this context, the concept of the zero-energy home (ZEH) is really a misnomer because in practical terms we envision "near zero net energy homes." These have on-site power generation and are tied to the community power grid. They use grid power as needed and feed excess on-site generation to the grid when available through net-metering (i.e., running the electric meter backwards). What we are seeing today are "near-zero net energy homes," and some cities are gaining experience in this aspect of Whole Community Energy. For example, the Sacramento Municipal Utility District (SMUD) has used solar photovoltaic (PV) arrays on building rooftops for several years to add to its power capacity by 10 MW. SMUD's program is discussed in Chapter 18 (Figure 18.30 shows a solar subdivision in Sacramento).

The ZEH movement has been advanced by DOE's Building America (BA) program, a research and technology development effort in partnership with builders and manufacturers, including Building Science Corporation. BA's goal is to reduce residential energy use by 40% to 70% through advanced efficiency technologies and on-site generation. By 2007, more than 40,000 homes had been built in Building America projects across the country, incorporating some of the new technologies.

Building America's objectives are illustrated in Figure 8.15:

- Reduce energy demand by 70% from a benchmark design by 2020 by incorporating Whole Building efficiency.
- Increase on-site generation to make up the 30% remaining.

Figure 8.16 shows a BA affordable house in Denver's Habitat for Humanity program that incorporates several technologies and reduces net energy use to 40% of the benchmark house.

California is taking the first steps toward implementing zero net energy buildings into incentives and codes. The Califonia Solar Initiative (CSI) is a 10-year, $3.3 billion program to achieve one million solar PV roofs. As part of CSI, the New Solar Homes Partnership engages residential builders in developing solar homes. To qualify for some of the program's $400 million over 10 years, builders' homes must exceed Title 24 efficiency standards by 15% (35% encouraged) and be targeted to single family, low-income, and multifamily housing.

figure 8.15   DOE's Building America Program Goal: Zero Net Energy Home

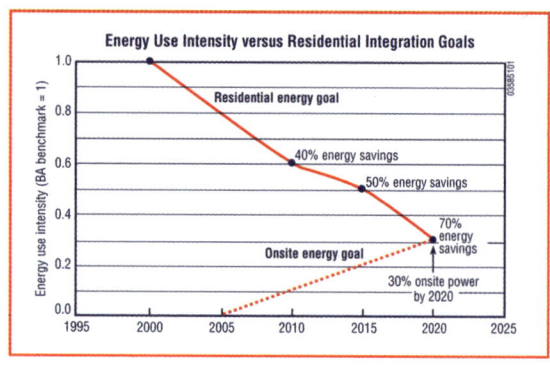
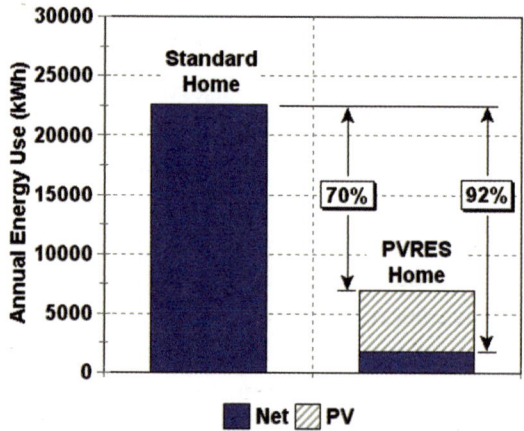

Goals include by 2020, 70% reduction in building energy demand with remaining 30% from on-site generation. On-site generation is most effective when on-site demand is minimized.

figure 8.16 Denver Habitat for Humanity House: Affordable Efficiency and On-Site Generation

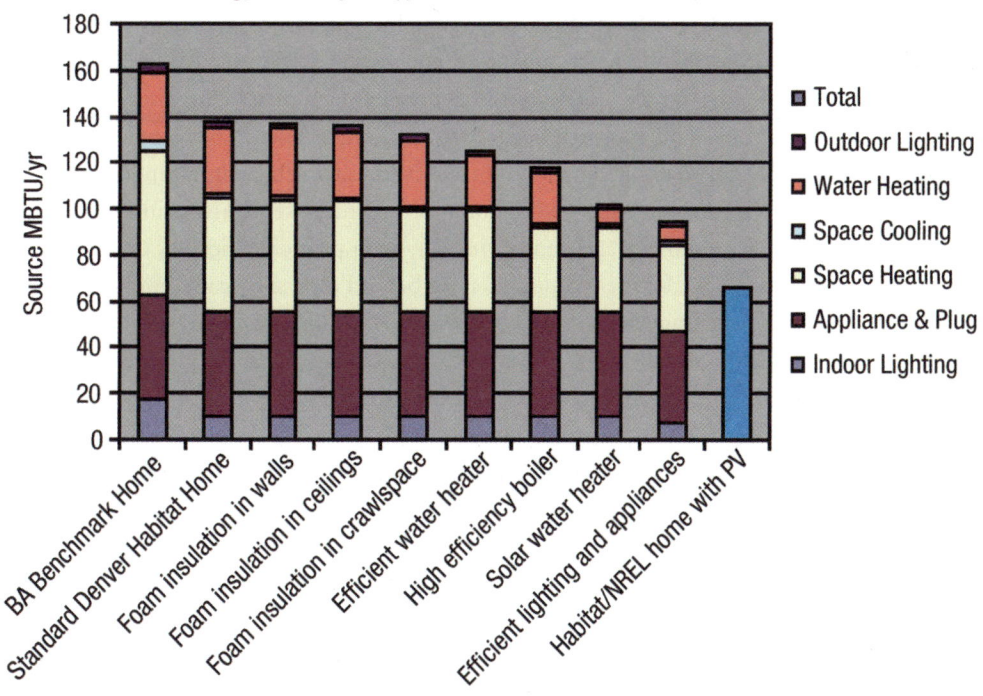

The house demonstrates that efficiency and on-site generation can be incorporated into affordable housing.

Source: U.S. DOE, 2006

By October 2007, the program had received more than 1200 applications and had approved about half of them totaling 1.2 megaWatts (MW).

On-site generation from rooftop photovoltaic systems may be finding their way into California building codes. In 2007, California passed SB 1 calling for the CEC to examine if PV should be required in future building codes. The California PUC also ordered utilities to prepare "next generation" efficiency plans so that all new buildings will be zero net energy by 2020 (residential) and 2030 (commercial).

8.7 Summary

Although these goals and model developments provide a demonstration and vision for zero net energy buildings, the mainstream building industry lags far behind. This is why codes and standards are important to raise the bar for the mainstream. But housing and building markets are also important. Consumer interest, builder and designer innovations, and Green Building ratings help fuel that market, and according to our virtuous cycle described in Figure 8.1, codes and standards may follow.

We have seen that the approach to energy in buildings has evolved considerably in the last forty years from an emphasis on the building envelope to a Whole Building approach including building systems, appliances and lighting, and more recently to a Whole Building Life-Cycle approach considering embodied energy and environmental and health effects. Further evolution to a Whole Community Energy approach may be necessary to achieve sustainable communities. This approach begins with building efficiency but extends to on-site generation of electricity and to building location and land use for efficient transportation. Except for some innovative developments, Whole Community Energy remains a vision, not yet a reality.

The most direct and penetrating way to implement efficiency on a large scale is through building codes. But these codes often lag behind technological innovation and best practices, and building rating systems such as ENERGY STAR and LEED often reflect these improvements before codes do. As a result of the growing Green Building movement, these rating systems are having an increasing influence on consumer choice. Driven by improved technology, we are seeing continual improvement in ratings systems and codes, and at increasing frequency. These improvements are necessary to change our patterns of energy use in buildings to greater sustainability.

One improvement is incorporating electricity efficiency in ratings and codes. These Whole Building Energy considerations are critical because electricity for lighting, appliances, and equipment amounts to about half of the primary energy in buildings. State and federal appliance efficiency standards have been very effective in saving energy and money.

Embodied energy and associated environmental effects are important considerations in the Whole Building Life-Cycle approach. Embodied energy can be as much as 10%–20% of the life-cycle energy of a building depending on its lifetime and operating efficiency. Research is showing that wood products have far less embodied energy than steel and concrete, and have the added benefit of sequestering potential GHG carbon in building materials. Much more research is needed, and results will then need to be translated into building practices and codes to fully implement the Whole Building Life-Cycle approach.

Some of these expanding considerations are addressed in building rating systems and Green Building programs. EPA's ENERGY STAR homes have begun to take a Whole Building approach by incorporating lighting and appliance efficiency. USGBC's LEED program goes further by including life-cycle considerations of materials, indoor air quality, and environmental impacts. The growing interest in LEED certification and new LEED protocols under development bode well for greater penetration of measures for Whole Building Life-Cycle and Whole Community Energy.

LEED includes on-site generation in its protocol, and as we improve operating efficiency and lower energy demand of buildings, we can use on-site generation to move toward zero net energy buildings. DOE's Building America program is advancing this vision by partnering with builders to demonstrate advanced technologies.

These trends move us toward Whole Community Energy. We explore further the distributed generation and transportation components of this vision in subsequent chapters.

Sustainable Electricity

CHAPTER 9

Centralized Electric Power Systems

9.1 Introduction

Those of us lucky enough to live in the industrialized world take for granted the truly remarkable power, literally at our fingertips, that we can access by simply flipping a switch. The lights go on, the air-conditioning keeps us cool, our food stays fresh in the refrigerator, the Internet gives us access to a treasure-house of information, our TVs entertain us. About the only time many of us pay any attention at all to the electric power grid that provides these amazing benefits is when it suddenly goes down and we're left sitting in the dark, in a building that begins to get too hot or too cold, worried about food spoiling in the fridge, and wondering how the next episode in our favorite TV series is going to come out.

The electricity infrastructure providing power to North America includes more than 275,000 miles of high-voltage transmission lines and 1000 gigawatts of generating capacity to serve a customer base of over 300 million people. Although the cost of constructing this infrastructure has been staggering—well over $1 trillion—its value is incalculable. Without it we could not even imagine having a modern economy. Managing that investment is a complex technical challenge that requires real-time control and coordination of tens of thousands of power plants to move electricity across a vast network of transmission lines and distribution networks to meet the exact, constantly varying, power demands of those customers.

Although this book is mostly concerned with alternatives to large, centralized power systems, we need to have some understanding of how these conventional systems work. This chapter explores the history of the utility industry; the physics and engineering that go into the generation, transmission, and distribution of electric power; and some of the regulatory issues involved in the buying and selling of electric power. In the next chapter, the alternative model of a grid based on smaller-scale, decentralized energy systems will be explored.

9.2 Electromagnetism: The Technology behind Electric Power

In the early nineteenth century, scientists such as Hans Christian Oersted, James Clerk Maxwell, and Michael Faraday began to explore the wonders of electromagnetism. Their explanations of how electricity and magnetism interact made possible the development of electrical generators and motors—inventions that have transformed the world.

Early experiments demonstrated that a voltage (originally called an *electromotive force*, or *emf*) could be created in an electrical conductor by moving it through a magnetic field as shown in Figure 9.1a. Clever engineering based on that phenomenon led to the development of direct-current dynamos, and later to alternating-current generators. The opposite effect was also observed; that is, if current flows through a wire located in a magnetic field, the wire will experience a force that wants to move the wire as shown in Figure 9.1b. This is the fundamental principle by which electric motors are able to convert electric current into mechanical power.

Notice the inherent symmetry of the two key electromagnetic phenomena. Moving a wire through a magnetic field causes a current to flow, whereas sending current through a wire in a magnetic field creates a force that wants to move the wire. If this suggests to you that a single device could be built that could act as a generator if you applied force to it, or act as a motor if you put current into it, you would be absolutely right. In fact, the electric motors in today's hybrid-electric vehicles do exactly that. In normal operation the electric motor helps power the car, but when the brakes are engaged, the motor acts as a generator, slowing the car by converting the vehicle's kinetic energy into electrical current that recharges the vehicle's battery system.

A key to the development of electromechanical machines, such as motors and generators, was finding a way to create the required magnetic fields. The first *electromagnet* is credited to a British inventor, William Sturgeon, who, in 1825, demonstrated that a magnetic field could be created by sending current through a number of turns of wire wrapped around a horseshoe-shaped piece of iron. With that, the stage was set for the development of generators and motors.

The first practical direct-current (dc) motor/generator, called a dynamo, was developed by a Belgian, Zénobe Gramme. His device, shown in Figure 9.2, consisted of a ring of iron

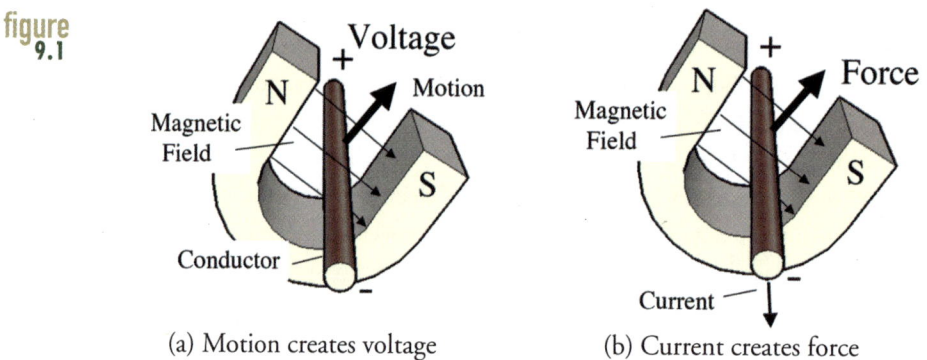

figure 9.1

(a) Motion creates voltage   (b) Current creates force

The voltage created when an electrical conductor moves through a magnetic field is the basis for generators (a). The force created when current is passed through a conductor in a magnetic field is the basis for electric motors (b).

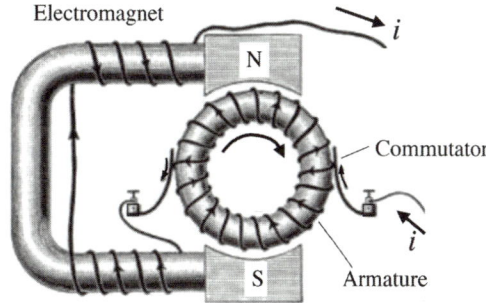

figure 9.2

Electromagnet

Commutator

Armature

Gramme's "electromotor" could operate as a motor or as a generator.

(the *armature*) wrapped with wire, which was set up to spin within a stationary magnetic field. The magnetic field was based on Sturgeon's electromagnet. The key to Gramme's invention was his method of delivering dc power to and from the armature using contacts (called a *commutator*) that rubbed against the rotating armature windings. Gramme startled the world with his machines at a Vienna Exposition in 1873. Using one dynamo to generate electricity, he was able to power another, operating as a motor, three-quarters of a mile away. The potential to generate power at one location and transmit it through wires to a distant location, where it could do useful work, stimulated imaginations everywhere. An enthusiastic American writer, Henry Adams, even proclaimed the dynamo as "a moral force" comparable to the European cathedrals in a 1900 essay called "The Dynamo and the Virgin."

9.3 Creating the Modern Electric Utility: Edison, Westinghouse, and Insull

Motors and generators quickly found application in factories; however, the first major electric power market developed around the need for illumination. Although many others had worked on the concept of electrically heating a filament to create light, it was Thomas Alva Edison who, in 1879, created the first workable incandescent lamp. Simultaneously he launched the Edison Electric Light Company, which was a full-service illumination company that provided not only the electricity but also the lightbulbs themselves. In 1882, his company began distributing power primarily for lights, but also for electric motors, from his Pearl Street Station in Manhattan. This was to become the first investor-owned utility in the nation.

There was a fatal flaw in Edison's electric utility. Edison's system was based on direct current, which he preferred in part because it provided flicker-free light but also because it enables easier speed control of dc motors. The downside of Edison's dc, however, was that it was generated at low voltage for safety, but at low voltages it is very difficult to move much power from one place to another without incurring unacceptably high losses in the power lines connecting the generators to the loads. Edison's customers, therefore, had to be located within just a mile or two of a generating station, which meant power stations were beginning to be located every few blocks around the city.

9.3.1 The Important Role of Transformers

To understand the difficulty that Edison faced, we need to review briefly some of the electricity concepts introduced in Chapter 4. As presented there, the power delivered by power lines is equal to the voltage of the lines times the current they deliver ($P = vi$). For example, suppose you want to transport 100,000 watts of power on a transmission line. Consider the choice between delivering 100 amps at 1000 volts (100 A × 1000 V = 100,000 W) versus delivering 10 amps at 10,000 volts (10 A × 10,000 V = 100,000 W). Is there any advantage to one over the other?

Recall that power losses in wires are equal to the square of the current times the resistance of the lines (i^2R losses). Suppose the connecting wires in this example have a resistance of 2 ohms, then the i^2R line losses at 100 A and 1000 V would be

$$\text{Line losses @ 1000 V} = 100^2 \times 2 = 20{,}000 \text{ W}$$

and the line losses at 10 A and 10,000 V would be

$$\text{Line losses @ 10,000 V} = 10^2 \times 2 = 200 \text{ W}$$

Notice that by increasing the voltage by a factor of 10, the line losses decrease by a factor of 100! At the lower voltage, 20% of the power would be lost in the connecting wires, whereas at the higher voltage line losses would be only 0.2%. Minimizing line losses is why modern transmission lines operate at such high voltages—some as high as 765 kV. Of course, such high voltages must be reduced to much lower levels for safe use in our homes and offices.

In Edison's day, the only way to change voltages conveniently was to take advantage of an 1883 invention called the *transformer*, which, unfortunately for Edison, works only on alternating current (ac). As shown in Figure 9.3, a simple transformer consists of an iron core with two sets of windings. As shown, the primary side of the transformer has N_1 turns of wire carrying current i_1, and the secondary side has N_2 turns carrying i_2. The change in voltage from the primary side to the secondary side is equal to the turns ratio, N_2/N_1.

9.3.2 The Battle between Edison and Westinghouse

Edison's mistake was that he placed his bet on dc power, but dc was unable to take advantage of the reduction in line losses that transformers could provide by increasing the voltage as it goes onto transmission lines and then decreasing it back again to safe levels at the customer's facility. Meanwhile, George Westinghouse recognized the advantages of ac for transmitting

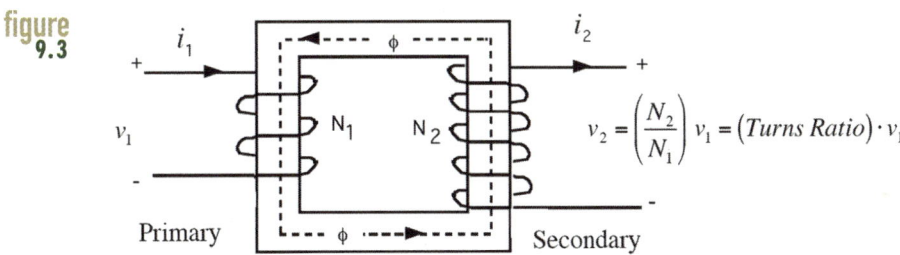

figure 9.3

Transformers are very important in power systems because they can step up voltages to help reduce transmission line losses, and then step them down again for safe use by customers.

power over great distances and so in 1886 he launched a competing company based on ac power, called the Westinghouse Electric Company. Within just a few years, Westinghouse was making significant inroads into Edison's electricity market, and a bizarre feud developed between these two industry giants. Rather than hedge his losses by developing a competing ac technology, Edison stuck with dc and launched a campaign to discredit ac by condemning its high voltages as a safety hazard. To make the point, Edison began demonstrating its lethality by coaxing animals, including dogs, cats, calves, and eventually even a horse, onto a metal plate wired to a 1000-volt ac generator and then electrocuting them in front of the local press (Penrose, 1994).

The advantages of high-voltage transmission, however, were overwhelming and Edison's insistence on dc eventually led to the disintegration of his electric utility enterprise. Through buyouts and mergers, Edison's various electricity interests were incorporated in 1892 into the General Electric company, which shifted its focus from being a utility to manufacturing electrical equipment and end-use devices for utilities and their customers.

One of the first demonstrations of the ability to use ac to deliver power over large distances occurred in 1891 when a 106-mile, 30,000-volt transmission line began to carry 75 kW of power between Lauffen and Frankfurt, Germany. The first transmission line in the United States went into operation in 1890 using 3.3-kV lines to connect a hydroelectric station on the Willamette River in Oregon to the city of Portland, 13 miles away. Meanwhile, the flicker problem for incandescent lamps with ac was resolved by trial and error with various frequencies until it was no longer a noticeable problem. Surprisingly, it wasn't until the 1930s that 60 Hz finally became the standard in the United States. Some countries had by then settled on 50 Hz, and even today, some countries, such as Japan, use both.

9.3.3 Insull Develops the Business Side of Utilities

Another important player in the evolution of electric utilities was Samuel Insull. He is credited with having developed the business side of utilities. It was his realization that the key to making money was to find ways to spread the high fixed costs of facilities over as many customers as possible. One way to do that was to aggressively market the advantages of electric power, especially for use during the daytime to complement what was then the dominant nighttime lighting load. In previous practice, separate generators were used for industrial facilities, street lighting, streetcars, and residential loads, but Insull's idea was to integrate the loads so that he could use the same expensive generation and transmission equipment on a more continuous basis to satisfy them all. Because operating costs were minimal, amortizing high fixed costs over more kilowatt-hour sales results in lower prices, which creates more demand. With controllable transmission line losses and attention to financing, Insull promoted rural electrification, further extending his customer base.

With more customers, more evenly balanced loads, and modest transmission losses, it made sense to build bigger power stations to take advantage of economies of scale, which also contributed to decreasing electricity prices and increasing profits. Large, centralized facilities with long transmission lines required tremendous capital investments. To raise such large sums, Insull introduced the idea of selling utility common stock to the public.

Insull also recognized the inefficiencies associated with multiple power companies competing for the same customers, with each building its own power plants and stringing its own wires up and down the streets. The risk of the monopoly alternative, of course, was that without customer choice, utilities would charge whatever they could get away with. To counter that criticism, he helped establish the concept of regulated monopolies with established franchise territories and prices controlled by *public utility commissions* (PUCs). The era of regulation had begun.

9.4 Electric Power Infrastructure: Generation

The electric power industry in the United States is truly immense, worth more than a trillion dollars with sales that exceed $300 billion each year. About 40% of total U.S. primary energy is used to generate electricity, with about 70% of that coming from the combustion of fossil fuels. All of that combustion is responsible for three-fourths of the country's emissions of sulfur oxides (SO_x), one-third of its carbon dioxide (CO_2) and nitrogen oxides (NO_x), and one-fourth of particulate matter and toxic heavy metals emissions.

How is all that power generated; how does it get from one part of the country to another; and how does it make its way up and down every street in town to get to our homes, businesses, and factories? We'll break these questions into two parts: In this section and the next, we'll look at the power plants themselves. Then, in Section 9.6, we'll examine the grid that transports and distributes power to customers.

Power plants come in a wide range of sizes, run on a variety of fuels, and utilize a number of different technologies to convert fuels into electricity. Most electricity today is generated in large, central stations with power capacities measured in hundreds or even thousands of megawatts (MW). A single, large nuclear power plant, for example, generates close to 1000 MW (also described as one gigawatt, 1 GW). The total generation capacity of the United States is equivalent to roughly 1000 such power plants. At the other extreme are small-scale, distributed-generation technologies, such as fuel cells and microturbines, with rated capacities of several kilowatts. These will be described in the next chapter of this book.

Coal is the dominant fuel, accounting for 52% of all power plant input energy; nuclear is 21%; natural gas, 15%; and renewables (especially hydro and geothermal), 9%. Notice that petroleum is a very minor fuel in the electricity sector, only about 3%, almost all of which is residual fuel oil—literally the bottom of the barrel—that has little value for anything else.

The distribution of power plants based on fuel type is very uneven as Figure 9.4 suggests. The Pacific Northwest generates most of its power at large hydroelectric facilities owned by the federal government. Coal is predominant in the midwestern and southern states, especially Ohio, West Virginia, Kentucky, and Tennessee. The states of Texas, Louisiana, Oklahoma, and California derive significant fractions from natural gas, whereas what little oil-fired generation there is tends to be in Florida and New York.

Most large power plants, whether they are fueled by coal, natural gas, or even nuclear fission, use heat to boil water, which creates high-temperature, high-pressure steam. The steam expands as it passes through a steam turbine, which in turn powers a generator. These large steam plants tend to be *base-load* plants, which means they operate more or less continuously,

figure 9.4   **Energy Sources for Electricity Generation by Region**

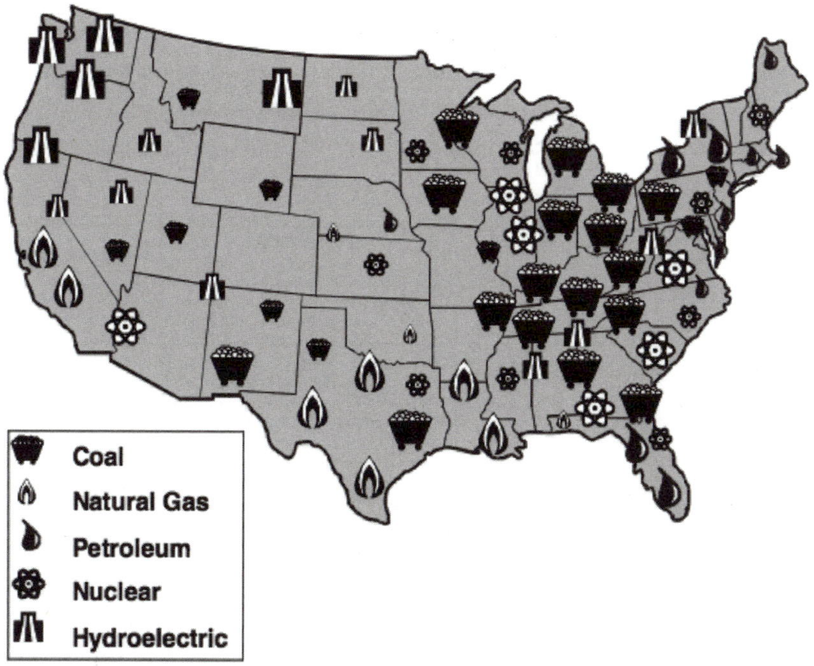

Coal
Natural Gas
Petroleum
Nuclear
Hydroelectric

Each large icon represents about 10 GW of capacity; small ones about 5 GW.

SOURCE: from *The Changing Structure of the Electric Power Industry 2000: An Update* (EIA 2000)

twenty-four hours a day, at relatively constant output. Base-load plants tend to be expensive to build, but cheap to operate, so they are economically most efficient when running as much of the time as possible. But, the mix of power demands for houses, commercial buildings, and industrial facilities, which a utility must supply, varies throughout the day. Figure 9.5 shows an example in which demand shows peaks during the day and valleys at night, along with reduced power demands on weekends.

The implications of Figure 9.5 are extremely important. It shows that power companies must have enough generating capacity to meet the highest peak demand (and then some to allow an adequate reserve margin), but many of those power plants will operate only part of the time during the day and be shut down at night. These are called *intermediate-load* and *peaking power plants* (or, just, *peakers*). Given this operating pattern, peakers tend to be plants that are cheap to build, but expensive to operate, which is just the reverse of large, base-load plants.

9.4.1 Conventional Coal-Fired, Steam Power Plants

More than half of all of the electrical power generated in the United States is from pulverized-coal, steam power plants similar to the one shown in Figure 9.6. Finely pulverized coal is burned in a boiler, which, as the name implies, boils water to make high-temperature, high-pressure steam. Steam expanding in the blades of the turbine causes the turbine shaft to rotate, which spins the armature of a generator to make electric power. A transformer increases the generator output to the voltage needed for efficient power delivery on the high-voltage transmission lines. Back at the turbine, the expanded steam is condensed back to the liquid state by passing it over a heat exchanger carrying cooling water, usually taken from a local river, lake, or ocean. The condensed steam is returned to the boiler to continue the cycle.

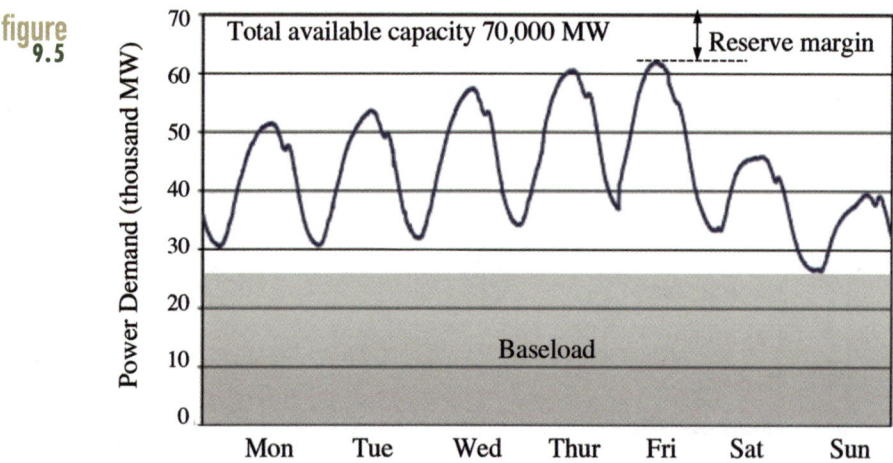

figure 9.5

Power demand in the summer usually peaks in the afternoon. Base-load plants operate at nearly constant power all day long, whereas the output of intermediate and peaking power plants is adjusted to track the daily load pattern.

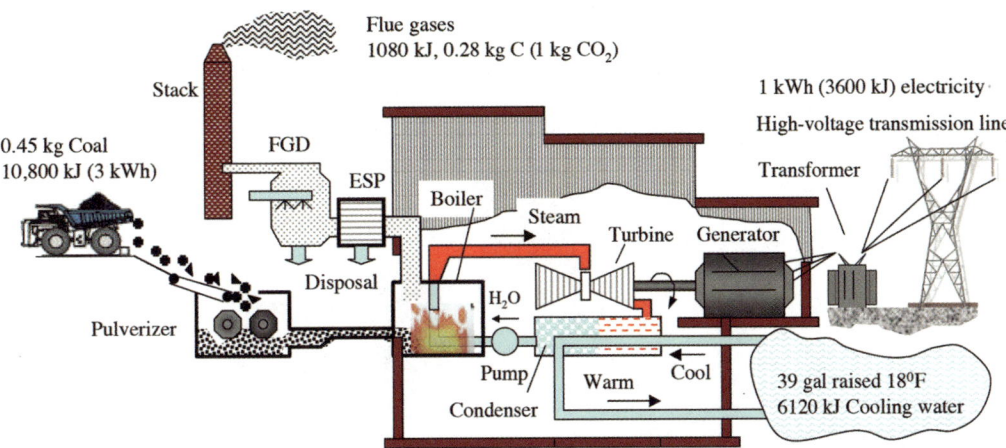

figure 9.6 **Essential Features in a Conventional Coal-Fired, Steam-Turbine Power Plant**

Numerical values correspond to a 33.3%-efficient plant generating 1 kWh of electricity. Carbon emissions are based on 24 MJ/kg coal with 62% carbon. Emission controls to be explained later include an electrostatic precipitator (ESP) and flue-gas desulfurization (FGD) system.

You might ask, Why bother to cool the steam in the condenser when you're just going to heat it up again to make new steam? That seems like a waste of energy: heating it, cooling it, and heating it back up all over again. There are several ways to think about this. For starters, we need to create a large pressure difference across the turbine to make it spin efficiently, which means we've got to get that spent steam out of the turbine to make room for the incoming steam. You might suggest simply exhausting the steam to the atmosphere, but that would waste a lot of water. Moreover, turbines are easily damaged by impurities in the steam, which means highly purified water has to be used to protect the blades, so we would spend a lot of money on water purification if we use the water only once. We avoid both of those problems by condensing the steam and reusing it. Also, by condensing the steam we create a slight vacuum on the exhaust side of the turbine, which helps create the higher pressure difference across the turbine mentioned above. Finally, in the next chapter we will learn that the maximum possible efficiency of heat engines like this depends on how cold the working fluid becomes in its cycle as well as how hot it gets. So cooling water and condensers are an essential part of the system.

A typical coal plant converts only about one-third the energy in its fuel into the desired output—that is, electricity. About 85% of the remaining two-thirds of the fuel's energy leaves the plant in the form of waste heat in the cooling water. The remainder is lost out the stack. Figure 9.6 includes an energy balance for such a plant along with an estimate of the carbon emissions and cooling water requirements. To generate one kilowatt-hour of electricity, about 1 pound of coal and 39 gallons of cooling water are required and about 1 kg of CO_2 will be emitted into the atmosphere.

The cooling water demands of a large 1000 MW power plant are enormous. Approximately 1 billion gallons of water per day are withdrawn, passed through the condenser, and returned to the source typically 10°C warmer than its initial temperature. When a source of

figure
9.7   **Cooling Towers on the John Amos Coal-Fired Power Plant in West Virginia**

SOURCE: U.S. Army Corps of Engineers

cooling water is not conveniently located relative to the plant, large cooling towers are often used, such as the ones shown in Figure 9.7. A portion of the cooling water sprayed into these towers evaporates, transferring heat directly into the atmosphere, leaving the remaining water cool enough to return to the condenser.

Almost two-thirds of the energy put into a conventional power plant ends up in cooling water, but it is at such a low temperature that it is fairly useless. This is one of the principle disadvantages of centralized power generation: there is just not much you can do with a lot of lukewarm water, especially when it is likely to be miles from any potential application. Conversely, it is one of the principle advantages of small-scale, decentralized systems that can generate electricity at the site of the end user who might be able to put that waste heat to work. Such *combined-heat-and-power* systems (CHP) will be explored in the next chapter.

9.4.2 Flue-Gas Emission Controls

Power plants, especially those that burn coal, emit a number of toxic pollutants, including oxides of sulfur (SO_x), oxides of nitrogen (NO_x), and particulate matter (as well as the main culprit responsible for global warming, CO_2). Figure 9.6 shows some of the emission-control devices that can help remove such pollutants from the flue gases. Flue gas from the boiler is

SIDEBAR 9.1

Using Fly Ash to Reduce Carbon Emissions from Cement Production

A surprisingly large fraction of global carbon emissions (about 6% to 7%) is attributable to the production of cement. Cement, mixed with water, is the binding agent that holds together sand and gravel aggregate to make concrete. To produce 1 ton of cement requires about 2 tons of limestone and clay or sand along with a lot of heat. In the process, about 1 ton of CO_2 is liberated, partly from the fuel needed to supply that heat and partly from the chemical reactions taking place (calcination):

$$\underbrace{\text{Limestone (CaCO}_3) + \text{Clay or Sand (Si)}}_{\text{2 tons material}} + \overbrace{\text{Heat}}^{\substack{6 \times 10^6 \text{ Btu} \\ 1500°\text{C}}} \rightarrow \overbrace{\text{Cement}}^{1 \text{ ton}} + \overbrace{CO_2}^{1 \text{ ton}}$$

$$\left. \begin{array}{llll} \text{Calcination:} & CaCO_3 & \rightarrow & CaO + CO_2 \\ \text{Energy:} & C_xH_y + O_2 & \rightarrow & H_2O + CO_2 \end{array} \right\} \text{Roughly half and half}$$

Fly ash from power plants can replace cement on a one-for-one basis, meaning that for every ton of fly ash used in concrete, a bit more than 1 ton of CO_2 emissions will be saved. The concrete that results, even when well over half of the cement has been replaced with fly ash, has been shown to be stronger and more durable than ordinary concrete. Saving carbon emissions as well as avoiding disposal costs of fly ash is gaining attention, but even so, less than 10% of the 650 million tons generated annually is currently being recycled this way.

often sent to an electrostatic precipitator (ESP), which adds a charge to the particulates in the gas stream so they can be attracted to electrodes that collect this fly ash. Fly ash is normally buried, but it has a much more useful application as a replacement for cement in concrete (see Sidebar 9.1). Next, a flue-gas desulfurization system (FGD, or *scrubber*), sprays a limestone slurry over the flue gases, precipitating the sulfur to form a thick calcium sulfite sludge that must be dewatered and either buried in landfills or reprocessed into useful gypsum.

Not shown in Figure 9.6 are emission controls for nitrogen oxides, NO_x. Nitrogen oxides have two sources. Thermal NO_x is created when high temperatures oxidize the nitrogen, N_2, in air. Fuel NO_x results from nitrogen impurities in fossil fuels. Some NO_x emissions reductions have been based on careful control of the combustion process rather than with external devices such as scrubbers and precipitators. More recently, *selective catalytic reduction* (SCR) technology has proven effective. The SCR in a coal station is similar to the catalytic converters used in cars to control emissions. Before exhaust gases enter the smokestack, they pass through the SCR where anhydrous ammonia reacts with nitrogen oxide and converts it to nitrogen and water.

Flue-gas emission controls are not only very expensive, accounting for upwards of 40% of the capital cost of a new coal plant, but they also drain off about 5% of the power generated by the plant, which lowers overall efficiency.

9.4.3 Combustion Gas Turbines

Natural gas as a fuel for power plants has many environmental advantages over the coal-fired power plants just described. It burns cleaner and it is much less carbon intensive. Rather than boiling water to make steam, most gas plants use a turbine similar to that of a jet engine. As shown in Figure 9.8, a simple gas turbine consists of three major components: a compressor, a combustion chamber, and a power turbine. In the compressor, air is drawn in, compressed, and accelerated to several hundreds of miles per hour as it enters the combustion chamber. In the combustion chamber, a steady stream of fuel (usually natural gas) is injected and ignited, creating a high-pressure, high-temperature gas stream that expands through the turbine blades. The expanding hot gases spin the turbine and are then exhausted to the atmosphere. The compressor and turbine share a connecting shaft, so that a portion, typically more than half, of the rotational energy created by the spinning turbine is used to power the compressor. That shaft is also connected to the generator, which produces the desired electrical power output.

Gas turbines have long been used in industrial applications and as such were designed strictly as stationary power systems. These industrial gas turbines tend to be large machines made with heavy, thick materials whose thermal capacitance and moment of inertia reduce their ability to adjust quickly to changing loads. These workhorses tend to have relatively low efficiencies in the 20% to 30% range.

A newer style of gas turbine takes advantage of the billions of dollars of development work that has gone into designing lightweight, compact engines for jet aircraft. The thin, light, super-alloy materials used in these *aeroderivative turbines* enable fast starts and quick acceleration, so they easily adjust to rapid load changes and numerous start-up/shut-down events. Their small size makes it easy to fabricate the complete unit in the factory and ship it to a site, thereby reducing field installation time and cost. Their 30% to 40% efficiency makes them typically more efficient than their industrial counterparts.

figure 9.8   A Simple-Cycle, Natural-Gas-Fired Gas Turbine

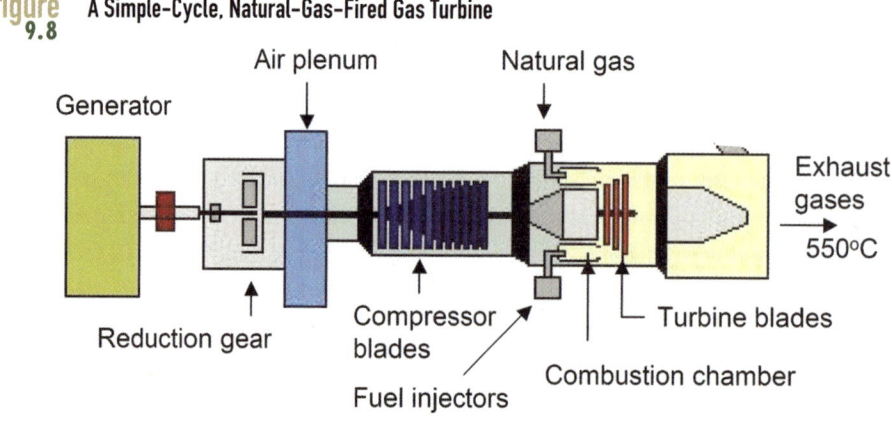

9.4.4 Combined-Cycle Power Plants

Notice the temperature of the gases exhausted into the atmosphere in the simple-cycle gas turbine shown in Figure 9.8 is over 500°C. Clearly that is a tremendous waste of high-quality heat that could be captured and put to good use. One way to do so is to pass those hot gases through a heat exchanger to boil water and make steam. The heat exchanger is called a heat-recovery-steam-generator (HRSG) and the resulting steam can be put to work in a number of applications, including industrial process heat or water and space heating for buildings. Of course, this combined heat-and-power (CHP) operation is viable only if the gas turbine is located very close to the site where its waste heat can be utilized.

Why not use the steam generated in an HRSG to power a second-stage steam turbine to generate more electricity? That is precisely what is done in a new generation of high-efficiency natural-gas-fired power plants called *combined-cycle* plants. An example of how a gas turbine can be coupled with a steam turbine is shown in Figure 9.9. Working together, such combined-cycle plants have achieved fuel-to-electricity efficiencies approaching 60%.

9.4.5 Integrated Gasification Combined-Cycle (IGCC) Power Plants

With combined-cycle plants achieving such high efficiencies, and with natural gas being an inherently cleaner fuel, the trend in the United States has been away from building new coal-fired power plants. Coal, however, is a much more abundant fuel than natural gas, but in its

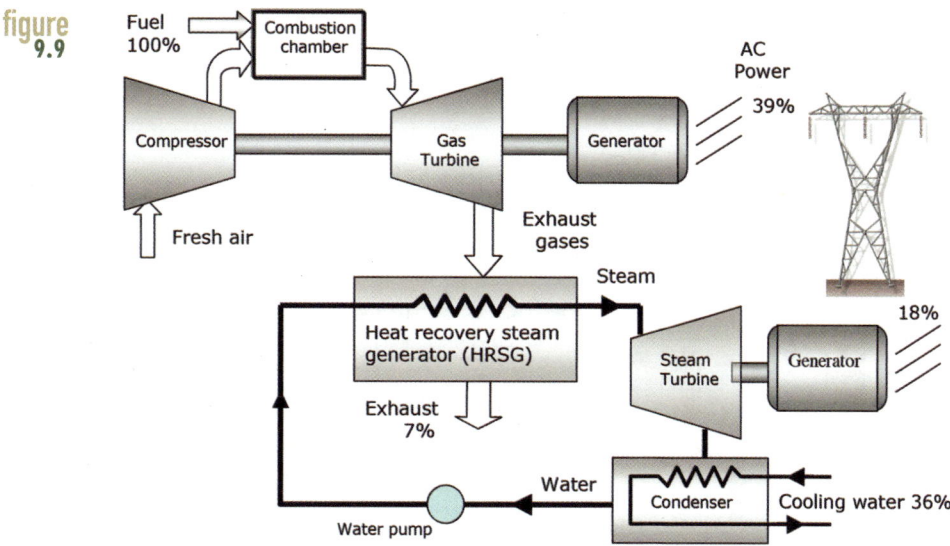

figure
9.9

Combined-cycle power plants have achieved efficiencies approaching 60%.

conventional, solid form, it cannot be used in a gas turbine. Erosion and corrosion of turbine blades due to impurities in coal would quickly ruin a gas turbine. However, coal can be processed to convert it into a synthetic gas, which can be burned in what is called an *integrated gasification, combined-cycle* (IGCC) power plant.

Gas derived from coal, called "town gas," was popular in the late 1800s before the discovery of large deposits of natural gas. One hundred years later, coal's air pollution problems prompted the refinement of technologies for coal gasification. Several gasification processes have been developed, primarily in the Great Plains Gasification Plant in Beulah, North Dakota, in the 1970s and later in the 100 MW Cool Water project near Barstow, California, in the 1980s. These early experimental facilities established the technical foundations for future, more commercially viable, IGCC plants.

As shown in Figure 9.10, the essence of an IGCC consists of bringing a coal-water slurry into contact with steam to form a fuel gas consisting mostly of carbon monoxide (CO) and hydrogen (H_2). The fuel gas is cleaned up, removing most of the particulates, mercury, and sulfur, and then burned in the gas turbine. Air used in the combustion process is first separated into nitrogen and oxygen. The nitrogen is used to cool the gas turbine and the oxygen is mixed with the gasified coal, which helps increase combustion efficiency. Despite energy losses in the gasification process, by taking advantage of combined-cycle power generation an IGCC can burn coal with an overall thermal efficiency of around 45%. This is considerably higher efficiency than conventional pulverized coal plants, but still far below the best 60%-efficient combined-cycle natural-gas plants.

IGCC plants are more expensive than pulverized coal plants and they have trouble competing economically with natural-gas-fired combined-cycle plants. As of 2007, there were only four IGCC plants in the world—two in Europe and two in the United States. One of the U.S. plants is located on the Wabash River in Indiana; the other is a newer, state-of-the-art plant near Tampa, Florida. Several others are in the planning stage, including a proposed 1200 MW plant that American Electric Power (AEP) hopes to build somewhere along the Ohio River. Rising and

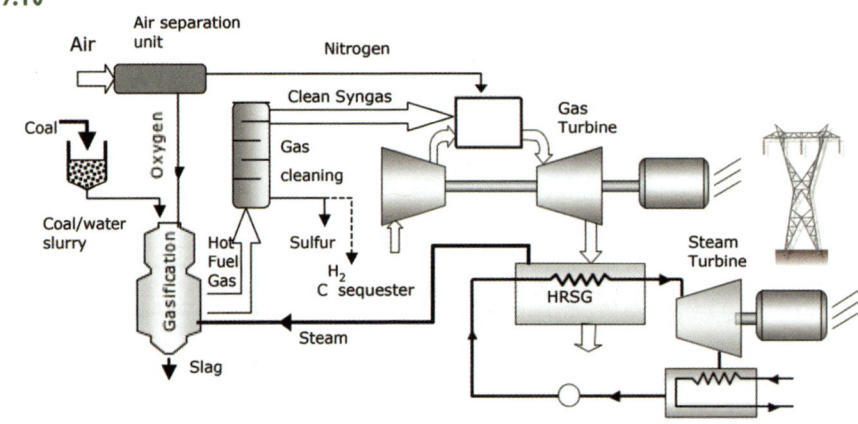

figure 9.10   An Integrated Gasification, Combined-Cycle (IGCC) Power Plant

uncertain future natural gas prices, coupled with the potential to remove carbon in the syngas before it is burned, have reignited interest in IGCC plants. They enjoy significant political support and so are likely to grow in importance in our future electricity mix.

9.4.6 IGCC with Carbon Sequestration

Coal is the most carbon-intense fossil fuel, and it is also the most abundant. If we continue to rely on coal to satisfy most of the world's growing electric power demands, and if we don't control its carbon emissions, our global climate future is indeed precarious.

Table 9.1 presents estimates of the energy content of the world's remaining fossil-fuel resources divided into conventional sources of the type now being exploited, as well as unconventional sources such as oil shale, tar sands, and heavy crude that might be developed in the future. Also shown is an average value of the *carbon intensity* of each fuel (carbon per unit of energy). The product of the resource base times the carbon intensity gives us an estimate of the carbon that would be released if the entire resource were to be consumed. As can be seen, the total carbon content of coal is triple the combined carbon content of oil and gas.

The final column in Table 9.1 converts the carbon emission potential of each fuel into the added CO_2 that would accumulate if half of the carbon emitted remains in the atmosphere (that ratio is called the *airborne fraction*). By these estimates, burning all of the world's coal could triple the CO_2 concentration in the atmosphere from the current 380 ppm to 1144 ppm (380 + 764).

The most promising way to control carbon emissions from coal-fired power plants is based on the IGCC design shown in Figure 9.10. By first converting coal to a syngas in an IGCC plant, it is possible to design the gas-cleaning stage in such a way that the carbon in the fuel gas can be extracted before combustion. By doing so, it may be possible to build "clean coal" power plants that would be able to take advantage of the relative abundance of coal without contributing to global warming. If a carbon sequestration technology could be developed to store that carbon in perpetuity, it may be possible to envision a future with carbon-free, high-efficiency, coal-fired power plants capable of supplying clean electricity for several centuries into the future.

table 9.1 Fossil-fuel resource estimates, low-heating-value (LHV) carbon intensity, and potential additions to global atmospheric CO_2 if the resource is totally consumed and half of its carbon remains in the atmosphere. Burning all of our coal could triple atmospheric CO_2.

Fuel	Conventional resources (Exajoules)	Unconventional resources (Exajoules)	Total resource base (Exajoules)	Carbon intensity (Gton C/EJ)	Carbon potential (Gton C)	Add'l CO_2 at 50% AF (ppm CO_2)
Natural gas	9,200	26,900	36,100	0.0150	542	128
Petroleum	8,500	16,100	24,600	0.0200	492	116
Coal	25,200	100,300	125,500	0.0258	3,238	764

SOURCE: based on Nakicenovic, 1996

The key, of course, is finding a way to store all of that carbon, essentially forever. At present there are some carbon sequestration processes underway, but those involve capturing CO_2 and injecting it into oil fields to enhance oil recovery. The injected CO_2 helps push more oil out of the source rock far below the surface. Of course, when that oil is burned, more CO_2 is added back into the atmosphere.

More promising is permanent CO_2 storage in geologic formations such as deep brine aquifers. Such formations consist of highly porous rock, similar to those containing oil and gas, but without the hydrocarbons that produced our fossil fuels. Instead, they are filled with water containing high concentrations of salts dissolved out of the surrounding rocks. When these formations are capped with impermeable rock they might be viable for CO_2 sequestration for the indefinite future.

9.4.7 Nuclear Power

Nuclear power has had a rocky history, leading it from its glory days in the 1970s as a technology thought to be "too cheap to meter," to a technology that in the 1980s some characterized as "too expensive to matter." The truth is probably somewhere in the middle. It does have the advantage of being a carbon-free source of electric power, so it is beginning to enjoy a resurgence of interest. Whether a new generation of cheaper, safer reactors can overcome public misgivings over where to bury radioactive wastes and how to keep plutonium from falling into the wrong hands, remains to be seen.

The essence of nuclear reactor technology is basically the same simple steam cycle described for fossil fueled power plants. The main difference is the heat is created by nuclear reactions (see section 4.7.2 of this book) instead of fossil fuel combustion.

Light water reactors. Water in a reactor core not only acts as the working fluid, it also serves as a *moderator* to slow down neutrons ejected when uranium fissions. In *light water reactors* (LWRs), ordinary water is used as the moderator. Figure 9.11 illustrates the two principle types of LWRs. *Boiling water reactors* (BWRs) make steam by boiling water within the reactor core itself, whereas in *pressurized water reactors* (PWRs) a separate heat exchanger, called a steam generator, is used. PWRs are more complicated, but they can operate at higher temperatures than BWRs and hence are somewhat more efficient. PWRs can be somewhat safer because a fuel leak would not pass any radioactive contaminants into the turbine and condenser. Both types of reactors are used in the United States, but the majority are PWRs.

Heavy water reactors. Reactors commonly used in Canada use heavy water; that is, water in which some of the hydrogen atoms are replaced with deuterium (hydrogen with an added neutron). The deuterium in heavy water is more effective in slowing down neutrons than ordinary hydrogen. The advantage in these Canadian deuterium reactors (commonly called CANDU) is that ordinary uranium as mined, which contains only 0.7% of the fissile isotope U-235, can be used without the enrichment that LWRs require.

figure 9.11   The Two Types of Light Water Reactors Commonly Used in the United States

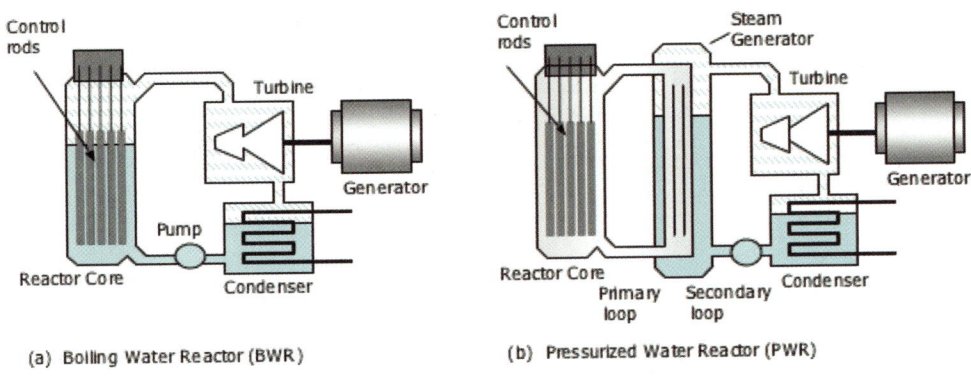

(a) Boiling Water Reactor (BWR)   (b) Pressurized Water Reactor (PWR)

High-temperature, gas-cooled reactors (HTGR). HTGRs use helium as the reactor core coolant rather than water, and, in some designs, it is helium itself that drives the turbine. These reactors operate at considerably higher temperatures than conventional water-moderated reactors, which means their efficiencies can be higher—upwards of 45% rather than the 33% that typifies LWRs.

There are two HTGR concepts under development: the Prismatic Fuel Modular Reactor (GT-MHR) based on German technology and the Modular Pebble Bed Reactor (MPBR), which is being developed in South Africa. Both are based on microspheres of fuel, but differ in how they are configured in the reactor. The MPBR incorporates the fuel microspheres in carbon-coated balls ("pebbles") roughly 2 inches in diameter. One reactor will contain close to half a million such balls. A potential advantage of pebble reactors is that they could be refueled continuously by adding new balls and withdrawing spent-fuel balls without having to shut down the reactor.

The nuclear fuel "cycle." The costs and concerns for nuclear fission are not confined to the reactor itself. Figure 9.12 shows current practice from mining and processing of uranium ores, to enrichment that raises the concentration of U-235, to fuel fabrication and shipment to reactors. Highly radioactive spent fuel removed from reactors these days sits on-site in short-term storage facilities while we await a longer-term storage solution such as the underground federal repository planned for Yucca Mountain, Nevada. Eventually, after forty years or so, the reactor reaches the end of its useful lifetime. At that point, it will have to be decommissioned, and its radioactive components will also have to be transported to a secure disposal site.

Reactor wastes contain not only fission fragments formed during the reactions, which tend to have half-lives measured in decades, but also include some radionuclides with very long half-lives. Of major concern is plutonium, which has a half-life of 24,390 years. Only a few percent of the uranium atoms in reactor fuel are the fissile isotope U-235, whereas essentially all of the rest are U-238, which does not fission. Uranium-238 can, however, capture a neutron and be transformed into plutonium as the following reactions suggest:

Eq. 9.1
$$^{238}_{92}U + n \rightarrow \,^{239}_{92}U \xrightarrow{\beta} \,^{239}_{93}Np \xrightarrow{\beta} \,^{239}_{94}Pu$$

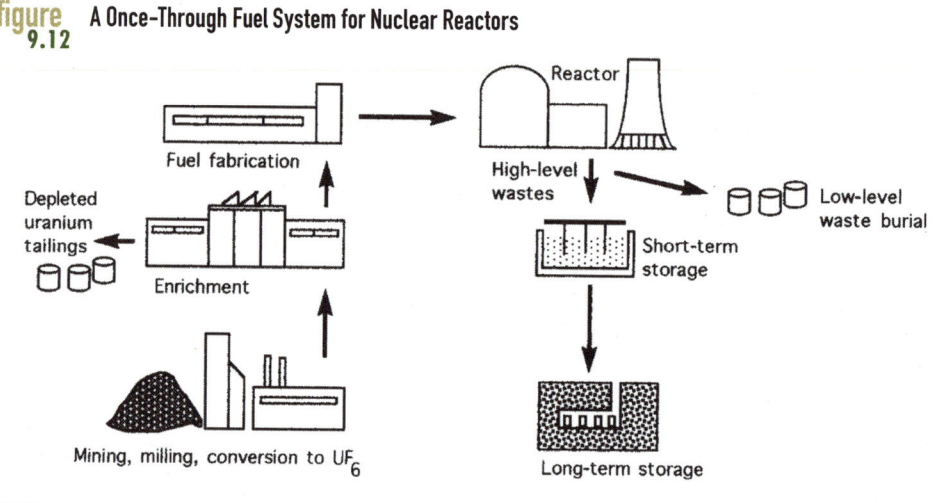

figure 9.12 **A Once-Through Fuel System for Nuclear Reactors**

This plutonium, along with several other long-lived radionuclides, makes nuclear wastes dangerously radioactive for tens of thousands of years, which greatly increases the difficulty of providing safe disposal. Removing the plutonium from nuclear wastes before disposal has been proposed as a way to shorten the decay period, but that introduces another problem. Plutonium not only is radioactive and highly toxic, it is also the critical ingredient in the manufacture of nuclear weapons. A single nuclear reactor produces enough plutonium each year to make dozens of small atomic bombs and some have argued that if the plutonium is separated from nuclear wastes the potential of illicit diversions for such weapons would cause an unacceptable risk.

On the other hand, the plutonium is a fissile material, which, if separated from the wastes, can be used as a reactor fuel. Indeed, France, Japan, Russia, and the United Kingdom have reprocessing plants in operation to capture and reuse that plutonium. In the United States, however, Presidents Ford and Carter considered the proliferation risk too high and commercial reprocessing of wastes has ever since not been allowed. Moreover, a recent major study of nuclear power at MIT recommends that reprocessing continue to not be pursued (Sidebar 9.2). Figure 9.13 shows the nuclear fuel cycle including the controversial reprocessing steps.

9.5 Economics of Centralized Power Plants

With such a range of generation technologies to choose from, how should a utility, or society in general, make decisions about which to use? An economic analysis is usually the basis for comparison. Costs of construction, fuel, operations and maintenance (O&M), and financing are crucial factors. Some of these can be straightforward engineering and accounting estimates and others, such as the future cost of fuel and whether there will be a carbon tax and if so, how much and when, require something akin to a crystal ball. Even if these cost estimates can be agreed upon, there are other costs, called *externalities*, that society must bear

SIDEBAR

SIDEBAR 9.2

The Interdisciplinary MIT Study on the Future of Nuclear Power

In July 2003, a distinguished team of researchers from the Massachusetts Institute of Technology and Harvard released one of the most comprehensive, interdisciplinary studies ever conducted on the future of nuclear energy. Their findings and recommendations were heavily influenced by the need for carbon-free sources of power.

They believe there are only four realistic options for reducing CO_2 emissions from electricity generation in the next few decades, and that all four need to be pursued:

1. Increased efficiency in generation and use
2. Expanded use of renewable energy sources such as wind, solar, and geothermal
3. Carbon sequestration, especially from coal-fired power plants
4. Increased use of nuclear power

They identify four unresolved problems that limit today's prospects for nuclear power:

1. *Costs:* Nuclear power has higher overall lifetime costs compared to coal and combined-cycle natural gas, at least in the absence of a carbon tax. To be competitive, capital costs, operations and maintenance costs, construct time, and financing costs all need to be reduced.
2. *Safety:* The Three Mile Island and Chernobyl nuclear accidents, growing concern for the security of nuclear facilities from terrorist attack, and risks associated with transporting nuclear materials have adversely affected public perception.
3. *Proliferation:* Spread of nuclear weapons either by theft when wastes are reprocessed to extract plutonium, or by misuse of nuclear technologies for weapons production by countries developing their own reactor programs.
4. *Waste:* Unresolved challenges in long-term management of radioactive wastes.

They recommend that priority be given to the deployment of the once-through fuel cycle, rather than the development of the more expensive, greater proliferation risk, closed-fuel-cycle technology involving reprocessing.

that are not usually included in such calculations, such as health care and other costs of the pollution produced. Other complicating factors include the vulnerability we expose ourselves to with large, centralized power plants, transmission lines, pipelines, and other infrastructure that may fail due to natural disasters, such as hurricanes and earthquakes, or less-natural ones, due to terrorism or war.

9.5.1 Cost Per Kilowatt-Hour

In concept, figuring out the cost of electricity from a power plant is simple. Just figure out the annual cost of owning and operating the power plant and divide that by the annual number of kilowatt-hours of electricity generated.

figure 9.13   Nuclear Fuel Cycle with Reprocessing

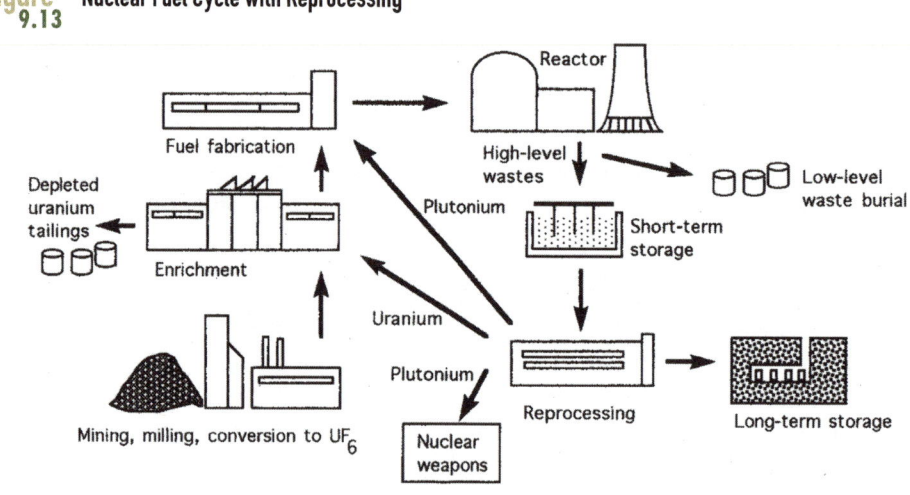

In its simplest form, the cost of electricity from a power plant can be expressed in terms of *fixed costs* that are incurred whether or not the power plant ever runs, and *variable costs* that depend on how much energy the plant actually generates. Fixed costs include money that has to be spent every year to pay for construction financing, return on debt and equity, insurance, taxes, depreciation, and routine O&M needed whether or not the plant is operated. Variable costs associated with running the plant consist mostly of fuel costs and production-related O&M. When both fixed and variable costs are expressed as $/year amounts, they can be combined to determine the annual cost of owning and operating the power plant.

Fixed costs can be annualized by multiplying the capital cost of the plant by a quantity known as the *fixed charge rate* (FCR):

Eq. 9.2   **Annual fixed costs ($/yr) = Capital cost ($) × FCR (%/yr)**

The FCR typically ranges between 11% and 18% per year, depending mostly on the cost of capital, which in turn is a function of current economic conditions, and the perceived risk and O&M for a particular technology.

Variable costs depend on the power plant efficiency, the price of fuel, operations-related O&M, and how much the plant actually runs.

Power plant efficiency in the United States is often described in terms of a *heat rate*, which is the number of Btus needed to generate 1 kWh of electricity; the smaller the heat rate, the higher the efficiency. For example, a new pulverized coal plant may have a heat rate of about 9300 Btu/kWh whereas an advanced combined-cycle natural gas plant heat rate can be down around 6000 Btu/kWh.

Fuel costs in these calculations should be levelized to account for the varying price of fuel over the life of the power plant. This, of course, gets pretty tricky. As an indication of the volatility of fuel prices, consider the twenty-year variation in natural gas prices shown in Figure 9.14.

figure
9.14 **Twenty Years of Natural Gas Prices at Henry Hub, Louisiana**

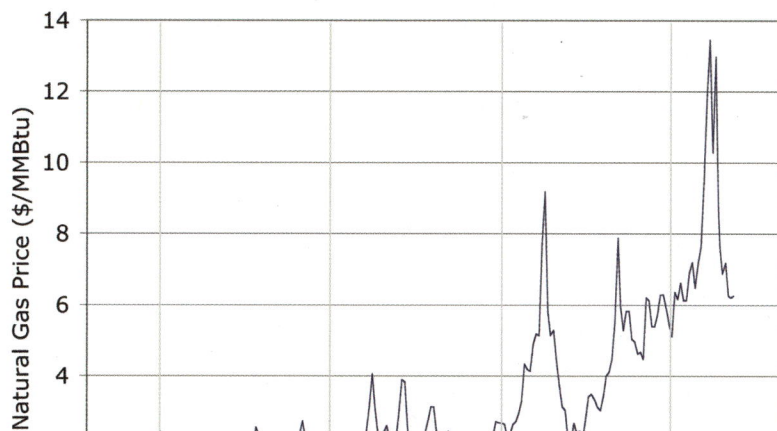

The energy delivered by a power plant can be described by its *rated power*, P_R, which is the power it delivers when operating at full capacity, and its *capacity factor* (CF), which is the ratio of the actual energy delivered by a power plant to the energy that would have been delivered if the plant ran continuously at full rated power. Assuming rated power in kW, annual energy in kWh, and 24 hours/day × 365 days/year = 8760 hours in a year, the annual energy delivered by a power plant is given by

Eq. 9.3 $$\text{Annual energy (kWh/yr)} = P_R \text{ (kW)} \times 8760 \text{ hr/yr} \times \text{CF}$$

Solution Box 9.1 shows how to combine costs and energy production to estimate the average cost of electricity generated.

If we repeat the calculations shown in Solution Box 9.1, while varying the capacity factor and the cost of fuel, we can easily derive the graph shown in Figure 9.15. At 2007 prices for natural gas at around $7/MMBtu, with a high capacity factor, electricity from an NGCC can cost a little less than 6¢/kWh. If the trends shown in Figure 9.14 are any indication, however, a levelized cost of natural gas of more like $10/MMBtu seems likely and the cost of electricity from an NGCC plant rises above 8¢/kWh.

9.5.2 Comparison of Costs for Generation Technologies

Some technologies, such as coal and nuclear plants, tend to be expensive to build and cheap to operate, so they make sense only if they run almost all of the time. Others, such as gas

SOLUTION BOX 9.1

The Cost of Electricity for a Natural Gas, Combined-Cycle Plant

What is the cost of energy (COE) for a natural gas, combined-cycle (NGCC) power plant using the following cost factors?

Capital cost:	$500/kW
Fixed charge rate:	14.8%/year
Average heat rate:	7000 Btu/kWh
Levelized fuel/O&M cost:	$7.00 per million Btu ($7.00/MMBtu)
Capacity factor:	0.85

Solution:

For simplicity, let's assume the rated power of the plant is just 1 kW:

$$\text{Annualized capital cost} = 1 \text{ kW} \times \$500/\text{kW} \times 0.148/\text{yr} = \$72.40/\text{yr}$$
$$\text{Annual energy produced} = 1 \text{ kW} \times 8760 \text{ hr/yr} \times 0.85 = 7446 \text{ kWh/yr}$$
$$\text{Fuel/O\&M} = 7446 \text{ kWh/yr} \times 7000 \text{ Btu/kWh} \times \$7/10^6 \text{ Btu} = \$364.85/\text{yr}$$
$$\text{Total annual costs} = \$72.40 + \$364.85 = \$437.25$$
$$\text{Cost of electricity} = \frac{437.25/\text{yr}}{7446 \text{ kWh/yr}} = \$0.059/\text{kWh} = 5.9 ¢/\text{kWh}$$

(Notice that the final cost of electricity does not depend on having chosen a 1 kW plant.)

turbines, are just the opposite: cheap to build and expensive to operate, so they are better used as peakers (recall Figure 9.5). An economically efficient power system will include a mix of power plant types appropriate to the variation in power demand from day to day and from month to month.

An example of the cost of energy for four types of power plants is shown in Figure 9.16. Costs for simple combustion turbine (CT), a pulverized coal plant (coal), a natural gas combined-cycle (NGCC), and an estimate for the cost of a new nuclear power plant, are compared. As can be seen, for this example CT is the least expensive option as long as it runs fewer than about 1500 hours/year (CF < 0.17), which means it is very appropriate for a peaking power plant that operates only a few hours each day. The coal plant is most cost-effective when it runs at least 3700 hours/year (CF > 0.38), which means it should be a base-load plant operating almost all of the time. The NGCC is least expensive when it runs between 1500 and 3700 hours/year, which means it is most appropriate as an intermediate-load power plant.

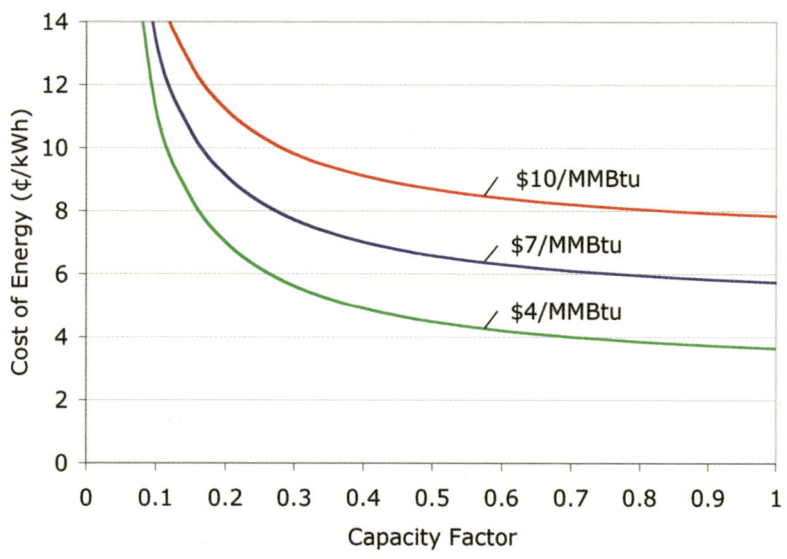

figure 9.15   Sensitivity Analysis for the Natural Gas, Combined–Cycle (NGCC) Power Plant of Solution Box 9.1

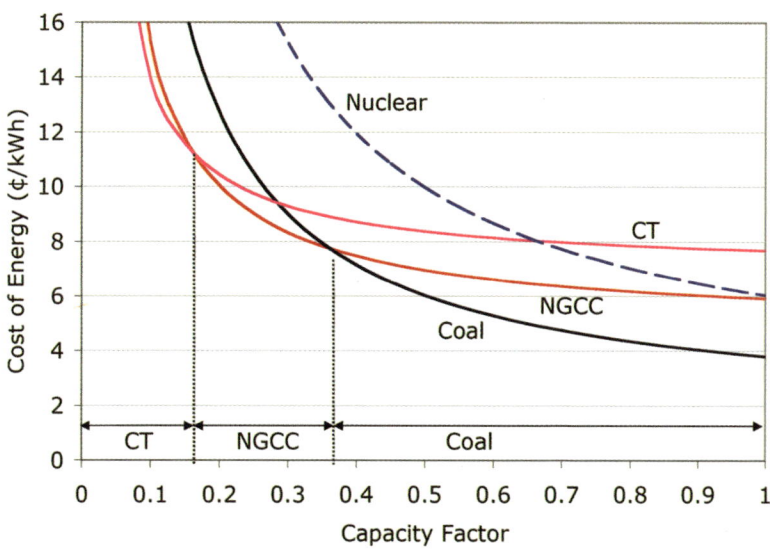

figure 9.16

The least expensive power technology depends on its capacity factor. For this example, CT is the least expensive for CF < 0.17; coal is least for CF > 0.38; in between NGCC is the least expensive option. Assumptions: FCR = 15%, gas @ $7.00/MMBtu; CT $400/kW, 10,000 Btu/kWh; NGCC $600/kW, 7000 Btu/kWh; Coal $1300/kW, 9300 Btu/kWh, $1.70/MMBtu; Nuclear $2300/kW, 10,500 Btu/kWh, $2.00/MMBtu.

Under the assumptions given in Figure 9.16, nuclear power is not competitive. However, if society institutes a carbon tax that penalizes coal and gas plants, the situation could change. Figure 9.17 shows the implications of adding a cost of carbon emissions to the coal, NGCC, and nuclear plants analyzed in Figure 9.16. With carbon emissions costing more than \$28 per metric ton of CO_2, nuclear power would be the least expensive option; above \$38/ton, coal becomes the most expensive option.

9.6 Electric Power Infrastructure: Transmission and Distribution

Power plants generate electricity and transmission lines and distribution (T&D) systems carry it to customers. Figure 9.18 provides a simple schematic of a complete system consisting of generating stations with transformers to bump up voltages to the high values needed for efficient transmission. High-voltage transmission lines carry bulk power tens or hundreds of miles away to major load centers. Distribution substations drop voltages to levels suitable for local power lines to deliver power to every factory, business, and home that needs it.

Although the emphasis in this chapter thus far has been on generation of electricity, those costs are often less than half of the total utility bill you are likely to receive for your home. In fact, for the last twenty years or so, utilities have spent more on transmission and distribution than on generation.

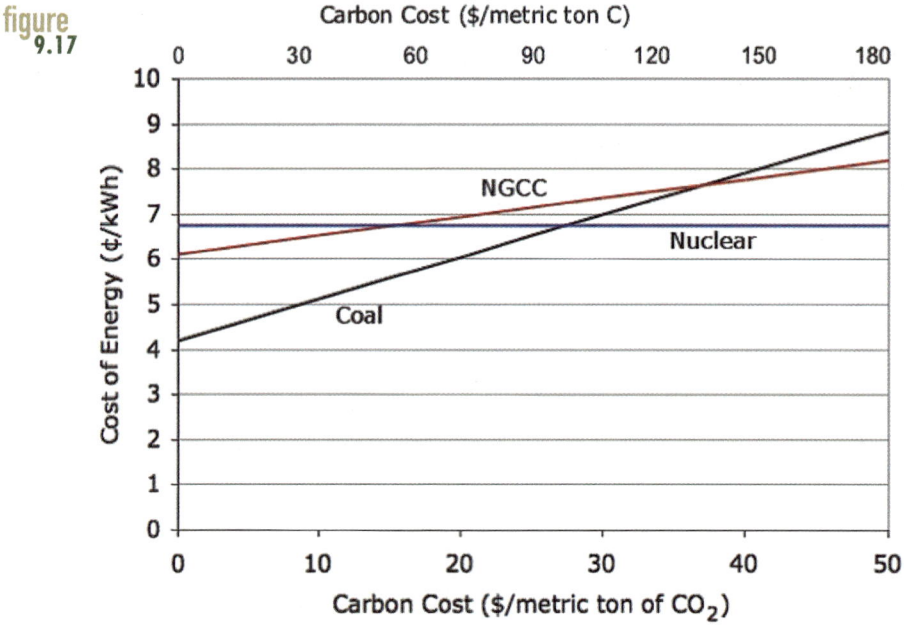

figure 9.17

In this analysis, nuclear power would be the least expensive option if carbon were priced above \$28 per metric ton of CO_2. Assumptions are as given in Figure 9.16 with all three plants operating at an 85% capacity factor.

figure
9.18

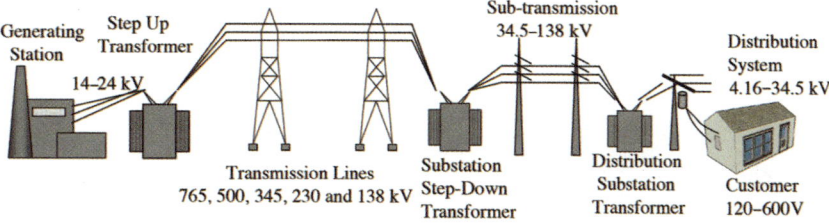

Generation, transmission, and distribution systems convert fuels to electricity and deliver it tens or hundreds of miles away to utility customers.

9.6.1 The North American Power Grid

The system in Figure 9.18 suggests a rather linear system with one straight path from sources to loads. In reality, there are multiple paths that electric currents can take to get from generators to end users. Transmission lines are interconnected at switching stations and substations, with lower-voltage, "sub-transmission" lines and distribution feeders extending into every part of the system. The vast array of transmission and distribution lines is called a power "grid." Within a grid, it is impossible to know which path electricity will take as it flows at nearly the speed of light, seeking out the path of least resistance, to get from generator to load.

Figure 9.19 shows a map of the basic structure of the North American power grid including transmission lines and the 140 or so major control-area dispatch centers. Because it is uneconomical to store significant quantities of electricity, at every instant in time the power generated within a dispatch area must equal the power demanded by loads. It is the job of these dispatch centers to constantly juggle the output of their generation facilities to provide just the right amount of power to meet that moment's demand.

Figure 9.19 suggests the North American power grid is one giant interconnected machine, but it actually consists of three separate interconnected grids: the Western Interconnection, the Texas Interconnection, and the Eastern Interconnection (Figure 9.20). Within each of these interconnection zones, everything is synchronized so that voltages, frequencies, phase angles, and currents are locked together into a single enormous ac circuit. Interconnections between the grids are made using the high-voltage dc (HVDC) links shown in Figure 9.19. These links consist of *rectifiers* that convert ac to dc, a connecting HVDC transmission line, and *inverters* that convert dc back to ac. The advantage of a dc link is that problems associated with exactly matching ac frequency, phase, and voltages from one interconnect to another are eliminated in dc. HVDC links can also connect various parts of a single grid, as is the case with the 6000 MW Pacific Intertie between the Pacific Northwest and southern California. Quite often national grids of neighboring countries are linked this way as well.

figure
9.19

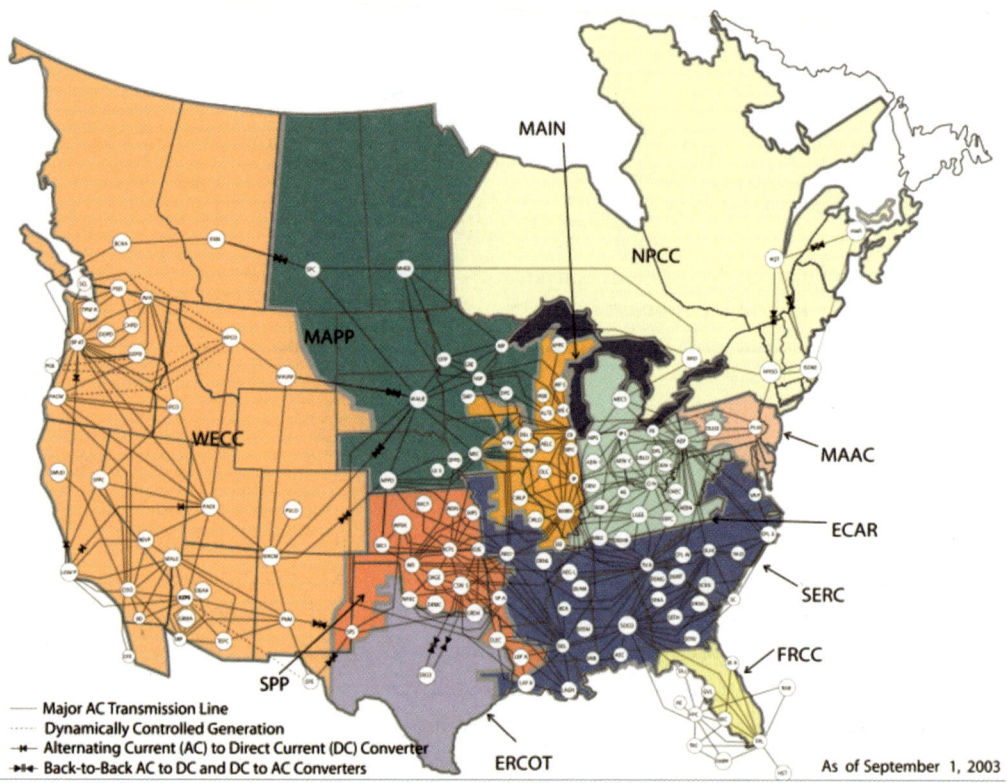

The North American power grid, showing major transmission lines, control-area dispatch centers, and the areas covered by regional electricity reliability councils.

SOURCE: U.S.–Canada Power System Outage Task Force, 2004

9.6.2 Grid Stability

Most blackouts are short term, such as when someone pays more attention to his cell phone than his driving, hits a power pole, and our lights go out for a few minutes. Some are predictable, as in the rolling blackouts that California had to endure during its ill-fated experiment with deregulation in 2000–2001. Some are longer term, as in the great blackout that hit the Midwest and Northeast parts of the United States, as well as Ontario, Canada, in August 2003. That blackout caused 50 million people to be without power, some for as long as four days, and cost the United States roughly $4 billion to $10 billion.

The organization that takes responsibility for overall grid reliability has traditionally been the North American Electric Reliability Council (NERC). NERC is a nonprofit corporation made up of members of the ten regional reliability councils shown in Figure 9.20. The blackout of 2003, coupled with growing concern for terrorism, motivated Congress to pay

figure
9.20

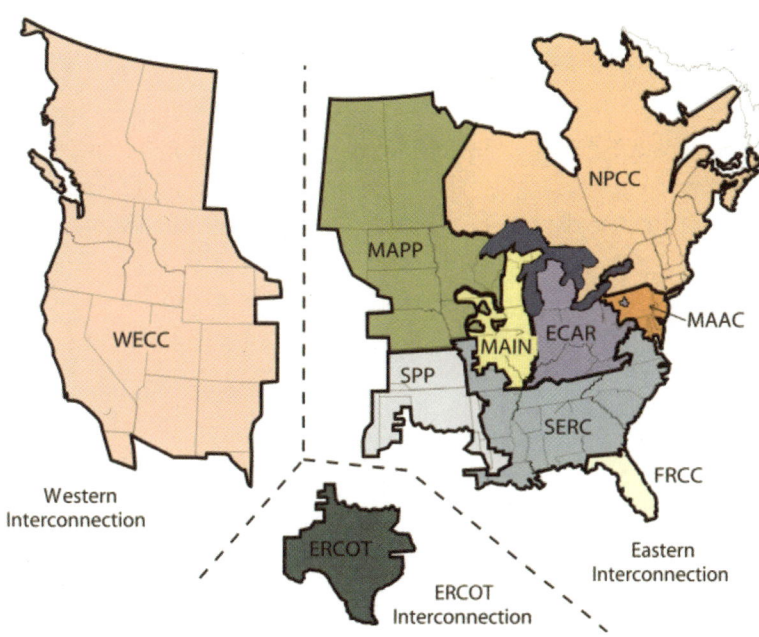

The North American power grid actually consists of three separate grids: The Western Inter-connection, the Texas (ERCOT) Interconnection, and the Eastern Interconnection. These are connected to each other using relatively small HVDC links.

more attention to the grid and to include in the Energy Policy Act of 2005 the creation of an *Electricity Reliability Organization* (ERO) that will likely augment or even replace NERC.

During normal operation, the grid responds to slight imbalances in supply and demand by adjusting the voltage and/or frequency of the nominal 60-hertz voltage (it increases the frequency and/or raises the voltage when generation exceeds demand). Small variations are routine; however, large deviations in frequency can cause the rotational speed of generators to fluctuate, leading to vibrations that can damage turbine blades and other equipment. Significant imbalances can lead to automatic shutdowns of portions of the grid, which can affect thousands of people. When parts of the grid shut down, especially when that occurs without warning, power that surges around the outage can potentially overload other parts of the grid causing those sections to go down as well.

Most often, major blackouts occur when the grid is running at near capacity, which for most of the United States occurs during the hottest days of summer when the demand for air-conditioning is at its highest. Perhaps surprisingly, one of the most common triggers for blackouts on those hot days results from insufficient attention having been paid to simple management of tree growth within transmission-line rights-of-way (see Sidebar 9.3). When lines get hot, they expand. When they expand, they sag more and are more likely to short out

SIDEBAR

SIDEBAR 9.3

Why Do Trees Sometimes Cause Blackouts on Hot Days?

Several major blackouts have been caused by transmission lines shorting out when contacting trees in the right-of-way—especially on hot summer days. Why is that?

1. Peak power demands increase i^2R heating of lines causing them to expand and sag.
2. On hot days, air does not cool the lines as much, increasing the temperature sag.

3. Low wind speeds may not adequately cool the lines, raising line temperature and sag.
4. Vegetation grows during the summer, increasing the chance of line contact.
5. Power outage in one place may increase current in another line, increasing its sag.

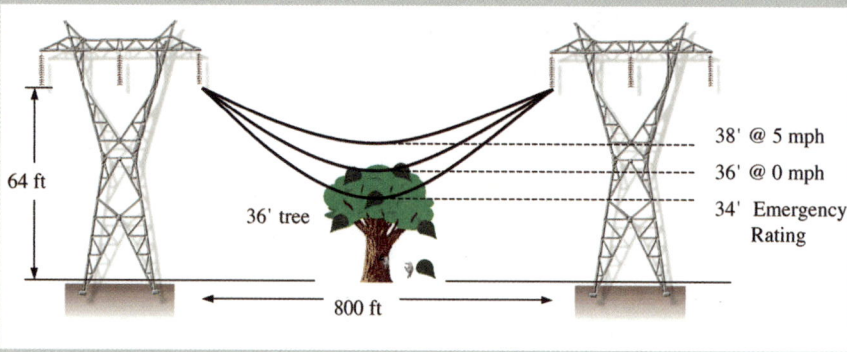

64 ft

36' tree

800 ft

38' @ 5 mph
36' @ 0 mph
34' Emergency Rating

in trees. Lines, if allowed to stay stretched, acquire a permanent stretch in them, so there are emergency limits to how much they are allowed to stretch and for how long.

9.7 Evolving Regulation of Electric Power

Samuel Insull shaped what became the modern electric utility by bringing into being the concepts of regulated utilities with monopoly franchises. In exchange for the right to be the only provider of electric power within a designated service territory, utilities accepted the obligation to serve the public by providing reliable service to every customer at rates that were to be determined by public utility commissions (PUCs). The economies of scale that went with increasingly large steam power plants led to an industry based on centralized generation coupled with a complex infrastructure of transmission lines and distribution facilities.

At the end of the twentieth century, however, the benefits of Insull's model began to unravel. Economies of scale had played out and big customers wanted direct access to power that could now be generated by new, smaller turbines at a lower price than grid power. Focus also shifted to the customer's side of the meter when it was realized that it was cheaper and faster to help customers save energy than for utilities to build more power plants. And, finally, growing awareness of the environmental benefits of renewable energy systems led to pressure to find ways to encourage their use.

9.7.1 The Public Utility Holding Company Act of 1935 (PUHCA)

In the early part of the twentieth century, as enormous amounts of money were being made, utility companies began to merge and grow into larger conglomerates. A popular corporate form emerged, called a *utility holding company*. A holding company is a financial shell that exercises management control of one or more companies through ownership of their stock. Holding companies began to purchase each other and by 1929, sixteen holding companies controlled 80% of the U.S. electricity market, with just three of them owning 45% of the total.

With so few entities having so much control, it should have come as no surprise that financial abuses would emerge. Holding companies formed pyramids with other holding companies, each owning stocks in subsequent layers of holding companies. An actual operating utility at the bottom found itself directed by layers of holding companies above it, with each layer demanding its own profits. At one point, these pyramids were sometimes ten layers thick. When the stock market crashed in 1929, the resulting Great Depression drove many holding companies into bankruptcy causing investors to lose fortunes. Insull became somewhat of a scapegoat for the whole financial fiasco associated with holding companies and he fled the country amidst charges of mail fraud, embezzlement, and bankruptcy violations, charges of which he was later cleared.

In response to these abuses, Congress created the *Public Utility Holding Company Act of 1935* (PUHCA) to regulate the gas and electric industries and prevent holding company excesses from reoccurring. Many holding companies were dissolved, their geographic size was limited, and the remaining ones came under control of the newly created Securities and Exchange Commission (SEC).

Although PUHCA has been an effective deterrent to previous holding-company financial abuses, recent changes in utility regulatory structures, with their goal of increasing competition, led many to say it had outlived its usefulness. The main issue was a provision of PUHCA that restricted holding companies to business within a single integrated utility, which is a major deterrent to the modern pressure to allow wholesale wheeling of power from one region in the country to another. As a result, Congress repealed PUHCA in its Energy Policy Act of 2005.

9.7.2 The Public Utility Regulatory Policies Act of 1978 (PURPA)

With the country in shock from the oil crisis of 1973 and with the economies of scale associated with ever larger power plants having pretty much played out, the country was drawn toward energy efficiency; renewable energy systems; and new, small, inexpensive gas turbines.

To encourage these systems, President Carter signed the Public Utility Regulatory Policies Act of 1978 (PURPA).

The two key provisions of PURPA relate to allowing *independent power producers* (IPPs), under certain restricted conditions, to connect their generators to the utility-owned grid. IPPs, for example, were often customers trying to generate some of their own power on-site. Prior to PURPA, utilities could refuse service to such customers, which meant they would have to provide all of their own power, all of the time, including their own redundant, backup power systems. That reality virtually eliminated the possibility of using efficient, economical, on-site power production to offset a portion of a customer's load.

PURPA not only allowed grid interconnection but it also required utilities to purchase electricity from certain *qualifying facilities* (QFs) at a "just and reasonable price." The purchase price of QF electricity was to be based on what it would have cost the utility to generate the power itself or to purchase it on the open market (referred to as the *avoided cost*). This provision stimulated the construction of numerous renewable energy facilities, especially in California, because PURPA guaranteed a market, at a good price, for any electricity generated.

PURPA not only gave birth to the renewable energy industry, it also clearly demonstrated that small, on-site generation could deliver power at considerably lower cost than the retail rates charged by utilities. Competition had begun.

9.7.3 The Energy Policy Act of 1992 (EPAct)

The Energy Policy Act of 1992 (EPAct) created even more competition in the electricity generation market by opening the grid to more than just the QFs identified in PURPA. A new category of access was granted to *exempt wholesale generators* (EWGs), which can be of any size, using any fuel and any generation technology, without the restrictions and ownership constraints that PURPA and PUHCA imposed. EPAct allows EWGs to generate electricity in one location and sell it anywhere else in the country using someone else's transmission system to wheel their power from one location to another. The key restriction of an EWG is that it deals exclusively with the *wholesale wheeling* of power from the generator to a buyer, usually a regulated utility, who is not the final retail customer who uses that power.

9.7.4 FERC's Order 888 (1996)

While the 1992 EPAct allowed independent power producers (IPPs) to gain access to the transmission grid, problems arose during periods when the transmission lines were being used to near capacity. In these and other circumstances, the investor-owned utilities (IOUs) that owned the lines favored their own generators, and IPPs were often denied access. In addition, the regulatory process administered by the Federal Energy Regulatory Commission

(FERC) was initially cumbersome and inefficient. To eliminate such deterrents, FERC issued Order 888 in 1996, which had as a principle goal the elimination of anticompetitive practices in transmission services by requiring IOUs to publish nondiscriminatory tariffs that applied to all generators.

9.7.5 The Emergence of Competitive Markets

Prior to PURPA, the accepted method of regulation was based on monopoly franchises; vertically integrated utilities that owned some or all of their own generation, transmission, and distribution facilities; and consumer protections based on strict control of rates and utility profits. In the final decades of the twentieth century, however, the successful deregulation of other traditional monopolies such as telecommunications, airlines, and the natural gas industry, provided evidence that introducing competition in the electric power industry might also work. Although the disadvantages of multiple systems of wires to transmit and distribute power continue to suggest they be administered as regulated monopolies, there is no inherent reason why there shouldn't be competition between generators who want to put power onto those wires. The whole thrust of PURPA, Order 888, and EPAct 1992 was to begin the opening up of that grid to allow generators to compete for customers, in the hopes of driving down costs and prices.

The emergence of small, less capital-intensive power plants helped independent power producers get into the power generation business. By the early 1990s the cost of electricity generated by IPPs was far less than the average price of power charged by regulated utilities. With EPAct opening the grid, large customers began to imagine how much better off they would be if they could just bypass the regulated utility monopolies and purchase power directly from those small, less expensive units. Large customers, with the wherewithal, threatened to leave the grid entirely and generate their own electricity, whereas others, when allowed, began to take advantage of retail wheeling to purchase power directly from IPPs.

Not only did small power plants become more cost-effective, independent power producers found themselves with a considerable advantage over traditional regulated utilities. Even though utilities and IPPs had equal access to new, less expensive generation, the utilities had huge investments in their existing facilities so the addition of a few low-cost new turbines had almost no impact on their overall average cost of generation.

To help utilities successfully compete with IPPs in the emerging competitive marketplace, FERC included in Order 888 a provision to allow utilities to speed up the recovery of costs on power plants that were no longer cost-effective, also known as *stranded assets*. The argument was based on the idea that when utilities built those expensive power plants, they did so under a regulatory regime that allowed cost recovery of all prudent investments. To be fair, and to help assure utility support for a new competitive market, FERC believed it was appropriate to allow utilities to recover those stranded-asset costs even if that might delay the emergence of a competitive market.

9.7.6 California's Attempt to Restructure

In the 1990s, California's electric rates were among the highest in the nation—especially for its industrial customers—which led to an effort to try to reduce electricity prices by introducing competition among generation sources. In 1996, the California Legislature passed Assembly Bill 1890. AB 1890 had a number of provisions, but the critical ones included the following:

1. To reduce their control of the market, the three major investor-owned utilities (IOUs), Pacific Gas and Electric (PG&E), Southern California Edison (SCE), and San Diego Gas and Electric (SDG&E), which accounted for three-fourths of California's supply, were required to sell off most of their generation assets. About 40% of California's installed capacity was sold off to a handful of independent power producers including Mirant, Reliant, Williams, Dynergy, and AES. The thought was that new players who purchased these generators would compete to sell their power; thereby, lowering prices.

2. All customers would be given a choice of electricity suppliers. For a period of about four years, large customers who stayed with the IOUs would have their rates frozen at 1996 levels, and small customers would see a 10% reduction. Individual rate payers could choose non-IOU providers if they wanted to, and this "customer choice" was touted as a special advantage of deregulation. Some providers, for example, offered elevated percentages of their power from wind, solar, and other environmentally friendly sources as "green power."

3. Utilities would purchase wholesale power on the market, which, due to competition, was supposed to be comparatively inexpensive. The hope was that with their retail rates frozen at relatively high 1996 levels, and with dropping wholesale prices in the new competitive market, there would be extra profits left over that could be used to pay off those costly stranded assets—mostly nuclear power plants.

4. The competitive process was set up so that each day there would be an auction in which generators would submit bids indicating how much power they were willing to sell the next day and at what price. A new entity, called the California Power Exchange (CalPX) selected enough bids to meet the projected demand. All of those successful bidders were paid the same amount, equal to the highest accepted bid. Any provider who bid too high would not sell power the next day. So if a generator bid $10/MWh (1¢/kWh) and the market clearing price was $40/MWh, that generator would get to sell power at the full $40 level. This was supposed to encourage generators to bid low so they would be assured of the ability to sell power the next day.

On paper, it all sounds pretty good, doesn't it? Competition would cause electricity prices to go down and customers could choose providers based on whatever criteria they liked, including environmental values. As wholesale power prices dropped, utilities with high, fixed retail rates could make enough extra money to pay off old debts and start fresh.

For two years, up until May 2000, the new electricity market seemed to be working with wholesale prices averaging about $30/Mwh (3¢/kWh). Then, in the summer of 2000, it all began to unravel (Figure 9.21). In August 2000, the wholesale price was five times higher than it had been in the same month in 1999. During a few days in January 2001, when demand is traditionally low and prices normally drop, the wholesale price spiked to the astronomical level of $1500/Mwh. By the end of 2000, Californians had paid $33.5 billion for electricity, nearly five times the $7.5 billion spent in 1999. In just the first month and a half of 2001, they spent as much as they had in all of 1999.

What went wrong? Factors that contributed to the crisis included higher-than-normal natural gas prices, a drought that reduced the availability of imported electricity from the Pacific Northwest, reduced efforts by California utilities to pursue customer energy-efficiency programs in the deregulated environment, and, some argue, insufficient new plant construction. But, when California had to endure rolling blackouts in January 2001, a month when demand is far below the summer peaks and utilities normally have abundant excess capacity, it became clear that none of the above arguments was adequate. Clearly, the IPPs had discovered they could make a lot more money manipulating the market, in part by withholding supplies, than by honestly competing with each other.

The energy crisis finally began to ease by the summer of 2001 after the Federal Energy Regulatory Commission (FERC) finally stepped in and instituted price caps on wholesale power, the governor began to negotiate long-term contracts, and the state's aggressive energy-conservation efforts began to pay off. Those conservation programs, for example, are credited with cutting the June 2001, California energy demand by 14% compared with the previous June.

In March 2003, FERC issued a statement concluding that California's electricity and natural gas prices were driven higher because of widespread manipulation and misconduct by Enron and more than thirty other energy companies during the 2000–2001 energy crisis.

figure 9.21 **California Wholesale Electricity Prices during the Crisis of 2000–2001**

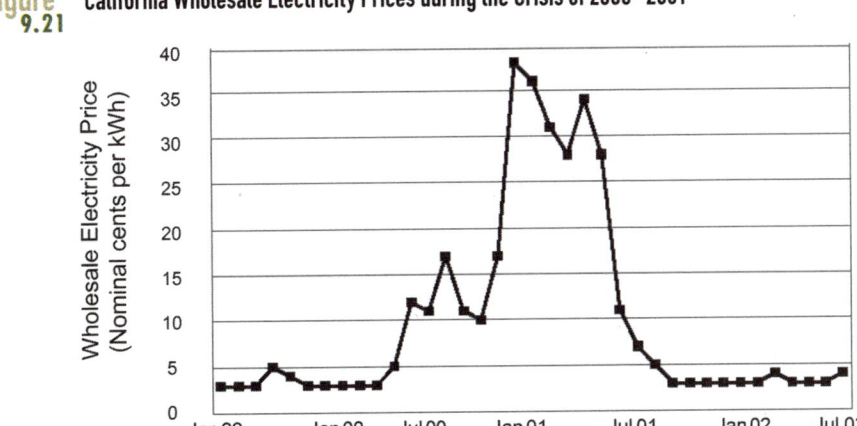

SOURCE: from Bachrach, et al., 2003

In 2004, audiotapes were released that included Enron manipulators joking about stealing money from those dumb grandmothers in California. By 2005, Dynergy, Duke, Mirant, Williams, and Reliant had settled claims with California totaling $2.1 billion—a small fraction of the estimated $71 billion that the crisis is estimated to have cost the state.

Although the momentum of the 1990s toward restructuring was shaken by the California experience, the basic arguments in favor of a more competitive electric power industry remain attractive. Analysis of the failure there has guided the restructuring that continues in a number of other states.

9.8 Summary

This chapter has attempted to describe the history of electric utilities and how they are regulated as well as the main centralized generation facilities that send power over our vast and intricate transmission and distribution system. Techniques were presented to evaluate the economics of different types of power plants and how sensitive those cost numbers are to the assumptions made. The focus has been on the utility's side of the meter; that is, on wholesale markets. This background provides the context for the next chapter in which an alternative model based on small-scale energy systems, often located on the customer's side of the meter, will be presented.

CHAPTER 10

Distributed Energy Resources

In the last chapter, the inefficiencies associated with large central stations sending power over hundreds or thousands of miles of transmission and distribution (T&D) lines were described. An alternative, or perhaps a complementary, model is emerging that integrates what are often referred to as *distributed energy resources* (DER) into the supply-and-demand equation that governs the electricity network. Distributed energy resources include the following:

- *distributed-generation* (DG) facilities typically located at, or very near, the end user,
- *demand-side management* (DSM) programs to help customers reduce or shift their loads,
- *electricity storage* systems and, potentially,
- *vehicle-to-grid* (V2G) systems in which cars can be both suppliers and users of power.

Figure 10.1 illustrates some of these distributed energy resources installed in industrial, residential, and commercial facilities, as well as in utility distribution substations to provide grid support.

10.1 Distributed Generation (DG)

Small, modular power plants located close to loads are referred to as *distributed generation*. Any number of technologies may be utilized, including gas turbines, wind turbines, photovoltaic systems, fuel cells, microturbines, reciprocating engine-generators, micro-hydroelectric plants, systems fueled by biomass, and even electric-drive vehicles connected to the grid when parked. Details about these technologies are included in subsequent sections of this chapter as well as in the following two chapters.

Oftentimes distributed-generation plants may be owned by customers who may be motivated by environmental values, the potential to reduce costs, increased reliability, or sometimes to improve power quality for sensitive electronic loads. On-site generation has several efficiency advantages including reduced power-line losses as well as the possibility of capturing waste heat in *combined-heat-and-power* (CHP) systems. Burning a fuel to generate

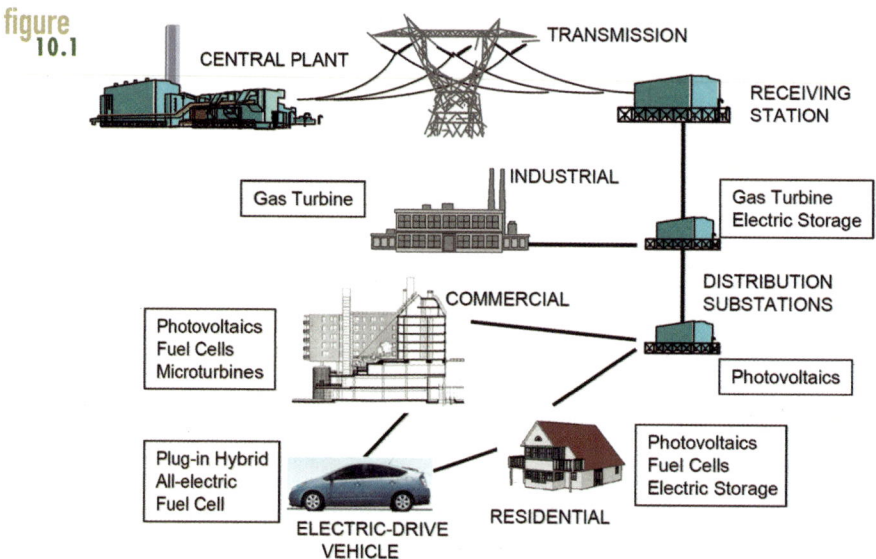

figure 10.1

Distributed-generation technologies and electricity storage systems can be owned and operated by customers or the utilities that serve them.

electricity and then capturing and utilizing waste heat can reduce primary energy demand by 40% to 50% compared to separate grid electricity and fuel-fired boilers. Those efficiency gains translate into greenhouse gas reductions as well. Figure 10.2 makes that point by comparing carbon emissions from conventional plants with emissions from simple-cycle and combined-cycle plants designed to provide combined-heat-and-power CHP. As shown there, replacing a conventional coal-fired power plant with a natural-gas-fired combined-cycle CHP plant reduces carbon emissions by over 80%.

Utilities may also derive benefits from generation facilities located closer to loads, no matter who owns them, by deferring otherwise needed upgrades to their T&D systems. In any grid system, thousands of kilowatts are wasted in the process of delivering large currents over long distances, all the while squeezing those amps through bottlenecks and sending them around and around in various distribution loops. Properly positioned DG can greatly reduce system congestion and curtail waste. When rapid load growth in a particular part of the network threatens to exceed the capacity of the system, it can be quicker and cheaper to strategically inject power into the grid with DG than to upgrade the entire T&D system to meet the new demand. The potential savings of grid support can often help justify DG investments.

Small distributed-generation systems can also provide considerable value by allowing utilities to more accurately match generation growth to load growth. Consider the difference between building a single large power plant that will meet anticipated load growth for many years versus building a series of smaller units that will each, perhaps, satisfy growth for only a year or two. Planning, permitting, and building the large plant requires a longer lead time

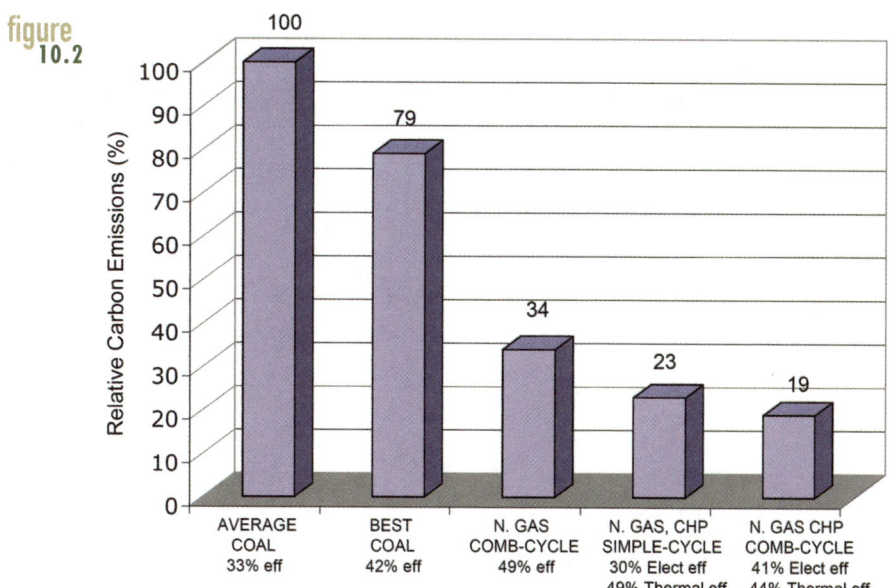

Relative carbon emissions for various power plants compared to those of the average coal plant. CHP units assume thermal output replaces an 83%-efficient gas boiler.

before any power is generated, all while capital investments are tied up without earning revenues. Moreover, once operational, for some period of time the larger plant will provide more power generation capacity than is needed to satisfy load growth. Idle capacity costs money without generating income. Swisher (2002) explores this *option value* of smaller generation facilities with some surprising results. For example, consider the option of building one plant that takes three years to build, capable of supplying load growth for eight years, versus eight smaller plants, each with one-year lead time and each capable of supplying one year's worth of growth. Using a 10% discount factor, Swisher's analysis indicates that the smaller plants could each cost 50% more per kilowatt of capacity than the large one and still be equally cost-effective. For example, if the capital cost of a big plant is $1000/kW, then a series of smaller ones coming on line as needed could cost $1500/kW and still have the same present value cost.

Of the approximately 250 GW (1 GW = 10^9 watts) of currently installed distributed generation in the United States (defined as generation of less than 50 MW), more than 80% is provided by internal combustion engines tied to generators. Air quality permitting constraints, however, often limit the ability to use these resources for anything other than emergency standby power. In fact, only about 10% of these distributed generation systems are even interconnected with the grid. That suggests an enormous resource—equivalent to about one-quarter of the entire 1000 GW of grid capacity—is just sitting there for all but a few hours each year. This could all change with cleaner, quieter, easier-to-permit, emerging technologies such as the fuel cell systems to be described later in this chapter. Imagine the reliability and efficiency implications if those reciprocating engines were replaced with grid-connected fuel cells providing power and heat whether the grid is operating or not. Moreover, for many high-tech industries, their power quality and absolute reliability would be highly valued and could be worth much more than the energy provided.

10.2 Demand-Side Management (DSM)

Utility programs designed to control energy consumption on the customer's side of the meter are known as *demand-side management*. DSM can be broadly defined to include the following:

1. *Conservation/energy efficiency* programs that have the effect of reducing consumption during most or all hours of the day.
2. *Load management programs* that have the effect of reducing peak demand through conservation or by shifting the demand to nonpeak hours.
3. *Fuel substitution programs* that encourage a customer to replace electricity with another energy source. For example, replacing a standard air conditioner with one based on absorption cooling can shift the load from electricity to natural gas.

Reducing energy demand can save customers money and it can reduce the environmental impacts of providing that energy. So, how can a utility influence the energy use of its customers, and why should it want to do so?

For DSM programs to succeed, customers and utilities must be provided with incentives to participate. Customers can be encouraged to reduce, or shift, loads through a combination of rebates, rate structures, and smarter meters. Many utilities offer rebates to residential customers who purchase energy-efficient appliances and to commercial customers for more efficient lighting systems and other performance improvements in their buildings.

The price of electricity will also certainly affect the demand. However, as explained in the next section, today's rate structures and electric meters are pretty dumb. They usually don't differentiate between using power in the middle of a hot summer afternoon, when generation is most costly, and at 3:00 a.m. when both demand and the cost of generation are low. We need to have smarter meters that will act as two-way energy and information portals, coupled with real-time pricing. With proper price signals, customers could avoid using power when it is most costly, which would lower their bills while simultaneously benefiting utilities by deferring the need for added generation or expansion of expensive T&D infrastructure.

The key to getting utilities to want to help customers save energy is *decoupling* sales from profits. The usual procedure by which utility rates are established is based on utilities providing evidence for costs and expected demand to their public utilities commission (PUC). The PUC in turn sets rates so that the utility can earn a reasonable return on their investments. The problem with this procedure is it encourages utilities to sell more electricity than they predicted, and penalizes them if they sell less.

For example, suppose a utility that predicts sales of 10 billion kWh needs $400 million to cover costs with a reasonable profit. That means they need to be allowed to charge 4¢/kWh

($400 \times 10^6/10 \times 10^9$ kWh). Suppose, however, that the marginal cost of running their system just a little longer to sell one additional kWh beyond their prediction is only 1¢/kWh. That means if they sell that extra kWh, their revenues would be 4¢ and their costs would be 1¢, so they would make an extra profit of 3¢ (hence the incentive to sell more electricity). Conversely, if they sell one less kWh than their prediction, they lose 4¢ in revenue and save 1¢ in costs, for a net loss of 3¢ (hence the incentive to discourage conservation).

To avoid this conundrum, a few states have begun to decouple utility sales from profits by introducing something called the *electric rate adjustment mechanism* (ERAM). ERAM simply incorporates any revenue collected above or below the forecasted amount into the following year's authorized revenues, thereby eliminating incentives to sell more kWh and eliminating disincentives to reduce kWh sales. Although ERAM, or something similar, is necessary for utility DSM programs, it has been found to not be sufficient. In addition, utilities need to be allowed to recover the costs of running their DSM programs and, on top of that, incentives need to be provided to allow utilities to make more profits helping their customers save energy than they would have made through generation.

10.3 Electricity Storage

Until recently, the only method of providing energy storage for electric utilities was based on pumping water uphill into a reservoir at night, when demand is low, and then running the water back down again through turbines during the day when demand is high. These pumped-storage systems are difficult to site, relatively inefficient, and very costly.

New storage options being developed include battery systems, flywheels, ultra-capacitors, compressed air energy storage, and reversible electrolyzer/fuel cells with hydrogen storage. As a load management technique, storage allows utilities to shave their power demand peaks and fill the valleys. Some of these storage systems would not only help the grid by providing power during times of peak demand but they may also provide what are called *ancillary services* to the grid, including quick-response "spinning" reserves, voltage stability, frequency regulation, and reactive power support.

Another important attribute of these energy storage devices is their potential ability to "firm up" the output of "intermittent" solar and wind energy systems. Intermittent sources aren't available all of the time, making it difficult for a utility dispatcher to be sure they will be there when needed. By providing some amount of energy storage, these intermittent sources can become more *dispatchable*, which means they can be scheduled for use at the planner's convenience—an attribute that greatly increases their value.

10.3.1 Stationary Battery Storage

Batteries have advantages over hydroelectric pumped storage or compressed air storage systems because they don't require special nearby geographical features to make them viable. On

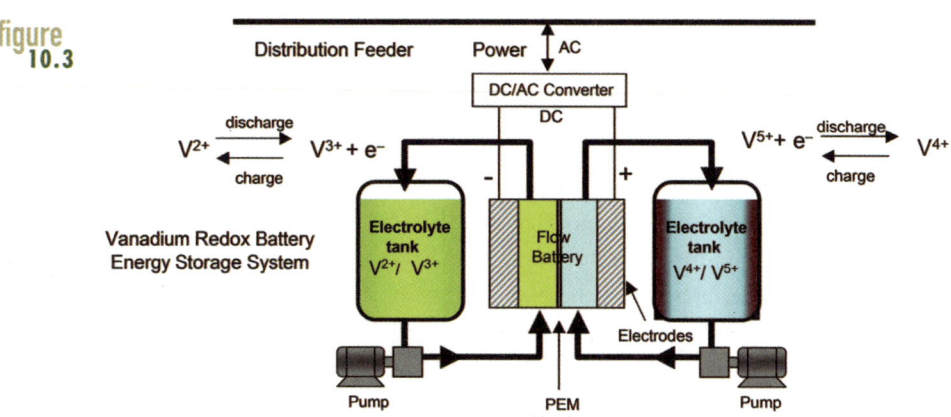

figure 10.3

A vanadium redox battery energy storage system seems well suited to stationary applications such as power line voltage and power support.

a per-kWh-of-storage basis, the most important attributes of battery systems include cost, weight, volume, and cycle-lifetime. For stationary grid-support systems, weight and volume are less important, but for vehicles they are critical.

An example of a promising battery for stationary applications is the vanadium redox battery energy storage system (VRB-ESS) shown in Figure 10.3 and Sidebar 10.1. The VRB-ESS is an example of what is called a *flow battery*, so named because electrolytes stored in large plastic tanks are continuously circulated through the battery cells where the reactions actually take place. Increasing the volume of the tanks increases energy stored, whereas increasing the number of cell stacks increases the power that the system can deliver.

An example of another promising storage technology is the sodium-sulfur 1.2 MW, 7.2 MWh battery storage, grid support system installed in 2006 at an Appalachian Power Company substation near Charleston, West Virginia. The installation is expected to delay the need for facility equipment upgrades by six or seven years.

10.3.2 Batteries for Vehicles: Vehicle-to-Grid (V2G) Systems

Battery-powered electric vehicles have been under development for quite some time, but their limited range has always made them more of an oddity than a mainstream technology. The weak link has always been their batteries, which have tended to be bulky and heavy. Meanwhile, research into small, lightweight batteries for laptops and cell phones has resulted in new battery technologies that can be used not only for portable electronic devices but also for electric vehicles.

The most promising battery technology for vehicles is based on an electrochemical reaction involving very lightweight lithium ions. As shown in Figure 10.4, lithium-ion batteries

SIDEBAR 10.1

A Duracell in the Desert: The Castle Valley VRB Project

By the turn of the twenty-first century, PacifiCorp's 209-mile-long, 25 kV, feeder line that snakes its way through southeastern Utah had reached its capacity. New service connects were being denied and existing customers toward the end of the line were subject to periodic low-voltage episodes that dimmed lights and sometimes burned out motors that were unable to start properly. Rather than upgrade the entire line, in 2004 PacifiCorp installed a 2 MWh battery system close to the load center of the distribution line, partway down the feeder as shown below.

By charging the batteries at night when power demand is low, and discharging them into the feeder during the peak hours of the day, the substation no longer has to supply all of the power to the far end of the line during peak demand periods and avoids the need to upgrade the whole system.

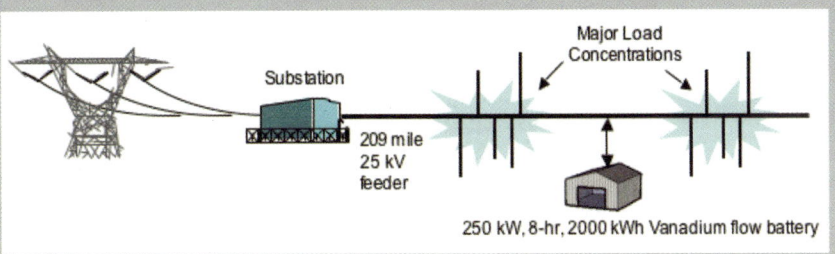

figure
10.4

Energy Density of Existing, Conventional Batteries

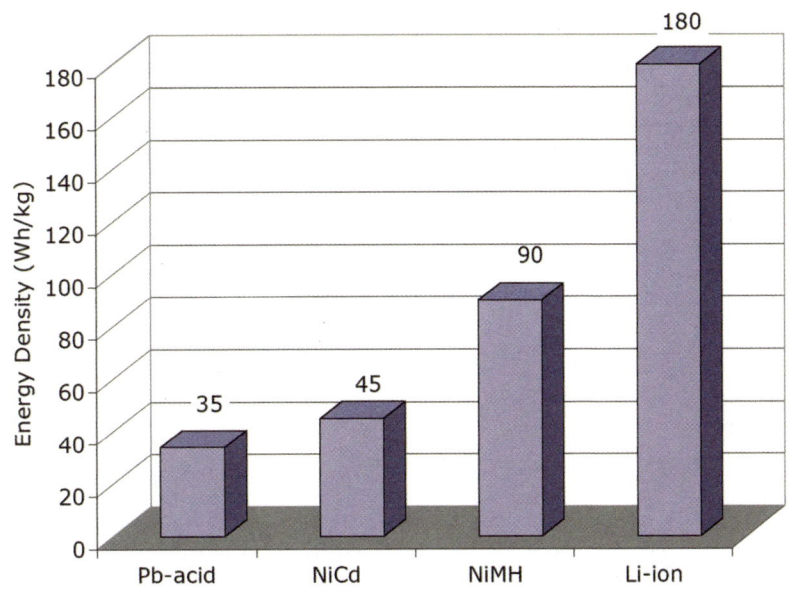

have a kWh/kg energy density roughly double that of nickel-metal-hydride (NiMH), quadruple the density of nickel-cadmium (NiCd), and five times that of conventional lead-acid batteries. The current generation of lithium-ion batteries has an energy density of around 180 Wh/kg, which is high by conventional standards, but is still far below their theoretical limit.

With the development of new, lightweight, powerful batteries, there is renewed interest in battery-powered vehicles and with that comes new attention to possible interactions between vehicles and the grid.

Two versions of electric-drive vehicles are emerging. One is referred to as a *plug-in hybrid electric vehicle* (PHEV); the other is an *all-battery-powered electric vehicle* (BEV). A PHEV is basically a conventional hybrid-electric vehicle with extra batteries added to extend the number of miles that can be driven on electricity before the backup internal combustion engine has to kick in. A battery pack capable of providing 30 miles per day of electric drive could reduce the average U.S. light-vehicle gasoline consumption by half. The advantages of electric-drive vehicles include helping to reduce our dependence on oil, which is of enormous importance as a national security issue; reducing emissions of pollutants that cause urban smog; and reducing overall climate-changing carbon emissions.

For either PHEVs or BEVs, the batteries would likely be charged from the grid at night, when demand and prices are low, and discharged during the day while the vehicle is in use. Figure 10.5 shows a simple analysis of what this "valley-filling" might look like. It has been estimated that the idle capacity of the U.S. grid could supply nearly three-fourths of the energy needs of today's light-vehicle fleet without adding generation or transmission and distribution

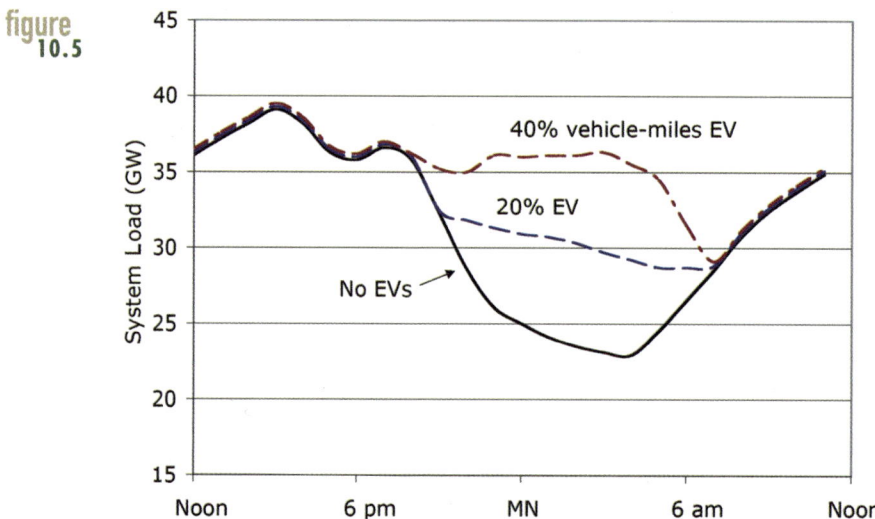

Typical electricity system load in California showing impacts of night charging for electric vehicles with fraction of total vehicle miles provided by electricity as a parameter. Based on 3 miles/kWh and 280 billion vehicle-miles/yr.

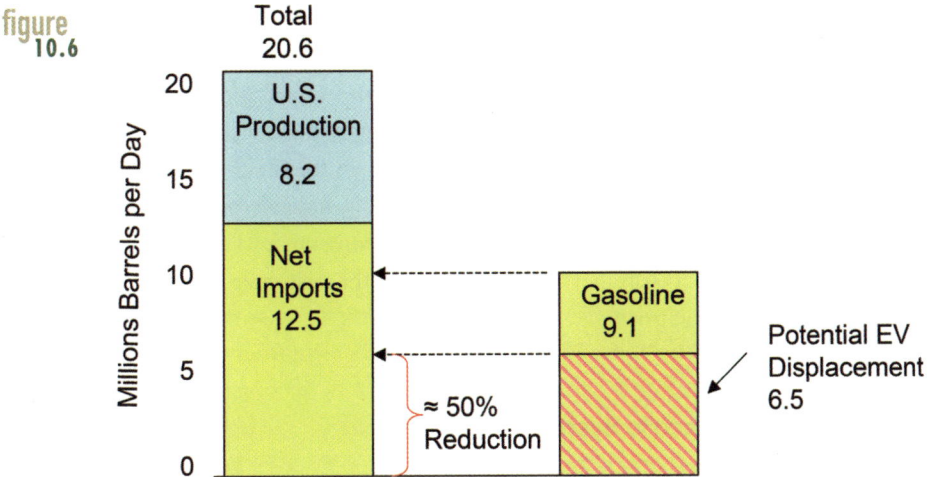

Imported oil could be reduced by about half with electric vehicles powered by idle grid capacity.

SOURCE: based on Kintner-Meyer, 2007

capacity (Kintner-Meyer, 2007). As Figure 10.6 shows, at that rate of EV use, about half of all U.S. oil imports could be eliminated by electricity generated from idle grid capacity.

It is interesting to contemplate the implications of EVs not only as consumers of electric power, but also as sources of power for the grid. Realize that a typical car is used only about one hour per day, so it is parked somewhere almost all of the time. Plugged into an ordinary home electric outlet, an EV could potentially sell or buy about 15 kW of power to or from the grid. Scaling that up, if one-third of U.S. cars were grid-connected EVs they would have the potential to deliver as much power as the entire U.S. electrical grid. Such V2G systems wouldn't be used for base-load power because it is much cheaper to provide that with conventional generators, but properly organized and controlled V2G could provide large amounts of quick-response, high-value electric services for the grid.

The most likely initial application of V2G would be to help utilities fine-tune the balance between generation and constantly fluctuating loads. To keep generators from slowing down as loads increase, grid operators need to be able to purchase relatively small amounts of quick-response power. And to keep them from speeding up when loads decrease they need to be able to quickly dump power somewhere as well. Fleets of EVs connected to the grid could easily provide these *automatic generation control* services (also known as *regulation* services), which are worth literally billions of dollars per year. These fluctuations happen twenty-four hours a day, which suggests EVs could be used during the day and still be able to sell regulation services at night when they aren't being driven. After accounting for hook-up fees and accelerated battery replacements, Kempton, et al. (2005) estimate that an aggregated fleet of EVs selling regulation services while parked could earn net revenues on the order of $2500 per vehicle per year.

10.4 The View from the Customer's Side of the Meter

There is an important economic advantage associated with technologies that generate power on the customer's side of the meter. By doing so, these small, distributed-generation systems compete against the full retail price of electricity, rather than the much lower wholesale price paid by grid operators for electricity from power plants. Wholesale prices on spot markets typically run a few cents per kWh, but by the time transmission and distribution costs and other charges are added to a customer's bill the retail rate the customer pays can easily be two to three times the wholesale price.

10.4.1 Your Home's Electric System

Utility distribution lines running through your neighborhood probably carry 4160 volts (4.16 kV), but some carry as much as 34.5 kV. Obviously, those voltages are far too dangerous for household use, so a transformer, usually mounted on pole nearby, drops those voltages to the nominal 120 V/240 V used in homes. Three wires come from that transformer to your electric meter, which dutifully records the accumulating kilowatt-hours of energy being delivered (Figure 10.7).

The wall outlets in homes in the United States provide 60-hertz (Hz) ac power at a nominal voltage of about 120 volts (actual voltages are usually in the range of 110 V to 120 V; for a brief primer on ac voltages, see Sidebar 10.2). Such voltages are sufficient for typical, low-power applications such as lighting, electronics, toasters, and refrigerators. For appliances that require higher power, such as an electric clothes dryer or an electric stove or space heater, special outlets provide power at a nominal 240 V. Running high-power equipment on 240 V rather than 120 V cuts current in half, which cuts the i^2R heating of wires by three-fourths, allowing normal wire gauges to be used without causing undue fire danger.

figure 10.7   **Residential 120 V/240 V Electrical System**

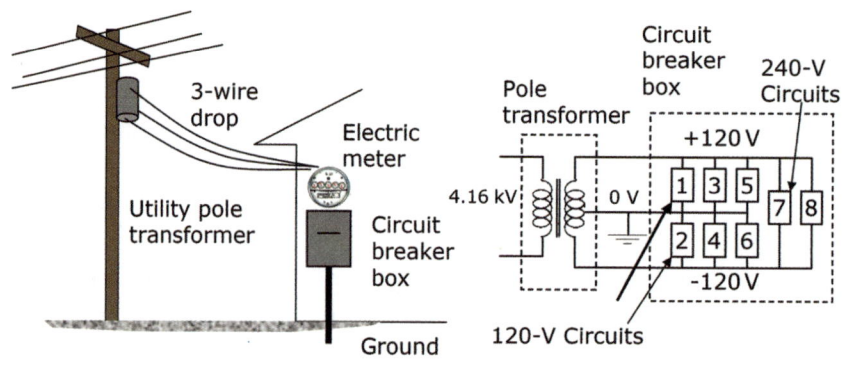

SIDEBAR 10.2

Some Background on AC Voltages and Currents

When voltages are nice, steady dc, it is intuitively obvious what we mean when we say, for example, that "this is a 9-V battery." But what does it mean to say the voltage at our wall outlet is 120-V ac? Because it is ac, the voltage is constantly changing, so what does the "120" refer to? It isn't the average value, because the average of a sinusoidally varying quantity is zero. It could mean the maximum value, or amplitude, but it doesn't. It is, instead, a quantity called the *root-mean-squared*, or *rms*, value of the voltage. The rms value of a quantity is the square *R*oot of the *M*ean value of the *S*quare of the quantity. The idea behind using rms values of voltages and currents for ac circuits is that when you do so the relationships developed in Section 4.8.4 for power dissipated in a resistive load work just as well in ac as they do in dc. Thus, for example, an rms voltage of 120 V across the 240-ohm filament resistance in a lightbulb will dissipate $P = V^2/R = 120^2/240 = 60$ watts.

A 120-V rms sinusoidal voltage with a frequency of 60 Hz can be concisely written as follows:

$$v = 120\sqrt{2}\,\sin(2\pi \cdot 60t)$$

where v = voltage measured in volts

t = time measured in seconds

If you look carefully, you'll see where the "120 volts" we are used to appears as well as where the 60 Hz frequency pops up. Suppose we want to find the rms value of the voltage in above equation. From the definition of rms we get

$$V_{rms} = \sqrt{(v^2)_{avg}} = \sqrt{\left[(120\sqrt{2})^2 \sin^2(2\pi \cdot 60t)\right]_{avg}}$$

$$= 120\sqrt{2}\sqrt{\left[\sin^2(2\pi \cdot 60t)\right]_{avg}}$$

As the following figure indicates, the average value of the square of a sinusoid is $\frac{1}{2}$.

So, the rms value of voltage is simply

$$V_{rms} = 120\sqrt{2} \cdot \sqrt{\frac{1}{2}} = 120 \text{ volts}$$

That's what we hoped it would come out to be.

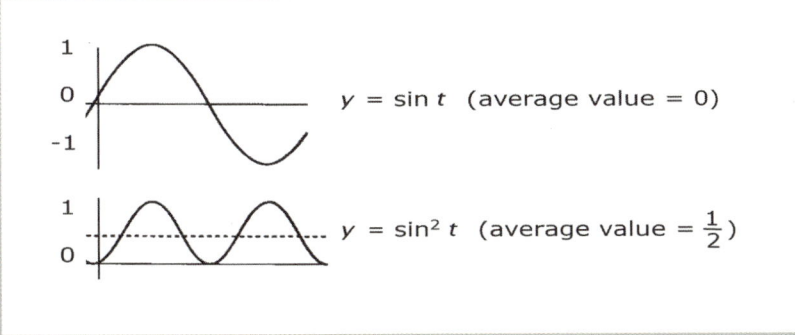

Your home's electric meter not only measures kWh of energy, it can also be used to measure power (the rate at which energy is being used). This can be a handy feature if you want to know how many watts various household appliances use when turned on. Sidebar 10.3 shows how to measure both power and energy with your meter.

SIDEBAR 10.3

Reading Your Electric Meter

Most electric meters on homes have a series of dials that indicate kilowatt-hours of energy. The meters are usually read once a month and the difference between readings is the total kWh used during that month. Read the meter from left to right, but be careful when an individual dial points directly at a number. For example, the meter below reads 44949 kWh.

By timing the rate at which the horizontal disk spins in your meter you can also determine the power (watts or kW) used at any time in the house. The key is the Kh factor shown on the meter face, which indicates the watt-hours of energy per revolution of the disk. For example, if this disk rotates at 1 revolution per minute, the rate of power consumption would be

$$P = 1 \text{ rev/min} \times 60 \text{ min/h} \times 7.2 \text{ Wh/rev}$$
$$= 432 \text{ watts}$$

kWh dials

Rotating disk

7.2 Wh/Revolution

10.4.2 Three-Phase Power for Larger Customers

Utility transmission and distribution systems almost always transmit power using three-phase voltages and currents. What that means is that each of the three incoming lines carries an ac voltage that is 120 degrees out of phase with the other two lines (Figure 10.8). Three-phase power not only utilizes transmission lines more efficiently, it also results in smoother motor and generator operation because the total power delivered by three-phase systems is a constant, not a function of time (power is proportional to the square of voltage). In comparison, single-phase power varies constantly as shown in Figure 10.8a.

Large buildings are often served with 480-V, three-phase power using the four-wire scheme shown in Figure 10.9. The fourth wire is a ground line that is supposed to carry very little current. The line-to-line voltage (called the line voltage) is 480 V. The voltage between any phase line and ground (called the phase voltage) is 277 V. Quite often, lighting circuits in buildings run on the 277-V phase voltage.

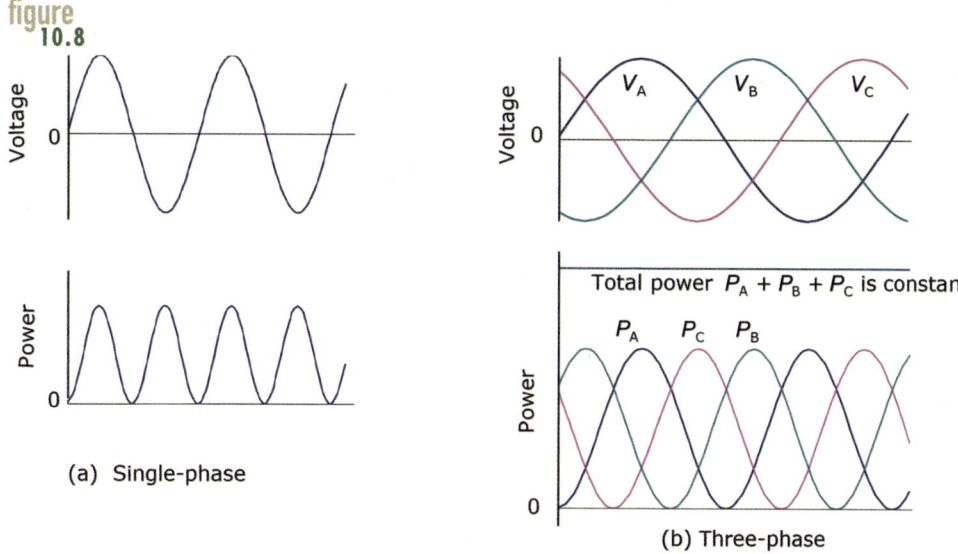

(a) Single-phase

(b) Three-phase

Power is proportional to the square of voltage. (a) In a single-phase system instantaneous power varies sinusoidally. (b) For three-phase systems, total power delivered is constant, leading to smoother motor and generator operation.

10.4.3 Utility Rate Schedules

Investment decisions made by customers to reduce their own loads, or perhaps to generate some of their own power, are critically dependent on how much they have to pay for electricity when they buy it from their local utility. Obviously, if electricity costs 15¢/kWh it is a lot easier

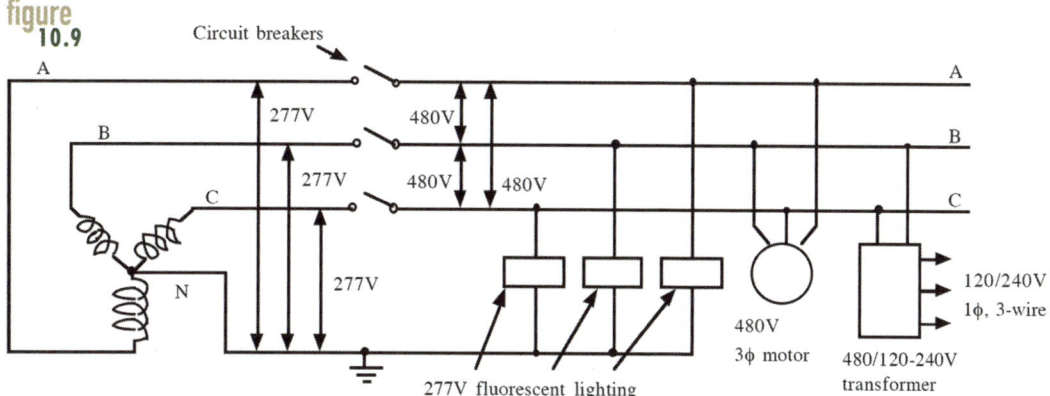

A three-phase, 480-V, large-building wiring system that provides 480-V, 277-V, 240-V, and 120-V service to various loads. The three windings shown represent the secondary side of the building's three-phase transformer.

to justify installing a photovoltaic system or purchasing a more energy-efficient refrigerator than if the local rate is only 6¢/kWh. It is handy to know roughly how much a kilowatt-hour costs in your area, but in reality utility rate structures are often quite complicated.

Utility rates vary within an individual class of customers depending on how much energy is purchased, the season of the year, and sometimes even the time of day that it is used. Large industrial customers pay a lot less per kWh than most businesses, and businesses usually pay less than residential customers. Some of this discrepancy has to do with the cost of providing service. It is considerably cheaper to provide a lot of power through a single distribution line to a big factory with a steady, predictable demand, and then send that factory a single bill each month than it is to supply a residential area with power lines, transformers, power drops, meters, and bills to every single house.

Most customers are given a choice of rate structures by their local utility. Figuring out which rate structure to choose can be a tricky task. The following sections attempt to unravel some of these complexities.

Inverted block rates. An example residential rate schedule is shown in Table 10.1. Notice it includes four tiers based on monthly kWh consumed, and notice the rates increase with increasing demand. This is an example of what is called an *inverted block rate* structure designed to discourage increasing consumption (not that many years ago, however, utilities typically used declining block rates with prices that decreased with usage to encourage their customers to buy more energy).

For example, using the rate structure given in Table 10.1, a customer using 600 kWh would be charged:

$$255 \times \$0.1143 + (331 - 255) \times \$0.1299 + (510 - 331) \times$$
$$\$0.1781 + (600 - 510) \times \$0.2194 = \$90.64$$

If this customer were to buy an energy-efficient refrigerator, for example, which saves 60 kWh/month, the monthly bill would drop by 60 kWh × \$0.2194 = \$13.16/month. The very same refrigerator, however, would save only about half as much for a more frugal customer whose demand is always within Tier 1 (60 kWh × \$0.1143/kWh = \$6.86/month). The financial incentive to reduce energy demand is thus higher for customers who use larger amounts of energy.

Time-of-use (TOU) rates. Another rate schedule incorporates *time-of-use* (TOU) charges that reflect the increased cost of generation during periods when power demand is the high-

table 10.1 Example Residential Inverted Block Rate Structure

kWh	Tier 1:0–255	Tier 2:256–331	Tier 3:332–510	Tier 4: > 510 kWh
¢/kWh	11.43	12.99	17.81	21.94

est. An example of a simple residential TOU rate schedule is given in Table 10.2. In this rate structure, prices are very high in the late afternoon in the summer. If you need a lot of air-conditioning, you probably wouldn't want to choose this TOU schedule. But, if you had a photovoltaic array and could imagine selling electricity to the utility during the day at 29.4¢/kWh and then buying it back at night at 8.7¢/kWh, this could be a very attractive rate structure in which to enroll. The hours during which the winter peak demand period occurs, by the way, reflect increased heating demand on those cold winter mornings and evenings.

Demand charges. Rate schedules for commercial and industrial customers usually include *demand charges* ($/month per kW of peak demand) as well as energy charges ($/kWh). These demand charges are a monthly fee based on the half-hour of time during the month at which the customer's power demand (kW) is the highest. Demand charges reward customers who have relatively constant power demands with no large peaks. Table 10.3 shows three different rate schedules for commercial buildings served by a utility in South Carolina. For customers who routinely use less than 7500 kWh/month (about the same as ten average homes), there is just a standard energy charge of 6.56¢/kWh. For customers who use more than that amount, a demand charge is included. Those customers can choose between a standard demand charge of $11.85/month-kW or one that is based on time of use. Choosing the best rate structure for a given customer requires a careful analysis of the expected month-by-month demand profile as the example in Solution Box 10.1 illustrates.

Real-time pricing (RTP). Although time-of-use rates attempt to capture the true cost of utility service, they are relatively crude because they only differentiate between relatively large blocks of time (peak, partial-peak, and off-peak, for example) and they typically acknowledge only two seasons: summer and non-summer. The ideal rate structure would be one based

table 10.2   An Example of Time of Use (TOU) Residential Rates

	SUMMER: May to October		WINTER: November to April	
On peak	2–8 p.m.	29.4¢/kWh	7–10 a.m. 5–8 p.m.	11.5¢/kWh
Off peak	All other times	8.7¢/kWh	All other times	9.0¢/kWh

table 10.3   Commercial Rate Schedule for a South Carolina Utility, 2005

	General Service < 7500 kWh/mo	Medium Service > 7500 kWh/mo	Medium Service TOU Rate
Energy charge	6.56¢/kWh	2.60¢/kWh	2.32¢/kWh
Demand charge	—	$11.85/mo-kW	$13.20/mo-kW on peak* $3.87/mo-kW off peak

*On peak is 1:00 p.m. to 10:00 p.m., May to October; and 6:00 a.m. to 10:00 a.m., November to April.

SOLUTION BOX 10.1

Choosing the Right Rate Structure with Photovoltaics

A small office building that uses 40,000 kWh/month during the summer has a peak demand of 100 kW. A photovoltaic system is being proposed that will provide a supply peak of 80 kW along with 20,000 kWh/month of energy. The before and after load curves are shown in the following diagrams.

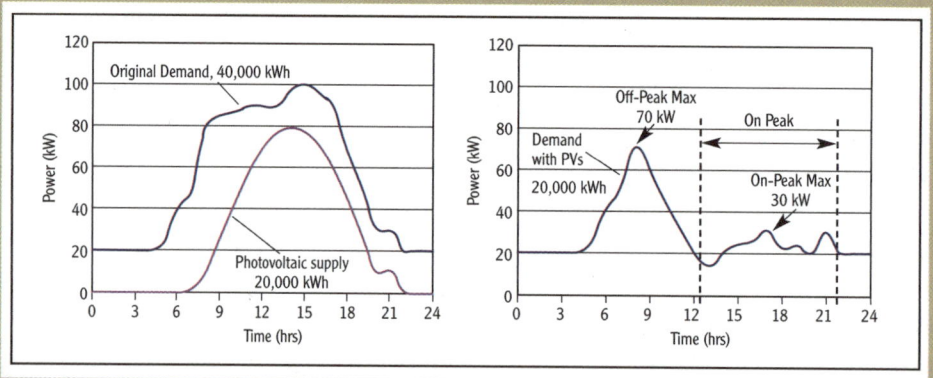

a. Using the standard Medium Service rate structure in Table 10.3, what is the original monthly utility bill?
b. With the PVs in place, which rate structure should the customer use to provide the maximum dollar savings?

on real-time pricing (RTP) in which the true cost of energy is reflected in rates that change throughout the day, each and every day. With RTP, for example, there would be no demand charges, just energy charges that might vary, for example, on an hourly basis.

Some utilities now offer one-day-ahead, hour-by-hour, real-time pricing. When customers know that tomorrow afternoon the price of electricity will be high, they can implement appropriate measures to respond to that high price. With the price of electricity more accurately reflecting the real, almost instantaneous, cost of power, it is hoped that market forces will encourage the most efficient management of demand.

10.4.4 Net Metering

For many industrial customers who generate some of their own power on-site, whatever they generate will be used on-site at the instant it is generated. In these circumstances the grid is there to supply any additional power to the site at the moment it is needed. Metering is conventional

Solution:

a. Without the PVs, the non-TOU (time of use) Medium Service rate from Table 10.3 results in a monthly bill of

$$40,000 \text{ kWh} \times \$0.026/\text{kWh} + 100 \text{ kW} \times \$11.85/\text{kW} = \$2225$$

Notice the demand charge ($1185) is more than the cost of energy ($1040).

b. With the 80 kW photovoltaic array, the highest on-peak demand drops from 100 kW to 30 kW, but the highest off-peak demand is now 70 kW. The monthly energy has been reduced from 40,000 kWh to 20,000 kWh. The three rate choices result in the following bills:

The General Service rate bill would be

$$20,000 \text{ kWh} \times \$0.0656/\text{kWh} = \$1312/\text{mo}$$

The Medium Service without the time-of-use (TOU) rate would be

$$20,000 \text{ kWh} \times \$0.026/\text{kWh} + 70 \text{ kW} \times \$11.85/\text{kW} = \$1350/\text{mo}$$

For the Medium Service with TOU rate, the bill would be

$$20,000 \text{ kWh} \times \$0.0232/\text{kWh} + 30 \text{ kW} \times \$13.20/\text{kW} + 70 \text{ kW} \times \$3.87/\text{kW} = \$1131$$

So, the TOU rate is the least expensive. The PVs save $2225 – $1131 = $1094/month.

because power flow is always in one direction—from the grid to the customer. However, if a distributed-generation system is designed to sell power to the grid some of the time as well as to purchase it back again at another time, the metering becomes more complicated.

For "Qualifying Facilities" (QFs) under the Public Utility Regulatory Policies Act of 1978 (see Section 9.7.2 of this book for a description of QFs and PURPA), the utility is required to purchase power at what is called the "avoided cost," which is meant to be what it would have cost the utility if it had generated the power. This is in essence the wholesale cost of generation. On the other hand, any power purchased by the QF would be priced at the much higher retail rate. Under PURPA, self-generators sell at the low wholesale price and purchase at the higher retail price. To keep track of the different rates for buying and selling power, a two-meter approach is often used with ratcheted meters that record flows in only one direction (Figure 10.10a). Thus, for example, a QF may sell power to the utility for say 2¢/kWh and they might purchase it back again for perhaps 9¢/kWh. Clearly, this isn't a very lucrative deal unless the power plant owner almost always sells power to the utility and rarely buys any back.

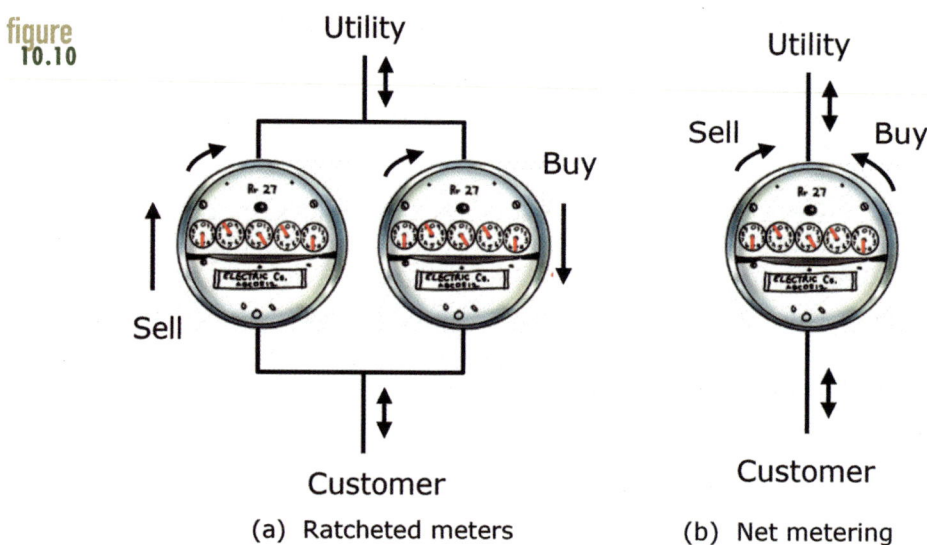

(a) Ratcheted meters allow different prices for selling and buying electricity. (b) With net metering a single meter runs in either direction.

A more advantageous contract for customers who generate some of their own power uses only one meter that spins in both directions. When the customer buys power from the grid, the meter goes in the forward direction, but when the customer generates more power than is needed at that moment, the meter spins backward (Figure 10.10b). Thus both the sale and purchase of electricity are at the retail rate. This is known as *net metering*. Most, but not all, states and jurisdictions have developed laws allowing this net metering approach to customer-utility interconnections. At the end of each month's billing period, the meter is read and the customer pays for the net amount of energy purchased. If more energy is sold than purchased, a credit is entered into that customer's account.

Suppose you install a photovoltaic system and then sign up for net metering with the TOU rates shown in Table 10.2. During those clear summer days, the meter credits energy sold to the utility at 29.4¢/kWh, while at night you can buy it back again for only 8.7¢/kWh. This really enhances the economics of the PV system. With a large PV system and sunny weather you could imagine receiving big checks from the utility. Most net metering contracts, however, are designed to tally up the books at the end of each year in such a way as to prevent a net amount of cash to flow from the utility to you.

10.5 Heat Engines and the Carnot Efficiency Limit

As you recall, the first law of thermodynamics states that energy is conserved in any process. This is a wonderful bookkeeping tool that helps us analyze energy processes to be sure we know where every bit of it came from and where it went. What the first law doesn't do, however, is keep track of what happens to the quality of energy as it moves about in various systems. Energy quality relates to the ability to do useful things with the energy. Electricity is very high-quality energy. You can power your computer, drive your electric car, or heat your coffee with it. An equal amount of energy contained in some lukewarm water can't do much more than warm your hands.

This suggests that there is a hierarchy of energy forms, with some being "better" than others. Electricity and mechanical energy (doing work) are of the highest quality. In theory, we could go back and forth between electricity and mechanical work with 100% conversion efficiency. Thermal energy is of much lower quality. Moreover, the quality of thermal energy relates to its temperature. High-temperature heat can boil water to run a steam turbine, so it is more useful than lower-temperature heat. The relationship between energy quality and temperature is tied to the notion of entropy, which was briefly introduced in Chapter 4.

10.5.1 Entropy

One way to think about entropy is that it is a measure of molecular disorder, or molecular randomness. At one end of the entropy scale is a pure crystalline substance at absolute zero temperature. Because every atom is locked into a predictable place, in perfect order, its entropy is defined to be zero. The same substance, as a liquid, has more randomness and higher entropy. And as a gas, it has even more. For example, when we burn a piece of wood, there is more entropy in the randomness of the gaseous end products than in the solid log we started with, so the entropy of the universe increases. Notice there is no such thing as a law of conservation of entropy. In fact, quite the opposite is the case. Everything we do, except for perfect, frictionless work, increases the entropy of the universe.

In any isolated system in which the total energy is fixed, any process that occurs will either increase the entropy of the system or, at best, break even. This requirement is actually one way to express the second law of thermodynamics. Moreover, the only way to break even is to consider work as an idealized, reversible process that allows us to convert back and forth from one mechanical form of energy (e.g., kinetic energy) to another (e.g., potential energy) without heating anything (e.g., no friction). In fact, processes can be analyzed in terms of just these two components: heat transfer and work. Heat transfer is accompanied by increased entropy, but work is entropy-free.

The concept of ever-increasing entropy is enormously important. It can be used to explain why heat flows naturally from warm objects to cold ones, and not the other way around. It can be used to explain why heat can't be converted entirely into electricity, yet the opposite is possible. We'll use it in the next section to determine the maximum theoretical efficiency of certain kinds of power plants.

10.5.2 Heat Engines

Over 90% of U.S. power plants convert heat, usually from a boiler or a nuclear reaction, into work done by a rotating turbine/generator shaft. And they reject waste heat into the

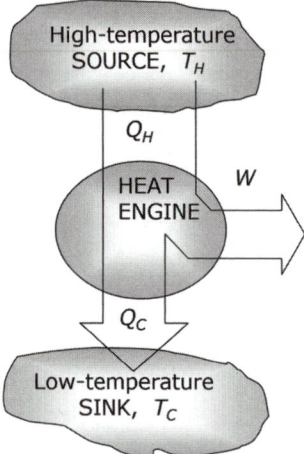

figure 10.11

A heat engine converts some of the heat extracted from a high-temperature reservoir into work, rejecting the rest into a low-temperature sink.

environment. Such *heat engines* can be described as consisting of some kind of device that takes thermal energy Q_H from a high-temperature source, converts some of it into work W, and rejects the remainder Q_C into a low-temperature sink (Figure 10.11).

As shown in Figure 10.11, not all of the energy taken from the high-temperature source ends up being converted to work (and then into electricity, if this is a power plant). If you think this must have something to do with entropy, you're exactly right.

The most efficient heat engine that could possibly operate between hot and cold thermal reservoirs was first described back in the 1820s by the French engineer, Sadi Carnot. To sketch out the basis for his famous equation, which links the maximum possible efficiency to the temperatures of the two reservoirs, we begin with a simple expression for the efficiency of the heat engine shown in Figure 10.11.

Eq. 10.1
$$\text{Efficiency } \eta = \frac{\text{Work done}}{\text{Heat input}} = \frac{W}{Q_H}$$

Using the first law, we can substitute $W = Q_H - Q_C$ into Equation 10.1 to get

Eq. 10.2
$$\text{Efficiency } \eta = \frac{Q_H - Q_C}{Q_H} = 1 - \frac{Q_C}{Q_H}$$

For our heat engine, when heat transfer Q occurs at constant temperature T the entropy transferred ΔS is defined as

Eq. 10.3
$$\Delta S = \frac{Q}{T}$$

where T = an absolute temperature measured using either the Kelvin
   ($K = °C + 273$) or Rankine ($R = °F + 460$) scale

Notice entropy goes down as temperature goes up. We know that high-temperature heat is more useful than the same amount at lower temperature, which reminds us that entropy is not such a good thing. Less is better!

Because the work done by our heat engine is entropy-free, the only entropy transfers are those associated with taking heat Q_H out of the hot reservoir and dumping Q_C into the cold reservoir. In the unachievable, best-of-all-possible worlds, there would be no increase in entropy of the system, which would mean the entropy gained by the cold reservoir would equal the entropy lost by the hot one:

Eq. 10.4
$$\frac{Q_C}{T_C} = \frac{Q_H}{T_H}$$

Substituting Equation 10.4 into Equation 10.2 gives the maximum possible efficiency of a heat engine:

Eq. 10.5
$$\eta_{\max} = 1 - \frac{Q_C}{Q_H} = 1 - \frac{T_C}{T_H}$$

This is the classical result described by Carnot and it bears his name. Notice that the only way the efficiency could be 100% is to have an infinite source temperature or a thermal sink at absolute zero. Neither of these is possible, so there will always be waste heat rejected into the environment.

For example, consider the Carnot efficiency limit applied to a conventional steam-electric power plant. A reasonable estimate of T_H might be the temperature of the steam from the boiler, which is typically around 600°C. For T_C we might use a typical condenser operating temperature of around 30°C. Using these values in Equation 10.5, and remembering to convert temperatures to the absolute scale, gives

Eq. 10.6 $$\text{Carnot efficiency} = 1 - \frac{T_C}{T_H} = 1 - \frac{(30 + 273)}{(600 + 273)} = 0.65 = 65\%$$

Actual steam power plants have considerably lower efficiency, with the most efficient ones approaching 40%.

10.6 Combined-Heat-and-Power (CHP) Systems

The above analysis of heat engines assumes the only output that is useful is the work or electricity produced by the engine. Although there is nothing wrong with that analysis, it ignores the potential benefits associated with capturing and using the heat rejected by the engine. Such waste heat can be used for water heating, space heating of buildings, or process heat for industry. It can also be used to cool buildings.

Distributed generation technologies include some that produce usable waste heat as well as electricity (e.g., combustion turbines and fuel cells), and some that don't (e.g., photovoltaics, micro-hydro, and wind turbines). For those that do, one of the advantages of having generation located close to loads is the possibility of capturing that waste heat and putting it to good use. In the past, systems that produce both electricity and useful thermal energy in a sequential process from a single source of fuel were referred to as *cogeneration* systems; now they are more often called CHP systems.

A representative example of the efficiency benefit associated with CHP systems is shown in Figure 10.12. In this example, a CHP system with 100 units of energy input delivers 35 units of electricity and 50 units of useful waste heat, for an overall efficiency of 85%. If the same 35 units of electricity were to be delivered from a 30%-efficient utility grid, 117 units of fuel would need to be burned. And, to provide the 50 units of heat in a separate fuel-fired, 80%-efficient boiler, another 63 units of fuel would be burned. The total in a separate grid-and-boiler system is 180 units of fuel compared to the 100 units needed in the CHP system. In other words, the CHP system results in an overall energy savings of 44%.

Sorting out the economics of a CHP system can be a bit tricky. For example, we might like to be able to characterize the cost of electricity from the system separately from the cost of heat. A common approach is based on adding the amortized capital cost of the CHP unit to the fuel cost to get an annual cost of heat and power. We can then subtract the cost of the heating fuel that we don't need to purchase because we have a CHP system, and use the remainder as an indicator of the cost of electricity. Solution Box 10.2 illustrates the procedure.

figure 10.12

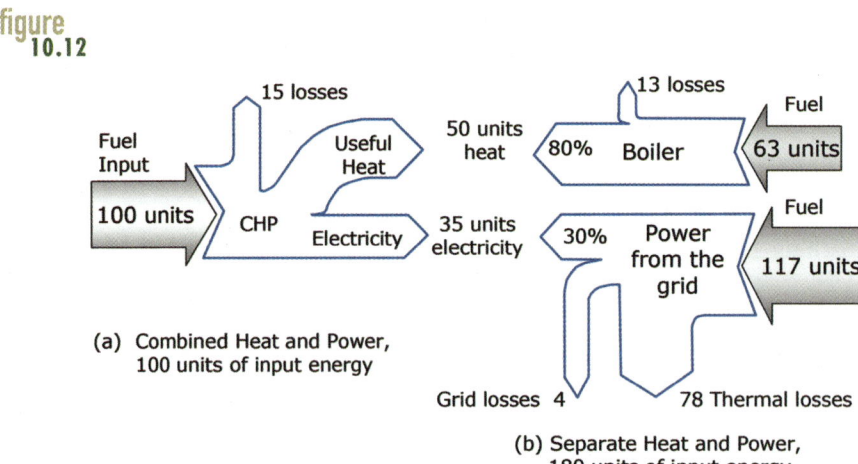

(a) Combined Heat and Power, 100 units of input energy

(b) Separate Heat and Power, 180 units of input energy

In this example, the CHP system needs only 100 units of fuel to deliver the same electricity and useful heat as 180 units from a combination of grid electricity and a separate boiler, resulting in an overall efficiency improvement of 44%.

Economics of a Combined-Heat-and-Power System

A 10-kW CHP system has an electrical efficiency of 40% and a thermal efficiency of 40%. Its $30,000 cost is paid for with an 8%, twenty-year loan having annual payments of $3055/year. Its heat output displaces gas costing $10 per million Btu that would have been burned in an existing 85%-efficient boiler. If the CHP system has a capacity factor (CF) of 0.90 (in essence it runs only 90% of the time during which it delivers the full 10 kW), find its cost of electricity.

Solution:

Because the electrical efficiency is 40%, it takes 25 kW of heat to produce its 10 kW of electrical power. That is, it takes 25 kW × 3412 Btu/kWh = 85,300 Btu/hr of fuel. Forty percent of that fuel is captured waste heat (0.40 × 85,300 = 34,120 Btu/hr) that the 85%-efficient boiler doesn't need to provide. That means cogeneration saves 34,120 Btu/hr/0.85 = 40,140 Btu/hr of boiler fuel.

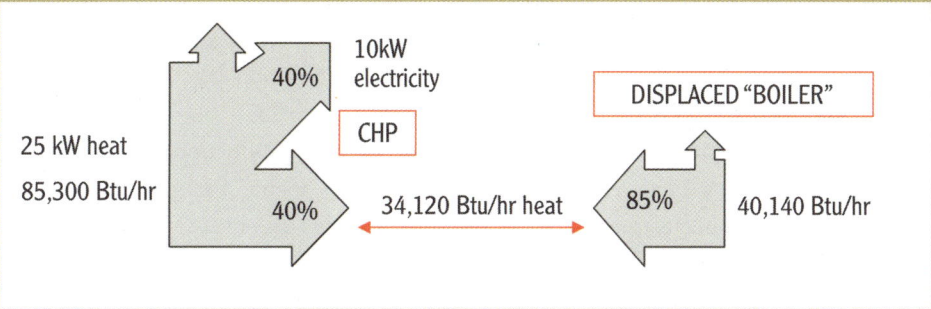

To find the cost of fuel attributed to generation of electricity, we'll subtract the cost of displaced boiler fuel from the total fuel cost:

Fuel for electricity = (85,300 − 40,125) Btu/hr × $10/10^6 Btu = $0.452/hr

That translates to $0.452/hr to deliver 10 kW, which is $0.0452/kWh. With a CF of 0.90, the system will deliver

10 kW × 8760 hr/yr × 0.90 = 78,840 kWh/yr

Adding the $3055/yr cost of the loan to the annual net fuel for electricity yields

$$\text{Cost of electricity} = \frac{\$3055/\text{yr}}{78,840 \text{ kWh/yr}} + \$0.0452/\text{kWh} = \$0.084/\text{kWh}$$

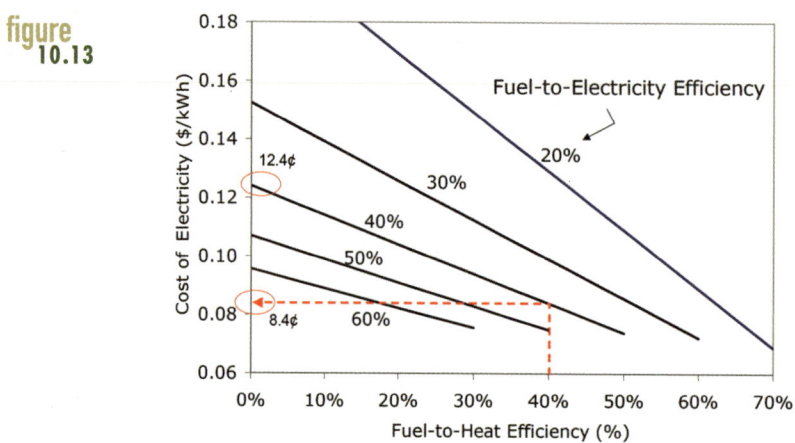

figure
10.13

The cost of electricity from a CHP system is reduced when waste heat is captured. For the example in Solution Box 10.2, electricity drops from 12.4¢/kWh to 8.4¢/kWh with 40%-efficient heat recovery. Assumptions: Capital cost, $3/Watt; fuel cost, $10/MMBtu; twenty-year, 8% loan; boiler efficiency, 85%; CF, 90%.

Notice how the fuel saved by heat recovery in the above example helps subsidize the cost of electricity generated. Without heat recovery, the cost of electricity would have been 12.4¢/kWh, but with that recovery it is 8.4¢/kWh—a savings of about one-third. Figure 10.13 shows how the cost of electricity found in Solution Box 10.2 varies as the electrical and thermal efficiencies for this CHP system are varied.

Quite often, systems designed primarily for electrical production are characterized using "dollars per watt" and electrical efficiency measures. Figure 10.14 provides a quick indication of the cost of electricity from a CHP system using these measures. For example, it indicates that a 35%-electrical-efficiency CHP fuel cell system could cost as much as $5 per watt and still be competitive with 11¢/kWh electricity from the grid.

10.7 Microturbines

Microturbines are small combustion turbines with individual units that generate roughly 500 watts to several-hundred kilowatts of electricity along with sizeable amounts of waste heat. They tend to be fueled with conventional natural gas, but also may operate with lower-quality biogas from landfills or wastewater treatment plants, propane, hydrogen, or diesel. The technology is scaleable, which means multiple units can be ganged together for larger loads.

Whereas the electrical efficiency of microturbines is relatively low at around 25%–30%, when waste heat is captured and utilized their overall efficiency can be well above 80%. The economic advantage of these turbines is closely tied to the value of that heat, which tends

figure 10.14

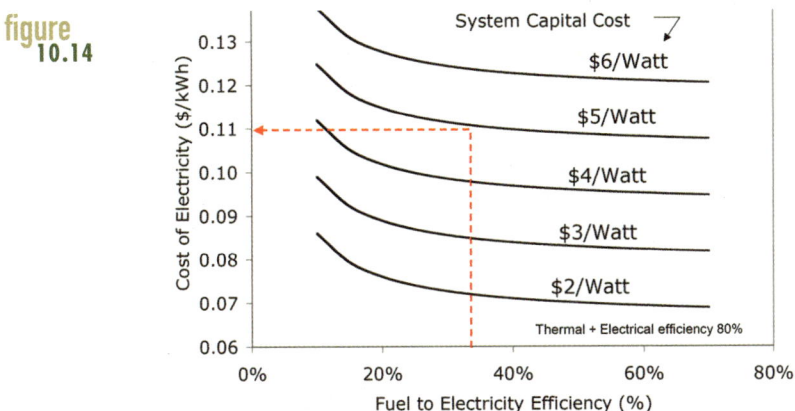

Cost of electricity from a CHP system with overall efficiency of 80%. The example indicates that a $5/watt system with 35% electrical efficiency would generate electricity at a cost of $0.11/kWh. Assumptions: Fuel cost = $10/MMBtu; twenty-year, 8% loan; 85% boiler efficiency; 90% CF.

to be relatively low-temperature 50°C–80°C hot water from a heat-recovery unit. Good candidates for microturbines include applications that can use as much of the waste heat as possible, year-round. Water heating loads in hotels and apartment houses, and swimming pools if they are heated all year long, are good examples. Figure 10.15 shows an example of a microturbine system that incorporates a heat recuperator to boost turbine efficiency and a heat exchanger to capture waste heat.

figure 10.15

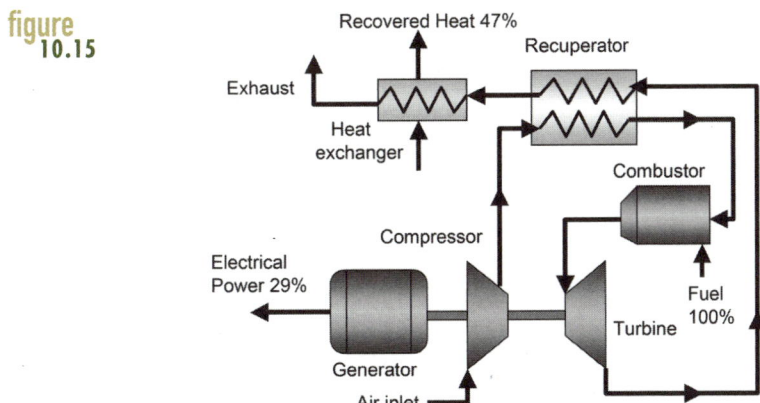

Schematic diagram of a microturbine with a recuperator to improve turbine efficiency and a heat exchanger to capture waste heat. Representative LHV efficiencies are shown.

10.8 Reciprocating Engines

Distributed generation today is dominated by reciprocating engines—that is, piston-driven, internal combustion engines connected to constant-speed ac generators. They are used primarily for standby power, which means they don't run very often, so they are rarely set up to take advantage of their cogeneration potential. They are readily available in sizes that range from less than 1 kW to over 6 MW and can be designed for a number of fuels, including gasoline, diesel, natural gas, kerosene, propane, fuel oil, alcohol, digester gas, and hydrogen. Because emissions are a significant issue, many of these reciprocating engines use clean-burning natural gas.

The heart of these systems is a rather conventional four-stroke engine, very much like the one in your car. Their combustion can be based on spark-ignition (Otto cycle) or compression ignition (diesel-cycle) with fuel-to-horsepower conversion efficiencies that can exceed 40%. When used in combined-heat-and-power systems, most of the heat recovery is from the engine's cooling water along with some from the engine exhaust. Figure 10.16 shows an example of a CHP engine with an overall efficiency of 85%.

The U.S. Department of Energy, along with Caterpillar, Cummins, Waukesha Engines, and a number of universities, has initiated an Advanced Reciprocating Engines Systems (ARES) program to help improve the efficiency of natural-gas engines while lowering emissions. Their goals include 50% electrical efficiency, and 80%⁺ cogeneration efficiency, with NO_x emissions below 0.1 g per horsepower, at a capital cost of less than \$450/kW.

Although reciprocating engines boast high CHP efficiencies, their NO_x emissions may prove to be problematic for areas with significant smog problems. Moreover, maintenance and lifetime issues suggest they may not be appropriate for continuous operation. To get a feeling for that constraint, think about servicing your car every 10,000 miles and replacing the engine at 200,000 miles. At a leisurely 20 mph you would service your engine after every 500 hours of use and you would replace it after only 10,000 hours. If you were generating electricity with that engine constantly running, you'd be changing the oil every three weeks and replacing the engine after barely fourteen months of use.

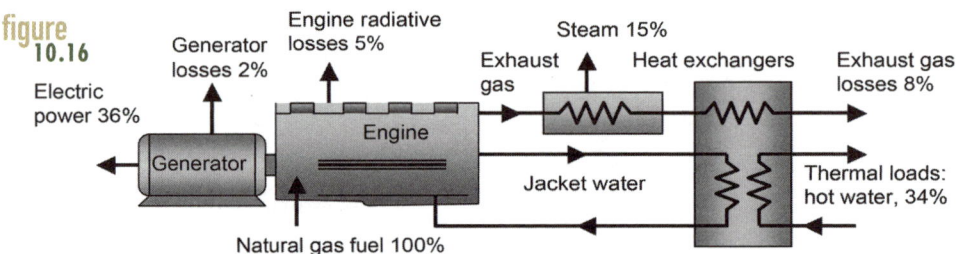

figure 10.16

Example energy balance for a reciprocating engine 85%-efficient cogeneration system.

10.9 Fuel Cells

I believe that water will one day be employed as a fuel, that hydrogen and oxygen which constitutes it, used singly or together, will furnish an inexhaustible source of heat and light.

Jules Verne, *Mysterious Island*, 1874

The portion of the above quote in which Jules Verne predicts the joining of hydrogen and oxygen to provide a source of heat and light is a remarkably accurate description of how a fuel cell actually works. It misses the mark, however, when it suggests that water itself will be the fuel. Whereas water can be electrolyzed to yield hydrogen and oxygen, the energy to do so far exceeds the energy that would be returned when these gases are recombined in a fuel cell. It is more accurate to consider a fuel cell as a sort of battery that delivers electricity (and heat) as long as it receives a continuous supply of energy-rich fuel (usually hydrogen).

There are several potential advantages of fuel cells compared to other energy-conversion systems. Because they convert chemical energy directly into electrical power, they avoid the traditional intermediate steps of converting fuel into heat, heat into mechanical motion, and mechanical energy into electricity. Because they are not heat engines, they are not constrained by the Carnot limits and fuel-to-electricity efficiencies as high as 65% are likely, which gives fuel cells the potential to be roughly twice as efficient as the average central power station operating today.

Moreover, the usual combustion products such as SO_x, CO, particulates, and assorted partially burned hydrocarbons, are totally eliminated. For most fuel cells, NO_x emissions are either eliminated or greatly reduced. If the original fuel is a hydrocarbon, such as natural gas (which is usually the case right now), there will still be carbon emissions, but if fuel cells are powered by hydrogen obtained by electrolysis of water using renewable energy systems such as wind, hydroelectric, or photovoltaics, they have no greenhouse gas emissions at all.

Because fuel cells are vibration-free, quiet, and emit little pollution, they can be located very close to their loads—for example, in the basement of a building. Being close to their loads not only avoids transmission and distribution system losses, but their waste heat can be used to cogenerate electricity and useful heat for applications such as space heating, hot water, and even air-conditioning. As a combined-heat-and power system, fuel cells can have overall efficiencies of over 80%.

10.9.1 Basic Proton-Exchange-Membrane (PEM) Cell and Stack

There are many variations on the basic fuel cell concept, but a common configuration looks something like Figure 10.17. As shown there, a single cell consists of two porous gas diffusion electrodes separated by an electrolyte. It is the choice of electrolyte that distinguishes one fuel cell type from another.

The electrolyte in Figure 10.17 consists of a thin membrane that is capable of conducting positive ions (protons) but not electrons or neutral gases. This particular membrane

figure 10.17   **Basic Configuration for a PEM Cell**

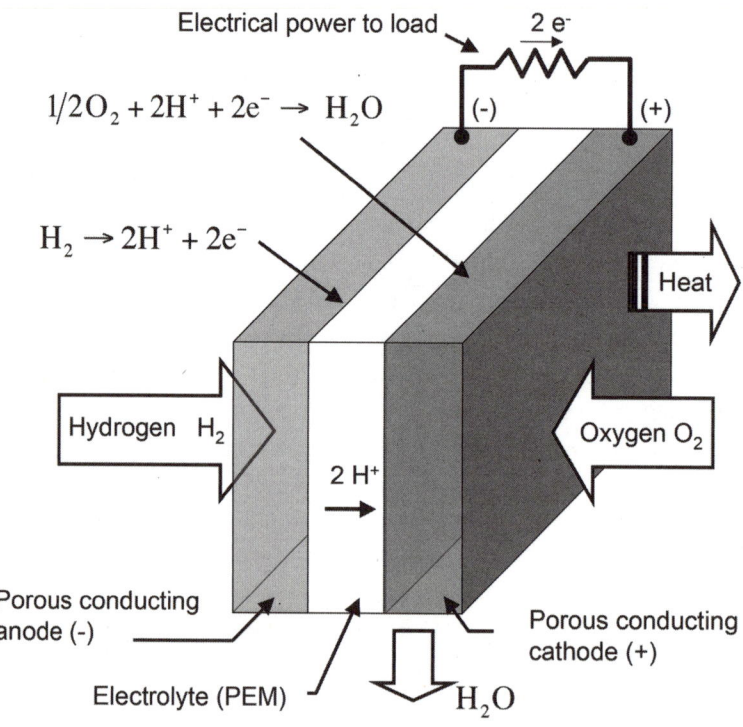

Electrical power to load

$2\,e^-$

$1/2O_2 + 2H^+ + 2e^- \rightarrow H_2O$

(-)   (+)

$H_2 \rightarrow 2H^+ + 2e^-$

Heat

Hydrogen H_2

$2\,H^+$

Oxygen O_2

Porous conducting anode (-)

Porous conducting cathode (+)

Electrolyte (PEM)

H_2O

passes protons, so it is usually called a *proton-exchange-membrane*, though sometimes it is called a *polymer electrolyte membrane*—in either case, it is usually referred to as a *PEM* cell. PEM cells were originally known as *solid polymer electrolyte* (SPE) cells, but that descriptor is less common these days.

Hydrogen (H_2) introduced on one side of a PEM cell (the anode) is encouraged by an expensive platinum catalyst to dissociate into protons and electrons as follows:

Eq. 10.7 $H_2 \leftrightarrow 2H^+ + 2e^-$

With a high concentration of protons accumulating in the anode, there is a tendency for some to diffuse through the membrane to the cathode, leaving the electrons behind. The accumulating negative charge on the anode and positive charge on the cathode creates an open-circuit voltage on the order of 1 volt across the cell. When connected to a load, electrons flow from anode to cathode delivering dc electrical power to the load.

The two half-cell reactions occurring in a PEM cell are as follows:

Eq. 10.8   Anode: $H_2 \rightarrow 2H^+ + 2e^-$

Eq. 10.9   Cathode: $\frac{1}{2}O_2 + 2H^+ + 2e^- \rightarrow H_2O$

which results in a net reaction that looks like the simple combustion of hydrogen to produce energy and water:

Eq. 10.10   Combined: $H_2 + \frac{1}{2}O_2 \rightarrow H_2O + energy$

Although an entropy analysis is beyond the scope of this book, it can be shown that at least 17% of the energy liberated in Equation 10.10 must be heat, which limits the hydrogen-to-electricity HHV efficiency of a PEM cell to a maximum of 83% (e.g., Masters, 2004).

Under normal loads, the voltage across a single cell is only about 0.5 volts, so to create decent voltages many cells are connected in series into a multicell stack as shown in Figure 10.18.

As shown in Figure 10.19, a complete fuel cell system typically consists of three major components: a reformer to convert fuel—usually natural gas—to hydrogen, the fuel cell stack itself, and a power conditioning unit that converts the dc output of the fuel cell into ac power with the right frequency and voltage.

10.9.2 Hydrogen Production

Oxygen for the PEM cell is readily available from ordinary air; the real fuel problem is providing the needed hydrogen. Realize that hydrogen is not an energy source. It is, like electricity, a high-quality energy carrier that is not naturally available in the environment. It must be manufactured, which means an energy investment must be made to create the desired hydrogen fuel.

The main technologies currently in use for hydrogen production are steam reforming of methane (MSR), partial oxidation of hydrocarbons (POX), and electrolysis of water. In

figure 10.18

A fuel cell stack consists of multiple cells arranged in series to increase the voltage.

Source: Ballard Corporation

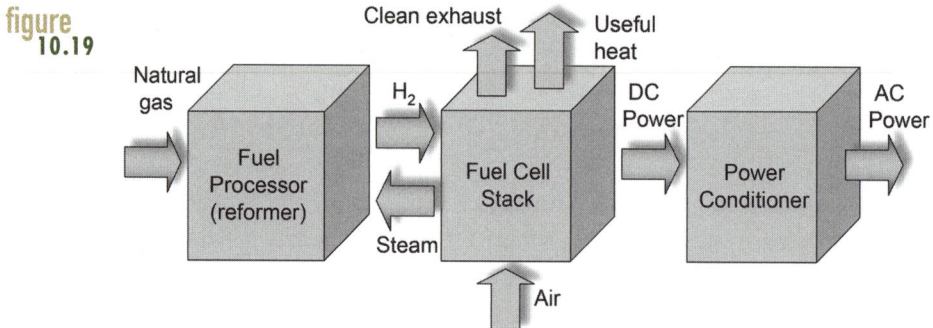

figure
10.19

A complete fuel cell system consists of a reformer to produce hydrogen-rich fuel, the fuel cell stack itself, and a power conditioner to convert dc to ac.

the last chapter (Section 9.4.5), we saw how future integrated gasification combined-cycle (IGCC) power plants might use pyrolysis of coal to produce hydrogen. More exotic methods for the future include photocatalytic, photoelectrochemical, or biological production of hydrogen using sunlight as the energy source.

Most hydrogen production today uses the MSR process in which a mixture of steam and natural gas is passed through a catalyst at very high temperature producing a synthesis gas, or syngas, made up mostly of CO and H_2. Syngas in this form could be used as a fuel directly in certain high-temperature fuel cells; however, it is incompatible with PEM cells because the CO poisons their anode catalysts. To upgrade syngas, a second-stage, shift reaction is usually incorporated that converts the CO to CO_2. The overall efficiency of MSR production of hydrogen is typically 75%–80%, but higher levels may be achievable.

The other promising approach, called *electrolysis,* is the reverse of the conventional fuel cell. In this process, electrical current is forced through the cell, which breaks apart water molecules into protons, electrons, and molecular oxygen. As shown in Figure 10.20a, the protons pass through the membrane, recombine with electrons from the power source, and form molecular hydrogen, which can be captured, compressed, and used as a fuel. The overall efficiency from electricity into compressed hydrogen is in the range of 70%–75%.

The concept of a reversible fuel cell is intriguing. We could imagine such cells operating in one direction to convert electricity to hydrogen and then being reversed to convert that hydrogen back into electricity. This is especially interesting in the context of intermittent, renewable sources of electricity such as wind or photovoltaics. As suggested in Figure 10.20b, such a system could yield an inexhaustible, pollution-free source of electricity, available whenever it is needed. Unfortunately, the roundtrip efficiency, from electricity to hydrogen to electricity would only be around 20%–30% using these low-temperature PEM cells. With higher-temperature fuel cells, the efficiency might be improved considerably. But, with rapid improvements in battery technology having roundtrip efficiencies above 90%, the reversible fuel cell concept will have tough competition.

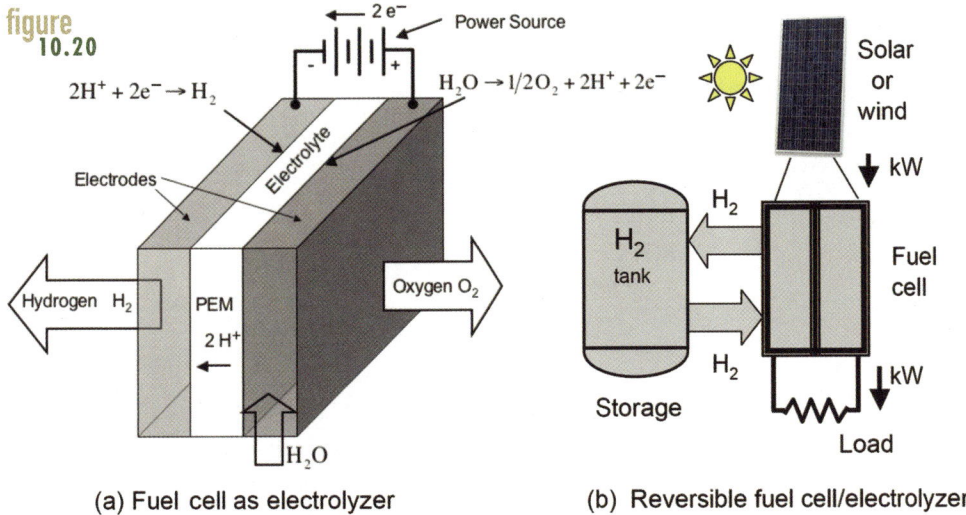

figure 10.20

$2H^+ + 2e^- \rightarrow H_2$ $\leftarrow 2e^-$ Power Source $H_2O \rightarrow 1/2O_2 + 2H^+ + 2e^-$

Electrolyte

Electrodes

Hydrogen H_2 PEM $2H^+$ $\leftarrow$ Oxygen O_2

H_2O

(a) Fuel cell as electrolyzer

Solar or wind $\downarrow$ kW

H_2 tank H_2 Fuel cell

H_2 $\downarrow$ kW

Storage   Load

(b) Reversible fuel cell/electrolyzer

One way to provide reliable electric power from a renewable source such as wind or solar is with a reversible fuel cell with hydrogen storage.

10.9.3 Other Promising Fuel Cell Technologies

Most of the recent work on fuel cells has been directed toward their possible use in vehicles, for which polymer electrolyte cells seem the most promising. Because vehicles aren't actually driven for much of the time, the limited lifetime of PEM membranes is less of an issue than it would be for stationary applications in which continuous operation might be desirable. And, because their operating temperature is relatively low, they can turned on and off without waiting long for them to come up to temperature.

A number of other fuel cell technologies have been developed and some of them threaten the current dominance of PEM cells. Alkaline cells using potassium-hydroxide electrolytes were developed for the Apollo and Space Shuttle programs, but they don't tolerate exposure to CO_2 and are really appropriate only for space applications. Another technology based on a phosphoric acid electrolyte actually reached the marketplace in the early 1990s, but its high cost is still limiting its acceptance. Direct methanol fuel cells (DMFC) use the same polymer electrolytes as PEM cells, but they have the significant advantage of being able to utilize the liquid fuel, methanol (CH_3OH), which is more convenient and portable than gaseous hydrogen. There is considerable interest in using DMFCs powered by little capsules of fuel as replacements for batteries in portable electronic devices such as cell phones or laptops.

Molten-carbonate fuel cells (MCFC) and solid-oxide fuel cells (SOFC) are competing technologies that operate at much higher temperatures than any of the other technologies, which makes them very attractive for combined-heat-and-power applications. Moreover, their high temperature allows for self-reforming of natural gas into hydrogen-rich fuel.

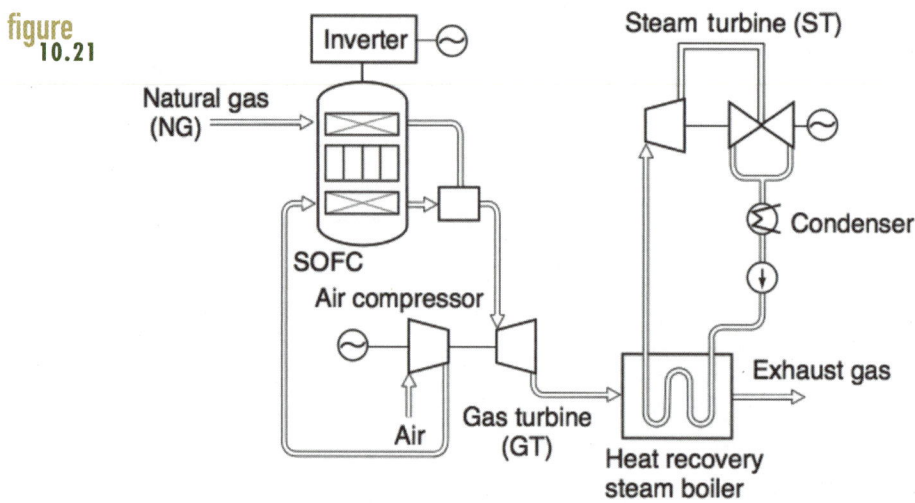

figure 10.21

A hybrid solid-oxide-fuel-cell/combined-cycle power system could have a 70% LHV efficiency.

SOURCE: Yoshida, Mitsubishi Heavy Industries, 2003

Fuel-to-electrical efficiencies near 60% are likely and when coupled with a combined-cycle steam and gas turbine (Figure 10.21) their electrical efficiency may approach an astonishing 70%.

A summary of the major characteristics of the most promising fuel cell technologies is included in Table 10.4.

10.10 Stirling Engines

One of the most promising approaches to new CHP systems is based on an old concept, called a Stirling engine, that is just now being rediscovered and refined. Stirling engines have

table 10.4   Essential Characteristics of the Most Promising Fuel Cell Technologies

Characteristic	Solid polymer (PEM)	Solid Oxide (SOFC)	Molten Carbonate (MCFC)
Electrolyte	Proton exchange membrane	Solid oxide ZrO_2–Y_2O_3	Molten carbonates (Li, K, Na)
Charge carrier	H^+	O^{2-}	CO_3^{2-}
Operating temperature	80°C	1000°C	650°C
Methane reformer?	Yes	No	Maybe
Applications	Portable, transportation	Stationary power	Stationary power
Cogeneration	Hot water, space heating, cooling	Steam	Steam
Electrical efficiency (LHV)	40%	60%	50%–60%
		(higher with combined cycle)	(higher with combined cycle)

a long and interesting history. They were invented in 1816 by a Scottish minister, Robert Stirling, who apparently was motivated in part by concern for the safety of his parishioners who were at risk from poor-quality steam engines that had a tendency, in those days, to violently and unexpectedly explode. His engines had no such problem because they were designed to operate at relatively low pressures. Stirling engines were reasonably popular in the last decades of the nineteenth century, but by the early 1900s advances in steam and spark-ignition engines, with higher efficiencies and greater versatility, pretty much eliminated them from the marketplace. At the beginning of the twenty-first century, however, there is a resurgence of interest in them both as small, quiet, somewhat efficient, combined-heat-and-power systems and also as electricity-generating, parabolic-dish solar energy systems.

Stirling engines are very different from the internal-combustion engine in your car. There are no valves opening and closing, no fuel explosions (and hence very little noise or vibrations), and no high-pressure exhaust propelled out a tailpipe. They are classic heat engines in that all they require is a temperature difference between a hot source and a cold sink. In fact, you can buy small demonstration models that can spin a rotor using just the heat of your hand, or a hot cup of coffee, and ambient air for the cold sink (http://www.stirlingengine.com).

An example Stirling engine is shown in Figure 10.22. As shown, a continuous source of heat (in this case a flame) and cooling (in this case cold water) provide the needed temperature difference. A gas, which could be air but is more likely to be nitrogen, helium, or hydrogen,

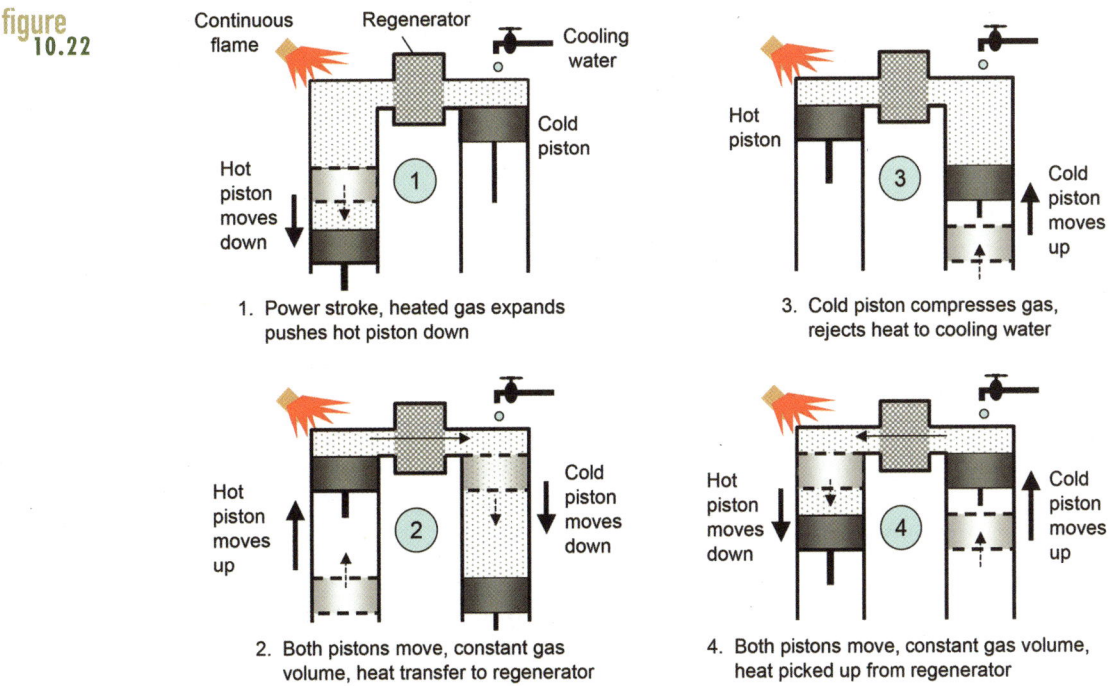

figure 10.22

A simplified example of the four stages of operation for a Stirling engine.

is kept in a confined space. The heating, cooling, compression and expansion of the gas are what drives the engine. In this case, two pistons move up and down in their cylinders in a four-step cycle. The pistons are connected to a crankshaft that converts the up-and-down motion of the cylinders into a rotating shaft.

There are two sides to the engine—a hot side and a cold side—separated by a "short-term" thermal energy storage device called a regenerator. The regenerator is designed to allow easy flow of the working gas back and forth between the hot side and the cold side of the engine. It may be just a wire or ceramic mesh or some other kind of porous plug with sufficient mass to allow it to maintain a good thermal gradient from one face to the other.

As shown, the cycle begins with a power stroke caused by the expansion of the confined working gas as the gas is heated by the flame. This is the only part of the cycle that actually does any work. The crankshaft then moves the hot piston up and the cold piston down in a constant-volume process that transfers heat into the regenerator. With most of the gas now on the cold side of the engine, the cold piston rises and compresses that gas, but the heat created is transferred to the cooling water. In the fourth stage, both pistons move the cold gas through the regenerator where it picks up some heat, and the cycle is ready to begin again. Thermo-dynamically, this is a very efficient cycle with the potential to achieve efficiencies approaching the Carnot limit. In reality, although they are not that good, they can be more efficient than almost any other practical heat engine operating between the same temperature difference.

An example of a commercially available product based on Stirling engine technology is a dishwasher-size WhisperGen 1.2 kW micro-CHP unit. Overall efficiency is claimed to be around 90% (roughly 15% electrical and 75% thermal efficiency). Similar units have been built as battery-charging, stand-alone systems for such exotic applications as submarines and private yachts where their quiet operation is highly valued. In Chapter 11, the use of Stirling engines powered by solar energy will be described.

10.11 Summary

The economies of scale that motivated the era of ever-larger power stations have begun to play themselves out. By 2000, traditional vertically integrated utilities that owned everything from generation to transmission to distribution found themselves competing against new independent power producers (IPPs) who were better able to capitalize on the economic advantages of smaller, cheaper generation technologies. Meanwhile, the customer's side of the meter was discovered and the potential to save energy cheaper than it could be generated began to be tapped. In addition, utilities began to find they could save money by delaying the need to upgrade transmission and distribution systems to meet growing loads by strategi-cally siting small, distributed-generation systems close to their customers. And, finally, the world's growing concern for emissions of climate-changing greenhouse gases has motivated the search for ways to improve the efficiency of both generation and end uses of electricity.

The technologies and policies driving these dramatic changes have been the subject of this chapter. Distributed energy resources are finally gaining traction in the planning and implementation of our nation's energy systems. Demand-side management programs to reduce loads or shift them away from times of peak demand; distributed generation systems located on either the utility side or customer side of the meter; new storage systems both stationary and, potentially, in future electric-drive vehicles; and combined-heat-and-power systems that can significantly cut energy demand will contribute to a lower-cost, more sustainable, energy future.

In the next two chapters, we will look at several of the most promising renewable energy technologies for electricity generation. Some of these, including most prominently photovoltaics, are located primarily on the customer's side of the meter. Others, especially wind turbines, are usually owned by independent power producers who sell power into the grid on the utility side of the meter.

CHAPTER 11

Photovoltaic Systems

In Chapter 7, techniques for capturing and using solar energy to provide heat for domestic hot water and for space heating of buildings were introduced. By and large, these involve relatively simple, low-cost technologies and design concepts that can greatly reduce the need for conventional fuels at little or no extra cost. Indeed, simple passive solar design ideas such as careful building orientation, overhangs, and thermal mass have been effectively utilized throughout human history.

Relatively new on the scene are "rooftop" photovoltaic (PV) systems that convert sunlight directly into electricity—the highest quality, most versatile form of energy. Because buildings use almost three-fourths of U.S. electricity, the potential to meet some of that demand with PVs is intriguing. As will be discussed in this chapter, there are enough rooftops with appropriate solar exposure for PVs to supply over one-third of today's U.S. total electricity demand. Costs are still too high to make a dent in that potential without significant subsidies, but with rapidly increasing demand, and corresponding decreases in cost, we may well be able to wean ourselves of that necessity within the next decade or so.

In the next chapter, we will explore other solar energy systems for electricity generation, including concentrating parabolic troughs, Stirling engine dish systems, and, most importantly, wind turbines. The key advantage that building-integrated PVs have over these more centralized energy systems is that the electricity produced with PVs competes against the relatively expensive retail price of electricity, which homeowners and building operators pay, rather than the much lower wholesale price that large systems have to meet to in order to sell bulk electricity into the grid.

11.1 Introduction to Photovoltaics

Back in 1839, a 19-year-old French physicist, Edmund Becquerel, was able to cause a voltage to appear when he illuminated a metal electrode immersed in a weak electrolyte solution. That was the first known observation of what is now referred to as the photoelectric effect. Albert Einstein, in 1904, was the first to provide a theoretical explanation for the phenomenon, which led to his Nobel Prize in Physics in 1921. Then in 1916, in what has turned out to be a cornerstone of modern electronics in general, and PVs in particular, a Polish chemist, Jan Czochralski, developed a technique for fabricating pure single-crystal materials. His approach led to the modern-day Czochralski method for growing perfect crystals of silicon, the most commonly used PV materials today.

The first practical PV devices for power generation were developed as part of the space program in the late 1950s, for which their high cost was much less important than their low weight and high reliability. By the late 1980s, PVs began to be used in more mundane applications where utility power lines were not a cost-effective option; these uses included off-shore buoys, highway lights, signs and emergency call boxes, rural water pumping, and small off-grid home systems. By the end of the twentieth century, however, as PV costs declined and efficiencies increased, it has been grid-connected, rooftop systems that have dominated sales. As shown in Figure 11.1, the global rate of production of PV modules has been growing at close to 40% per year. Japanese and European manufacturers provide half of that total, whereas the United States manufactures less than 10%. A milestone of sorts was reached in 2006 when the 2.5 GW of PV production used more tons of silicon than the entire micro-electronics industry.

Annual installations of PVs in the last few years have been predominantly in three countries: Germany, Japan, and the United States, with Germany alone accounting for almost 40% of the total. Germany's aggressive Renewable Energy Law of 2000, which enables PV

figure 11.1   **Global Production of Photovoltaics**

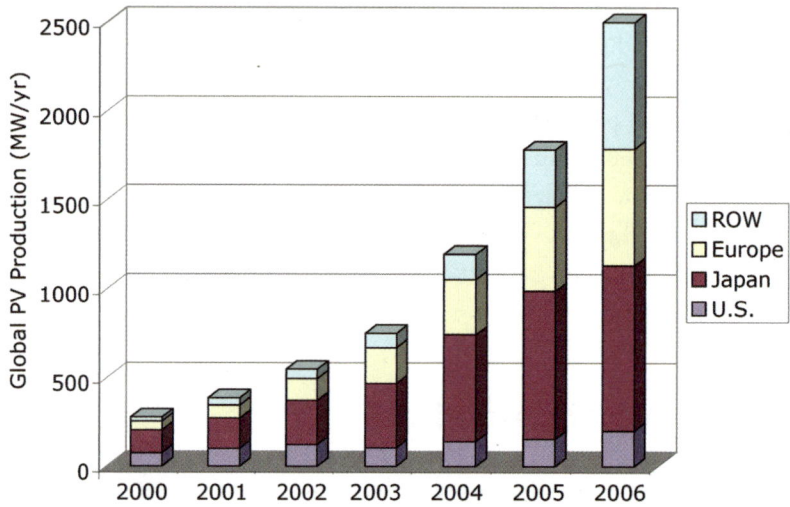

Japanese and European manufacturers provide half of the total, whereas the United States and the rest of the world (ROW) provide the other half.

generators to sell their electricity for more than 50¢/kWh, has pushed that country to the forefront of global PV sales in spite of having only about half the annual average solar radiation of a typical site in California.

11.2 Basic Semiconductor Physics

Einstein's revolutionary hypothesis that led to his Nobel Prize was that in certain circumstances light could be considered to consist of discrete particles, called photons, each carrying an amount of energy proportional to its frequency. Photons with high enough frequency can cause electrons in PV materials to break free of the atoms to which they are normally bound. If a nearby electric field is provided, those electrons can be swept toward a metallic contact where they can emerge as an electric current.

Although there are a number of promising PV materials under development, the starting point for almost all of the world's current PV devices, as well as almost all semiconductors used in electronic circuits, is pure crystalline silicon. Silicon has fourteen protons in its nucleus and fourteen orbital electrons. For all intents and purposes, the only electrons that matter are the four valence electrons in the outer orbit, so it is common practice to draw a silicon atom as if it has a +4 charge on its nucleus and four, tightly held, electrons that form covalent bonds with nearby silicon atoms as shown in Figure 11.2.

11.2.1 Hole-Electron Pairs

Crystalline silicon at absolute zero temperature would be a perfect electrical insulator. All of its electrons would be so tightly bound to their nuclei that none would be available to carry current, which means it would be useless as an electrical component. For an electron

figure
11.2

(a) tetrahedral

(b) 2-D version

Crystalline silicon forms a three-dimensional tetrahedral structure (a), but it is easier to draw it as a two-dimensional flat array (b).

to do us some good, it must be able to contribute to current flow. To do so, it must acquire enough energy, called the *band-gap energy*, to free itself from its covalent bonds. Heating it up will free some electrons, but not very many. Exposing silicon to sunlight, however, can allow photons to provide the energy needed to free those electrons. Those photons must have at least as much energy as the band gap, which for silicon means that the wavelength of an incoming photon must be less than 1.11 microns (millionths of a meter). Solution Box 11.1 helps clarify the relationship between photon energy and wavelengths.

When a negatively charged electron leaves its nucleus, it also leaves behind a net positive charge, called a *hole*, associated with that nucleus. As suggested in Figure 11.3, if an electron from an adjacent silicon atom slides into that hole, the positive charge will appear to move. Imagine a room with every seat occupied. If someone (the electron) gets up to stretch his legs, an empty seat is created (a hole). Someone else may like that seat better and move into it, leaving her seat behind. The empty seat appears to move, just as a hole in silicon appears to move when a valence electron slips into a nearby hole. In a PV device, the trick is to get the electron to move away from the hole before the two have a chance to recombine. That is done by cleverly creating an internal electric field within the PV device that pushes holes in one direction and electrons in the other. The accumulating charge on opposite sides of the cell creates a voltage. Hook this up to a load and you have a solar-powered source of electricity.

11.2.2 The *p-n* Junction

Sunlight falling on a hunk of crystalline silicon will create hole-electron pairs; that is, negatively charged free electrons and positively charged holes. Both are capable of contributing to current flow. That's a great start to creating a device to convert sunlight into electricity. However if that is all you do, those electrons will quickly fall back into nearby holes and nothing will have been accomplished. To avoid recombination of holes and electrons, an internal electric field must be created within the device to separate the two charge carriers, sending holes toward one end of the device and electrons toward the other.

To create the needed electric field, two regions are established within the crystal. On one side of the dividing line separating the regions, pure (intrinsic) silicon is purposely contaminated with very small amounts of an element having five electrons in its outer orbit, such as phosphorus. Only about one phosphorus atom per 1000 silicon atoms is a typical amount of doping. When a pentavalent atom such as phosphorus forms covalent bonds with nearby silicon atoms, there is a leftover electron that is so loosely bound to its nucleus that it easily drifts off and becomes a free electron that can roam around the crystal. This side of the cell is referred to as being *n*-type material because there are now a fair number of free, negatively charged electrons that can move about. Meanwhile, the original +5 nucleus that the electron left behind becomes an immobile positive charge embedded in the crystal as shown in Figure 11.4(a).

On the other side of the device, about one atom of some trivalent element is added, such as boron, per 10 million atoms of silicon. When a trivalent atom forms covalent bonds in the crystal, it quickly grabs a fourth electron from a nearby silicon atom, creating an immobile negative

---

SOLUTION BOX 11.1

Light as Photons and
Light as a Wave Phenomenon

Light can be described as a continuous wave phenomenon characterized by wavelengths and frequencies, or it can be described as discrete packets of energy called photons. The relationship between the two is described by the following:

Eq. 11.1
$$E = hv = \frac{hc}{\lambda}$$

where E = the energy of a photon (J)

h = Planck's constant (6.626×10^{-34} J-s)

c = the speed of light (3×10^8 m/s)

v = the frequency (Hz)

λ = wavelength (m)

Because the energy of a photon is so low, it is often expressed in the more convenient units of electron-volts (eV), where 1 eV = 1.6×10^{-19} J. For our purposes, we can roll these various constants into a simple relationship between eV and wavelength:

Eq. 11.2
$$E\,(eV) = \frac{1.2424 \times 10^{-6}}{\lambda\,(m)}$$

Notice the inverse relationship between wavelength and energy. Short wavelength radiation has more energy per photon than long wavelength radiation.

For example, find the maximum wavelength with sufficient energy to send an electron into the conduction band for silicon, which has a band gap of 1.12 eV.

Solution:

A photon must have at least 1.12 eV to free an electron from its nucleus. In terms of wavelengths, that means the wavelength must be no more than

$$\lambda\,(m) = \frac{1.2424 \times 10^{-6}}{1.12\,eV} = 1.11 \times 10^{-6}\,m = 1.11\,\mu m$$

As we shall see soon, the band gap of a PV material and the wavelengths in the incoming solar spectrum limit the maximum theoretical efficiency of cells.

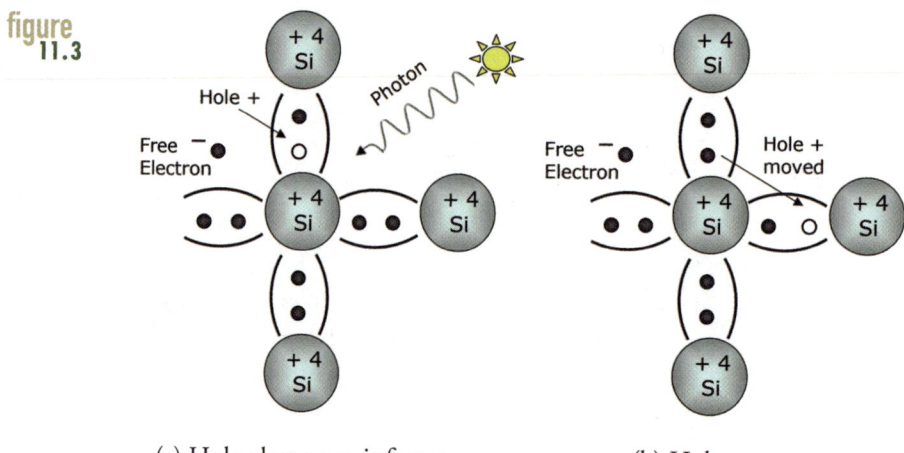

(a) Hole-electron pair forms   (b) Hole moves

(a) A photon with sufficient energy can create a hole-electron pair. (b) A valence electron from an adjacent nucleus can slide into the hole, which gives the appearance of the positively charged hole moving. Both holes and electrons can move about in the crystal, so both can contribute to current flow.

charge in the vicinity of the +3 nucleus. Meanwhile, the silicon atom that lost an electron leaves behind a nice, movable, positively charged hole, as is suggested in Figure 11.4b. The crystal on this side of the device is called *p*-type because it has an abundance of positively charged carriers.

Both *n*-type and *p*-type materials have charged mobile carriers, which greatly increases the electrical conductivity. They're not as conductive as metals, but they are a lot more so than the original intrinsic silicon. Hence the name, *semiconductors*.

Now imagine what happens when some *n*-type material is put next to some *p*-type material, forming a *p-n* junction. With such a concentration of free electrons on the *n*-side

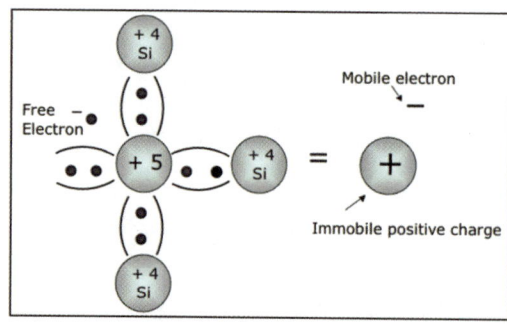

(a) Representation of *n*-type material

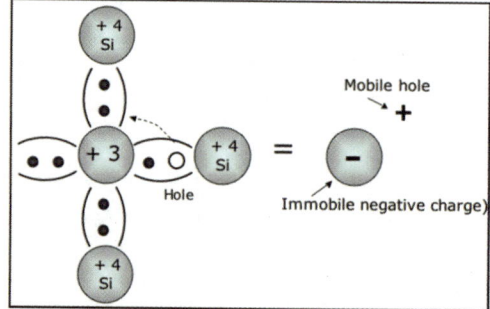

(b) Representation of *p*-type material

An *n*-type material consists of immobile positive charges with mobile electrons whereas *p*-type materials have fixed negative charges and mobile, positively charged holes.

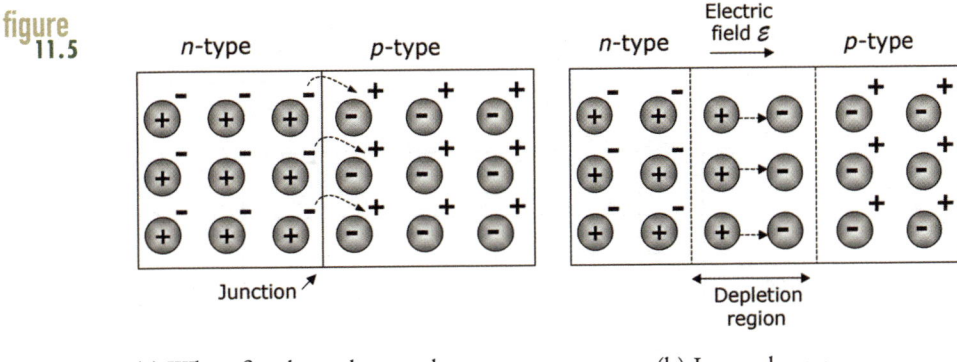

figure 11.5

(a) When first brought together

(b) In steady-state

When *p*-type and *n*-type semiconductors are brought together (a), electrons diffuse from the *n*-region into the *p*-region filling holes and creating immobile charges on each side of the junction. The electric field created by those fixed charges opposes further diffusion, keeping holes on the *p*-side and electrons on the *n*-side (b).

---

of the junction, and hardly any on the other, there will be a tendency for those mobile electrons to drift over to check out the action on the *p*-side. As they cross over, those electrons leave behind immobile positive charges in the *n*-side. And, as they cross over, they will find themselves falling into holes on the *p*-side, creating immobile negative charges on that side of the junction. These immobile charged atoms in the *p*- and *n*-regions create an electric field that works against the continued movement of electrons across the junction. Almost instantaneously, the electric field reaches a level sufficient to stop any further diffusion of holes and electrons across the junction. Figure 11.5 shows the resulting stalemate.

11.2.3 A Complete Solar Cell

We just about have everything we need now to understand how a PV cell works In essence, a PV cell is just a *p-n* junction that we will expose to sunlight. Photons with sufficient energy create hole-electron pairs. If those holes and electrons reach the vicinity of the depletion region, the electric field sweeps the electrons into the *n*-region and the holes into the *p*-region. This creates a voltage across the cell. When a load is connected to the cell, electrons will flow from the *n*-region through the load and return to the *p*-region. Power is delivered to the load as long as the sun shines on the cell. Figure 11.6 summarizes the whole thing.

11.3 Photovoltaic Efficiency

Now that we have a sense of how PVs work, we can begin to address a key question in terms of their performance. That is, what fraction of the sunlight hitting a PV will be collected and

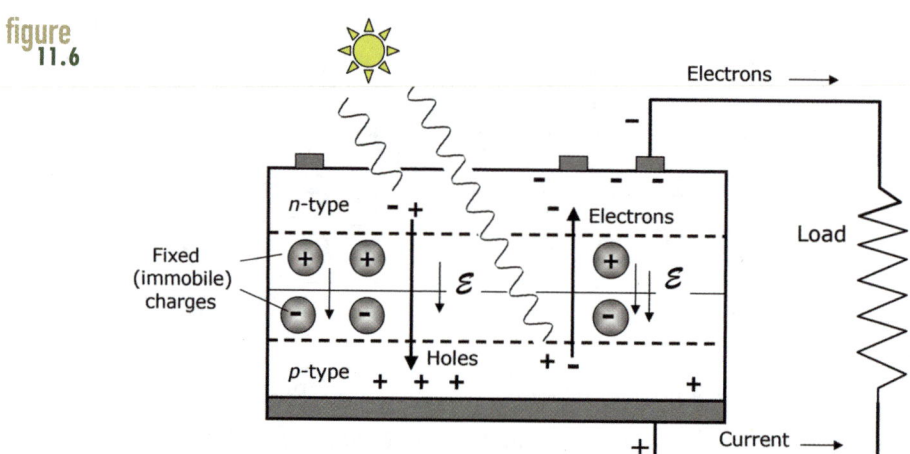

figure
11.6

When photons create hole-electron pairs near the junction, the electric field in the depletion region sweeps holes into the *p*-side and electrons into the *n*-side of the cell. As shown, electron flow is clockwise through the load; conventional current is in the other direction.

transformed into electrical power? The efficiency of a PV device depends on a number of factors, including the following:

1. The band-gap constraint, which has to do with some photons not having enough energy to create hole-electron pairs, whereas others have more energy than is needed to do so.

2. Photons that are not absorbed by the cell either because they are reflected off the face of the cell, or because they pass right through the cell, or because they are blocked by the metal conductors that collect current from the top of the cell.

3. Recombination of holes and electrons before they can be separated by the junction's electric field.

4. Internal resistance within the cell, which dissipates power.

5. Environmental effects such as temperature (which lowers efficiency) and the spectral distribution of sunlight striking the device, which varies depending on sun angles and sky clarity.

11.3.1 The Solar Spectrum

As was described in Chapter 4, the surface of the sun emits radiant energy with spectral characteristics that closely match that of a 5800 K blackbody. Some of those photons are absorbed by various constituents in the Earth's atmosphere so that by the time sunlight reaches the Earth's surface its spectrum is significantly distorted. The amount of sunlight reaching the Earth and its spectral characteristics depend on the *air mass ratio*, which is a measure of the amount of air the rays have to pass through before reaching the Earth's surface. With the

figure
11.7

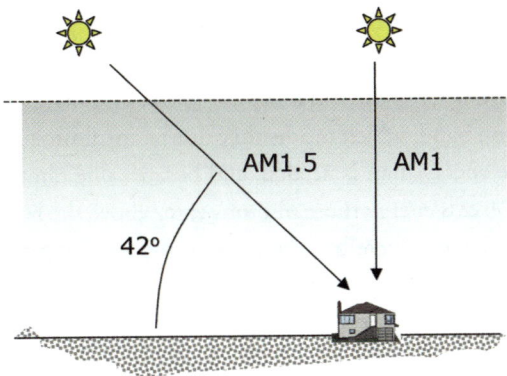

PV performance measures usually assume sunlight passes through 1.5 times as much air before it reaches the Earth's surface (designated as AM1.5) as it would if the sun were directly overhead (AM1). AM1.5 is equivalent to the sun being about 42 degrees above the horizon.

sun directly overhead, the air mass ratio is defined to be 1, which is designated as AM1. It is standard practice to evaluate PV performance under AM1.5 conditions; that is, with the sun passing through an amount of air equivalent to 1.5 times as much as when the sun is directly overhead (Figure 11.7). Sunlight passing through clear AM1.5 skies has the spectral distribution shown in Figure 11.8.

figure
11.8   **The Clear-Sky Solar Spectrum at AM1.5**

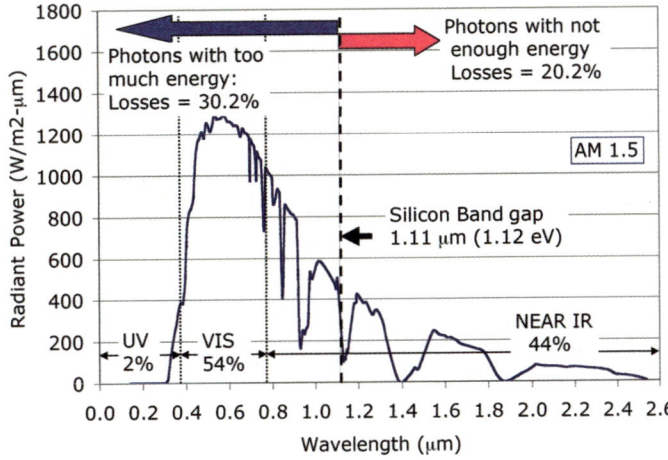

For silicon, over half of the incoming solar energy is wasted because photons either don't have enough energy or they have more than is needed to create hole-electron pairs.

SOURCE: based on ERDA/NASA, 1977

11.3.2 Band Gap Impact on Photovoltaic Efficiency

Some photons coming from the sun don't have sufficient energy to cause an electron to jump into the conduction band, which means they don't contribute to the generation of electricity. Others have more energy than is needed, and because one photon can create only one hole-electron pair, any excess energy those photons carry above the band-gap energy is also wasted. How well a PV will work, therefore, depends on the wavelengths of the energy arriving from the sun as well as the band gap of the cells.

For silicon, the band gap is 1.12 electron-volts (eV), corresponding to a wavelength of 1.11 μm (see Solution Box 11.1 for the relationship between eV and wavelength). As shown in Figure 11.8, at AM1.5 about 20.2% of the available solar energy has wavelengths above 1.11 μm so those are not absorbed and cannot create hole-electron pairs. They simply pass right through the silicon. Wavelengths shorter than that are absorbed, but they have excess energy that simply heats the crystal and wastes another 30.2% of the sun's energy. Between the two, more than half of the sun's energy doesn't get turned into electricity.

Even this simple analysis can provide some insights into the importance of the band gap for various potential PV materials. Think of band gap as an indicator of the voltage that a cell will produce, and the number of hole-electron pairs created as current that will be delivered. And remember that power is the product of voltage and current. A high band gap material will deliver more voltage but less current than one with a lower band gap. And the other way around: lower band gap means lower voltage but higher current. So clearly there is a trade-off when it comes to power delivered, which is after all what we are after.

As Figure 11.9 indicates, there is an optimum band gap of about 1.4 eV (which means the maximum wavelength to create current is 0.89 μm). Also shown in the figure are various promising materials for PVs, including conventional silicon; a non-crystalline amorphous silicon (a-Si); cadmium telluride (CdTe); gallium arsenide (GaAs); copper gallium diselenide ($CuGaSe_2$); copper indium diselenide, or CIS cells, ($CuInSe_2$); and CIS cells with added gallium, $Cu(In,Ga)Se_2$, known as CIGS cells. Notice how the addition of gallium to CIS cells shifts the band gap a little closer toward the optimum.

11.3.3 Single-Junction PV Efficiency under Laboratory Conditions

A careful thermodynamic analysis of the maximum possible PV efficiency for single band gap materials under unconcentrated solar irradiance yields an upper limit of about 31% (Shockley and Queisser, 1961). Under laboratory conditions with cells not yet built into PV modules, the best solar cells now have efficiencies of about 25%.

Figure 11.10 shows an example of a single-crystal silicon PV that has been designed to maximize efficiency. Several features are worth commenting on. Notice there are no wire contacts on the front surface, which avoids the blockage of sunlight that those surface conductors normally cause. Instead, the contacts, both positive and negative, are positioned on the underside of the device. Also note the top surface is textured in such a way as to help any reflected sunlight to bounce down into the cell rather than away from it. Finally, a reflective surface on the underside

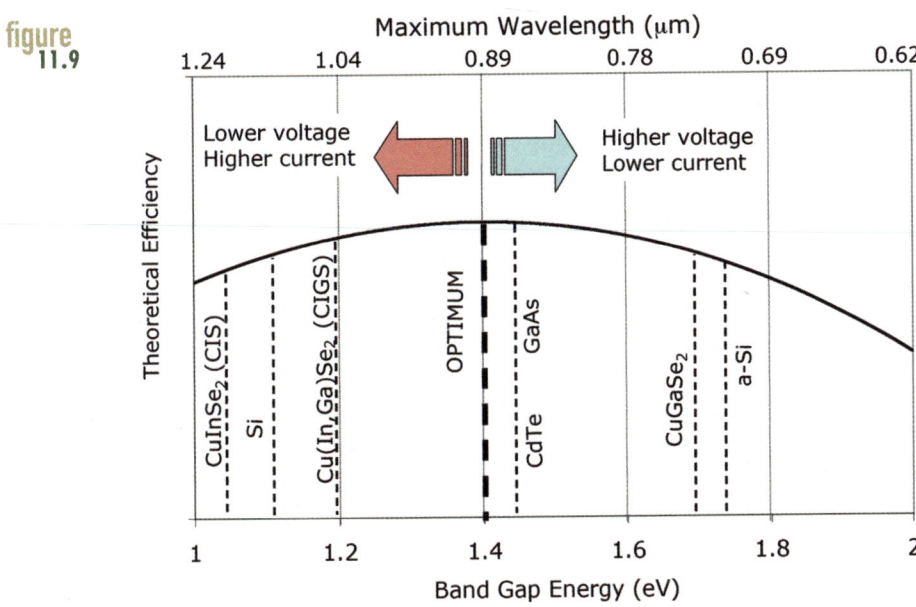

Increasing the band gap increases cell voltage but reduces current, and vice versa. The product of voltage and current is power, which means there is an optimum band gap at which maximum power will be produced.

of the cell helps bounce photons back into the cell that otherwise might pass completely through the photovoltaic. This reflective feature allows the cell thickness to be reduced as well. With less silicon in each cell, the embodied energy needed to produce them is also reduced.

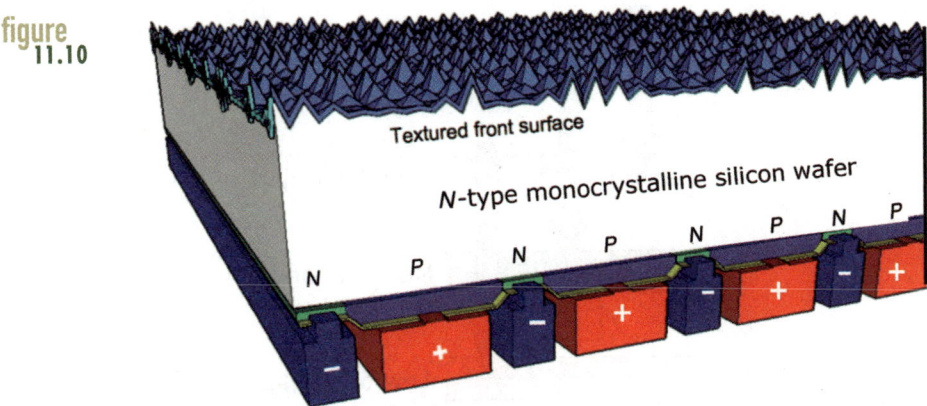

Putting the electrical contacts on the underside of the cell and texturing the surface to bounce reflected sunlight into the cell, helps boost efficiency of this device to over 20%.

Source: courtesy of SunPower Corp.

11.3.4 Multijunction (Tandem) Cells to Increase Efficiency

By carefully manipulating various alloys in some types of solar cells, the band gap can be adjusted to increase the cells' ability to capture different portions of the incoming solar spectrum. For example, the band gap of CIGS cells can be tuned to anything from about 1.04 eV to 1.68 eV. This suggests that perhaps a clever way to significantly increase PV efficiencies is to fabricate them with multiple *p-n* junctions, one on top of the other, with each tweaked to capture different wavelengths of solar energy. These multijunction, or tandem, cells are very promising. In fact, with an infinite combination of band gaps in a solar cell, the theoretical efficiency could be as high as 66%.

In multijunction cells, the uppermost junction is designed to have a high band gap so it will capture short-wavelength photons. Longer-wavelength photons are not absorbed so they pass right through to the next level. Subsequent junctions capture photons with longer and longer wavelengths (Figure 11.11).

11.4 Photovoltaic Fabrication

There are a number of different ways to fabricate PVs. One way to characterize these technologies is by the thickness of the cells. Relatively speaking, crystalline silicon cells tend to be very thick—on the order of 150–250 μm (micrometers), which is a bit thicker than a human hair. Most of today's commercially available PVs are these thick, crystalline silicon cells. An alternative approach to PV fabrication is based on thin films of semiconductor material, where "thin"

figure 11.11 An Example of a Triple-Junction (Tandem) Cell

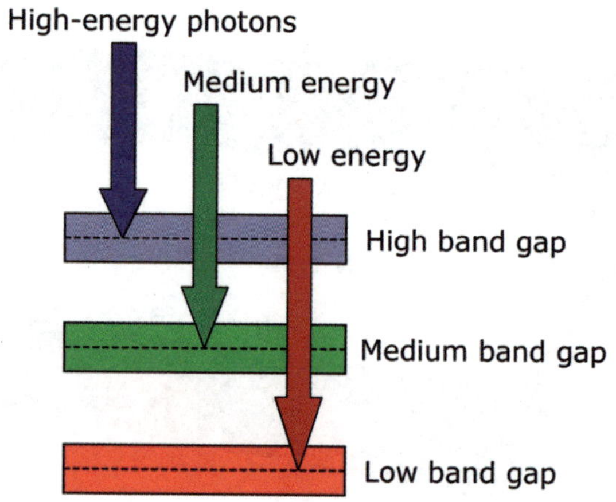

Higher-energy photons (shorter wavelengths) are captured in the upper junction. Longer wavelength photons may be captured in subsequent junctions.

means something like 1–10 µm. Thin-film cells require much less semiconductor material and they are potentially easier to manufacture, so they offer the tantalizing promise of eventually being cheaper than crystalline silicon.

11.4.1 Crystalline Silicon Cells (*x*-Si)

Most solar cells are made of silicon, which is the second most abundant element in the Earth's crust. In its natural state it is usually in the form of silicon dioxide (SiO_2), or silica. That SiO_2 is first purified into a 99.9999% pure polysilicon form, which looks like the shiny rock-like material shown in Figure 11.12a.

The most commonly used technique for forming single-crystal silicon is based on the *Czochralski* (pronounced check-ralski), or CZ, method in which a small seed of solid, crystalline silicon about the size of a pencil is dipped into a molten vat of polysilicon. A tiny amount of *n*- or *p*-type material is added to the vat to dope the silicon one way or the other. As the seed crystal is slowly withdrawn from the vat, molten silicon atoms bond with atoms in the crystal and then solidify (freeze) in place. The result is a large cylindrical ingot of single-crystal silicon perhaps a meter or so long and 15 to 20 cm in diameter. As shown in Figure 11.9, the ingot is then sliced into wafers, the top layer of which is then doped in the other direction to create the necessary *p-n* junction.

Other, less expensive, approaches to manufacture crystalline silicon solar cells are providing tough competition for the CZ method. Single-crystal silicon can also be grown as a long, continuous ribbon from a silicon melt. Molten silicon can also be poured into a mold and allowed to solidify into a massive rectangular block that can be sliced into silicon wafers. The resulting wafers turn out to have a less well organized multicrystalline structure made up of regions of single-crystal silicon separated by grain boundaries. They are easy to

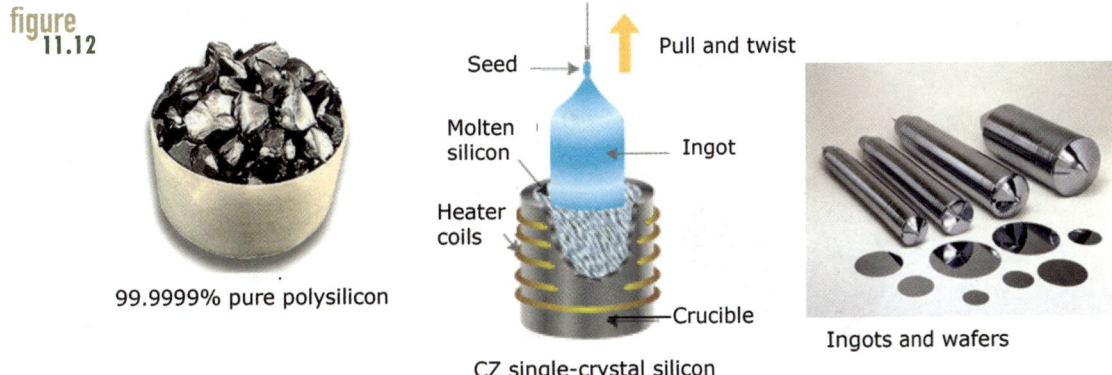

figure 11.12

99.9999% pure polysilicon

Seed — Pull and twist

Molten silicon — Ingot

Heater coils

Crucible

CZ single-crystal silicon

Ingots and wafers

The CZ method for growing single-crystal silicon begins with polysilicon, which is melted in a heated crucible. A seed crystal drawn from the crucible forms an ingot, which is then sliced into wafers.

recognize when looking at multicrystalline PVs because light reflects slightly differently from each single-crystal region of the cell.

11.4.2 Thin-Film, Organic, and Nano-Solar Technologies

Although crystal-silicon cells dominate the current PV marketplace, a number of thin film technologies are beginning to receive considerable research and development funding and might one day shift the balance. At present these technologies tend to be less efficient, which means larger areas and area-related costs, but they hold the promise of ultimately being cheaper per watt generated.

Examples of thin-film cells and their maximum laboratory efficiencies as of 2006 include those made from amorphous silicon (a-Si, 13.1%), cadmium telluride (CdTe, 16.7%), and copper-indium-gallium diselinide (CIGS, 19.3%). Many of these thin-film technologies have their *n*-layers made from different materials than their *p*-layers and are thus referred to as *heterojunction* cells (as opposed to silicon *homojunction* cells). CIGS cells, for example, may have an *n*-layer made of cadmium and zinc sulfide (CdZn)S and their *p*-layer of copper-indium-gallium diselenide, $Cu(In,Ga)Se_2$.

Some of these emerging PV technologies offer the tantalizing promise of very low cost roll-to-roll production methods, similar to the way newspapers are printed and photographic films are manufactured. In one process, CIGS semiconductors are vacuum coated onto a film; another process applies CIGS materials using an ink printing method.

New organic and nano cells are being developed that could be printed or sprayed as an ultra-thin layer of semiconductor onto a roll of plastic. One of the most promising of these new devices is known as the Dye-Sensitized Solar Cell (DSSC), or Graetzel cell, after its Swiss inventor Michael Graetzel (Figure 11.13). The Graetzel cell uses an organic dye injected into nanocrystalline titanium-dioxide (TiO_2), a white pigment commonly used in sunscreen, toothpaste, and paint. The dyes absorb sunlight and create pairs of charged hole-electron pairs using a principle similar to photosynthesis. These photoreactive materials

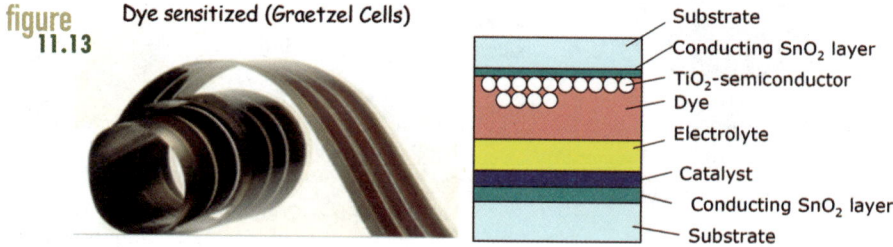

figure 11.13

Dye sensitized (Graetzel Cells)

- Substrate
- Conducting SnO_2 layer
- TiO_2-semiconductor
- Dye
- Electrolyte
- Catalyst
- Conducting SnO_2 layer
- Substrate

Organic and nano PVs can be printed onto flexible plastic rolls. Cross section is for a dye-sensitized, or Graetzel, solar cell.

Source: from Konarka Web site

can be printed onto flexible plastic materials using a continuous roll-to-roll manufacturing process similar to the way newspapers are printed. Different colors of dye can be used to absorb different wavelengths of light allowing a range of colored, semitransparent flexible films to be incorporated into windows, roofing materials, and a variety of portable electronic devices. They can even be invisible when tuned to collect just the near-infrared portion of the spectrum. Someday, the tent you take backpacking may generate its own electricity.

11.5 From Laboratory Cells to Commercial Modules

Because individual solar cells produce only about 0.5 V, it is a rare application for which just a single cell is of much use. Instead, the basic building block for PV applications is a *module* consisting of a number of preconnected cells in series, all encased in tough, weather-resistant packages. Multiple modules, in turn, can be wired together in series to increase voltages even more and in parallel to increase current. Recall power is the product of voltage and current. A simple figure illustrating the cell-to-module-to-array concept is shown in Figure 11.14.

As Figure 11.15 indicates, the overall efficiency of actual commercially available modules is never quite as good as the best cells being developed in laboratories. Some of that difference is related to packaging and some is just the difference between something someone can build in the lab versus a commercial product that can be warranted to perform as advertised for twenty years or more in the field.

11.6 Grid-Connected Photovoltaic Systems

Whereas much of the early attention was directed toward remote, off-grid systems, the vast majority of current and future PV sales are projected to be for residential and commercial

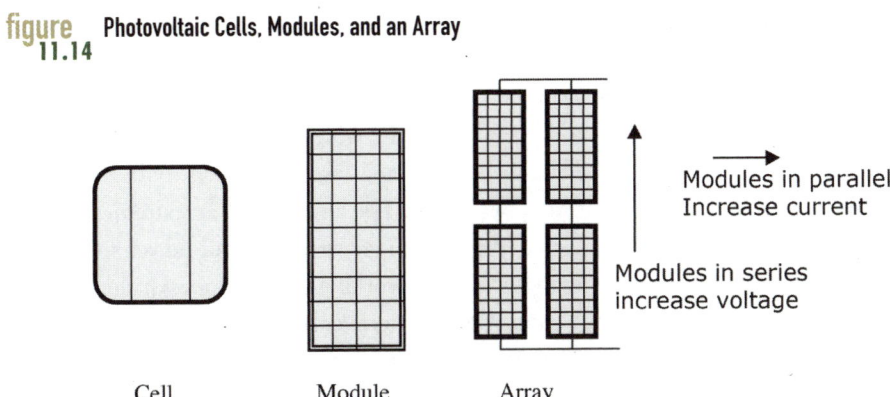

figure 11.14 Photovoltaic Cells, Modules, and an Array

Cell   Module   Array

Modules in parallel
Increase current

Modules in series
increase voltage

Modules can be connected in series to increase voltage and in parallel to increase current, the product of which is power.

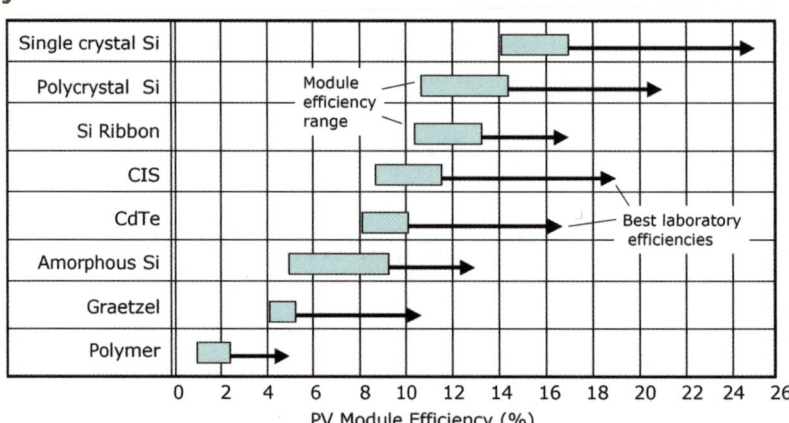

figure 11.15   Efficiencies of Various Photovoltaic Technologies

The rectangles show the range of large-module efficiencies. The arrows indicate laboratory cells.

rooftop systems connected to the local utility grid. The primary advantage to being able to hook up to the grid is the ability to use it as your energy storage system. As suggested in Figure 11.16, a grid-connected system offers the opportunity to sell to the local utility any excess electricity your PVs might generate during the day, running your meter in one direction, and then buying electricity back from the grid at night or at other times when your load exceeds your solar supply, running your meter in the other direction. With time-of-use (TOU) rates, it is even possible to sell your electricity for a higher price during those hot, clear summer days when it is more valuable, and then buy it back at night at a cheaper rate (see, for example, Section 10.5.2). Most states now allow net metering, but with the usual proviso that at the end of the year, when the books are tallied, you cannot sell more kWh to the utility than you purchased back again, and you cannot make a net dollar profit on your sales either.

The power conditioning unit (PCU) shown in Figure 11.16 serves several purposes. Because PVs generate direct-current (dc) power and your house needs ac, the main function of the PCU is to convert dc into ac using an electronic device called an *inverter*. The PCU also will include a *maximum power point tracker* (MPPT), which helps optimize the electrical output of the PVs, a set of protective circuit breakers and fuses, and circuitry to disconnect the PV system from the grid if the utility loses power. The latter serves an extremely important safety function by ensuring that your PVs don't inadvertently send power to the grid during a power outage when utility workers may be working on the lines. Grid-connected PV systems can be designed with some battery storage to cover those power outages, but most customers elect not to include that added level of system complexity.

11.6.1 Various DC and AC Power Ratings

PV modules are rated under standard laboratory test conditions (STC) that include a solar irradiance of 1 kW/m² (called "1-sun"), a cell temperature of 25°C, and an air mass ratio

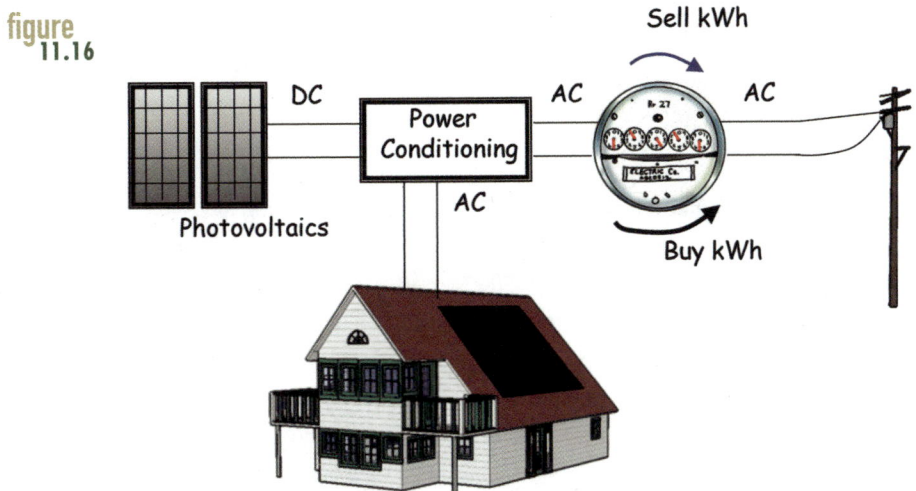

figure 11.16

Grid-connected systems allow you to spin your electric meter backward when your PVs generate more power than you need, in essence using the grid for energy storage.

of 1.5 (AM1.5). Under laboratory conditions, module outputs are thus often referred to as "watts (STC)" or "peak-watts." Up on your rooftop, modules are subject to very different conditions and their outputs will vary significantly from the STC rated power that the manufacturer specifies. You don't always have 1-sun of insolation, modules get dirty, and most importantly, modules are very temperature sensitive, with most losing about 0.5% of their power for each degree Celsius of increased cell temperature. Because cells are typically 20°–30°C hotter than the surrounding air, unless it is very cold outside, or it is not a very sunny day, they will usually be much hotter than the 25°C at which they are rated.

Based on extensive data collected in the field under a program called PVUSA, another rating system has emerged that attempts to more accurately specify dc output of PV modules. The PVUSA rating system is based on exposure to 0.8 kW/m² of solar irradiation, an ambient temperature of 20°C, and a wind speed of 1 m/s. Module dc output under these PVUSA test conditions is referred to as DC,PTC. When the DC,PTC power is multiplied by the efficiency of the dc-to-ac inverter, the result is an ac power rating based on PVUSA conditions referred to as AC,PTC. In some states, financial incentives are based somewhat on these PVUSA assumptions.

Finally, when all system losses are lumped together, including temperature effects, dirt, electrical mismatch of modules, and dc-to-ac inverter inefficiencies, the actual ac power delivered at 1-sun, call it P_{AC}, can be represented as the following product:

Eq. 11.3 $$P_{AC} = P_{DC,STC} \times \text{(de-rating factor)}$$

where $P_{DC,STC}$ = the dc power of the array under standard test conditions

The de-rating factor is based on the sum of all of the system losses just mentioned.

SIDEBAR 11.1

How Many Watts Do You Have?

The need to be quite clear about whether a system is specified in terms of its dc rated power, its actual ac output, or its output under some government- or utility-defined rebate program is absolutely essential, and is unfortunately, all too often overlooked. For example, system costs and rebates are often stated in $/W terms. But are those DC,STC watts, expected ac watts for real systems, or ac watts defined by the rebate program?

Figure 11.17 shows three ways of quoting the cost of a 1 kW (DC,STC) system with an installed cost of $7000, which is roughly the 2006 average cost of PV systems in California before rebates and tax credits (Wiser, 2006). For systems defined under California's rebate programs, the de-rating factor is based on the buyer's choice of modules and inverter and is typically about 0.84. In actual field conditions, however, including dirt, mismatch, and other factors, a more likely de-rating factor to predict actual performance is about 0.75. As shown, this system could be described as costing $7/W, $8.33/W, or $9.20/W depending on the definition of what kind of watts are being quoted.

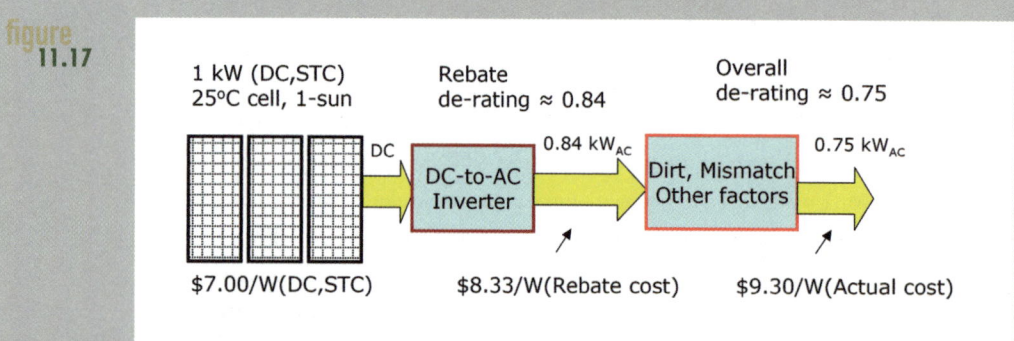

figure 11.17

Three different power outputs for a PV system with representative values based on 1 kW dc at standard test conditions. Also shown are corresponding ways to express the $/watt cost of system having a pre-rebate cost of $7/watt (DC,STC).

A typical PV system de-rating factor system is about 75%, which means an array typically delivers only about three-fourths of the manufacturer's DC,STC rated power. Sidebar 11.1 points out the importance of keeping clear the distinction between various dc and ac power ratings.

11.6.2 Annual Insolation is What Matters for Grid-Connected Photovoltaics

Predicting PV performance is a matter of combining the dc (STC) rated power of the array, an estimate of the overall de-rating factor, and the local insolation at the site. The National

Renewable Energy Labs has produced a wonderful publication called the *Solar Radiation Data Manual for Flat-Plate and Concentrating Collectors,* which provides solar insolation data for a number of cities across the United States. The full report can be downloaded from the Internet at http://rredc.nrel.gov/solar/pubs/redbook/.

For illustrative purposes, Table 11.1 provides some of NREL's insolation data. For fixed-orientation south-facing collectors, the tilt angles are designated as L +15°, L, and L –15°, where L is your local latitude. The table also includes data for the following tracking arrays: single-axis, north-south oriented, horizontal trackers; single-axis trackers for arrays with fixed tilt equal to the local latitude (called a polar axis since the axis points toward the north star *polaris*); and double-axis trackers that always face directly into the sun. The improvement of single-axis trackers over the best, fixed arrays is significant—often more than 30% better. It is interesting to note, however, that two-axis trackers, which always point the PVs directly into the sun, are not much better than single-axis trackers with tilt equal to the local latitude.

Sample monthly insolation data for Boulder, Colorado, for various fixed collector orientations, are plotted in Figure 11.18. If we were designing a stand-alone system with battery storage, those month-to-month variations would be very important and we would probably pick a collector tilt angle that provides relatively uniform insolation throughout the year. A tilt equal to your local latitude plus about 15° would be pretty ideal for that application because it evens out monthly variations, which reduces the amount of battery storage required.

For grid-connected systems, however, it is the average annual insolation that matters the most. If you generate more than you need in the summer, you just sell the excess to the utility and buy it back again in the winter. In fact, designing your system to emphasize generation in the summer can be a good economic strategy because that is when grid electricity prices are usually the highest.

table 11.1   Annual Average Insolation in kWh/m²-day for South-Facing Surfaces and Various Tracking Collectors

Tilt Angle	Tucson, AZ	Los Angeles, CA	Sacramento, CA	San Francisco, CA	Boulder, CO	Miami, FL	Atlanta, GA	Honolulu, HI	Chicago, IL	Boston, MA	Minneapolis, MN	Raleigh, NC	Albuquerque, NM	Cleveland, OH	Medford, OR	Austin, TX	Roanoke, VA	Seattle, WA
Horizontal	5.7	4.9	4.9	4.7	4.6	4.8	4.6	5.4	3.9	3.9	3.9	4.4	5.6	3.8	4.4	4.9	4.2	3.3
L – 15	6.3	5.5	5.5	5.3	5.4	5.1	5.0	5.5	4.4	4.5	4.6	4.9	6.3	4.2	4.9	5.2	4.8	3.8
Lat	6.5	5.6	5.5	5.4	5.5	5.2	5.1	5.7	4.4	4.6	4.6	5.0	6.4	4.1	4.9	5.3	4.8	3.7
L + 15	6.3	5.4	5.2	5.1	5.3	5.1	4.9	5.5	4.1	4.4	4.4	4.8	6.2	3.9	4.5	5.1	4.6	3.5
90	3.9	3.5	3.4	3.4	3.8	3.0	3.1	2.9	3.0	3.2	3.3	3.2	4.1	2.7	3.1	3.1	3.2	2.6
1-Axis (Horiz) N,S	8.1	6.4	6.8	6.3	6.4	6.2	6.0	7.2	5.1	5.2	5.3	5.7	7.8	4.8	6.0	6.4	5.6	4.3
1-Axis (Latitude)	8.7	7.0	7.4	6.9	7.2	6.5	6.4	7.4	5.5	5.7	6.0	6.2	8.5	5.1	6.5	6.7	6.1	4.7
2-Axis	9.0	7.2	7.6	7.1	7.4	6.7	6.6	7.7	5.7	5.9	6.2	6.4	8.8	5.3	6.7	7.0	6.3	4.9

Source: data from NREL, 1994

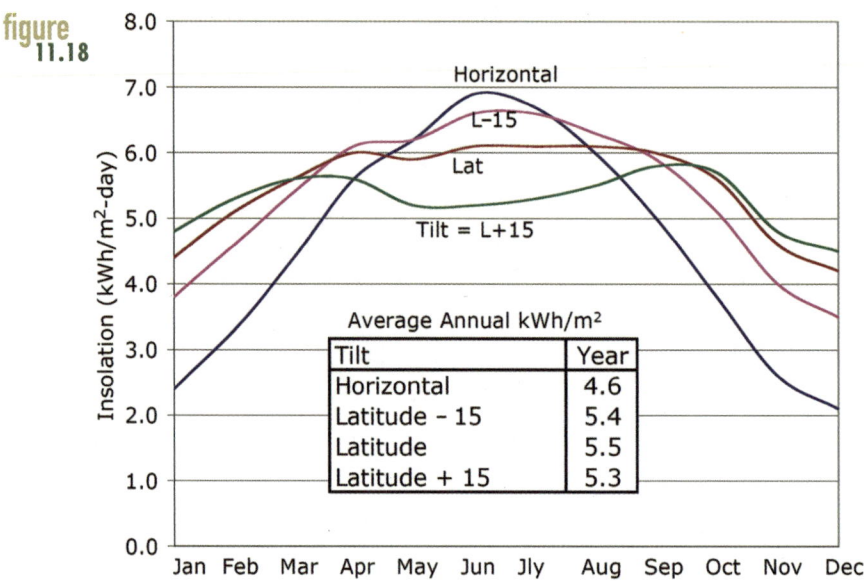

figure 11.18

Tilt	Year
Horizontal	4.6
Latitude – 15	5.4
Latitude	5.5
Latitude + 15	5.3

Average Annual kWh/m²

Monthly and annual average insolation for Denver for various south-facing collector tilt angles. On an annual basis, insolation varies only a few percent for a wide range of tilt angles.

SOURCE: based on NREL, 1994

As shown in Figure 11.18, for quite a range of south-facing collector tilt angles—all the way from an L –15° to an L +15° tilt—the Denver total average daily insolation hardly changes at all (5.3-to-5.5 kWh/m²-day). Indeed, even a collector lying flat on a horizontal roof loses only about 16% compared to the ideal tilt angle (tilt = latitude). Although these conclusions were noted for Denver, they are reasonably true for most locations in the United States.

11.6.3 The "Peak-Hours" Approach to Sizing a Grid-Connected PV System

Because the grid provides the backup source of electricity, how large a PV system you choose is mostly a matter of how much you can afford and the available area of appropriately oriented roof having good solar exposure.

The "peak-hours" approach makes a very simple, and reasonable translation between average daily insolation expressed in kWh/m²-day and the number of equivalent hours of full sun. Because 1-sun of insolation is defined as 1 kW-m², we can think of an insolation of say 5.4 kWh/m²-day as being the same as 5.4 hours per day of 1 kW/m² sun, or 5.4 hours of "peak sun." So, we can write

Eq. 11.4 $\text{Energy (kWh/yr)} = P_{AC}(\text{kW}) \times (\text{hr/day of 1-sun}) \times 365 \text{ day/yr}$

where P_{AC} = the ac power produced by the array when exposed to 1-sun of insolation

SOLUTION BOX 11.2

Sizing a PV Array for San Francisco

How many kW (DC,STC) would be needed to deliver 3600 kWh/yr (300 kWh/mo) to a home in San Francisco? Assume south-facing collectors tipped up at an L −15° angle (23° for San Francisco at latitude 38°) and use a de-rating factor of 0.75. At 17% efficiency, how big would the PV array be?

Solution:

Table 11.1 indicates an L −15° tilt exposes the array to 5.3 kWh/m²-day of insolation (5.3 hours of 1-sun). Equation 11.5 suggests we will need

$$P_{DC,STC}(kW) = \frac{3600 \text{ kWh/yr}}{0.75 \times 5.3 \text{ hr/day} \times 365 \text{ day/yr}} = 2.48 \text{ kW}$$

Using Equation 11.6 with 17% collector efficiency, we can find the area needed for this array:

$$A(m^2) = \frac{P_{DC,STC}(kW)}{1 \text{ kW/m}^2 \cdot \eta} = \frac{2.48}{0.17} = 14.6 \text{ m}^2 \ (157 \text{ ft}^2)$$

Modules are rated according to their dc output under standard test conditions (DC,STC), so we can insert Equation 11.3 into Equation 11.4 and get the following simple sizing equation:

Eq. 11.5   Energy (kWh/yr) = $P_{DC,STC}$(kW) × (de-rating) × (hr/day of 1-sun) × 365 day/yr

We would also like to know how big an array must be to deliver the energy found in Equation 11.5. To do that, we need to know the PV efficiency under standard test conditions, η, which is easy to obtain from manufacturer specifications. The area required can then be calculated as follows:

Eq. 11.6 $P_{DC,STC}$(kW) = 1 kW/m² insolation × A (m²) × η

The area, A, found in Equation 11.6 is in square meters. The conversion to square feet is 1 m² = 10.76 ft². Solution Box 11.2 illustrates the use of these sizing relationships.

One simple way to make quick estimates of annual energy production from a PV array as a function of the average insolation at the site is to use the sizing estimator provided in Figure 11.19. For example, a good site with 6 kWh/m²-day of insolation can provide about 1600 kWh/yr of delivered ac electricity per kilowatt of rated DC,STC power.

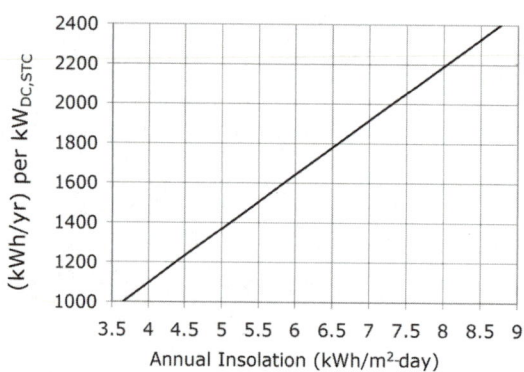

Simple sizing estimator from annual insolation to energy generated (kWh/yr) per kW of DC,STC rated power. Assumes a de-rating factor of 0.75.

Using PV efficiency and local insolation as parameters, we can use the graph shown in Figure 11.20 to help estimate array area requirements. For example, with 14%-efficient collectors in a site with 5.5 hours of full sun, one square foot of collector could deliver about 20 kWh/yr of ac electricity.

Solution Box 11.3 provides an intriguing estimate that suggests the total potential for rooftop PV systems is sufficient to supply over one-third of all of U.S. electricity demand.

11.7 Economics of Photovoltaics

The manufacturer's price of PV modules has followed a fairly steady and consistent decrease over time. As shown in Figure 11.21, when plotted on a log-log scale versus cumulative production, a nearly straight line results. Projecting that line into the future suggests that

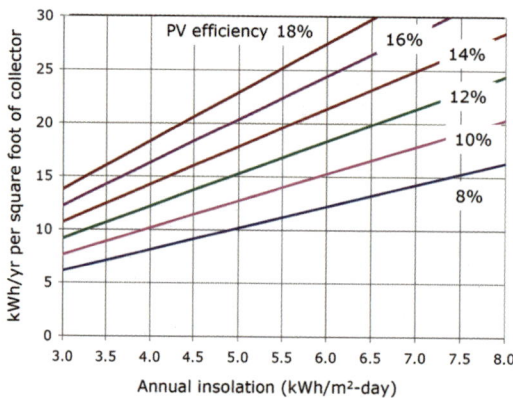

Annual energy production from a PV array per square foot of collector using efficiency as a parameter. Assumes a de-rating factor of 0.75.

SOLUTION BOX 11.3

The Total Potential for Rooftop PV in the United States

A study of the roof area in the United States potentially available for PVs estimates 3.5 billion square meters of residential rooftop area and 2.9 billion square meters of commercial roof area (Chaudhari, 2004). These estimates account for roof orientation, shading, and structural issues. Assuming 17%-efficient collectors, an average annual solar exposure of 5 kWh/m²-day, and a de-rating factor of 0.75, find the annual energy that could be delivered if that entire available space is utilized.

Solution:

The total area of 6.4 billion square meters would allow an installed capacity of

$$P_{DC,STC} = 6.4 \times 10^9 \text{ m}^2 \times 1 \text{ kW/m}^2 \times 0.17 = 1088 \times 10^6 \text{ kW}$$

With 5 kWh/m²-day of insolation (equivalent to 5 hrs of 1 kW/m² sun), and using the 0.75 de-rating factor, the energy that could be delivered would be

Annual energy = 1088×10^6 kW $\times$ 0.75 $\times$ 5 hr/day $\times$ 365 day/yr = 1490 billion kWh/yr

The total net output of all U.S. power plants in 2005 was 4340 billion kWh, so PVs could supply just over one-third of the entire demand. In fact, if we include transmission losses from traditional power plants to end users, which are avoided by on-site generation, this full build-out of PVs would be sufficient to supply half of the total electricity demand of all U.S. buildings.

---

when cumulative production reaches about 100,000 MW, modules might cost as little as $1/Watt (dc). At that price, subsidies would not be necessary. Integrating the subsidy using the projected experience curve shown leads to a conclusion that the total subsidy needed before PV systems would be cost-effective is on the order of $25 billion (Swanson, 2004).

But, modules are only one part of the cost of PV systems. Inverters, installation fees, and other *balance of systems* (BOS) costs add to the total. In one study of these costs, modules were 62% of the total installed cost of residential PV systems (Figure 11.22).

11.7.1 Amortizing Costs

Because the capital cost of a PV system pretty much covers the cost of the next several decades of electricity delivered, we need some way to amortize that cost into something that can be

figure 11.21

Historical and projected manufacturer's price of PV modules suggests with cumulative production of 100,000 MW, costs might drop to $1/watt(dc).

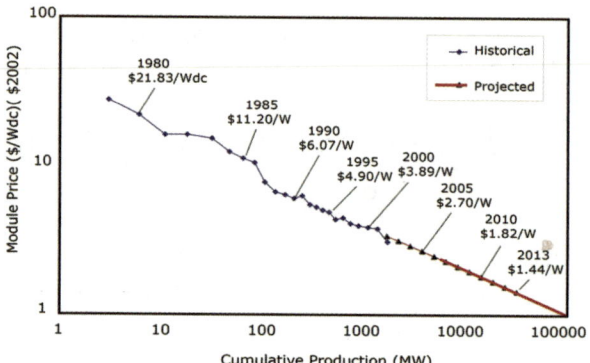

SOURCE: based on Swanson, 2004

compared to the usual cents/kWh found on your electric bill. A convenient way to do so is based on imagining that you take out a loan to pay for the system, which results in annual loan payments that can be divided by the annual kWh generated.

Recall the Capital Recovery Factor, CRF(i,n) introduced in Chapter 4. If I take out a loan of P dollars at interest rate i with a loan term of n years, my annual payments will be

Eq. 11.7
$$A = P \cdot \mathrm{CRF}(i,n) \text{ where } \mathrm{CRF}(i,n) = \frac{i(1 + i)^n}{(1 + i)^n - 1}$$

A short set of capital recovery factors is given in Table 11.2.

If Equation 11.7 is combined with the annual electricity delivered in Equation 11.5 we can find the cost of PV-generated electricity. Solution Box 11.4 illustrates the approach.

Eq. 11.8
$$\$/\mathrm{kWh} = \frac{\$/\mathrm{yr}}{\mathrm{kWh/yr}} = \frac{P \cdot \mathrm{CRF}(i,n)}{P_{\mathrm{DC,STC}}(\mathrm{kW}) \cdot (\text{de-rating}) \cdot (\mathrm{hr/day \ of \ 1 - sun}) \cdot 365 \ \mathrm{day/yr}}$$

Figure 11.23 shows the impact of PV capital cost on annual cost of electricity produced, using the same assumptions that were incorporated in the example in Solution Box 11.4.

figure 11.22

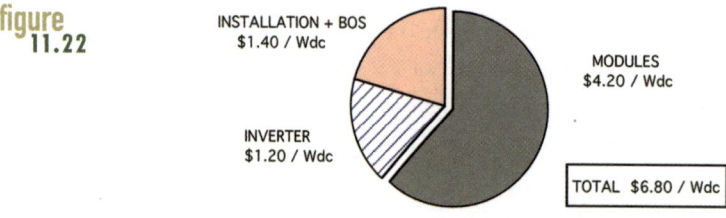

Average installed cost for 625 residential PV systems installed between 1994 and 2000.

SOURCE: Chang, 2000

**table
11.2** Capital Recovery Factors, CRF (*i,n*)

Term (yr)	Interest rate (% yr)*				
	5.0%	5.5%	6.0%	6.5%	7.0%
15	0.09634	0.09963	0.10296	0.10635	0.10979
20	0.08024	0.08368	0.08718	0.09076	0.09439
30	0.06505	0.06881	0.07265	0.07658	0.08059

* Units are per year.

11.7.2 Rebates and Tax Credits

The 34¢/kWh found in Solution Box 11.4 doesn't make a very good case for investing in a PV system. There are, however, two other factors that can help close the deal. One is government and utility incentive programs. The other is tax-deductible interest on home loans.

As of 2007, there is a federal tax credit of 30% for residential and commercial PV systems (unlimited on commercial, but a maximum of $2000 on residential). There are also many state and utility incentive programs. These rebates and tax credits can work together to help close the cost gap between utility electricity and your PV electricity (Solution Box 11.5).

**figure
11.23**

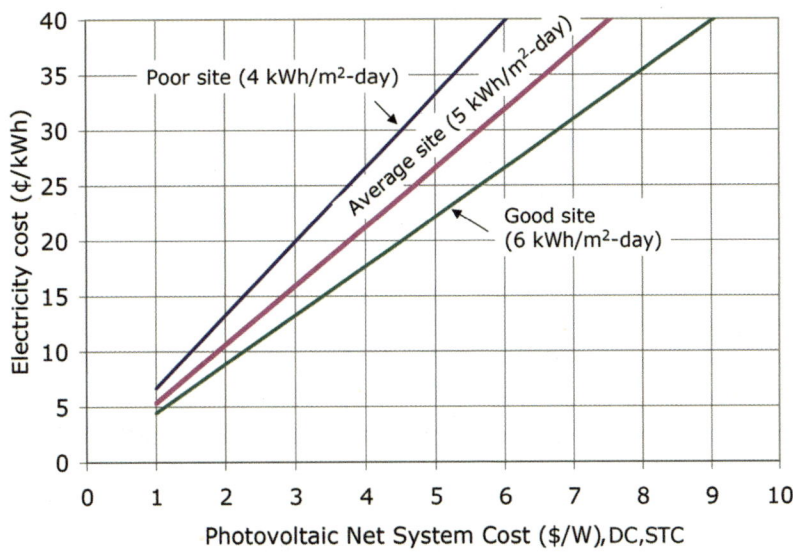

Showing the impact of net system cost $/W(DC,STC) (after tax credits and rebates) on annualized electricity cost. Assumptions: de-rating = 0.75, 6%, 30-year loan, no deduction for mortgage interest.

SOLUTION

SOLUTION BOX 11.4

Cost of Electricity from PVs

Suppose you install a 2 kW(DC,STC) array (also referred to as 2 kW peak, or 2kWp) in Los Angeles that costs $7/W(DC,STC). If you borrow the money at 6% interest on a 30-year loan, find the cost of electricity generated if it is installed on a south-facing roof with tilt = L −15° and we assume a de-rating factor of 0.75.

Solution:

From Table 11.1, the average annual insolation is 5.5 kWh/m²-day (5.5 hr/day @ 1-sun), so the system generates 2 kW × 0.75 × 5.5 hr/day × 365 day/yr = 3011 kWh/yr.

The system costs $14,000. From Table 11.2, CRF(6%,30yr) = 0.07265/yr, so the annual cost of the loan is $14,000 × 0.07265/yr = $1017/yr. Combining the $/yr and kWh/yr, or just using Equation 11.8, we find the cost of electricity to be

$$\$/kWh = \frac{\$1017/yr}{3011 \text{ kWh/yr}} = \$0.338/kWh$$

11.7.3 Closing the Deal with Tax-Deductible Home-Loan Interest

If the cost of a PV system is part of a home-equity loan, or part of your home mortgage, the interest on your loan is tax deductible. The value to you of a tax deduction depends on your marginal tax bracket (MTB). For instance, if you are in a 28% tax bracket, every dollar of deductions (e.g., charity, interest on home loans) saves you $0.28 on your taxes.

Eq. 11.7   **Tax savings on interest = Unpaid balance × Interest rate × MTB**

In the first years of your mortgage, almost all of the loan payment you make is paying for interest on the original balance of the loan. A good approximation, then, is that in those early years, the value of tax-deductible interest is just the original principal times the interest rate times your marginal tax bracket. The example in Solution Box 11.6 shows how this tax deduction on loan interest helps the overall economics of PVs.

In some states with relatively high utility rates, such as California, the 17.6¢/kWh cost of PV electricity is potentially attractive. This is especially true for customers who use a lot of kilowatt-hours, for whom the marginal cost of electricity can be extremely high. Table 11.3 shows the 2007 residential rate structure for a major California utility in which high-demand customers may be paying as much as 37¢/kWh.

PV Economics with a Rebate, Tax Credit, and Tax Deduction

Suppose we continue the example in Solution Box 11.5, in which a 2 kWp, $14,000 system without incentives generates electricity costing $0.34/kWh. Let's assume we are eligible for the 30% (max $2000) federal tax credit and a state rebate of $2.80/W based on a de-rating factor of 0.84. Find the cost of electricity.

Solution:

The system cost after applying the state rebate is therefore

$$\text{System cost} = \$14{,}000 - 2000W \times 0.84 \times \$2.80/W = \$9296$$

The federal tax credit is 30% of the price you pay after other rebates, so it is

$$\textbf{Tax credit} = \textbf{Minimum (0.30} \times \$9296 = \$2789, \textbf{ or } \$2000) = \$2000$$

As a credit, this actually reduces your taxes by $2000 making the system cost now

$$\textbf{After-tax-credit and rebate system cost} = \$9296 - \$2000 = \$7296$$

Borrowing that on a loan with CRF(6%,30) = 0.07265 gives you annual loan payments of

$$A = \$7296 \times 0.07265 = \$530.05/\text{yr}$$

The system still generates 2 kW × 0.75 × 5.5 hr/day × 365 day/yr = 3011 kWh/yr. So our cost per kWh is now

$$\$/\text{kWh} = \frac{\$530.05/\text{yr}}{3011\ \text{kWh/yr}} = \$0.176/\text{kWh}$$

That's better but still more than most people pay for their utility-generated electricity.

11.8 Stand-Alone Photovoltaic Systems

Stand-alone PV systems are much harder to design than those connected to the grid. Without the grid to provide energy storage, stand-alone systems are much more dependent on month-by-month load estimates, collector tilt angles, and solar availability. They are also more complicated since they involve battery storage and (usually) backup generators,

SOLUTION

SOLUTION BOX 11.6

Including Tax-Deductible Interest

Continuing the previous two examples, we have a system that costs $7296 after tax credits and rebates. Now let's add the impact of that 6% loan interest. Let's assume you are doing well and are in the 28% marginal tax bracket (MTB). Now find the cost of PV electricity.

Solution:

Using Equation 11.9, in the first year your loan balance is the full amount you borrowed; that is, $7296.

Tax reduction for interest (year 1) = $7296 × 0.06 × 0.28 = $122.57

The cost of your loan in the first year is now the loan payment of $530.05 minus the tax savings on interest of $122.57, which is a net cost for the PV system of $407.48. The cost of PV electricity to generate those 3011 kWh/yr is therefore

$$\$/kWh = \frac{\$407.48/yr}{3011 \ kWh/yr} = \$0.135/kWh$$

which means they cost more, require more maintenance, and are considerably less reliable. On the other hand, stand-alone systems don't have to compete economically with relatively cheap utility power. When your site is miles from the nearest power lines, it can cost more to bring power to the site than to buy a PV system. And, those who live with these systems have a personal stake in their operation and maintenance and are much more likely to value the electricity produced. After all, compared to no electricity at all—the situation faced by a couple of billion people on this planet—having even a small amount of PV electricity, maybe just enough to power a light for a few hours in the evening, can dramatically change lives.

table 11.3   **Residential Rate Structures in California, 2007**

Customers who use large amounts of electricity have added motivation to use PV Power.
Compare these prices with the $0.135/kWh cost of PV electricity derived in Solution Box 11.6.

Baseline Rate	$0.11430 per kWh
101–130% of Baseline	$0.12989
131–200% of Baseline	$0.22986
201–300% of Baseline	$0.32227
Over 300% of Baseline	$0.37070

A complete stand-alone system has a number of components, including the PV array, some batteries for energy storage, and an ac fuel-fired generator for backup power (Figure 11.24). Other components include a charge controller to keep from overcharging the batteries; a charger, which converts ac to dc to let the generator charge up the batteries when necessary; and an inverter, which converts dc from the batteries into ac for ac loads. Depending on the design, some loads may take dc directly from the batteries, thereby eliminating inverter losses, whereas others may be normal ac loads served by either the battery/inverter or directly from the ac generator.

Although the complete design of these systems is beyond the scope of this book (for such an analysis see, for example, Masters, 2004), we can illustrate the most important considerations.

Because every kilowatt-hour supplied by these systems is costly, the starting point for any design is a careful analysis of the loads that need to be supplied. Table 11.4 presents some examples of the power requirements for a variety of such loads. For most of these devices, energy is just the product of power and the number of hours the device is in use. For thermostatically controlled devices, such as refrigerators and freezers, the table provides daily energy estimates. An emerging concern is devices, such as TVs and satellite receivers, that use standby power even when they appear to be turned off, along with an array of other devices, such as portable phones, battery chargers, and so on, that are also constantly sucking up small amounts of power (look around the room and see how many little green and red lights you spot). For many entertainment systems, as much as two-thirds of their energy consumption occurs when they are not even being used.

Solution Box 11.7 shows an example estimate of a modest household demand and a first pass at sizing a PV array.

The example PV sizing for a stand-alone system for the cabin in Solution Box 11.7 is merely a first-cut at the design. To do a more careful job, we would have to trade off the number of days of battery storage we might want to provide with the number of hours of backup generator use that we could tolerate.

figure 11.24   A Versatile Stand–Alone PV System Capable of Providing DC or AC Power

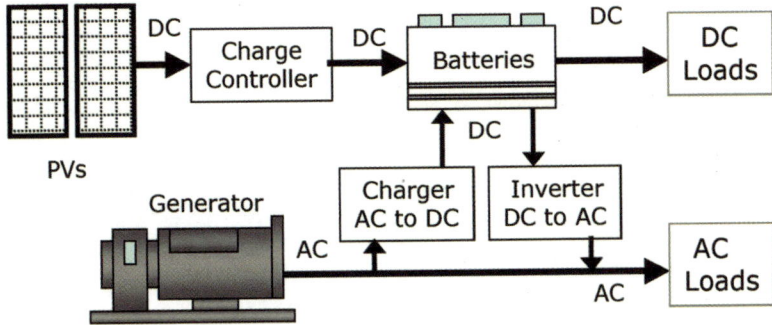

table 11.4   **Approximate Power Requirements of Typical Loads***

Kitchen Appliances	DEMAND
Refrigerator: AC ENERGY STAR, 14 cu. ft.	1080 Wh/day
Refrigerator: AC ENERGY STAR, 22 cu. ft.	1250 Wh/day
Refrigerator: DC Sun Frost, 12 cu. ft.	560 Wh/day
Freezer: AC 7.5 cu. ft.	540 Wh/day
Electric range (small burner)	1250 W
Microwave oven	750–1100 W
Toaster	800–1400 W

General Houshold	
Clothes washer: vertical axis	500 W
Clothes washer: horizontal axis	250 W
Dryer (gas)	500 W
Vacuum cleaner	1000–1400 W
Furnace fan: 1/3 hp	700 W
Ceiling fan	65–175 W
Whole house fan	240–750 W
Air conditioner: window, 10,000 Btu/hr	1200 W
Heater (portable)	1000–1500 W
Compact fluorescent lamp (60 W equivalent)	16 W
Electric clock	4 W

Consumer Electronics	
TV: Conventional CRT (active/standby)	4.5 W/in. / 4 W
TV: LCD per inch (active/standby)	3.6 W/in. / 10 W
TV: Plasma per inch (active/standby)	7.6 W/in. / 20 W
TV: Rear projector microdisplay per inch (active/standby)	3.7 W/in. / 16 W
Analog cable box (active/standby)	12/11 W
Satellite receiver (active/standby)	17/16 W
VCR (active/standby)	17/5.9 W
Component stereo (active/standby)	44/3 W
Compact stereo (active/standby)	22/9.8 W
Clock radio (active/standby)	2.0/1.7 W
Computer, desktop (active/idle/standby)	125/80/2.2 W
Laptop computer	20 W
Ink-jet printer	35 W
Laser printer	900 W
Xbox 360	160 W

(*continued*)

Shop

Circular saw, 7 1/4"	900 W
Table saw, 10-inch	1800 W
Hand drill, 1/4"	250 W

Water Pumping

Centrifugal pump: 36 VDC, 50-ft @ 10 gpm	450 W
Submersible pump: 48 VDC, 300-ft @ 1.5 gpm	180 W
DC pump (house pressure system)	100 Wh/day

* Some include estimates of standby power as well as power when in use.

SOLUTION BOX 11.7

PVs for a Modest Cabin

Find the monthly energy demand for a cabin with a 22 cu. ft refrigerator, five 16 W compact fluorescent lamps used 6 hr/day, a 20 in. LCD TV used 4 hr/day, a 1000 W microwave used 6 min/day, and a 300-ft well that supplies 150 gallons of water per day (1.5 gpm, 180 W). If the average sun available is 5 kWh/m²-day (5 hr of full sun), use a de-rating factor of 0.75 and a PV efficiency of 14% to size a PV system to meet this load.

Solution:

From Table 11.4, we can figure out the daily energy demand:

refrigerator:		= 1250 Wh/day
CFLs:	5 × 16 W × 6 hr/day	= 480 Wh/day
LCD TV:	3.6 W/in × 20 in × 4 hr/day	= 288 Wh/day
LCD TV standby:	10 W × 20 hr/day	= 200 Wh/day
microwave:	1000 W × 6 min/60 min/hr	= 100 Wh/day
well:	150 gal/day /1.5 gal/min/60 min/hr × 180 W	= 300 Wh/day
		Total: 2618 Wh/day

Notice the energy required by this TV when it isn't turned on is almost as much as when it is. This user might consider a power strip to really turn that thing off when not in use. From Equation 11.5, a first-cut at the rated power needed for this PV array would be

$$P_{DC,STC} = \frac{2618 \text{ Wh/day}}{(5 \text{ hr/day @ 1-sun}) \times 0.75} = 698 \text{ W} \approx 0.7 \text{ kW}$$

A 14%-efficient 0.7 kW$_{DC,STC}$ PV array would have an area of

$$\frac{0.7 \text{ kW}}{0.14 \text{ kW/m}^2} = 5 \text{ m}^2$$

11.9 Summary

This chapter has presented quite a range of PV topics, from the basic physics that describes how they work, to the influence of collector orientation, to array sizing and evaluation of system economics. The emphasis has been on PV systems located on the customer's side of the meter, in which case they compete against the retail price of electricity. In sunny locations, with existing tax credits, utility rebates, and tax-deductible interest on loans, PV systems are cost-effective, especially for customers who use lots of power and for whom the marginal cost of utility electricity is high.

Sales of PV modules have been increasing at roughly 40% each year, which is a phenomenal rate of growth. But, they are starting from a very small base and total installed capacity is still just a few gigawatts. In the next chapter we will look at bigger renewable energy systems located on the utility side of the meter, including the very rapid growth of large wind-turbine power plants around the globe.

CHAPTER 12

Large-Scale Renewables: Wind and Solar

12.1 Renewable Electric Power Systems

Locating generation on the customer's side of the meter has a significant economic advantage: it allows photovoltaic (PV) and other self-generation systems to compete against the retail price of electricity. It is much harder for renewables to compete against conventional central power stations in wholesale markets because wholesale prices are often one-half to one-third of retail rates. The most advanced, and cost-effective, renewable electricity generation systems are wind turbines with economics that can compete, toe-to-toe, with other large central stations at the wholesale level. Much of this chapter is devoted to these systems.

In addition to wind systems, concentrating solar power (CSP) systems are also making inroads into the marketplace for central power stations. These CSP systems convert sunlight into heat, which is then used to drive a heat engine coupled to an electrical generator. These solar-thermal systems include reflective parabolic troughs that concentrate sunlight onto a focal line containing a circulating heat-transfer fluid used to generate steam to run a rather conventional steam-cycle power plant. The largest such system, some 354 MW, has been operating in the desert in southern California for more than two decades. More recently, an even larger solar-dish array has been proposed that will use arrays of parabolic dish concentrators to focus heat onto electricity-generating Stirling engines of the sort described in Chapter 10.

12.2 Historical Development of Wind Power

Wind has been utilized as a source of power for thousands of years for such tasks as propelling sailing ships, grinding grain, pumping water, and powering factory machinery. The world's first wind turbine used to generate electricity was built in 1891 by a Danish inventor and school principal, Poul la Cour. It is especially interesting to note that la Cour experimented with electrolysis to produce hydrogen for gas lights in his schoolhouse (records include reference to a number of windows that had to be replaced as a result of his tinkering). In that regard, we could say that he was one hundred years ahead of his time—the concept of using renewable sources of electricity to electrolyze water to power fuel cells has re-emerged as an intriguing possibility for the twenty-first century.

In the United States, the classic multibladed, water-pumping wind turbines used to be ubiquitous across the Great Plains. Indeed, it can be argued that they enabled the crucial first steps in the expansion of farming, ranching, and human settlements into that vast stretch of relatively arid country. Those turbines were ideal for water pumping because their multiblade design produces high torque even at low wind speeds—just what is needed to overcome the friction and weight of that heavy pumping rod that moves up and down in the well. The strong winds in those Great Plains states also stimulated the development of small wind-electric systems for rural areas not yet served by the electricity grid. Hundreds of thousands of these fast spinning, two- and three-bladed turbines used to dot the landscape in the 1930s and 1940s, but they disappeared as soon as the more reliable and economic utility grid spread across the landscape. The role that wind once played in the economic development of these windy states seems now on the verge of being repeated as wind farms begin to create employment, tax revenues, and hefty royalty payments that are providing a much-needed jolt to many local economies.

The oil shocks of the 1970s, which heightened awareness of our energy problems, coupled with substantial financial and regulatory incentives for alternative energy systems, stimulated a renewal of interest in wind power in the United States. California became the proving ground for dozens of manufacturers who installed thousands of new wind turbines in the Altamont Pass region just east of San Francisco, the Tehachipi Pass near Barstow, and San Gorgonio Pass just north of Palm Springs. Many of these early machines did not perform very well, and their very location in mountain passes often put them directly in the path of migrating birds. Their location coupled with their small-diameter, high-speed blades, created the image that these were lethal "bird cuisinarts." When lucrative tax incentives were terminated in the mid-1980s, the U.S. wind industry nearly collapsed as well.

Lack of interest in wind in the United States was reflected in the lackluster growth in installed capacity in the decade between 1988 and 1998. By contrast, between 2000 and 2006 it grew at an average annual rate of 23%, reaching a total of just under 14,000 MW in 2007—enough to satisfy the entire electricity demand of 3.5 million homes. The 2500 MW of new wind projects brought on line in 2006 was 19% of that year's rated-power additions to the U.S. grid and represented a $3.7 billion investment in new installations (Wiser, et al., 2007). As shown in Figure 12.1, these later years have been characterized by a boom-and-bust cycle of construction, which reflects the on-again, off-again, short-term extensions of a federal production tax credit (PTC) that as of 2007 provides a ten-year, $0.019/kWh incentive for wind projects.

The U.S. wind industry stalled from the late 1980s to the mid-1990s, but wind technology development continued unabated in Europe—especially in Denmark, Germany, and Spain—and they entered the market with larger, less costly, more efficient, more reliable turbines that created the global sales boom in the late 1990s. By the turn of the century it was possible to make the case that wind, in good locations, was as cheap as any other source of electricity (Jacobson and Masters, 2001). Figure 12.2 presents a comparison of the average price of wind power in recent years with the range of wholesale prices paid for bulk power in the United States. On a cumulative basis, wind has consistently been priced at or below the low end of the wholesale power price range.

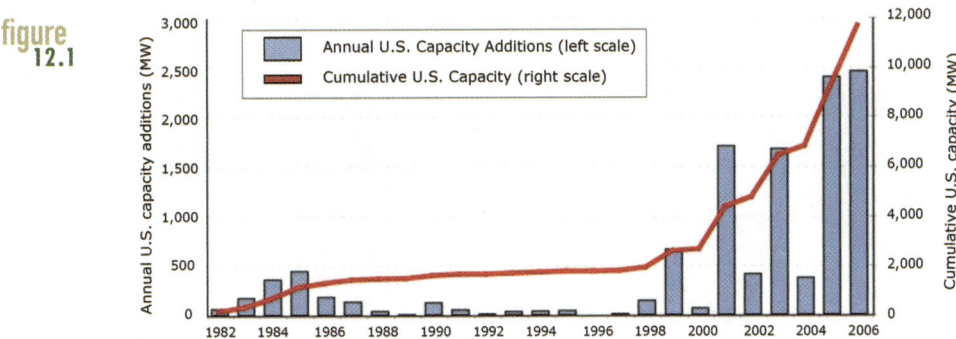

Installed capacity of wind turbines in the United States and net annual additions. Irregular net additions illustrate the impact of unpredictable tax credits.

SOURCE: Wiser, et al., 2007

Globally, the cumulative installed capacity of wind turbines grew at a fairly steady rate of 25% per year between 1996 and 2006, reaching just over 74,000 MW in 2006. As shown in Figure 12.3, the country with the largest total installed wind capacity in 2006 was Germany (28%) followed by Spain and the United States, each with 16%. The largest increases in capacity in 2006 were in those same three countries, with the United States in the lead at 2.4 GW of added capacity. As an interesting comparison, the new installed capacity of wind around the globe in 2006 (15 GW) was about 6 times the increase of photovoltaic installations (2.5 GW), but the rate of growth of PVs was higher (40% versus 26%).

12.3 The Wind Resource

How much power and energy is available in the wind? To help answer that question, the National Renewable Energy Laboratory (NREL) defines the wind power classification scheme

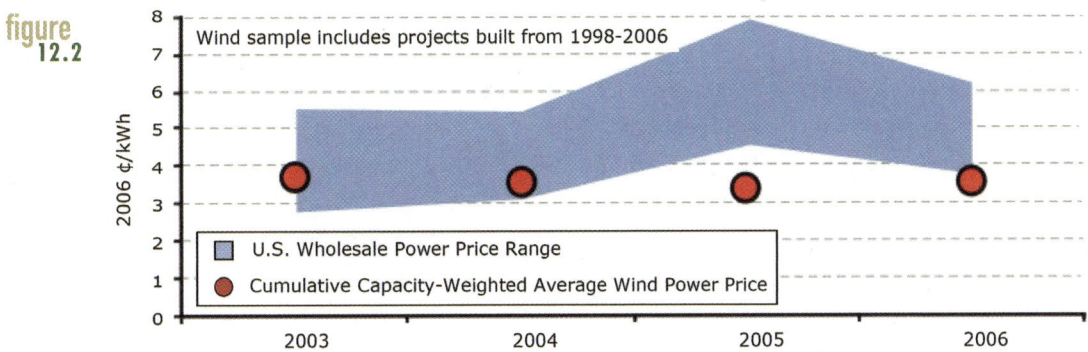

The cumulative capacity-weighted average wind price in the United States compared with conventional generation wholesale power prices.

SOURCE: Wiser, et al., 2007

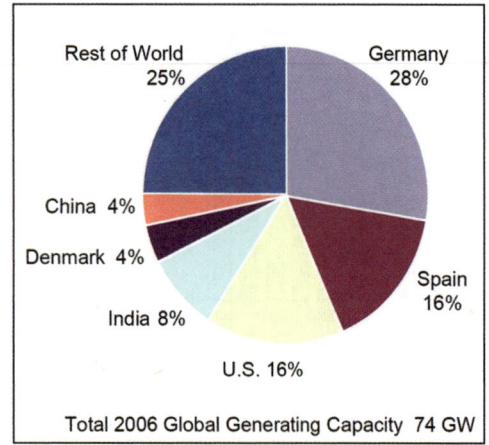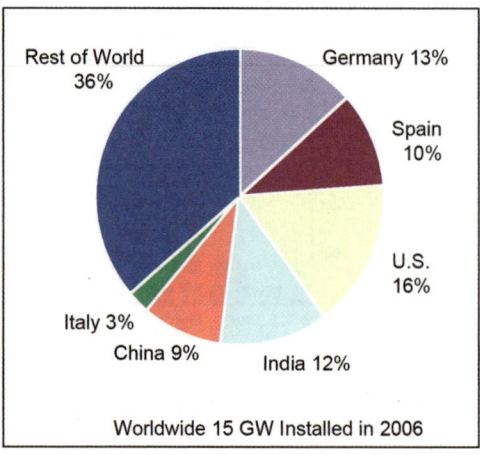

Worldwide installed capacity of wind turbines and net annual additions. Installed capacity grew at 25% per year for the decade from 1996 to 2006.

Source: data from GWEC, 2006

table 12.1 — Standard Wind Power Classification Scheme*

Wind Power Classification	Resource Potential	Wind Power Density (W/m²)	Average Wind Speed (m/s)	Average Wind Speed (mph)
2	Marginal	200–300	5.6–6.4	12.5–14.3
3	Fair	300–400	6.4–7.0	14.3–15.7
4	Good	400–500	7.0–7.5	15.7–16.8
5	Excellent	500–600	7.5–8.0	16.8–17.9
6	Outstanding	600–800	8.0–8.8	17.9–19.7
7	Superb	> 800	> 8.8	> 19.7

* Average wind speeds for each category are based on Rayleigh statistics (see Sidebar 12.1).

shown in Table 12.1. For example, Class-4 winds (referred to as "Good") have between 400 and 500 watts of power per square meter of cross-sectional area, which correlates to an average wind speed of 15.7 to 16.8 mph and is often thought of as the threshold of economic viability for wind power.

Wind quality varies with geography, so NREL has developed a continuously improving series of national and regional maps that apply this wind power classification scheme, such as the one shown in Figure 12.4. These have traditionally been based on wind evaluations at an assumed elevation of 50 meters, which was roughly the hub height for turbines at the time the maps were first made. With wind turbines getting ever larger, mounted on taller and taller towers, new maps are being developed that show the resource at 80 meters, which is closer to the current hub height of large turbines (Archer and Jacobson, 2005). At that higher elevation, wind

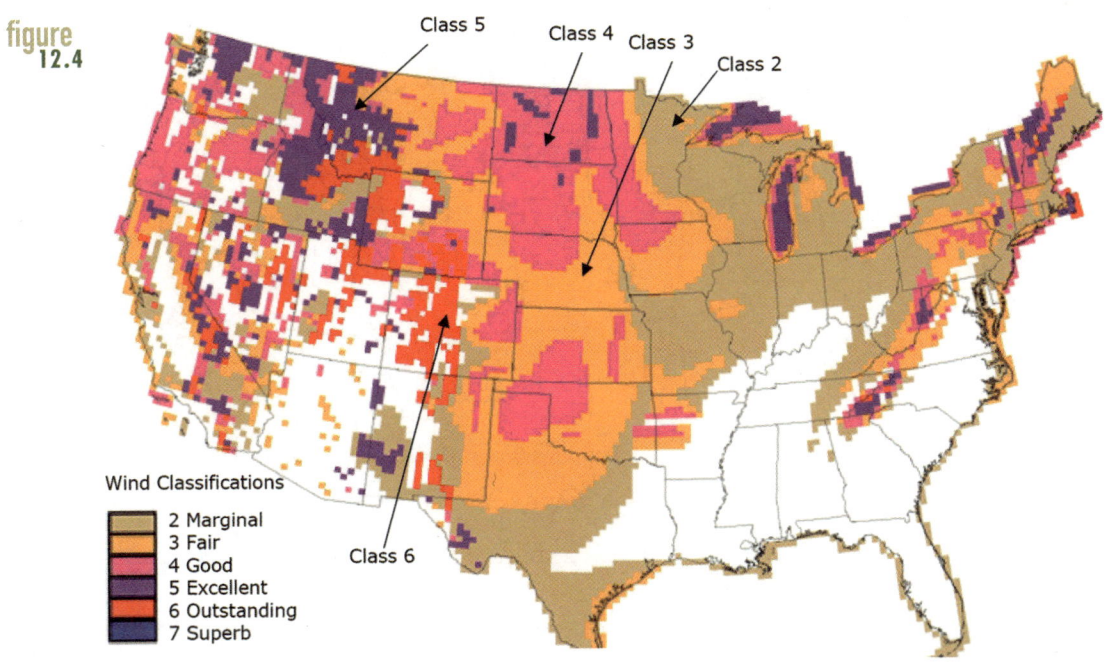

U.S. wind resources organized by wind classifications at an elevation of 50 meters.

Source: NREL, 1987

speeds increase enough to often bump the wind regime upward by one classification (e.g., from class 3 to class 4), making available an even larger wind resource base than previously estimated.

Wind maps such as those shown in Figure 12.4 can be used as a starting point for estimates of the electrical energy that wind turbines can potentially deliver in a given region. To make such estimates, significant land use questions must be evaluated. Flat grazing lands would be easy to develop, and the impacts on current usage of such lands would be minimal. On the other hand, developing sites in heavily forested areas or along mountain ridges, for example, would be much more difficult and environmentally damaging. Urban areas and highly sensitive lands such as national parks also need to be excluded from consideration. Economic viability of remaining areas will often be closely tied to the proximity to transmission lines with available capacity as well as load centers near enough to take advantage of the available power.

Land-use constraints play a major role in siting wind power systems. Estimates of their effect on U.S. wind energy potential have been made by the Pacific Northwest Laboratory. In one assessment the exploitable wind resource at 50 meters was estimated to be 16,700 billion kWh/yr with no land-use restriction, but decreased to 4600 billion kWh/yr under the most severe constraints (Elliott, 1991). By comparison, the total electricity generated by all power plants in the United States in 2005 was 4340 billion kWh, which suggests the wind resource is theoretically sufficient to meet the entire U.S. demand. That, of course, seems highly unlikely because wind doesn't always blow at exactly the right time and the cost of energy storage to buffer the mismatch between instantaneous supply with current demand

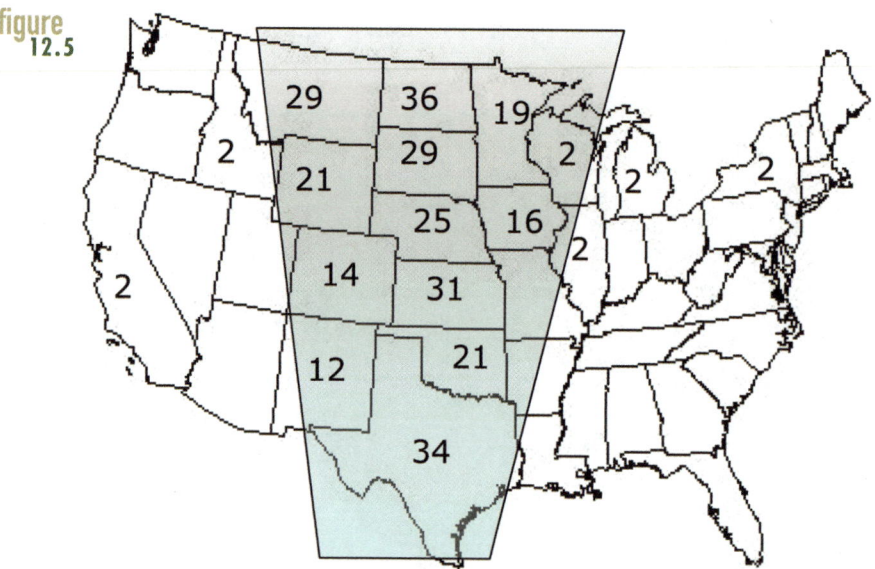

figure 12.5

Estimated percentage of U.S. electricity demand that could be met with the wind resource in the windiest states. North Dakota or Texas alone could theoretically supply more than one-third of national demand. Based on data from Elliott (1991) for winds of class 3 or better at 50 meters.

seems prohibitive. The fraction of U.S. demand that could actually be met with wind power in the future is highly uncertain.

Another important difficulty in supplying a high fraction of U.S. power by wind is the disparity between where the wind blows and where the major load centers are located. As Figure 12.5 suggests, the best winds tend to be located along a wedge of states from Montana and Minnesota in the north, down to Texas in the south. The wind resource in one state alone, North Dakota, is thought to be sufficient to supply over one-third of all U.S. electricity. Some have dubbed the Great Plains region "the Saudi Arabia of wind" because of the great, untapped potential. California, which ranks seventeenth in terms of wind resource, used to have more installed capacity than any other state. However, in 2006, Texas, which has abundant wind resources, overtook California as the number one state (Figure 12.6).

12.4 Wind Turbine Technology

Wind turbines are characterized by the axis about which the blades rotate, the number of blades, and whether they face into the wind or away from it. Figure 12.7 illustrates these distinctions.

The only vertical-axis machine that has had any commercial success is the Darrieus rotor, named after its inventor, French engineer G. M. Darrieus, who first developed the turbines in the 1920s. The shape of the blades is that which would result from holding a

figure
12.6

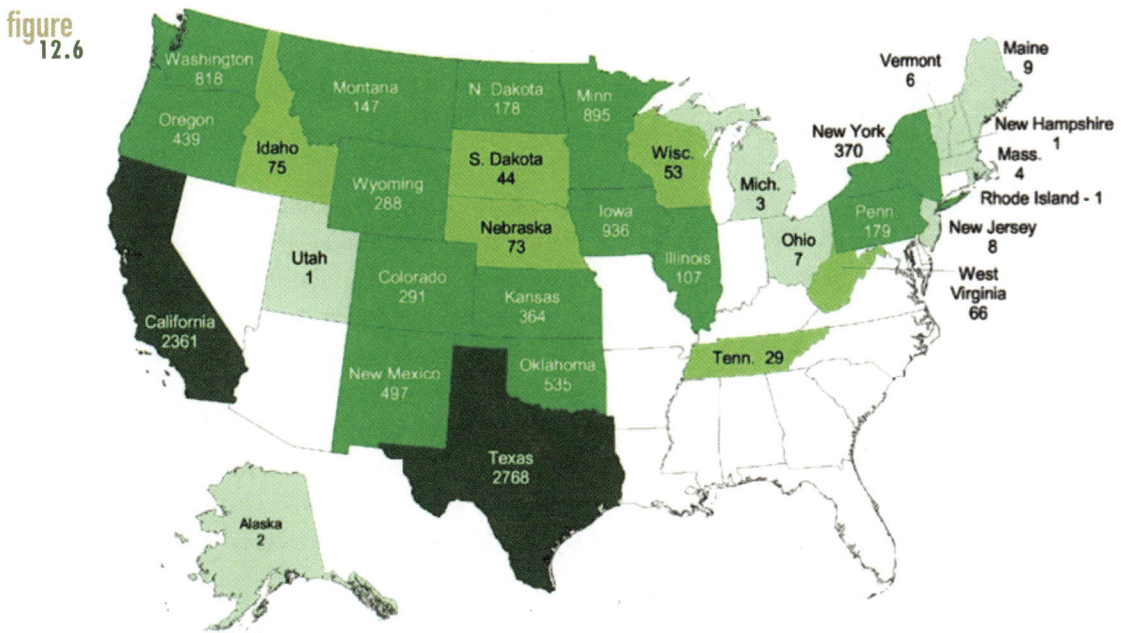

Distribution of the 11,600 MW of installed wind turbine capacity by state as of January 2007.

SOURCE: National Renewable Energy Laboratory

rope at both ends and spinning it around a vertical axis, giving it a look not unlike a giant eggbeater. There are a number of potential advantages of vertical-axis machines over their horizontal-axis counterparts. They always point into the wind, which eliminates the need for special yaw (left-right directional) controls. The heavy machinery contained in the *nacelle*

figure
12.7

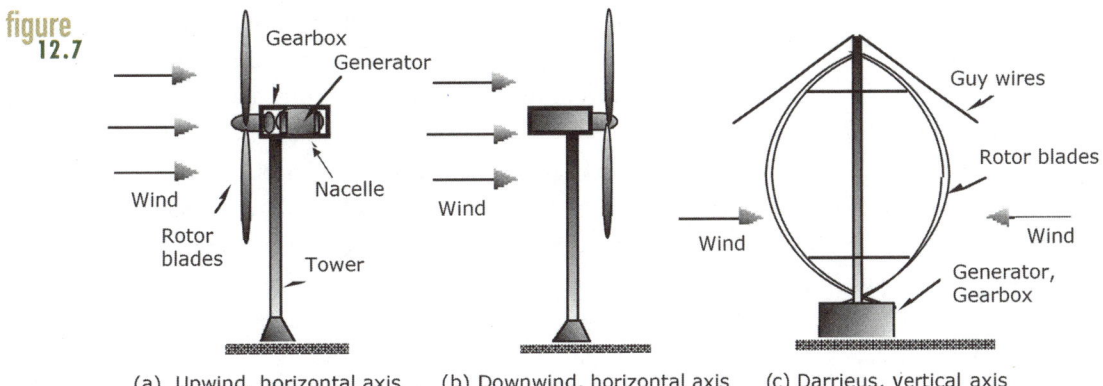

Horizontal-axis wind turbines are either upwind machines (a) or downwind machines (b). Vertical-axis wind turbines accept the wind from any direction (c). Most turbines these days are three-bladed, horizontal-axis, upwind machines.

(the housing around the generator, gearbox, etc.) is at ground level so the support structure for the turbine doesn't need to be nearly as strong. Moreover, the blades in a Darrieus rotor are almost always in pure tension, which means that they can be relatively lightweight and inexpensive because they don't have to handle the constant flexing associated with blades on horizontal-axis machines. On the other hand, the best winds are higher up, which means these low-to-the-ground rotors don't have nearly the wind resource to tap into compared to their horizontal counterparts, which tend to be mounted on tall towers.

Vertical-axis machines do remain intriguing, but the industry has pretty much abandoned the concept and virtually all turbines are now horizontal-axis machines. The next key design question is whether to make them upwind machines, in which the wind hits the blades before it reaches the tower, or the other way around, as downwind machines. Downwind machines have the advantage of automatic yaw control as the wind pushes the blades away from the tower, but the wind-shadowing effect of having the tower in front of the blades causes blade flexing every time the blades swing behind the tower. This flexing not only has the potential to cause blade failure due to fatigue, but it also reduces power output while increasing blade noise. The result is that the wind industry has adopted horizontal-axis, upwind machines as the standard.

The remaining question is how many blades the turbine should have. The classic multi-bladed farm windmill has a large area of rotor blades facing into the wind, which provides the high torque needed for simple water pumping. They don't spin very fast, so the turbulence caused by one blade on the following blade is relatively unimportant. For electricity generation, however, the tip speed of the blades is very high and the turbulence caused by one blade on another can significantly reduce overall efficiency, which suggests the fewer the number of blades, the better. Most new turbines have two or three blades. Three-bladed rotors run smoother than their two-bladed counterparts because the impact of tower interference as well as the variation of wind speed with height are more evenly transferred from rotors to drive shaft. Most wind turbines now have three blades.

The number of blades affects the overall rotor efficiency as a function of a quantity called the *tip-speed ratio* (TSR), as illustrated in Figure 12.8. The tip-speed ratio is the speed at which the outer tip of the blade is rotating divided by the wind speed. For the American multiblade windmill so common in the 1930s and 1940s, the optimum TSR is less than one, whereas for a typical three-bladed wind turbine the tip speed is about four times the speed of the wind. The maximum theoretical efficiency of a rotor is the *Betz limit*. Albert Betz was a German physicist who in 1919 showed that the maximum possible rotor efficiency occurs when the blades slow the wind by two-thirds, which results in a maximum possible efficiency of 59.3%. Solution Box 12.1 illustrates the use of TSR to estimate the rpm of the rotor.

Figure 12.9 shows an artist's rendition of the inner workings of the 3.6-MW wind turbine manufactured by GE. The main components inside the nacelle are the generator, gearbox, and yaw drive system. The three-bladed rotor hub includes pitch drive mechanisms to vary the pitch of the blades to control speed.

Turbine manufacturers have exploited the economies of scale that come with building larger and larger turbines. Most turbines built before 2000 were rated at less than 1 MW

SOLUTION

SOLUTION BOX 12.1

How Fast Does the Turbine Spin?

One of the first impressions you are likely to have when you see a modern wind turbine is how slowly it seems to turn. Imagine a three-bladed turbine with a blade diameter of 102 meters that generates 3.6 MW of power when exposed to 14 m/s wind speeds. If we assume a tip-speed ratio of 4, estimate how fast the turbine spins.

Solution:

Begin with the definition of TSR.

$$\text{Tip-speed-ratio (TSR)} = \frac{\text{Rotor tip speed}}{\text{Wind speed}} = \frac{(\text{rev/min}) \times \pi D \ (\text{m/rev})}{V_w \ (\text{m/s}) \times 60 \ (\text{s/min})}$$

$$\text{rev/min} = \frac{60 \, V_w}{\pi D} = \frac{60 \ (\text{s/min}) \times 14 \ (\text{m/s}) \times 4}{\pi \times 102 \ (\text{m/rev})} = 10.5 \ \text{rpm}$$

That is about 5.7 seconds per revolution. Although that looks very slow, the tip of the blades would be moving at 4×14 m/s = 56 m/s, which is about 125 miles per hour.

figure
12.8

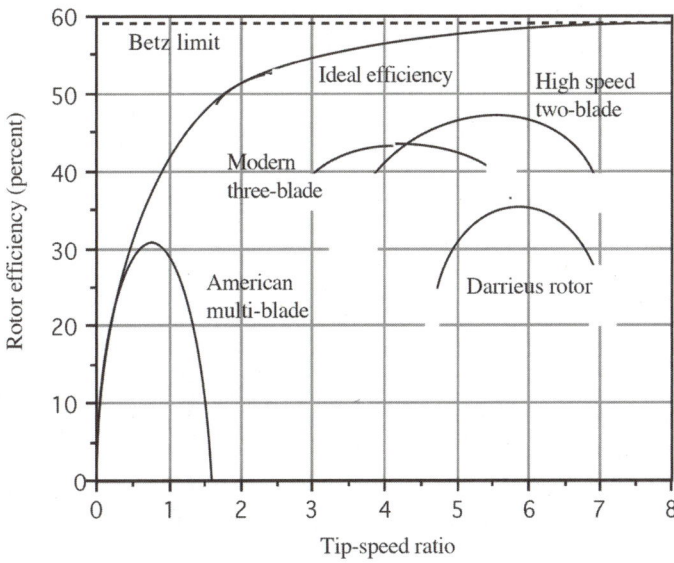

Rotor efficiency depends on the number of blades and tip-speed ratio. The theoretical maximum, called the Betz limit, is 59.3%.

figure 12.9 An Inside View of the 102-meter, 3.6-MW GE Wind Turbine

1. Nacelle
2. Heat Exchanger
3. Generator
4. Control Panel
5. Main Frame
6. Impact Noise Insulation
7. Hydraulic Parking Brake
8. Gearbox
9. Impact Noise Insulation
10. Yaw Drive
11. Yaw Drive
12. Rotor Shaft
13. Oil Cooler
14. Pitch Drive
15. Rotor Hub
16. Nose Cone

each; those are now considered small machines. New turbines are several megawatts each and some under development are as large as 4.5 MW. Figure 12.12 shows how big these new machines are becoming. The 4.5-MW turbine being developed by Vestas Wind Systems for the offshore market will have a blade diameter of 120 meters (longer than a football field), and by the time it is placed on its tower the top of the sweep of its blades will be over 160 meters (roughly the height of a fifty-story building).

12.5 Energy from the Wind

Just how much power is in the wind and how much can we imagine the turbine being able to extract? The answers, of course, depend on how fast the wind is blowing, the swept area of the blades, and the detailed characteristics of the turbine. The starting point is the power in the wind itself.

figure 12.10

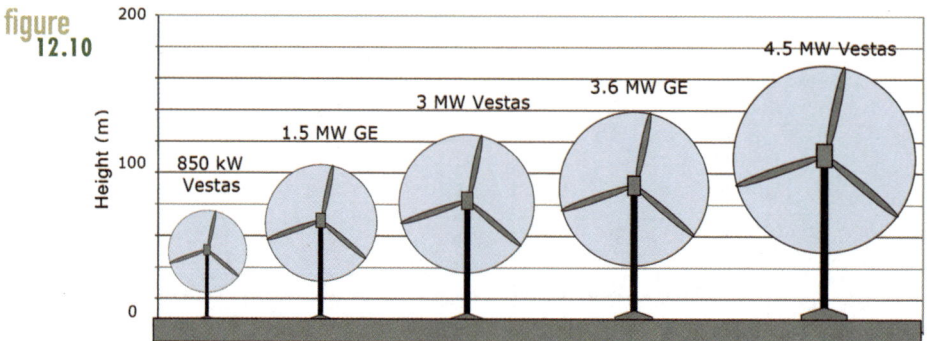

Showing the evolution of turbine sizes. The 4.5-MW Vestas turbine will be as tall as a fifty-story building.

12.5.1 Instantaneous Power in the Wind

Consider a "piece" of air with mass m moving at a speed v. Its kinetic energy (K.E.) is given by the familiar relationship described in Section 4.3 of this book:

Eq. 12.1

$$m \quad \xrightarrow{\ v\ } \quad \text{K.E.} = \frac{1}{2} mv^2$$

Because power is energy per unit of time, the power represented by a mass of air moving at velocity v through area A will be

Eq. 12.2

$$\dot{m} \quad \xrightarrow{A \ \ v} \quad \text{Power through area } A = \frac{\text{Energy}}{\text{Time}} = \frac{1}{2}\left(\frac{\text{Mass}}{\text{Time}}\right)v^2$$

The mass flow rate $\dot{m}$, through area A, is the product of air density ρ (which is a function of temperature and atmospheric pressure), wind speed v, and cross-sectional area A.

Eq. 12.3

$$\dot{m} = \frac{\text{mass passing through } A}{\text{time}} = \rho A v$$

Combining Equation 12.2 with 12.3 gives us an important relationship:

Eq. 12.4

$$P_{\text{w}} = \frac{1}{2}\rho A v^3$$

In S.I. units,

   P_{w} = power in the wind (watts)
   ρ = air density, which is 1.225 kg/m³ at 15°C and 1 atm
   A = cross-sectional area through which the wind blows (m²)
   v = wind speed (m/s); note: 1 m/s = 2.237 mph

The important thing to note from Equation 12.4 is that power in the wind increases as the cube of wind speed. That is, doubling the wind speed increases the power eight-fold. Another way to look at it is, for example, 1 hour of wind blowing at 20 mph carries as much energy as 8 hours at 10 mph, or 64 hours (2.7 days) of wind blowing at 5 mph. What really matters for a wind turbine is its ability to capture those faster winds. In fact, most big turbines aren't even turned on for low-speed winds. Another thing to note from Equation 12.4 is that power goes up as the swept area increases, which means a doubling of the blade diameter increases the available power by a factor of four.

SOLUTION BOX 12.2

Average Power in the Wind—Be Careful!

Suppose the wind blows for 10 hours at 8 m/s and 10 hours at 4 m/s. What would be the total energy and average power per square meter of area over those 20 hours?

Solution:

Applying Equation 12.4 to each wind regime:

$$\text{Energy} = \tfrac{1}{2}\,\rho v^3(\text{W/m}^2) \times \Delta t\,(\text{hr})$$

$$\text{Energy (10 hr @ 8 m/s)} = 0.5 \times 1.225 \times 8^3 \times 10 = 3136\,\text{Wh/m}^2$$
$$\text{Energy (10 hr @ 4 m/s)} = 0.5 \times 1.225 \times 4^3 \times 10 = 392\,\text{Wh/m}^2$$
$$\text{Total} = 3136 + 392 = 3528\,\text{Wh/m}^2$$

Notice how insignificant the energy contributed by those low-speed, 4 m/s winds is. The average power over those 20 hours is 3528 Wh/20 hr = 176.4 Wh/m².

Suppose we had simply plugged the average wind speed of 6 m/s into Equation 12.4. What would we have gotten for average power?

$$\text{Average Power} = \tfrac{1}{2}\,\rho\big(v\big)^3 = 0.5 \times 1.225 \times 6^3 = 132.3\,\text{W/m}^2$$

Our 132.4 W/m² estimate using average wind speed in Equation 12.4 is 25% lower than the correct answer of 176.4 W/m².

12.5.2 Average Power in the Wind

Although Equation 12.4 correctly provides the instantaneous power in the wind, the nonlinear relationship between wind speed and power tells us we need to be cautious about using it to estimate the average power in winds that have variable speeds. Just plugging the average wind speed into Equation 12.4 will underestimate the average power by a significant amount (as we shall see later, the error may be close to 50%).

Even the very simple example shown in Solution Box 12.2 shows the need to have some idea of the distribution of wind speeds at a site if we want to estimate the average power or total energy that a wind turbine will produce. For a real wind project, a lot of data need to be collected over a considerable period of time to try to determine the typical number of hours each year that the wind will blow at 1 m/s, 2 m/s . . . and so forth. From this data, an analysis similar to that shown in Solution Box 12.2 can be worked out.

There are some shortcuts that are often taken, however, the most common of which begins with a simple estimate of the average wind speed at the site. Average wind speed is easy to measure using an inexpensive anemometer, which is used as a "wind odometer" to measure miles of wind that pass by as indicated by the number of revolutions of the spinning anemometer cups. Dividing by the hours it took to record those miles gives an average wind speed in miles per hour. Coupling average wind speed with an assumption about the distribution of wind speeds about that average enables us to find the average power in the wind. The mathematics is a little tricky, so we're putting the analysis in Sidebar 12.1 for those who are interested in such details. Later we will summarize the conclusions.

12.5.3 Energy from a Turbine Using Average Power in the Wind

If we can make some assumptions about the efficiency of a wind turbine, we can quickly estimate the power and energy that will be delivered if we assume Rayleigh winds with some average wind speed. Table 12.2 uses Equation 12.8 to assemble a convenient conversion from average wind speed to average power in the wind.

Although wind turbine efficiencies vary depending on the wind regime in which they are placed, in good winds they tend to operate with an overall efficiency of somewhere between 25% and 35%. Those state-by-state estimates of wind energy potential shown in Figure 12.5, for example, assumed average turbine efficiencies of 25%, which is on the low side for today's modern turbines.

To illustrate this simple procedure, suppose we want to estimate the energy delivered from a 30%-efficient 2000 kW wind turbine with 80-meter blades if it is located in an area with an average wind speed of 7 m/s (the beginning edge of class-4 winds). From Table 12.2, the average power in the wind is 401 W/m² so the average power that the turbine would deliver would be

$$P_{avg} = \eta_{Turbine} \times A_{Rotor} \, (\text{m}^2) \times P_{Wind} \, (\text{W/m}^2) = 0.30 \times \frac{\pi}{4} \times 80^2 \times 401 = 604{,}693 \, \text{W}$$

Over a year's time (8760 hours), the output would be about

$$\text{Energy delivered} = 604.7 \, \text{kW} \times 8760 \, \text{hr/yr} = 5.3 \times 10^6 \, \text{kWh/yr}$$

12.5.4 Wind Turbine Capacity Factors

All power plants, whether they be nuclear, hydroelectric, coal, or whatever, have a *rated power* output, P_R, which tells us how many kilowatts or megawatts they deliver when running at full power. A conventional power plant may operate at full, or near-full, output most of the time, but that is not the case for wind turbines because they are so dependent on available winds. This means, for example, that a 100-MW base-load, coal plant will be likely to deliver many more

SIDEBAR 12.1

Rayleigh Statistics

The distribution of wind speeds at a site is often assumed to follow a Rayleigh probability density function described by the following equation:

Eq. 12.5
$$f(v) = \frac{\pi v}{2(v_{avg})^2} \exp\left[-\frac{\pi}{4}\left(\frac{v}{v_{avg}}\right)^2\right]$$

where v_{avg} = average wind speed

Figure 12.11 shows what this looks like.

What we would like to find is the average power in the wind. That is, we want

Eq. 12.6
$$P_{avg} = \left(\frac{1}{2}\rho A v^3\right)_{avg} = \frac{1}{2}\rho A \cdot (v^3)_{avg}$$

Using some notions from statistics, we can evaluate Equation 12.6 with the following:

Eq. 12.7
$$P_{avg} = \left(\frac{1}{2}\rho A v^3\right)_{avg} = \frac{1}{2}\rho A \cdot (v^3)_{avg}$$
$$= \frac{1}{2}\rho A \int_0^\infty v^3 f(v)\,dv$$

If we plug the Rayleigh probability density function given in Equation 12.5 into Equation 12.7 and do some fancy calculus, we get the following interesting result:

Eq. 12.8
$$P_{avg} = \frac{6}{\pi} \cdot \frac{1}{2}\rho A \cdot (v^3)_{avg} = 1.91 \cdot \frac{1}{2}\rho A\,(v^3)_{avg}$$

That is, if we just plug the average wind speed into the usual equation for power in the wind and then multiply the result by $6/\pi = 1.91$ we get the average power in the wind if the wind distribution follows Rayleigh statistics.

For example, in Table 12.1 the threshold of class-4 winds is an average wind speed of 7.0 m/s that supposedly creates an average power in the wind of 400 W/m². We can test that combination using Equation 12.8:

Eq. 12.9
$$P_{avg}\,(@v_{avg} = 7\text{ m/s}) = \frac{6}{\pi}\left(\frac{1}{2} \times 1.225 \times 7^3\right)$$
$$= 401\text{ W/m}^2$$

which pretty closely agrees with the wind classification table.

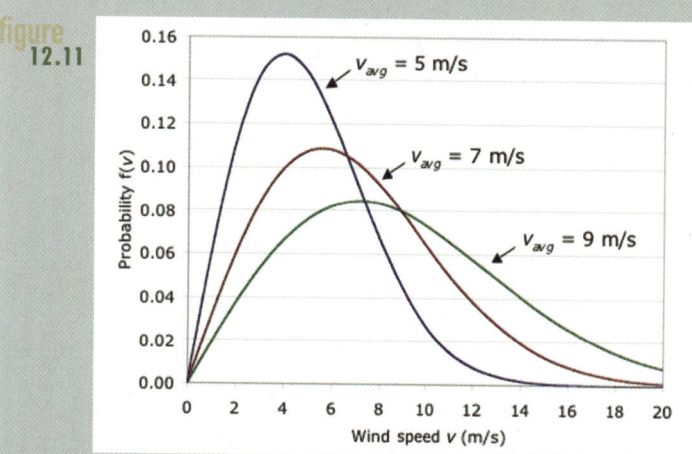

figure 12.11

The Rayleigh probability density function for varying average wind speeds.

table 12.2	Average Power in the Wind Assuming Rayleigh Statistics	
Average Wind Speed (m/s)	Average Wind Speed (mph)	Average Power in Wind (W/m²)
3	6.7	32
4	8.9	75
5	11.2	146
6	13.4	253
7	15.7	401
8	17.9	599
9	20.1	853

kWh per year than a 100-MW wind farm—perhaps as much as three times as much! Therefore, we have to be very careful comparing the two to be sure we aren't overstating the case for wind.

If a power plant operated at full power all 8760 hours per year (24 hr/day × 365 day/yr), its annual energy output would be P_R (kW) × 8760 hr/yr. Because power plants don't operate at full output all of the time, we can describe their annual output using an overall average capacity factor (CF) such that

Eq. 12.9   **Annual energy (kWh/yr) = P_R (kW) × 8760 hr/yr × CF**

For example the 2000 kW wind turbine in the previous example delivered 5.3×10^6 kWh/yr, which means its CF would be

$$CF = \frac{5.3 \times 10^6 \text{ kWh/yr}}{2{,}000 \text{ kW} \times 8760 \text{ hr/yr}} = 0.302 = 30.2\%$$

That's a pretty typical CF for modern turbines operating on the edge between class-3 and class-4 winds. Most wind plants installed today are in class-4 and class-5 sites, resulting in CFs of roughly 30%–40%, with some as high as 45% (Figure 12.12). For comparison, base-load coal plants often operate with capacity factors of 80%–90%.

Wind turbine CFs are affected both by the wind regime and the turbine's power curve, which is a graph of its power output as a function of wind speed. An example of an idealized power curve is shown in Figure 12.13. For winds below the *cut-in wind speed*, V_C, the turbine isn't even turned on because the power that would be generated isn't enough to offset generator losses. Above the cut-in wind speed, the power output climbs rapidly, more or less as the cube of wind speed, until it reaches the point at which the generator is delivering as much power as it can, namely its rated power P_R. The *rated wind speed*, V_R, that goes with the rated power isn't a very clearly defined number for real turbines because the power curve is usually somewhat rounded as shown in the figure. Above the rated wind speed, the pitch of the turbine blades is adjusted to shed some of the wind to keep from overpowering the generator. Finally, at some point, called the *furling* or *cut-out wind speed*, V_F, the winds are just too high and too dangerous, so the turbine shuts down.

2006 Capacity Factors for U.S. Wind Farms by Date of Installation

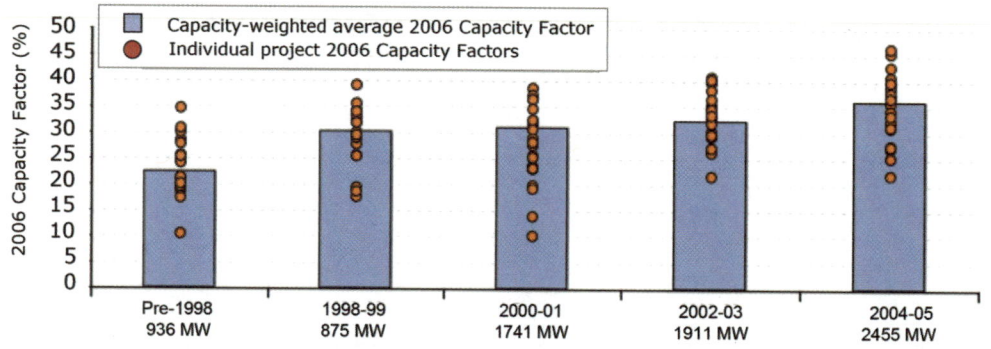

SOURCE: Wiser, et al., 2007

figure 12.13   **An Example Wind–Turbine Power Curve**

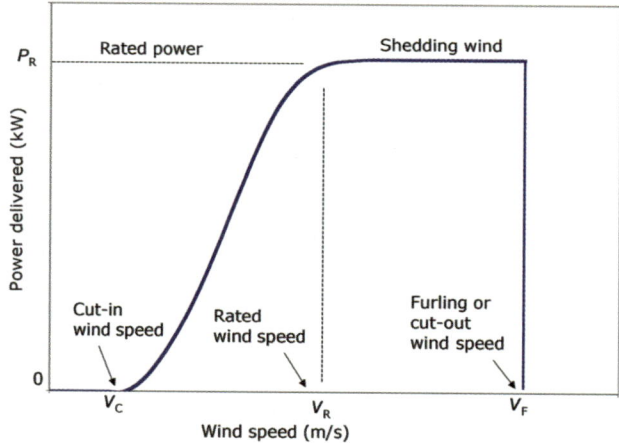

No power is generated below V_C. Above V_R the output remains relatively constant at P_R. Above V_F the turbine is shut down.

The CF for a wind turbine depends on the interaction between its power curve and the distribution of wind speeds to which it is exposed. In fact, the CF, as a function of average wind speed, will have a similar shape to the turbine's power curve, as is suggested in Figure 12.14. The CF will be very low if a lot of the wind is below the cut-in wind speed. On the other hand, when a lot of the wind is above the rated wind speed the CF saturates, and in fact, can begin to decrease when winds include some above the furling wind speed.

The most interesting aspect of Figure 12.14 is the fact that within the range of average wind speeds (e.g., about 4–9 m/s) to which a turbine is likely to be exposed, the CF is quite linear. This may seem quite surprising because earlier we talked about how power in the wind

figure
12.14

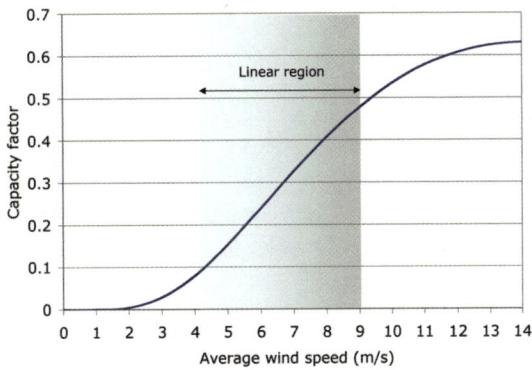

For turbines in good sites (e.g., average wind in the 4–9 m/s range), capacity factors are fairly linear with increasing average wind speeds.

itself increases as the cube of wind speed. However, once we include the characteristics of actual turbines, which include discarding low-speed winds and shedding much of the available power at higher wind speeds, the energy delivered by an actual turbine seems to increase in a straight-line fashion with increasing average wind speed. In fact, based on correlating CFs for a number of actual turbines with simple turbine characteristics, the following very handy relationship has been derived (Masters, 2004).

Eq. 12.10
$$CF = 0.087 V_{avg} - \frac{P_R}{D^2}$$

where CF = turbine capacity factor
V_{avg} = average wind speed (m/s) assuming Rayleigh statistics
P_R = rated power of the turbine (kW)
D = rotor diameter (m)

Be aware that Equation 12.10 is a simple correlation that was not derived from fundamental principles, so the units don't cancel. To make it work, you must use the units specified above. The example in Solution Box 12.3 illustrates its use.

12.6 Economics of Wind Power

One of the key advantages of newer wind turbines is the reduction in capital cost per kW of installed capacity as economies of scale kick in, Reasons for this reduction include the following:

- Cost of a rotor is roughly proportional to its diameter, but power delivered is proportional to diameter squared.

SOLUTION

SOLUTION BOX 12.3

Estimating the Energy Delivered by a Wind Turbine

Let's return to the example of a 2-MW, 80-meter wind turbine in Rayleigh winds with average wind speed equal to 7 m/s. Use Equation 12.10 to estimate its CF and annual electricity production.

Solution:

From Equation 12.10 the capacity factor is

$$CF = 0.087 V_{avg} - \frac{P_R}{D^2} = 0.087 \times 7 - \frac{2000}{80^2} = 0.297$$

From Equation 12.9 the annual energy delivered is estimated to be

$$Energy = CF \times P_R \times 8760 = 0.297 \times 2000 \text{ kW} \times 8760 \text{ hr/yr} = 5.2 \times 10^6 \text{ kWh/yr}$$

Pretty simple.

- Taller towers reach into higher winds, which increases energy faster than tower cost.
- Labor to assemble the components for larger machines is not that much higher than for small ones.
- Planning, permitting, site preparation, and installation costs don't increase much when size increases.
- Servicing large turbines isn't much different from servicing small ones and newer turbines are designed to need less servicing in the first place.

As a result, capital costs for U.S. projects dropped by about 85% in the last two decades to about $1200/kW. Since then, prices have begun to rise slightly in part due to the rising cost of materials (somewhat due to the surge in all types of construction in China), a weakening dollar compared to the Euro (because a significant fraction of turbines are imported), and a shortage of turbines (due to explosive growth in demand). Offshore installations at about $1600/kW are more expensive, but the strength and consistency of winds there can offset that initial cost disadvantage.

There are a great many financial considerations that go into an actual analysis of a wind project, including things such as loan interest rates, the desired return on any equity investment in the plant, tax advantages associated with accelerated depreciation of capital equipment, property taxes and income taxes, and various incentives provided by government agencies and utilities. A special tax incentive, called the *production tax credit* (PTC), has been an especially important and somewhat problematic factor in the economics of wind. The PTC was enacted in 1992 to provide a 1.5¢/kWh, tax credit that would be inflation adjusted over the years (1.9¢/kWh in 2006). Although the credit does provide a significant financial incentive, Congress has several times allowed the credit to expire, causing the boom-and-bust cycle in construction shown in Figure 12.1. Just before each expiration, construction booms as the industry tries to get plants installed in time to take advantage of the credit, after which the industry virtually shuts down while it waits for the next renewal.

To find a levelized cost estimate for energy delivered by a wind turbine, we need to divide annual costs by annual energy delivered. We just learned how to find annual energy, and in Section 9.5.1 we described annual costs in terms of a fixed charge rate on capital along with annual operations and maintenance (O&M) costs. The example in Solution Box 12.4 illustrates how these factors combine to determine the cost of energy (COE).

The 5.5¢/kWh cost of energy from the example wind plant analyzed in Solution Box 12.4 is actually somewhat on the high side. When accelerated depreciation of the capital investment is included, most wind projects are able to make profits while selling electricity on the wholesale market for around 4¢/kWh. For comparison purposes, consider the cost of a natural-gas-fired efficient, combined-cycle power plant. Assuming a heat rate of 7500 Btu/kWh and natural gas at the 2006 price of around $7 per million Btu, the cost of fuel alone would be over 5¢/kWh. The levelized cost of new coal-fired power plants is around 4¢/kWh. The bottom line suggests that new wind plants in good locations compare favorably against any other new generation source. Moreover, wind avoids the risk of future increases in prices for conventional fuels and likely future carbon taxes, which adds a valuable "hedge value" to wind.

Other economic attributes add value to wind plants. Because turbines are compatible with traditional farming and ranching, land lease revenues of thousands of dollars per year per turbine can provide a real income boost to farmers and ranchers, which can make the difference between just getting by and prosperity. Wind farms also add a new and significant contribution to the local tax base, which helps fund local schools, hospitals, and all the other services that county governments provide. Construction of wind farms and ongoing mainte-nance provides local jobs, while adding significantly to the local tax base.

12.7 Environmental Impacts of Wind

Wind turbines have many environmental benefits as well as some negative attributes. On the positive side of course, wind turbines generate electricity without the CO_2, SO_2, NO_x, particu-late matter, and mercury air pollutants that conventional power plants emit in great quantities. The American Wind Energy Association (AWEA) estimates that the development of just 10% of the wind potential in the ten windiest states in the United States would provide more than

Simple Levelized Cost of Wind Energy

Suppose a 2-MW, 80-m turbine is set up in an area with an average wind speed of 7 m/s. Suppose the complete system cost of $2.4 million ($1200/kW) is amortized using a 14%/yr fixed charge rate (FCR). Annual O&M costs are estimated to be $48,000/yr (2% of capital costs). An additional royalty revenue of $5000/yr is to be paid to the local rancher on whose land the turbine sits. Finally, the system is eligible for a PTC of $0.019/kWh. Find the COE.

Solution:

First, we'll amortize the capital cost using Equation 9.4.

$$\text{Annual fixed costs (\$/yr)} = \text{Capital cost (\$)} \times \text{FCR (\%/yr)}$$
$$= \$2.4 \times 10^6 \times 0.14/\text{yr} = \$336,000/\text{yr}$$

Add in the O&M and royalty payments to get an annual cost of the project

$$A = \$336,000 + \$48,000 + \$5,000 = \$389,000/\text{yr}$$

These are the same turbine and wind conditions already analyzed in Solution Box 12.3, which we calculated would deliver 5.2 million kWh/yr. Dividing annual cost by annual energy gives

$$\text{Cost of electricity} = \frac{\$389,000/\text{yr}}{5.2 \times 10^6 \text{ kWh/yr}} = \$0.075/\text{kWh}$$

Reducing this by the production tax credit yields a final

$$\text{COE of } \$0.075 - \$0.019 = \$0.055/\text{kWh}$$

enough energy to displace emissions from the nation's coal-fired power plants and eliminate the nation's major source of acid rain, reduce total U.S. emissions of CO_2 by almost one-third, and help contain the spread of asthma and other respiratory diseases aggravated or caused by air pollution in this country. In addition, wind turbines don't need water for cooling and hence can be located in arid areas without using up that precious resource. Whereas a nuclear or coal-fired power plant consumes through evaporation more than 500 gallons of water per MWh, a wind turbine requires about one gallon (mostly for cleaning turbine blades).

Much of the negative impact of wind turbines has been associated with avian collisions—especially with birds in the Altamont Wind Resource Area about 50 miles east of San Francisco and with bats in Appalachia. These two locations in particular seem to stand apart from most of the potential wind sites in the rest of the country. The northern California wind farms are located along ridges and in mountain passes through which the winds blow and the birds fly. The high concentration of turbines (more than 5000 of them); the fact that many of them are older, smaller machines with blades that spin much faster and reach closer to the ground; coupled with older, lattice towers that provide convenient resting and nesting spots for birds, all contribute to the high mortality rate. Moreover, the year-round abundance of prey in the area leads to an unacceptably high mortality rate of birds the public cares most about—raptors. California, with about 20% of U.S. wind turbine installed capacity, has been estimated to be the site of over 90% of turbine-caused raptor deaths. These circumstances are not common, and wind farms with more sensitive siting and taller, slower turbines have greatly reduced bird death rates.

An interesting comparison of bird deaths caused by wind turbines with other causes of mortality is provided in Figure 12.15. Turbines are thought to cause on the order of 30,000 bird deaths per year; feral and domestic cats kill 100 million each year; and bird collisions with buildings over 500 million.

On the other side of the country, a wind farm in the mountains of eastern West Virginia and another in Pennsylvania have been implicated in the deaths of a large number of bats. Although bats are not perceived by the public in the same way that eagles and hawks are, their deaths are worrisome in part because they have lower reproductive rates than birds and their populations are more vulnerable to blade impacts. A 2005 Government Accounting Office survey indicates that bat death rates may be more of a problem in the Appalachian Mountains,

figure 12.15 Estimated Bird Fatalities Caused by Wind Turbines and Other Lethal Encounters

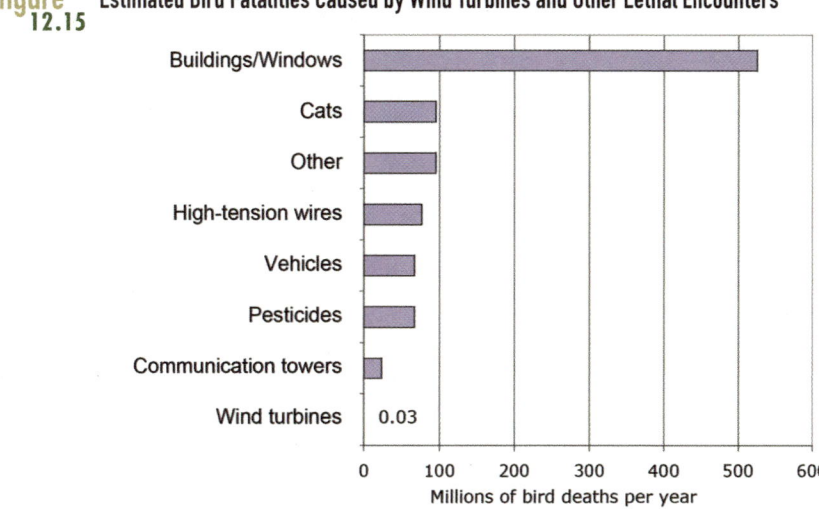

Source: based on Erickson, et al., 2002

where annual fatality rates may be as high as 38 bats per turbine compared with other areas where the numbers range from 0 to 4.3 deaths per year.

Another environmental concern is noise. Again, newer turbines are much better than older ones. Most now are upwind machines, which avoids the thumping sound caused when downwind blades flex as they pass behind their towers. And newer blades have improved aerodynamic characteristics that make them much quieter. For most of us who might stop along the highway to look at a wind farm, turbine noise is insignificant in comparison to the sound of cars and trucks whizzing by. Standing next to a turbine, the sound is a gentle "whup, whup, whup" that is masked in large part by the sound of the wind itself. And a few hundred meters away, turbine sound levels have been compared to that encountered in the reading room of a library.

Perhaps the most troublesome environmental impact of wind turbines is in the eye of the beholder. For some, wind farms are a blight on the landscape, comparable to oil derricks, whereas for others they are welcomed as an elegant, fascinating symbol of a modern, pollution-free energy future. As Martin Pasqualetti (2004) points out, negative attitudes often change dramatically a few years after turbines are installed. In Palm Springs, for example, after thousands of turbines were installed in San Gorgonio Pass just north of the city limits, the community reacted with alarm at the perception that turbines were destroying the aesthetic appeal of the very thing that drew tourists to the area's fancy resorts in the first place. A few years later, influenced somewhat by the financial windfall that local tax revenues were producing, attitudes abruptly changed and wind farms began to pop up in advertisements and postcards as a major tourist attraction.

The major test case for wind aesthetics is being played out in Nantucket Sound where a proposed $500–$700 million Cape Wind project is being proposed that would include 130, 400-foot-tall turbines as close as 5 miles from the shore. The project has split the community with some taking the side of preserving an astonishingly beautiful marine environment (i.e., their view), with others describing it as a classic example of a "Not in My Back Yard" (NIMBY) objection. Perhaps, as Pasqualetti suggests, wind developers should consider a fresh approach and avoid such conflicts, at least for the time being, by focusing their attention on regions where wind would be welcomed as a boon to the local economy—replacing NIMBYs with PIMBYs ("Please, in My Back Yard").

The bottom line for wind power systems is their bottom line. They are essentially pollution-free sources of low-cost electricity that can compete with any other source of electricity.

12.8 Concentrating Solar Power (CSP) Technologies

Although wind systems are the fastest growing renewable energy technology for large-scale generation, there are other solar energy systems that are beginning to enjoy a resurgence of attention as potential electricity sources for wholesale markets. CSP technologies convert sunlight into thermal energy, which is transformed into mechanical power in the form of a rotating shaft that spins a generator, that delivers electrical power. In spite of all of these energy transformations, CSP systems can transform sunlight into electricity with efficiencies that range from about 10 to 20%, which puts them in the same range as photovoltaics.

At the heart of a CSP solar-thermal system is a heat engine (see Section 10.5.2) that takes heat from a high-temperature source, converts some of it to work, and rejects the rest to a cold-temperature sink. Their maximum possible efficiency is therefore related to how hot the source is and how cold the sink is. With the environment providing the cold-temperature sink, the maximum efficiency of these heat engines depends on how hot the sun can make the high-temperature source. Without concentration, sunlight cannot provide high enough temperatures to make the thermodynamic efficiency worth pursuing, but with concentration, it is an entirely different story. Three successfully demonstrated approaches to concentrating sunlight for heat engines have been pursued with some vigor: linear parabolic trough systems, parabolic dish systems with Stirling engines, and heliostats (mirrors) reflecting sunlight onto a heat-exchange boiler located at the top of a tower.

12.8.1 Parabolic Trough Systems

As of 2006, the world's largest solar power plant was a 354-MW parabolic-trough facility located in the Mojave Desert near Barstow, California, called the Solar Electric Generating Station (SEGS). For comparison, the largest wind plant in the United States—the 300-MW Stateline wind farm located along the border between Washington and Oregon—is just a bit smaller.

The SEGS plant consists of nine large arrays made up of rows of parabolic-shaped mirrors that reflect and concentrate sunlight onto linear receivers located along the foci of the parabolas. The receivers, or heat collection elements (HCE), consist of a stainless steel absorber tube surrounded by a glass envelope with a vacuum drawn between the two to reduce heat losses. A heat

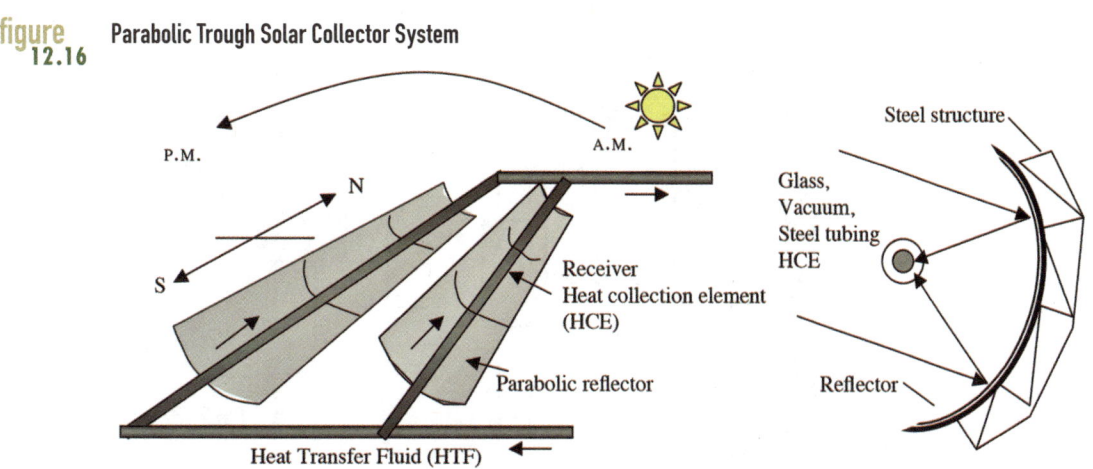

figure 12.16 **Parabolic Trough Solar Collector System**

transfer fluid (HTF) circulates through the receivers, delivering the collected solar energy to a conventional steam turbine/generator. The SEGS collectors, with over 2 million square meters of surface area, run along a north-south axis, and they rotate from east-to-west to track the sun throughout the day. Figure 12.16 illustrates the parabolic trough concept.

The first SEGS array, a 13.4-MW unit, was built in 1985, and the final plant, which produces 80 MW, was completed in 1991. Continuous operation of the last of these units over the past fifteen years has demonstrated their reliability. The overall cost of generation has been around $0.12/kWh, which is about double the cost of conventional thermal power plants.

Whereas solar trough technology has languished for nearly fifteen years, there has been a recent resurgence of interest in constructing a new generation of these power plants. Figure 12.17 provides a simple schematic of their basic design. Solargenix Energy (formerly Duke Solar Energy) has started two solar trough plants, one in Arizona and one in Nevada. The Arizona Public Service Company's 1-MW Saguaro power plant, located about 30 miles north of Tucson, uses sunlight to heat mineral oil to approximately 300°C (Figure 12.18). A heat exchanger transfers that heat to a Rankine-cycle turbine system that is unusual in that it uses a hydrocarbon, pentane, as the working fluid rather than steam. The plant generates approximately 2000 MWh per year, giving it a 23% capacity factor. The $106 million Nevada Solar One plant is a 64-MW facility covering 320 acres of desert land in the Eldorado Valley of Boulder City. It uses a more conventional, steam cycle to generate power.

12.8.2 Solar Dish/Stirling Power Systems

Solar dish/Stirling systems use a concentrator made up of multiple mirrors that approximate a parabolic dish (Figure 12.19). The dish tracks the sun and focuses it onto a thermal receiver, which energizes the hot side of a Stirling engine (Stirling engines were described in Section 10.10). The cold side of the engine is maintained using a water-cooled, fan-augmented radiator system similar to that used in an ordinary automobile. Being a closed system, very little

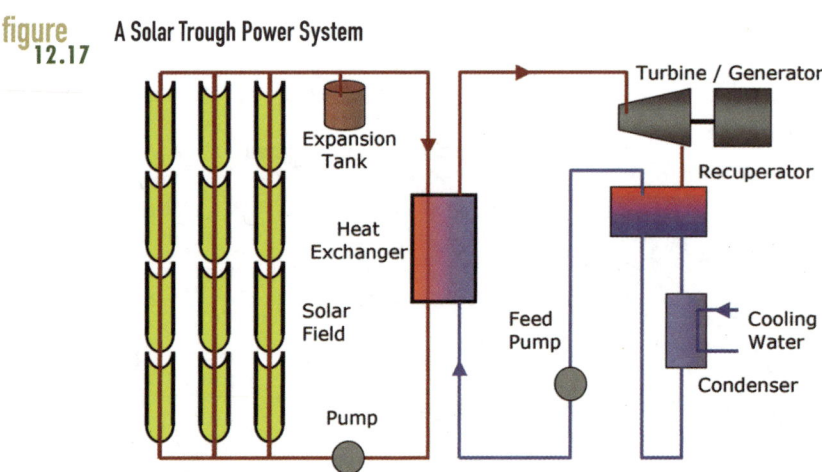

figure 12.17   A Solar Trough Power System

figure
12.18

The parabolic trough and generator for the APS 1-MW Saguaro plant outside of Tucson.

make-up water is required, which means they are ideally suited to sunny desert areas where water constraints are a serious issue. This gives them a significant advantage over other CSP technologies that use turbines and condensers, which require a constant source of cooling water. Moreover, many rapidly growing load centers, including those in southern California, Arizona, and Nevada, are located close to these sunny deserts, so transmission distances could be short. These, of course, are areas where air-conditioning on those hot, sunny days drives up daytime power demands, which is just when these systems are performing at their best.

Dish/Stirling systems are on the verge of commercial reality. At the end of 2005, the California Public Utilities Commission (CPUC) approved plans for two very large solar dish systems to be installed in the deserts of southern California. The current design is based on a set of roughly ninety mirrors forming a dish with a diameter of 38 feet, which focuses sunlight onto a 25-kW Stirling engine/generator (Figure 12.19). These systems have been shown to have overall efficiencies, from sunlight to electric power, of well over 20%, with capacity factors exceeding 25% in sunny desert climates.

A Phoenix-based company, Solar Energy Systems, will own and operate the plants and sell power to Southern California Edison (SCE) and San Diego Gas and Electric (SDG&E). The plan is to build a 20,000-dish, 500-MW plant for SCE in the Mojave Desert near Victorville, and a 300-MW, 12,000-dish system for SDG&E covering about 3 square miles in the Imperial Valley, near Calexico. The first 1-MW increment for the SCE system is expected to be on line by the end of 2008.

figure 12.19   **The SES 25–kW Dish/Stirling Solar Energy System**

Source: courtesy of Solar Energy Systems

With no air emissions and very little water requirement (mostly just to keep the mirrors clean), permitting should be relatively fast. With short construction times, easy permitting, and relatively small, 25-kW modular sizes, which simplifies financing, the time from project design to delivered power can be very short—on the order of just a year or two. Short lead times, and a modular design philosophy, mean that capacity can be added incrementally and power can be delivered while the project continues to expand. The key to the economic viability of these systems is the potential for mass-production economies of scale to bring costs down to the point where power will be produced for less than 10¢/kWh.

figure
12.20

The 11-MW PS10 plant in Spain consists of 624 large (120 m²) heliostat mirrors that focus sunlight onto a boiler at the top of a 115-m tall tower.

SOURCE: Solucar Energia, S.A.

12.8.3 Solar Central Receiver Systems

Another approach to concentrating sunlight for solar thermal power plants is based on a system of computer-controlled mirrors, called heliostats, which bounce sunlight onto a receiver mounted on top of a tower. These are properly described as *central receiver systems*, but they are more often referred to simply as "power towers."

The first significant attempt to generate power using this concept was a 10-MW facility called *Solar One*, built near Barstow, California. Solar One consisted of 1800 heliostats that focused sunlight onto a receiver mounted at the top of a 90-m tower. Water pumped up to the receiver was turned into steam that was brought back down to power a steam turbine/generator. Steam could also be diverted to a large, thermal storage tank filled with oil, rock, and sand to test the potential for continued power generation during marginal solar conditions or after the sun had set. This system was operational from 1982 to 1988, after which it was dismantled. Parts of Solar One were used in a next-generation system, called Solar Two, which operated at the same site until 1999.

At present there are no power towers generating power in the United States, but an 11-MW plant called PS10 (Planta Solar 10) was completed in 2007 near Seville in Spain (Figure 12.20). It is the first of a series of concentrating solar plants planned for Spain, with a total capacity of 300 MW expected by 2013.

12.9 Summary

Large-scale renewable energy systems capable of competing directly with conventional fossil fuel and nuclear power stations have begun to make inroads into the wholesale power markets. The most viable of these are wind turbine systems, which are currently experiencing growth rates of installed capacity of over 25% per year. Global wind resources are abundant and environmental impacts are minimal, but transmission constraints, dealing with wind intermittency, and aesthetic concerns are issues that will need to be resolved if these growth rates are to continue.

CSP systems are also beginning to gain some traction as carbon-free sources of electric power. Some of these systems have been on line for decades, but their costs have been just a little too high for them compete with fossil fuels. As prices of fossil fuels rise and as the likelihood of carbon taxes or other forms of carbon constraints emerges, these technologies may well become significant generation sources in the coming decades.

Sustainable Transportation and Land Use

CHAPTER 13

Transportation Energy and Efficient Vehicles

Transportation is the fastest growing use of global energy and petroleum and a major source of GHG emissions. If we hope to arrest energy demand growth, petroleum dependency, and global warming, we must deal with energy in transportation. Next time you fill up at the gas station, think about these stunning facts:

- Transportation and oil go hand in hand. Transportation relies almost exclusively (96%) on oil, and it uses more than two-thirds of the petroleum products consumed in the United States and more than half of world oil.
- Transportation produces one-third of all U.S. carbon dioxide emissions and is the primary source of urban air pollution.

If the oil-intensive U.S. patterns of transportation, dominated by personal vehicles, are adopted by developing countries, such as China, pressure will continue on oil markets, GHG emissions, and urban air pollution. The line in Figure 13.1 shows the growth in U.S. vehicles per 1000 people since 1900, and the 2002 levels for different countries and regions. For example, in 2002 Canada had the same ratio that the United States had in 1972. As of 2004, the United States, Canada, and Western Europe vehicles per 1000 people have stabilized, but this indicator is growing in all other regions of the world except Africa. There are about 800 million vehicles in the world today. If trends continue, that number could grow to 3.25 billion by 2050, led by China and India, each of which now has a middle-class population exceeding the total U.S. population. Each has a growing auto industry.

Solving energy problems posed by oil and carbon emissions requires more sustainable patterns of transportation energy use. The key factors that drive transportation energy and related air emissions are as follows:

- **Vehicle energy intensity,** measured by efficiency or economy; for example, miles per gallon (mpg)
- **Fuel type,** for example, petroleum-based gasoline or diesel; alternative fossil-fuel natural gas or coal-liquids; renewable biofuels, ethanol, or biodiesel; or electricity

figure 13.1 Growth of U.S. Vehicles per 1000 People, 1900–2002, with 2002 Values for Selected Countries and Regions

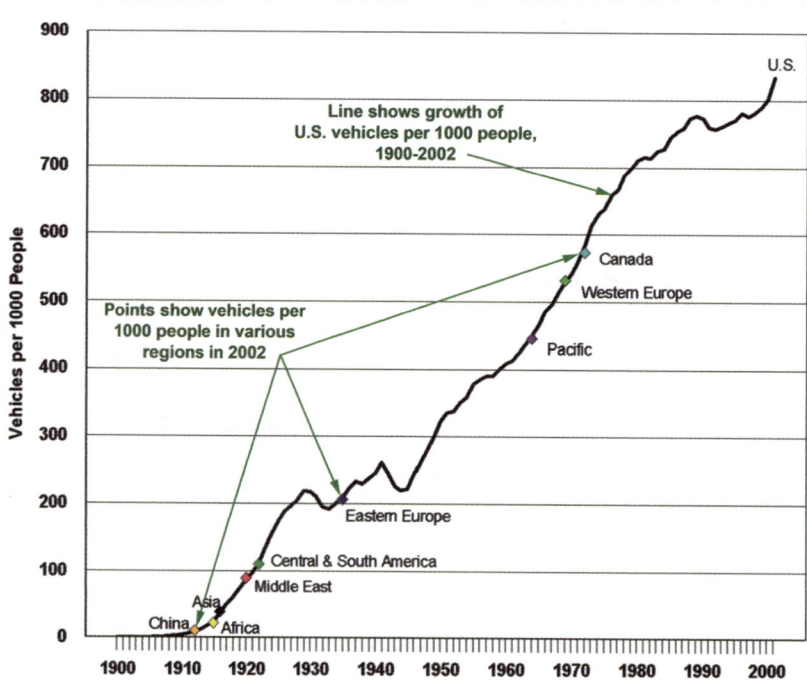

Source: U.S. DOE, 2006

- **Vehicle miles traveled** (VMT), affected by load factor (people per vehicle), distance of travel, and use of other modes (e.g., transit, walking)

U.S. EIA 2007 Annual Energy Outlook to the year 2030 forecasts continued growth in petroleum-based fuels and VMT, small changes in vehicle efficiency, and replacement of some petroleum with alternative fuels, mostly from coal-to-liquids, oil shale, and natural gas-to-liquids (70%) and biofuels (30%). However, other studies show greater opportunities for efficiency and biofuels. For example, the Natural Resources Defense Council (NRDC) estimates that vehicle efficiency, reduction of VMT through smart growth land-use practices, and use of biofuels could eliminate the need for vehicle gasoline by 2050 (see Figure 13.2).

This section of the book describes in greater detail transportation energy use patterns and discusses the roles that vehicle efficiency, modes of transport, alternative and renewable fuels, transit, and land-use and spatial development patterns, can play in transitioning to more sustainable transportation. After reviewing data on transportation energy use, this chapter focuses on vehicle technologies. In the United States and increasingly in the rest of the world, we have a fixation on personal vehicles fueled by petroleum, and any solution must improve vehicle efficiency and replace petroleum with alternative fuels.

figure 13.2 NRDC Estimate of Potential Savings in Gasoline and Transportation Oil Demand, 2005–2050

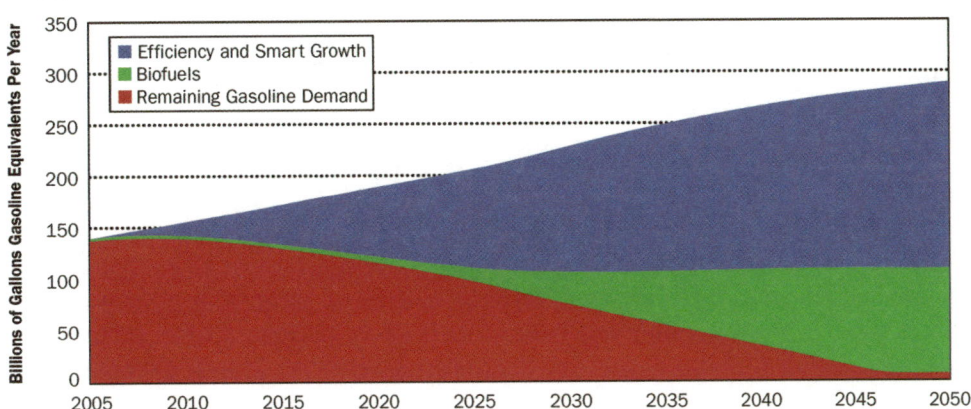

(a) Gasoline savings.

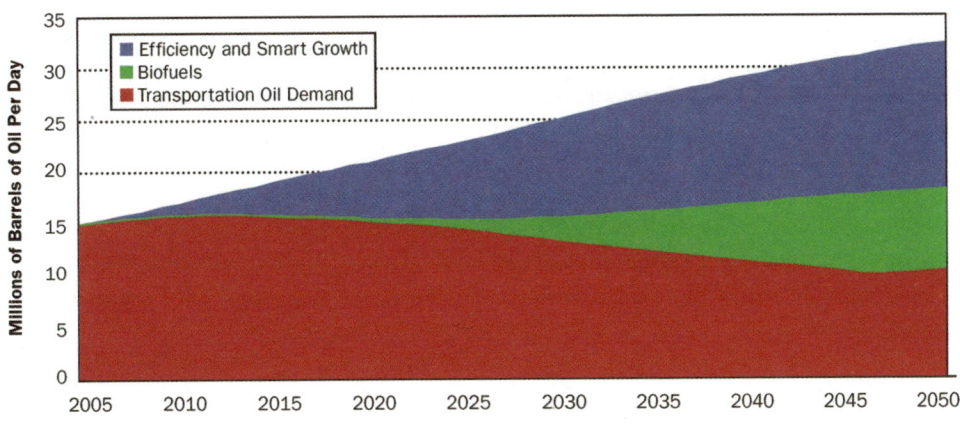

(b) Oil savings.

Potential savings come from: increased use of biofuels, vehicle efficiency, and smart growth, or land-use changes that reduce vehicle miles traveled.

SOURCE: Greene, 2004. Used with permission of NRDC.

Chapter 14 discusses the development of biofuels and other alternative fuels, including cellulosic ethanol and biodiesel from algae. Efficient and alternative fueled vehicles can help reduce per-mile energy and environmental impacts of transportation, but increasing vehicle miles traveled (VMT) can offset those improvements. VMT and land-use trends that drive them are discussed in Chapter 15. That latter chapter also applies our concept of Whole Community Energy to transportation. In the Whole Community approach, we can plan and develop building location and land use to reduce travel distances and enhance transit and pedestrian modes to reduce VMT. In addition, as introduced in Chapter 10, electrification of vehicles can make excellent use of distributed renewable electricity in communities.

13.1 Energy Use in Transportation

Before discussing vehicle technologies and efficiency, it is useful to review patterns of transportation energy use. This can help us to see where we should focus our efforts to reduce oil use and carbon emissions.

13.1.1 Global Transportation Energy Use

Global transportation consumes more than 20% of world energy and it comes almost entirely from oil. It is growing faster than other uses of energy, and U.S. EIA expects transportation energy to increase by 2.5% per year through 2025 when it would be *double* the levels of 2000. Developed countries, including the United States (represented by the OECD nations in Figure 13.3) consumed 56% of global transportation energy in 2002, but they are expected to grow slowly (at 1.3% per year) compared to developing countries (non-OECD) led by China and India. This is where nearly two-thirds of the growth in transportation energy is expected to occur, at an astounding rate of 4.4% per year.

Will we have enough oil and the atmospheric capacity to absorb the carbon associated with these trends? We are already feeling the economic and environmental effects of transportation dependency on oil. It is hard to imagine that we can sustain the expected growth of transportation energy if it follows current patterns of oil use. We must either slow the growth of transportation energy, change to less oil- and carbon-dependent patterns of transportation, or both.

figure 13.3   International Transportation Energy Outlook by Region, 2004–2030

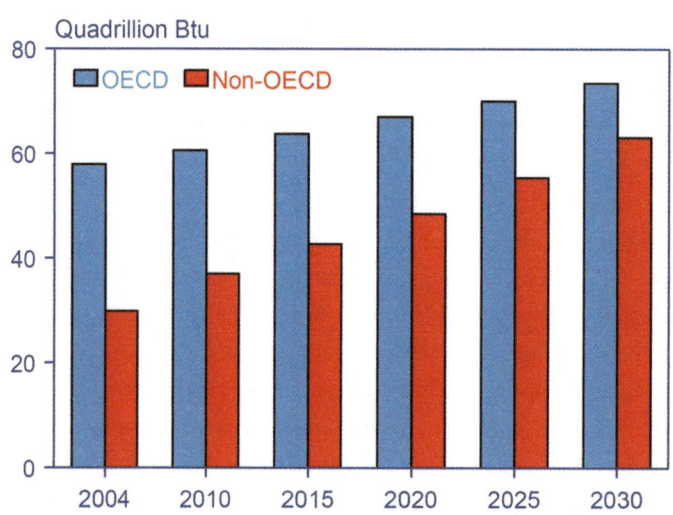

13.1.2 Transportation Energy Use in the United States

In Chapter 1 we saw that transportation is the fastest growing and most oil-dependent energy-consuming sector in the United States (see Figures 1.5[c] and 1.13). Our goal is to arrest these trends: this section provides more detail to help identify opportunities for improved efficiency.

13.1.2.1 Transportation Modes and Energy Use

We use various modes to transport people and goods, including highway vehicles, airplanes, rail, waterways, and pipelines. Let's not forget pedestrian and bicycle transport, but these do not use fuel energy. As shown in Table 13.1, highway (81%) dominates non-highway (19%) transport, and light vehicles (including autos, pickups, and sport-utility vehicles [SUVs]) dominate highway transport with 61% of total transportation energy. Heavy trucks consume 19%, and other modes are small in comparison: air transport (8%), waterborne (5%), pipeline (4%), and rail (2.4%).

Figure 13.4 shows the growth of petroleum use by these modes of transportation to 2004 and projections to 2030. These government projections are an extension of current trends, but are we destined to continue these current patterns of petroleum use for transportation? They are probably not attainable, because they require doubling our current imports of oil, even with large-scale production of nonconventional liquid fuels from coal, biofuels,

table 13.1 U.S. Transportation Energy Use by Mode, 2004

	Trillion Btu	% of Total Transport Energy	1000 Barrels/Day Crude Oil Equivalent
HIGHWAY	21,945	81.1	11,242
Light vehicles	17,217	63.6	8820
Automobiles	9330	34.5	4780
Light trucks	7861	29.0	4027
Motorcycles	25	0.1	13
Buses	193	0.7	99
Medium/heavy trucks	4535	16.8	2323
NON-HIGHWAY	5121	18.9	2623
Air	2348	8.7	1203
Water	1300	4.8	666
Pipeline	822	3.0	421
Rail	659	2.4	338
TOTAL Hwy + Non-Hwy	27,066	100.0	13,866

Source: U.S. DOE, 2007

figure
13.4   Growth of Transportation Petroleum Use by Mode, 1970–2004, with U.S. EIA Projections to 2030

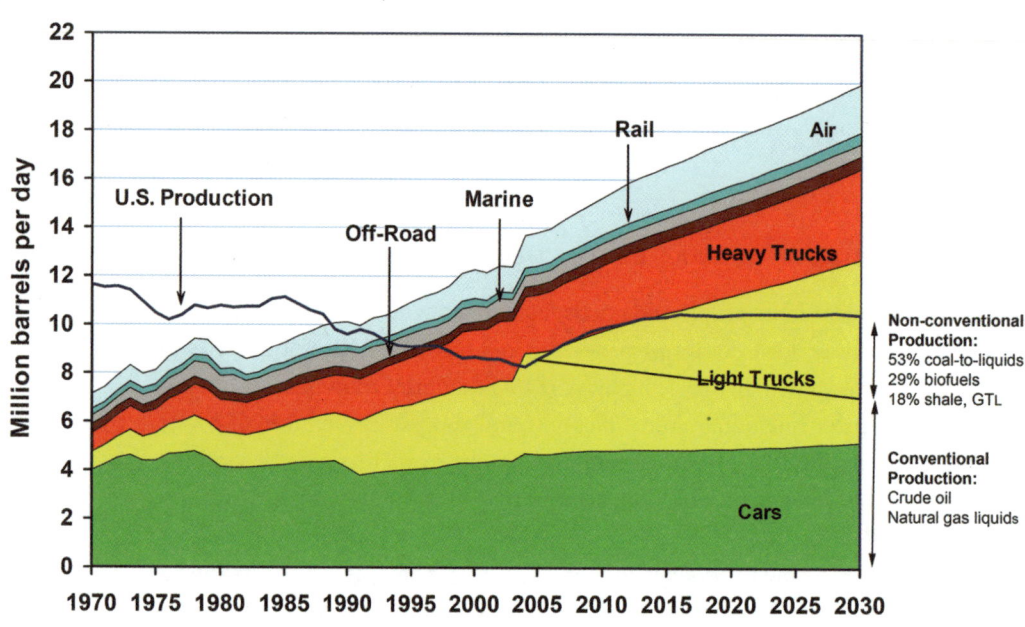

EIA foresees declining domestic petroleum production offset by new sources of nonconventional liquid fuels, especially coal-to-liquids (CTL). Most current and projected growth comes from light and heavy trucks.

Source: U.S. DOE, 2006

and oil shale. We currently don't produce any commercial liquid fuels from coal or oil shale, and doubling imports may have significant economic, trade balance, and geopolitical consequences. On the other hand, other scenarios for transportation energy take advantage of opportunities for efficiency, modal change, and non-fossil alternative fuels.

What do we want from transportation? Our objective is not to consume energy but to move around ourselves and the materials we want. And we want to do that with timeliness, flexibility, convenience, comfort, safety, style, and performance (vroom, vroom). Modes of transport and vehicle markets have shifted to maximize these objectives. And as travel distances have increased due to growing (and sprawling) metropolitan areas, traditional pedestrian and transit passenger travel is less practical. Passenger travel has shifted more to single-occupancy vehicles (SOV) to maximize flexibility, convenience, and style. Passenger vehicles have increased in size to maximize comfort, perceived safety, and style.

Intercity passenger travel in the United States has shifted from rail and car to air, again for added convenience and timeliness. Freight has shifted from rail to heavy trucks as markets have had to become more responsive to customer expectations for timeliness and convenience.

Energy use has increased not only because of modal shift and increased transportation miles but also because of lower load factors. **Load factor** is the quantity per vehicle, such as passenger per vehicle or freight tons per vehicle. Modal shift, load factors, and vehicle miles traveled, as well as market and user preferences, are issues transportation planners have to accommodate in planning and developing transportation systems. Below, we look a little more closely at highway and freight transport before discussing the largest energy user, passenger transportation.

13.1.2.2 Highway Transportation Energy

We saw in Table 13.1 that highway miles traveled are by far the biggest energy user (81%). Highway energy use doubled from 1970 to 2005 (Table 13.2). Overall, highway energy increased 1.9% per year from 1995 to 2005, with the largest growth coming from light trucks (pickups and SUVs; 3.6%/year growth). Consumer preference created a shift from autos to light trucks. Manufacturers encouraged this shift with advertising because bigger vehicles meant bigger profits. As fuel prices rose in 2005–2006, the market fell for larger SUVs, but as gas prices returned to $2 per gallon, the market rebounded. Freight transport has shifted from rail and water to heavy trucks, spurred by market preference for faster delivery. We will see how these shifts have reduced the efficiency of our transportation energy as measured by energy per passenger-mile and energy per freight-ton-mile (see Tables 13.3 and 13.6).

Clearly, highway passenger miles are the biggest energy use in U.S. transportation (> 61%) and we will spend much of this chapter discussing passenger transport and vehicles. Before doing so, the next section looks at the opportunities for increasing energy efficiency of freight transport.

13.1.2.3 Freight Transportation Energy

Passenger travel amounts to about 75% of transportation energy, and freight movement uses about 25%. Table 13.3 compares the three major modes of freight shipments: trucks, water-

table 13.2 Highway Transportation Energy Consumption by Mode, 1970–2005 (trillion Btu)

Year	Autos	Light Trucks	Light Vehicles Subtotal	Motor-Cycles	Buses	Heavy Trucks	Highway Subtotal	Total Transportation
1970	8479	1539	10,018	7	129	1553	11,707	15,402
1980	8800	2975	11,774	26	143	2686	14,629	18,937
1990	8688	4451	13,139	24	167	3334	16,663	21,598
2000	9100	6607	15,707	26	208	4819	20,760	26,268
2005	9140	8108	17,248	27	191	4577	22,043	27,385
Average Annual Percentage Change								
1970–2005	0.2%	4.9%	1.6%	3.9%	1.1%	3.1%	1.8%	1.7%
1995–2005	0.7%	3.6%	2.0%	0.8%	0.4%	1.5%	1.9%	1.6%

Source: U.S. DOE, 2007

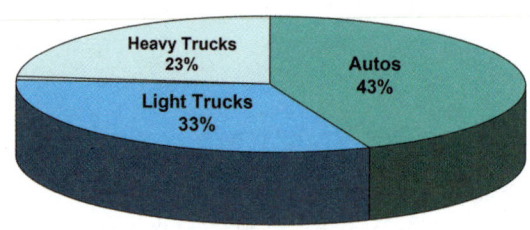

figure 13.5 Highway Transportation Energy, 2005

SOURCE: U.S. DOE, 2006

borne, and Class I railroads. Class I includes the seven major railroads that handle 92% of the nation's rail freight. In 2002, 44% of the Class I tonnage and 21% of revenues came from rail shipments of coal.

Whereas heavy truck vehicle miles and energy use has grown during the last three decades by 3.1% per year (doubled every 20 years), more efficient domestic waterborne ton-miles have declined in the last decade, although foreign waterborne commerce has increased steadily. Rail freight has also had little growth in tonnage over the last three decades but has increased at 2.4%/year since the mid-1990s. The trend from water and rail to trucks is disturbing for energy use, because trucks are seven times less energy efficient than rail and waterborne transport on a Btu/ton-mile basis (Figure 13.6).

13.1.2.4 Freight Transport Efficiency

Heavy trucks consume nearly 19% of total U.S. transportation energy, and their use has grown considerably (Tables 13.1 and 13.2). Freight has moved from rail to trucks because of the flexibility and effectiveness of time-to-delivery. Increasing amounts are moved by air to improve time-to-delivery even more. Consumers have become used to overnight delivery and this has enhanced our economy, but at a cost of more energy for materials transport.

table 13.3 Intercity Freight Movement and Energy Use in the United States, 2005

	Trucks (2003)	Waterborne Commerce	Class I Railroads
Ton-miles (billions)	1051	591	1696
Tons shipped (millions)	4122	1029	1899
Energy intensity (Btu/ton-mi)	3476	514	337
Energy use (trillion Btu)	3653	304	571

SOURCE: U.S. DOE, 2006, 2007

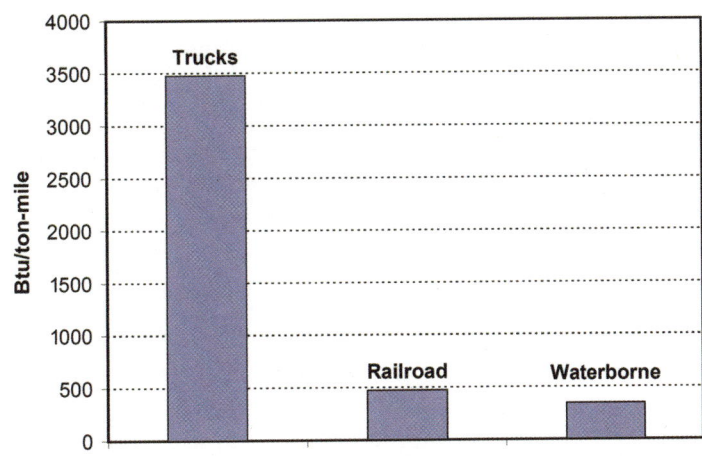

figure 13.6 Freight Energy Intensity, 2003

SOURCE: U.S. DOE, 2006

But there are opportunities for improving the energy efficiency of freight movement. Efficiency studies have demonstrated opportunities for efficiency improvements from aerodynamics and other design changes, hybrid electric drives, idle reductions and other operations improvements, modal shift to rail, and increased use of B-20 biodiesel. Dierkers' (2005) estimates of energy savings as indicated by reduction in CO_2 emissions are given in Table 13.4.

table 13.4 U.S. Freight Truck Potential Efficiency Opportunities to 2025 (Dierkers, 2005)

Option	Technical Potential Fuel and CO_2 Reduction	Additional Penetration	Total Savings	Total MMTCO$_2$ Reduction
Technology Measures				44.4
Aerodynamics	10.9%	50%	5.4%	
Tire base/inflation	3.2	50%	1.6%	
Weight reduction	1.8	50%	0.9%	
Low-friction lubricant	1.5	50%	0.8%	
Drive Train				
Hybrid technology	25%	25%	6.3%	33.3
Operations				50.4
Idle reduction	8.9%	50%	4.5%	
Speed reduction	13.6%	25%	3.4%	
Driver training	3.8%	50%	1.9%	
Modal Shift				
Shift to rail	80%	10%	8%	42.7
Fuel Options				
Biodiesel 20% (B-20)	20%	50%	6.9%	36.8

SOURCE: Dierkers, 2005

13.1.3 Passenger Miles Traveled and Energy Intensity

13.1.3.1 International Variation in Passenger Miles, Modes, and Energy Intensity

The patterns of transportation passenger miles vary considerably throughout the world. Kenworthy (2003) conducted a study of urban passenger transport systems in eighty-four global cities. He grouped the cities into five "higher income" regions and six "lower income" regions. He argued that "automobile cities" such as those in the United States are more vulnerable during the transition to a post-petroleum world.

His results in Table 13.5 reveal that the major factors associated with high energy and CO_2 emissions from passenger transportation do *not* include the wealth of cities. One key factor underlying automobile dependency and energy use is the extent and quality of the public transportation system, especially the amount of service provided by rail. Urban density is also a key factor, as higher densities are associated with higher levels of transit use and non-motorized transport (walking and bicycling), and lower densities are associated with lower levels. Lower densities and reduced transit use are associated with more freeways and parking requirements.

The high U.S. levels of energy use per passenger mile and passenger CO_2 per capita result from travel distances, patterns of land use (urban density), automobile dependency, and low levels of rail transit use. Western Europe and high-income Asia, with comparable levels of urban wealth, have much higher densities and public rail transport use, while lower passenger energy and CO_2 emissions. Cities in lower-income regions have a much higher percentage of public transport passenger miles, but little of this is by efficient rail. The table's 1995 data for China do not reflect the recent changes going on there. Vehicle registrations grew by more than 8% per year in both China and India from 1992 to 2002, and by more than 19% in China from 2004 to 2005. Both China and India (lower income Asia) have much to do to approach the indicators of the higher income countries, but they appear well on their way.

13.1.3.2 U.S. Passenger Miles, Miles, Load Factors, and Energy Intensity

Passenger miles make up most of highway transportation but also include a significant portion of air and about 16% of rail transportation energy. Indicators of energy use for different modes of passenger travel are given in Table 13.6. Automobiles and personal pickups and SUVs dominate in number of vehicles, passenger miles, and energy use (Figure 13.7).

Energy efficiency is given by energy intensity measures of Btu per passenger mile (Btu/p-m) and Btu per vehicle mile. Vanpools and motorcycles have the best Btu/p-m (1294–2270), and autos, personal trucks, SUVs, buses, and overall rail have comparable Btu/p-m (3496–4329). Energy intensity depends not only on vehicle efficiency (Btu per vehicle mile) but also on load factor or persons per vehicle (P/v). The overall efficiency of autos and personal trucks is constrained by their low load factors (1.5–1.7 P/v). The potential efficiency of bus transit is also

table 13.5 Urban Passenger Transport Indicators in Major Global Cities by Region, 1995

Indicator	Units	Higher Income				Lower Income			
		USA*	ANZ	WEU	HIA	EEU	LAM	LIA	CHN
Number of cities in group sample		10	5	32	6	3	3	8	3
Average population of cities	million people	5.7	2.0	2.2	11.0	1.3	7.9	9.7	7.2
Urban density	persons/ha	15	15	55	150	53	75	204	146
Metropolitan GDP/cap†	US$/person	$31K	$20K	$32K	$32K	$6	$5	$4	$2
Length freeway per person	m/person	0.16	0.13	0.08	0.02	0.03	0.003	0.015	0.003
Parking spaces/1000 CBD jobs		555	505	261	105	75	90	127	17
Total reserved public transport routes per urban hectare	m/ha	0.8	3.4	9.5	5.9	10.7	1.2	2.5	0.3
Passenger cars/1000 persons		587	575	414	210	332	202	105	26
Passenger car passenger-km/cap	p-km/person	18,155	11,387	6,202	3.614	2,907	2,862	1,855	814
Private passenger vehicles/road km	units/km	99	73	182	144	169	144	236	117
Average road network speed	km/hr	49	44	33	29	31	32	22	19
Public transport seat-km service/cap	seat km/person	1,557	3,628	4,213	4,995	4,170	4,481	2,699	1,171
% public transport seat-km rail	%	48%	68%	62%	46%	59%	7%	15%	4%
Ratio public/private transport speed	km/hr	0.58	0.75	0.79	1.04	0.71	0.60	0.81	0.73
% non-motorized mode trips	%	8	16	31	28	26	31	32	65
% motorized public mode trips	%	3	5	19	30	47	34	32	19
% motorized private mode trips	%	89	79	50	42	27	35	36	16
% motorized pass-km on public	%	3	8	19	46	53	48	41	55
Private passenger energy/cap	1000 MJ/person	60.0	29.6	15.6	9.6	6.7	7.3	5.5	2.5
Private passenger energy/$GDP	MJ/$1000	1913	1497	489	303	1119	1477	1471	1055
Public transport energy/capita	MJ/person	809	795	1118	1423	1242	2158	1112	419
Energy per private pass-km	MJ/p-km	3.25	2.56	2.49	2.33	2.35	2.27	1.78	1.69
Energy per public pass-km	MJ/p-km	2.13	0.92	0.83	0.48	0.40	0.76	0.64	0.28
Overall energy per pass-km	MJ/p-km	3.20	2.43	2.17	1.40	1.31	1.60	1.20	0.87
Passenger CO_2 emissions/capita	kg/person	4405	2226	1269	825	694	678	509	213

* USA = United States; ANZ = Australia-New Zealand; WEU = Western Europe; HIA = High Income Asia; EEU = Eastern Europe; LAM = Latin America; LIA = Low Income Asia; CHN = China

† Abbreviations: cap = capita; CBD = central business district; GDP = gross domestic product; ha = hectare; hr = hour; kg = kilogram; km = kilometer; MJ = mega-joules, million joules; pass = passenger

SOURCE: Kenworthy, 2003

limited by a low average load factor (8.7 P/v), but commuter rail has a better load factor (32.9) and a lower Btu/p-m (2569; Table 13.6 and Figure 13.8).

13.1.4 Overview

Transportation is a critical sector of energy use because of its dominant use of petroleum and because it is a major source of air pollution and GHG emissions. Transport may conjure

table 13.6 Passenger Travel and Energy Use, 2004

				Energy Intensity			
	Number of Vehicles (thousands)	Vehicle-miles (millions)	Passenger-miles (millions)	Load Factor (persons/vehicle)	Btu per vehicle-mile	Btu per passenger-mile	Energy use (trillion Btu)
Automobiles	136,431	1,699,890	2,668,827	1.57	5489	3496	9331
Personal trucks	80,818	859,902	1,479,031	1.72	7447	4329	6403
Motorcycles	5768	10,122	11,134	1.1	2500	2272	25
Demand response	37	890	930	1	14,952	14,301	13
Vanpool	6	85	541	6.4	8,226	1294	0.7
Bus—Transit	78	2435	21,262	8.7	38,275	4318	93
Air certificated	NA	6071	548,629	90.4	357,750	3959	2,172
Rail	19	1313	31,160	23.7	70,694	2978	93
Intercity	<1	308	5511	17.9	51,948	2760	15
Transit	13	710	15,930	22.4	70,170	2750	44
Commuter	6	295	9719	32.9	91,525	2569	25

Source: U.S. DOE, 2007

up visions of the movie *Trains, Planes, and Automobiles,* but 60% of transportation energy is used just for the latter: highway passenger miles in light vehicles. Compared to the rest of the world, passenger travel in U.S. cities is much more energy intensive because of its greater dependence on automobile use and fewer rail transit options.

figure 13.7 U.S. Passenger Travel Energy Use, 2003

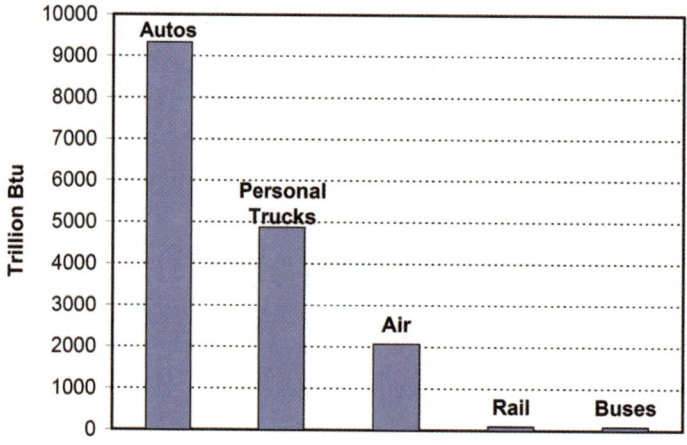

Source: U.S. DOE, 2006

figure 13.8   U.S. Passenger Travel Energy Intensity, 2003

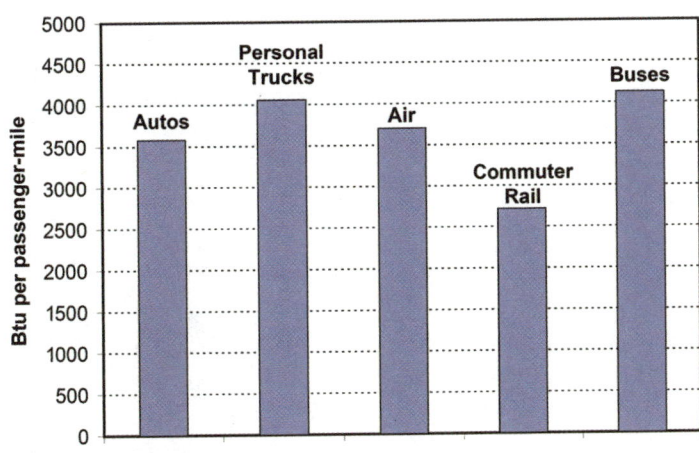

SOURCE: U.S. DOE, 2006

Shifts from smaller to larger vehicles, from rail to truck freight, and from higher to lower load factors create additional challenges as we try to improve transportation efficiency. But the most important factor affecting energy use is light vehicles, and the next section discusses progress and opportunities for their improved efficiency and reduced emissions.

13.2 Highway Passenger Vehicle Technologies, Efficiency, and Emissions

13.2.1 Commercially Available Vehicle Types and Technologies

Before discussing vehicle efficiency and environmental emissions, we need to introduce vehicle types, technologies, and fuels. Nearly all vehicles currently use internal combustion engines (ICE), but new technologies are being introduced to the market. Because of constraints on oil and carbon, future vehicles may turn to electrification, through currently popular hybrid electric vehicles (HEV), all-electric vehicles (EV), and/or fuel cell technologies. We look at emerging technologies in Section 13.2.4.

13.2.1.1 Internal Combustion Engines

Because of their fascination with automobiles, most people are familiar with the workings of internal combustion engines. In these engines, fuel is burned in the engine itself as opposed to in an external combustion engine such as the steam engine. Most cars use the classic **Otto cycle gasoline engine** that takes a mixture of gas and air into a cylinder, compresses it with

figure 13.9 Otto Cycle Gasoline Engine Cylinder, Piston, Valves, and Spark Plug

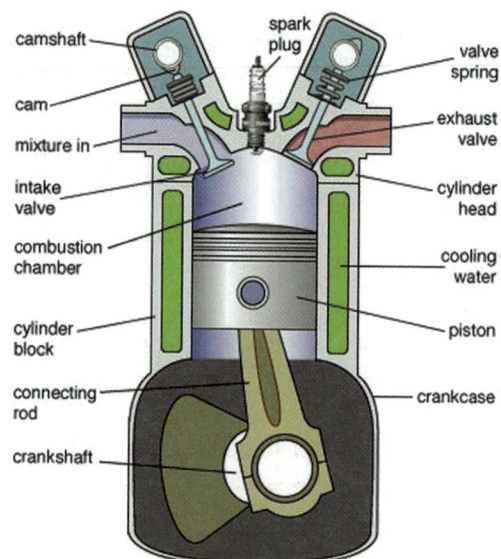

Fuel-air mixture is injected through the intake valve into cylinder at downstroke of piston, then is ignited on upstroke by spark, driving piston down. Exhaust gases are expelled through exhaust valve on next upstroke of piston.

a piston, and ignites it with a spark. The ignition explosion drives the piston down, turning the crankshaft into mechanical rotational motion and that energy is transferred to the rotating wheels (see Figure 13.9).

The **diesel engine** also operates with cylinders, pistons, and rotating crankshaft, but differs from the Otto gasoline engine. It takes air into the cylinder, compresses it, then injects distillate or diesel fuel. The higher compression ratio (piston downstroke to upstroke volume) of the diesel engine (about 15:1 or 20:1 compared to 8:1 or 10:1 for the Otto cycle) heats the compressed air hot enough to ignite the fuel without a spark, driving the piston downward and turning the crankshaft. Diesel vehicles tend to operate more efficiently than gasoline engines and are a mainstay in heavy-duty vehicles. Despite their higher efficiency, diesel engines have been plagued by higher particulate emissions than gasoline engines and have not penetrated the light vehicle market in the United States. However, recent advances by Daimler-Chrysler, Volkswagen, and others in "clean diesel" technologies have spurred sales of diesel cars especially in Europe (see Section 13.2.3).

The efficiency of internal combustion engines varies considerably because of tradeoffs between power performance, engine longevity, compression ratio, and controls of air pollution of exhaust emissions. Engines can operate as hot as 1000 K, giving a maximum thermal efficiency of about 70% assuming ambient temperature of 300 K. Efficiency losses occur as a result of exhaust and water heat losses; friction losses in motor, drive train, and braking; and in-vehicle energy (e.g., lights, air-conditioning). Overall efficiency of fuel energy to transport energy for a typical vehicle is about 20%–25%, meaning only one-fifth to one-fourth of the fuel energy is converted to energy of motion. And we thought electric power plants were inefficient. Of course, of greater meaning to users is the fuel efficiency rating given as miles per gallon of fuel, which we discuss in the next section.

13.2.1.2 Flex-Fuel Vehicles (FFV)

Otto cycle engines run primarily on gasoline, but they can also operate on non-petroleum fuels including compressed natural gas (CNG), liquid petroleum gas (LPG or propane), hydrogen, methanol, and ethanol. Generally, methanol and ethanol are blended with gasoline at various mixtures (e.g., M10 is 10% methanol, E85 is 85% ethanol).

"Flex fuel" Otto cycle engines can run on gasoline or alcohol blends up to M- or E85. Maintaining 15% gasoline helps with cold starts and requires only simple engine modification. The only difference between an FFV and a gasoline engine is an oxygen sensor that measures the amount of ethanol in the fuel at any time, provides this information to the onboard computer, which then adjusts the fuel injector to maximize efficiency and performance. The cost is less than $100, and FFVs generally cost the same as gasoline-only versions of the same model. E85 has a higher octane rating than gasoline so it enhances engine wear and performance, but it has lower energy content so it achieves only about three-fourths of the miles per gallon compared to straight gasoline. This is usually offset by a lower cost. Where E85 is available, gasoline price averages about 15% more per gallon than that of the E85 fuel.

There are sixty-two flex-fuel vehicle models on the market for 2008 in the United States, including mostly larger vehicles and light trucks made by Ford, GM, and Daimler-Chrysler. The primary motivation for manufacturing these vehicles is a credit these companies get on meeting the CAFE efficiency standards. There are about 6 million FFVs on the road in the United States today, and about 700,000 new FFVs have been sold each year. GM announced plans to double FFVs sold, and Toyota announced its plans to market FFVs in the United States by 2008. Brazil ramped up its sales of FFVs from 4% to 70% in just three years.

Although there are many FFVs on the road and more being brought to market, as of 2007, there were only 1200 E85 filling stations out of the 170,000 gas stations in the United States. One-quarter of them were in Minnesota. As a result, most flex-fuel vehicle owners are just filling up with gasoline.

Diesel engines use diesel (distillate) fuel, but they can also run on synthetic diesel derived from vegetable or animal oils, so-called biodiesel. Like alcohol blends, biodiesel is usually mixed with petroleum diesel at a variety of mixtures, from B-2 (2% biodiesel) to B-10, B-20, and B-100. We will see in Chapter 14 that biodiesel production has grown significantly in Europe. The U.S. market, which was essentially nonexistent in 2000, grew from 25 million gallons (Mgal) in 2004 to an estimated 450 Mgal in 2007. Wow! Chapter 14 discusses the significant prospects for bringing more ethanol and other biofuels to market.

13.2.1.3 Hybrid Electric Vehicles (HEV)

One of the most significant recent advances in vehicle technology is the hybrid electric vehicle (HEV). The HEV has an Otto cycle gasoline engine like conventional vehicles but it also has an electric motor drive that works in tandem with the gasoline engine (Figure 13.10). The electric motor is run by a battery bank that is charged by the engine when excess engine power is available. There are three variations of hybrid drivetrains.

- In the **Series** drivetrain (Figure 13.10[a]), the gasoline engine simply charges the batteries and the electric motor drives the car. The Chevy Volt under development uses this technology.
- In the **Parallel** drivetrain (Figure 13.10[b]), the electric and gasoline motors work together to drive the wheels and are coordinated by computer controls and transmission. Honda uses this drivetrain in its Integrated Motor Assist technology in the Civic and Accord hybrids.
- In the **Parallel-Series** drivetrain (Figure 13.10[c]), the electric motor and the gasoline engine operate independently as a dual drivetrain, so that the gas engine can operate at near optimum efficiency and the electric motor can drive the vehicle on its own. Toyota's Synergy Drive uses this technology.

Most hybrids have **regenerative braking** that helps charge the battery. The electric motor normally helps power the car, but when the brakes are engaged, the motor acts as a generator, slowing the car by converting the vehicle's kinetic energy into electrical current that recharges the batteries. Only the Series and Parallel-Series drivetrains can be converted to plug-in hybrids (Section 13.3.1) because they have independent electric drive.

HEVs are the most efficient vehicles sold today. The 2007 Prius with EPA rating of 60 mph (city) and 51 mpg (highway) is the most efficient midsize car, the Honda Civic Hybrid (49 city, 51 hwy) is the most efficient compact car, and the Ford Escape Hybrid is the most efficient SUV (36 city, 31 hwy). EPA changed the way it measures fuel economy ratings for 2008 to reflect faster speeds and acceleration, air conditioner use, and colder outside temperatures. The 2008 Prius is still the leader at 48 mpg (city), 45 mpg (highway).

The market for hybrid vehicles has grown rapidly. For 2008 there are seventeen hybrid models on the market and more are expected. As shown in Table 13.7, U.S. HEV sales, less than 10,000 in 2000, increased to more than 300,000 in 2007, or 2% of all sales. Toyota had a 76 percent share of the hybrid market in 2006, with the Prius commanding 43 percent of all sales.

Future sales of hybrid vehicles will depend on consumer choice for low-impact vehicles, government incentives, and especially gas prices. Estimates range from 5%–6% of U.S. car sales by 2010 (Oak Ridge National Lab) to 30% of sales in 2030 (ExxonMobil). Although these forecasts vary and nobody really knows for sure, one thing is clear: HEVs have begun to capture the market and that market will grow. An important factor in the growth of the HEV market is that current technology is compatible with emerging technologies, especially

**figure
13.10**   Basic Components of Series, Parallel, and Parallel–Series Hybrid Electric Vehicles

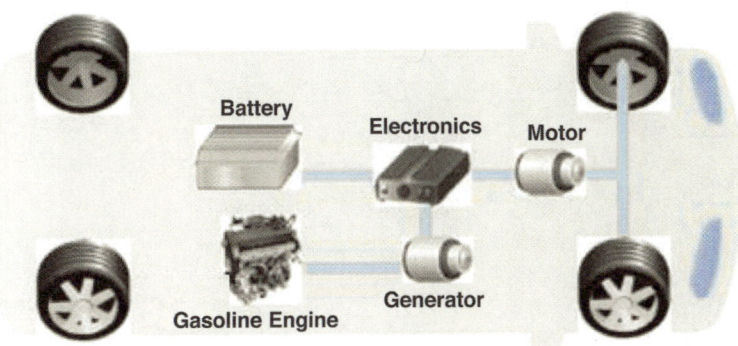

(a) Series hybrid has only electric drive motor and gas engine simply drives generator to charge the battery (e.g., Chevy Volt under development).

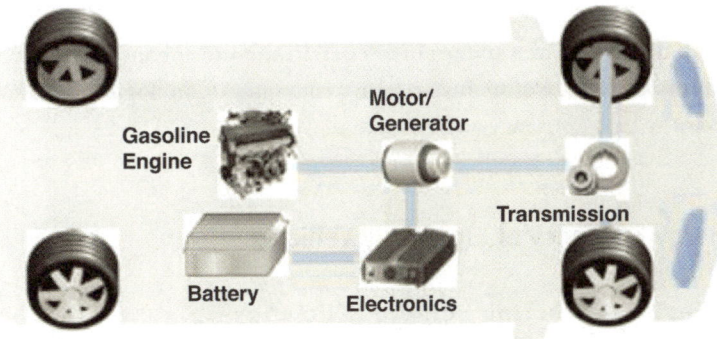

(b) Parallel hybrid has electric drive motor that assists gas-engine drive (e.g., Honda).

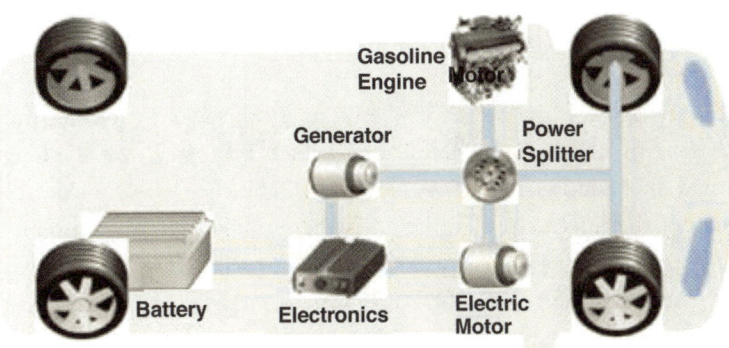

(c) Parallel-Series hybrid has independent gasoline drive and electric drive motors (e.g., Toyota).

SOURCE: Union of Concerned Scientists, 2007

table 13.7 U.S. Hybrid Electric Vehicle Sales, 2000–2007

Year	Number Sold
2000	9350
2001	20,287
2002	35,000
2003	47,525
2004	88,000
2005	210,000
2006	268,000
2007	330,000[p]

p = preliminary

flex-fuel options and, for series and parallel-series drivetrains, enhanced electrification through plug-in hybrids (PHEV). Before investigating emerging technologies the following sections review current vehicle efficiency and emissions.

13.2.2 Energy Efficiency of Light Duty Vehicles

13.2.2.1 Factors Affecting Vehicle Efficiency

Technology, fuel prices, public policy, and consumer choice influence the average efficiency of light duty vehicles. Consumer choice in the United States for larger (and less efficient) passenger vehicles such as SUVs increased considerably in the last two decades. Large SUV sales began to decline in 2005, but despite higher fuel prices, other SUVs, vans, and pickups have increased, and "car" sales dropped below 50% of light vehicle sales for the first time in 2006 (Figure 13.11).

Fuel prices affect travel behavior (discretionary travel and choice of transport mode) over the short term, and consumer choice for vehicle efficiency over the long term. Prices have increased dramatically since 2004 (Table 5.4), and they vary considerably around the world. Much of the international variation is the result of differing gasoline taxes. Figure 13.12 compares mid-2007 gasoline prices and taxes for selected countries in U.S. dollars per gallon. The gasoline portion of the total price is fairly consistent, but taxes vary considerably from $0.38/gallon in the United States to $4.68/gallon in the United Kingdom, where total price exceeded $7.00/gallon.

The lower cost of fuel in the United States has contributed to the patterns of transportation energy, vehicle efficiency, and modal choice relative to other developed countries as demonstrated by the data in Table 13.5. Fuel taxes are an example of public policy that indirectly affects vehicle efficiency and behavior through market forces.

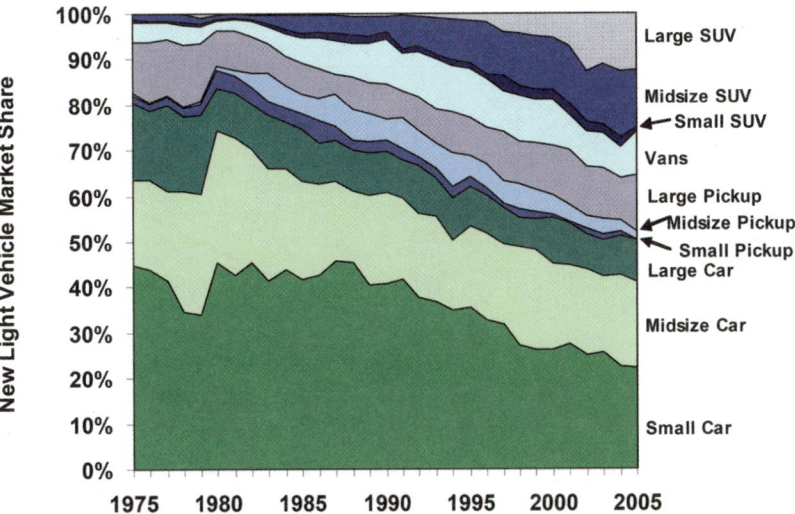

figure 13.11 Changing Market Share for Light Vehicles in the United States, 1976–2005

Market for large SUVs began to slow in 2005–2006, but other SUVs vans and large pickups increased their share despite higher gas prices.

SOURCE: U.S. DOE, 2007

13.2.2.2 U.S. CAFE Standards and Efficiency Trends

Regulatory standards are a more direct policy tool that affects vehicle efficiency. The United States first adopted federal auto efficiency standards in the 1975 Energy Conservation Policy Act, which mandated doubling average 1974 new auto fuel efficiency to 27.5 mpg by 1985. The average efficiency of all cars sold by each manufacturer must meet the Corporate

figure 13.12 Gasoline Price and Taxes, Selected Countries, 2007

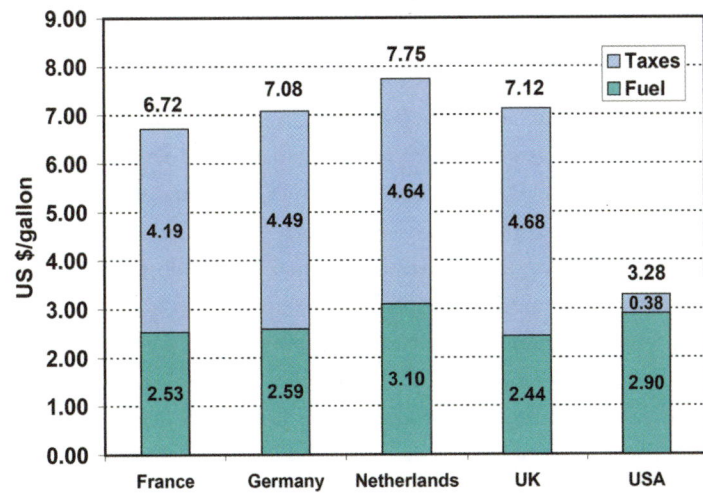

SOURCE: U.S. EIA, 2007

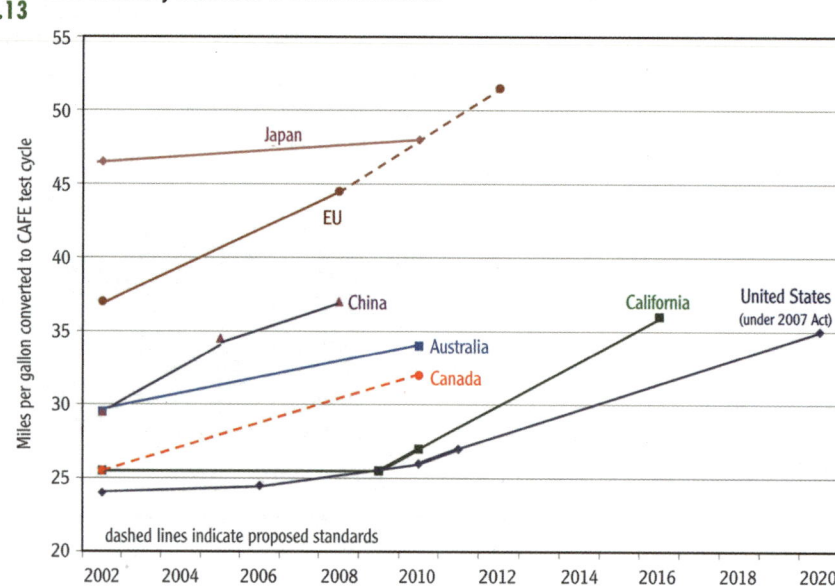

figure 13.13   **Auto Efficiency Standards in Various Countries**

Includes proposed standards, all converted to U.S. CAFE test cycle.

SOURCE: CEC, 2007; data from Pew Center on Global Climate Change, Comparison of Passenger Vehicle Fuel Economy and Greenhouse Gas Emission Standards Around the World, December 2004

Average Fuel Economy (CAFE) standard. The program is administered by the National Highway Transportation Safety Administration (NHTSA), which has authority to raise or lower the standard. It did lower the standard to 26 mpg in 1986–1989 but raised it back to 27.5 mpg in 1990. Figure 13.13 shows that the U.S. vehicle efficiency standards lag behind the standards of many other countries, including China and especially Japan and the EU. California vehicle efficiency implied by its proposed GHG emission standards is also shown, as is the 35 mpg U.S. standard for all light vehicles by 2020 adopted in the 2007 Energy Act.

Congress did not set a target for light trucks but gave NHTSA authority to set a standard at the "maximum feasible" level. Light trucks and SUVs now make up more than half of the passenger vehicle market. In 2006, NHTSA announced higher standards for 2008–2011 rising to 24 mpg in 2011, including the largest SUVs (8500–10,000 lb gross vehicle weight rating [GVWR]) that had not been regulated before. The agency estimates the new standard will save 10.7 billion gallons of fuel. However, in November 2007, the federal 9th Circuit Court of Appeals in San Francisco ruled in a suit brought by eleven states that NHTSA's new light truck/SUV standard was inadequate considering the criteria on which they were to be based and ordered NHTSA to conduct a full environmental impact statement of the standard.

If manufacturers' annual sales do not comply with the standards, they pay a penalty. The penalty is $55 per mpg under the target value per vehicle sold. To determine the penalty the company calculates the average mpg of its sales weighted by volume, subtracts this average from the standard, and multiplies this difference by $55 and the total sales volume. Through 2004, manufacturers paid more than $590 million in CAFE civil penalties. Most European manufacturers regularly pay CAFE civil penalties ranging from less than $1 million to more than $20 million annually. Asian manufacturers have never paid a civil penalty. Automakers get a credit of 0.9 mpg for flex-fueled vehicles, and U.S. makers have taken advantage of this provision.

Have total sales matched the standards? They have but not by much. Table 13.8 gives the automobile standards and average new fleet estimates for autos and autos/light trucks combined for various years. Figure 13.14 plots those values for passenger cars, light trucks, and overall average. New car fleet average and the standards themselves stagnated after 1985. In fact, the overall combined *new* car and light truck average efficiency was lower in 2004 than it was in the mid-1980s. By the way, the new EPA fuel economy testing procedures for 2008 vehicles (that reduced the fuel economy displayed on windshield stickers compared to prior years) does not affect compliance with the CAFE standards because the two tests are different.

These data are for new vehicles, but what about vehicles on the road? On-the-road efficiency for all vehicles increased in 1980 to 1990 from 13 to 17 mpg as new vehicles meeting the standards replaced older, less efficient ones. But on-the-road average stagnated after 1990 because less efficient light trucks and SUVs have a bigger market share. In 2004, on-the-road efficiency for all vehicles was 17.1 mpg (only up from 16.9 in 1991); for autos, it was 22.4 (up from 21.1); for light trucks, 16.2 (down from 17.3); and for heavy trucks, 6.7 (up from 6.0). No wonder people are upset over higher gas prices.

How would improvement in vehicle fuel economy affect our concerns about oil and carbon? We'll see in the next section that improvement in fuel economy directly reduces CO_2 emissions. What about oil imports? Solution Box 13.1 calculates the effect of average light vehicle efficiency on oil consumption and imports. Increasing average efficiency from 22 to 32 mpg would reduce vehicle oil use by 31% and imports by 20%. Increasing to an HEV-equivalent 42 mpg would cut vehicle oil use by half and imports by 30%. If average

table 13.8 Automobile Corporate Average Fuel Economy Standards* versus Sales-Weighted Fuel Economy Estimates, 1978–2004

| Model Year | CAFE Standards | Passenger Cars CAFE Estimates | | | CAFE Estimates Autos and Light Trucks Combined |
		Domestic	Import	Combined	
1978	18.0	18.7	27.3	19.9	19.9
1985	27.5	26.3	31.5	27.6	25.4
1995	27.5	27.7	30.3	28.6	24.9
2004	27.5	29.3	29.3	29.3	24.7

* Standards are in miles per gallon.
SOURCE: U.S. Department of Transportation, NHTSA, "Summary of Fuel Economy Performance," Washington, DC, March 2004.

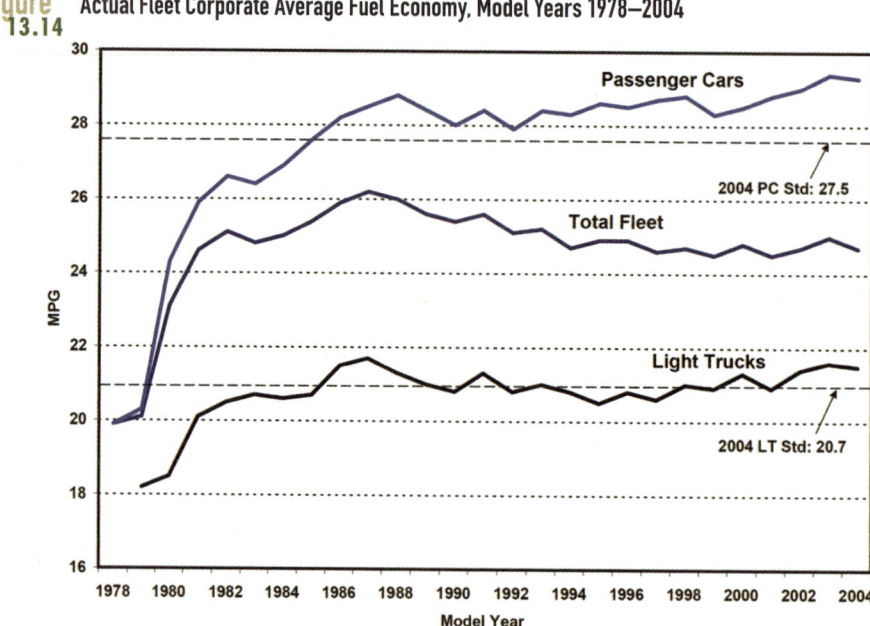

figure 13.14 Actual Fleet Corporate Average Fuel Economy, Model Years 1978–2004

Average light vehicle sold in the United States in 2004 had less than 25 mpg.

Source: NHTSA, 2004

efficiency were as high as emerging technologies plug-in HEV (100 mpg) and flex-fuel PHEV (500 mpg gasoline; see Section 13.2.4), oil imports would drop by 50% and 62% respectively. See also Figure 10.6.

13.2.3 Vehicle Air Emissions: Criteria Air Pollutants and GHG Emissions of Current Vehicles

A great peripheral benefit of higher vehicle efficiency is that it reduces GHG emissions and urban air pollution. Transportation vehicles contribute about half of the nation's NO_x and volatile organic compounds (VOC). These two pollutants combine to form photochemical smog, which is measured by atmospheric ozone (O_3), our most serious urban air pollution problem (see Figure 2.17). The 1970 Clean Air Act (CAA) mandated significant reductions in regulated, "criteria" air pollutants. Congress boldly called for a 90% reduction in vehicle emissions by 1975. Vehicle manufacturers said it couldn't be done, but with a two-year extension, and the development of the catalytic converter, they did it. The CAA also let California, given its severe smog problem and its large auto market, develop its own stricter emissions standards, and gave other states the option to adopt California's regulations. For such standards, California must file a petition to EPA for a waiver from federal preemption under the

SOLUTION BOX 13.1

Passenger Vehicle Efficiency and Oil Imports

If current on-the-road efficiency of light passenger vehicles of 22 mpg were improved to 32 mpg (assuming the same vehicle miles traveled), what quantity and percent of current oil imports could we avoid? What if efficiency were improved to 42 mpg, to 100 mpg (PHEV), to 500 mpg (FF-PHEV) of gasoline?

Current data show that the United States is importing 13.5 million barrels of oil per day (MMbbl/d) and 8.7 MMbbl/d of oil products are used to fuel light passenger vehicles. Let's calculate the fuel it would take for 22 miles under our different mpg scenarios.

Gallons used for 22 miles @ 22 mpg = 22 mi/22 mpg = 1 gal

Gallons used for 22 miles @ 32 mpg = 22 mi/32 mpg = 0.688 gal or

(1 − 0.688)100 or 31% less than 22 mpg

Gallons used for 22 miles @ 42 mpg = 22 mi/42 mpg = 0.524 gal or

48% less than 22 mpg

Gallons used for 22 miles @ 100 mpg = 22 mi/100 mpg = 0.22 gal or

78% less than 22 mpg

Gallons used for 22 miles @ 500 mpg = 22 mi/500 mpg = 0.044 gal or

96% less than 22 mpg

Reduction of oil imports @ 32 mpg = (8.7 MMbbl/d)(0.312) = 2.71 MMbbl/d or

2.71/13.5 = 20% less imports

Reduction of oil imports @ 42 mpg = (8.7 MMbbl/d)(0.476) = 4.14 MMbbl/d or

4.14/13.5 = 31% less imports

Reduction of oil imports @ 100 mpg = (8.7 MMbbl/d)(0.78) = 6.79 MMbbl/d or

6.79/13.5 = 50% less imports

Reduction of oil imports @ 500 mpg = (8.7 MMbbl/d)(0.956) = 8.32 MMbbl/d or

8.32/13.5 = 62% less imports

SOLUTION

CAA. EPA always granted such petitions (until 2007 as discussed below), and California has established stricter emission standards. Several other states have adopted them.

The situation with GHG and CO_2 emissions is not as positive. Transportation vehicles contribute about one-third of the CO_2 emissions in the United States, but EPA has chosen not to include CO_2 among regulated pollutants. Two prominent court cases could affect how vehicle CO_2 emissions are addressed. In the first, *Massachusetts v. EPA*, twelve states and other parties claim that EPA must regulate CO_2 emissions from vehicles. Lower courts gave mixed opinions but generally sided with EPA. In April 2007, the U.S. Supreme Court ruled in a 5 to 4 decision that GHG are air pollutants and remanded the case to the Circuit Court to determine how EPA should regulate them.

The second case relates to California's 2002 Pavley statute (AB 1493) to reduce vehicle CO_2 emissions beginning in 2009 and achieving 30% reduction by 2016. Reducing vehicle CO_2 emissions can be done by higher efficiency, use of lower or non-carbon fuels or biofuels, or new technology. California petitioned EPA for waiver from federal preemption under the CAA, and EPA denied the petition in late 2007 after passage of new federal vehicle efficiency standards for 2020, prompting California to file suit. The *Massachusetts v. EPA* ruling will likely affect California's suit. So will a September 2007 Vermont Federal District Court decision. In *Green Mountain Chrysler Plymouth Dodge Jeep v. Crombie*, the judge rejected all of the industry's claims challenging the validity of the California standards which Vermont is poised to adopt. Two other cases are pending in Rhode island and California.

13.2.3.1 Vehicle Emission Rates and Standards

Let's look at current emission rates for different vehicles. Table 13.9 gives emission rates in grams per vehicle mile traveled (g/vmt) for vehicles on the road (2nd through 4th columns), federal and state emissions standards for new vehicles (5th through 7th columns), and the Toyota Prius, the lowest emission vehicle on the market (8th column) that far exceeds existing standards.

The last two rows give EPA's Air Pollution (AP) score and GHG emission score that it uses to rate "green vehicles." A maximum AP score of 10 is a zero-emission vehicle (ZEV). To qualify for EPA's **SmartWay** class of green vehicles, a car or light truck must have a minimum AP score of 6, a minimum GHG score of 6, and a combined score of 13.

Vehicle emission standards are a bit complicated because of the number of regulated pollutants, the variety of vehicle types, and the categories of emission reduction (e.g., from low-emission vehicle [LEV] to partial zero-emission vehicle [PZEV]). We should know the basic framework of the standards given in Table 13.10 that shows various Bin number categories of emission reduction, comparable California standards, and applicable AP scores.

We should also know that the maximum U.S. emission rates for 2004–2008 passenger vehicles are given as EPA Tier 2, Bin 10 (in Tables 13.9 and 13.10), and that California's LEV II emissions standards, adopted by several other states, are the most stringent in the country, comparable to Bin 5.

How do U.S. standards compare with those of other countries? The United States has been a world leader in pollution control, but some countries now exceed U.S. standards. For example, Table 13.11 gives European Union emission standards for passenger vehicles. They have specific standards for diesel and gasoline fueled vehicles because diesel is a popular and growing fuel for cars there. Standards are given in both g/km and g/mi, the latter so that they can be compared to U.S. standards. The Euro 5 standards proposed for mid-2008 gasoline cars are much more stringent than the U.S. Tier 2 Bin 10 standards.

What about GHG emissions? As discussed above, the United States does not yet regulate CO_2 emissions as an air pollutant, but EPA must address this issue under

table 13.9   Emission Rates for Light Vehicles, Per Vehicle Mile Traveled

	Emission Factors (gs/vmt)						
	Vehicles on the Road			**New Vehicles**			
	Average	Passenger Car	Light Truck	EPA Tier 1 2003 std	EPA Tier 2 Bin 10 2004+ std	CA/NE States LEV II std	2005 Prius
NO_x	1.54	1.39	1.81	0.60	0.60	0.07	0.009
PM-10	0.07			0.10	0.08	0.01	0.01
PM-2.5	0.05						
SO_2	0.09						
CO	23.4	20.9	27.7	4.2	4.2	4.2	1.0
VOC/NMOG	3.06	2.8	3.5	0.31	0.156	0.09	0.004
NH_3	0.09						
CO_2 (lb/mi)	1.0	0.92	1.15	0.9*	0.9*	**	< 0.45
CH_4	0.08						
N_2O	0.03						
Gasoline (gal/mi)		0.047	0.058		0.036		0.018
EPA AP Score	0	0	0	0	1	5	9.5
EPA GHG Score	4	5	2	5	5	**	10

* There is no federal standard for CO_2 emissions, but 0.9 is based on CAFE fuel economy standards.

** California has promulgated a CO_2 emissions standard, and, if approved, 16 other states will also adopt it.

CA/NE states LEV II is California low-emission vehicle standards also adopted by northeastern states.

Pollutants: NO_x = nitrogen oxides; PM = particulate matter; SO_2 = sulfur dioxide; CO = carbon monoxide; VOC/NMOG = non-methane organic gases; NH_3 = ammonia; CO_2 = carbon dioxide; CH_4 = methane; N_2O = nitrous oxide

Source: U.S. DOE, Transportation Energy Data Book; EPA Green Vehicle; fueleconomy.gov; EPA, 2005

direction from the *Massachusetts v. EPA* Supreme Court decision and other pending suits over its denial of California's vehicle GHG emission standard. EPA does recognize that vehicles cause one-third of our GHG emissions and the agency created a scale for the vehicle GHG score based on fuel economy and fuel type to inform consumers of the impact of their vehicle purchases. Shown in Table 13.12, the scale penalizes diesel fuel vehicles with higher minimum fuel economy relative to gasoline vehicles, whereas alternative fuel (E85, compressed natural gas, and liquid petroleum gas) vehicles are credited with lower minimum miles per gallon. EPA provides an online interactive **Green Vehicle Guide** that allows users to find AP and GHG scores for any vehicle on the market. See http://www.epa.gov/green-vehicles/.

Solution Box 13.2 gives an example that shows how we can use these emission rates to assess air pollution and GHG emission impacts of vehicles and driving patterns.

table 13.10   **U.S. EPA Federal Light Duty Vehicle Emissions Standards for Air Pollutants, Tier 2**

Standard	Model Year	Vehicle Types	Emission Limits at Full Useful Life (100,000–120,000 miles) Maximum Allowed Grams per Mile					Air Pollution Score and California Standard Category	
			NO$_x$	NMOG	CO	PM	HCHO	CA Std Cat	AP Score
Bin 1	2004+	LDV, LLDT, HLDT, MDPV	0.00	0.000	0.0	0.0	0.0	ZEV	10
–	2004+	LDV, LLDT	0.009	0.004	1.0	0.01	0.004	PZEV	9.5
Bin 2	2004+	LDV, LLDT, HLDT, MDPV	0.02	0.010	2.1	0.01	0.004	SULEV II	9
Bin 3	2004+	LDV, LLDT, HLDT, MDPV	0.03	0.055	2.1	0.01	0.011	—	8
Bin 4	2004+	LDV, LLDT, HLDT, MDPV	0.04	0.070	2.1	0.01	0.011	ULEV II	7
Bin 5	2004+	LDV, LLDT, HLDT, MDPV	0.07	0.090	4.2	0.01	0.018	LEV II	6
Bin 6	2004+	LDV, LLDT, HLDT, MDPV	0.10	0.090	4.2	0.01	0.018	LEV II option 1	5
Bin 7	2004+	LDV, LLDT, HLDT, MDPV	0.15	0.090	4.2	0.02	0.018	–	4
Bin 8a	2004+	LDV, LLDT, HLDT, MDPV	0.20	0.125	4.2	0.02	0.018	–	3
Bin 8b	2004–2008	HLDT, MDPV	0.20	0.156	4.2	0.02	0.018	SULEV LT	3
Bin 9a	2004–2006	LDV, LLDT	0.30	0.090	4.2	0.06	0.018	–	2
Bin 9b	2004–2006	LDT2	0.30	0.130	4.2	0.06	0.018	–	2
Bin 9c	2004–2008	HLDT, MDPV	0.30	0.180	4.2	0.06	0.018	ULEVII HT	2
Bin 10a	2004–2006	LDV, LLDT	0.60	0.156	4.2	0.08	0.018	–	1
Bin 10b	2004–2008	HLDT, MDPV	0.60	0.230	6.4	0.08	0.027	LEV II HT	1
Bin 10c	2004–2008	LDT4, MDPV	0.60	0.280	6.4	0.08	0.027	–	1
Bin 11	2004–2008	MDPV	0.90	0.280	7.3	0.12	0.032	–	0

Vehicle type: L = light; D = duty; V = vehicle; T = truck; M = medium; H = heavy; P = passenger

CA Cat: E = emission; V = vehicle; Z = zero, P = partial; S = super; U = ultra; L = low; LT = light truck; HT = heavy truck

Pollutants: NMOG = non-methane organic gases; HCHO = formaldehyde; NO$_x$ = nitrogen oxides; CO = carbon monoxide; PM = particulate matter

SOURCE: EPA, 2005

table 13.11	European Emission Standards and Clean Diesel, g/km (g/mi)					
Tier	Date	CO	HC	HC + NO_x	NO_x	PM
Diesel						
Euro 4	2005.01	0.50 (0.80)	-	0.30 (0.48)	0.25 (0.40)	0.025 (0.40)
Euro 5	mid-2008	0.50 (0.80)	-	0.25 (0.40)	0.20 (0.32)	0.005 (0.008)
Gasoline						
Euro 4	2005.01	1.0 (1.6)	0.10 (0.16)	-	0.08 (0.13)	-
Euro 5	mid-2008	1.0 (1.6)	0.075 (0.12)	-	0.06 (0.10)	0.005 (0.008)

13.2.3.2 Vehicle Emission Control Technologies

We have made impressive progress in reducing vehicle emissions as a result of innovative technology. Emissions come from two primary sources: exhaust emissions from fuel combustion and evaporative emissions fuel vapors from refueling, the tank, and the engine. Exhaust emissions include the full range of pollutants given in Table 13.9, whereas evaporative emissions are primarily the volatile organic compounds (VOC).

In gasoline engines, the evaporative emissions are easier to control with a canister system shown in Figure 13.15(b). Vapors from the tank and the engine flow to a holding

table 13.12	U.S. EPA Vehicle Information Program: Greenhouse Gas Score					
		Minimum Fuel Economy: Combined mpg				
Greenhouse Gas Score	Max. lbs CO_2/mile	Gasoline	Diesel	E85	LPG	CNG
10	0.45	44	50	31	28	33
9	0.54	36	41	26	23	27
8	0.64	30	35	22	20	23
7	0.74	26	30	19	17	20
6	0.84	23	27	17	15	18
5	0.94	21	24	15	14	16
4	1.04	19	22	14	13	14
3	1.14	17	20	13	12	13
2	1.24	16	18	12	11	12
1	1.34	15	17	11	10	11
0	> 1.34	< 15	< 17	< 11	< 10	< 11

E85 = 85% ethanol fuel; LPG = liquid petroleum gas; CNG = compressed natural gas
SOURCE: EPA, 2005

Calculating Vehicle Emissions

When I bought my 2005 Prius, my neighbor bought a 2005 4WD Ford Explorer. We both drive our vehicles about 1000 miles per month. How do the vehicles compare in fuel use, fuel cost (at \$3/gal), air pollution emissions, and CO_2 emissions?

Solution:

On the EPA Green Vehicles Web site, I look up Air Pollution and GHG scores for the two vehicles. The Prius has values of 9.5 and 10, and the Explorer has values of 2 and 2, respectively. Tables 13.11 and 13.12 give emission rates for these scores. The rates in grams or pounds per mile can be multiplied by 12,000 miles per year to give annual emissions.

For example, the Explorer NO_x emission rate is taken from Bin 9a in Table 13.12 as 0.30 g/mi, and the CO_2 emissions rate is taken from Table 13.13 as 1.24 lb/mi.

Explorer annual NO_x emissions = 0.30 g/mi × 12,000 mi/yr = 3600 g/y = 3.6 kg/yr

Explorer annual CO_2 emissions = 1.24 lb/mi × 12,000 mi/yr
= 14,880 lb/yr × 1/2.2 kg/lb = 6763 kg/yr

The full results are given in the table below.

	AP Score	GHG Score	Mpg	CO_2 (kg/yr)	NO_x (kg/yr)	NMOG (kg/yr)	CO (kg/yr)	Gas (gal/yr)	Gas Cost (\$/yr)
Explorer	2	2	16	6763	3.6	1.10	50.4	750	2250
Prius	9.5	10	55	2182	0.1	0.05	12.0	218	654

Not only do I enjoy economic benefits from a gas bill 70% less than my neighbor's, but I take pleasure in living more lightly on the planet: my Prius emits 68% less CO_2 and 95% less urban air pollutants than his Explorer.

canister and then are burned in the engine. Exhaust emissions are more complicated. In gasoline engines, it is difficult to control both NO_x and volatile hydrocarbons and CO. NO_x is basically burnt air (air is 78 percent nitrogen) and results from lean fuel mixtures with more air. A richer mixture produces less NO_x but more volatile hydrocarbons and CO from incomplete combustion. The invention of the catalytic converter (CC) helped solve this conundrum. As shown in Figure 13.15(a), the engine can be run rich to control NO_x and then remaining unburned volatile hydrocarbons and CO are combusted in the CC. An oxygen sensor in the exhaust stream can provide data to fine-tune the fuel mixture to optimize the operation.

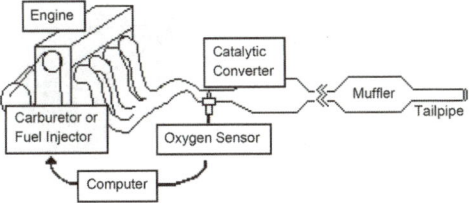

figure 13.15   **Basic Controls for Exhaust and Evaporative Emissions**

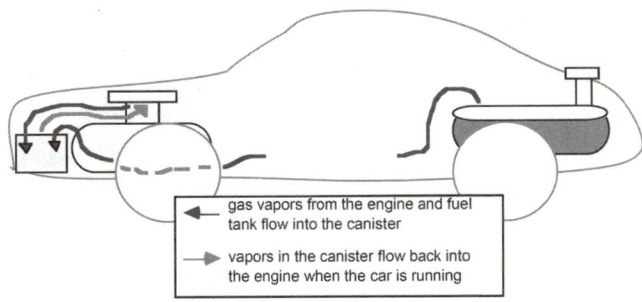

(a) Typical catalyst system for exhaust emissions.

(b) Typical canister system for evaporative emissions.

SOURCE: U.S. EPA, 1994

Diesel engines offer additional challenges because of particulate and sulfur emissions. Emissions from heavy diesel trucks have been blamed for public health effects. EPA has initiated a National Clean Diesel Campaign to reduce diesel emissions. For light vehicles, diesel engines are more energy efficient, and if emissions could be controlled, diesel could be a more effective alternative to our energy and emissions challenge. As mentioned earlier, the Europeans seem to be banking on "clean diesel." Volkswagen and Daimler-Chrysler use "BlueTec" technology combined with low-sulfur diesel fuel to meet clean-diesel Euro 5 standards. Figure 13.16 shows that after the diesel exhaust passes through a particle filter, the BlueTec system adds urea to the exhaust from a separate "AdBlue" tank. The urea releases ammonia that converts NO_x in harmless N_2. The system also uses a conventional catalytic converter (SCR-Kat in the figure). These companies hope that clean diesel can find a strong U.S. market.

13.3 Emerging Vehicle Technologies

13.3.1 Plug-In Hybrid Electric Vehicles (PHEV)

Hybrid electric vehicles can improve efficiency considerably, but they are still dependent on gasoline for energy. Flex-fuel hybrids are under development, and should be on the market soon. Another option is to enhance the battery capacity of a series or parallel-series hybrid

figure 13.16 BlueTec System for Clean Diesel

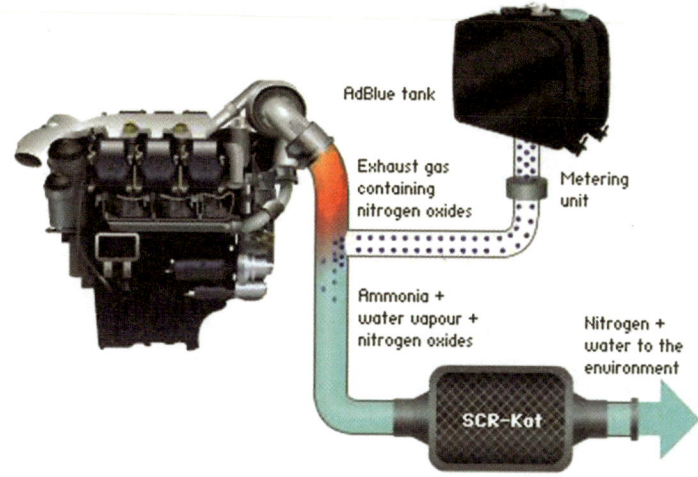

Source: Daimler-Chrysler

(Figure 13.10), use grid power to assist in charging the battery, and make greater use of the electric motor drive. These so-called plug-in hybrids (PHEV) offer certain advantages:

1. With greater use of the electric drive, the vehicle uses less gasoline per mile than conventional HEVs. Plug-in Prius retrofits can easily achieve 100 mpg.
2. With a battery capacity of 5 kWh and all-electric travel at 200 Wh/mile, a fully charged PHEV has a range of 25 miles at a cost of 50¢, about one-fourth the cost of an efficient gasoline car and about one-third the cost of a hybrid car (see Solution Box 13.3). Half the cars on the road today drive 25 miles/day or less.
3. With use of the electric drive in city driving, the PHEV is a zero-emission vehicle (ZEV) that can reduce emissions and improve urban air quality.
4. PHEV can be easily adapted to flex-fuel option with the added advantage of further offsetting petroleum use with E85.

Of course, PHEV batteries must be charged with grid power, and we know that electricity is high-value energy that is usually inefficient to produce and has its own environmental impacts. Are we just trading in one energy problem for another? Isn't electricity the most expensive form of energy we have? Well, PHEVs have significant benefits for the power grid:

1. PHEVs can be charged by grid power at night during off-peak hours when grid capacity is idle and base-load power is available. At off-peak rates, this power can be very inexpensive. As we saw in Figure 10.5, 40% of California's auto VMT could be met by night-charged electricity without needing additional power plant capacity.
2. PHEVs can be charged by excess power from rooftop photovoltaics (your garage roof is your filling station [Solution Box 13.4]), wind power, or other renewable

electricity. They may offer a significant opportunity for our growing wind electric capacity. As discussed in Chapter 12, one disadvantage of wind power is that it is intermittent and cannot be programmed to meet the peak demands when the grid needs the power the most. A fleet of PHEV (and/or all-electric or battery electric vehicles [BEV]) could provide a ready market for grid wind power whenever it is produced.

3. As we discussed in Chapter 10 (section 10.3.2), a large fleet of PHEV and BEV enables a vehicle-to-grid (V2G) power system, where batteries in electric vehicles (charged primarily at night) can provide a bank of electricity storage for the grid when they are parked and plugged in at parking ramps and lots during the day when peak power is needed.

Although PHEVs are not yet available in new vehicles, they are likely to be on the market soon. Daimler-Chrysler, Ford, Toyota, and others are actively developing plug-in hybrids. GM's Chevy Volt is a concept plug-in series hybrid with a small gasoline motor that simply keeps the batteries charged for long trips. There is already a fledgling market for retrofitting HEV to PHEV, which indicates that converting or adapting existing parallel-series hybrids would be straightforward, if not inexpensive at the moment. *EDrive Systems* and *Hymotion* are two firms that are offering plug-in retrofit packages for Prius and Ford Escape hybrids (Figure 13.17).

Both use lithium ion (Li-ion) batteries either as replacement or in addition to the standard nickel metal hydride (NiMH) batteries. The keys to effective batteries for both PHEV and BEV

figure 13.17 **Hymotion L5 Lithium Power Specifications**

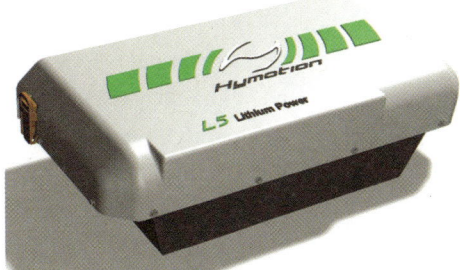

(a) Plug-in Hybrid Priuses being tested at Argonne National Lab.

(b) *Hymotion*™ retrofit lithium-ion battery package for HEV to convert them to PHEV.

Battery Type:	Lithium Polymer
Energy:	5kWh
Charge temperature range:	-10 deg C to 35 deg C
Operation temperature range:	-20 deg C to 45 deg C
Charge Voltage:	120V/240V (15A circuit)
Charge Time:	5.5hrs/4.0hrs
Weight:	72.5 Kg

Source: Argonne National Lab, 2007; Hymotion, Inc., 2007

SOLUTION

SOLUTION BOX 13.3

Electric-Drive Vehicles: Gas-Equivalent "Price per Gallon" and CO_2 Emissions

What are the cost and CO_2 per 25 miles of a grid supplied plug-in vehicle vs. a gasoline supplied 37.5 mpg vehicle?

Solution:

Assumptions:

- Electric drive: 200 Wh/mi = 5 mi/kWh
  = 37.5 mi/7.5 kWh
- Gasoline drive: 37.5 mi/gal
- CO_2 emissions
  - Electricity: 1.4 lb/kWh (U.S. average, see Table 5.7)
  - Gasoline, auto: 37.5 mpg = 0.54 lb/mi (Table 13.13)

Gasoline cost per 37.5 mi = 299¢/gal = 299¢/37.5 mi
Electricity cost = 10¢/ kWh × 7.5 kWh/37.5 mi = 75¢/37.5 mi

CO_2 emissions gasoline = 37.5 mi × 0.54 lb-CO_2/mi = 20 lb CO_2
CO_2 emissions electric = 37.5 mi = 7.5 kWh × 1.4 lb/kWh = 10 lb CO_2

are low cost and low weight. The laptop computer industry has helped advance lithium ion battery technology to reduce both weight per Wh and cost. It is expected that increased production will lead to further technical improvements and cost and weight reductions (see Figure 10.4).

Solution Box 13.3 demonstrates the potential economic and environmental benefits of plug-in electric drive vehicles. The figure from *EPRI Journal* (Sanna, 2005) gives a gas-equivalent price of 75¢ per "gallon" for plug-in electric vehicles. The box gives assumptions and calculations for this price to hold up. Assuming gas at $2.99/gal and electricity at 10¢/kWh, an electric-drive vehicle would operate at only **one-fourth the cost** of an efficient 37.5 mpg gasoline car. Because there would be no CO_2 emissions from the vehicle tailpipe, it would have **half the CO_2** emissions as gasoline vehicles, even when assuming the plug-in electricity comes from average U.S. power plants (i.e., 52% coal).

Plug-in electric drive vehicles can take advantage of renewable solar and wind power. These intermittent sources crave a storage system, and a fleet of vehicle batteries could provide it. One vision of the future would have our garage rooftops turned into solar recharging stations. Solution Box 13.4 shows that a south-facing garage roof size photovoltaic array (150–225 ft² depending on location) is sufficient to charge a PHEV (or all-electric vehicle)

SOLUTION BOX 13.4

Sizing a Rooftop PV Array to Charge a Plug-In Hybrid

How much roof area dedicated to a PV system does it take to produce equivalent electricity for a PHEV or BEV?

Solution:

It depends where you live, how much sun you get, and how much you drive. For example, Table 11.1 tells us that Atlanta has an annual average of five hours of full sun per day (or average insolation of 5.0 kWh/m²/day) on a stationary south-facing collector at Lat –15° tilt. To produce 45 miles of driving per day at 200 Wh/mi requires 45 mi/day × 200 Wh/mi = 9000 Wh = 9 kWh/day.

Using the method in Solution Box 11.2, assuming a 0.75 de–rating factor,

$$\text{PV } P_{DC,STC} = \frac{9 \text{ kWh/day}}{0.75 \times 5.0 \text{ hr/day}} = 2.4 \text{ kW}$$

Assuming 14% efficiency and a cost of \$4/W $P_{DC,STC}$ (current California price with rebates),

$$\text{Area (ft}^2) = \frac{P_{DC,STC} \times 10.75 \text{ ft}^2}{1 \text{ kW/m}^2 \times \text{m}^2} = \frac{2.4 \times 10.75}{0.14} = 184 \text{ ft}^2$$
$$\text{Cost} = \$4/\text{W} \times 2400 \text{ W} = \$9600$$

The table shows the rooftop area and cost needed for such a photovoltaic array in different cities with the assumptions given.

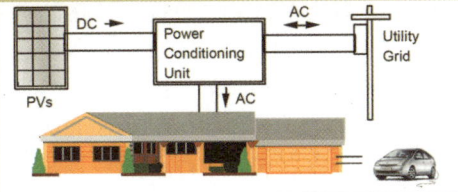

CITY	Hr/day 1-sun	kW STC	AREA (ft²)	PV (\$)
Atlanta	5.0	2.40	184	\$ 9,600
Boston	4.5	2.67	205	\$ 10,667
Boulder	5.4	2.22	171	\$ 8,889
Los Angeles	5.5	2.18	168	\$ 8,727
Madison	4.1	2.93	225	\$ 11,707
Phoenix	6.4	1.88	144	\$ 7,500

Small 1-car garage 14'x22' = 310 ft²

ASSUMPTIONS:
- 9 kWh/day from grid
- PV kW STC-to-Grid AC efficiency 75%
- PV Solar-to-kW efficiency 14%
- PV south-facing, Lat-15° tilt
- PV @ \$4/W dc STC

Area and Cost of Photovoltaic Array for a Plug-In Hybrid

for travel of about 45 miles per day. Of course, such a system would be grid-connected so that the PV system would mostly feed the grid during day and the grid would charge the vehicle at night. A utility might pay more for on-peak power than it would sell off-peak power, so the PV garage with a time-of-day meter would allow users to "buy low, sell high."

13.3.2 Flex-Fuel Plug-In Hybrid Electric Vehicle (FF-PHEV)

As we noted earlier, both flex-fuel vehicles and PHEVs offer significant advantages by themselves. When combined, they offer dramatic possibilities. PHEVs can use a flex-fuel engine that can run on gasoline or E85 ethanol blend. This would further reduce CO_2 emissions and reduce oil use and imports to the equivalent of 500–700 mpg of gasoline. Now we're talking! Solution Box 13.1 showed that such a FF-PHEV would use 95% less gasoline per mile than an average car on the road today.

The technology for the FF-PHEV is readily available today, and a flex-fuel option should be available on PHEVs when they hit the market in a year or two. When they do, the benefits of FF-PHEV will be constrained by ethanol production and the limited number of E85 filling stations, except in Minnesota, Iowa, and Illinois. However, as discussed in Chapter 14, these limitations may change.

13.3.3 Battery Electric Vehicles (BEV)

If plug-in systems and electric drive motors are so good for PHEV, why not just skip the hybrid part and go all electric? The battery all-electric vehicle (BEV) is not new; in fact, BEVs outnumbered gasoline cars 10 to 1 in the 1890s. But it has had fits and starts ever since and has never really captured a market. Some of the big automakers have developed concept cars and some sales, especially after 1990 when California mandated that 10% of cars sold there be ZEV by 2003. GM began producing its EV-1 in 1996 and Toyota sold the RAV4-EV in 2002–2003. However, California weakened its ZEV mandate in 2003, allowing credits for non-ZEV vehicles, and the ready market for BEVs dried up; GM and Toyota stopped production of their EVs that year.

The biggest constraints to BEVs have been cost and weight of batteries and slow recharge times that work against the U.S. driving culture to "fill 'er up" and go. But recall from Chapter 10 (Section 10.3.2), advances in Li-ion batteries have achieved an energy density of 180 Wh/kg, double that of NiMH batteries, with prospects for even higher densities and lower weight (Figure 10.4).

There has been renewed interest in BEVs in recent years as a result of surging gasoline prices and growing concern about carbon emissions. In 2007, GM announced it will get back into the BEV market by 2010: its plug-in Volt is actually a series HEV because it will have a small gasoline engine to charge batteries on long trips. But by then it may lag behind a number of entrepreneurial start-up companies which are developing high-performance BEVs using advanced motors, batteries, and controls. Although these vehicles are now at the (very) high end of the market, the experience gained is likely to spread to more affordable production models.

Leading these companies is Tesla Motors, which unveiled its Tesla Roadster EV in 2006 to high acclaim (Figure 13.18). The stylish, high-performance two-seater boasts 0 to 60 mph in 4 seconds, 250 miles per charge at about 1¢ per mile, the equivalent of 135 mpg, and one-third the carbon emissions of a Prius. The fuel economy and emissions are based on well-to-wheel studies assuming natural gas combined-cycle electricity generation (see Section 13.4). The two-gear transmission, watermelon-sized 70-pound motor, power electronics module that can control over 200 kW during peak acceleration, and the modular battery energy storage system (ESS), make this one of the simplest vehicle technologies.

The heart of the vehicle is the ESS. It consists of 6800 lithium-ion (Li-ion) cells, each just a bit larger than a AA battery, making a 1000 lb (450 kg) battery bank. The ESS can be fully charged in 3.5 hours, usually overnight. It has a 250-mile range, and is designed for 500 full charge-discharge cycles. Thus the ESS is estimated to last 125,000 miles before replacement. The 250-mile range, easy recharging, and high performance output of the ESS helps the Tesla Roadster stand apart from other BEVs. So does its price—at $92,000 it is targeted at the high-end sports car market. But the lessons Tesla Motors is likely to learn in coming years may have lasting effects on the broader vehicle market.

figure 13.18   **Tesla Roadster Electric Vehicle**

High performance (0 to 60 mph in 4 seconds); 250-mile battery range from $3\frac{1}{2}$ hour charge; 1¢ per mile; 135 mpg energy equivalent.

SOURCE: Tesla Motors www.tesla.com

13.3.4 The Hypercar™: Ultralight Composite Materials and Vehicle Efficiency

Independent of propulsion system, all vehicles can be made more efficient if they were lighter. Conventional vehicles consume 7–8 units of fuel energy to deliver 1 unit of power to the wheels; the ratio for hybrid vehicles is 3–5 to 1. Thus, reducing the power needed at the wheels can result in 7–8 times the fuel savings. Power at the wheels is needed to overcome drag, rolling resistance, and weight, with weight requiring two-thirds to three-fourths of the fuel consumption of a typical midsize sedan. Amory Lovins maintains that "contrary to folklore, it's more important to make a car light and low-drag than to make its engine more efficient or change its fuel" (RMI, 2004). And of course, these measures of lightness, low-drag, engine efficiency, and renewable fuel are not mutually exclusive.

In 1999, Lovins founded *Hypercar, Inc.* to support the transition of the auto industry to higher efficiency vehicles. The key element of its design concept, incorporated in its 2000 concept car *Revolution*, was the use of ultralight carbon composite materials (Figure 13.19). These materials have been used in some auto body parts and other lightweight applications such as airplanes and high-performance vehicles, but they have long been considered too expensive to replace steel in typical cars. Although they are ultralight, carbon composites can also be ultrastrong, so there is no sacrifice in safety.

Lovins suggests that the concept car would triple the efficiency of a comparable steel car. Because of its reduced weight, the concept car could use a smaller, lighter engine, and use efficient engine systems such as hybrids or fuel cells much more effectively. Lovins argues that the lessons of ultralight materials can be easily transferred to other non-automotive markets once prices drop.

figure 13.19 **Hypercar *Revolution***

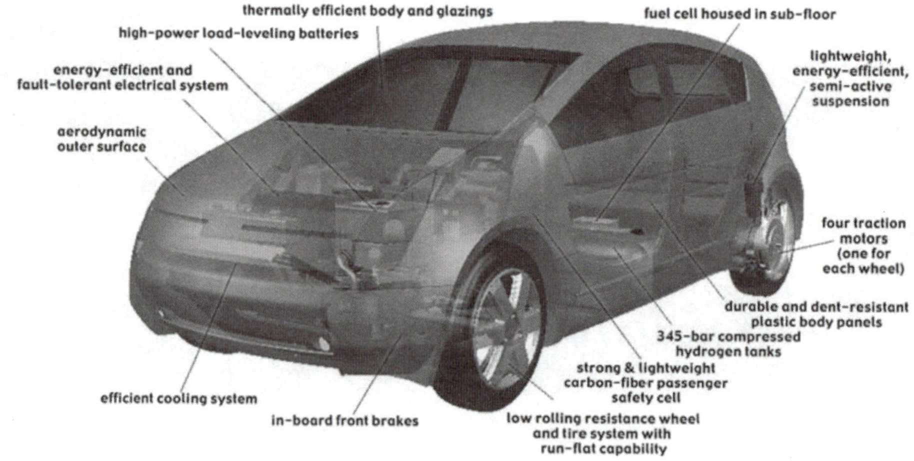

Ultralight composite materials, aerodynamic design, and fuel cell electric drive create an efficient, oil-free car.

SOURCE: Lovins, 2004. Used with permission.

figure
13.20

Fiberforge carbon-fiber panels reduce vehicle weight and improve efficiency.

SOURCE: Fiberforge, Inc. Used with permission.

The key to this revolution in the vehicle industry is the improvement in cost of manufacturing the composite materials. As a result, since 2002, Lovins has focused not on vehicle design, but on composite materials manufacturing processes to reduce costs. He co-founded Fiberforge, Inc., in Glenwood Springs, Colorado, to perfect high-volume thermoforming of advanced composites (Figure 13.20). In 2007, Fiberforge was named a Technology Pioneer by the World Economic Forum.

13.3.5 Fuel Cell Electric Vehicles (FCEV)

We introduced fuel cell technology in Chapter 10. Considerable attention has been given to fuel cells because of their efficient means of converting hydrogen fuel to electricity and doing so without pollution. In transportation they can offset use of petroleum and of all fossil fuels if the hydrogen can be produced from renewables. Most hydrogen is now produced from natural gas. Lovins' vision for the Hypercar assumes it to ultimately run on hydrogen fuel cells.

The motor drive of a fuel cell vehicle is an electric motor like that in a BEV, PHEV, or HEV. The difference is that the battery bank that drives the motor is charged by the fuel cell, not by a gasoline engine-driven generator or the grid, although a plug-in FCEV is an option. As discussed in Section 10.9.1, a single fuel cell has a small voltage, so multiple cells are stacked as shown in Figure 10.18 and in Figure 13.21(b). Honda has taken the lead in FCEV development by announcing in 2006 its intent to produce its next-generation

figure 13.21 Honda Fuel Cell FCX

(a) Honda FCX concept car

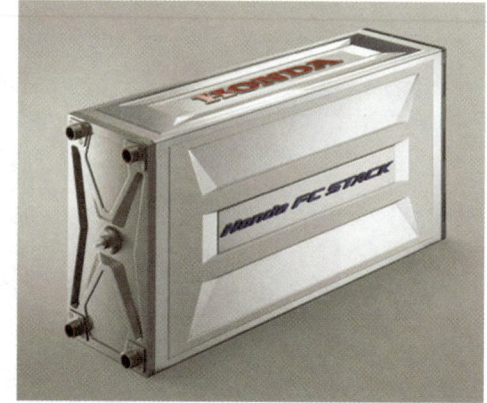

(b) Honda FC stack

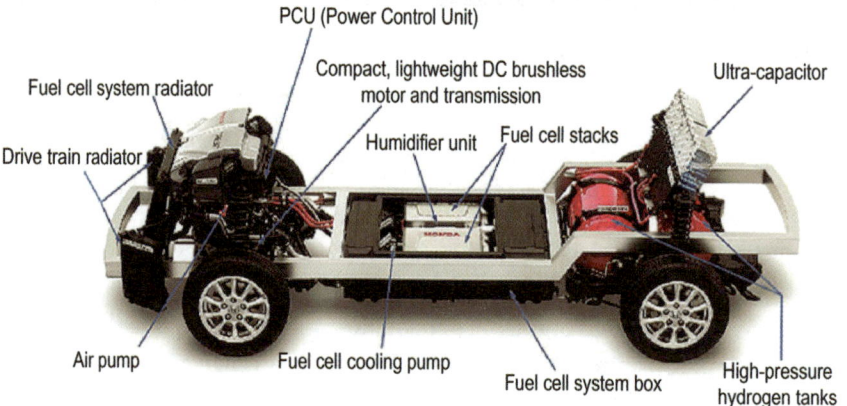

(c) Honda FCX components and configuration

Fuel cell stacks on low-floor platform, power control unit, motor/transmission up front, ultra-capacitor electric storage, high-pressure hydrogen tanks

SOURCE: courtesy Honda Motors, Inc., 2007

FCX for the commercial market in three to four years. Figure 13.21(a) shows Honda's FCX concept vehicle, and Figure 13.21(c) shows the configuration of system components, including the fuel cell stack on a low-floor platform, power control unit and motor/ transmission up front, and high-pressure hydrogen tanks and ultra-capacitor (battery) storage in the back.

There are complications in bringing FCEVs to market. They include developing an inexpensive, small, and lightweight fuel cell; the energy source and production of hydrogen fuel; improved hydrogen storage methods; and especially the infrastructure to deliver that fuel. All of these will take many years, and in the end hydrogen fuel cell vehicles will likely be expensive to buy and operate (see Section 3.3.4).

figure
13.22 **Honda Home Energy Station**

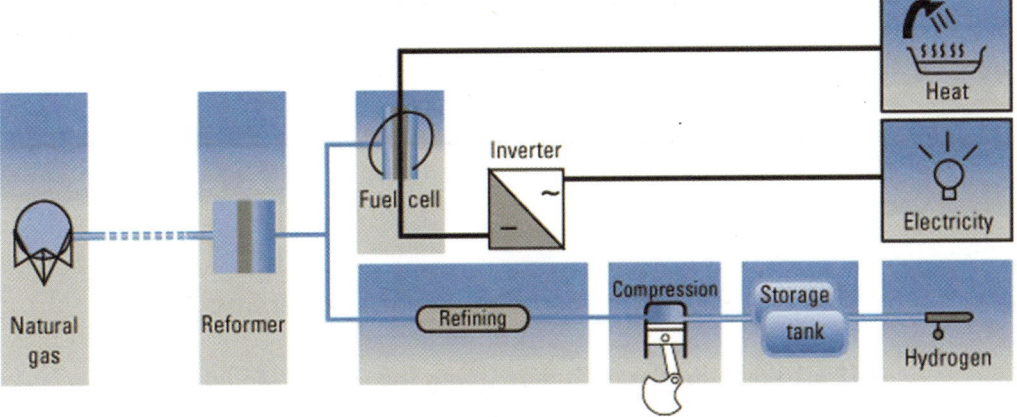

Reforms natural gas to hydrogen for use in residential fuel cell for home heat and electricity and for use in FCEV.

SOURCE: courtesy Honda motors, Inc., 2007

Understanding that marketing its FCX vehicle requires fuel options, in 2001 Honda developed an on-site solar PV electrolysis hydrogen producer at its Torrance, California, North American headquarters. In 2005, it developed a Home Energy Station that reforms natural gas to hydrogen and produces heat and electricity for home use and hydrogen for a FCEV (see Figure 13.22).

Although hydrogen fuel cell vehicles have been touted as the vehicles of the future and the center of a future hydrogen economy, EV, PHEV, and FF-PHEV have greater immediate promise. Consider the efficiency comparison of an FCEV and an EV if fuel cells rely only on hydrogen electrolysis from grid power and the EV relies on the same power to charge its batteries. Figure 13.23 shows that it would be more efficient to forget the fuel cell and use the power to charge the EV. The EV is $3\frac{1}{2}$ times as efficient, assuming 40% fuel cell efficiency (the EPA rating of the Honda FCX is 49 mi/kg H_2 or 37% efficient). We compare more vehicle technologies in the next section.

figure 13.23   Grid–to–Motor Efficiency for BEV versus FCEV with Electrolysis

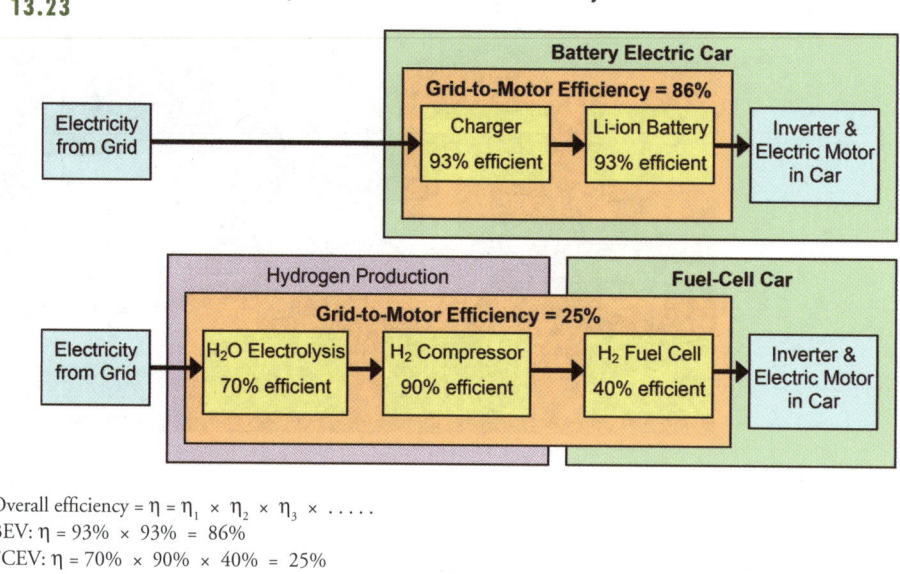

Overall efficiency = $\eta = \eta_1 \times \eta_2 \times \eta_3 \times \dots$
BEV: $\eta = 93\% \times 93\% = 86\%$
FCEV: $\eta = 70\% \times 90\% \times 40\% = 25\%$

SOURCE: Eberhard and Tarpenning, 2006. Used with permission of Tesla Motors, Inc.

13.4 Well-to-Wheel Studies of Vehicle Technologies

13.4.1 Well-to-Wheel Studies Using the GREET Model

As we begin to choose what technologies and fuels to develop for transportation in our increasingly carbon-rich and oil-poor world, the choice is complicated by many options, impacts, and life-cycle considerations. To assist with the comparative analysis to inform decisions, in 1995 Argonne National Lab began developing a life-cycle model called GREET—the **G**reenhouse gas, **R**egulated **E**missions, **E**nergy use in **T**ransportation. The model has been used for Well-to-Tank assessments of different fuel options and Tank-to-Wheels assessments of drive train and technology options (Figure 13.24). Combining the assessments gives a Well-to-Wheel (WTW) analysis, for which GREET is now the tool of choice. The "cradle-to-grave" concept of life-cycle assessment, introduced in Chapter 5 and applied to buildings in Chapter 8, is the basis for WTW analysis. As with cradle-to-grave studies, we can focus on one part of or the entire process. GREET now deals with the fuel energy process, but, as discussed under limitations of WTW studies, its developers are currently adding assessment of the vehicle production cycle to assess embodied energy and related impacts.

Most of the studies have included a variety of vehicle technologies, including gasoline and diesel ICE, gas and diesel hybrid, and fuel cell and fuel cell hybrid. Few have included the full range of technologies discussed in this chapter. Figures 13.25 and 13.26 show results of some of the studies. Figure 13.25 from the California Energy Commission's Full Fuel Cycle Assessment

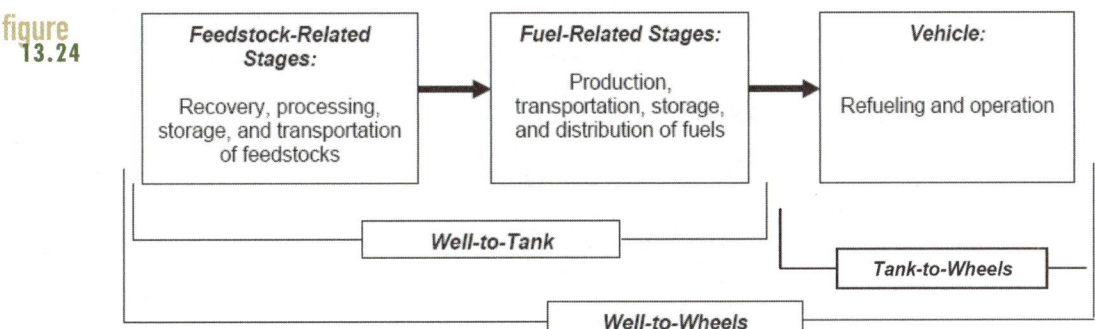

figure
13.24

GREET model combines well-to-tank assessment of the fuel cycle and tank-to-wheels assessment of drive train technologies. The model will be adding cradle-to-grave assessment of the vehicle manufacturing cycle.

Source: adapted from Weiss et al., 2000

(2007) compares alternative fuels (CNG, biofuel blends, plug-in hybrids, and hydrogen fuel cells) to reformulated gasoline (RFG3) for petroleum and GHG emission savings. The cellulosic E85, plug-in hybrid, and fuel cell hydrogen from biomass and on-site steam methane reforming (SMR) options have the greatest savings. Figure 13.26 from Wang (2005) shows the typical results of many of these studies: that there is likely to be a path from conventional technologies to hybrids to hydrogen fuel cell vehicles (see also Demirdoven et al., 2004).

figure
13.25

California Energy Commission Well-to-Wheels Assessment of Vehicle–Fuel Options

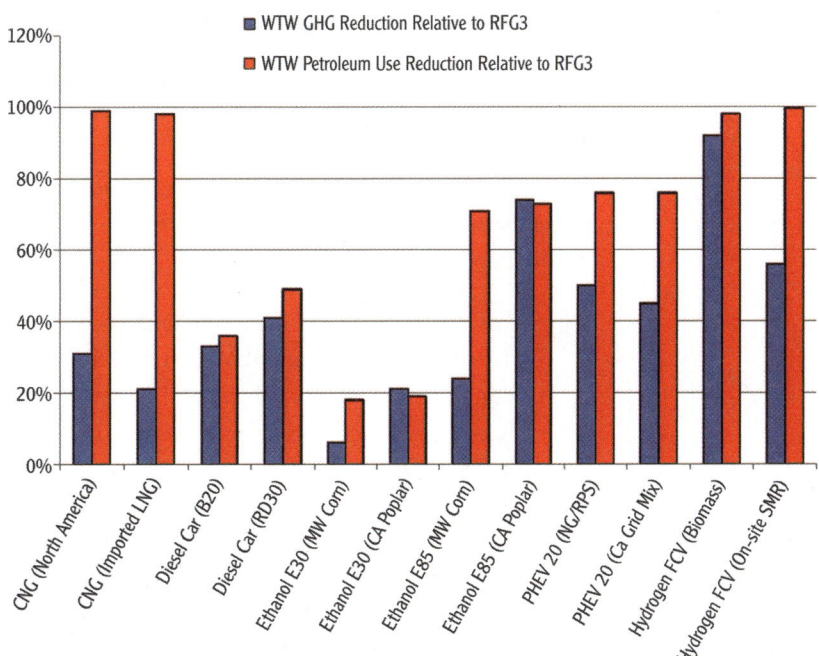

CEC compared compressed natural gas (CNG), biofuel blends, lug-in hybrids, and hydrogen fuel cell vehicles for savings of petroleum and GHG emissions compared to a standard reformulated gasoline (RFG3) vehicle.

Source: CEC, 2007

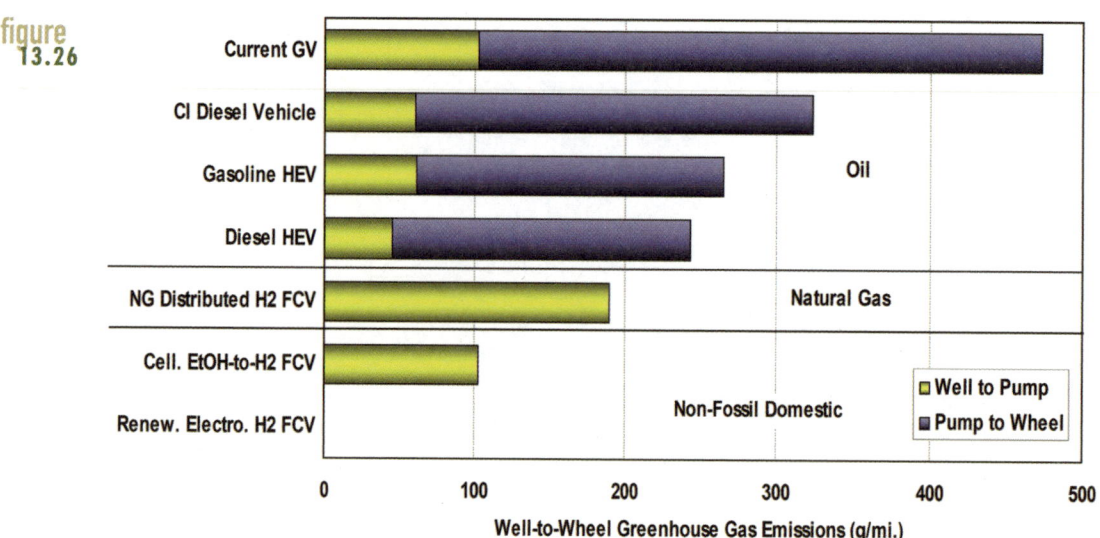

SOURCE: Wang, 2005

GREET analysis implies a path from conventional technologies to hybrids then to fuel cells to achieve zero GHG emissions if H_2 is generated from renewable energy.

We are not so sure of this path. When the GREET model is applied to some different combinations of technologies and fuels, including plug-in hybrids, flex-fuel plug-in hybrids, and all-electric vehicles charged by renewable electricity, the assumption of a fuel cell vehicle future may change. Let's try our own back-of-the-envelope WTW comparison of these technologies.

13.4.2 A Simple WTW Assessment of Current and Emerging Vehicles

Given the previous discussion on vehicle technology, efficiency, and emissions, we conducted a comparison of conventional, HEV, PHEV, EV, and FCEV cars for urban uses. The six vehicles are as follows:

- Gasoline vehicle (GV) based on Ford Focus
- Hybrid electric vehicle (HEV) based on Toyota Prius
- Plug-in hybrid electric vehicle (PHEV) based on plug-in "Prius Plus"
- Flex-fuel plug-in hybrid electric vehicle (FF-PHEV) based on flex-fuel plug-in "Prius Plus"
- Battery electric vehicle (BEV) based on Tesla Roadster
- Prototype fuel cell electric vehicle (FCEV) based on Honda FCX

The assessment considers WTW energy, gasoline, cost, and carbon emissions for each vehicle. It is broken into two parts: the tank-to-wheel (TTW) efficiency for each vehicle type, and the well-to-tank (WTT) efficiency for each fuel type. The calculations are not too hard, but the most important part is setting and stating assumptions. That way everyone knows

table 13.13   Fuel Assumptions: WTT Energy, Efficiency, and Cost

Well-to-Tank	WTT Energy				WTT Efficiency	Cost per		CO_2	
Gasoline	150	1000 Btu/gal	42	MJ/l	80.0%	$2.50	gal	11.2	kg/gal
E85	74	1000 Btu/gal	21	MJ/l	80%-g 125%-e	$2.20	gal	7.85	kg/gal
Electric-grid	11.4	1000 Btu/kWh	8.3	MJ/kWh	30.6%	$0.10	kWh	0.70	kg/kWh
Electric-ngcc	7.8	1000 Btu/kWh	12	MJ/kWh	43.7%	$0.10	kWh	0.43	kg/kWh
H_2-reform	224	1000 Btu/kg	237	MJ/kg	60.0%	$3.00	kg H_2	13	kg/kgH
H_2-el-grid	629	1000 Btu/kg	663	MJ/kg	21.4%	$3.00	kg H_2	38	kg/kgH
H_2-el-ngcc	440	1000 Btu/kg	464	MJ/kg	30.6%	$3.00	kg H_2	23	kg/kgH

what the results are based on, and if you want to change the assumptions, it is easy to do, especially if you build a spreadsheet for the analysis.

For our assessment, the assumed drive cycle is 15,600 mi/year, with 6500 highway (42%) and 9100 city (58%). Assumptions on vehicle TTW and fuel WTT efficiencies, fuel price, and CO_2 emission rates are in Tables 13.13 and 13.14. For the electric drives and H_2 electrolysis, two electricity sources are included: average U.S. grid (35% efficient generation) and natural gas combined-cycle (50% efficient generation). Electricity also incurs fuel extraction and processing and transmission efficiencies.

In Solution Box 13.5, we step through a sample of the calculations for the GV, FF-HEV, and BEV to show the process. You might try some of the other fuel-vehicle scenarios on your own.

The overall WTW results are given in Table 13.15 and Figures 13.27–13.30. The FF-HEV and the BEV are the winners with the lowest WTW energy per mile, and lowest CO_2 emissions. Like the BEV, the fuel cell vehicle has zero gasoline use (good for petroleum reduction) but has high WTW energy and carbon emissions especially for the electrolysis options.

table 13.14   Vehicle Assumptions: TTW Efficiency

	Vehicle	Basis	TTW Efficiency
Gas	Ford Focus	24 mpg	
HEV	Prius	49 mpg	
PHEV	Prius Plus	48 mpg (highway)	5 mi/kWh (city)
FFPEV	FF Prius Plus	40 mpg (highway)	5 mi/kWh (city)
BEV	Tesla Roadster	5.6 mi/kWh	
FCEV	Honda FCX	57 mi/kg H_2	(49–66 mi/kg H_2)

SOLUTION BOX 13.5

WTW Calculations for Gasoline, FF-HEV, and BEV

Given the assumptions in Tables 13.13 and 13.14, what are the WTW energy per mile and annual CO_2 emissions for the gasoline vehicle, the FF-HEV and the BEV both with NGCC electricity?

1. **Gasoline Vehicle (GV):**

   Fuel WTT: The energy value of gasoline is 120,000 Btu/gal. WTT efficiency for gasoline is estimated at 80%, that is about 20% of crude oil energy is consumed in production of crude, processing to gasoline, and transport to filling station. The WTT energy is 120,000/0.8 or 150,000 Btu/gal. CO_2 of production and combustion is given as 11.2 kg CO_2/gal.

$$\frac{\text{WTW energy}}{\text{mi}} = \text{WTT} \times \text{TTW} = \frac{150 \times 10^3 \text{ Btu} \times \text{gal}}{\text{gal} \quad 24 \text{ mi}} = \frac{6.3 \times 10^3 \text{ Btu}}{\text{mi}} = \frac{4.1 \text{ MJ}}{\text{km}}$$

$$CO_2 \text{ emissions} = \frac{15,600 \text{ mi}}{\text{yr}} \times \frac{11.2 \text{ kg } CO_2}{\text{gal}} \times \frac{\text{gal}}{24 \text{ mi}} = \frac{7.3 \times 10^3 \text{ kg } CO_2}{\text{yr}}$$

2. **Flex-Fuel Plug-In Hybrid Electric Vehicle (FF-PHEV) with natural gas combined-cycle electricity:**

   Fuel WTT: This is a little more complicated because we have two fuels to deal with: E85 (85% ethanol, 15% gasoline) and electricity. So we figure out the WTT energy for each and weight them by their use in the driving cycle (42% highway E85, 58% city electric). Ethanol WTT accounts for the fossil fuel energy needed to grow and process corn into ethanol (not the solar energy that grows the corn). Wang, 2005, estimates this at 56%–79% of the energy value of the ethanol (Figure 5.4), we will conservatively use 80%. For NGCC electricity, we assume 95% well-to-power plant, 50% generation, and 92% transmission efficiency for a total of 43.7%.

$$\frac{\text{WTW energy}}{\text{mi}} = \text{WTT} + \text{TTW} = 0.42 \times \frac{74 \times 10^3 \text{ Btu}}{\text{gal}} \times \frac{\text{gal}}{40 \text{ mi}} +$$

$$0.58 \times \frac{7.8 \times 10^3 \text{ Btu}}{\text{kWh}} \times \frac{\text{kWh}}{5 \text{ mi}} = (777 + 905) \frac{\text{Btu}}{\text{mi}} = 1.7 \times 10^3 \frac{\text{Btu}}{\text{mi}} = 1.1 \frac{\text{MJ}}{\text{mi}}$$

$$CO_2 \text{ emissions} = \frac{16,500 \text{ mi}}{\text{yr}} \times \frac{\text{gal}}{40 \text{ mi}} \times \frac{7.85 \text{ kg } CO_2}{\text{gal}} +$$

$$\frac{9100 \text{ mi}}{\text{yr}} \times \frac{\text{kWh}}{5 \text{ mi}} \times \frac{0.43 \text{ kg } CO_2}{\text{kWh}} = 2.0 \times 10^3 \frac{\text{kg } CO_2}{\text{yr}}$$

3. **Battery Electric Vehicle (BEV) with average grid electricity:**

Fuel WTT: We don't have the complications of multiple fuels here but we assume average grid electricity, so the WTT efficiency and CO_2 emission rates are different from the case above. For grid electricity, we assume 95% well/mine-to-power plant, 35% generation, and 92% transmission efficiency for a total of 30.6% (Table 13.13).

$$\frac{\text{WTW energy}}{\text{mi}} = \text{WTT} + \text{TTW} = \frac{11.4 \times 10^3 \text{ Btu}}{\text{kWh}} \times \frac{\text{kWh}}{5.6 \text{ mi}} = 2.0 \times 10^3 \frac{\text{Btu}}{\text{mi}} = 1.3 \frac{\text{MJ}}{\text{km}}$$

$$CO_2 \text{ emissions} = \frac{15,600 \text{ mi}}{\text{yr}} \times \frac{0.7 \text{ kg } CO_2}{\text{kWh}} \times \frac{\text{kWh}}{5.6 \text{ mi}} = 2.0 \times 10^3 \frac{\text{kg } CO_2}{\text{yr}}$$

table 13.15 WTW Results: Gasoline, Cost, Energy, and CO_2 Emissions

WTW Results	Gasoline gal/yr	Cost $/yr	Energy 1000 Btu/mi*	Energy MJ/km*	CO_2 1000 kg/yr*
GV	650	1625	**6.3**	**4.1**	**7.3**
HEV	318	796	3.1	2.0	3.6
PHEV-grid	135	521	2.6	1.7	2.8
PHEV-ngcc	135	521	2.2	1.4	2.3
FFPEV-grid	24	540	2.1	1.4	2.5
FFPEV-ngcc	24	540	**1.7**	**1.1**	**2.0**
BEV-grid	0	279	**2.0**	**1.3**	**2.0**
BEV-ngcc	0	279	1.4	0.9	1.2
FCEV-reform	0	821	3.9	2.6	3.6
FCEV-el-grid	0	821	11.0	7.2	10.4
FCEV-el-ngcc	0	821	7.7	5.1	6.3

* Bold values are results from Solution Box 13.5.

figure 13.27 WTW Energy Consumed per Mile

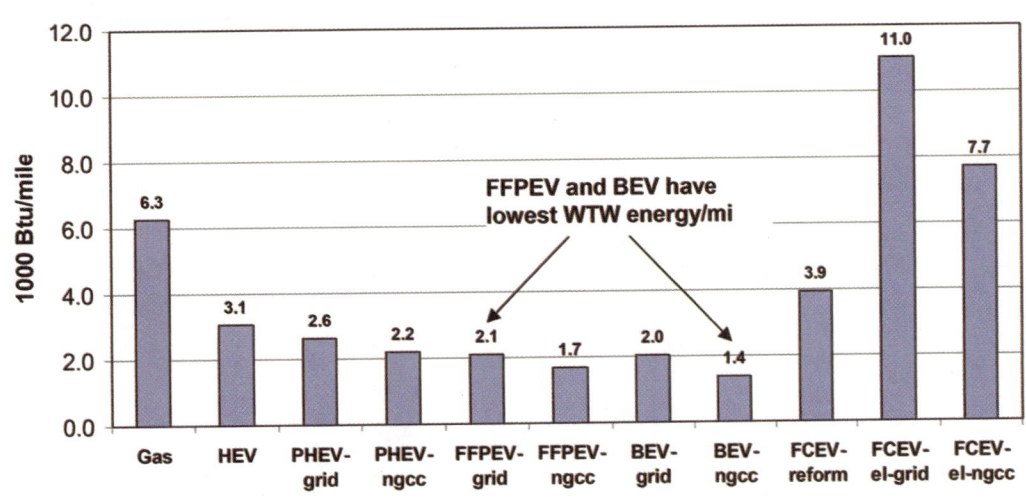

figure
13.28   WTW Annual CO$_2$ Emissions

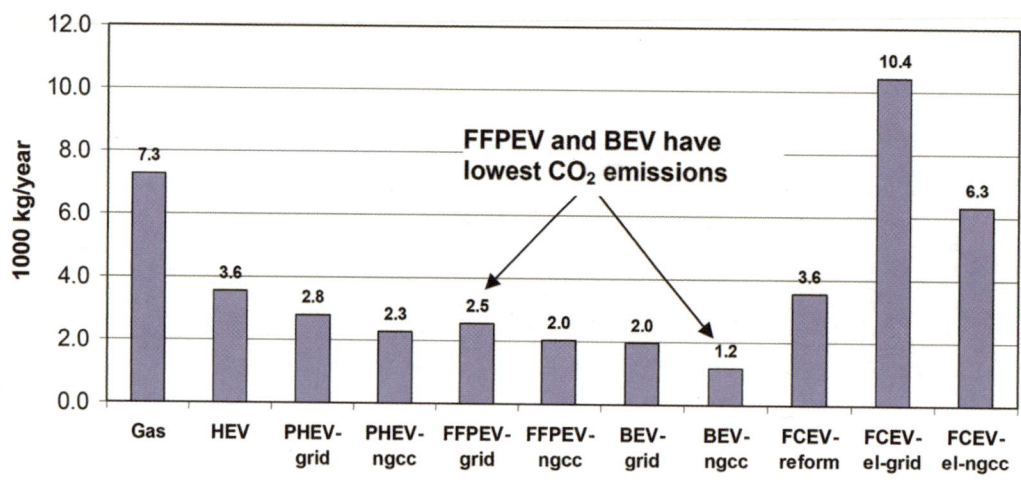

13.4.3 Limitations of Well-to-Wheel Analysis

Well-to-wheel analysis aims to compare a diverse mix of vehicles and fuels in a common framework. This is a great step forward in energy analysis, but the state of the art is not complete. For example, our analysis did not consider vehicle cost and life-cycle energy and emissions for vehicle manufacturing. We were comparing a $15,000 Ford Focus to a $92,000 Tesla Roadster and assessed fuel cost only. Perhaps more importantly, we did not consider the embodied energy in the vehicles. If they were similar technologies, we could assume

figure
13.29   WTW Annual Gasoline Consumption

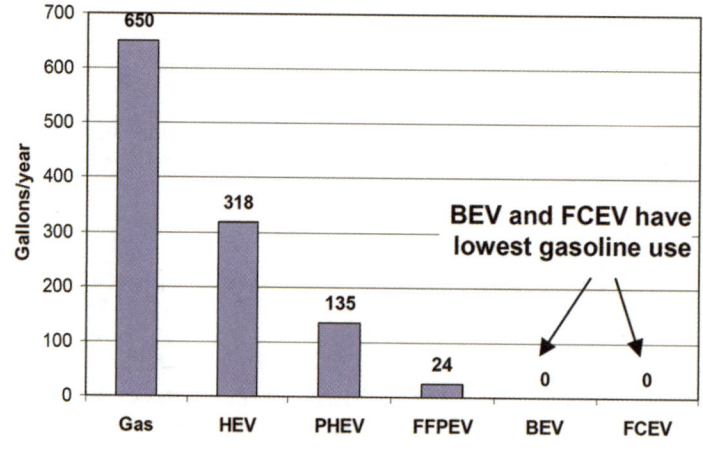

figure
13.30 **WTW Annual Fuel Cost**

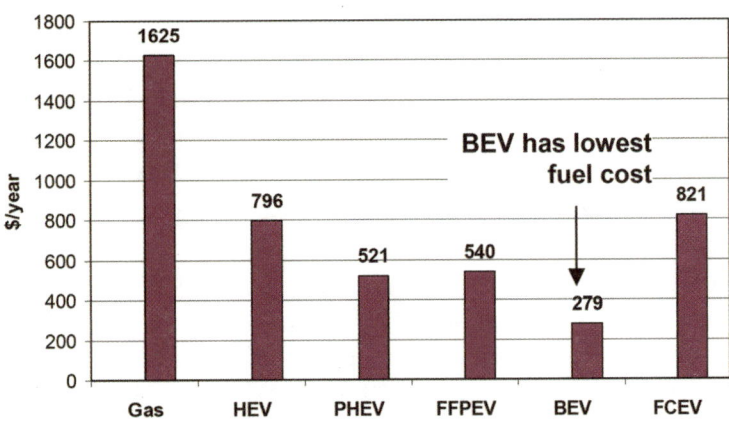

comparable embodied energy, but they are not. For example, what is the embodied energy (and related emissions) of manufacturing a steel-intensive Ford Focus to a Li-ion battery-intensive PHEV or BEV? The answer to this question certainly would affect the results on life-cycle energy and emissions.

Even the "gold standard" WTW GREET model has dealt only with the fuel energy process. However, the model's creators recognize this limitation and are expanding the model (GREET 2 Series, GREET 2.7) to include the vehicle production cycle to assess embodied energy, GHG emissions, air emissions, and other impacts of the materials and production cycle of vehicle manufacture, decommissioning, recycling, and disposal.

13.5 Summary

Transportation is a critical energy sector because of its growing demand, its dependency on oil, and its contribution to urban air pollution and GHG emissions. If we are to adequately address our oil and carbon problems, transportation is the first place to start. We need a three-prong solution:

- Improve vehicle efficiency.
- Increase use of alternative fuels to replace petroleum fuels.
- Reduce vehicle miles traveled.

This chapter has addressed the first of these responses. The next chapter focuses on alternative fuels, and Chapter 15 discusses means of mitigating VMT.

Technology is the key to improving vehicle efficiency (i.e., mpg). It also is important for reducing energy intensity (e.g., energy per passenger mile, energy per freight ton-mile), but energy intensity is also affected by modal shifts and load factors of both passengers and freight. The world's and especially the nation's fascination with the automobile makes vehicle technology and efficiency the subject of broad consumer and public interest. Improvements in the traditional internal combustion engine and ancillary emission control equipment have succeeded in reducing emissions per vehicle mile and increasing efficiency. But these improvements have been offset by increased vehicle miles traveled and vehicle size and weight. Further, the efficiency of average new light vehicles in the U.S. market has not improved much in twenty years because the federal government has chosen not to increase fuel efficiency standards, the consumer market has turned to less efficient larger light trucks and SUVs, and fuel prices have remained relatively low compared to other countries.

Higher fuel prices and new federal efficiency standards adopted in 2007 for 2020 will help push technology development to create more efficient vehicle options for consumers. The most promising commercial development is the growing market for efficient hybrid electric vehicles. Simple enhancements of this technology, adding extra batteries and a plug-in option, as well as using a flex-fuel engine that can operate on gasoline or E85 blend, provide the best mid-term and perhaps long-term option to improve vehicle efficiency, reduce carbon emissions, and reduce oil use. This technology currently looks more promising than other options including hydrogen fuel cells.

CHAPTER 14

Biofuels, Biomass, and Other Alternative Fuels

We don't need oil, and we definitely don't need hydrogen for our cars and light trucks. We don't need new engines, new fuel distribution and storage, and we don't need a lot of money or time to do this. Through three simple inexpensive policy changes we can kick start the transition and reassure investors that there is a long-term market for ethanol, not subject to price manipulation by the oil-producing countries. . . . And this is not an alternative fuel option. It can replace all our oil imports and become the center of our transportation fuels economy. The other impediment, the various politically powerful interest groups also seem to be well aligned. Other objections, like land use, environmental impact and energy balance can be overcome.

Vinod Khosla, April 2006

These words sound like those of a utopian environmentalist, but they come from one of the world's most successful venture capitalists. They would have sounded like a pipe dream as recently as mid-2005, but in 2006 and 2007, similar versions were coming from a variety of unlikely sources, including energy companies, farm groups, environmental organizations, Madison Avenue, and the White House. Combined with increased efficiency offered by lighter hybrid vehicles with plug-in options, these groups argue that biofuels can take a big bite out of oil imports and overall consumption, improving urban air quality, and reducing GHG emissions. Global biofuel production grew by 22% per year in 2005 and 2006. But not all are so optimistic about biofuels. Debate continues about the net energy of biofuels production, the effect of biofuels from food crops on food prices, the prospects for non-food crop biofuels, and the subsidies and import tariffs that affect biofuels markets.

This chapter explores the options for alternative transportation fuels to replace oil. Some of these options are linked to the propulsion technologies discussed in the last chapter, especially flex-fuel vehicles that can use 85% ethanol (E85) or biodiesel blends, and electric drive motors using batteries charged by hybrid engines, plug-in grid, on-site generation, or fuel cells. This chapter gives an overview of alternative transportation fuels and then presents in more detail the opportunities and constraints for fuel ethanol, biodiesel, other biomass energy, as well as natural gas and hydrogen.

14.1 Introduction to Alternative Transportation Fuels

We learned in the last chapter that petroleum fuels 96% of the transportation energy in the United States. Because of the current and long-term problems with oil, there has been considerable talk of switching to alternative fuels since the 1974 oil embargo. But there has been little progress because petroleum products have remained relatively cheap (until recently) and special interests have helped protect the status quo. Today, we see increased attention to alternative fuels, and we do have some experience in several alternatives to help us decide how to proceed.

Let's review the several types of alternative fuels available for use in transportation vehicles:

- **Alternative fossil fuels** include liquefied petroleum gas (LPG), liquefied natural gas (LNG), and compressed natural gas (CNG). Together these were the largest source of energy for alternative fueled vehicles in the United States in 2005 (Table 14.1). These vehicles are modified to use these fuels. Some are "dual-fuel" vehicles that can use both gasoline and the alternative, but unlike flex-fuel vehicles they need two separate fuel-handling systems. Because they have much lower urban air pollutant emissions than gasoline and diesel vehicles, they are primarily used to fuel buses and other large vehicles in metropolitan areas not meeting air quality standards. Once looked upon as a viable option for alternative fuels, price volatility is a growing concern as these fuels tend to follow oil prices.
- **Electricity** used in electric vehicles amounted to less than 0.1% of alternative fuel in 2005, but it has grown considerably since 1995 (Table 14.1). Plug-in, electric, and fuel cell vehicles could greatly increase the use of electricity for transportation.
- **Alcohol fuels** include ethanol and methanol, and these are blended with gasoline in different proportions. E85 is 85% ethanol and 15% gasoline, while E10 or "gasohol" is 10% ethanol and 90% gasoline. Methanol from natural gas was used until the mid-1990s, but little is used today (Table 14.1). Alcohol has less energy content than gasoline; E85 has 27% less energy per gallon, but mileage comparison tests have shown that E85 has 5%–12% less mpg than gasoline. E85 is growing quickly because an increasing number of flex-fuel vehicles are available, and some areas, especially the midwestern states, have expanded the availability of E85. Still most ethanol fuel use has been in E10 gasohol as an oxygenate additive to gasoline to reduce carbon monoxide emissions.
- **Biodiesel** is a distillate or diesel fuel replacement made from biomass oils including vegetable oils, such as soybean or rapeseed; waste vegetable oils; or algae. It is blended with petroleum diesel in blends that range from B-2 to B-100. Still a very small source of fuel, U.S. biodiesel production grew eightfold from 2004 to 2006 and the number of fueling stations quadrupled.

table 14.1 Alternative Fuel and Oxygenate Consumption, 1995–2005

Alternative fuel	1995*	2005*	2005 %
Liquefied petroleum gas	232,701	188,171	5.9%
Compressed natural gas	35,162	166,878	5.3%
Liquefied natural gas	2759	22,409	0.7%
M85	2023	0	0.0%
M100	2150	0	0.0%
E85	190	38,074	1.2%
E95	995	0	0.0%
Electricity	663	5,219	0.0%
Subtotal	**278,121**	**420,776**	**13.2%**
Oxygenates			
MTBE	2,693,407	1,654,500	–
Ethanol in gasohol	**934,615**	**2,756,663**	**86.8%**

* Values for 1995 and 2005 are thousand gasoline-equivalent gallons.
Source: U.S. DOE, Alternative Fuel Database, 2007

- **Other alternative fuels:** Other fuels have more limited availability or are under development. Hydrogen re-formed from natural gas for use in fuel cells and coal-to-liquids, natural gas-to-liquids, and fuels from unconventional oil sands and shales may contribute to future transportation fuels, but they have greater economic and environmental obstacles to overcome than do biofuels or electricity.

Table 14.1 gives the gasoline-equivalent gallons consumed of different alternative fuels in 1995 and 2005. Ethanol amounted to 87% of all alternative fuels in 2005, just about all as an oxygenate additive for gasoline. Oxygenates were mandated in gasoline to reduce carbon monoxide emissions under the Clean Air Act amendments of 1990. Use of methyl tertiary butyl ether (MTBE) oxygenate grew rapidly until the late 1990s when concerns grew over its water pollution problems. California and New York, with a combined 40% of total MTBE consumption, banned its use after January 1, 2004. As a result the consumption trends shown in Table 14.1 have continued. Daily MTBE use has dropped significantly from 2004 to 2006, whereas ethanol use in gasohol increased 30% over the same period. In 2006, EPA lifted the requirement for oxygenate additives.

Although increased use of ethanol in gasohol displaces some gasoline, ethanol's real potential is in E85. Table 14.1 shows significant growth from 1995 to 2005 and this growth continues. The leading state, Minnesota, saw sales of E85 increase from 2.6 million gal (Mgal) in 2004 to 21 Mgal in 2007. Still only 6% of the state's filling stations offer the fuel, and only 5% of the state's vehicles are flex-fuel.

A key issue in the widespread adoption of any alternative fuel is the development of infrastructure for its delivery. Table 14.2 lists the number of alternative-fuel filling stations in the top sixteen states, as well as national totals in 2007 and previous years. With its CNG and

table 14.2   Alternative-Fuel Filling Stations, July 2007

State	CNG	E85	LPG	ELEC	LNG	BD	H_2	ALL
California	184	3	232	379	28	34	23	883
Texas	16	29	556	1	2	45	0	649
Minnesota	1	306	31	0	0	2	0	340
Illinois	14	146	64	0	0	12	0	236
Missouri	7	60	82	0	0	48	0	197
Michigan	13	44	79	0	0	16	2	154
Pennsylvania	29	11	78	0	0	35	1	154
South Carolina	5	46	29	1	0	67	0	148
Indiana	14	84	33	0	0	11	0	142
Colorado	21	26	67	2	0	24	0	140
Ohio	11	34	68	0	0	21	0	134
Oklahoma	51	1	71	0	0	7	0	130
North Carolina	10	9	54	0	0	56	0	129
Wisconsin	16	60	47	0	0	4	0	127
Washington	13	6	56	0	0	32	0	107
Iowa	0	68	24	0	0	13	0	105
Total 2007	**727**	**1166**	**2459**	**444**	**35**	**705**	**31**	**5567**
Total 2006	**732**	**762**	**2619**	**465**	**37**	**459**	**17**	**5091**
Total 2003	**1035**	**188**	**3966**	**830**	**62**	**142**	**7**	**6230**
Total 2000	**1217**	**113**	**3268**	**558**	**44**	**2**	**0**	**5205**

Source: U.S. DOE, Alternative Fuel Database, 2007

electric refilling stations, California has the largest number, but Minnesota leads in E85 stations, Texas leads in LPG, and South Carolina leads in biodiesel filling stations. Whereas most other alternative-fuel stations have dropped in number, E85 and biodiesel stations have increased sixfold from 2003 to 2007. Still the number of alternative-fuel stations is small compared to the 170,000 gasoline filling stations across the country.

Price is also a major issue in the adoption of alternative fuel, and the price of alternative fuels varies among fuels. They are affected by the price of gasoline as shown by the data for 2005 to 2007 in Table 14.3. Although CNG and E85 are the lowest-priced fuels per gallon in Table 14.3, they are usually hard to find. On an energy equivalent basis, E85's price is not as competitive. CNG requires a dedicated or dual-fuel vehicle, and E85 requires a flex-fuel vehicle.

table
14.3   **Prices by Fuel, September 2005 to June 2007**

	Sept 05 per gal	Feb 06 per gal	June 06 per gal	Oct 06 per gal	Mar 07 per gal	Jun 07 per gal	Jun 07 per gg/de	Jun 07 per MBtu
Gasoline	$2.77	$2.23	$2.84	$2.22	$2.30	$3.03	$3.04	$26.25
Diesel	$2.81	$2.56	$2.98	$2.62	$2.63	$2.96	$2.65	$22.98
CNG	$2.12	$1.99	$1.90	$1.77	$1.94	$2.09	$2.10	$18.18
Propane (LPG)	$2.56	$1.98	$2.08	$2.33	$2.62	$2.58	$3.57	$30.91
Ethanol (E85)	$2.41	$1.98	$2.43	$2.11	$2.10	$2.63	$3.72	$32.21
Biodiesel (B-2–B-5)	$2.81	$2.46	$2.67	$2.75	$2.75	$2.84	$2.55	$22.09
Biodiesel (B-20)	$2.91	$2.64	$2.67	$2.66	$2.53	$2.96	$2.70	$23.43
Biodiesel (B-99–B-100)	$3.40	$3.23	$3.76	$3.31	$3.31	$3.27	$3.22	$27.89

SOURCE: U.S. DOE, Alternative Fuel Price Report, 2007

14.2 Prospects and Potential for Biomass Fuels

Have you thanked a green plant today? Of course, we know that through the miracle of photosynthesis, plants are able to absorb solar energy and atmospheric carbon, providing not only the food and materials for all living things, but possibly a significant answer to our economy's energy needs and our environment's need for carbon sequestration. **Biomass** is vegetative and animal waste organic matter that can be converted into useful energy. It includes **solid fuel** like wood and plant residues that can be burned directly for thermal energy or power production. It can be converted to **liquid biofuels** like ethanol and biodiesel that can substitute for gasoline and diesel fuel, or to **gaseous fuels** like methane and **biogas** that can be burned for thermal energy or used in gas turbines to produce electric power.

Figure 14.1 shows the classic carbon cycle with an emphasis on bioenergy. With the sun's radiant energy and atmospheric carbon dioxide (and other nutrients), plants produce physical biomass, which stores the sun's energy. The biomass can be processed in a variety of ways and converted to solid, liquid, and gaseous fuels that can be used for vehicle transport, heat, and power generation, and/or for biomaterials that can be used for building materials, paper, and other products. Processing in a biorefinery usually emits some of the biomass carbon as CO_2 and may require some fossil fuels that also emit CO_2. The end-use combustion of biomass energy also emits CO_2 as the carbonaceous materials are oxidized; and all of this CO_2 ends up in the atmosphere. But biomass combustion is generally considered to be **greenhouse-gas-neutral** because it is part of the contemporary carbon cycle—its carbon recently came from the atmosphere and its carbon emissions are in balance with subsequent absorption by revegetation. Biomaterials can sequester carbon in building materials and other products or can be recycled back to reprocessing for subsequent use.

Biomass energy is the largest source of renewable energy in the world today, and the primary fuel for cooking and heating for nearly half of the world's population. In Africa,

figure 14.1   **The Biomass Energy Carbon Cycle**

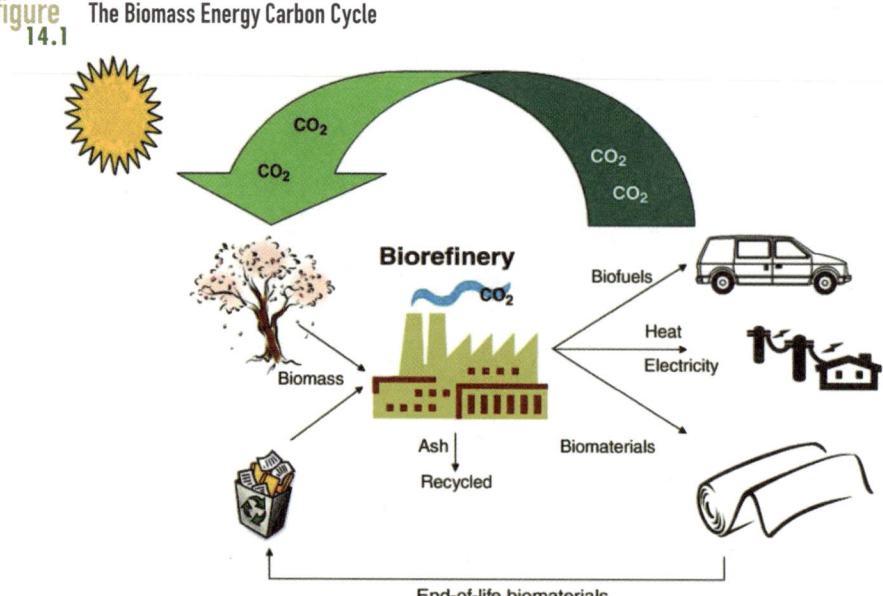

Biomass is stored solar energy and atmospheric carbon that can be processed in biorefineries and converted to usable fuels and biomaterials. Combustion releases carbon back to the atmosphere and materials can be recycled. Operating in the contemporary (rather than fossil) carbon cycle, biomass energy combustion is "carbon neutral."

SOURCE: from Ragauskas, et al., *Science 311*:484–489 (2006). Reprinted with permission from AAAS.

biomass contributes about one-half of primary energy; in Asia about one-fourth. Of course, this is mostly traditional firewood and charcoal and is not included in commercial energy markets that we use to monitor energy data. If traditional biomass were accounted for, it probably would make up about 10%–15% of global energy use.

But this extensive use of biomass among the world's poor is also an indicator of poverty. Potentially productive time must be used simply gathering fuelwood and charcoal. Indoor and urban air pollution from wood and charcoal cookstoves is a major health hazard. Both charcoal production and traditional stoves are extremely inefficient, so most of the hard-earned energy is lost. Improvements in stove technology, like the $2 Kenya Ceramic Jiko charcoal stove, can greatly improve efficiency and air quality.

Many think development in poorer countries will come with modern energy systems, but these need not be fossil-fuel based. Wind and photovoltaic power are promising sources in such contexts, but so too is biomass energy as long as it transitions from traditional to more modern applications, such as biomass power plants, fuel ethanol, biodiesel, and biogas from methane digestion.

In the United States, biomass is the largest source of renewable energy today. In 2006, just 7% of U.S. energy came from renewables, and biomass contributed 48% of that, followed closely by hydroelectric power (42%). Hydro's annual contribution fluctuates with precipitation. About half of biomass energy currently comes from residues and pulping liquors in the forest products industry; about 18% from urban wastes and construction residues; about 18% from fuelwood for residential woodstoves and electricity generation; and about 10% from liquid biofuels, mostly ethanol. About 190 million dry tons (Mdt) of forest and agricultural biomass are used for energy. The fastest growing forms of biomass energy are fuel ethanol and biodiesel, liquid products that directly replace petroleum products in transportation. World production of ethanol is growing by more than 20% per year and U.S. production grew by 24% in 2006 to 4.86 billion gallons (Bgal) and another 42% in 2007 to about 7 Bgal. Biodiesel volume is smaller, but its growth rate is even faster. U.S. production increased by 18 times from 25 Mgal in 2004 to an estimated 450 Mgal in 2007. This recent growth in biofuels is not reflected in the 2004 data in Figure 14.2. Before turning specifically to these biofuels, we look at overall potential for biomass.

14.2.1 U.S. Biomass Energy Potential

A number of studies have recently assessed the potential biomass energy production in the United States (e.g., U.S. DOE, 2002; Greene and Mugica, 2005). Perhaps the most comprehensive recent study prepared at Oak Ridge National Laboratory (Perlack, et al., 2005),

figure 14.2 U.S. Biomass Resource Consumption, 2004

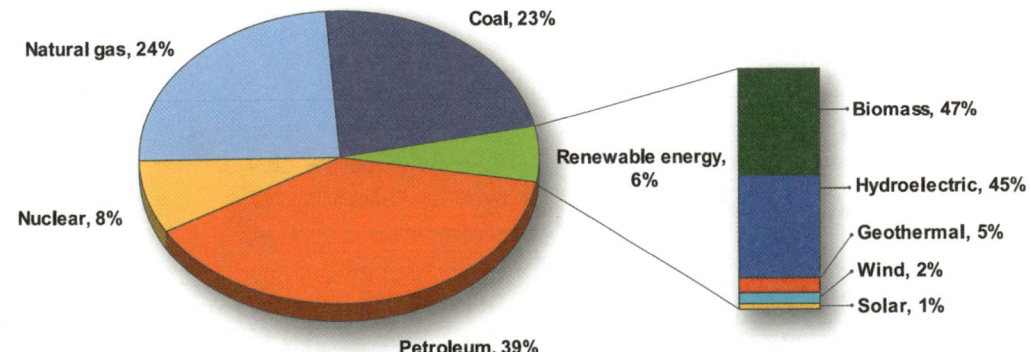

Biomass Consumption	Million dry tons/year
Forest products industry	
Wood residues	44
Pulping liquors	52
Urban wood and food & other process residues	35
Fuelwood (residential/commercial & electric utilities)	35
Biofuels	18
Bioproducts	6
Total	**190**

Source: Perlack, et al., 2005; Oak Ridge National Laboratory

assessed the potential of forest and agriculture land resources to sustainably supply biofuels to offset petroleum consumption in the United States. The study found that more than 1.3 billion dry tons (Bdt) per year of biomass energy materials could be produced, which could replace more than one-third of the nation's current petroleum consumption. This includes 370 Mdt on forest lands and 1 Bdt from agricultural lands. A key issue is balancing the need for energy with that for other agricultural products. According to ORNL, these production targets could be achieved while meeting expected food, feed, fiber, and export demands, as well as needs for environmental conservation.

To achieve this potential, we would have to rely on woody, cellulosic fiber from fields and forests. Figure 14.3 shows that forestlands' potential of 370 Mdt includes the 140 Mdt currently used residues generated in the manufacture and use of various forest products and wood for residential space heating. Removal of logging and other residues and fuel treatment thinning are not being fully utilized and can sustainably provide more than 120 Mdt annually. These residues can be recovered from commercial harvest and land-clearing operations, and fuel treatment thinning can be done in conjunction with forest fire hazard mitigation and forest health projects.

There is much greater potential available from agricultural lands. Figure 14.4 shows that farmland could provide 1 Bdt of sustainably collectable biomass without impacting food, feed, and export demands. This includes 87 Mdt from grains for biofuels and another 87 Mdt from animal manures, process residues, and other residues generated in the consumption food products.

figure 14.3   U.S. Forestland Biomass Energy Potential

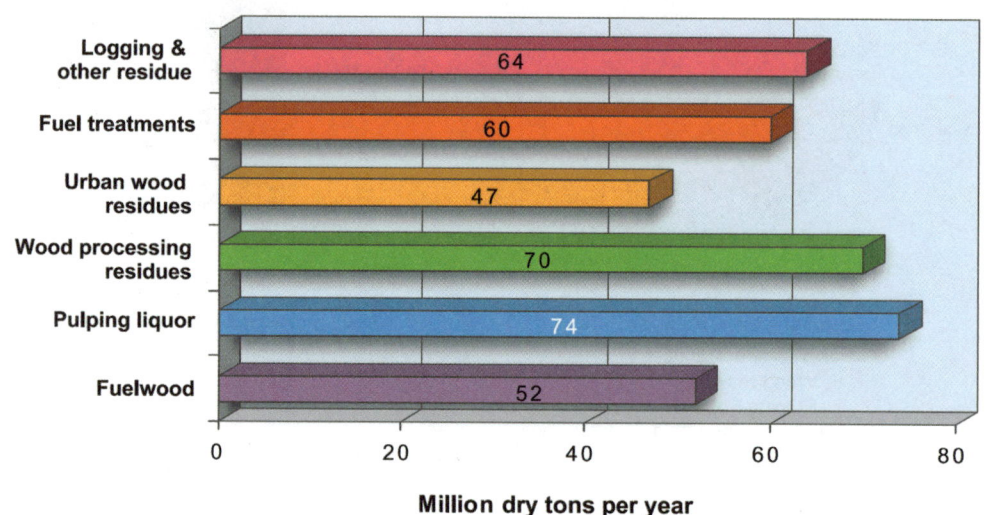

Fuelwood, pulping liquors, and processing residues are mostly existing uses. Others represent new potential.

Source: Perlack, et al., 2005; Oak Ridge National Laboratory

figure 14.4 U.S. Agricultural Land Biomass Energy Potential

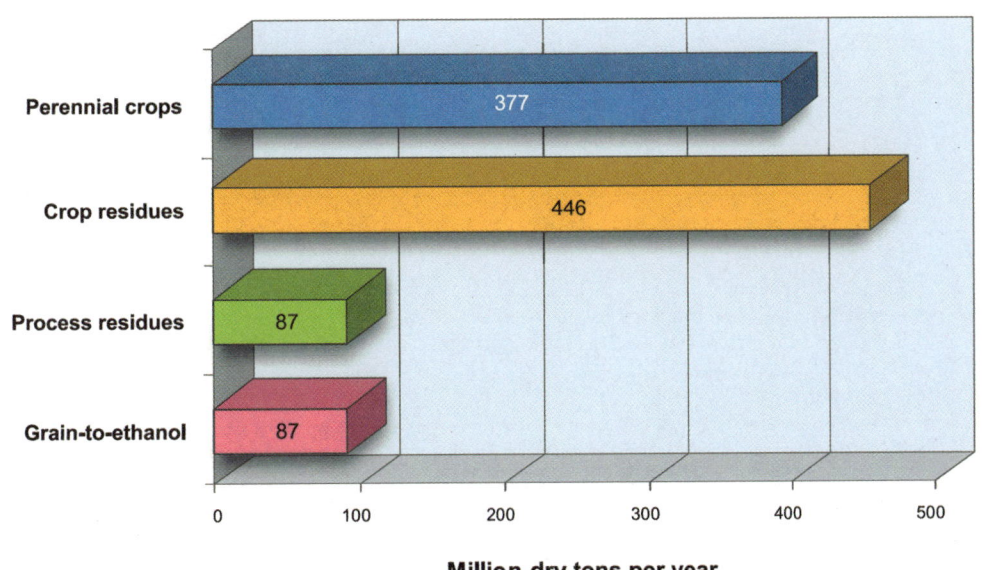

Million dry tons per year

Grain-to-ethanol and process residues are modest increases above current land use, enhanced by improved yields. Significant new potential comes from perennial grasses/woody crops and residues.

SOURCE: Perlack, et al., 2005; Oak Ridge National Laboratory

The increase in grains for biofuels over current levels is largely due to expected increase in yields and not a diversion from food and export markets. However, the biggest potential increase in agricultural biomass is not from grain crops but from crop residues (446 Mdt) and perennial grasses and woody crops (377 Mdt), or 82% of the total potential.

However, the job of achieving this potential will not be an easy one. It will require

- Increasing yields of corn, wheat, and other small grains by 50%
- Doubling residue-to-grain ratios for soybeans
- Developing much more efficient residue harvesting equipment
- Managing active cropland with no-till cultivation
- Growing perennial crops dedicated to bioenergy purposes on 55 million acres of pasture, cropland, and idle cropland (including Conservation Reserve Program [CRP] lands)
- Using for bioenergy excess animal manure not applied on-farm for soil improvement
- Using a larger fraction of other secondary and tertiary residues for bioenergy
- Developing a large-scale biorefinery industry (Perlack, et al., 2005)

It is important to understand the scale of change necessary to pull off this billion ton growth in biomass energy. Table 14.4 gives the assumptions and scenarios of the ORNL study. All scenarios assume the same agricultural land harvested or reserved as today,

table 14.4   **Agricultural Lands Biomass Energy Potential, Three Scenarios**

Crop	Acres	Product Yield	Residue Yield	Total Mass	Total Residue	Residue Sustainably Removable	Grains for Bioenergy	Secondary Residues	Total Sustainable Biomass
	Million Dry Tons/Acre/Year			Million Dry Tons/Year					
						Baseline			
Corn	69	3.3	3.3	450	225	75	14	6	95
Other crops	244			608	267	38	1	0	39
CRP grasses	25	2	0	51	0	0	0	0	0
Pasture	68	1.5	0	101	0	0	0	0	0
Perennials	0	0	0	0	0	0	0	0	0
Other	42			23	58.5	0	0	60	60
Total	**448**			1233	550	113	14.6	66	194
Moderate Yield without Land Use						**Change***			
Corn	77	**4.1**	**4.1**	**626**	**313**	**170**	**47**	**8**	**225**
Other crops	235			689	308	**67**	9	1	**98**
CRP grasses	25	2	0	51	0	**25**	0	0	**25**
Pasture	68	1.5	0	101	0	0	0	0	0
Perennials	0	0	0	0	0	0	0	0	0
Other	43			23	73	22	0	75	75
Total	**448**			1490	694	284	**56**	84	423
High Yield with Land Use						**Change***			
Corn	77	**4.9**	**4.9**	**751**	**376**	**256**	**75**	**12**	**343**
Other crops	218			760	359	**173**	12	0	**186**
CRP grasses	15	2	0	31	0	15	0	0	15
Pasture	43	1.5	0	64	0	0	0	0	0
Perennials	**55**	**0.6**	**7.4**	**440**	**409**	**368**	**0**	**0**	**368**
Other	40			62	83	11	0	75	86
Total	**448**			2108	1227	823	87	87	998

* Major changes in bold.
SOURCE: Perlack, et al., 2005; Oak Ridge National Laboratory

448 million acres. The top portion of the table gives the "baseline" of biomass currently available: 194 Mdt. The second portion of the table gives the scenario assuming moderate increase in yields (+30%), no perennials, and no land use change, resulting in 423 Mdt. Finally, the bottom portion gives the high-yield (+50%), perennials, and land-use-change scenario, which yields 998 Mdt. The land use changes involve transfer of some pasture, CRP grasses, and hay acreage to perennials, such as switchgrass.

Of this total potential of 998 Mdt, less than 9% would come from grain crops (mostly corn) and 37% would come from perennial grasses such as switchgrass, 26% from corn stover, and 26% from other agricultural residues, which all could be removed "sustainably" or with minimal impact on soils and waters. Indeed, the future potential of biomass energy depends on our ability to tap these crop residues, grasses, and woody corps, and to convert

figure 14.5   Perennial Grasses and Crop Residues Are the Key to Increased Fuel Ethanol Production

(a) Miscanthus

(b) Switchgrass

(c) Corn stover bales in storage

Current U.S. ethanol production uses grain crops like corn, but future production growth must come from crop residues (like the bales of corn stover above [c]) and from perennial grass energy crops (like miscanthus [a] and switchgrass [b]), which combined constitute 82% of ORNL's estimated biomass energy potential.

SOURCES: a: Miscanthus: photo by Patrick Schmitz, S. Long lab, University of Illinois–UC; b: Switchgrass: photo by Warren Gretz, DOE/NREL; c: Corn stover bails in storage: D. Glasser, NREL, Corn Storer, Approaching its Real Worth, 1999.

them into fuel ethanol economically (Figure 14.5). This is because biofuels from grain crops such as corn and soybeans will never be able to make a significant dent in our oil consumption without impacting food supply and price.

Where in the United States would a biomass energy industry be located? Crop-based biomass would be centered in the country's agricultural heartland, as shown in Figure 14.6 from NREL (2005). The map shows the location of the biomass resource yield (t/km²/yr)

figure 14.6   Biomass Resources of the United States

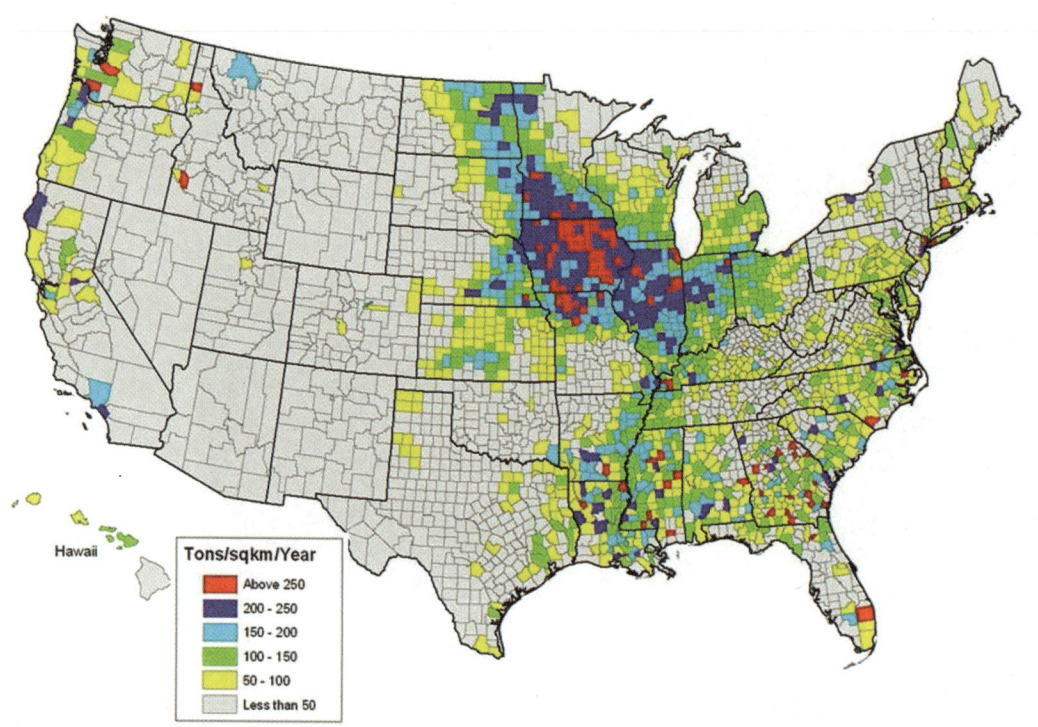

Resources in tons/sqkm/yr, including dedicated energy crops, residues, municipal wastes.

Source: NREL, 2005

corresponding to the baseline case, including agricultural and woodland residues, municipal wastes, and dedicated energy crops. But conversion to perennial grass-based biomass would broaden the geographic source area beyond the heartland, especially to the southern states that have longer growing seasons. Figure 14.7, from Greene (2004), projects perennial switchgrass production (in 1000 tons per county) in 2030.

14.3 Fuel Ethanol

Humans have been making alcohol for six millennia. Yes, early humans liked beer too. The same fermentation process using microorganisms, mostly yeasts, to convert sugars into alcohol in beer is used to produce fuel ethanol. Another similarity is the current popularity of both alcoholic brews. Fuel ethanol is a replacement for gasoline, can be produced domestically, and has far lower net GHG emissions than gasoline. With higher gas prices, interest has grown among investors, government policy makers, and energy analysts in greatly expanding the production capacity of ethanol in the United States and other countries. World production

figure 14.7   Potential Perennial Switchgrass Production in 2025 at $40/dry ton

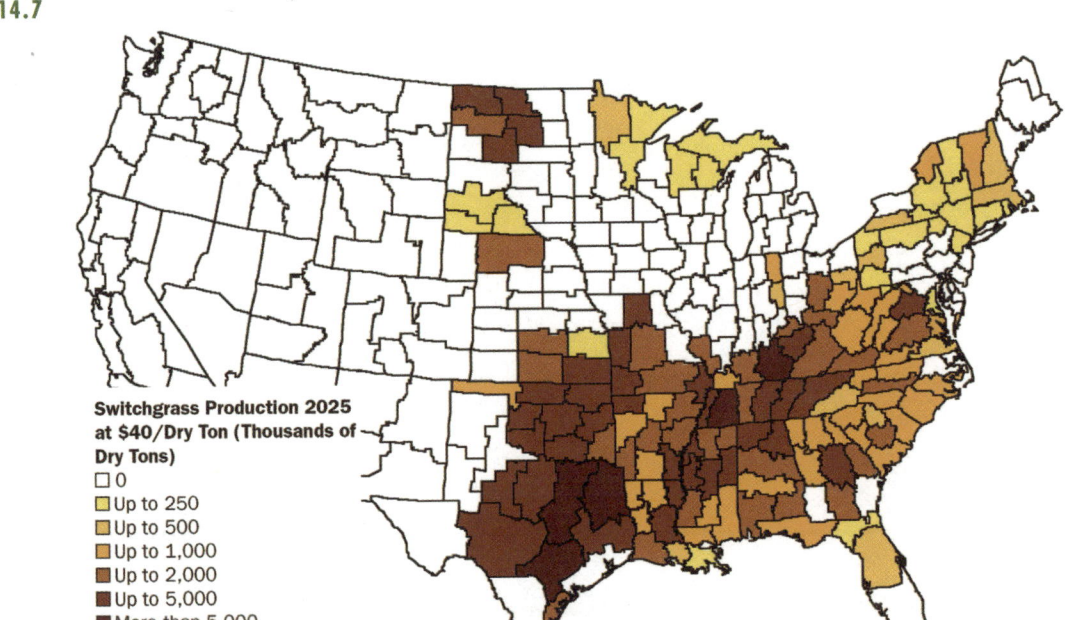

SOURCE: Greene, 2004

of fuel ethanol grew from 4.6 Bgal in 2000 to 13.5 Bgal in 2006, or about 22% per year (Figure 14.8). Brazil had been the world leader in ethanol production until 2005 when the United States surpassed it.

figure 14.8   World Fuel Ethanol Production, 1995–2006

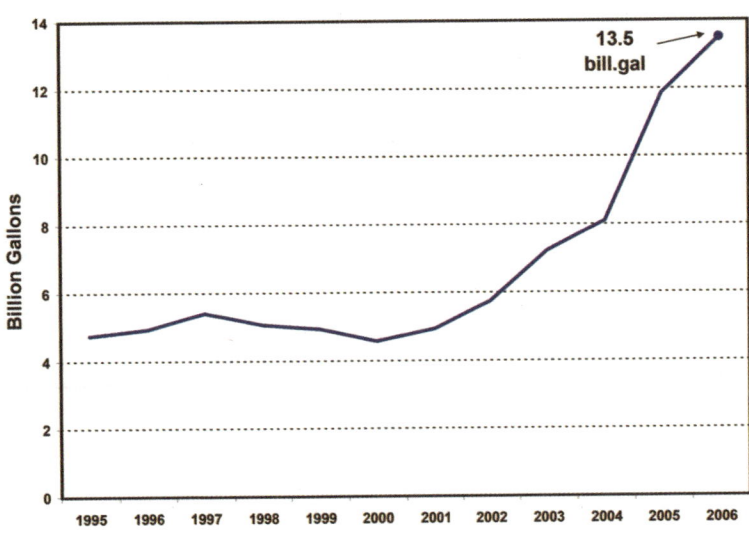

figure 14.9   U.S. Ethanol Production, 1980–2006, and 2005 RFS for 2006–2012 (million gallons)

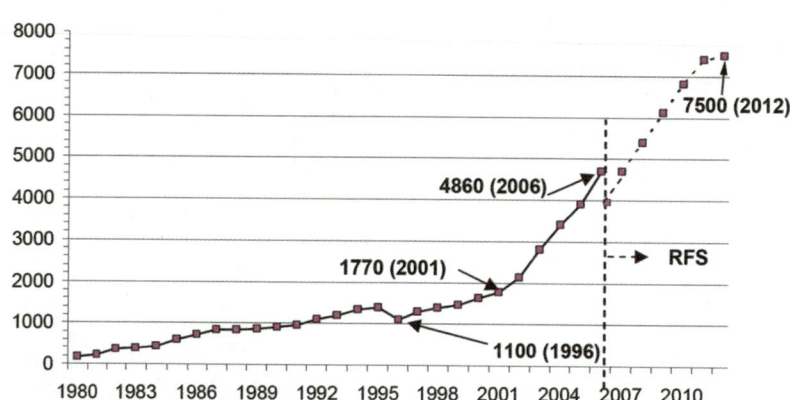

The RFS enacted by the 2005 EPAct was exceeded in its first year (2006). The 2007 act's RFS calls for 35 Bgal by 2022.

The explosive growth of ethanol production in the United States is shown in Figure 14.9. Production grew by 23% per year, nearly tripling from 2001 to 4.86 Bgal in 2006. Production in 2007 is on target to hit 7 Bgal.

14.3.1 U.S. Ethanol Production Capabilities

This growth is expected to continue, driven in part by the national **Renewable Fuels Standard** (RFS) enacted by the 2005 and 2007 Energy Policy Acts. The 2005 EPAct's RFS requires fuel suppliers to include in their fuel supply a minimum amount of renewable fuels gradually increasing from 4.0 Bgal in 2006 (or 2.7% of total fuel supply) increasing by 0.7 Bgal each year to 7.4 Bgal in 2011 and 7.5 Bgal in 2012 (see Figure 14.9). The 2012 standard is equivalent to a savings of 80,000 bbl/day of petroleum. The act also provided a $1 billion loan guarantee program for 80% non-recourse loan guarantees for the first four plants up to maximum of $250 million/plant. And it provided a blender's production tax credit of $0.51/gal ethanol through 2008. Several states have their own RFS and other incentives for ethanol (see Sidebar 14.1 and Chapter 18).

The 2005 EPAct's RFS provides a minimum production schedule that assures investors and producers of a guaranteed market, and with this assurance the fledgling industry may find its wings. It already has. In its first year, the industry exceeded the standard by 22% in 2006 when it outproduced the 2007 level a year ahead of schedule, and in 2007 production may hit 7 Bgal, exceeding the 2010 RFS. In mid-2007, President Bush called for expanding production at a much faster rate to 35 Bgal by 2017, or 20% of gasoline needs by 2017, his so-called 20-in-10 goal. By the end of 2007, Congress passed and Bush signed the expanded RFS requirement for 35 Bgal/yr by 2022 (see Chapter 17).

With new production capacity under development, this pace of growth may well occur. Table 14.5 shows the fall 2007 ethanol production capacity in the leading states, and Figure 14.10

figure 14.10 Location of Existing and New Fuel Ethanol Biorefineries, October 2007

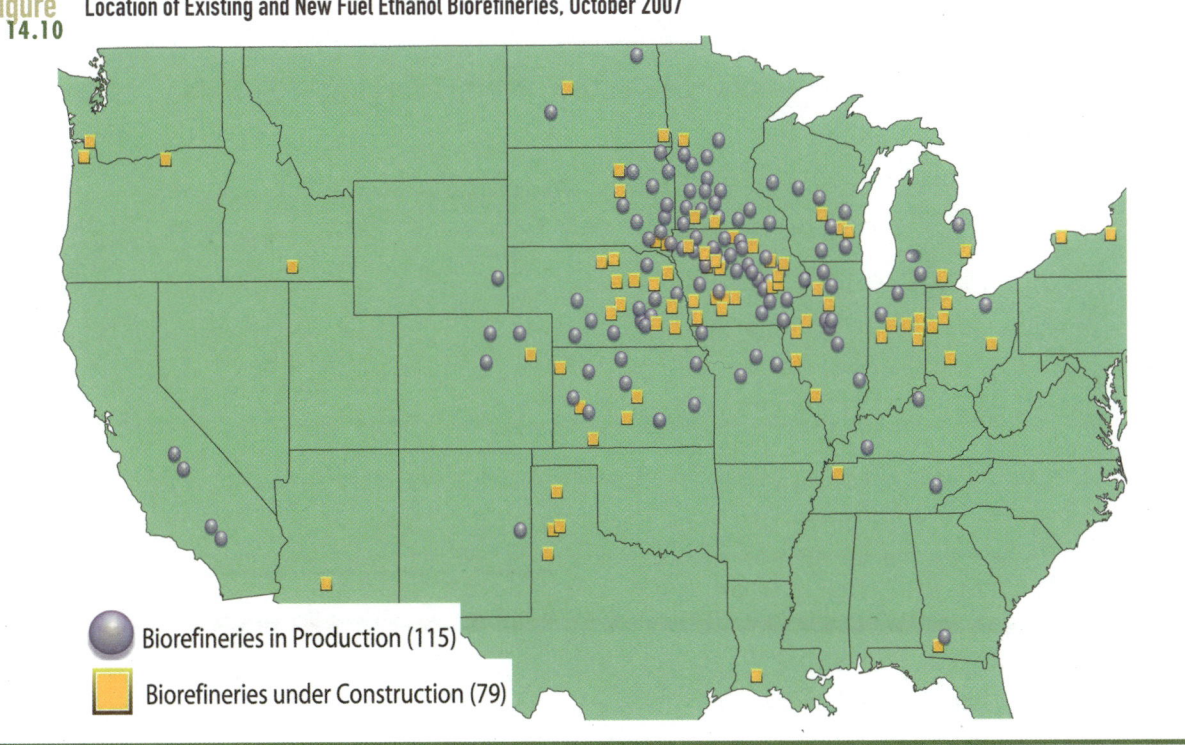

● Biorefineries in Production (115)

▢ Biorefineries under Construction (79)

Source: RFA, 2007

table 14.5 Ethanol Production Capacity* by State, 2007

	Online, 10/07	Under Construction	Total 10/07
Iowa	1863	1495	3358
Nebraska	1018	728	1746
Illinois	881	291	1172
South Dakota	607	378	985
Minnesota	605	498	1102
Indiana	292	556	848
Wisconsin	278	220	498
Kansas	213	295	508
Michigan	214	50	264
Missouri	186	na	186
North Dakota	123	210	333
Ohio	0	529	529
Texas	0	355	355
New York	0	164	164
Other	138	688	826
Total 10/07	**7023**	**6452**	**13,475**
Total 4/06	**4486**	**2049**	**6715**
Total 1/05			**4398**

* Million gallons per year (Mg/yr).
Source: RFA, 2007

SIDEBAR 14.1

Minnesota: State Promoter of Ethanol Production and E85 Sales

In 1997, Minnesota took a mandate for gasohol sales that applied to the cities only part of the year and extended the mandate to an annual statewide requirement. All gasoline sold in Minnesota was to be E10. There was 97% penetration by 2005. Sales of E85 and the number of stations selling E85 are growing rapidly. In 2007, 21 million gallons were sold, eight times the 2004 sales. The number of fueling stations has grown from 85 in 2003 to 312 in 2007. With ethanol production capacity of almost 600 Mgal/yr in mid-2006, Minnesota is an ethanol exporter. The $1.5 billion industry employs thousands, and the state looks to expand both production and marketing. In 2005, the state approved an **E20 Renewable Fuel Standard,** requiring all gasoline sold in the state to average 20% ethanol by 2010. With its mandatory minimum E10 for all gasoline and rising sales of E85, the state may achieve the standard without any blending to E20. The RFS aims to provide more certainty to investors in the state's ethanol industry.

Minnesota E85 Sales and Stations

	E85 Sales 1000 gal	E85 Stations
2000	301	56
2001	694	65
2002	1244	70
2003	2179	85
2004	2606	101
2005	8085	175
2006	17,934	291
2007	21,400	312

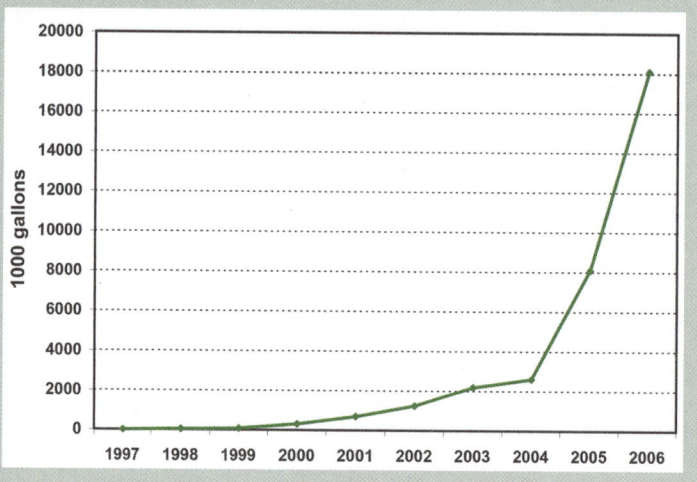

gives their location. Operating and planned capacity doubled between April 2006 and October 2007. In 2007, there were 115 ethanol plants with a total production capacity of 7 Bgal/yr and 79 new plants and expansions under construction, for a total of 13.5 Bgal/yr capacity. There is a strong industrial cluster for ethanol in the Iowa, Illinois, Nebraska, South Dakota, and Minnesota region, with nearly 65% of existing and new capacity, but other states are also adding capacity. Sidebar 14.1 highlights Minnesota's ethanol industry.

This explosive growth is impressive, but 2006 production is still a drop in the bucket compared to U.S. gasoline use (3.5%), domestic oil production (6.2%), and oil imports (2.6%). Currently, ethanol is made primarily from corn in the United States, and it is used primarily as an oxygenate additive in E10 gasohol. To make a greater impact on oil and gasoline markets, three changes in ethanol production and application are needed:

1. Ethanol capacity must continue to grow rapidly to offset a greater proportion of gasoline use.
2. The fuel ethanol market must expand from oxygenate additive to E85 fuel. This requires more E85 production and fueling stations and more flex-fuel vehicles that can use the fuel.
3. Ethanol production must transition from corn-based raw material to cellulosic residues and grasses. This requires production of perennial grasses and residues on the scale of Table 14.4 and further development and commercialization of enzymatic hydrolysis (saccharification) for large-scale conversion of cellulose to ethanol.

Although these developments will take time and investment, many see them as easier, quicker, and less risky than increasing dependence on imported oil or other alternative fuel options such as coal-to-liquids or hydrogen. Before addressing these challenges, let's first look at how ethanol is produced.

The trick in ethanol production is turning biomass materials into readily fermentable sugars. For some feed stocks, such as sugarcane used in Brazil, this production process is straightforward. For others, such as corn and especially cellulosic material, it is more complicated.

The process for four types of feedstock is illustrated in Figure 14.11. The main difference is in the feedstock conversion to sugars. Sugar crops are easy to convert. Grain crops,

figure 14.11 Ethanol Production Steps by Feedstock and Conversion Method

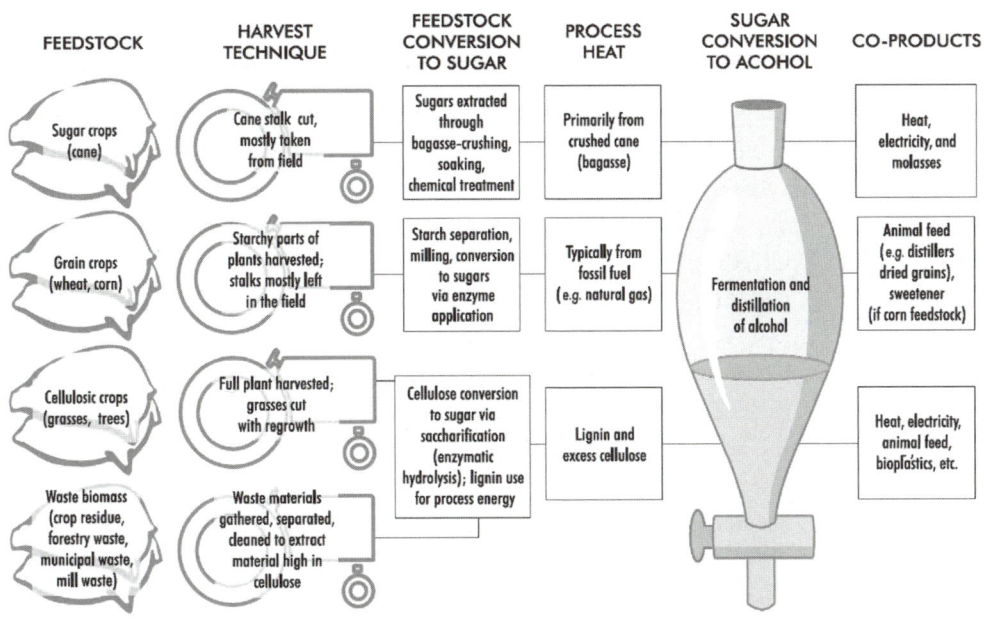

Source: IEA, *Biofuels for Transport: An International Perspective,* 2004

such as corn or wheat, require milling, starch separation, and enzymatic reaction. Conversion of cellulosic materials, such as grasses, crop residues, paper wastes, and wood, to fermentable sugars is the most complex. Their sugars are locked in complex carbohydrates (polysaccharides). To free the sugars from the woody lignin requires enzymatic hydrolysis or **saccharization.**

The key to these processes is the **enzymes.** Enzymes are active proteins that catalyze microorganism action to break down complex carbohydrates ultimately into simple sugars. The ease, time, control, and ultimately the cost of these processes depend on effective and often expensive enzymes.

All these processes require energy. Most grain crop conversions use fossil fuels for this process, whereas sugar crop and cellulosic material conversions have residual biomass materials that can be used for process heat, and this enhances the overall energy efficiency of the process.

14.3.2 Ethanol from Corn

Most ethanol from corn uses a dry milling process in which liquefied corn starch is produced by heating ground cornmeal with water and enzymes. The basic process looks like this:

$$Corn \rightarrow Starch + H_2O + enzymes + heat \rightarrow Sugars + yeast \rightarrow$$
$$(56\ lb)\quad (32\ lb)\qquad\qquad\qquad (36\ lb)$$
$$Ethanol\ (C_2H_5OH) + CO_2 + heat + DDGS$$
$$(18\ lb)\qquad\qquad (18\ lb)\qquad (17\ lb)$$

where the bold italic terms = inputs to the process
DDGS = dried distillers grain solids that have a very high feed value

About 82% of ethanol production in the United States is done using a dry milling process and about 18% is from wet milling. The overall dry milling production process for corn includes the following steps:

1. **Grinding** to the consistency of coarse flour in a hammer mill or roller mill.
2. **Cooking:** Ground corn is mixed with water and two enzymes at high temperature (>120°F) and pressure (10–40 psig), then held at about 180°F–195°F for 4–8 hours. Considerable energy is used in this process and it generally comes from fossil fuels. One enzyme, **alpha amylase**, chemically liquefies the starch polymers into shorter strings, and the other, **gluco amylase** chemically "saccharifies" the short strings into sugars (mostly glucose [$C_6H_{12}O_6$]).
3. **Fermentation:** Sugar "mash" is put in tanks with large amounts of yeast that convert the simple sugars into ethanol, CO_2, and heat.
4. **Distillation:** More heat is added to boil off the ethanol, which is then condensed, separating it from non-fermentable constituents and water.

5. **Dehydration:** The 190 proof (95%) ethanol from distillation still has 5% water, which is removed by more distillation or drying columns.

6. **Fate of non-fermentables:** Because residue materials have feed value, further processing by centrifuge to 25%–40% solids (wet distillers grains with solubles) or by additional drying to about 90% solids (dry distillers grains or DDGS) adds productive and energy value (Kohl, 2003).

14.3.3 Ethanol from Cellulose

As we learned in section 14.2.1, the biggest potential for ethanol production is not from corn but from cellulosic materials, including

- Crop residues such as corn stover (the stock and husk) and other materials now left in the field
- Perennial grasses such as switchgrass that can be harvested and regrown rapidly without cultivation
- Fast-growing trees such as poplar and willow
- Municipal and other wastes that are high in cellulosic fibers

But the process for converting these materials into ethanol is more complex than it is for corn. They are made up of lignin, hemicellulose, and cellulose. Cellulose molecules are made up of long chains of glucose molecules similar to those of corn but they are encapsulated in lignin. Hemicellulose has long chains of 6-carbon sugars like glucose (called hexose sugars) but also 5-carbon sugars (called pentoses), and these vary depending on the plant. Special enzymes and microorganisms are needed to break down and ferment these different sugars.

figure 14.12 Cellulose Ethanol Process

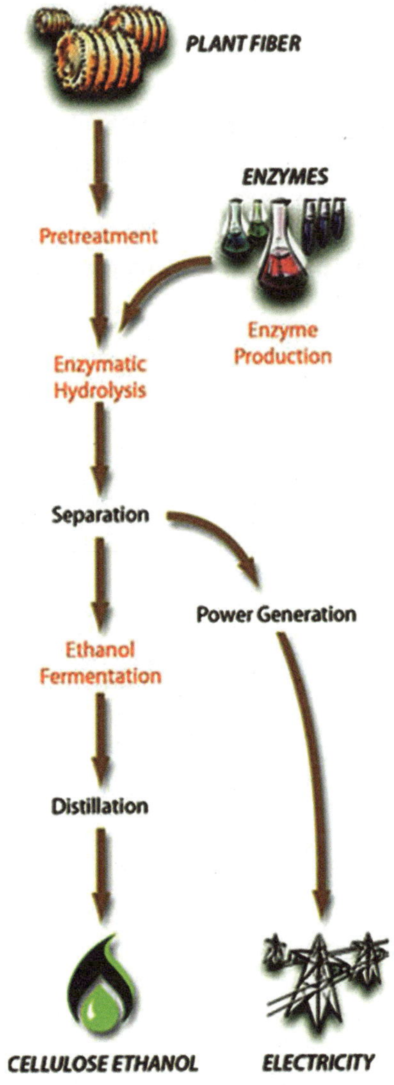

PLANT FIBER

Pretreatment

ENZYMES

Enzyme Production

Enzymatic Hydrolysis

Separation

Power Generation

Ethanol Fermentation

Distillation

CELLULOSE ETHANOL   ELECTRICITY

SOURCE: Passmore, 2006

Acid hydrolysis and enzymatic hydrolysis are two options for freeing the fermentable sugars from the complex polysaccharides in the cellulose; enzymatic hydrolysis is more promising. The overall process has the following steps:

1. **Pretreatment:** To break down the lignin sheath surrounding the cellulose, options include dilute acid, steam explosion, ammonia fiber explosion (AMFE), or organic solvent processes. AMFE uses ammonia under moderate heat and pressure to disrupt biomass components, and it appears to be most promising (Greer, 2005).

2. **Enzymatic hydrolysis or saccharization:** This process converts cellulose to sugars. The key barrier to cost-effective cellulosic ethanol is the high cost of enzymes, but enzyme producers in the biotechnology industry are driving down costs. Iogen, Inc., Arkenol Holdings, Genencor International, Novozymes Biotech, and other companies are active in enzyme development and other processes.

3. **Waste separation:** The lignin must be separated from the fermentable materials. It has high energy value comparable to coal and can be used to generate electricity and heat to power the operation and even feed the grid. This is a significant net energy savings for this process.

4. **Fermentation, distillation, dehydration, and the production of residue feed** materials are basically the same as in the corn-ethanol process.

Commercializing Cellulosic Enthanol: President Bush's 20-in-10 Initiative

In 2007, President Bush announced the goal of making cellulosic ethanol cost competitive with gasoline by 2012 and producing 35 Bgal per year by 2017. This, in conjunction with increased auto efficiency, could reduce gasoline consumption by **20% in 10 years.** In December 2007, this initiative was codified by Congress and extended as a 35 Bgal RFS by 2022. To facilitate this development, under the authority of the 2005 Energy Policy Act, in 2007 DOE announced grants to six companies to build six cellulosic plants in the next four years. The $385 million in grants will leverage private funds for a total investment of more than $1.2 billion. The six companies are as follows:

- Abengoa Bioenergy Biomass of Kansas (11.4 Mg/yr ethanol plus net electricity from corn stover, wheat straw, switchgrass)
- ALICO, Inc. (13.9 Mg/yr ethanol plus net electricity from biomass waste)
- Bluefire Ethanol, Inc. (19 Mg/yr ethanol from green and wood waste)
- Broin Companies (125 Mg/yr ethanol, 25% from cellulosic corn fiber, cobs, and stalks)
- Iogen Biorefinery Partners, LLC (18 Mg/yr ethanol from agricultural residues including wheat straw, barley straw, corn stover, and switchgrass)
- Range Fuels (40 Mg/yr ethanol plus 9 Mg/yr methanol from woody residues and crops)

figure 14.13   Iogen Corporation's Demonstration Cellulose-to-Ethanol Plant in Ottawa

SOURCE: Iogen Corp.

Iogen has a proprietary enzyme used in its 260,000 gal/yr wheat straw-to-ethanol plant in Ottawa, Canada. Its U.S. partnership has attracted funding from Shell Oil, Goldman Sachs, and other investors for its Idaho commercial scale facility.

14.3.4 Ethanol Conversion Efficiencies

The conversion efficiency of biomass energy to ethanol depends on the material. NREL has a biomass feedstock and composition database for several energy crops, residues, and waste. The composition includes C-5 and C-6 polymeric sugars, which determine the theoretical ethanol yield from the material. NREL also has a convenient online calculator that computes the yield for different compositions. See http://www.eere.energy.gov/biomass/ethanol_yield_calculator.html. Table 14.6 gives the ethanol yield for several materials.

Table 14.7 gives current and prospective conversion efficiencies of biomass to ethanol for corn and switchgrass, as well as projected yields per acre. **Gasoline equivalent gallons** (gge) per dry ton and per acre are also given. The projected increase yields (dry tons per acre [dt/ac]) are taken from Perlack, et al., (2005) for corn and from Greene (2004) for switchgrass. Projections of increased conversion efficiency for switchgrass (gallons ethanol per dry ton [gal/dt]) come from Greene (2004). Included in the last two lines are energy credits from coproduction of biofuels (lignin) for generation of power and heat.

Solution Box 14.1 looks at these yields and efficiencies and the ORNL estimate of biomass potential (1.3 billion tons per year [Bdt/yr]) in the context of the 2007 RFS for 2022 (35 Bgal/yr), 2007 production (7 Bg/yr), and current petroleum data. As shown in Table 14.8, this is twenty-one times current ethanol production, three times the 2022 RFS,

table 14.6	Theoretical Ethanol Yield for Feedstock			
Material	**Yield (gal/dt)**		**Material**	**Yield (gal/dt)**
Corn grain	124.4		Switchgrass	105.0
Corn stover	113.0		Hardwood sawdust	100.8
Rice straw	109.9		Mixed paper	116.2

SOURCE: NREL Biomass Feedstock Database

one and one-fourth times current domestic oil production, three-fourths of current gasoline consumption, and half of current oil imports.

The immense potential for cellulosic ethanol becomes clear when musing about dedicating production to energy grasses and crops. Solution Box 14.2 considers dedicating all of South Dakota's current farmland to energy grasses and crops. Given expected increases in yields and process efficiencies, South Dakota could become comparable to a member of OPEC in supply of liquid fuel.

14.3.5 Ethanol Net Energy and Greenhouse Gas Emissions

We introduced the debate over ethanol net energy in Chapter 5. The issue was popularized by Cornell University's David Pimental, who has maintained for more than a decade that it takes more energy to produce corn-based ethanol than you get out of it. Others, like USDA's Hosein Shapouri, disputed his claims, and a battle of competing net energy studies ensued. It became clear that the results were a function of the assumptions and data used (see Figures 5.3 and 5.4). While their studies tried to measure total net energy, Michael Wang from Argonne National Lab helped by changing the question: if we care

table 14.7	Current and Prospective Annual Yields and Conversion Efficiencies of Biomass-to-Ethanol			
		Biomass to Ethanol Efficiency		
Feedstock and Assumptions	**Yield (dt/ac)**	**gal/dt**	**gge/dt**	**gge/ac**
Corn (2005)	3.3	124	83	274
Corn, increased yield	4.9	124	83	407
Switchgrass (SG; 2005)	5	50	33	165
SG + improved conversion efficiency (IC)	5	105	69	345
SG + IC + biofuel co-production (BC)	5	117	77	385
SG + IC + BC + increased yield	12.4	117	77	955

* dt/ac = dry tons per acre; gal/st = gallons per dry ton; gge/dt = gallon gasoline efficiency per dry ton; gge/ac = gasoline equivalent per acre
SOURCES: Perlack, et al., 2005; Greene 2004

SOLUTION BOX 14.1

Putting Ethanol Production Potential in Context

How would the potential ethanol production measure up to current ethanol production, the 2022 RFS, current gasoline consumption, domestic petroleum production, and U.S. oil imports?

Solution:

Let's assume an overall conversion efficiency of 117 gal/dt (77 gge/dt) for energy crops, crop residues, and perennial grasses, and a 0.66 factor for gasoline equivalent.

$$\text{Biomass production (tons/yr)} \times \text{fuel yield (gal/dt)} =$$
$$\text{fuel ethanol (gal/yr)} \times 0.66 = \text{gasoline equivalent (gge/yr)}$$
$$1.3 \text{ Bdt/yr} \times 117 \text{ gal/dt} = 150 \text{ Bgal/yr} \times 0.66 \text{ gge/gal} = 100 \text{ Bgge/yr}$$

How does this potential compare to 2007 ethanol production?

$$2007 \text{ production} = 7 \text{ Bgal/yr} \times 0.66 \text{ gge/gal} = 4.7 \text{ Bgge/yr}$$
$$\frac{\text{Ethanol potential}}{2007 \text{ production}} = \frac{100}{4.7} = 21$$

table 14.8   Potential Annual U.S. Ethanol Production vs. 2005 Petroleum Indicators

Indicator	Annual Value	Potential Ethanol Indicator
U.S. potential ethanol production	100.0 Bgge	1
U.S. 2007 ethanol production	4.7 Bgge	21
RFS for 2022	30.0 Bgge	3.3
2005 U.S. gasoline consumption	138.6 Bgal	0.72
2005 U.S. crude oil production	78.5 Bgal	1.27
2005 U.S. petroleum net imports	189.4 Bgal	5.3

SOURCES: Perlack, et al., 2005; Greene 2004

about oil consumption and carbon emissions, not just energy, then shouldn't we be measuring the relative impact of fuel ethanol on those factors? His studies showed corn-based ethanol could displace a considerable amount of petroleum with far less fossil fuel inputs (see Figure 5.3).

SOLUTION

Let's Make South Dakota Comparable to an OPEC Member

If South Dakota were to dedicate its 44 million farm acres to energy crops, crop residues, and perennial grasses, what would its annual yield of biomass be at current and expected crop and ethanol yields?

Solution:

Let's assume a mix of corn, residues, and perennial grasses that yields about 4 tons per acre today, and perhaps 10 tons per acre later with greater use of residues and grasses and higher yields. Let's assume an average ethanol yield of 60 gal/dt today and 100 gal/dt tomorrow with more efficient ethanol and energy recovery (after Khosla [2006], and Ceres Company).

South Dakota Biofuel Potential

	Today	Tomorrow
Farm acres	44 million ac	44 million ac
Tons/acre	4 dt/ac	10 dt/ac
Ethanol yield, gal/dt	60 gal/dt	100 gal/dt
1000 bbl ethanol/day	689	2870
1000 bbl gas equiv/day	455	1894

OPEC Oil Production (1000 bbl/day)

Saudi Arabia	9400
Iran	3900
Kuwait	2600
Venezuela	2500
UAE	2500
Nigeria	2200
South Dakota	**1894**
Iraq	1700
Libya	1650

We also discussed a study published in *Science* by the University of California-Berkeley Energy Resources Group (ERG). Using their ERG Biofuels Analysis Meta-Model (EBAMM), the study reviewed six previous net energy and life-cycle studies of ethanol production as well as its own three ethanol scenarios: Ethanol Today, CO_2 Intensive Ethanol, and Cellulosic Ethanol. Their net fossil energy and GHG emissions results were given in Figure 5.4. In Figure 14.14, the energy flows in mega-Joules (MJ) inputs/MJ-fuel and GHG emissions are compared for gasoline (including its petroleum feedstock) and the three ethanol options. Cellulosic ethanol has by far the lowest energy inputs, even negative for coal because of net generation of electricity from lignin by-products, and only 12%–14% of the GHG emissions of gasoline and the other ethanol scenarios (Farrell, et al., 2006).

figure 14.14

Energy flow (MJ input/MJ fuel), GHG emissions (kg CO_2/MJ fuel) from gasoline production and three ethanol scenarios: Ethanol Today, CO_2 Intensive Ethanol, and Cellulosic Ethanol. Cellulosic ethanol requires the fewest energy inputs (0.10 MJ/MJ) and produces the lowest net GHG (11 kg CO_2/MJ).

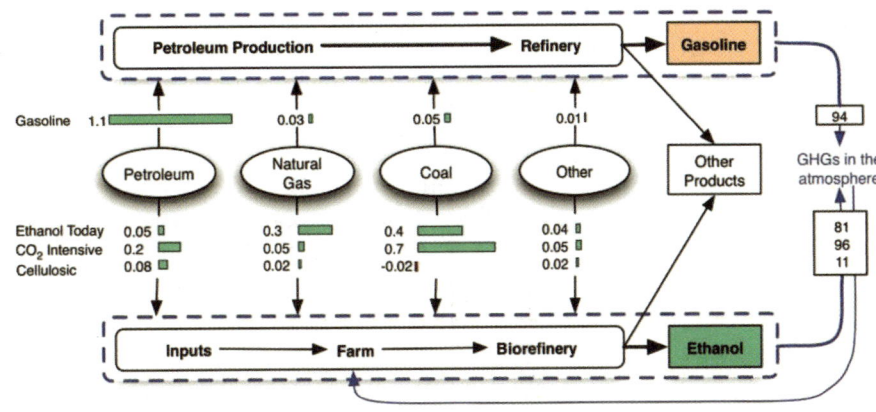

SOURCE: From Farrell, et al., *Science 311*:506–508 (2006). Reprinted with permission from AAAS.

The UC-Berkeley study prompted a lively online debate among critics and advocates of ethanol fuel (see *Science* online), but the message of the study remained clear:

- Corn-based ethanol has considerable petroleum-saving benefits, and small net-energy and GHG emission benefits compared to gasoline.
- Cellulosic ethanol has clear advantages in net energy, petroleum savings, and GHG emission reduction over gasoline and corn ethanol.

14.3.6 Food Crops versus Energy Crops

We can't look at energy production on farms in isolation from food and other agricultural production. In fact, the boom in ethanol production in the United States (Figure 14.9) began to affect the market and price for corn in 2006 and 2007. As shown in Figure 14.15, ethanol use has grown from 7% of U.S. corn supply in 2001 to 27% in 2007. This has prompted world food experts such as Lester Brown (2006) to raise concerns about the effect of corn-based ethanol on food markets. Brown fears that this continuing trend, especially with the growing ethanol production capacity (see Table 14.5), will raise prices and affect exports. The U.S. crop is a major factor in global food markets and a safety net for poor production years in other countries. Indeed, corn prices in the nine primary U.S. markets averaged 66% higher in 2007 than in 2005.

The implications of this changing market are unclear. Farmers welcome the higher corn prices, which have not yet affected food prices significantly in the United States. But at some point, and perhaps soon, the competition between corn for food and feed and corn for fuel will have adverse effects on prices and markets. Significant growth in ethanol fuel production from corn will ultimately be limited by these effects. The growth in ethanol needed to offset petroleum use will have to rely on cellulosic ethanol.

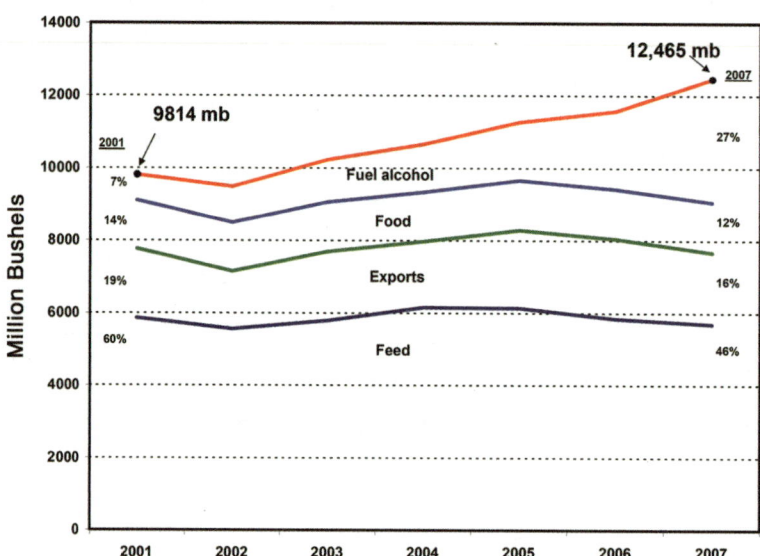

figure 14.15 Changing U.S. Corn Market, 2001–2007

Ethanol market has grown from 7% to an expected 27% share of the corn crop, more than the corn export market.

SOURCE: USDA, ERS Yearbook, 2007

14.3.7 Urban Air Quality and Other Environmental Effects of Ethanol

Although gasohol was originally developed to provide oxygen to gasoline to help reduce emissions of carbon monoxide in urban areas, E10 gasohol has a tendency for slightly greater emissions of nitrogen oxides (NO_x) that contribute to the formation of urban smog and ozone. To address this and other urban air quality issues with ethanol, Argonne Lab along with Dartmouth and Princeton universities, conducted a major well-to-wheel (WTW) study of biofuel options (Wu, Wu, and Wang, 2005).

The study compared six biofuel multi-product production options, focusing on cellulosic feedstocks with various forms of combined-heat-and-power (CHP) generation using lignin by-products, including gas turbine combined-cycle (GTCC) and steam power. Some options included Fischer-Tropsch diesel and dimethyl ether production.

All options showed benefits compared to gasoline and diesel fuel in reducing fossil fuels, petroleum, CO_2 emissions, and urban air pollutants, especially SO_x and surprisingly NO_x. The study concluded: "From a multiple-production perspective, for each unit of biomass, the (ethanol/GTCC) option that co-produces cellulosic ethanol from a consolidated biological process and power from advanced GTCC is the most promising option in that it displaces the greatest amount of fossil fuel and ranks at the top in overall energy and emission benefits among the six options" (Wu, Wu, and Wang, 2005).

Experience with E10 gasohol has shown slight increases in NO_x emissions, but the situation with E85 appears to be different. A 2004 Minnesota study of emissions of E85, E10, and non-ethanol fueled vehicles sheds some light on NO_x emissions. The study tested tailpipe emissions from a flex-fuel Ford Explorer with various fuels, three different brands of E10 (mandatory in Minnesota), non-ethanol gasoline (purchased in Wisconsin), and E85. All ethanol fuel mixes had less total hydrocarbons (THC) emissions than straight gasoline, but two of the three E10 brands had higher NO_x emissions than the non-ethanol gasoline. The E85 test had considerably lower emissions of both pollutants. The air quality benefits of E85, even for NO_x, have also been shown in other studies.

However, Jacobsen (2007) issued a recent caution about the urban air pollution impacts of large-scale use of ethanol due to volatilized ethanol, which is a strong photochemical agent. His work indicates there is alwaays more to know about the effects of potentially beneficial solutions.

Regarding other environmental effects, cellulosic biomass, especially perennial grasses such as switchgrass, has soil and water and carbon sequestration benefits. The perennial grasses are harvested for product but regrow without cultivation or planting. As Figure 14.16 shows, the root system grows deep, holding soil together to reduce erosion and sequestering carbon in its root materials. Perennials provide wildlife habitat, water retention, and wind erosion control, relative to intensive agricultural practices needed for energy crops.

figure 14.16 Soil, Carbon, and Habitat Benefits of Perennial Switchgrass Grown as an Energy Crop

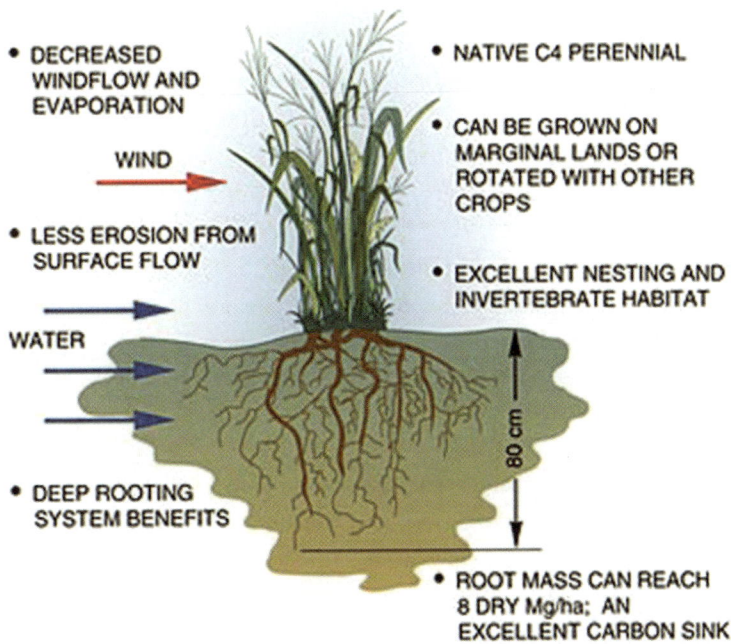

- DECREASED WINDFLOW AND EVAPORATION
- NATIVE C4 PERENNIAL
- CAN BE GROWN ON MARGINAL LANDS OR ROTATED WITH OTHER CROPS

WIND

- LESS EROSION FROM SURFACE FLOW
- EXCELLENT NESTING AND INVERTEBRATE HABITAT

WATER

80 cm

- DEEP ROOTING SYSTEM BENEFITS
- ROOT MASS CAN REACH 8 DRY Mg/ha; AN EXCELLENT CARBON SINK

SOURCE: Oak Ridge National Laboratory: ORNL-DWG-93-M-8892

These land conservation values make perennial grasses a natural productive use on the nation's CRP lands. CRP currently pays farmers $1.5 billion per year to keep 34 million acres out of production. Most of these lands are highly erodible and not suitable for intensive cultivation but ideally suited for perennial production. Growth and harvest of perennial energy crops on these lands are compatible with the objectives of the CRP program, would contribute to farm income, and would save or shift farm subsidies from keeping land idle to putting it into biofuel production.

14.3.8 Achieving Ethanol's Potential

As we have shown, ethanol is now primarily used as an additive in gasohol. If ethanol production capacity, efficiency, and cost improvements meet expected potential, the biofuel can make a significant impact on gasoline consumption, oil imports, and GHG emissions in the years ahead. The market will likely be E85, the 85% ethanol–15% gasoline blend that requires minor changes in existing engines. This requires significant expansion of the agricultural and biorefinery industries dedicated to producing the biomass and processing it to ethanol, as well as the infrastructure to deliver it. This expansion has begun as shown in production (Figure 14.9) and capacity (Table 14.5).

Many are voicing the battle cry for rapid expansion of biofuels, especially E85. These include not only the usual suspects, such as environmental and farm groups, but also not-so-usual ones, such as politicians from all persuasions and notable investors and venture capitalists. Figure 14.17 shows U.S. production through 2006 (4.9 Bgal) and four future visions and goals. They include the following:

a. The 2005 Energy Policy Act Renewable Fuels Standard (RFS): 7.5 Bgal/yr in 2012. At current trends this production will be achieved by 2008 or 2009.
b. President Bush's "20 in 10" goal to achieve ethanol production of 20% of gasoline in 10 years, by 2017. The 2007 Energy Policy Act requires this 35 Bgal/yr by 2022.
c. The increasingly popular "25 × '25" EPT goal to achieve 25% of electric power and transportation fuel energy from renewables by 2025. This goal has been endorsed by half the states and congressional resolutions (see Chapter 3).
d. The vision of Vinod Khosla, the venture capitalist whose words began this chapter. He maintains that ethanol can grow to 150 Bgal/yr by 2031 and displace three-fourths of our gasoline use.

Even if we can expand ethanol production toward these goals by building more biorefineries and dedicating more energy crops and residues, we still need to bring this ethanol to market. As shown in Table 14.2, in 2007 there were only 1100 filling stations that had E85, but the number is growing. More than one-quarter of these were in Minnesota. Stations now generally use the same gasoline tanks and pumps for E85 and regular gasoline. Infrastructure issues seemed to be minor until Underwriters Laboratory (UL) decided in late 2006 not to list gasoline pumps for E85 because of suspected corrosion problems. This created uncertainty about E85 infrastructure and only eighty E85 stations were added between

figure
14.17   Visions of U.S. Ethanol Production

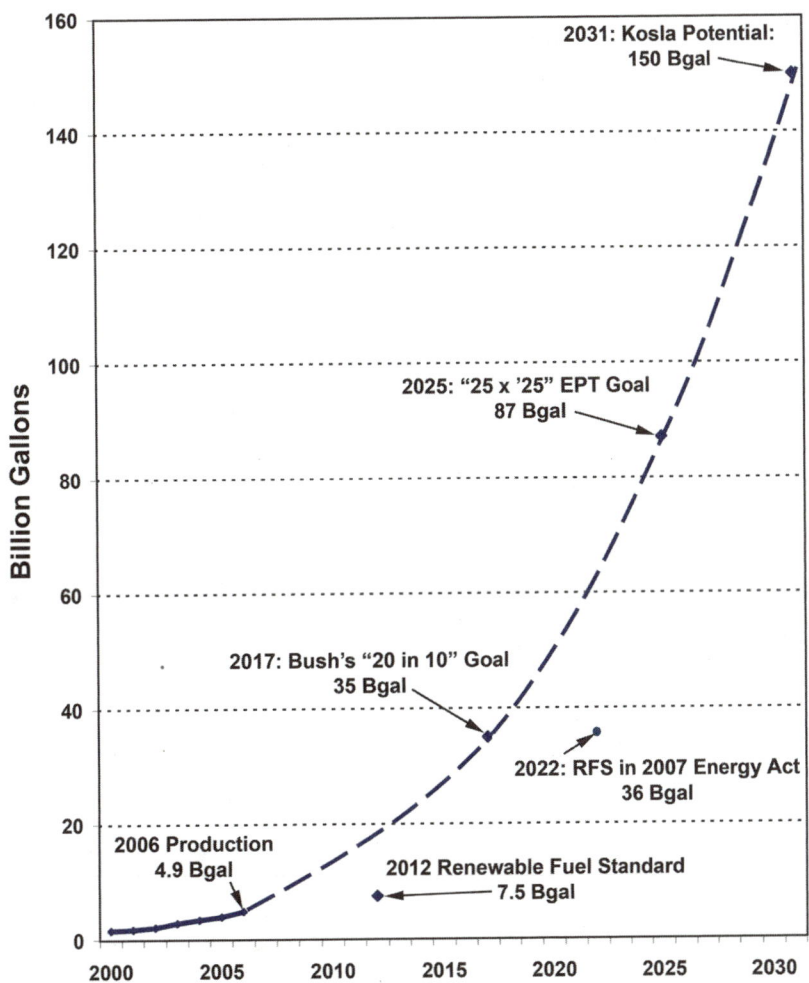

Bush's "20 in 10" goal far exceeds the 2005 act's RFS for 2012 and is the same quantity as the 2007 act's RFS for 2022. It is on track with "25 × '25" goal and even Khosla's ambitious goal of 150 Bgal by 2031.

December 2006 and October 2007. But in October 2007, UL completed an exhaustive research effort and announced accepted safety protocol that is expected to clear the way for a huge expansion of E85 stations in 2008.

Minnesota is providing good lessons for the rest of the country in ethanol fueling (see Sidebar 14.1), but more states and more retailers outside the upper Midwest need to provide access to E85 as the market for ethanol and flex-fuel vehicles grows.

In fact, the vehicle market has already outstripped the fuel delivery infrastructure. There are more vehicles that can use E85 than can find a filling station to get it. As discussed in Chapter 13, there are 6 million flex-fuel vehicles (FFV) on the road today that can use both straight gasoline and E85 seamlessly. Thirty-six models of FFV are on the market, almost all made by the U.S. Big Three automakers.

Why would they be making so many if there are so few opportunities to use E85? Well, manufacturers are given a 0.9 mpg credit on their CAFE fuel economy standards for flex-fuel, and they have taken full advantage. No surprise then that the FFV models on the market are mostly large SUVs. See the current list at http://www.fueleconomy.gov/feg/byfueltype.htm.

Although this tactic by automakers has not led to much additional ethanol use, it has put a large FFV fleet on the road, it has given automakers valuable experience in the technology (which is simple and essentially adds no cost), and it positions the U.S. industry for greater sales of FFV. In November 2005, Senators Lugar (R-IN), Harkin (D-IA), and Obama (D-IL) proposed the bi-partisan Fuel Security and Consumer Choice Act, which would require all vehicles marketed in the United States to be FFV within ten years.

In summary, cellulosic ethanol provides an opportunity to produce a majority of our vehicle gasoline consumption and greatly reduce our oil imports within twenty-five years, while helping revitalize rural economies across America. The ethanol fuel can be marketed as E85 in existing infrastructure and used in flex-fuel vehicles that U.S. auto companies have been making for years at no additional cost. With flex-fuel plug-in electric hybrid vehicles, expected on the market soon, consumers have the choice of E85, gasoline, or electric fuel, with the prospect of using zero-urban-emission electric drive in the city and 85% ethanol on the highway, further reducing gasoline use, oil imports, and GHG emissions.

14.4 Biodiesel

Biodiesel is a fuel that comprises mono-alkyl esters of long-chain fatty acids derived from vegetable oils or animal fats. It offers a biofuel option for diesel vehicles. As discussed in Chapter 13, diesel engines power heavy trucks and equipment. But their use in passenger vehicles is rebounding because of their inherent energy efficiency and improvements made in emissions reductions. Europe has made a strong commitment to diesel cars, and there is likely to be a growing market for clean diesel in the United States.

In fact, like ethanol, biodiesel production has been booming in 2005–2007. Although worldwide biodiesel had only about 8% of the volume production as ethanol in 2005, it grew 67% that year. About 90% of 2005 world biodiesel was in Europe (see Figure 14.18). Why Europe? The European Union (EU) has committed to a renewable fuel standard that 5.75% of its diesel fuel come from biofuels by 2010. Germany leads this effort—its production of 2.0 Mt in 2006 hit its 2010 target four years early. Italy had the next highest capacity at 0.6 Mt. European countries primarily use rapeseed oil as a feedstock.

14.4.1 U.S. Biodiesel Production Capabilities

Production is also growing rapidly in the United States—from 25 Mgal in 2004 to 75 Mgal in 2005 to 250 Mgal in 2006 to an estimated 450 Mgal in 2007. Filling stations selling biodiesel

figure 14.18   World Biodiesel Capacity, 1995–2006

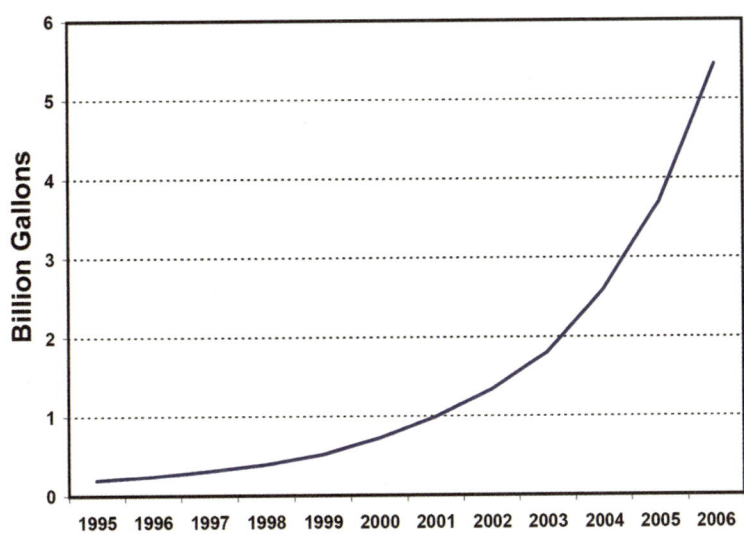

increased from 142 in 2003 to 705 in 2007, with South Carolina leading the nation with 67 (Table 14.2). Other states and communities are encouraging biodiesel development. Minnesota adopted a "blend specific" RFS for requiring all diesel sold in the state to be 2% biodiesel or B-2. It became effective when enough production capacity was developed in the state; when two 30 Mgal/yr facilities were built in 2005, the rule became effective in September 2005. Other states, such as Washington, have adopted "volumetric" RFS like the federal RFS. It has prompted development of production plants. Seattle Biodiesel built a 5 Mgal/yr production plant in downtown Seattle in 2005. Imperium Renewables, parent company of Seattle Biodiesel, is building a 100 Mgal/yr plant in Grays Harbor, Washington.

In the United States, soybean and seed oils are the primary feedstock, although waste vegetable oils are also used. One appeal of biodiesel is that conversion from raw or waste oils is a fairly simple "backyard" technology that can be done at small scale. But one disadvantage is that to be commercial, resulting biodiesel must meet the same strict standard that all diesel fuel must meet, the American Society of Testing and Materials' ASTM D 6751–02. If the product meets this standard it can be used at a range of blends, from B-2 to B-100. Most dealers and manufacturers of diesel engines will honor engine warranties up to B-5, but B-20 is increasingly becoming the blend of choice.

Meeting this standard and dealing with some of the by-products like glycerol, have plagued some small producers. But "backyard" production will not impact our petroleum problem, and only large-scale development will likely have sufficient economies of scale to produce a quality product at competitive prices.

Biodiesel production capacity is growing rapidly in the United States. Figure 14.19 shows the existing production plants and those under construction or expansion in mid-2007. By early 2008, capacity totaled 2240 Mgal/yr. Capacity does not equal production because the plants will not operate at 100% all year. Much of this capacity was just coming on line in 2007, so actual production in 2007 was about 450 Mgal for the year. Nineteen of these plants have BQ-9000 certification, a voluntary and cooperative program of the National Biodiesel Accreditation Board. Capacity under construction or expansion, shown in Figure 14.19(b), totals an additional 1230 Mgal/yr.

figure 14.19   Existing and New U.S. Biodiesel Production Capacity, 2007

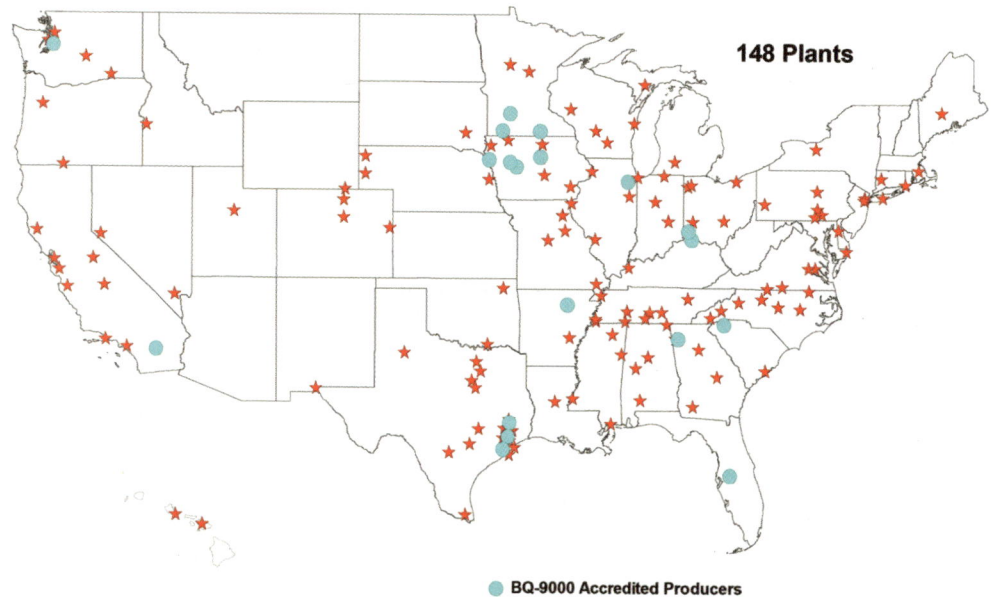

(a) Biodiesel production capacity in June 2007 totaling 1.39 Bgal/yr. BQ-9000 are accredited producers.

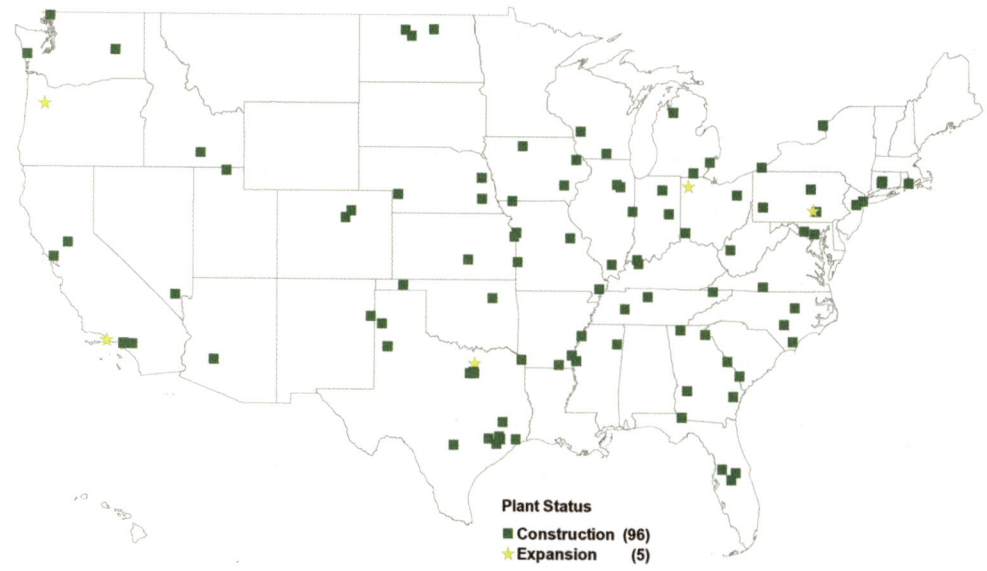

(b) Biodiesel production capacity under construction or expansion in 2007 totaling 1.89 Bgal/yr.

Source: NBB, 2007

figure 14.20 Transesterification Process for Producing Biodiesel from Vegetable or Animal Oils

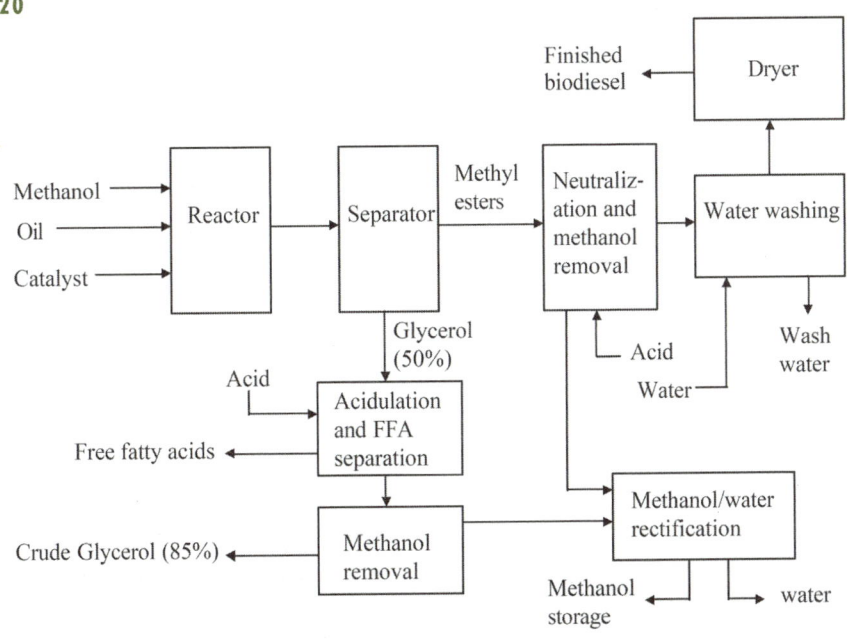

SOURCE: Jon Van Gerpen, Biodiesel Production and Fuel Quality, University of Idaho

14.4.2 Biodiesel Production Technologies

Biodiesel is produced through a process called **alkali-catalyzed transesterification,** illustrated in the schematic in Figure 14.20. Oils along with alcohol (usually methanol) and catalysts are placed into a reactor and then into a separator to divide methyl esters from glycerol. The methyl esters are neutralized and methanol is removed, then they are washed and dried to produce finished biodiesel. The remainder of the process deals with by-products glycerol and methanol, both of which have marketable value if they can be purified. Methanol can be recycled in the process.

The two primary challenges in biodiesel production are meeting the high ASTM D 6751 standard for diesel fuel and converting the potentially hazardous by-products glycerol and methanol into marketable quality. The former requires tight quality control of the methanol removal, water washing, and drying process steps. The latter conversion through methanol and free fatty acid separation is just as important as the production process for commercial effectiveness.

14.4.3 Biodiesel from Algae

Algae is to the future of biodiesel as cellulose is to the future of fuel ethanol. The primary feedstock for existing and planned capacity in the United States is soybean oil, although other

table 14.9	Biodiesel Yield Estimates for Various Sources	
Source	Yield (gal/ac)
Corn	15–20
Soybeans	40–50
Safflower	80–90
Sunflower	100–110
Rapeseed	110–130
Palm oil	625–650
Microalgae	5000–15,000

Source: NREL, 1998

seed oils and some waste oil are used. In Europe, rapeseed oil is the most popular source. These are energy crops that have alternative productive uses, so large-scale biodiesel production from these feedstocks will encounter the same competition with food products as will corn-based ethanol.

Microalgae, including diatoms and green algae, may provide an answer for biodiesel source material. Algae can produce at least thirty times the amount of oil per area of land as terrestrial oil seed crops, because of their abundance; proliferation; high oil content; ideal structure for photosynthesis; and ideal access to nutrients, water, and CO_2 through their aqueous suspension. Table 14.9 gives an estimate of biodiesel yields from various crop seeds (15–650 gal/ac) and microalgae (5000–15,000 gal/ac).

U.S. DOE funded an Aquatic Species Research program at the National Renewable Energy Laboratory from 1978 to 1996. The program funding was small, peaking at $2.5 million in 1984, then declining to closeout in 1996. Although the program ended, it provided basic research for the renewed current interest in algae as a source for biofuels. One concept developed through the program was an algae growth system enhanced by CO_2 capture from fossil fuel power plant emissions. Figure 14.21 shows a conceptual diagram of a series of algae growth lagoons with inputs of water, nutrients, CO_2, and sunlight, and outputs of algae for biofuel production.

This design is now being further developed by GreenFuel Technologies Corp. (GFT), which adapted the concept to the 20 MW cogeneration plant at MIT. The demonstration showed an 82% reduction of CO_2 emissions on sunny days and 50% on cloudy days, and also an 85% reduction in NO_x emissions. The GFT system does not use lagoons, but 3-meter long, 10–20 cm diameter polycarbonate tubes tilted to the sun like evacuated tube solar collectors. Flue gases are introduced at the bottom and bubble up through the algae medium. On a daily basis, 10%–30% of the algae are removed. GFT estimates biodiesel yields of 5000–10,000 gal/ac and comparable yields of ethanol. Others, such as Global Green Solutions, are claiming even higher yields of biodiesel from algae, with a similar design using thin-film membranes developed by Valcent Products. Its demonstration pilot plant was targeted for late 2007.

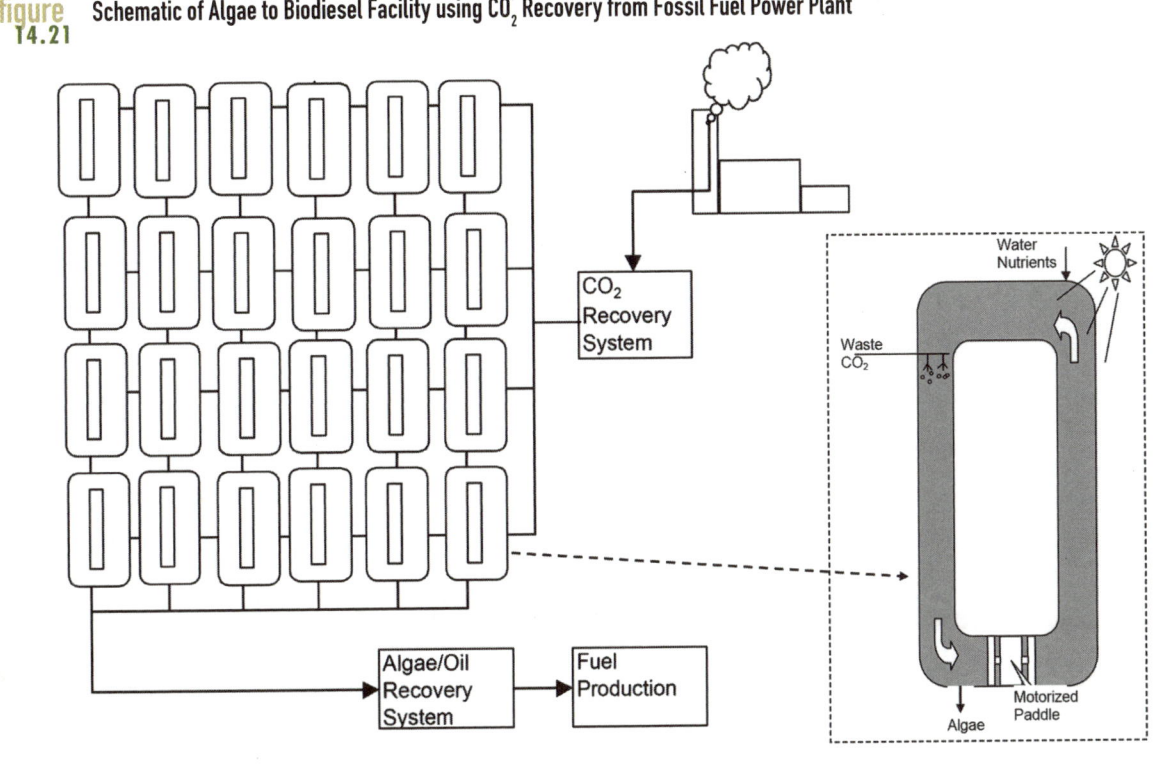

figure 14.21 Schematic of Algae to Biodiesel Facility using CO_2 Recovery from Fossil Fuel Power Plant

Source: NREL, 1998

14.4.4 Environmental and Life-Cycle Considerations of Biodiesel

Like all biofuels, the overall environmental impacts of biodiesel depend on what raw material is used, what land and production practices are used in growing it, what processes and controls are used to extract oils and convert them to biodiesel, and how the biodiesel is used. Although net energy for biodiesel from seed crops is not as well analyzed as ethanol, it appears to be positive. Like ethanol from corn, the real life-cycle advantages come from GHG emission reduction and petroleum savings. Life-cycle net energy, economic, and environmental benefits are likely to be much greater for biodiesel from algae than from seed crops, just as they are likely to be much greater for ethanol from cellulose than from corn.

Another significant life-cycle advantage of biodiesel and other biofuels is the reduction of CO_2 emissions and most criteria pollutants. When additional public policies for GHG emission reductions (such as a CO_2 cap and trade system) are enacted, they will translate into further economic advantages for biofuels.

Table 14.10 shows that B-20 and especially B-100, have significantly lower emissions of criteria pollutants except NO_x. NO_x continues to cloud the otherwise significant air quality benefits of biodiesel, and research continues to address this and other remaining air quality issues such as black carbon.

table 14.10	Emission Impacts from Biodiesel Compared to Conventional Diesel	
	B-100	B-20
Total unburned hydrocarbons	−67%	−20%
Carbon monoxide	−47%	−12%
Particulate matter	−48%	−12%
NO_x	+10%	+2%

Source: EPA, 2002

Agricultural practices for biomass production must minimize impacts on water and soils. For example, in December 2006, the *Wall Street Journal* reported in 2006 that the rush to develop biodiesel from palm oil in Indonesia has created an environmental disaster as huge areas are being burned on Borneo to clear land for the palm oil plantations (Barta and Spencer, 2006).

14.5 Other Biomass Energy and Emerging Biotechnologies

14.5.1 Other Biomass Energy

There are a number of other biomass energy sources. Although they are not directly related to transportation, this is a convenient place in the book to address them. Most of these sources are used for power generation, thermal uses, or both in CHP facilities. The biomass types include wood wastes and residues, municipal wastes, landfill methane recovery, and methane digestion from sewage sludge and agricultural animal wastes (Figure 14.22).

14.5.1.1 Biomass/Wood/Waste Heat and Power Generation

Figure 14.2 showed that 47% of 2004 renewable energy in the United States (or about 3% of total energy) came from biomass. About half of this comes from industrial forest products operations that burn wood residues and pulping liquors for heat and electricity. Of the 1000 wood-fired power plants, about two-thirds are primarily for industrial use and one-third are independent power producers generating electricity for sale.

As several states have adopted Renewable Portfolio Standards (RPS) for which wood-fired power plants qualify, there has been a growing interest in wood-fired electricity generation. In 2006, five such plants were planned in New England. Although such plants provide renewable fuel and GHG emission reduction benefits compared to fossil fuel plants, they are still combustion facilities requiring delivery of bulk solid fuel usually by truck. Local residents have not always embraced these plants in their communities.

Another type of biomass energy facility that has met with local concerns is municipal waste-to-energy plants. These accept raw or processed municipal wastes and burn them similar to old incinerators, except that they have energy recovery for steam heating or power gen-

**figure
14.22** **Biomass Sources for Combined Heat and Power**

(a) Wood chip CHP plant

(b) Wood residue

(c) MSW stream

(e) Landfill gas microturbines

(d) Landfill gas recovery

(f) Methane digester at sewage
treatment plant

Wood waste and residues and municipal solid waste can be burned in steam boilers for combined heat and power. Landfill gas and methane from sewage sludge or animal waste digestion can be burned in gas or microturbines, reciprocating engines, or Stirling engines for power and heat.

Source: Wisconsin Distributed Energy Collaborative; New Jersey Meadowlands; CB&I, Inc.

eration, and they have modern pollution control. They have an added benefit of reducing the volume of waste to be landfilled. In 2005, 13.6% of U.S. municipal solid waste (MSW) or 33.4 Mdt was combusted for energy recovery in 88 plants, 39 of which are in the northeast and 26 in the south. But the number of waste-to-energy plants, tonnage, and percent of wastes burned are all down since 2000. Table 14.11 shows the MSW generated in the United States and its fate from 1960 to 2005. Combustion with energy recovery peaked in 1990.

table 14.11   U.S. Municipal Solid Waste Generation and Fate, 1960–2005

Activity	1960	1970	1980	1990	2000	2003	2004	2005
Generation (million tons)	88	121	152	205	237	240	247	246
Recovery for recycling	6.4%	6.6%	9.6%	14.2%	22.2%	23.2%	23.1%	23.8%
Recovery for composting	Neg.	Neg.	Neg.	2.0%	6.9%	7.9%	8.3%	8.4%
Total materials recovery	6.4%	6.6%	9.6%	16.2%	29.1%	31.1%	31.4%	32.1%
Combustion with energy recovery	0.0%	0.3%	1.8%	14.5%	14.2%	14.0%	13.8%	13.6%
Discards to landfill other disposal	93.6%	93.1%	88.6%	69.3%	56.7%	54.9%	54.8%	54.3%

SOURCE: U.S. EPA, 2005

14.5.1.2 Landfill Methane Recovery and Methane Digestion

Most municipal wastes are disposed in landfills. Biomass in the waste, mostly paper, decomposes anaerobically (without oxygen) and produces methane (CH_4). This gas is trapped but ultimately seeps into the atmosphere. Many sealed landfills are converted to parks, and leaking methane can be toxic and hazardous to users. To avoid risk, landfills are normally vented to release methane, but methane is a powerful GHG, 23 times more powerful than an equal mass of CO_2. In some landfills the vented methane is flared off, burning the methane to CO_2. Landfill methane recovery, on the other hand, not only captures the methane but uses it to generate heat and power. Using gas turbines, microturbines, or reciprocating or Stirling engines, this gas can be converted to useful power with additional heat recovery.

Landfill gas (LFG) is a growing source of community power. Figure 14.23 shows the status of landfill gas projects in the United States. As of October 2006, there are 410 operational projects with a total 1080 MW capacity capturing 225 million cubic feet per day (mcfd) of methane and generating about 10 billion kWh of electricity. An additional 575 candidate landfills offer a potential for 1380 MW and 700 mcfd capture. The energy provided is good, but the conversion of methane to carbon dioxide is even better.

In a critical assessment of landfill gas (LFG) recovery, the Natural Resources Defense Council (2006) developed the following priorities:

1. **Avoid LFG by avoiding landfills.** The first priority is increased resource reduction and recycling. Biomass—especially paper—is easily recycled or composted. If there is no biomass in landfills, then there will be no LFG.
2. **Burn all LFG that is produced.** Even if we could close all landfills today, they would continue to produce LFG for years to come. Burning LFG in an engine, a turbine, or simply in a flare has tremendous benefits by reducing toxicity and reducing greenhouse

figure 14.23 Location of Operational and Candidate Landfill Gas Recovery Projects

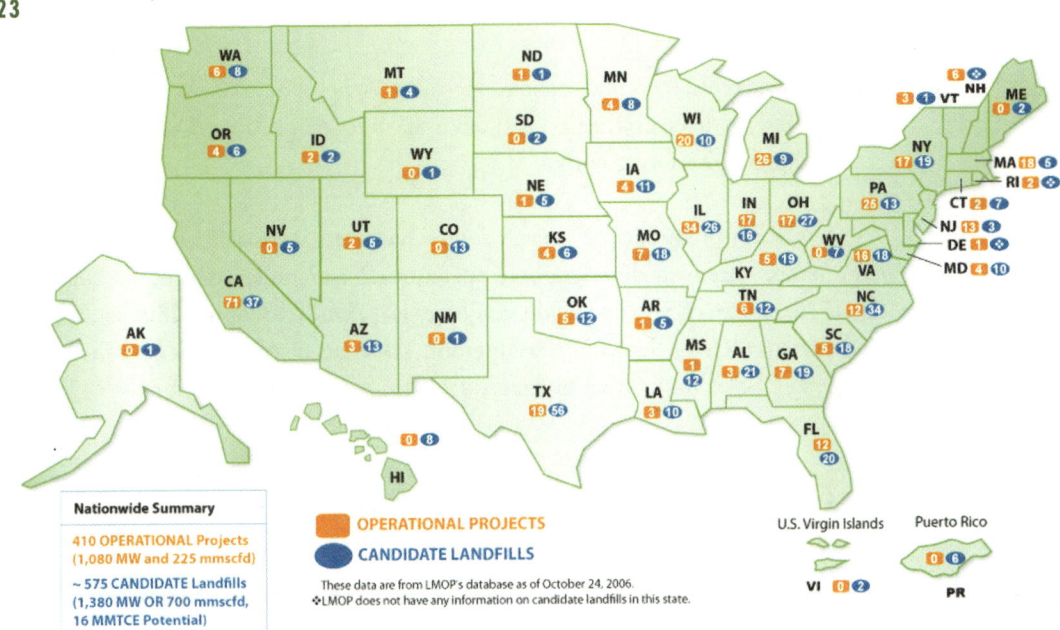

California and Illinois are the leading states.

Source: U.S. EPA LMOP, 2006

gases. Over 60% of LFG is generated at landfills with no collection system; and at landfills with collection systems, it is typical for at least 25% of LFG to escape.

3. **Use LFG for energy production.** The balance of benefits favor using LFG for energy.

Methane digestion. Usable methane can also be produced from organic wastes through anaerobic digestion processes. Although digestion has been used to stabilize sewage and animal wastes for decades in the United States and abroad, it has not been developed to its potential for usable energy production. Opportunities exist for additional methane digestion of municipal sewage sludges and especially concentrated animal facilities such as feedlots and poultry operations.

However, methane digestion involves a complex biological process, and a 63% failure rate of farm digesters installed in the 1970s has discouraged new farm applications. In addition to energy production, they also reduce odors and produce a more stable by-product, so as the costs of the conventional disposal methods increase for space or environmental reasons, and as fuel costs rise, methane digestion may increase in use for animal wastes.

Digestion is a two-stage process. First, acid-forming bacteria break down complex wastes into acids; then methane-forming bacteria convert these acids to methane. The resulting "biogas" is about 65%–70% methane. The process requires the right balance of these different bacteria, the right nutrient balance (carbon to nitrogen ratio), the right composition of waste and water, and the right temperature. The "active" ingredients of the wastes are their content of "volatile solids" (VS).

- Animals typically produce about one-third of their body weight in VS per day.
- Typical methane production is about 3–8 cubic feet per pound VS, higher for chickens and pigs, lowest for cows.

Digestion of municipal sewage sludge has been increasing in recent years. Higher performance systems include High Temperature Thermophilic-Mesophilic Digestion (e.g., Duluth, Minnesota), Separate Acid/Gas Phases (e.g., DuPage County, Illinois), and Extended Solids Retention (e.g., Spokane, Washington). These advanced systems have the advantage of producing Class A by-product sludges suitable for most land applications.

14.5.2 Emerging Biotechnologies for Energy

Photosynthesis in green plants is one of the miracles of life. So is human ingenuity. Advances in biotechnology research may create significant opportunities in capturing the sun's energy and transforming it into useful biomass energy more efficiently to increase energy yields. We have seen that common green algae and diatoms have potentially high yields of oils suitable for conversion to biodiesel, far more than terrestrial crops. Genetic engineering research can optimize algae production enhancing the efficiency of transfer of solar energy to biomass energy. Further advances may even use the photosynthetic process directly to produce hydrogen.

This exciting research aims to use the mechanisms of photosynthesis to extract hydrogen directly from water. As introduced in Chapter 4, the magic of photosynthesis is provided by a number of enzymes and nucleotides, such as ADP and ATP, which transport and accept electrons permitting a wide range of chemical reactions for the miracles of life.

Photobiological water splitting uses the natural enzymatic process of photosynthesis to split hydrogen gas directly from water. It uses bioengineered forms of green algae and cyanobacteria that consume water and produce hydrogen as a by-product. Bench lab experiments have collected hydrogen gas from beakers containing water-splitting algae. The biological process is complex, using an integrated system of hydrogen production with a combination of algae and photosynthetic and anaerobic bacteria. Although current processes are too slow for commercial application, this is a promising area of research.

14.6 Natural Gas and Hydrogen as Transportation Fuels

CNG has been a popular alternative urban transportation fuel (Tables 14.1 and 14.2) and natural gas has served as the energy source for what small amount of hydrogen is currently used. Future options for alternative transportation fuels include both synthetic liquids from natural gas and hydrogen.

14.6.1 Natural Gas as a Transportation Fuel

Two main options for natural gas are as CNG and natural gas derived synthetic liquid fuel. CNG has been a popular urban fuel for public fleets and buses, but its use is constrained by refueling needs and price uncertainties. As discussed in Chapter 2, natural gas supplies are believed to be greater than oil, and much natural gas is wasted in oil fields around the world for lack of easy transport to markets. Growth of the global market in LNG may help. LNG is simply cold, condensed natural gas (–163°C) that as a liquid occupies 1/600 the volume and can be shipped in tankers. At destination, the LNG is gasified by warming and fed into NG pipelines.

Another alternative is to convert natural gas to liquid (GTL) fuel at normal temperature and pressure so that it can replace gasoline and diesel fuel in transportation vehicles. With higher oil prices, there is renewed interest in GTL conversion. Most current ventures use the well-known Fischer-Tropsch (FT) method developed in the 1920s and used for decades in South Africa by Sasol, which has a capacity of 150,000 bbl/day of liquids from coal and natural gas. Modified versions of the Fischer-Tropsch technology (shown in Figure 14.24) are being developed in other countries.

However, dependency on natural gas for transportation fuel carries with it some of the same problems of petroleum. Although domestic natural gas supplies are greater than oil, they are limited, and future natural gas use in the United States will likely depend more on

figure 14.24   Schematic of Fischer–Tropsch Technology

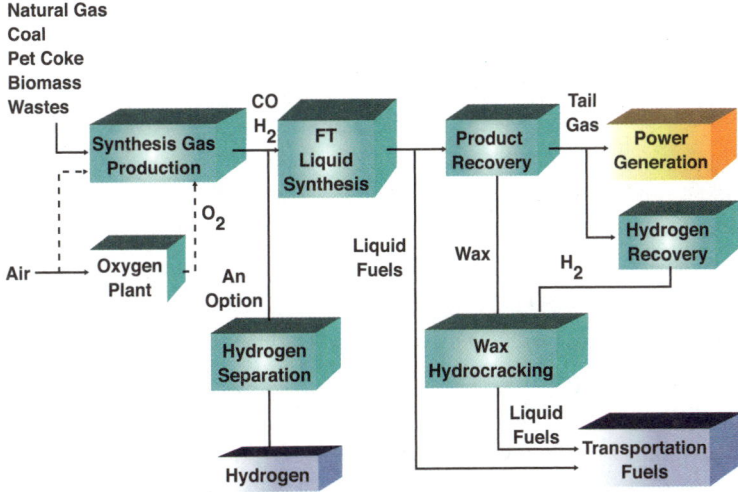

Fischer-Tropsch technology for converting coal, natural gas, or biomass into liquid transportation fuels.

SOURCE: William Harrison, U.S. DOD

imports, now 16% of supply (see Chapter 2). More importantly, natural gas prices follow the volatility of oil prices (see Table 5.4). Finally, natural gas is a source of fossil CO_2 emissions. Although its emissions are 25% less than gasoline, the FT process is only about 66% efficient, and it would exacerbate carbon emissions for transportation fuel. With natural gas feedstock, CO_2 emissions would be 14% more than gasoline; using coal feedstock for FT coal-to-liquids (CTL), emissions would be 100% more than gasoline.

14.6.2 Hydrogen as a Transportation Fuel

During the past five years, hydrogen has been touted as the answer to many of our energy problems—the perfect clean fuel that has multiple applications through direct combustion and conversion to electricity in fuel cells. Fuel cell technology in vehicles provides an opportunity to address petroleum and urban air pollution, and we introduced progress in the development of fuel cells and fuel cell vehicles in Chapters 10 and 13.

However, we know from Chapter 3 (Section 3.3.5) that hydrogen is just a storage medium and an energy source is needed to produce hydrogen. We need to consider life-cycle cost of hydrogen production—carbon-free, oil-free, and secure energy sources are the best options. Hydrogen can be extracted from water using electrolysis, in which electrical energy is used to break water into hydrogen and oxygen at about 70% efficiency from electrical energy to hydrogen energy. Electrolysis using carbon-free renewable wind and solar electricity is promising, but overall efficiency is small and it may be better to use the electricity directly.

Photobiological techniques discussed in the previous section are very exciting. So is **photoelectrochemical water splitting.** In this process, instead of using photosynthesizing algae and bacteria, hydrogen is produced from water using sunlight and specialized semiconductors. Different semiconductor materials work at different wavelengths of light and directly dissociate water molecules into hydrogen and oxygen. More research is needed to find the right materials and to collect the separated hydrogen.

But we are a long way from perfecting such a process, and until we do we are left with existing technologies of electrolysis or reforming hydrogen from natural gas or other fossil fuels, the most common method today. Reformation extracts hydrogen from natural gas at about 60% efficiency. Figure 13.22 showed a Honda prototype Home Energy Station that reforms natural gas to hydrogen, which can then be used in a fuel cell vehicle and a stationary home fuel cell. This looks pretty cool, but we saw in our well-to-wheel (WTW) assessment in Chapter 13 that reformed hydrogen in a fuel cell vehicle has 25% more energy use than a Prius hybrid without any reduction in CO_2 emissions over a hybrid (Figures 13.27 and 13.28).

Our WTW assessment also showed that the WTT efficiency of hydrogen electrolysis using fossil fuel steam-generated power is only 20%–30% depending on the type of generation. WTW energy use of fossil-steam electrolysis hydrogen fuel cell vehicles was by far the highest of all of the fuel-vehicle options, 22%–75% greater than a conventional gasoline vehicle depending on the type of generation. The hydrogen fuel cell vehicle using grid-average power for electrolysis also had the highest CO_2 emissions, 42% greater than the conventional gasoline car.

In addition to these life-cycle energy, economic, and carbon issues of hydrogen, there are technical and economic issues of storage, transport, and delivery of hydrogen to use, that pose significant barriers to what many have referred to as the "hydrogen economy." As far as vehicles are concerned, flex-fuel, plug-in hybrid electric vehicles may offer an easier, more cost-effective, and more energy efficient option than hydrogen fuel cell vehicles not only for the short term, but also for the long term.

14.7 Summary

In our quest to reduce oil use, carbon emissions, and energy demand growth, we must address transportation energy. Transportation consumes 68% of the oil and accounts for 32% of the carbon emissions in the United States today. More than 80% of that oil and carbon is attributable to highway vehicles, and three-fourths of that is from light cars, SUVs, and pickups. In Chapter 13, we introduced three approaches in our quest:

1. Improved vehicle efficiency through new designs and technologies
2. Alternative fuels to displace oil and reduce emissions
3. Reduction in vehicle miles traveled through better land use planning and increased use of commuter transit and other efficient transportation modes

Chapter 13 focused on vehicle efficiency and technology and Chapter 15 addresses vehicle miles traveled. This chapter has explored alternative fuels, especially biofuels. Electric and plug-in hybrid vehicles discussed in Chapter 13 may increase the use of electricity as an alternative "fuel." We saw that electric and plug-in hybrid vehicles can offer high performance, zero urban air emissions, and affordable per-mile costs, especially with overnight, off-peak battery charging. Figure 10.5 showed that if 40% of California vehicles were electric or plug-in hybrid, they could all be charged overnight with currently unused off-peak capacity. Advances in lightweight battery technology are likely to reduce cost and grow this market.

Biofuels, especially fuel ethanol produced from cellulosic crop residues such as corn stover, and perennial grasses such as switchgrass, may provide significant displacement of gasoline and reduction of GHG emissions. Biodiesel also has promise, but like ethanol, increased biodiesel volume depends on non-food biomass sources like microalgae. Recent studies by U.S. DOE, USDA, ORNL, and others estimate a domestic potential to grow 1.3 Bdt of biomass for energy, enough to displace 30% of our petroleum consumption and 60% of our gasoline consumption by 2030. This could be done without impacting domestic and export needs for food and fiber, while maintaining environmental land conservation and revitalizing rural economies.

The best bet for rapid expansion of biofuels is increasing production and marketing of E85, the blend of 85% ethanol and 15% gasoline. This requires

- Large development of biorefinery capacity
- Gearing up of production and recovery of cellulosic crop wastes and grasses

- Advances and cost reduction in enzymatic hydrolysis of cellulose
- Manufacture of more flex-fuel vehicles
- Increased availability of E85 fueling pumps at filling stations

None of these actions is difficult and the good news is that a coalition of diverse interests, including private investors, energy, and automobile companies, state and federal policy makers, and civil society organizations, are beginning to speak in one voice for these actions. RFS and other policies at the federal level and in several states help set the stage for significant private investment, the critical ingredient for rapid deployment.

The U.S. national RFS set by the 2005 Energy Policy Act of 7.5 Bgal of biofuels by 2012 was too modest, and in 2007 Congress changed it to 35 Bgal/yr by 2022. This is similar to President Bush's goal of "20 in 10," for a 20% biofuel contribution to vehicle fuel in ten years or by 2017. This would amount to about 35 Bgal of biofuels. The more ambitious "25 × '25" goal, endorsed by half of the states and resolutions in Congress, calls for 25% of transportation fuel from renewables by 2025, amounting to about 85 Bgal. And Vinod Khosla's vision of 150 Bgal/yr by 2031 takes the cake. Still this vision equals the production from ORNL's estimated biomass fuel potential of 1.3 Bdt per year.

Khosla (2006a) offers three simple policy recommendations to accelerate the movement toward this vision.

1. Require E85 distribution at 10% of all gas stations owned by those with more than fifty stations in the country.
2. Require 70% of all vehicles sold in the United States to be FFV within five years. Supply them all with yellow gas caps and give those caps to all who currently have FFV so they all know they can fill up at E85.
3. Establish a contingency tax on oil if it falls below $40/bbl to assure that will be the floor price. This assurance that oil will not undercut biofuels will spur needed investment. The tax revenues could be used to reduce oil price if it gets above $60 or $80/bbl.

CHAPTER 15

Whole Community Energy and Land Use

For buildings, energy considerations in technology, design, practice, ratings, and codes have evolved from a focus on the building envelope to greater attention to HVAC systems then to electricity use in what Don Aitken (1998) calls the Whole Building Energy. This "envelope + HVAC + electricity" approach is now represented in EPA ENERGY STAR criteria and some building codes. As discussed in Chapter 8, we are beginning to see this concept expand further in Green Building criteria, like LEED protocol, to include embodied energy of materials and life-cycle environmental and health effects.

Whole Community Energy extends these considerations even further to the role the building can play in community energy. Buildings not only consume energy, but they can also produce energy with on-site power generation for the building's needs and for the grid-connected community. In Chapters 10 and 11, we described distributed energy systems in which buildings' rooftop photovoltaics, combined-heat-and-power microturbines, and other on-site power systems can generate and feed excess power to the community grid, and do it economically through net metering. We also saw in Chapters 10 and 13 how grid-connected, on-site generation can feed the daytime grid in exchange for nighttime grid charging of batteries in electric and plug-in hybrid vehicles.

The Whole Community approach also addresses transportation energy by considering the building's location, its relationship to its site and neighborhood, and its connectivity to other neighborhoods and the region. In other words, efficient land use, brought about by compact, mixed-use, pedestrian, and transit-oriented development, and in-fill redevelopment, can reduce travel distances and enhance nonmotorized and transit modes of travel. Efficient land use in Whole Community Energy can reduce automobile dependency and vehicle miles traveled (VMT), saving energy, especially oil fuels. Green Building rating systems, such as LEED-H (homes) and LEED-ND (neighborhood development), are starting to incorporate Whole Community criteria, including on-site generation, efficient land use, and transport connectivity.

We learned in Chapter 13 that about half of our transportation energy in the United States is used in communities, especially in passenger travel in light vehicles, buses, and commuter rail to move us from our homes to work, school, commerce, and recreation.

This chapter first reviews community transportation patterns. It then focuses on transportation- and energy-efficient land use and how "smart growth" management can help arrest sprawling patterns of development and revitalize downtowns and inner suburbs, while reducing VMT and air pollutant and GHG emissions. The chapter concludes with a discussion of Whole Community Energy and land use, including solar access and the urban heat island, emerging "Green Community" development rating guidelines, some models of exemplary projects, and a vision of metropolitan development in the United States to accommodate the next 100 million Americans by 2040.

15.1 Community Transportation

15.1.1 Are U.S. Patterns an Anomaly or the Aspiration?

We saw in Table 13.6 the considerable difference in passenger transport indicators for cities in the United States and other countries. Table 15.1 highlights the comparison with western European countries. Despite having the same income per capita, the average European city is nearly four times as dense, has half the freeway length per person, and has half the parking spaces per 1000 jobs. The U.S. urban dwellers have three times the private car passenger miles per person, four times the passenger CO_2 emissions per person, and 1.5 times the energy use per passenger mile. Whereas Europeans use nonmotorized and transit modes for 50% of their travel, U.S. city residents use them for only 11%.

table 15.1 Urban Passenger Transport Indicators for the United States and Western Europe, 1995

Indicator	Units	United States	Western Europe
Number of cities in group sample		10	32
Average population of cities	million people	5.7	2.2
Urban density	persons/ha	15	55
Length freeway per person	mi/person	0.16	0.08
Parking spaces/1000 CBD jobs		555	261
Passenger cars/1000 persons		587	414
Passenger car passenger-km/cap	p-km/person	18,155	6202
Nonmotorized mode trips	%	8%	31%
Motorized public mode trips	%	3%	19%
Motorized private mode trips	%	89%	50%
Overall energy per pass-km	MJ/p-km	3.20	2.17
Passenger CO_2 emissions/capita	kg/person	4405	1269

Abbreviations: cap = capita; ha = hectare; kg = kilogram; km = kilometer; MJ = mega-joules, million joules; pass = passenger;
SOURCE: Kenworthy, 2003

What accounts for these differences? Certainly, culture, available land, and the price of motor fuel are influencing factors. Figure 13.13 showed that European gas prices are two to three times the price in the United States. Because of more limited land area, as well as their architectural and planning history, European cities are more dense, and this has facilitated the development and use of rail transit.

As poorer countries of the world develop, their passenger travel will increase. Figure 13.2 showed that transportation energy use in "emerging economies," including China and India, is expected to increase by 3.6% per year to 2025. To what extent will they develop transportation systems after the U.S. model based on personal vehicles or the European model based on public transit? And what impacts will much higher fuel prices expected in an oil- and carbon-constrained future, have on the U.S. patterns of travel? Let's look at the U.S. patterns more closely.

15.1.2 U.S. Vehicle Miles Traveled

Figure 15.1 shows the U.S. highway VMT through 2005, with projections to 2025. VMT have doubled since about 1980. Growth has been about 2.3% per year since 1990. No wonder our roads and freeways are congested.

Most of those miles are traveled within our communities, and most of those are for commuting to work, school, or commerce. Table 15.2 shows how travel to work has changed

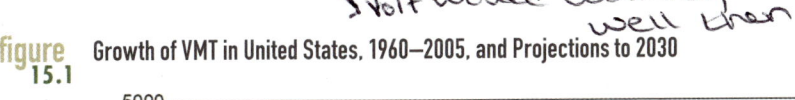

figure 15.1   Growth of VMT in United States, 1960–2005, and Projections to 2030

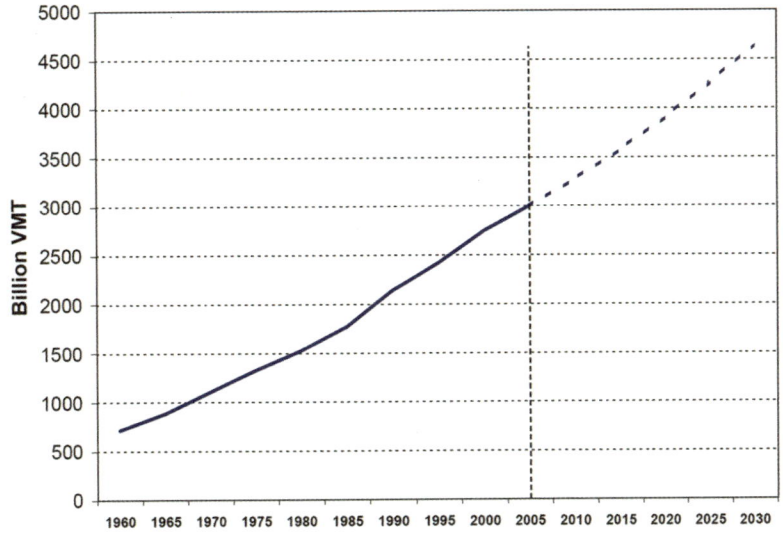

VMT is growing at an annual rate of 2.3%, doubling every thirty years.

SOURCE: Bureau of Transportation Statistics, 2006

table 15.2   **Means of Transportation to Work, 1980 and 2000 Census**

Means of Transportation	1980 Census Share	2000 Census Share
Private vehicle	84%	88%
Drove alone	64%	76%
Carpooled	20%	11%
Public transportation	6%	5%
Bus or trolley bus	4%	3%
Subway or elevated	2%	2%
Railroad	0.6%	0.5%
Taxicab	0.2%	0.2%
Motorcycle	0.4%	0.1%
Bicycle	0.5%	0.4%
Walked only	6%	3%
Other means	0.7%	0.9%
Worked at home	2%	3%
Total workers	96,617	127,437
Average travel time (min)	21.7	24.3

SOURCE: U.S. DOE, ORNL, Transportation Energy Data Book, 2007

between 1980 and 2000. In 2000, 76% of commuters drove alone to work, only 11% carpooled, and only 5% used public transportation.

15.1.2.1 Factors Affecting Vehicle Miles Traveled

Many factors affect VMT, including distance to work and other destinations, land use patterns, desire for convenience and flexibility, availability of alternative modes of transport, price of fuel, affluence, and culture, among others. Distance to work is affected by housing cost. Many home buyers will choose a longer commute so they can afford a better or bigger home. This choice is made more difficult as congestion and the price of fuel increase.

Land use development patterns also affect VMT. In single-use, sprawling suburban development, even the simplest travel requires the automobile. Compact, mixed-use development not only reduces travel distances but also provides the density and access to desired destinations that makes travel by walking, bicycle, and transit practical.

U.S. automobile culture. Perhaps the dominant factor affecting VMT and auto dependency in the United States is the automobile culture that has developed in the past century. One hundred years ago, transportation in the United States was dominated by physical activity (foot, bike, horse) and forms of public transport such as intracity streetcar and inter-city rail. Land use development patterns in urban areas were dense because they were based on walking distances from streetcar stops. Streetcar suburbs developed, but urban development was limited by the extent of the transit lines.

[handwritten: Europe is still like this.]

Then the advent of the automobile ushered in a major transformation of the physical and cultural character of the country. The automobile was first marketed in 1904, but it was Henry Ford's Model T, introduced in 1908, that brought the automobile to the masses. In less than twenty years, by 1927, 55% of American families had a car. The growth of the automobile led to the need for a national road system and within a decade of the 1921 Federal Highway Act, a two-lane road network linking population centers was largely complete. This incredible transformation revolutionized life in the United States, eliminating the filth and sanitation problems of horse-based transport, relieving overcrowding in cities, and providing better access to necessities for rural areas.

This great advance was slowed by the Great Depression and World War II, but after the war the influence of the automobile on the American landscape and culture was dramatic. The 1944 and 1956 Federal-Aid Highway Acts authorized 45,000 miles of the interstate highway system that was largely complete by 1991. The interstate and other road systems provided unprecedented access to outlying hinterlands, and urban dwellers seeking to flee the crime, grime, crowding, and later racial tensions of the city, found easy access via better roads and their beloved autos to the American dream in the increasingly sprawling suburbs.

This emphasis on auto-based transport and road systems discouraged investment in other modes of transport, especially transit but also pedestrian-oriented facilities. Transit could not compete with the flexibility and appeal of the car, and by 1960 only twelve cities still had electrified rail systems.

[handwritten: "Who killed the electric car?"]

The functional performance of the automobile provided a combination of flexibility, security, comfort, privacy, speed, independence, and access that other modes could not. But the automobile became more than just a means of transportation in American culture. It became a symbol of freedom and independence, social status, and identity: "You are what you drive." Gaining driving privileges became a rite of passage for youth and losing them became a devastating milestone of old age. The car culture has become a major source of consumerism and economic output not only in the United States but increasingly in other countries.

[handwritten: This could be a problem in the future.]

Although the price of fuel, efficient land use patterns, and availability of transit all influence VMT, it is the popular culture of the automobile that has the most lasting effect. It may also be the most difficult factor to change if VMT growth is to be slowed.

15.1.3 Public Transit Systems and Energy

Transit systems have long provided an alternative to private vehicles, especially for those who cannot afford to buy or operate one. Transit provides a critical option in dense cities that are short on road and parking capacity. European and some U.S. cities rely heavily on transit. With sufficient ridership (load factor), transit, especially rail, is the most efficient means of

U.S. Transit Ridership, 1900–2005

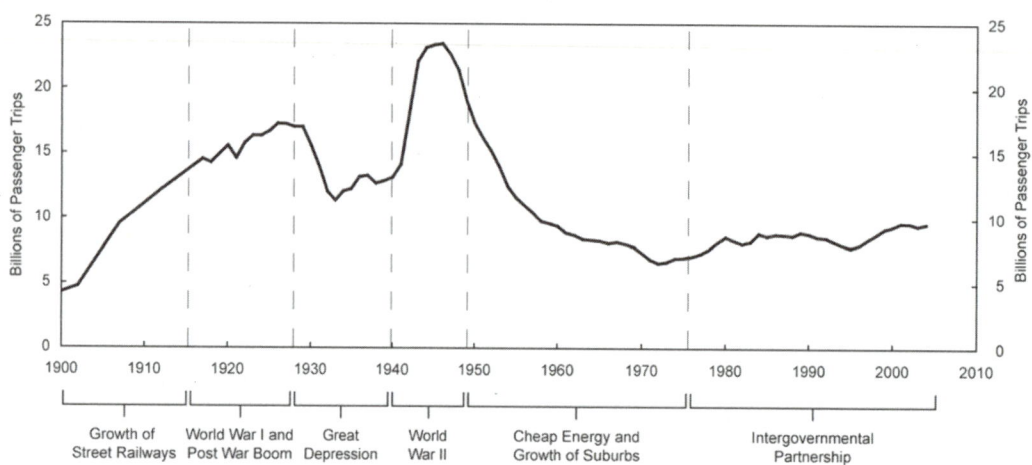

Cheap energy and growth of suburbs changed the transit culture of America. Intergovernmental partnership, especially federal funding, has helped increase ridership, albeit slowly.

SOURCE: APTA, 2006

moving people. Commuter rail transit, at a 2004 average 2569 Btu/passenger-mile (Table 13.6), is *40% more efficient* than cars and SUVs that have an average commuting load of 1.5–1.7 people per vehicle. With greater transit ridership and load factor, this difference would be even greater.

Figure 15.2 gives historic transit ridership in the United States. Since the mid 1970s, passenger trips have increased slowly from about 7 to 10 billion, but transit share has dropped. In the first half of the last century, transit was a critical part of our transportation system, when the percentage of passenger miles was very high. For commuters, it is now down to 5% (Table 15.2).

15.1.3.1 Types of Transit and Ridership

In 2004, there were about 50 billion passenger miles traveled (PMT) on transit in the United States, about 52% by rail and 44% by bus (Table 15.3). This is only about 1% of the 5000 billion VMT or 8000 billion PMT (assuming an average load of 1.6 people per vehicle). There are several types of transit systems, including heavy rail, commuter rail, and light rail (Figure 15.3); bus; paratransit (demand-response); and ferry. Among rail systems, Table 15.3 shows the highest riderships are on traditional rail systems in New York, Chicago, Boston, and Philadelphia, whereas systems in Washington, San Francisco, Los Angeles, and Atlanta have shown that newer systems are quite effective. Light rail systems have become popular because of their lower investment and easier development. Since the early 1990s, light rail

table 15.3 U.S. Rail Transit Ridership and Largest Systems, 2004

Commuter Rail	1000 Passenger Miles
Boston, MA	783,545
Chicago, IL	1,518,710
Los Angeles, CA	364,526
New York, NY (MTA-MNCR)	1,968,371
New York, NY (MTA-LIRR)	1,994,485
New York, NY (NJ TRANS)	1,890,460
Philadelphia, PA	433,572
Total commuter rail	**9,717,348**

Heavy Rail	1000 Passenger Miles
Atlanta, GA	455,359
Boston, MA	581,115
Chicago, IL	1,074,813
New York, NY	8,344,227
Philadelphia, PA	471,631
San Francisco, CA	1,228,433
Washington, DC	1,507,073
Total heavy rail	**14,354,281**

Light Rail	1000 Passenger Miles
Boston, MA	178,888
Dallas, TX	122,622
Los Angeles, CA	241,217
Portland, OR	181,760
San Diego, CA	170,376
San Francisco, CA	117,834
St. Louis, MO	127,210
Total light rail	**1,576,198**
Total rail	**25,647,827**
Bus transit	**21,376,973**
Other transit	**2,048,401**
Total transit	**49,073,201**

Source: APTA, 2006

figure 15.3 Types of Rail Transit Systems: Commuter, Heavy, Light

(a) San Francisco Commuter Rail

(b) New York Heavy Rail

(c) Minneapolis Light Rail

systems have been developed in Los Angeles, Portland (OR), Minneapolis, Denver, Kansas City, Dallas, and Salt Lake City.

15.1.3.2 Factors Affecting Transit Ridership

The most important factor affecting transit use is availability. But even where available, commuters weigh the advantages and disadvantages of transit versus vehicle commuting when making a choice. These factors include cost, time, convenience, perceived safety, privacy, hassles of parking and congestion, weather, and self-image. In some cases, these factors may favor automobile commuting, in other cases transit.

In recent years, transit passenger trips and passenger miles have increased faster than VMT. Increased fuel prices have been an important factor but so has improved transit availability and service. From 2003 to 2006 commuter rail ridership increased at more than 2% per year, heavy rail ridership at 2.6%, and light rail ridership at more than 6% per year. Since 1990, new light rail systems were installed in thirteen cities, and twenty-nine more have approved or proposed systems. In 2006, more than 130 miles of commuter rail routes and 150 miles of light rail routes were under construction.

15.1.3.3 Energy and Economic Savings from Public Transit in the United States

A recent American Public Transportation Association study (Bailey, 2007) estimated that a "public transit household" (two-adult, one-car household within three-fourths of a mile of public transportation) saves an average $6251 per year compared to an equivalent household with two cars and no access to public transit ("no-service household"; Table 15.4). These per household savings include the net savings in energy and fuel costs and the savings in avoiding a second car. They do not include savings on the cost of parking, which can be significant. For example, at $5/day, annual parking costs for a commuter are $1180.

These "public transit households" tend to drive less, walk more, and own fewer cars. The overall net energy savings of public transportation is estimated at 1.4 billion gallons (Bgal) of gasoline, the equivalent of 33.5 million barrels of oil. This is only 1% of our annual gasoline consumption.

The APTA study also modeled what improvement and expansion of the existing transit system would be needed to double the use of public transportation in the United States and double the energy savings to 2.8 Bgal gasoline per year. Figure 15.4(a) shows the 2001 distribution of households based on their proximity to public transportation: 51% are within three-fourths of a mile of a transit stop.

table 15.4   Savings for a "Public Transit Household" over a "No–Service Household"

	Public Transit Household	Non-Public Transit Household	Savings for Public Transit Household*
Annual fuel expenditures	$1705	$3104	$1399
Annual transit fare	$734	0	($734)
Vehicle cost	$5586	$11,172	$5586
Total	$8025	$14,276	$6251

* Does not include savings from parking.
Source: Bailey, 2007

Based on the recent experience of thirteen systems in seven cities, the study assumed that one-third of the increase in ridership could come from increased frequency and capacity of existing systems and two-thirds of the increase from expansion of high quality bus and rail routes. The expansion would require 11,700 additional miles of routes, 65% bus (7600 miles), and 35% rail (4100 miles). To put these numbers in perspective, there are about 6200 miles of transit rail lines in operation and another 4000 miles in various stages of planning and construction today.

Figure 15.4(b) illustrates the results of the expansion of routes: 64% of households would be within three-fourths of a mile of a transit stop. This assumes traditional development densities of about 2.4 units per buildable acre. However, the current practice and FTA rules for transportation system planning call for **transit-oriented development** (TOD) or greater density around transit stops (see 15.2.2). Figure 15.4(c) shows the increase in households within three-fourths of a mile of a transit stop if density of new development within that radius were increased from 2.4 to 4 units per buildable acre. Nearly three-fourths of the population would be within that radius.

15.2 Land Use, Transportation, and Energy

Vehicle miles traveled as well as transit opportunities depend on land use patterns. The last half of the twentieth century saw unprecedented growth in the United States. From 1950 to 2006, the population doubled to 300 million people, and massive highway construction, increased affluence, and the rise of the automobile freed those millions of people to flee the city for the suburbs. Decades later, however, those sprawling suburbs have spread out from

 **Household Distance to Transit**
figure 15.4

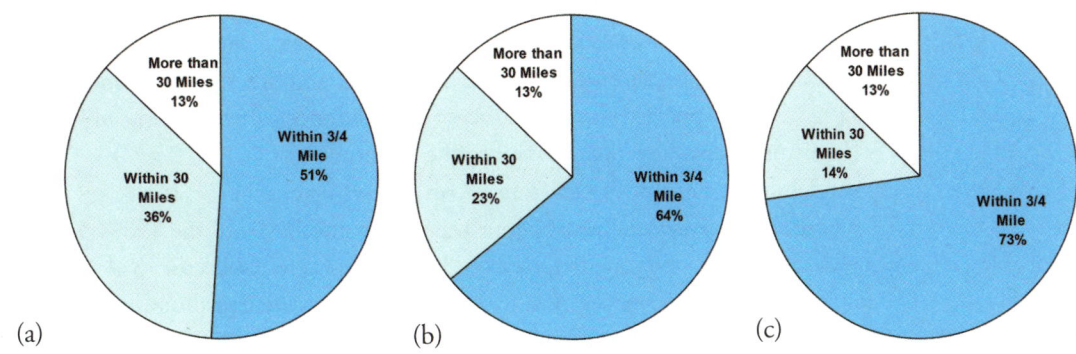

(a) Percent of U.S. households within distance of a public transportation stop, 2005. (b) Change in distribution with expansion of bus and rail routes by 11,700 miles. (c) Change in distribution if modest increase in transit-oriented development (TOD) density around new transit stops by 65% from 2.4 to 4 units per buildable acre.

SOURCE: APTA, 2007

the city, leaving behind vacant properties, consuming farmlands and natural areas in their wake, and isolating residents in auto-dependent developments with long, congested, and energy-intensive commutes. Some residents appear satisfied with suburban living, but many are increasingly unhappy with the dysfunctional social, environmental, and transportation problems of these land use patterns (especially when they fill up at the gas pump). But they often resign themselves to these problems as necessary evils of modern America.

Others believe there is a better way. Forward-thinking planners, developers, designers, and government officials are formulating new concepts of land use development. It is characterized by several related movements, including "sustainable communities," "new urbanism," "livable, walkable communities," "healthy communities," "brownfields redevelopment," and "smart growth." This section describes the transportation and energy problems related to land use and the means to arrest the sprawling nature of land development.

15.2.1 Auto-Dependent Urban and Ex-Urban Sprawl

Sprawl is land consumptive, dispersed, and auto-dependent land development patterns made up of homogeneous, segregated, automobile-dependent land uses. Sprawl's greatest (and perhaps only) triumph has been creation of the personal and family "private realm," be it home, yard, or personal car. But along with this private triumph has come a public or civic failure. Land uses have separated, and as people have become more segregated, they have retreated from a more public and interactive life to a more planned and structured life, from experiential communities of place to communities of interest. Sprawling development has diminished the visual and cultural diversity of communities, as suburban areas in all parts of the country now look the same, dominated by activity centers of big-box superstores, office parks, and homogeneous residential subdivisions.

The effects of these patterns of land use are illustrated in Figure 15.5. The top two maps show the Baltimore-Washington corridor in Maryland and the development from 1900 to 1960, which was concentrated in the urban cores and inner suburbs, and that from 1960 to 1997, which sprawled further out beyond the suburbs to what are called "exurban areas." The bottom maps show the effect on traffic density.

Land use has spread out from urban centers across the United States. Development land required per person grew by four times from the 1950s to the 1980s. In most sprawling development, everyone is forced to drive everywhere. Collector road designs and long commuting distances increase VMT, congestion, petroleum consumption, and air pollution. Sprawl consumes agricultural land, open space, and natural wildlife habitats and creates vast water-impacting impervious surfaces for subdivisions, shopping centers, roads, and parking lots. Local governments struggle financially to provide urban infrastructure and services in response to rapidly growing dispersed developments.

Sprawling development is enabled by government highway building and cheap fuel. But with government funds for ever more highway improvements diminishing and fuel prices increasing, the basic assumptions of these land use patterns are fading. From an energy

figure 15.5   Sprawl Development and Traffic Density in Maryland's Baltimore–Washington Corridor

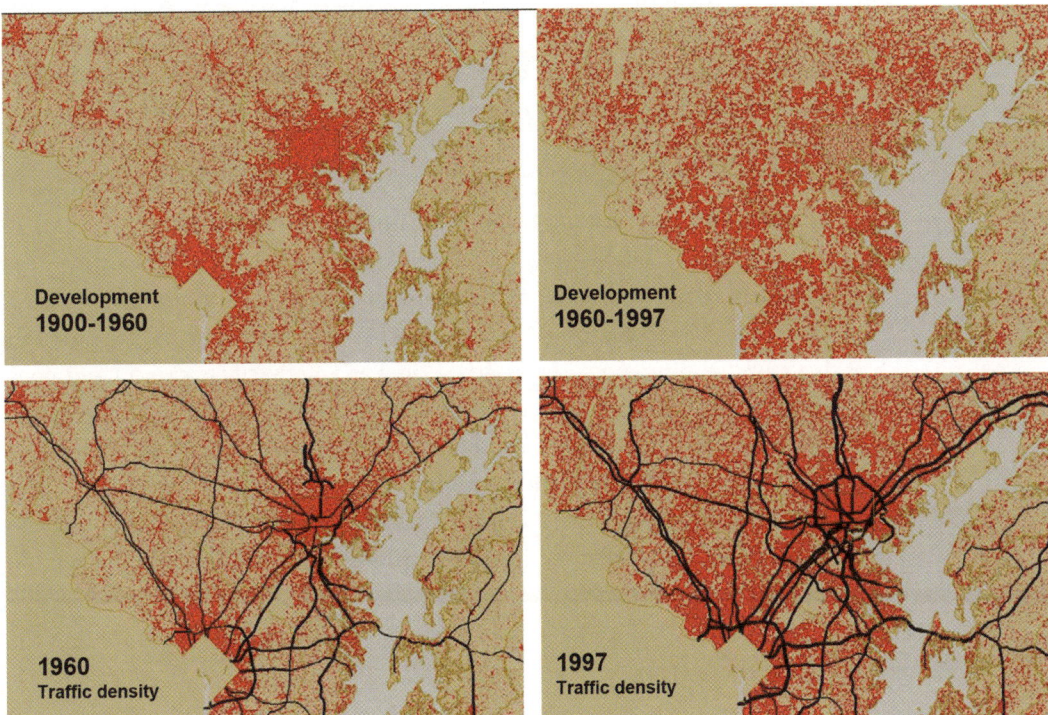

The red dots are dwellings, showing concentrated development from 1900 to 1960 and sprawling development from 1960 to 1997. Bottom maps show the difference in traffic density between 1960 and 1997, given by the breadth of the lines. Images like these generated public support for the Maryland Smart Growth program that aimed to concentrate development in areas of existing and planned infrastructure to revitalize existing communities and fund land conservation outside those areas.

Source: Frece, 2000

perspective, the low density of sprawl not only extends travel distance and often precludes efficient transit systems.

The consuming public states its preferences for the location, type, and amenities of residential living through its choices and purchasing power. The "American dream" has long been ownership of the large lot residence, usually in the suburbs. There is evidence that this preference is still alive and well, but there is also evidence that central city residential population is on the rise, along with a growing preference among some population segments for more dense, community-oriented living environments. Surveys are showing that sense of community is greater in traditional and pedestrian-oriented than in auto-oriented neighborhoods. This may bode well for the future, as discussed in Section 15.6.

Still, consumers can only choose from among the choices they are given. The land development industry, including developers, designers, investors, and contractors, provides the supply for this demand. Urban planners, local governments, community groups, and land trusts can help channel supply. Even though our somewhat dysfunctional land use patterns are the result of past efforts of those in the development industry and these patterns still dominate current markets, they are also our best hope for new and innovative designs that are more livable and sensitive to energy and environmental resources.

15.2.2 Land Use, Energy Use, and Carbon Emissions

What is the impact of land use on energy use and carbon emissions? Figure 15.6 is a well-cited figure from Jonathan Rose Companies comparing energy use for various housing types. It shows transportation energy (based on VMT and vehicle choice) and household energy based on dwelling heating efficiency (Btu/hr-deg-day), climate, and electrical energy use (kWh/yr). The typical suburban house has the greatest energy use because of high VMT. A "green" suburban household, assumed to have a higher efficiency house and vehicles, has 32% lower energy use. But it is still 15% higher than an average urban household because of higher VMT. The lowest energy users are green urban and especially green multifamily urban

 **figure 15.6**   Variation in Energy Use by Household Type

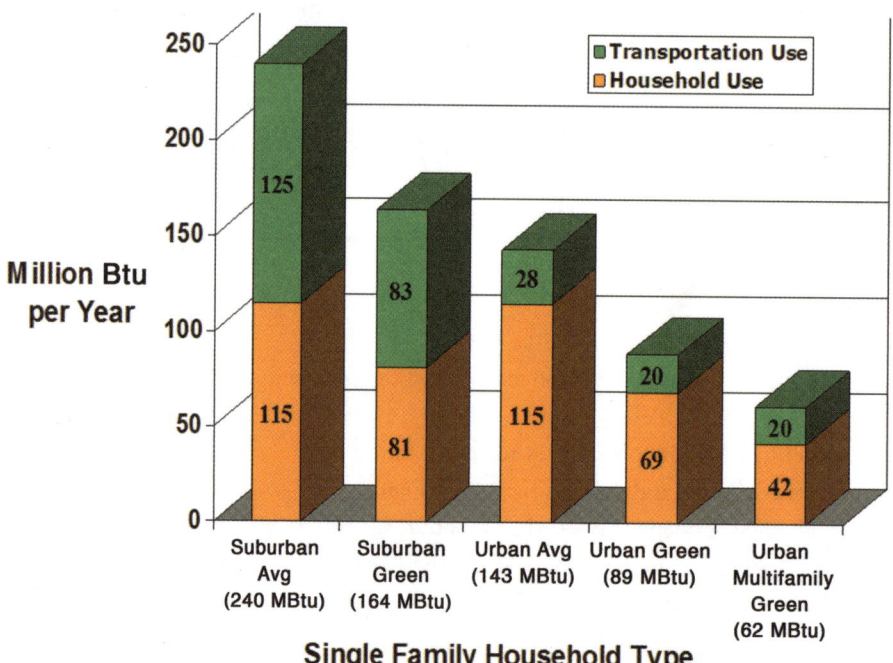

Suburban households have higher VMT than urban households and green households have higher MPG and more efficient houses than average.

households, which have both low house energy and very low transportation energy because of reduced automobile VMT and opportunity for public transit.

Table 15.5 calculates values for these variables that fit the results in Figure 15.6. The suburban average house has

- 125 MBtu/yr transportation energy consumed traveling 25,000 miles/yr @ 25 mpg
- 115 MBtu/yr household energy consumed in a 3000 sq. ft. house with a thermal index (TI) of 8 in an area with 3000 deg-days and that house uses 1050 kWh of electricity per month.

The "green" households have higher efficiency vehicles and lower TI. The urban households have lower VMT. Try to do the calculations or create a spreadsheet that allows you to vary the assumptions.

The higher VMT and lower density of suburban and exurban development create an interesting comparison of carbon emissions of suburban and urban dwellers. Figure 15.7(a) gives the annual tons of CO_2 emissions per sq. mile ($T\text{-}CO_2/\text{mi}^2\text{-yr}$) from VMT for the Chicago metropolitan area. This reinforces the traditional view that cities produce large amounts (about 75%) of GHGs. Figure 15.7(b) gives the emissions per household ($T\text{-}CO_2/\text{HH-yr}$) and the picture looks much different. This shows that city dwellers, because of their lower VMT, produce relatively low amounts of GHGs compared to suburban and rural households. The conclusion here is that denser urban and compact development patterns reduce VMT needs per person or household and reduce related energy and carbon emissions.

15.3 Land Use Design and Smart Growth Management

15.3.1 Energy-Efficient Land Use Design Principles

Beyond the neighborhood or subdivision scale, communities should fit together in a compact, mixed-use arrangement to foster efficient transportation, walkability, and livability. An important

table 15.5 Assumptions for Energy Use Results of Figure 15.6

	VMT/yr	Vehicle Mpg	Transport MBtu/yr	Thermal Index Btu/ft²-DD	House Size, ft²	Heating Energy* MBtu/yr	Electric kWh/mo	Electric kWh/yr	Electric MBtu/yr	Household MBtu/yr	Total Energy MBtu/yr
Suburban average	25,000	25	125	8	3000	72	1050	12,595	43	115	240
Suburban green	23,240	35	83	5	3000	45	879	10,545	36	81	164
Urban average	5600	25	28	8	3000	72	1050	12,595	43	115	143
Urban green	4800	30	20	5	2500	38	769	9227	32	69	89
Urban MF green**	4800	30	20	5	1500	23	476	5712	20	42	62

* Assumes 3000 degree-days.
** MF = multifamily.

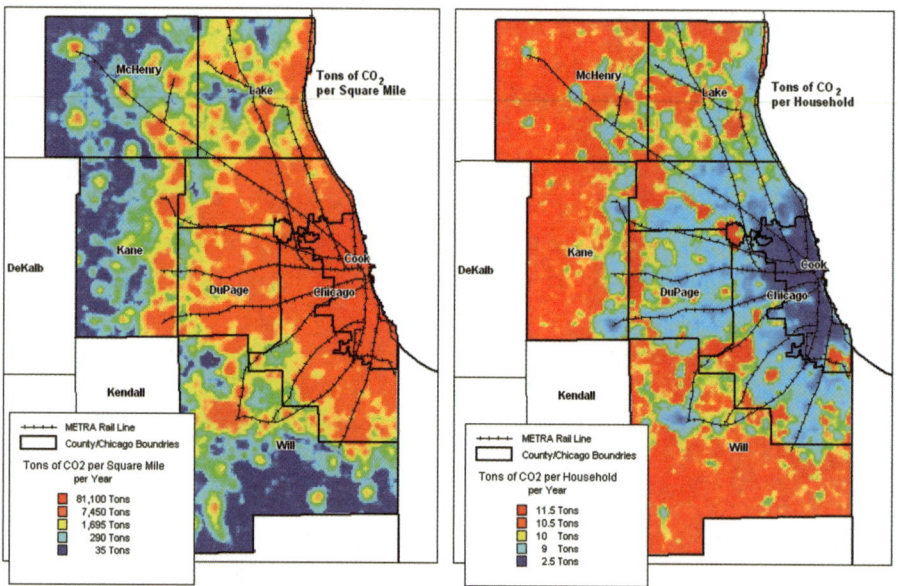

(a) Chicago Area CO$_2$ Emissions per Square Mile. Conclusion: Cities produce high amounts of GHGs.

(b) CO$_2$ Emissions per Household. Conclusion: City dwellers produce low amounts of GHGs.

SOURCE: Travel Matters, www.travelmatters.org

consideration in compact neighborhood and town design is reduced reliance on the automobile. Circulation within discrete towns should favor pedestrian and bicycle movement over vehicle transportation. "Walkability" has become a consistent feature of the various design approaches to community and neighborhood revitalization and is discussed further in the section on traditional neighborhoods.

Corbett's (1981) circulation schematic in Figure 15.8(a) shows a town center more easily accessible from the neighborhoods by foot than by car. Figure 15.8(b) shows a similar diagram from Van der Ryn and Calthorpe's (1984) sustainable community design. Their design includes mixed use, a discrete town center, and neighborhoods within a short walking radius. The town design could accommodate 5500 people within one-quarter mile of the town center, while still providing about 20% in open space.

Peter Calthorpe has championed the transportation aspects of "sustainable design," first through "pedestrian pockets," then transit orientation, finally through regional integration (Van der Ryn and Calthorpe, 1984; Calthorpe, 1993; Fulton and Calthorpe, 2001). His TOD is a mixed-use community within an average 2000-foot walking distance of a transit stop and core commercial area (Figure 15.9). The TOD is a key element of energy-efficient land use.

Each TOD is a "pedestrian pocket," with residential and public spaces within 2000 feet of the transit stop. Although the 2000-foot radius TOD is intended to be densely devel-

figure 15.8   **Neighborhood Circulation Patterns**

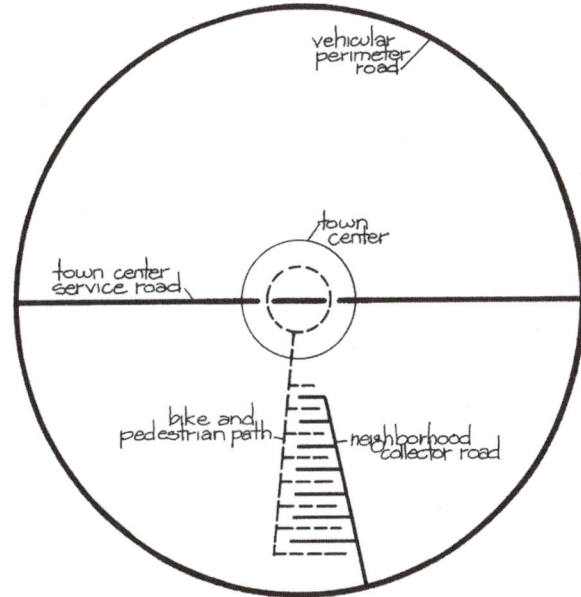

(a) Michael Corbett's simple neighborhood circulation favors bike and pedestrian movement over vehicles.

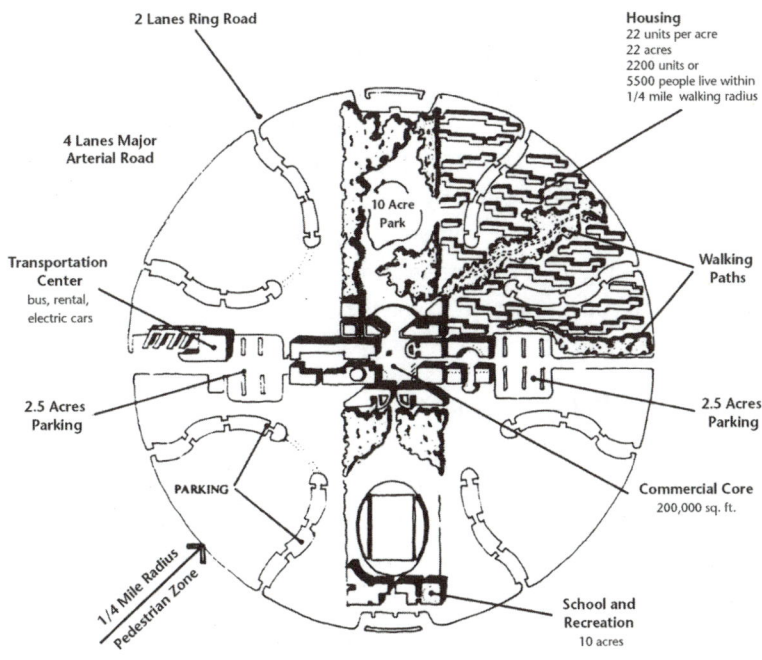

(b) Van der Ryn and Calthorpe's "sustainable community" design shows a similar circulation pattern.

SOURCES: Corbett, 1981; Van der Ryn and Calthorpe, 1984. Used with permission.

figure 15.9 Calthorpe's Transit-Oriented Development

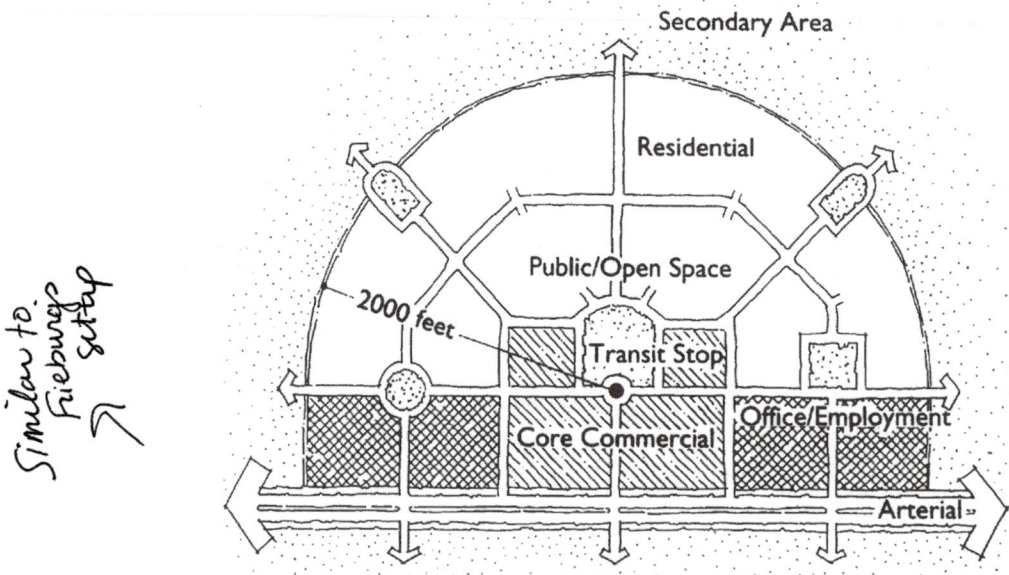

Similar to Freiburg setup → (handwritten annotation)

TOD concentrates development within walking distance of transit stops.

Source: Calthorpe, 1993. Used with permission.

oped, the design aims also to conserve riparian and other sensitive lands. TODs are linked to one another and to urban centers by bus or rail transit, characterized by the "urban network" (Figure 15.10). Calthorpe envisions regional development that ties a town or city center to its surrounding development area within an **urban growth boundary** (UGB) with TODs linked by rail and bus transit (see Figure 15.11).

15.3.2 Smart Growth Management

Government is charged with managing the land development industry. But urban planning and land use controls have too often been reactive to development proposals, rather than proactive in guiding development to more efficient land use. Fortunately, uncontrolled sprawl development has prompted many communities and states to adopt more aggressive growth controls to manage the impacts of development and facilitate the more efficient land use principles discussed above: compact, mixed-use, walkable, and transit-oriented designs. Management for "smart growth" emphasizes development in areas of existing infrastructure and de-emphasizes development in areas less suitable for development.

Growth management is defined as those government policies, plans, investments, incentives, and regulations to guide the type, amount, location, timing, and cost of development

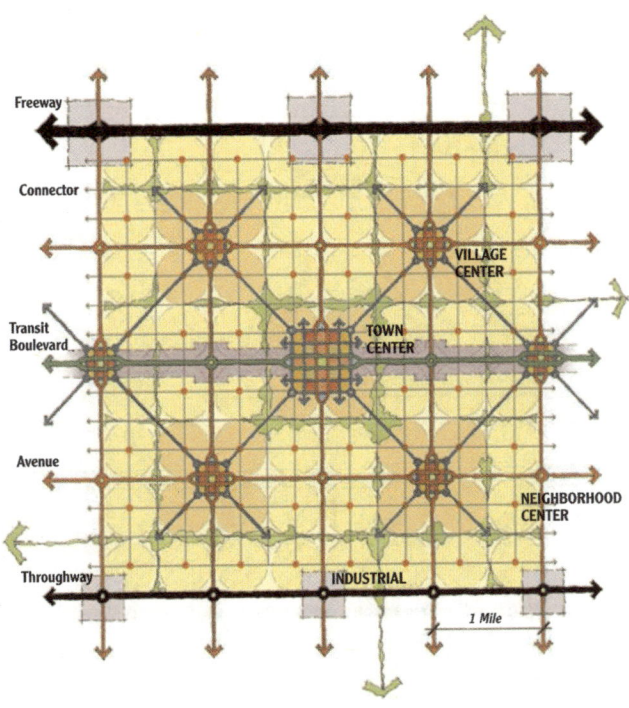

figure
15.10 **The Urban Network for Chicago Metropolis 2020: A Framework for Growth**

Peter Calthorpe's Urban Network provides interconnected centers and transit oriented developments with multiple types of transport: auto, transit, biking and walking. The road network reinforces access to walkable neighborhoods, urban town centers, and transit without cutting them off from local pedestrian movement.

Source: Calthorpe, 2002. Used with permission of Peter Calthrope.

to achieve a balance between the protection of the environment and the development to support growth, a responsible fit between development and necessary infrastructure, and enhanced quality of community life.

Smart growth emphasizes compact and mixed development in areas of existing infrastructure and de-emphasizes development outside existing infrastructure. By doing so, it aims to revitalize existing communities, preserve natural and agricultural resources, save the cost of new development infrastructure, and create spatial orientation and density that reduces travel distances and can support transit.

Managing urban growth and development should involve both local and state governments and take a regional interjurisdictional approach. Although an in-depth discussion of urban growth management is beyond the scope of this chapter, it is useful to introduce some of the tools involved, especially those that are important for energy-efficient land use.

Table 15.6 lists a number of planning, regulatory, and nonregulatory tools for smart growth management. Planning involves creating the technical and political basis for a

figure 15.11 Urban Growth Boundary

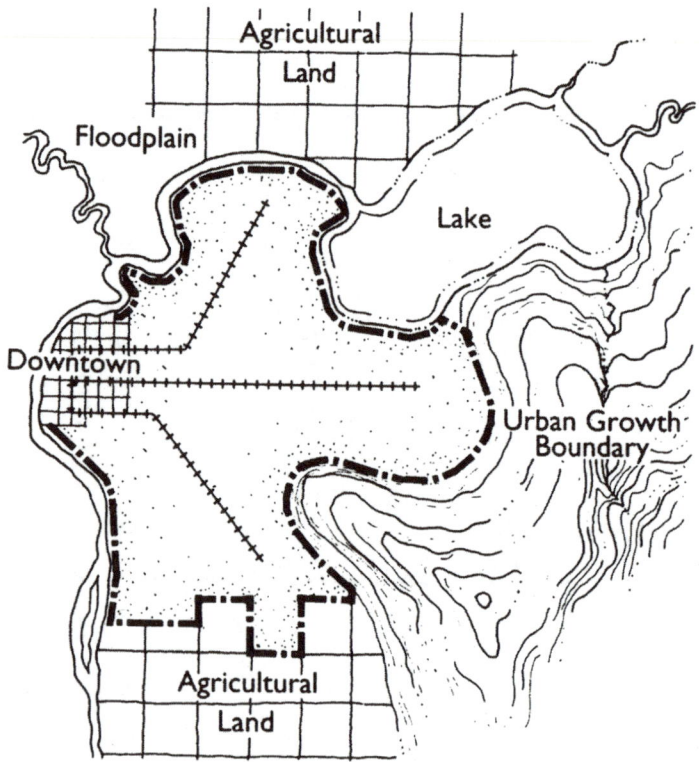

The urban growth boundary determines the extent of regional transit and TODs.

SOURCE: Peter Calthorpe, *The Next Metropolis: Ecology and the American Dream.* © 1993 Princeton Architectural Press. Used with permission.

community vision represented by the comprehensive or general plan. Nonregulatory tools include means of guiding where and when development occurs (a) by location and timing of infrastructure to support development, such as roads and water and sewer, or (b) by acquisition of land or the owners' development rights or conservation easements.

The most important smart growth management tools are the range of land use controls and regulations. Traditional zoning, subdivision regulations, and building codes are not enough to effect significant energy efficiency. Other land use ordinances have a greater effect on location, type, environmental protection, and energy efficiency of new development. We have already discussed the importance of building codes, especially those that incorporate green building standards. Cluster zoning can help enhance density and walkability. Performance zoning, point systems for phased development, and scorecard methods can incorporate energy-efficiency criteria.

Form-based codes (FBC) are a recent innovation that specify different building types for each face of the block (conventional zoning often assigns a single use/density category for

table 15.6 **Growth Management Tools**

Planning:

- Technical (energy/environment) and political basis for land use management
- The comprehensive or general plan, including energy chapter
- Functional plans: for example, energy plan, greenprint plan

Regulatory Tools:

- Zoning ordinance: use and density restrictions
- Subdivision regulations: rules for land division, solar access, and orientation
- *Building codes:* energy and green building requirements
- Variations on use and density restrictions (e.g., cluster/conservation zoning, parking standards that constrain availability and indirectly reduce VMT)
- *Form-based codes:* specify building types not uses, may address TOD
- Overlay districts: environmental zoning
- Performance/flexible zoning: performance criteria
- *Urban growth boundary:* contains most growth within designated boundary
- Transfer of development rights (TDR): allows greater density in development zone, compensation for no development in preservation zone
- Phased development: timing of development
- Development/design/construction standards and plan review: EIA, Smart Growth scorecard
- Environmental ordinances: for example, green building codes, storm water, tree preservation
- Development impact fees

Non-regulatory Tools:

- Land acquisition, conservation easements, purchase of development
- Infrastructure development: roads and sewers determine location of development
- Tax policies: use-value taxation, level-of-service areas

---

an entire block or group of blocks). Generally, the FBC incorporates design elements, allows mixed uses, encourages pedestrian opportunities, and can incorporate transit orientation.

The UGBs along with TODs are perhaps the most important elements of energy efficient land use. UGBs intend to contain development within a set boundary separating urban and rural land uses. UGBs were first used in Oregon's Land Conservation and Development program established in 1973. Portland has the best known UGB, shown in Figure 15.12 for 2040. For more than thirty years, it has led to community development within the boundary and farmland and environmental protection outside. It has created urban densities that can facilitate light rail transit and has made Portland one of America's most livable cities.

As a result of its UGB, Portland and its three-county region have created the density to support rail transit. In 1986, it completed the 15-mile eastern portion of Tri-Met MAX from Portland to Gresham and in 1998, the 18-mile western arm to Hillsboro. In addition, Tri-Met operates the bus transit system and Portland has a downtown trolley system, both of which are coordinated with MAX. Figure 15.13 shows the seventeen TODs that have been developed on the rail and trolley lines, and others on bus lines. It highlights three notable TODs and one under development.

figure 15.12   **Portland's 2040 Framework Plan**

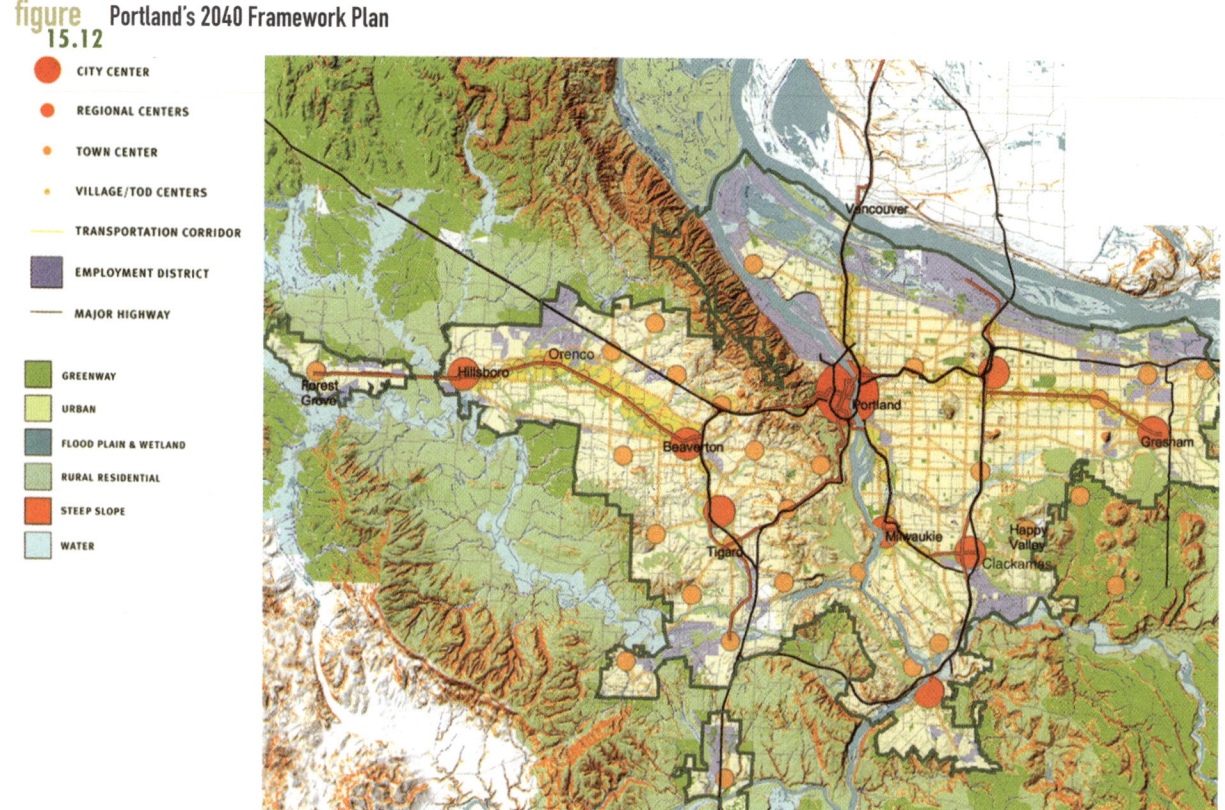

- ● CITY CENTER
- ● REGIONAL CENTERS
- ● TOWN CENTER
- · VILLAGE/TOD CENTERS
- — TRANSPORTATION CORRIDOR
- ▮ EMPLOYMENT DISTRICT
- — MAJOR HIGHWAY

- GREENWAY
- URBAN
- FLOOD PLAIN & WETLAND
- RURAL RESIDENTIAL
- STEEP SLOPE
- WATER

Shows open space elements, urban growth boundary, and metro centers, districts, and transit corridors.

Source: Calthorpe and Fulton, 2001. Used by permission of Island Press.

Oregon's regulatory UGB program prohibited dense development outside the UGB. After thirty years, this "UGB by regulation" met with voter opposition in 2004, but in 2007 voters reaffirmed the overall approach with modifications. Other states and localities have adopted UGBs, including Washington state and the Twin Cities (MN), among others.

Maryland has attempted to establish UGBs without regulation. The state provides financial support for development infrastructure only in Priority Funding Areas, essentially UGBs identified by localities based on existing and planned infrastructure. This incentive-based approach can slow the process of sprawl, but its effectiveness depends largely on how localities and the state choose to implement it.

15.3.3 Energy and Air Quality Benefits of Smart Growth

We have made significant progress in using vehicle technologies like catalytic converters to reduce vehicle emissions per mile (grams per mile [g/mi]) of criteria air pollutants including

figure 15.13 Portland's Tri-Met MAX Light Rail System

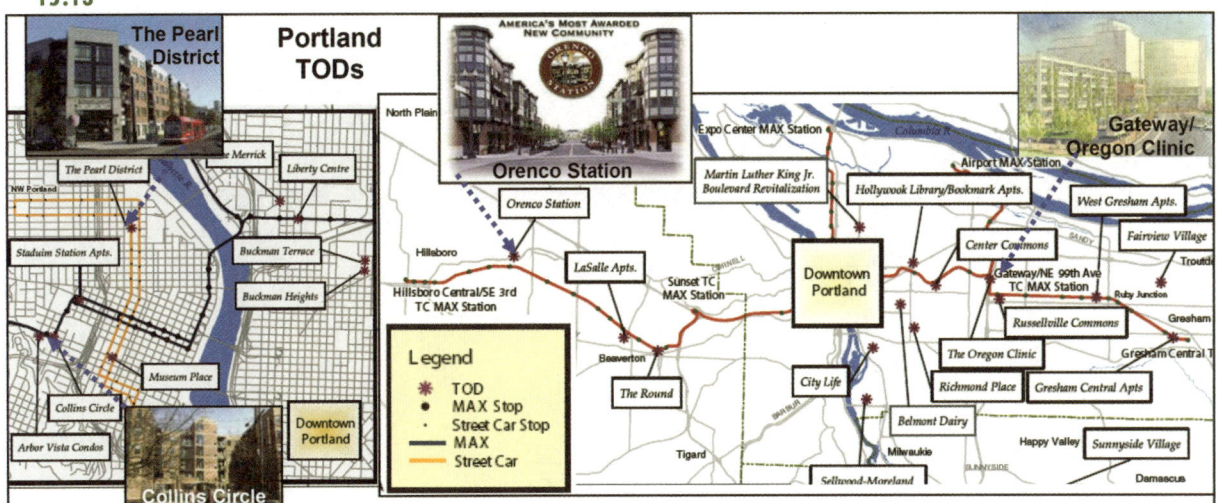

Shows transit-oriented developments, highlighting two in downtown, one on west side, and one on east side.

volatile organic compounds (VOC), oxides of nitrogen (NO$_x$), and carbon monoxide (CO). However, many of our cities still violate clean air standards because the reduced g/mi has been offset by the steady increase in VMT. Several cities have included land use measures to reduce VMT in their plans to achieve the clean air standards. In fact, the Clean Air Act is one of the few regulatory mechanisms used to address the runaway growth in VMT.

Brian Stone (2006) studied the effect of smart growth management on VMT and emissions reduction. Using Portland as a model of smart growth, he adapted its land use pattern to future growth plans of several Midwestern cities and modeled the effect on VMT and emissions.

Figure 15.16 shows the land use pattern of Columbus (OH) in 2000, continued business-as-usual (BAU) growth by 2050, and a smart growth (SG) density pattern in 2050. The SG 2050 scenario shows more concentrated density in the core city and inner suburbs, and less development outside. As a result, Stone's model calculated a 9% reduction in VMT and 7%–9% reduction in pollutant emissions compared to the BAU case. Other studies have estimated higher values for VMT reduction from smart growth strategies, including Puget Sound (10%–20%) and Sacramento (25%, see Chapter 18).

Smart growth management policies are part of indirect source control strategies in air quality improvement plans in areas of nonattainment with National Ambient Air Quality Standards (NAAQS). "Indirect sources," such as shopping centers and large residential subdivisions, are those that do not by themselves emit large amounts of pollution, but they stimulate more VMT and more pollutants. The California Air Resources Board (1995) recommends several land use measures to minimize VMT and emissions, including pedestrian facilities, increased density near transit stations and corridors, mixed development, infill

figure 15.14   Smart Growth 2050, Columbus, Ohio

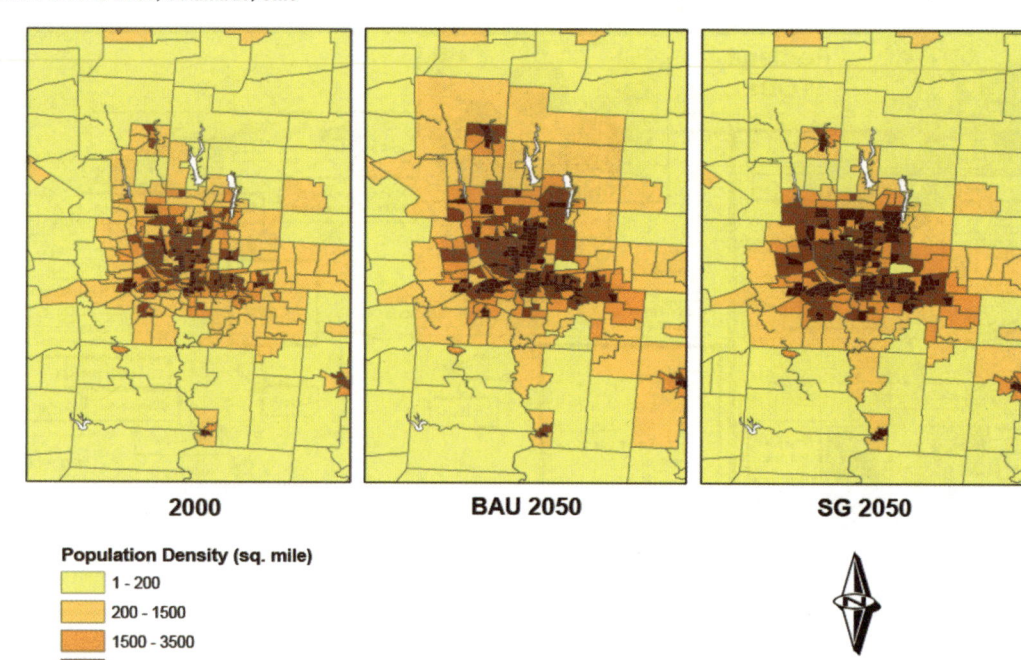

Development density in Columbus (OH) in 2000 and in 2050 under a "business-as-usual" (BAU) scenario and a smart growth (SG) scenario. The SG 2050 would have 9% less VMT and 7%–9% less pollutant emissions than the BAU scenario.

Source: Stone, 2006, 2007. Used with permission of Brian Stone.

and redevelopment, downtown revitalization including residential, interconnected street networks, and strategic location of parking.

Perhaps the best example of mitigation of indirect source impacts on VMT and emissions is California's San Joaquin Valley Air Pollution Control District (SJVAPCD). The District's Rule 9510, which became effective in March 2006, imposed impact fees on new developments for indirect air pollution impacts. The fees are reduced if mitigation measures are incorporated into the project.

Among the mitigation methods to reduce NO_x and PM10 are as follows:

- Proximity of project within a half-mile of Class 1 or Class 2 bike lanes or transit stop
- Higher density residential or intensity mixed use/retail commercial
- Residential density greater than seven dwellings per acre
- Bicycle storage
- Shower and locker facilities to encourage employees to bike or walk to work
- Complete, separate, safe, and convenient pedestrian sidewalks/paths, connecting multiple uses

- Parking pricing strategies, such as charging higher parking lot fees to low-occupancy vehicles
- Increase building energy efficiency rating above Title 24 requirements
- Implement an employee telecommuting policy
- Provide alternative transportation modes, such as "guaranteed ride home" or carpool matching
- Low emission delivery vehicles
- Reduce the number of wood fireplaces or wood stoves

Typical fees per dwelling unit for a 120-unit subdivision development on 24 acres:

- $780 with no mitigation
- $557 with typical on-site mitigation (increasing density to 5 units/acre, no wood stoves, improving building energy efficiency to 5% above code)
- $454 with additional mitigation to include typical measures plus 10% improved energy efficiency, sidewalks on both sides of street, 10% affordable housing

It is still too early to tell what effects this rule will have. But it has raised awareness of developers and citizens of the desirability of energy-efficient land use and development design.

The Center for Clean Air Policy (CCAP) has tried to educate states and communities about the air quality (and energy efficiency) benefits of VMT reduction through efficient land use. For this purpose it developed the Transportation Emissions Guidebook and a spreadsheet calculator model to assist users in analyzing potential VMT, energy, and air quality benefits of several land use options. Table 15.7 gives some of the "rule-of-thumb" VMT reduction

table 15.7 Potential VMT Savings from Various Measures

Measure	Savings	Scale
Transit-oriented development (TOD)	20%–30%	Site
Infill/brownfields	10%–50%	Site
Pedestrian-oriented (walkable) development (POD)	1%–10%	Area
Smart school siting (within walkable neighborhoods)	15%–50%	Site
Transit improvements	0.5 per 1% freq.	Area
Light rail	1%–2%	Corridor
Bicycle incentives (e.g., bikeways, transit bike racks)	1%–5%	Area
Road pricing (tolls)	1%–3%	Corridor
Commuter parking incentives (e.g., carpool parking rates)	5%–25%	Site
Green mortgages (reduced rates for Green buildings)	15%–50%	Household
Smart growth	3%–20%	Regional
Municipal parking program	15%–30%	Area

assumptions that are used as default values in the model, but these can be modified in the spreadsheet if more study specific information is known. The reduction values are applied to a specific scale ranging from site to regional scale. Check out the emissions guidebook calculator at http://www.ccap.org/guidebook.

15.4 Land Use and Whole Community Energy

Historically we have looked at energy in separate pieces. In fact, we have done that in this book: Section III focused on the Buildings piece, Section IV on Electricity, Section V on Transportation. But ultimately we need to integrate Whole Community Energy. This approach aims to think more broadly and see interconnections and opportunities across these primary energy sectors. Buildings are major consumers of energy that are connected at once by the electrical grid and by our transportation system. We have come to learn that buildings can be important sites for distributed electricity generation. And if we move toward electrification of our personal vehicles, such as plug-in hybrids, the building-as-power-source can generate daytime electricity to offset overnight grid power used to charge our car batteries.

But we have also learned that buildings, their density and distribution on the land, and their interconnected transportation system determine VMT, energy use, and related GHG and air pollutant emissions. We need to recognize these connections and begin to consider the opportunities of integrating energy and environmental considerations in building location and land use, pedestrian and transit orientation, on-site generation, in addition to efficient building envelope, equipment and appliances, and low-impact materials.

This section looks at that Whole Community Energy integration in emerging "green development" practices and what this may mean for future development in the United States as we plan how to accommodate the next 100 million Americans in the next thirty-five years. First, we introduce two additional considerations in energy and land use, solar access protection and the urban heat island.

15.4.1 Land Use and Solar Access

As we increase our use of building-integrated solar energy systems, such as passive solar heat, solar hot water heating (Chapter 7), and photovoltaic (PV) systems (Chapter 11), it is important we assure that these systems will have access to solar radiation. We need to site these systems well, using solar site surveys and other means, but we also need to prevent future shading of these collection systems from obstructions on adjacent sites, such as growing and new vegetation and new buildings. In addition, we need to accommodate the development of such systems in both new and existing buildings. This requires effective land use planning, subdivision design, and mechanisms to protect solar access.

figure 15.15 Layout of Solar Subdivision Village Homes in Davis, California

Source: Corbett, 1981. Used with permission.

Figure 15.15 shows Village Homes in Davis, California, one of the first solar subdivisions. East-west streets enable good solar orientation for houses, and spacing of houses and vegetative plantings maximize solar access to buildings' passive and active solar systems.

Some people living in subdivisions not specifically designed for solar access have encountered problems. In some cases, property owners installed a solar heating system, only to have an adjacent property owner build a house or addition or plant large trees that would shade the device. In other cases, property owners faced aesthetic or other restrictive covenants prohibiting such systems.

figure 15.16 Solar Skyspace

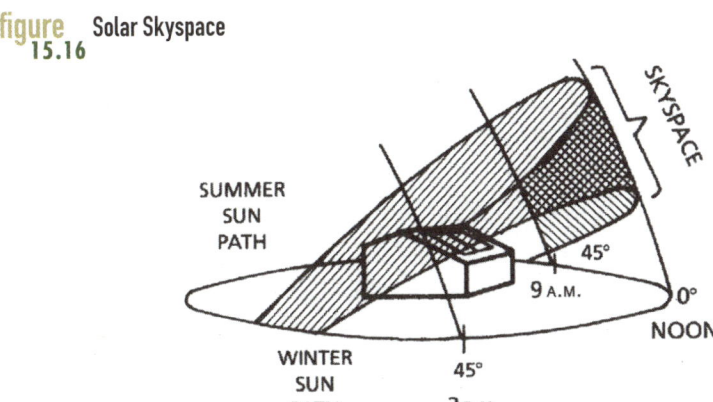

Source: Jaffe and Erley, 1980

table 15.8 **Recommended Skyspace Angles for December 21**

Latitude	a.m./p.m. Azimuth	a.m./p.m. Altitude	Noon Altitude	Insolation
25°	45°	25°	42°	76%
30°	45°	20°	37°	80%
35°	45°	16°	32°	85%
40°	45°	12°	27°	90%
45°+	50°	12°	22°	88%

Solar access laws in some states or localities have tried to address these "solar rights" problems. They have banned covenants and codes that would restrict or prohibit installation and operation of building-integrated solar energy systems. They have also allowed and encouraged neighboring property owners to use "solar easements" as formal agreements to protect solar access.

Others have gone further by establishing land use regulations to protect solar access. For example, California's Solar Shade Control Act of 1978, while encouraging the use of vegetative shading to reduce cooling energy demand, prohibits new vegetative growth from shading solar energy systems. The provision reads:

> After January 1, 1979, no person owning, or in control of a property shall allow a tree or shrub to be placed, or, if placed, to grow on such property, subsequent to the installation of a solar collector on the property of another so as to cast a shadow greater than 10% of the collector absorption area upon that solar collector surface on the property of another at any one time between the hours of 10 A.M. and 2 P.M., local standard time, [provided] . . . the location of a solar collector is . . . set back not less than 5 feet from the property line, and no less than 10 feet above the ground.

This provision indicates the need to consider time and distances because the geometry of solar access is important. We know about the path of the sun from solar site surveys in Chapter 7. The solar access "skyspace" that we want to keep free from obstructions is defined in Figure 15.16 and Table 15.8. We want to keep clear the sky from 45° east to 45° west of south (the south azimuth angle) and an altitude angle that depends on the latitude. For most U.S. latitudes, 80%–90% of December solar energy comes within that skyspace.

Whereas solar site surveys look at the sun's daily path in the sky from the position of the collector, shadow patterns look at the projections of shadows cast from the position of the obstruction. Figure 15.17 gives the shadow projection of a simple pole. These shadows depend on time of day, time of year, latitude, and orientation and slope. The equation for an obstruction's shadow length for the simple case of a level surface (slope = 0%) is given in the following:

$$S = \frac{H}{(\tan A)}$$

figure 15.17 Shadow Pattern of a Pole on Level Ground

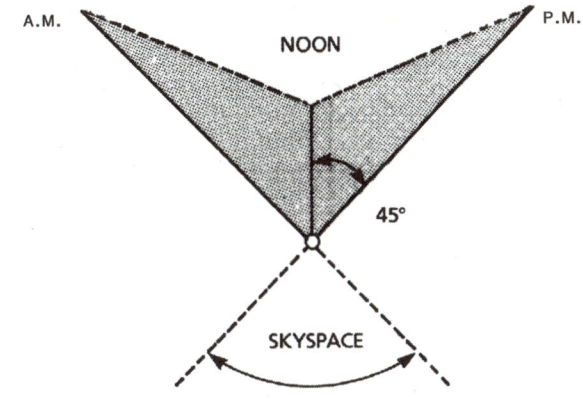

SOURCE: Jaffe and Erley, 1980

Of course, the direction of the shadow is opposite the direction of the incoming sun:

$$A_{z\,sh} = A_{z\,sun} \pm 180°$$

For a sloping surface, it is more complicated because the rise and fall of the land affects the length of the shadow:

$$S = \frac{H}{(\cos S_a)\{\tan A + (\tan S_a)[\cos (A_z - w)]\}}$$

where S = true shadow length, feet
H = height of obstruction, feet
A = solar altitude, degrees
S_p = projected shadow length, feet
$A_{z\,sun}$ = solar azimuth, degrees
$A_{z\,sh}$ = shadow azimuth, degrees
S_a = terrain slope angle, degrees
w = slope azimuth, degrees (see Figure 15.18)

Figure 15.19 shows typical shadow patterns of houses and trees in a subdivision, and Figure 15.20 shows how changing lot frontage from 75 to 60 feet and lot depth from 100 to 125 feet can improve solar access without increasing lot size.

One can predict how obstructions will shade an adjoining lot using the geometry of shadow patterns. Using these methods, some states and localities have adopted specific guidelines for development to provide and protect solar access. Some use the simple north shadow projection (S_p) to space buildings.

Others use a bulk plane approach to define a buildable envelope or volume on a site that will protect solar access on adjacent sites. The hypothetical wall approach shown in

**figure
15.18**   **Terms Used in Shadow Length Equations**

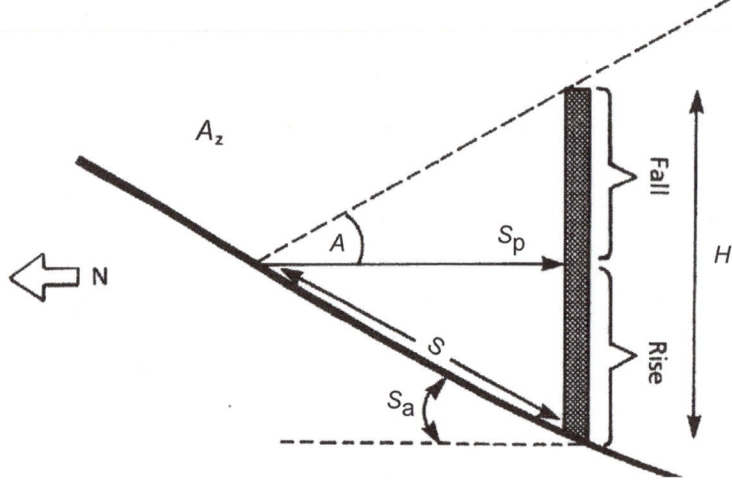

Source: Jaffe and Erley, 1980

Figure 15.21 uses a solar a.m./p.m. altitude angle and an acceptable amount of shading on the adjacent lot to define the height of a wall and the resulting bulk plane within which buildings and vegetation are allowed.

Figure 15.22 shows a "solar envelope" that reflects more accurately altitude angles throughout the day to define a three-dimensional buildable volume in which buildings and vegetation can reside without impacting solar access to adjoining sites.

**figure
15.19**   **Shadow Patterns in a Subdivision**

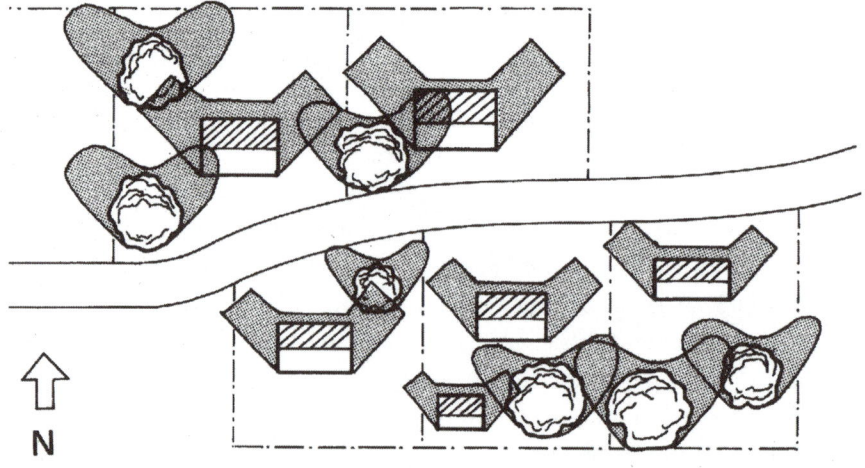

Source: Jaffe and Erley, 1980

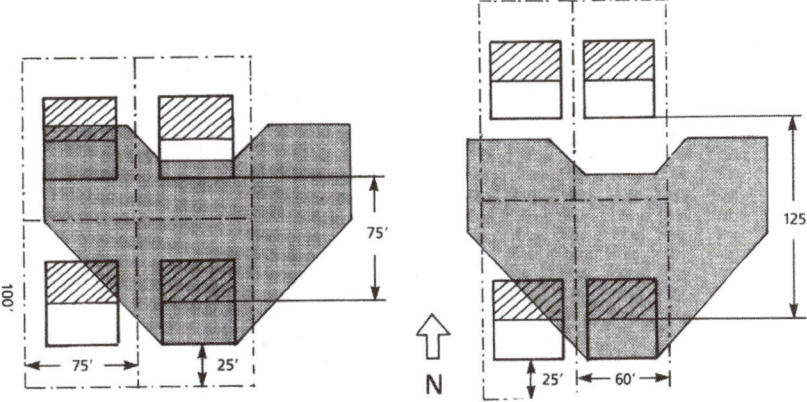

figure 15.20 **Adjustments to Lot Frontage and Depth Can Improve Solar Access**

By reducing frontage from 75' to 60', plan can accomodate 112' N. Shadow projection cast by 28' high buildings.

SOURCE: Jaffe and Erley, 1980

San Diego County (CA) enacted an innovative solar access approach in its subdivision regulations. It required developers to demonstrate that lots were laid out so that each lot had at least 100 square feet of rooftop solar access within its buildable area. Even if the houses built in the subdivision do not have solar systems, they were required to have solar access so

figure 15.21 **Hypothetical Wall Performance Standard for Protecting Solar Access on Adjacent Sites**

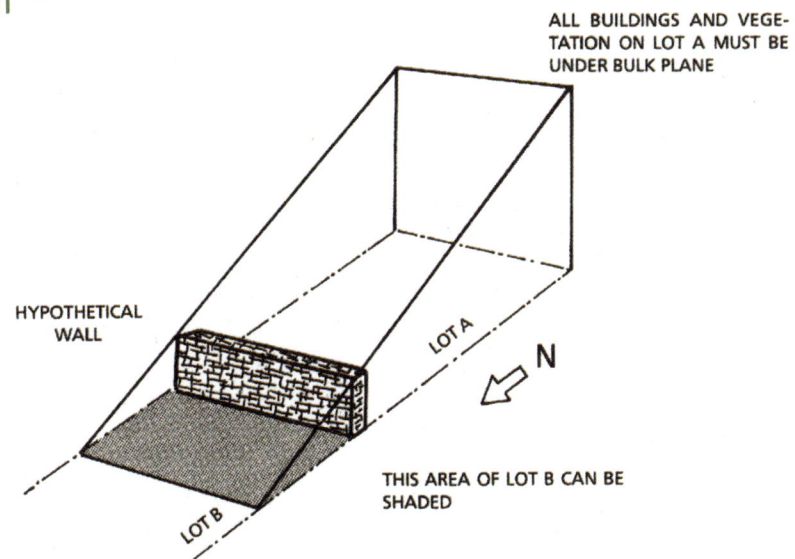

SOURCE: Jaffe and Erley, 1980

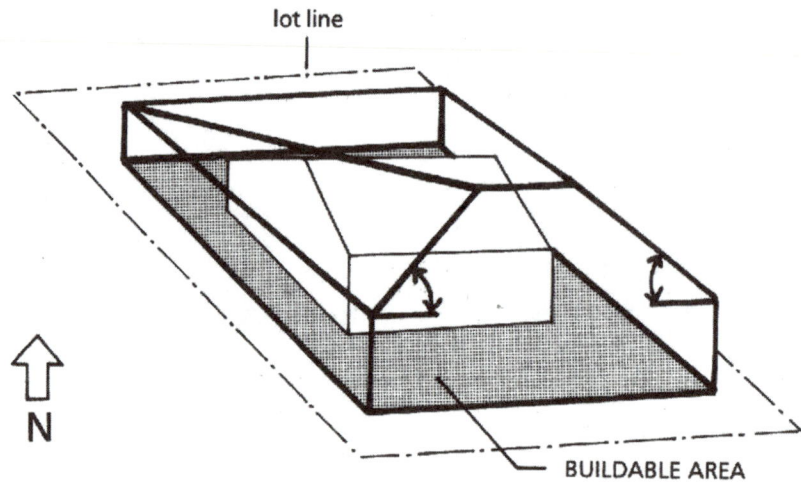

figure
15.22   **Solar Envelope Defines a Buildable Volume to Protect Solar Access on Adjacent Sites**

Source: Jaffe and Erley, 1980

systems could be added at a later date. The county provided a geometric template so that developers could easily access their plans. The ordinance reads:

> No tentative subdivision or parcel map received on or after October 1, 1979, shall be approved unless each lot within the subdivision can be demonstrated by the subdivider to have unobstructed access to sunlight to an area of not less than 100 square feet, falling in a horizontal plane 10 feet above the grade of the buildable area of the lot. The condition of unobstructed solar access shall be considered to be achieved when a specific area of not less than 100 square feet has an unobstructed skyview of the sun between azimuths of the sun at 45 degrees to the east and 45 degrees to the west of true south on December 21.

It is important to consider that solar access needs depend on solar application and location on-site. For example, access for winter solar heating is the most difficult to provide because the demand is concentrated when the sun is lowest in the sky and shadows are longest. Solar PVs and water heating have year-round needs. Solar swimming pool heating is needed in late spring to early fall so solar access is easiest because shadow lengths are short when the sun is higher in the sky. In addition, solar access to the roof is easier than to a south-facing passive solar wall because it is higher and less apt to be shaded.

15.4.2 Land Use and the Urban Heat Island

The urban heat island has been a phenomenon observed in all cities of the world. Because of reduced vegetative cover and more impervious surfaces, cities have a lower albedo (reflectivity) so that more short wavelength solar energy is absorbed and less is reflected. City streets, parking lots, and rooftops get hot and make the air above them hotter than the air above natural vegetation (Figure 15.23). In summer, urban heat islands increase air-conditioning demand and air pollution levels. Extreme heat is the number one weather-related killer in the United States; about 1100 people die from extreme heat each year, and this problem is exacerbated by the urban heat island and global warming. In an effort to show the differential effect of the urban heat island, Stone (2007) used historic meteorological data from 50 large metropolitan areas of the United States to show that the warming of urban areas has exceeded the warming of rural areas by an average 0.5°F.

Figure 15.23 shows that dense urban development has a greater heat island effect than suburban residential development. However, on a per household basis, larger lot suburban development has a higher heat island effect than more compact and dense development. This is a similar phenomenon we saw for VMT CO_2 emissions per household: Figure 15.7 showed that urban centers have higher CO_2 emissions per acre, but suburban areas have much higher emissions per household. In addition, suburban development that replaces tree canopy with grass has lower albedo and evapotranspiration and an increased thermal impact per household (Stone and Rodgers, 2001).

figure 15.23   Urban Heat Island

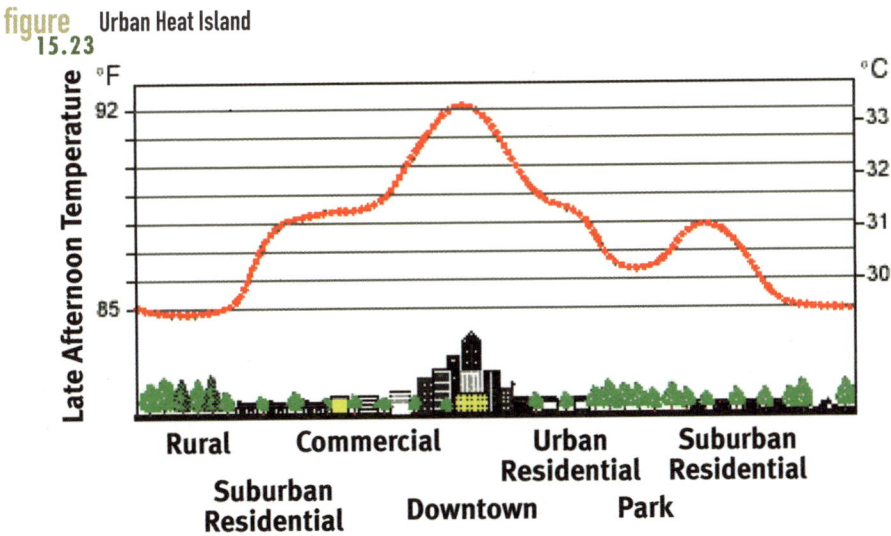

There are elevated summer temperatures in urban areas by as much as 7°F (3.5°C) above surrounding countryside. Suburban areas have less temperature increase but more thermal impact per household than denser urban centers.

Source: EPA, 1992

Denser compact development can mitigate the urban heat island effect on a per household basis, but additional measures are also needed. These include the following:

1. Retaining, protecting, and planting vegetation especially canopy trees
2. Utilizing "cool roofs" (Figures 7.17 and 7.18) and green roofs on buildings and "cool pavements" on roadways and parking lots to increase reflectivity and reduce absorbed summer solar radiation

Similar to cool roofs, "cool pavements" aim to increase the albedo of paved surfaces so they reflect more and absorb less of the sun's shorter wavelength radiation. Researchers at Lawrence Berkeley Lab tested various paved surfaces and showed that fresh asphalt had a reflectivity of only 5% and reached a temperature of 123°F on a summer day, whereas aged asphalt had a reflectivity of 10% and 115°F. A prototype cool asphalt coating increased the reflectivity to 50% and reduced summer day temperature to 90°F.

15.4.3 Whole Community Energy in Green Development: LEED-Neighborhood Development Guidelines

In Chapter 8, we discussed the adoption of innovative practices in buildings from new technology to codes and regulations (Figure 8.1). This process of adoption has been facilitated by guidance and rating systems such as ENERGY STAR and certification programs like LEED and ISO 14001. Not only do these programs educate consumers about the advantages of new technologies and designs, but they also provide a certification program for producers. This has greatly advanced the development of green buildings that have incorporated Whole Building Energy.

The evolution toward Whole Community Energy is being advanced by the U.S. Green Building Council's (USGBC) development of a LEED protocol for Neighborhood Development, so-called LEED-ND, which is moving LEED from Green Buildings to Green Communities. In conjunction with the Congress for New Urbanism (CNU) and the Natural Resources Defense Council (NRDC), USGBC has developed a draft protocol that began pilot testing in 2006, and after public comment in 2007, it is expected to be launched in 2008. Table 15.9 lists the main energy credit elements of the LEED-ND draft protocol. By achieving these elements, and having them certified by a licensed LEED inspector, developments accumulate points leading to LEED certification. Basic certification is obtained with 40% of total points, silver rating with 50%, gold rating with 60%, and platinum rating with 80% of total points. The major categories and their assigned points are:

- **Location Efficiency** is relative to employment, education, commerce, and opportunities for reduced auto use. This category includes 2 prerequisites, 7 credits, and 28 points or 25% of total points.

table 15.9   LEED–ND: Neighborhood Development: Energy-Related Credits

	# Credits	Points	% of Total Points
Location efficiency	7	28	25%
Reduced automobile dependence	2 to 6		
Environmental preservation	13		11%
Compact, complete, and connected neighborhoods	22	42	37%
Compact development	1 to 5		
Transit-oriented compactness	1		
Diversity of uses	1 to 3		
Comprehensively designed walkable streets	2		
Superior pedestrian experience	1 to 2		
Transit amenities	1		
Access to nearby communities	1		
Resource efficiency	17	25	22%
Certified green building	1 to 5		
Energy efficiency in buildings	1 to 3		
Heat island reduction	1		
Infrastructure energy efficiency	1		
On-site power generation	1		
On-site renewable energy sources	1		
Reuse of materials	1		
Recycled content	1		
Regionally provided materials	1		
Construction waste management	1		
Other	2	6	
TOTAL	48	114	100%
Total Energy-Related Credits:	*42*		*37%*

Certified: 46–56 (40%); **Silver:** 57–67 (50%); **Gold:** 68–90 (60%); **Platinum:** 91–114 (80%)

SOURCE: USGBC, 2005

- **Environmental Preservation** involves the protection of water, vegetation, and habitats. It includes 5 prerequisites, 11 credits, and 13 points (11% of total).
- **Compact, Complete, and Connected Neighborhoods** include efficient land use, such as compact, mixed-use, transit- and pedestrian-oriented development, and transit opportunities. It includes 3 prerequisites, 22 credits, and 42 points (37% of total).
- **Resource Efficiency** includes Whole Building, Life-Cycle Energy categories of Green building, efficiency, on-site generation, and low-impact materials. It provides up to 17 credits and 25 points (22% of total).

The energy elements of LEED-ND make up 37% of the total available points. They include the following:

- **Elements of Whole Building Energy:** Certified green building, energy efficiency in buildings, on-site renewable energy sources
- **Elements of Whole Building, Life-Cycle:** Certified green building, reuse of materials, recycled content, regionally provided materials
- **Elements of Whole Community Energy:** On-site power generation; infrastructure energy efficiency; reduced automobile dependence; compact, complete, and connected neighborhoods

Although LEED-ND certification will not be available until at least 2008, there are examples of community designs emerging that incorporate many of its elements. For example, there are many mixed-use and pedestrian and transit oriented developments featured on the CNU (www.CNU.org) and Calthorpe Associates (www.calthorpe.com) Web sites, and brownfield redevelopment success stories can be found at http://www.epa.gov/brownfields/success/success_scss.htm.

Figure 15.24 shows the design of a proposed suburban development, Covell Village in Davis, California. Although the 1864-unit, 422-acre development plan was approved by city council in 2005, it was voted down by citizen referendum due to its scale and conversion of farmland. Still, it is an excellent example of the next generation of dense, mixed-use, walkable, bikable, and transit-oriented suburban development. Although not formally scored on the LEED-ND scale, it would likely score well.

The project was designed by Village Homes' developer Michael Corbett and others, and it has some unique energy design elements:

- Buildings exceed CA Title 24 efficiency standards by 30%
- All single-family homes have 1 kW solar PV
- Eight miles of bike paths
- New transit line to downtown
- Compact/pedestrian design

In addition, it has several social and affordability benefits: two-thirds of units serve low- and moderate-income, mixed housing, co-housing, school site, civic land dedication, and senior care and hospice.

And the development includes many environmental features, such as a 772-acre farmland reserve, a 124-acre wetland habitat, 27 acres of parks, a 26-acre nature corridor, and 16 acres of greenbelts.

figure
15.24

Design for Covell Village in Davis, California, exemplifies Whole Community Energy.

SOURCE: Covell Village Partners

15.5 Planning for Whole Community Energy

Achieving the elements of Whole Community Energy, including green buildings, distributed generation, efficient land use, efficient vehicles, and effective public transit, requires strong federal and state energy policies and especially a strong local commitment. That local commitment takes the form of efficient building and land use regulations, municipal utilities and community choice of power, public transit, recycling programs, incentives and initiatives for community energy efficiency, community education, and community planning.

We will discuss some specific city and community energy policies in Chapter 18, but here it is useful to describe methods of developing energy plans. There was a strong interest in community energy planning in the late 1970s and early 1980s, but this field waned in the late 1980s as energy prices dropped and cities faced other priorities. Some cities have addressed energy in their comprehensive plans and others in their air quality improvement plans, such as the SJVAPCD discussed earlier.

A number of protocols and methods for energy planning were developed through this period and later, principal among them are PLACE³S and the clean air and climate protection planning approach of the International Council for Local Environment Initiatives (ICLEI).

15.5.1 PLACE³S: PLAnning for Community Energy, Economic, and Environmental Sustainability

PLACE³S is a planning method developed by the state energy offices in California, Oregon, and Washington in 1996. It uses energy accounting to evaluate the efficiency of land use, housing, and neighborhood design; transportation systems; and buildings and infrastructure. It follows a participatory planning process incorporating design concepts and quantitative measurement. The process has five steps:

1. **Start-up and Quantify Current Conditions:** Establish geographic scope, gather data on existing conditions, identify stakeholders, and begin participation process.
2. **Establish Business-as-Usual Alternative** given current land use policies and market trends that will serve as a baseline for comparing other alternatives.
3. **Analyze Alternatives** for redirecting growth and new transportation programs, including an Advanced Alternative that optimizes efficiency. Quantify the impacts of each alternative on energy efficiency and air pollutant and GHG emissions.
4. **Create Preferred Alternative** that may be a hybrid of the multiple alternatives developed in Step 3. Quantify expected levels of efficiency and emissions.
5. **Adopt, Implement, Monitor, and Revise:** While implementing the Preferred Alternative, monitor the efficiency and emissions effects, compare them to expectations, and revise the plan as necessary.

Figure 15.25 illustrates results of Steps 2–4, giving specifications and impacts of the Business as Usual future, the Advanced Alternative, and the Preferred Alternative.

figure 15.25 Illustration of PLACE³S Alternatives and Resulting Impacts on Energy, Cost, and Emissions

1. **BUSINESS-AS-USUAL:** Developer proposes to build on a 100-acre parcel at four units to the acre. The PLACE³S profile reveals the following:

    - Total development requirement: 100 acres
    - Open space reserved: 0 acres
    - Homeseekers served: 348
    - Transit feasibility: Poor, too few residents within walking distance of transit- good transit service not economically viable.
    - Local Street Connectivity: Poor, few streets provide direct access to transit.

    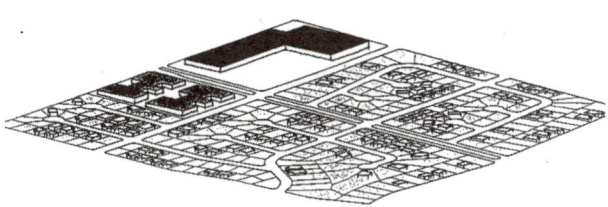

    Residential Units
    Single-family: 268
    Multi-family: 80

    Commercial Square Footage
    Retail: 65,000
    Other: 0

    - 175 MMBtu/person/yr
    - $2100/person/yr
    - 22 tons CO_2/person/yr

2. **ADVANCED ALTERNATIVE:** Community develops an alternative that doubles housing to meet projected need and doubles density to conserve resources, lower prices and preserve the environment. The PLACE³S profile reveals the following:

    - Total development requirement: 82 acres
    - Open space reserved: 18 acres
    - Homeseekers served: 770
    - Transit feasibility: Excellent, 95% of residents are within walking distance of transit
    - Vertical mixed uses in Activity Center
    - Local Street Connectivity: Excellent, streets provide direct access to transit, shopping and employment
    - Pavement minimization: skinny streets.

    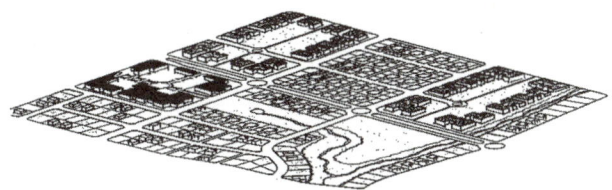

    Residential Units
    Single-family: 470
    Multi-family: 300

    Commercial Square Footage
    Retail: 35,000
    Other: 20,000

    - 125 MMBtu/person/yr
    - $1500/person/yr
    - 16 tons CO_2/person/yr

3. **PREFERRED ALTERNATIVE:** After assessing all alternatives in public meetings and negotiating trade-offs, the community removes some multi-family homes and open space, but agrees to a plan that is an improvement over the Business-As-Usual Alternative. The PLACE³S profile reveals the following:

    - Total development requirement: 85 acres
    - Open space reserved: 15 acres
    - Homeseekers served: 452
    - Transit feasibility: Good, density partially supports transit
    - Horizontal mixed uses in Activity Center
    - Local Street Connectivity: Good, most streets provide direct access to transit and shopping

    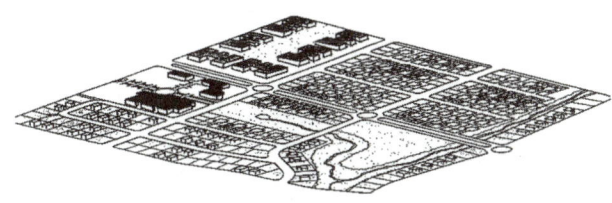

    Residential Units
    Single-family: 302
    Multi-family: 150

    Commercial Square Footage
    Retail: 45,000
    Other: 5,000

    - 140 MMBtu/person/yr;
    - $1900/person/yr
    - 19 tons CO_2/person/yr

Source: ODE, WSEO, CEC, 1996

The difficult part of PLACE³S is developing spatial alternatives of future development patterns, quantifying the energy use of various land uses, and matching energy production and distribution and transit systems to the alternative land uses. The method uses GIS and database software to facilitate analysis of large datasets and provide visual results. The California Energy Commission has been working to replace the PC version of the software with an Internet-based version, referred to as I-PLACE³S, which aims to simply the assessment. The PLACE³S method has been applied to planning efforts, most notably in the Sacramento Region and San Diego, California, and Eugene-Springfield, Oregon. The process does much to educate participants about the benefits of smart growth and the costs of sprawl.

15.5.2 ICLEI Process for Community Energy Inventories and Planning

ICLEI, founded in 1990, is an international association of local governments and associations that have made a commitment to sustainable development. In 2007, there were 800 member cities, towns, and counties. More than 250 U.S. cities and counties are involved in ICLEI's Cities for Climate Protection (CCP) Campaign, which has a five-milestone planning and implementation process:

1. Establish a baseline of energy use and GHG and air pollutant emissions data for both government operations and the community.
2. Set a target for emissions reductions, relative to current conditions and a "business-as-usual" projection (Figure 15.26).
3. Develop a local action plan, including energy codes, distributed energy, efficiency programs, land use planning, transportation planning, and solid waste management.
4. Implement the local action plan.
5. Measure results including energy use, VMT, waste stream and recycling, and related emissions.

To assist this process, ICLEI-CCP partnered with associations of air pollution control administrators and officials (STAPPA/ALAPCO) to develop a software package that quantifies baseline data, compares the impacts of alternatives, provides analytical evidence necessary for political and community support, and helps monitor progress. Participating cities throughout the country are using the software to create their baseline and assess alternative actions.

The baseline data are a critical first step, and the software provides user-friendly input screens to facilitate the process (Figure 15.27 in Sidebar 15.1). Data are input by sector for both the community as a whole and government operations. Input data come from an inventory of utility records, VMT data, and water, wastewater, and solid waste streams.

The brains of the software are the emission factors or coefficients (like those we discussed in Sections 5.5 and 13.2.3) that translate energy, VMT, and waste data into emissions.

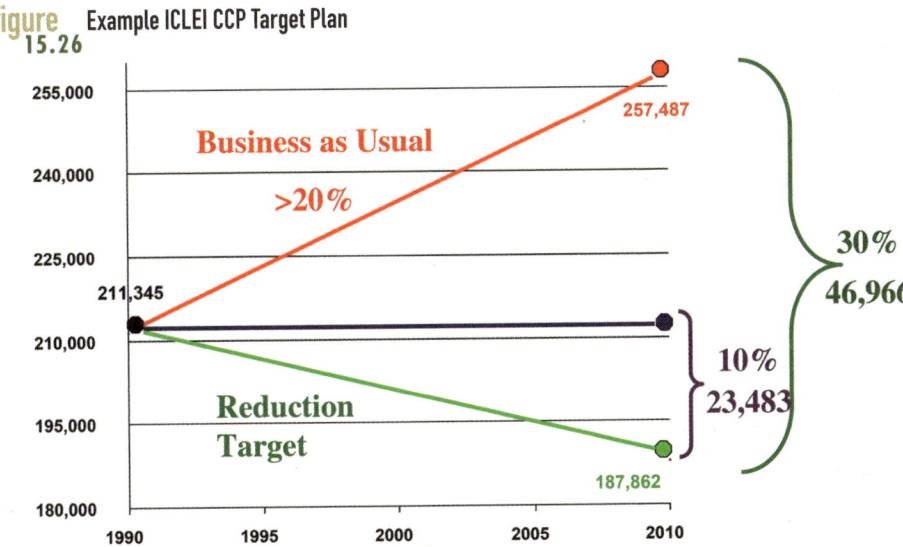

figure
15.26   **Example ICLEI CCP Target Plan**

Shows baseline today, business-as-usual future, and reduction target, which is 10% below today's GHG emissions and 30% below BAU.

Source: Burroughs, et al., 2006

Default coefficients are included in the software by region, but they can be overridden with more precise coefficients if they are known.

Once baseline data (and new coefficients if any) are input, the software calculates the emissions baseline and specific energy reduction measures can be input to assess their impact. Figure 15.28 in Sidebar 15.1 illustrates the input screen of a street light relamping program. With inputs of expected energy reduction and cost, the software calculates CO_2 and NO_x reduction and dollar savings. Figure 15.29 in Sidebar 15.1 illustrates an output pie-chart showing how the measures compare overall.

15.6 U.S.A. 2040: Land Use and Energy for the Next 100 Million Americans

Do visions of future U.S. development conform to this evolution toward Whole Community Energy? The U.S. population hit 300 million in 2006, and it is expected to hit 400 million in about thirty-five years. That 33% increase in population will spur an unprecedented building boom that Nelson (2006) estimates will require a $35 trillion investment in residential and nonresidential buildings and related public infrastructure. We will add the equivalent of our existing commercial building space to replace and expand commerce, and we will add the equivalent of half of our existing housing stock.

SIDEBAR 15.1

Sample Graphics from ICLEI's Climate Protection Protocol and Software

figure 15.27 ICLEI Software Sample Input Screen for Inventory Data

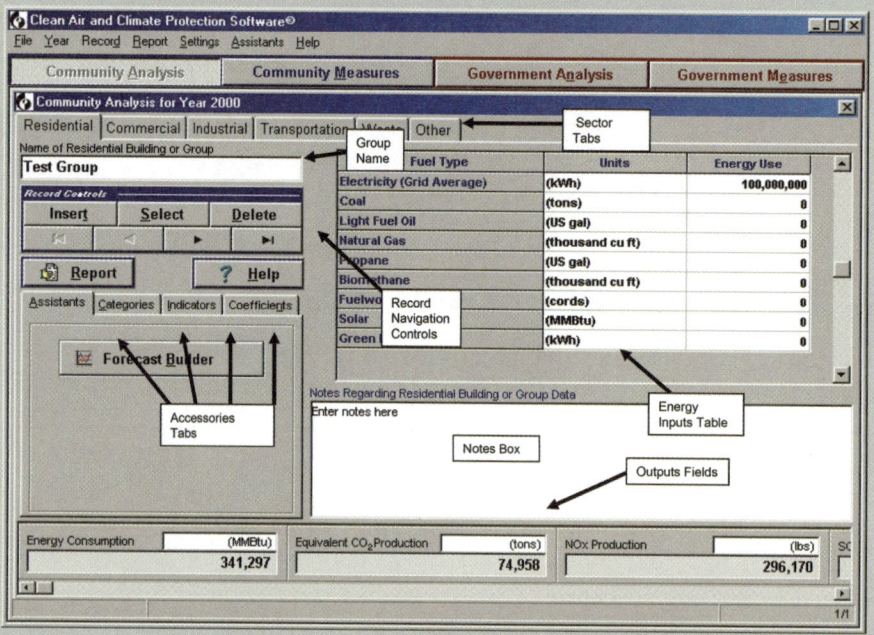

figure 15.28 ICLEI Software Sample Input Screen for Action Measure Data

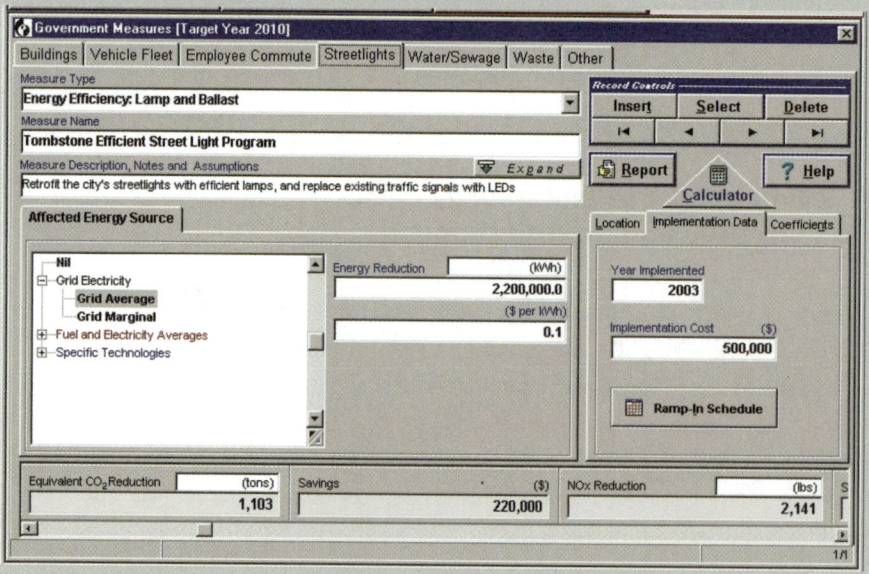

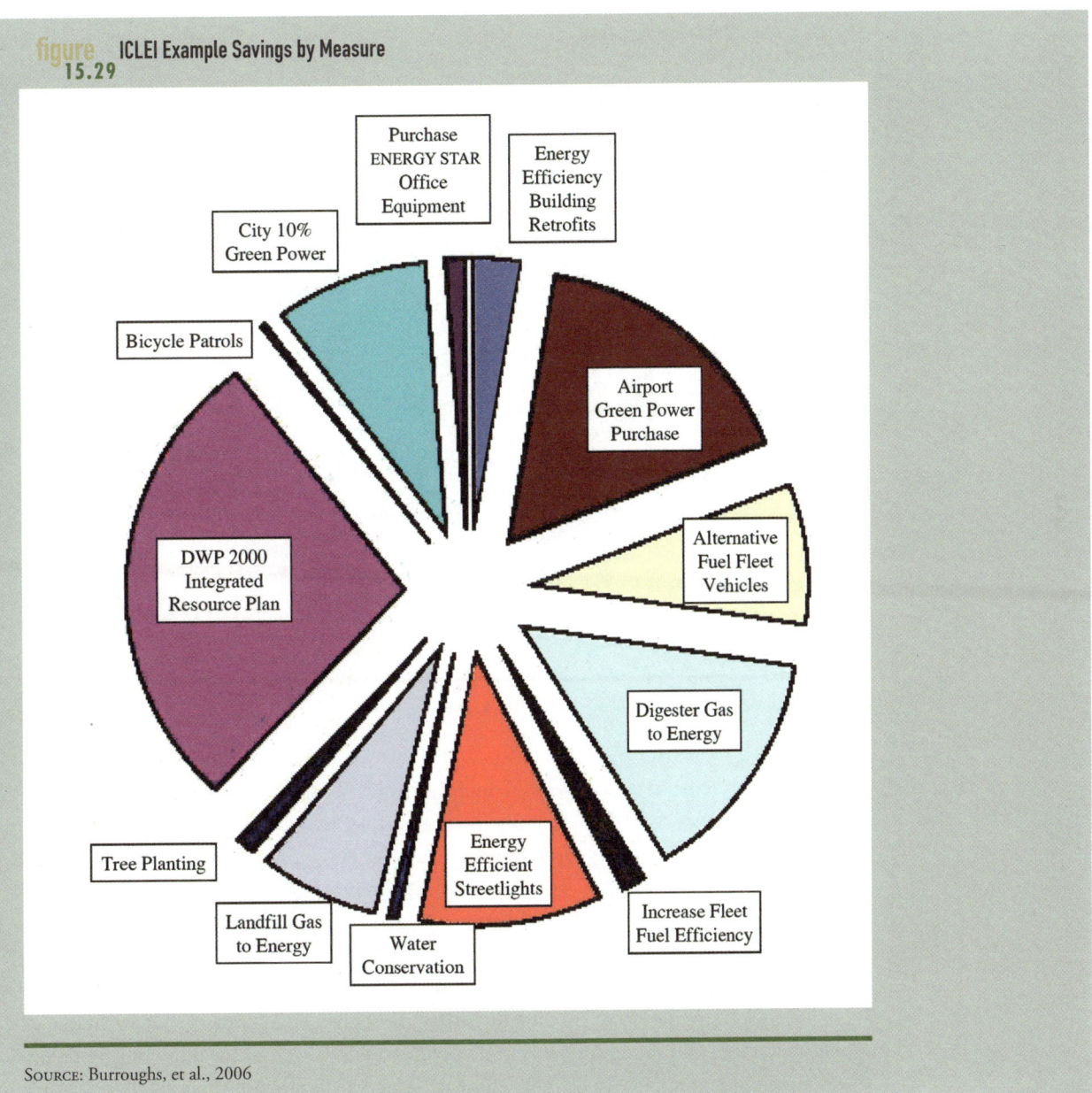

figure 15.29 ICLEI Example Savings by Measure

SOURCE: Burroughs, et al., 2006

This will take considerable energy as well as investment, but it also provides an opportunity to develop more sustainably, employing the Whole Community Energy concepts of green buildings, on-site generation, smart growth, improved transit, and compact, pedestrian- and transit-oriented developments.

Where will this development occur? Lang and Nelson (2007) suggest that it will occur largely where development is now concentrated, in "Megapolitan America." Figure 15.30 shows their map of megapolitan areas, which are defined as "two or more metropolitan areas

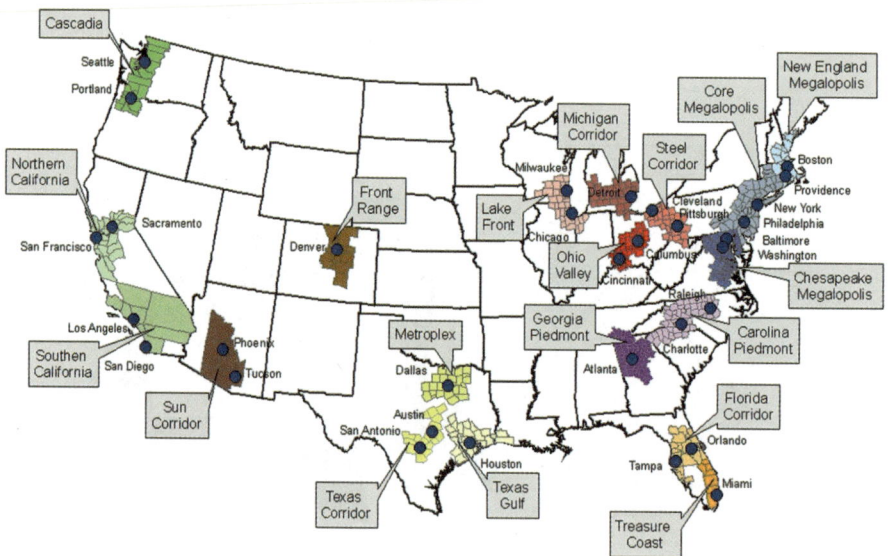

figure 15.30 Megapolitan America 2040

Lang and Nelson's image of America 2040, where "megapolitan" areas serve as interconnected development and employment zones.

SOURCE: Metropolitan Institute at Virginia Tech. Used with permission.

with anchor principal cities between 50 and 200 miles apart that will have an employment interchange measure (EIM) of 15% by 2040 based on projection." Other cities will also grow, but they foresee a more concentrated pattern of development rather than a dispersed pattern, driven by employment, cultural, social, economic, and environmental factors. They note that telecommuting, once thought to allow dispersed development because people were not tied to a place of work, has not materialized in a large way (only 4%–6% of all workers) and likely will not have a major effect on future development patterns.

What will this development look like? Because of changing demographics including an aging population and more nonconventional households as well as shifting housing preferences, development is likely to be more compact. The demand for attached, small lot, cluster, and other higher-density options is likely to outpace the demand for detached homes on large lots. Nelson and Lang foresee that metropolitan space of U.S.A. 2040 will be transformed by denser development, perhaps resembling European densities.

But most places beyond the "megas" will appear as they do today. "The image of a nation paved over from coast to coast is false. If anything, parts of the Great Plains and northern Rockies could be even less populated than now. . . . The unfinished business of settling the remaining frontier lands—the great national project of previous generations—may serve as a giant (open space) release valve in the collective conscience of twenty-first-century Americans who never fully warmed to the idea of being more built up than Old

Europe." This image of concentrated development bodes well for more sustainable Whole Community Energy.

15.7 Summary

Whole Community Energy integrates several components of sustainable energy. These include the following:

- Green buildings having efficient envelopes, HVAC systems, lighting, and appliances and minimizing embodied energy
- On-site generation in buildings and communities like solar PV, microturbines, combined heat and power district heating, wind systems, fuel cells, and landfill gas recovery that provide on-site demand and feed excess production to the grid
- Smart growth land use and infill development on vacant properties and brownfields in areas of existing infrastructure to revitalize communities, arrest sprawl, and reduce VMT
- Efficient and effective transit systems and transit-oriented development to reduce VMT
- Mixed-use, compact, and pedestrian-oriented development that promotes active living and reduced use of the automobile

This pattern of development can improve efficiency, increase renewable energy to replace oil and fossil fuels, reduce VMT, decrease CO_2 and air pollutant emissions, relieve traffic congestion, and create healthier and more livable communities.

Changing historic patterns of sprawling land use is a significant challenge in the United States and increasingly in other countries that are adopting U.S. land use patterns. But we see emerging innovative development design, smart growth land use policies, changing demographics, higher fuel prices, and shifts in consumer preference toward more compact development that enables efficient transit and freedom from the automobile. These trends need to provide the development model for the significant investment in efficient buildings, land use, and infrastructure to accommodate the growing population. Will it happen? Time will tell. And also telling will be government market transformation policies discussed in the next three chapters.

Energy Policy and Planning

CHAPTER 16

Market Transformation to Sustainable Energy

If we had every technology we needed to solve our energy problem, the job still wouldn't be finished. Even the best technologies don't always make it to the marketplace, or if they get to market, they don't win out in competition with lesser technologies. The market itself must be conditioned, or transformed, from oil and carbon-based fuels to more sustainable renewable sources and efficient use. The preceding chapters presented lots of technical and economic information about sustainable energy systems, and large-scale and rapid market transformation will fundamentally require economically competitive sustainable energy technologies— the **techno-economic solutions.** But market transformation depends on more than technical and economic feasibility. It depends on people and institutions making choices to change patterns of energy use, the subject of this chapter.

Technological and market forces and consumer choice are critical to market transformation, and government policies have a dramatic influence on both. Government market transformation programs and policies can, for example, spur the development and cost-effectiveness of new technologies, provide incentives for investment in sustainable energy, mandate efficiency improvements, regulate environmental protection that affects the relative cost of energy, and provide education and information that affect consumer choice. We call these the **policy solutions.**

Our choices are surely driven by technical and economic feasibility and policy mandates and incentives, but they are also affected by other factors including uncertainty, availability of products and investment capital, and personal and societal values. These values are affected by non-economic factors such as environmental protection, security, personal identity, and intergenerational equity, and they can influence not only consumer choice but also social movements that can accelerate market transformation. These values, choices, and movements are called the **social solutions.**

This part of the book turns from the technology of sustainable energy to market transformation and its policy and social dimensions. Chapter 17 looks at national energy transformation policies in the United States and other countries. Chapter 18 focuses on innovative state and local energy policies and programs in the United States. Before investigating these specific policies, this chapter introduces the key factors in the process of market transformation, and the role of techno-economic, policy, and social solutions.

The first section reviews some fundamentals of market transformation, including the effects of technology and market forces, market failure, and non-economic factors. The following three sections look more closely at the techno-economic solutions of technology innovation and cost-effectiveness, the policy solutions of market intervention, and the social solutions of consumer values, choice, and social movements.

16.1 Some Fundamentals of Market Transformation

We know that our global and U.S. patterns of energy use are not sustainable, and we need to transition to more sustainable non-carbon energy and efficiency before climate change and oil depletion inhibit our future options. In this section, we explore some of the theories and practicalities of energy market transformation. First, we briefly present the conceptual difference between technical, economic, and market potentials for emerging energy systems and efficiency measures. We then look at failures of the market that prevent or slow market transformation, and finally summarize nonmarket and noneconomic factors that influence market transformation.

16.1.1 Distinguishing Technical, Socio-Cultural, Economic, and Market Potential

It is often said that the amount of solar energy falling on the Earth in one day is more than the energy in the entire world oil reserve. But obviously this ultimate potential is not available because of a variety of logistical, technological, and thermodynamic constraints. Even within these constraints there is a vast "technical potential" for renewable energy and efficiency. We would like to achieve this technical potential, but we know there are nontechnical economic, social, and institutional barriers that limit our ability to develop this potential to transform from a carbon to non-carbon energy economy. On the road to energy market transformation, it is important to identify these barriers.

Energy and economic analysts define "potential" in different ways. Here, we adopt the following definitions that are highlighted in Figure 16.1 from Sathaye, et al. (2004), that shows market penetration of an energy technology (e.g., compact fluorescent lamps or photovoltaic [PV] systems) on the horizontal axis and the cost of energy or emissions saved by that technology on the vertical axis. **Market penetration** is the portion of the consumer market served by a product or technology.

- **Technical potential** is constrained by technical limits. It is the maximum amount of market penetration of a technology or effect (e.g., energy savings, GHG emission reduction) achieved over time if all technically feasible technologies were used in all relevant applications without regard to their cost or user acceptability. Technical potential expands by technological advancement.

figure 16.1   **Distinguishing Technical and Market Potential**

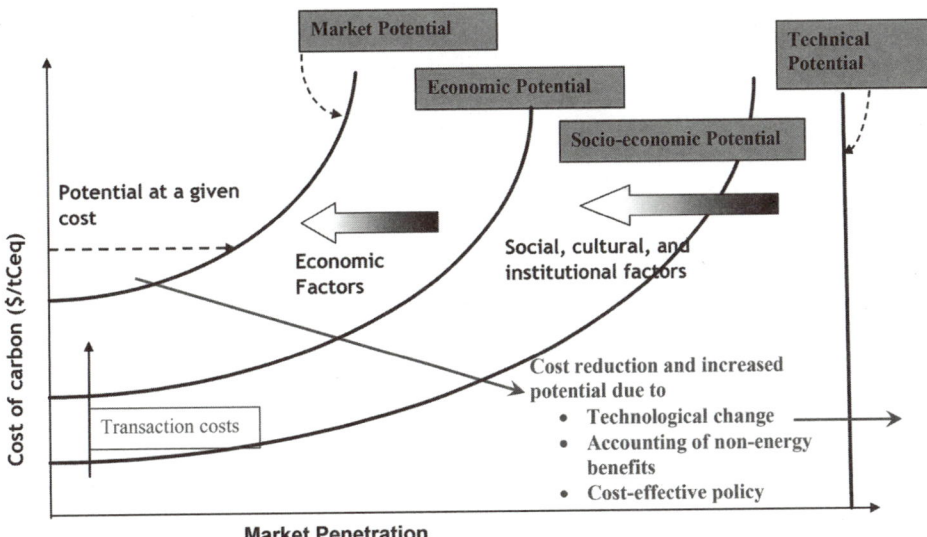

Market potential is less than technical potential because of social and economic factors, including transaction costs, but technological change, accounting for nonenergy and noneconomic benefits, and government policy can reduce these barriers to market penetration.

SOURCE: Sathaye, et al., 2004

- **Socio-cultural potential** is perhaps the most desirable level of market penetration from a societal point of view, because it assumes the elimination of market failures such as externalities (unpriced social costs) and transaction costs (consumer uncertainties and barriers). It is the maximum market penetration if all technologies were implemented that are cost-effective from a societal and cultural perspective, including noneconomic factors affecting consumer choice. Socio-cultural potential is constrained by limits imposed by social values and expands as cultural and consumer values for a technology change.
- **Economic potential** is the maximum amount of market penetration if all technologies are implemented that are economically cost-effective from consumers' point of view, assuming elimination of transaction costs. It is constrained not only by the technical and socio-cultural limits but also by economic limits including externalities. Economic potential expands as relative prices for a technology drop and shrinks as prices increase.
- **Market potential** is constrained by market limits including transaction costs. It is the amount of market penetration expected under forecast market conditions and consumer preference with no changes in policy. Market potential can expand as prices for this or competing technologies change and as policies regulate adoption and affect real prices and lower transaction costs.

So, reality falls short of technical potential because of consumer values (socio-cultural factors), externalities and false pricing (economic factors), and transaction costs (market factors). Public policy can affect all these barriers.

16.1.2 Market Failure: Transaction Costs and Externalities

One of the principal reasons that there is such a gap between market potential and socio-cultural potential is the failure of the free market to eliminate transaction costs of change and to internalize the externalities associated with all energy sources to level the economic playing field. Figure 16.1 shows that these market barriers (broad arrows pointed left) affect the levels of potential, and that transaction costs can drive up cost of energy saved. **Transaction costs** are the variety of barriers that confront consumers in making choices, thereby increasing the cost of those choices. They include poor and misinformation (e.g., consumers don't know about potential energy and cost savings), lack of access to capital (e.g., new choice requires investment and there is competition for precious cash), product unavailability, imperfect competition, and other hidden costs. Transaction costs are the main barrier between market and economic potentials. **Externalities** are external social costs such as pollution and GHG emissions that are not included in the cost of a technology such as coal-fired electricity. Externalities and consumer preference are the main barriers between economic and socio-cultural potentials.

The figure also shows that these barriers for renewable and efficient energy systems can be removed by cost reduction, technological advancement, government policies that reduce transaction and market costs and internalize externalities, and cultural values that consider nonenergy and noneconomic benefits and costs.

16.1.3 Noneconomic Factors and Market Transformation

Why do we buy the products we do? Cost of the product affects our ability to purchase it and which brand or model we choose ("the best deal"). We rarely compute the economic return of our purchases because the benefits (utility, convenience, entertainment, and pleasure) are hard to put in dollar terms, but we can easily recognize difference in quality or usefulness of consumer products. Generally, the choice to purchase renewable and efficient energy products and measures is somewhat different. These products provide the same basic functions or services as other conventional energy sources, and most consumers seek "the best deal" to provide these energy services.

So consumers often perceive that their "best deal" is the choice with the lowest initial cost. Well-informed, discriminating consumers of energy services will select the products and measures providing desired functions that have the *lowest life-cycle monetary costs*.

But there is more to consumer preference than simply monetary cost. Increasingly, even more discriminating consumers are selecting energy with the *lowest life-cycle sustainability costs*. Sometimes these latter choices will have greater monetary cost, but they meet the enlightened consumers' baseline of quality, and provide greater pleasure, as consumers know they are contributing to making society more sustainable.

For example, markets for "green" energy have developed for consumers willing to pay a bit more for environmentally friendly energy sources. These choices are made more rational by the improved means of life-cycle sustainability analysis, and certification systems like ISO 14000, LEED, and ENERGY STAR.

Kulakowski (1999) found that cost and payback alone do not determine energy investment decisions. In her study of institutional decisions to adopt energy-efficient technologies, she found that price and payback are very important and that higher initial costs of efficient technologies and artificially low energy prices (that do not internalize externalities) work against energy-efficient improvement. Transaction costs also impede adoption.

But she also found that *organizational structure, procedures, and culture* matter. Institutional culture and procedural rules influence availability of funds and decision making for investment in energy efficiency and new technology. And she found that *individual values and behavior* also matter. The personal values and commitment of individual consumers, institution or company leadership, and employees to the goals of energy efficiency and associated environmental benefits influence investment decisions. These values are examples of the social, cultural, and institutional factors shown in Figure 16.1 that define societal-economic-cultural potential.

However, price and monetary value still rule the roost. People face choices on how they invest their money. If we are to change our energy patterns on a large scale, renewable energy and efficiency must compete effectively against other investment choices. We discuss below the effect of price on market penetration and the importance of techno-economic solutions for market transformation.

16.2 The Techno-Economic Solutions

16.2.1 Technological Change and Diffusion of Innovation

How does the market behave as new technologies such as efficient and renewable energy systems come into commercial use? The process of diffusion and adoption of technology is the subject of considerable research in economics and marketing, and it is generally characterized by Everett Rogers' bell curve of adoption of a product or technology (Figure 16.2). The process is initiated by "innovators," the first 2.5% of users who are the risk takers. These are followed by "early adopters" (13.5%) and the "early majority" (34%), after which adopters follow a "bandwagon" effect; the rate of adoption declines with the "late majority" (34%) and finally the last-to-adopt "laggards."

Figure 16.3 illustrates the diffusion process and market penetration for a variety of U.S. consumer products and activities in the years after they were introduced. Note the rapid increase in use of electronic equipment. The difficulty in market analysis is to predict market penetration. So we need to be able to predict similar curves for hybrid vehicles, LED lighting systems, residential PVs, and biofuels, and what factors will influence that penetration.

figure 16.2   Model of Adoption Diffusion of Energy Efficient Technologies

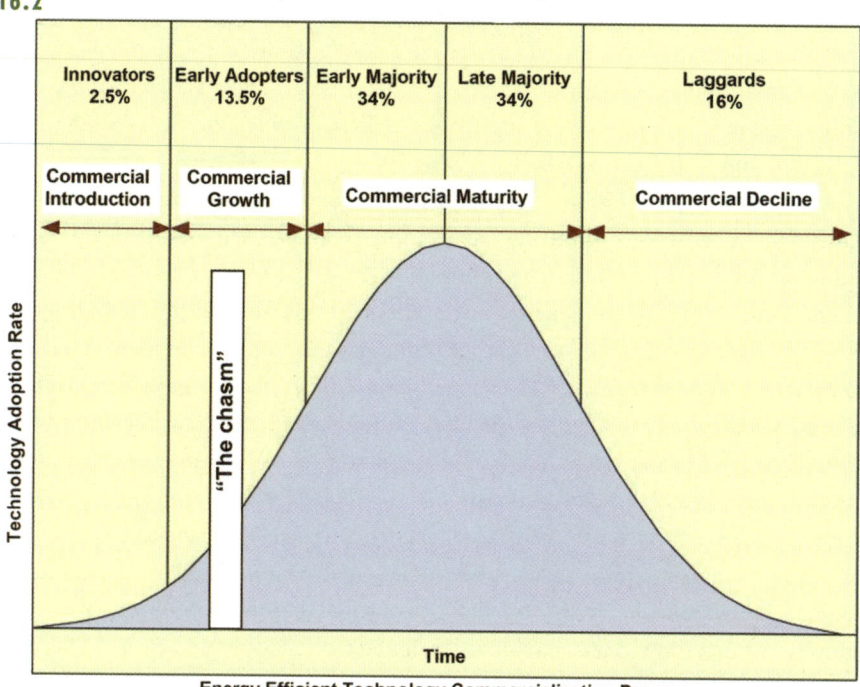

The technology adoption rate follows a bell-shaped curve from "innovators" and "early adopters" ultimately to "laggards." The difficult period in the process is overcoming "the chasm" in the early period of commercial growth.

Source: Jenkins, et al., 2004

16.2.2 Market Penetration and Simple Payback

Recall that market penetration is the portion of the consumer market served by a certain product or technology. Many market penetration models are based on simple payback period (SPP). As we know from Chapter 5, SPP is the number of years an investment will take to pay for itself from its returned savings:

$$\text{Simple payback period} = \frac{\text{Initial \$ cost}}{\text{Annual \$ savings}} = \frac{\text{Intial \$ cost}}{(\text{Annual energy savings})(\text{energy price})}$$

SPP is often a good predictor of market penetration because it is an intuitive measure of "a good deal" and financial return. Because it is understandable, it's a good predictor of purchasing behavior across products when decisions are based on energy cost savings. Some models assume a maximum eight-year SPP as a prerequisite for market penetration (Institute for Sustainable Energy, 2004). The U.S. DOE National Energy Modeling System (NEMS), which is used to

figure 16.3   **Market Penetration of Selected Consumer Products in the United States (showing years after they were introduced)**

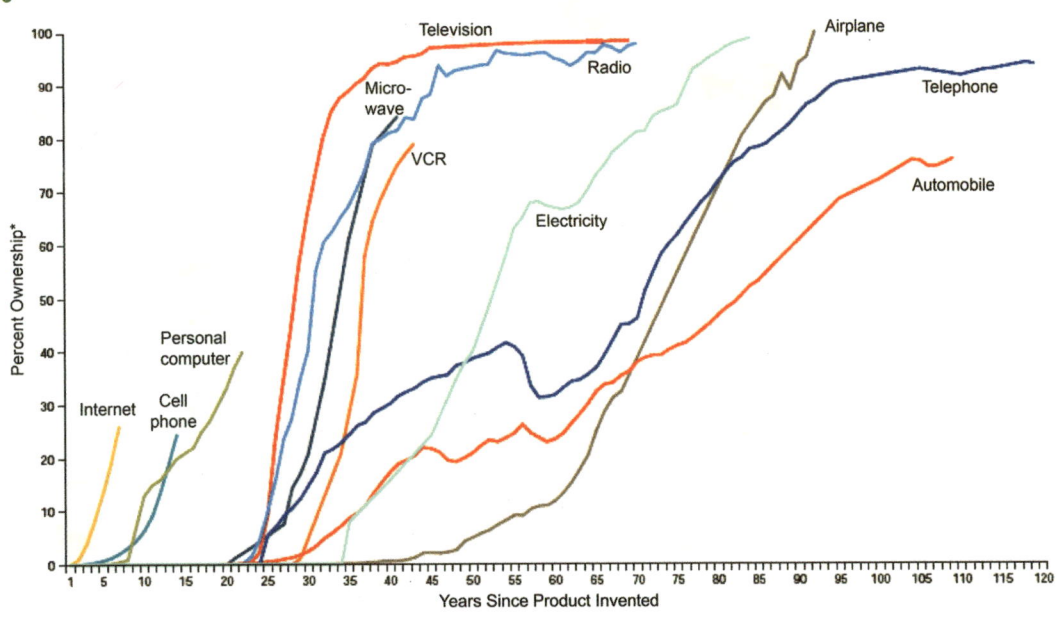

* Percent of households except airplanes (% relative to 1996), autos (% owned per adult > 16 yrs), cell phones (% phones owned per auto).
SOURCE: Federal Reserve Bank of Dallas, 1996 Annual Report

generate EIA's Annual Energy Outlook and other government projections, is more conservative and assumes market penetration into new construction based on SPPs given in Figure 16.4. This model has been used in several studies of distributed generation technologies in the building sector. The figure assumes that it takes some time for a new technology to "diffuse" into the market as adopters climb the learning curve and achieve ultimate penetration. Both the ultimate market share and the pace to get there are greater for longer SPP. The figure assumes that even with SPP as low as one year, the ultimate penetration rate in new construction may be only 30%.

16.2.3 The Price of Technology, the Experience Curve, and Learning Investments

The initial capital cost or price of a new technology or system is a critical factor in measures like SPP and the cost of conserved energy (CCE), which are important measures of performance and potential market penetration for a new technology. Basic microeconomics tells us that successful people, enterprises, and products do better as they operate and develop in competitive markets. Learning through market experience reduces price; reduced price then fuels additional demand and production; and more production experience further reduces price. The Learning Curve describes how marginal labor cost declines with cumulative production.

figure 16.4   **Market Penetration for Various Payback Periods**

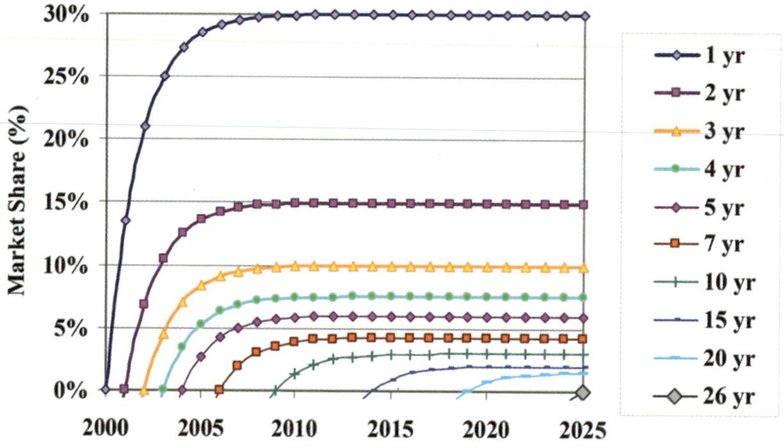

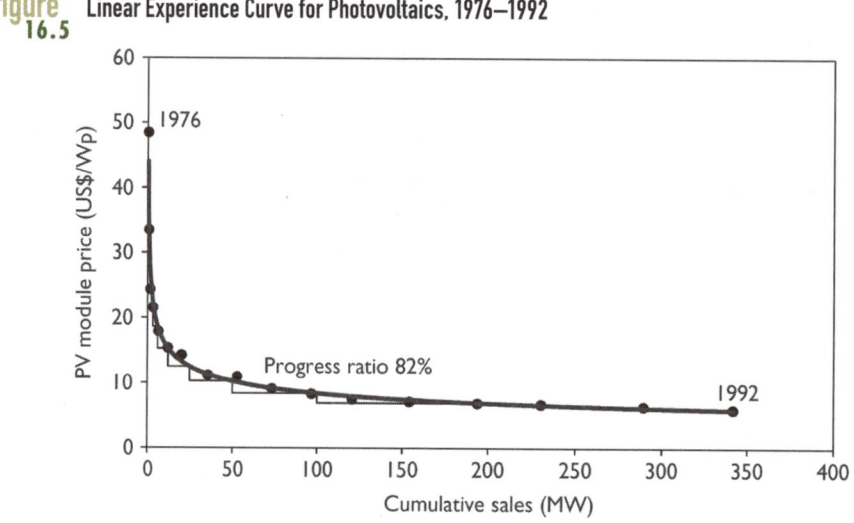

Source: LaCommare, et al., 2005

16.2.3.1 The Experience Curve

The Experience Curve describes how overall price declines with cumulative production. Bodde (1976) argues that the Experience Curve is very useful for gauging long-term trends and formulating long-range strategies for technology development. OECD/IEA (2000) suggests that it can be used to identify investments and public policy actions to advance renewable and efficient energy systems. Many other analysts have also applied it to energy systems (e.g., Duke and Kammen, 1999; Margolis, 2003; Buerskens, 2003; Swanson, 2004).

figure 16.5   **Linear Experience Curve for Photovoltaics, 1976–1992**

Source: OECD/IEA, 2000

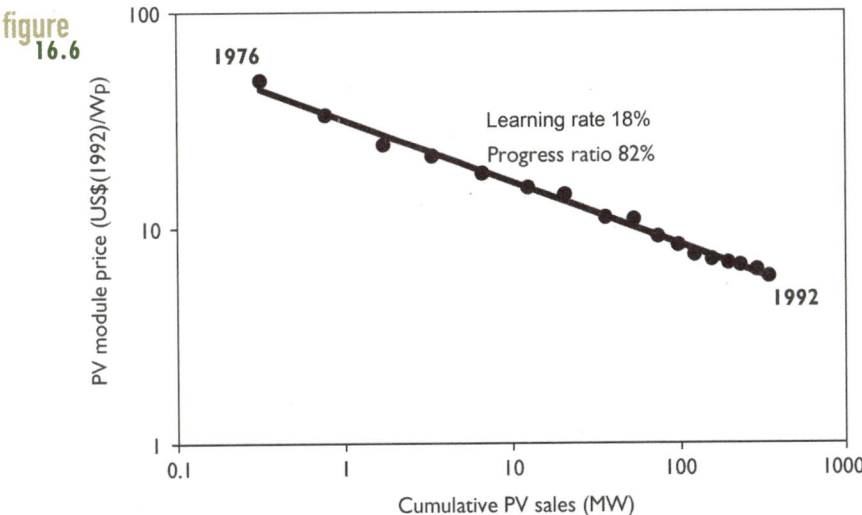

figure 16.6

The double logarithmic Experience Curve represents the relationship between cumulative production and price as a straight line, which defines the Progress Ratio.

SOURCE: OECD/IEA, 2000

The Experience Curve plots price versus cumulative sales. There is an obvious feedback loop that develops as technical advances and learning drive prices down, stimulating additional production, economies of scale, and research, which further drive prices down. Figure 16.5 gives price versus cumulative production for PV modules from 1976 to 1992 on a linear scale, meaning distances along the axes are directly proportional to *absolute* change in price or sales. The graph shows that large advances are made in the initial phases of development and they diminish quickly.

Figure 16.6 is a double-logarithmic graph of the same data; here, distances along the axes are directly proportional to the *relative* change in price or sales. This Experience Curve shows better that increasing levels of sales and production continue to affect price. The slope of this curve defines the **Progress Ratio (PR),** which measures how price will decline with every doubling of production sales. The Experience Curve is given by the following equations (Duke and Kammen, 1999):

$$P(t) = P(0) \times \left[\frac{q(t)}{q(0)} \right]^{-b}$$

where $P(t)$ = the average price at time t

$q(t)$ = the cumulative production at time t

b = the learning coefficient

$$PR = 2^{-b}$$

where PR = the Progress Ratio

In this case the PR is 82%, meaning that for each doubling of PV sales, price will be reduced to 0.82 of its previous level. The **Learning Rate** is $100 - PR$, or in this case $100 - 82 = 18$,

figure
16.7

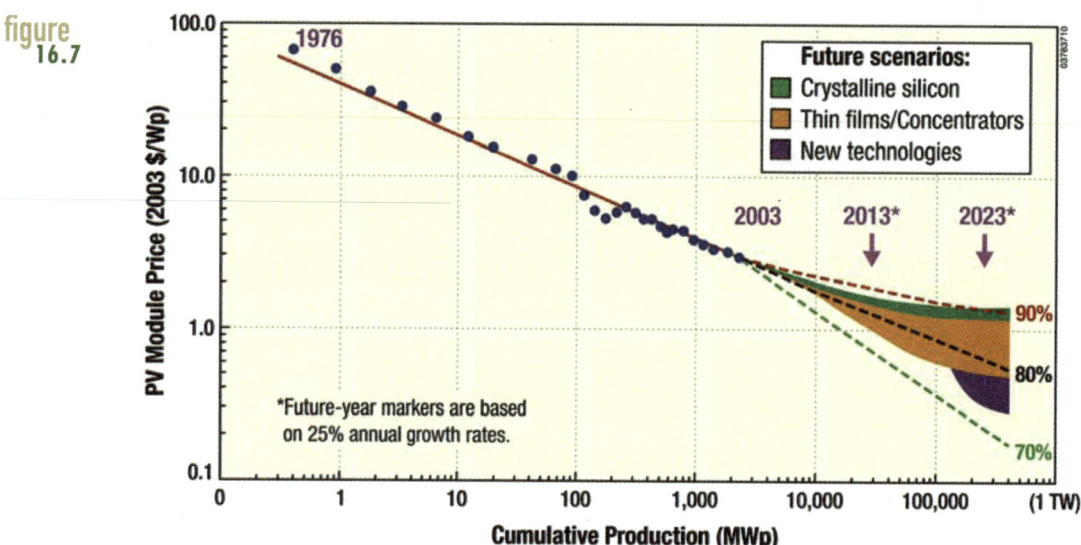

PV Experience Curve to 2003 with extension of current Progress Ratio of 80% (doubling of production will reduce price 20%) and *PR* scenarios of 70% and 90%.

Source: U.S. DOE, 2005

meaning each doubling of production will reduce price by 18%. The most successful technologies, such as semiconductors, have steeper Experience Curves, a smaller *PR*, and larger Learning Rates.

Figure 16.7 shows the Experience Curve for solar PV (see also Figure 11.21) with current learning rate of 80% extended and also 70% and 90% projections (U.S. DOE, 2005; Surek, 2005).

Figure 16.8 gives Experience Curves through 1995 for several power generation technologies in the European Union given in OECD/IEA (2000). More mature technologies have flatter curves. The natural gas combined-cycle (NGCC) option enjoyed the experience in the United States and elsewhere during the 1970s and 1980s.

16.2.3.2 Assessing Future Prospects and Estimating Learning Investments

Experience Curves can be interpreted in energy analysis and policy making, including assessing future prospects, estimating "learning investments," and formulating policy and other actions to accelerate learning rates.

The trend line of Experience Curves can be extended into the future to estimate production levels at which a technology may compete with others. Figure 16.9 shows that for PV with a *PR* of 80%, a break-even price with the fossil fuel alternative ($50¢/W_p$) could occur at cumulative production of about 200 GW assuming this *PR* continues. Trends for two other *PR*s are also shown.

The shaded area under the curve is equivalent to the total investment in dollars, the so-called **Learning Investment** necessary to reach the break-even point. For this graph, it is equal to about $60 billion. After these learning investments are made and the technology reaches break-even, they will be recovered as the technology continues to ride down the Experience Curve.

The curve shows the production (200 GW) and investment ($60 billion) needed to reach break-even for PV, but not when this will occur. This depends on the production growth rate of the technology to reach the break-even production level. For example, in

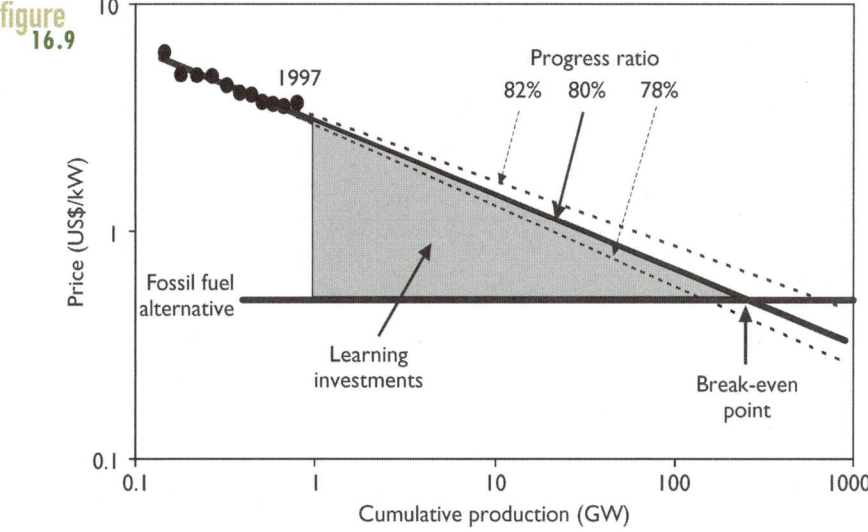

figure 16.8 Experience Curves for Different Generation Options in European Union

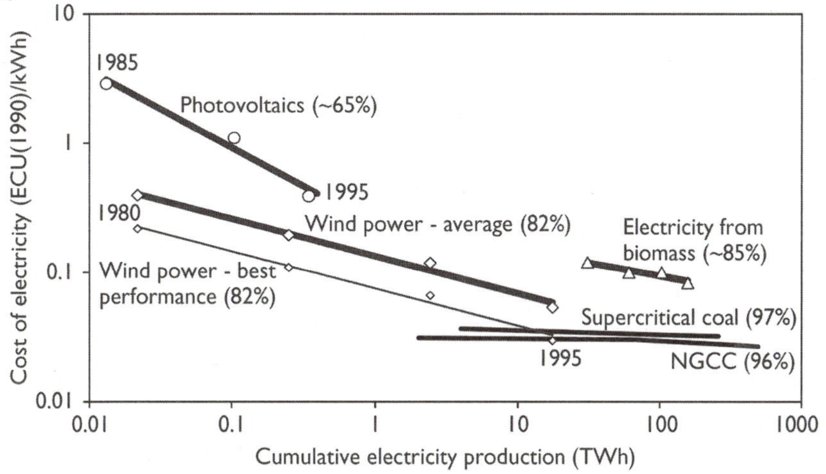

Source: OECD/IEA, 2000

figure 16.9

Extending the Experience Curve to break-even price identifies cumulative production necessary and Learning Investments needed to achieve it. The break-even production would decline for lower *PR* and increase for higher *PR*.

Source: OECD/IEA, 2000

2000, OECD/IEA estimated continuing the historic 15% growth of PV production would achieve production break-even with centralized power options in 2025; doubling that rate would move it up to 2015.

16.2.3.3 Energy Policy and Learning

The learning system that contributes to the Experience Curve phenomenon is influenced by investments for new discovery through research and development and production cost savings through economies of scale and improved efficiency as production grows. Price reduction creates more markets, which leads to greater production, which further reduces prices in a "virtuous" cycle (Duke and Kammen, 1999).

Although this is primarily driven by the engine of the free market, public policy can influence the learning system, the *PR*, and the rate of production growth. For example, government research and development funding can provide Learning Investment to developing technologies; government procurement programs can increase the rate of production that will drive the technology down the Experience Curve; and government incentives such as tax credits can reduce price and drive further production. These are what Duke and Kammen (1999) call **Market Transformation Programs.** The Experience Curve can identify Learning Investments or price reductions that are needed to achieve production objectives. We explore these policy solutions in the next section.

If society decides that some "breakaway" technological advance is necessary, the Experience Curve can be useful to identify the parameters of change. For example, Figure 16.10 gives a variation of the Experience Curve showing carbon intensity of the world economy (kg C/$GDP) versus cumulative $GDP. The Progress Ratio of 79% is extended to an

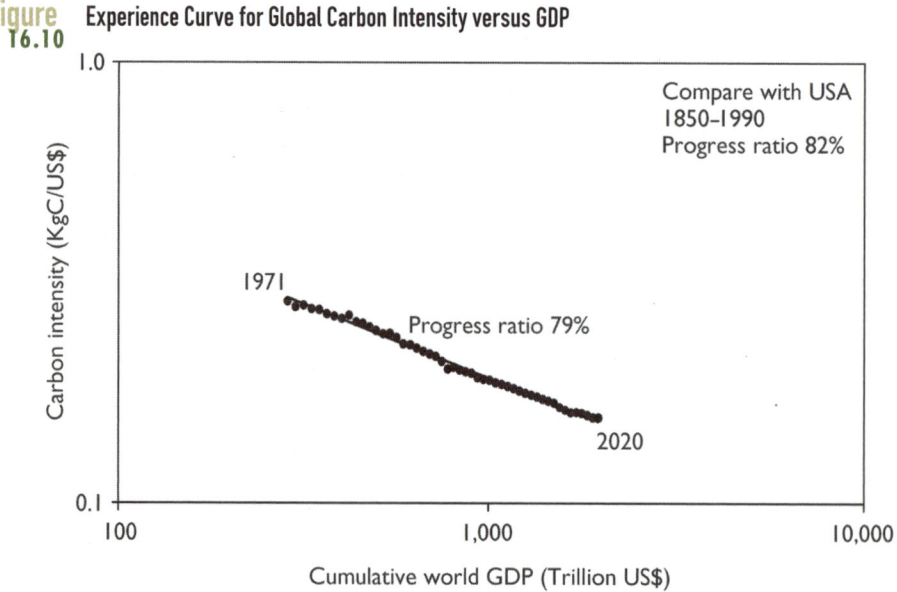

figure 16.10 Experience Curve for Global Carbon Intensity versus GDP

figure 16.11   The Necessary Breakaway Path to Stabilize CO_2 Emissions by 2050

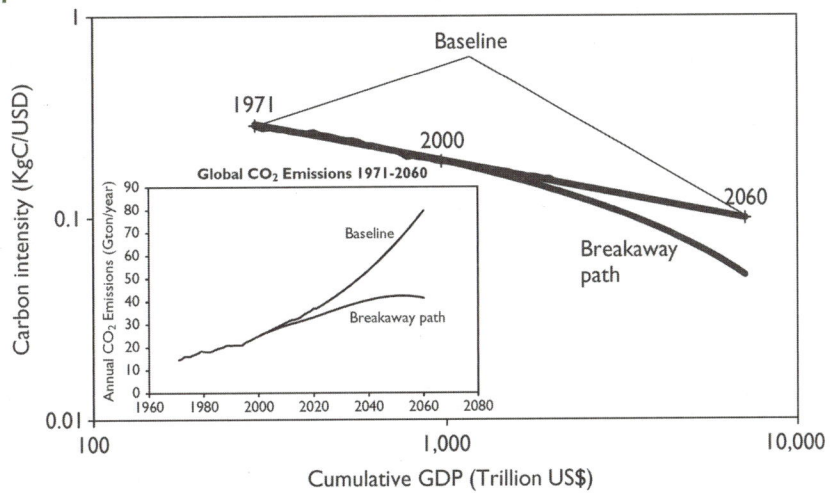

figure 16.12   The Learning Investments Needed in a Biomass/PV Technology Portfolio to Achieve that Breakaway

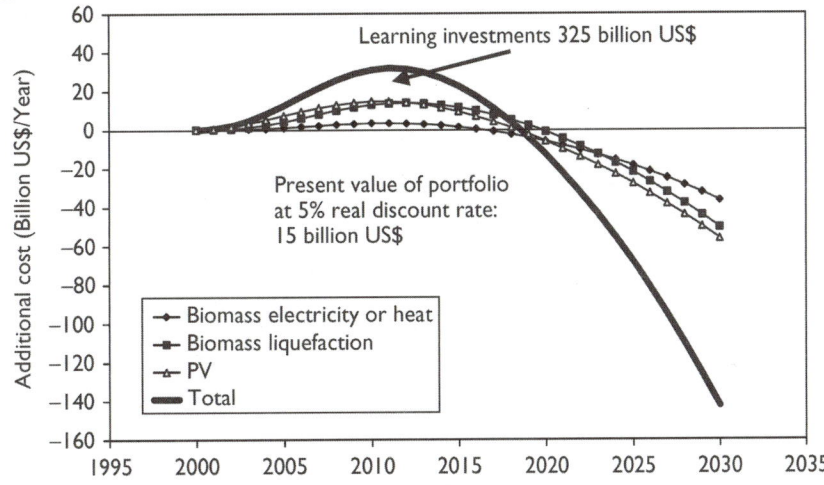

Experience Curves for these technologies indicate that $325 billion in investments by 2020 could achieve this result with substantial benefits thereafter, creating a Net Present Value of $15 billion at $d = 5\%$.

SOURCE: OECD/IEA, 2000

expected GDP in 2020. Figure 16.11 extends this curve to GDP level expected in 2060. Although the carbon intensity continues to go down, total carbon emissions would be four times 1990 levels. The figure also shows a breakaway path that would stabilize total emissions in 2050 (see inset). Converting this path to the Experience Curve, the resulting *PR* would need to be 50%.

This would require a steady increase in deployment of low/zero-carbon technologies. The OECD/IEA study provides one portfolio for achieving this deployment of equal parts biomass electricity, biomass liquid fuels, and PV technologies, shown in Figure 16.12. The learning investment of $325 billion by 2016 could provide a return on investment thereafter, and a positive present value of the portfolio of $15 billion.

16.3 The Policy Solutions

16.3.1 The Case for Market Intervention

There are at least three reasons for government intervention in energy markets to address the market failure and barriers introduced earlier in this chapter:

1. **Externalities:** Energy prices do not reflect the full range of costs and benefits associated with energy use, such as carbon emissions, urban air pollution, health and safety of coal miners, military costs to secure access to Middle East oil, and risks associated with nuclear safety and dependence on oil imports. Government intervention can internalize these external costs.
2. **Transaction costs:** Limited knowledge and information, poor access to capital, lack of availability of products, limited time, misplaced incentives and regulatory policies, and other market barriers inhibit investments in new technologies and efficiency. Government intervention can reduce these transaction costs.
3. **Poor future-orientation:** The market and consumers are today-oriented and give low priority to future energy problems especially when they are not felt today or are uncertain. Government intervention can make investments and help individuals and organizations make decisions that help themselves and society, today and for the future.

"For the future" is especially relevant to energy and sustainability. Some economists tell us not to worry about peak oil or even the impacts of global warming, because the market will make necessary adjustments to replace oil with other fuels and develop new sources to prevent impacts. But like most markets, energy markets are geared to today's economic forces in which demand and supply determine price, which in turn affects demand and supply. While we have "futures" markets for energy, especially oil, that future is usually only three to six months away. And markets can't by themselves correct large-scale, long-range problems such as climate impacts.

This system of the free market works very efficiently, *except* for the externality, transaction cost, and social welfare issues given above, and *except* for replacement costs beyond the short timeframe of futures markets. For limited nonrenewable conventional oil and natural gas, future replacement costs may be significant but current markets undervalue those costs and therefore underprice those fuels. Those low prices inhibit investment in replacement alternatives and improved efficiency of use. We saw significant increases in oil and gas prices in 2005–2007, but the market did not anticipate those increases and cannot respond fast enough with alternatives. And once we feel the full brunt of effects of global warming, it may be too late to reverse course. We need to plan ahead and we need to act more quickly than the market's slow pace of change.

Government intervention can "correct" these market imperfections; use the market to meet economic, environmental, and societal goals, including enhanced sustainability; and help the market plan for the future. Let's return to our model of commercial diffusion of technology, given in Figure 16.2, and see how government market transformation programs can help market forces in adoption of energy innovation. Figure 16.13 shows that energy

figure 16.13 Effect of Government Market Transformation Programs on Commercialization of Energy Efficient Technologies

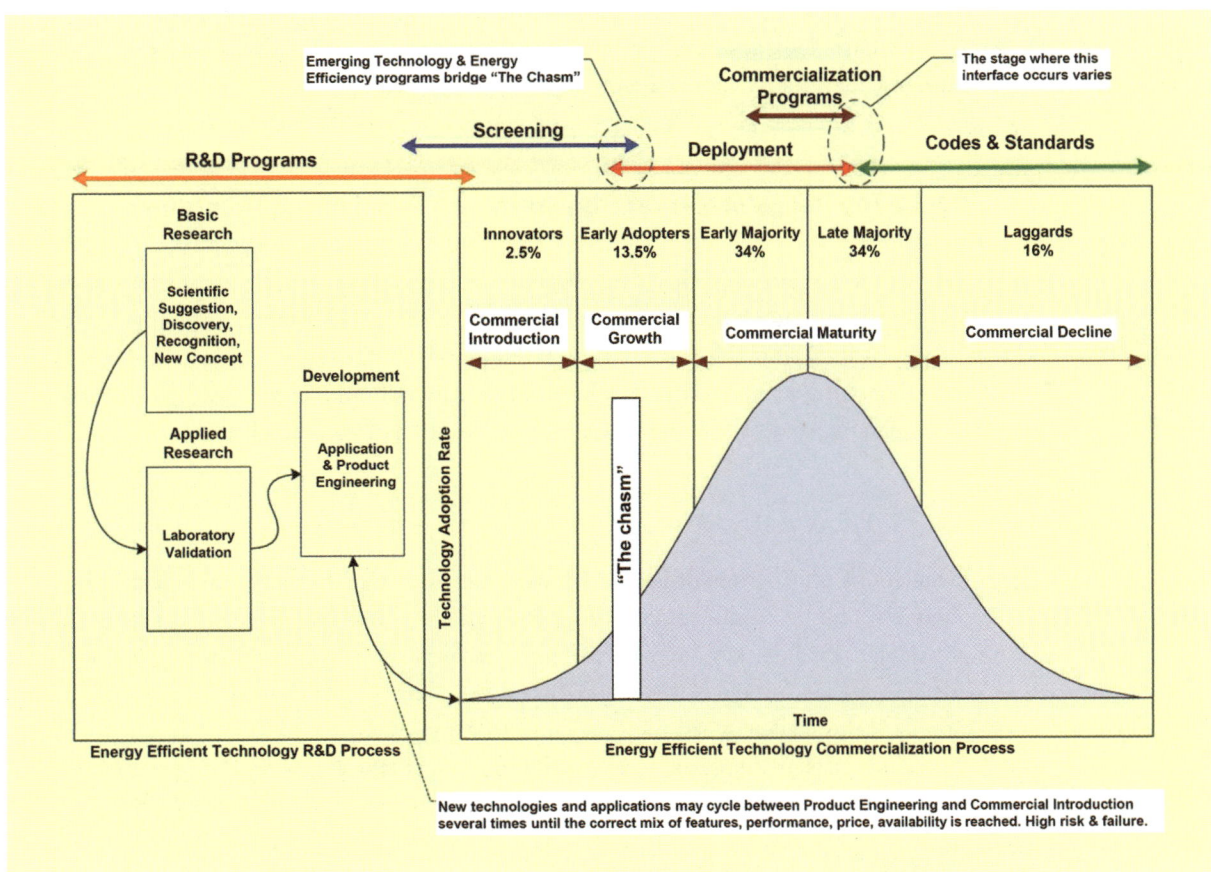

Research and development (R&D) funding spurs initial technology development and commercial introduction. Emerging technology and deployment programs help overcome the difficult "chasm" in the adoption process. Codes and standards finally push the remaining market into adoption.

SOURCE: Jenkins, et al., 2004

technology research and development programs can help fuel innovation. Energy efficiency programs, including improved information and incentives, can enhance commercialization and deployment, speeding the diffusion process. Ultimately codes and standards validate established technologies.

We will review a wide range of government market transformation programs in this section, but it is important to understand at the outset that enacting such policies is not straightforward. Government often has conflicting policy objectives, and there is constant political debate about the appropriate extent of market intervention and the specific industries and technologies to be advanced by policy initiatives. For example, some political interests aim to use policy to raise conventional energy prices to reflect external costs and create an incentive for more sustainable energy, but others fear that higher energy prices will slow economic growth with serious consequences. And of course, there is much at stake for different energy industries and other stakeholders who seek policies to protect or advantage their interests, so the policy process is further complicated by competing economic and political interests.

We will look at the politics of energy later in the chapter, but first, this section reviews various energy policy approaches to improve energy markets.

16.3.2 The Range of Market Transformation Policies and Programs

Market transformation policies and programs include a range of policy approaches using regulations, economic incentives and disincentives, learning investments, and direct assistance (Table 16.1). We provide below a general description of these approaches, and the following two chapters describe specific energy policy initiatives by the U.S. federal government and other national governments as well as U.S. state and local energy policies.

16.3.2.1 Regulations

Regulations provide one of the most direct means of market transformation because they require action by producers and consumers and are not solely dependent on market forces for change. Because they are mandatory, they achieve a high penetration rate close to 100% for new efficient products. Economic incentives affecting price and payback period cannot approach this market penetration as Figure 16.4 suggests.

Energy regulations can be grouped into product efficiency standards, production standards, utility and other energy industry regulation, and environmental regulation.

Product efficiency standards. We have introduced several product efficiency standards in previous chapters, including building codes, appliance efficiency standards, and vehicle efficiency standards. These regulations aim to transform markets where market forces are not sufficient to produce potential economic, environmental, or societal benefits. The potential market transformation and energy and economic savings associated with efficiency standards are significant because of near 100% market penetration.

table
16.1   Array of Energy Market Transformation Policies and Programs

Regulations

    Product efficiency standards

    Production standards

    Utility regulation and market reform

    Environmental regulations

    Price controls

Economic and Financial Measures

    Tax incentives and disincentives

    Financing assistance and risk insurance

    Research and development funding

    Procurement

    Energy assistance

Energy Planning and Information

    Energy planning

    Information and training

Capacity Building, Partnerships, and Voluntary Action

    Voluntary agreements and partnerships

    Capacity building and civil society

Product manufacturers often oppose stricter efficiency standards because of compliance costs, but if standards are applied equitably, they place the same requirements on all, and higher costs if any are passed on to all consumers. Some manufacturers who are early adopters producing efficient products may have a competitive advantage under stricter standards, but perhaps they should be so rewarded.

Although efficiency standards may have significant environmental benefits, the strongest case for them comes from the economic benefits to consumers. For example, cumulative net consumer savings to 2030 from energy cost savings of U.S. federal appliance standards enacted through 2007 are estimated at $250 billion (Nadel, et al., 2006; ASAP/ACEEE, 2008).

Production standards. Whereas product efficiency standards focus on the demand side of energy use, production standards focus on the supply side. They require a certain amount or percentage of supply to come from energy sources determined to be beneficial by public policy. Two production standards in use today are the Renewable Portfolio Standard (RPS) for electricity and the Renewable Fuels Standard (RFS) for vehicle fuel. Several states have adopted an RPS that requires each electric utility serving customers in the state to provide a certain percentage or amount of their marketed power from renewable sources by a certain date. Some states and the 2005 and 2007 federal energy policy acts include an RFS requiring that gasoline suppliers provide a minimum quantity or percentage of fuel from ethanol. For example, the federal RFS is 36 billion gallons by 2022, and the Minnesota RFS is 20% ethanol in gasoline fuel by 2010 (see Chapters 14, 17, and 18).

The primary purpose of the production standards is to establish a minimum market for renewable energy and thus greater certainty for developers and investors. Investments and greater production can help the industries move down the Experience Curve, lowering prices and growing their market penetration.

Utility regulation. Certain energy industries do not operate in competitive markets and are regulated to avoid abuse. The best examples are investor-owned electric and natural gas utilities that have designated service areas. Such utilities have operated as monopolies because consumers within the service area are essentially captured and have little choice. As discussed in Chapter 9, the Public Utility Holding Company Act (PUHCA) of 1938, the Public Utility Regulatory Policies Act (PURPA) of 1978, and other federal laws established guidelines for utility regulation mostly by state utility commissions. For decades, the rates, generating plants, transmission lines, services, and other practices of utilities have been subject to review and approval by state commissions. Because some utility operations, such as interstate transmission, cross state lines, federal law established the Federal Energy Regulatory Commission (FERC) to review and approve such operations.

From the late 1970s to the 1990s, many state commissions used their regulatory authority to encourage and mandate utility programs to enhance energy efficiency through demand-side management. The Energy Policy Act of 1992 established opportunities for restructuring of utility regulations and several states experimented with new regulatory structures that aimed to provide greater choice and competition in utility markets, which could lead to lower utility rates. Although California's restructuring failure (see Section 9.7.6) put a damper on several other states' attempts, consumers in most states now have greater choice in source of electricity, better access to renewable sources of power, and increasing opportunities for on-site generation through net-metering, than ever before as a result of utility market reforms.

Still, as discussed in Chapters 17 and 18, utility regulation remains a moving target. The Energy Policy Act of 2005 repealed PUHCA and amended PURPA. Although some states have moved more slowly toward further restructuring, others have moved forward by adopting RPS and greater consumer choice.

Environmental regulations. Environmental laws and regulations aiming to reduce the many environmental impacts of energy production, transport, and use have a significant effect on energy markets. The cost of compliance acts to reduce or internalize some of the externalities associated with energy options. For example, compliance with coal mine land reclamation regulations, miner safety and health laws, and air pollution control rules increases the cost of coal-generated electricity. The resulting higher price of coal power helps other, less environmental impacting power sources, such as wind, solar, and combined-cycle natural gas, compete with coal. Environmental laws affecting the cost of energy include air and water quality regulations, waste management controls, nuclear safety and fuel cycle management, energy facility siting requirements, and others.

Some regulations include a market component to enhance implementation. An **emission caps and trading** system is used in the U.S. Clean Air Act for control of sulfur oxides. Coal power plants are allocated caps on SO_x emissions; if they reduce emissions below the caps they can sell credits to other plants, which can use those credits in lieu of their own emissions reductions. The European Union (EU) uses a similar system for carbon emissions from industry (see Chapter 17), and such a carbon emissions control system is currently proposed in the U.S. Congress.

Energy price controls. Government has the authority to regulate wholesale and retail prices of energy. Indeed, utility regulation has essentially controlled electricity rates. But electricity price controls were at least partially blamed for California's electricity crisis of 2001. Most efforts in restructuring utility regulation have enabled more competition and integration of market forces into rate structures.

Government has used its authority over utilities to affect not only retail pricing but the rates utilities must pay non-utility generators supplying electricity to the grid. These can include a homeowner with rooftop PVs, a large windfarm, or an industry with a combined heat and power system. These so-called **buy-back** or **feed-in** rates will determine in large part the cost effectiveness of these on-site efficient and renewable electricity systems. Under the PURPA of 1978, these rates in the U.S. were to be based on the costs avoided by utilities for buying the on-site power. These rates were generally too low to provide an effective financial incentive for developing such systems. Most states now offer net metering for small-to-moderate on-site systems, essentially requiring utilities to buy back power at retail rates. In many European countries, especially Germany, feed-in rates are set well above retail, and this has led to an explosion of wind and solar systems that has made Germany the world leader in both (see Section 17.1.2.2).

Beyond utility rates, the evolving political climate for deregulation of markets has diminished government interest in direct control of consumer energy prices. This climate was affected by ineffectual efforts to regulate price of oil and natural gas. In 1971, the U.S. federal government had a complicated system of price controls on crude oil produced in this country, but by 1979 they were deemed ineffective and were repealed with an accompanying "windfall profits" tax on excess oil company profits from the higher prices that followed the repeal. Since that time, oil and natural gas prices have been determined largely by international markets. Recent record oil company profits in 2005–2007 resulting from record world oil prices have renewed proposals for windfall profits taxes, but they have not mustered sufficient political support for enactment.

16.3.2.2 Economic and Financial Measures

This is not to say that government policy does not aim to affect the cost of energy or energy systems. Government economic and financial measures are powerful policy tools used to affect investors, energy developers, and energy consumers. These measures can reduce financial risk, lower investment cost, fund development of new technology, and assist those hardest hit by the cost of energy. We can distinguish five basic types of economic and financial energy policies: tax policies, other financing and risk assistance, research and development funding, government procurement, and direct assistance.

Tax incentives and disincentives. Most individuals, firms, and investors are very sensitive to the taxes they pay, and energy tax policy can affect behavior of consumers, developers, and investors. There are different types of tax incentives and disincentives:

1. **Energy taxes and surcharges** increase the price of conventional energy, and higher prices can reduce demand and increase the value of energy saved by efficiency or alternative sources, and thus improve their SPP. An example of an energy tax is the excise tax on gasoline. In 2006 the average excise tax on gasoline in the United States was $0.39/gal, which has little effect on demand and energy saved by efficiency and conservation compared to the U.K. tax of $4.00/gal (see Figure 13.12).

   A surcharge per kWh electricity consumption is a common way state utility commissions have allowed utilities to generate revenues for demand-side efficiency programs.

   Broader taxes on energy, such as carbon or Btu tax, have been debated in the EU and the United States. For example, in 2007 Columbia Business School Dean and former Bush economic advisor Glenn Hubbard argued for a carbon tax, saying if you want to fuel innovation, you have to price it. Others argue that a carbon tax would be more effective and more easy to implement than a carbon cap and trade systems because it would apply to all energy markets at the point of sale including households and vehicles, whereas successful cap and trade systems have only been applied to large stationary sources. The fate of a carbon tax in the United States is uncertain because energy taxes are politically charged, and past federal proposals have been soundly voted down.

2. **Energy investment tax credits** aim to spur investment in qualified energy efficiency measures and production facilities by effectively lowering the initial cost by the value of the tax credit. For example, the 2005 Energy Policy Act provides a 30% tax credit for business investments in solar energy systems.

3. **Energy production tax credits** provide a direct incentive for the production of qualified energy sources. For example, producers of electricity from qualified renewable sources in the United States receive a tax credit of 1.8¢/kWh generated for commercial sale. Blenders of fuel ethanol receive a tax credit of 51¢/gal of ethanol blended for fuel sales.

4. **Energy research and development tax credits** are applied to expenditures on qualified energy research, removing some of the financial risk associated with such ventures.

5. **Energy investment and production deductions** on taxable income for investments provide an incentive similar to tax credits, but at a considerably lower rate. Solution Box 16.1 illustrates the different effects of an energy tax credit and tax deduction.

Financing assistance and risk insurance. Tax incentives can lower the initial cost of energy investments, but financing assistance can have a more direct effect in certain situations.

SOLUTION BOX 16.1

Comparing Energy Tax Credits and Deductions

The U.S. federal government wants to encourage consumers to buy energy-efficient, low-emission vehicles and provides tax incentives for the purchase of hybrid electric vehicles (HEV). In 2002–2005, the incentive was a $2000 federal tax *deduction* for the purchase of such a vehicle. The 2005 Energy Policy Act changed this to a tax *credit* of up to $3400 depending on the specific vehicle mileage and on its marketing success (i.e., the credit goes down as more such vehicles are sold). The Toyota Prius has an mpg rating to qualify for the full $3400 credit, but because of its marketing success in 2006, let's assume it will be eligible for a $2000 tax credit. How does the $2000 tax credit compare to the $2000 tax deduction in reducing the purchase price of a $20,000 Toyota Prius for a household with $100,000 taxable income and a 30% tax bracket?

Solution:

Purchase in 2005: $2000 Tax Deduction

Deduction is subtracted from income to which the tax-bracket percentage is applied:

Without deduction: tax = $100,000 × 0.30 = $30,000
With deduction: tax = ($100,000 − $2000) × 0.30 = $29,400
Tax savings = $30,000 − $29,400 = $600
Or tax savings = (tax deduction claimed) × (tax bracket %) = $2000 × 0.30 = $600

Purchase in 2006: $2000 Tax Credit

Credit is subtracted from tax obligation:

Without tax credit: tax = $100,000 × 0.30 = $30,000
With tax credit: tax = ($100,000 × 0.30) − $2000 = $28,000
Tax savings = $30,000 − $28,000 = $2000

The tax deduction lowers the purchase price by $600 or 3%, whereas the tax credit lowers the price by $2000 or 10%.

There are four types of government financing and insurance assistance, all of which may incur higher government administrative costs than tax credits:

1. **Low- or zero-interest loans:** To improve access to and reduce the cost of capital for energy investments by consumers, governments can offer, or direct utilities to offer, incentive financing for qualified energy systems or measures.

2. **Rebates:** Direct rebate of a portion of investment in qualified energy systems or measures. These are similar in effect to tax credits, but payment to consumer is more direct because it does not require filing a tax return.

3. **Feebates:** Rebates are paid out of government taxpayer funds or are rate-based by utilities and paid by all utility customers. Amory Lovins popularized the "feebate" that combines a fee or tax on consumers using high amounts of energy or purchasing inefficient products and a rebate for those consumers using less or buying efficient products. The fee builds a fund to pay for the rebate so the program is revenue neutral and does not cost taxpayers or utility customers.

4. **Loan guarantees:** Reduces the risk of investments by guaranteeing partial loan repayment if venture fails to meet certain return. They are generally applied to large industrial, high-risk ventures such as new nuclear reactors or synthetic fuel conversion plants.

5. **Risk insurance:** Government underwrites or provides insurance to reduce risk to ventures with high financial or safety risk. For example, the Price-Anderson Act, reauthorized in 2005, limits the liability to utilities for a nuclear accident at about $10 billion and provides a mechanism for the entire industry to share the damage cost to that amount, and for government to cover damages above that amount. Also, the 2005 Energy Policy Act authorized $2 billion in "regulatory risk insurance" to the nuclear industry to cover the cost of regulatory delays at six new reactors.

Research and development funding. Research and development (R&D) is critical for creating new commercial technologies for market transformation. This is especially important for energy technologies involving new energy sources, conversion systems, storage devices, and efficiency measures. Private funding of R&D is essential to advance energy technologies, but there is considerable risk in investments for long-term options. Therefore public government funding of R&D is important to support high-risk activities, to reduce risk for private investments, and to create incentives for additional private funding. If any one policy action is to prepare us for the energy future, it is R&D; it is our future.

Despite its importance and the considerable economic development potential of new energy technologies, both public and private funding of energy R&D in the United States has diminished considerably since the early 1980s (Figure 16.14). Kammen and others lament this "underinvestment" and call for an increase in public R&D investments of five to ten times the current levels (Kammen and Nemet, 2005; Margolis and Kammen, 1999).

Procurement. The government is a major consumer, and one way to stimulate market transformation is to create a dedicated market for sustainable energy technologies by requiring government to purchase them. Such requirements also help test the technologies and educate private consumers by example. To spur the alternative fueled vehicle (AFV) market, the 1992 Energy Policy Act required government vehicle fleets to include a large proportion of AFVs. Federal agencies were also required to purchase ENERGY STAR rated equipment. Government, or utilities under its direction, can also use bulk procurement of efficient lamps,

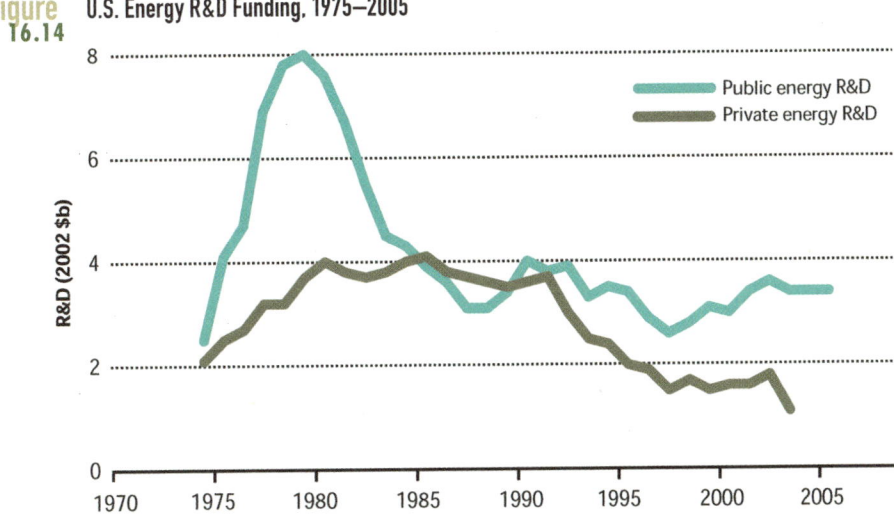

figure 16.14　**U.S. Energy R&D Funding, 1975–2005**

Kammen and Nemet (2005) argue that public energy R&D funding should increase by 5–10 times. Energy R&D as a percentage of total U.S. R&D has fallen from 10% to 2% since 1980.

SOURCE: Kammen and Nemet, 2005. Used with permission.

refrigerators, and other devices at reduced unit cost and use them to replace inefficient ones in selected consuming sectors (Geller, 2003).

Energy assistance programs. Energy costs add extra financial demands to the budgets of low-income consumers, especially when prices increase significantly. Low-income consumers are usually burdened with inefficient cars, housing, and appliances, which make matters worse. In response, government can complement social welfare programs with energy assistance.

Programs can provide financial assistance to eligible households to help pay utility bills, like the $5 billion per year U.S. federal Low-Income Home Energy Assistance Program (LIHEAP), or they can provide improvements in energy efficiency of eligible households, like the $500 million per year Weatherization Assistance Program (WAP). Whereas the former approach simply pays for fuel and electricity with no lasting return in efficiency improvements, the latter invests in housing energy efficiency that will continue to reduce energy bills in future years.

16.3.2.3 Energy Planning

Good energy decisions, be they consumer choices or government policies, require good information and good planning. Many have argued that our current energy problems are the result of poor planning. We simply have not prepared a strategic course of action to lead us to a sustainable future. In Chapter 3 we discussed the abysmal efforts at energy forecasting done in the past three decades. Forecasting is part of planning. But planning is broader and more normative and is simply defined as "figuring out what needs to be done and how to do it" through a process of problem solving. As John Friedmann says, it is "applying knowledge to action."

Government policy should direct careful, rational, iterative, and participatory planning to develop the most effective, efficient, and equitable actions to achieve energy sustainability. As applied problem solving, the planning process has the following basic steps:

1. **Let's scope out the problem and the process.** This can include identifying issues, stakeholders, and needs for data and information; developing scenarios; or articulating a desired future condition.
2. **Where are we now?** This includes baseline analysis of existing conditions, constraints, opportunities, objectives, and uncertainties.
3. **What can we do?** This step formulates alternative policies, projects, programs, designs, or other courses of action that might achieve objectives or a desired future condition.
4. **What should we do?** This assesses and evaluates the economic, environmental, and social effects of alternatives on objectives and future scenarios, and selects a course of action.
5. **Let's do it!** This is the implementation of the selected course of action, including post-implementation monitoring, evaluation, and modification if necessary.

Energy planning is conducted at all levels of government, by private companies, and by civil society organizations. Planning studies develop information and knowledge that can clarify uncertainties, articulate choices, and lead to better decisions.

Future energy is plagued by uncertainties, and this is the reason for the abysmal forecasting of the last three decades. As we discussed in Chapter 3, energy planning should not forecast "a future," but embrace uncertainty by formulating scenarios of possible futures and the conditions, consequences, and uncertainties related to them. We will review examples of energy planning at the national, state, and local levels in the next two chapters.

16.3.2.4 Capacity Building for Energy Action

Market transformation to sustainable energy requires action by everyone—government, energy companies, energy-consuming industry and commerce, civil society organizations, and individual consumers. Government policy can facilitate action through better information, voluntary agreements, partnerships, and capacity building of organizations and individuals.

Information and training. Inadequate and inaccurate information plagues planning and policy decisions. To improve information, government policies support research and analysis. For example, the U.S. Department of Energy's national laboratories and Energy Information Administration continuously support, develop, and disseminate new energy information to inform decisions.

In addition, market imperfections and transaction costs are driven by incomplete, unavailable, or incorrect information on available products, sources, costs, and benefits. Market transformation requires enhancing the quality of information for consumers, producers, and institutions. Government programs can develop and disseminate such information through product testing and labeling (e.g., EPA fuel economy ratings), certification programs (e.g., ENERGY STAR), and energy education and training.

Voluntary agreements and partnerships. Voluntary action can and must push market transformation beyond the limits of regulation and financial incentives. This involves countless participants from major industries to institutions to individual homeowners to make voluntary choices about their energy use. This voluntary approach is facilitated by the growing number of "green" or energy efficient and environmental protocols and certification systems such as ISO 14000 and LEED that help those taking voluntary action to make valid choices.

Government policy can also facilitate voluntary action through agreements and partnerships. Government-industry energy agreements have been very popular in Europe and have helped improve appliance efficiency and reduce auto CO_2 emissions (Geller, 2003).

Capacity building and civil society. Market transformation requires a knowledgeable public and the institutions to create and disseminate knowledge to the public. Government agencies, labs, and funding for energy studies contribute to this effort, but government cannot perform this task alone. It involves many participants in energy assessments, plans, and implementation, including K–12 schools, colleges and universities, energy research and demonstration centers, national public interest groups, and community organizations. Government programs can help build the capacity of these organizations through grant funding, technical assistance, and partnerships.

16.3.3 Pitfalls of Market Transformation Programs

There is considerable evidence of the benefits of government market transformation programs over the past thirty years, but there are also critics, many of whom argue that estimates of energy savings from efficiency programs are inflated and that leads to overinvestment in them. In a study done for the International Energy Agency, Geller and Attali (2005) provide a review of these critiques and draw on the literature of experience in IEA member countries to learn from them. The following list illustrates the pitfalls of energy efficiency programs identified by critics as well as Geller and Attali's responses.

1. **The "rebound effect" will erode energy savings.** The rebound effect is the increase in demand for energy services when the cost of service goes down because of efficiency improvements. If I make my house more efficient, I can turn up the winter thermostat and pay the same as before. My car is more efficient, so I'll drive more vehicle miles.

The rebound effect is real, but it is smaller than critics claim, and there are benefits associated with the greater services provided.

2. **The economy-wide effect will also erode energy savings.** Efficiency improvements can lower demand, which can reduce energy prices, which in turn can lead to economic growth and greater energy use. Research has shown that this effect is small (1%–2% of energy savings), and there are benefits to the economy.

3. **Most energy savings would happen anyway due to technical advances or rising energy prices.** This is true, but these "autonomous efficiency improvements" are slow and incomplete.

4. **Discount rates used to justify energy efficiency policies and programs are too low.** Critics suggest using "consumer purchase" discount rates of about 20%, but there is a good theoretical case for using "implicit" discount rates in evaluating government programs, in the range of 4%–8%, and even lower if the objective is for long-term benefit like GHG emission abatement.

5. **Rate- or taxpayer-funded energy efficiency programs are an unfair subsidy that hurts non-participants and low-income households.** Program participants do benefit more than non-participants, but carefully designed and administered programs should benefit all customers with lower rates than would otherwise be the case, and all society with less air emissions and greater energy security. Most programs dedicate a large share of program resources to low-income households.

6. **Energy efficiency programs are much less effective than their proponents claim.** It is important to use empirical data when evaluating energy efficiency programs.

7. **The market failures frequently used to justify energy efficiency programs are mostly a myth.** Externalities and transaction costs are well documented.

8. **Energy savings are impossible to meter and too difficult to estimate accurately.** Although savings are difficult to measure, there has been great progress in monitoring and evaluation methods for "before and after" assessment and estimation of "free riders" and net savings.

9. **Energy efficiency is a failure because energy use has been increasing.** Energy use has increased but not as fast as it would have without government market intervention programs. Figure 16.15 shows actual energy use and estimated energy use without programs for eleven OECD countries. Figures 1.8 and 1.9 give a similar assessment of U.S. energy use.

16.4 The Social Solutions

Some, like Lovins, et al. (2004), argue that the *techno-economic solutions* of efficiency, renewables, and new clean and safe fossil and nuclear technologies, along with economic market forces, will lead us to more sustainable energy patterns. Others, like Geller (2003) point out that market forces acting alone are too slow, and we need to accelerate the transition to sustainable energy through government *policy solutions*.

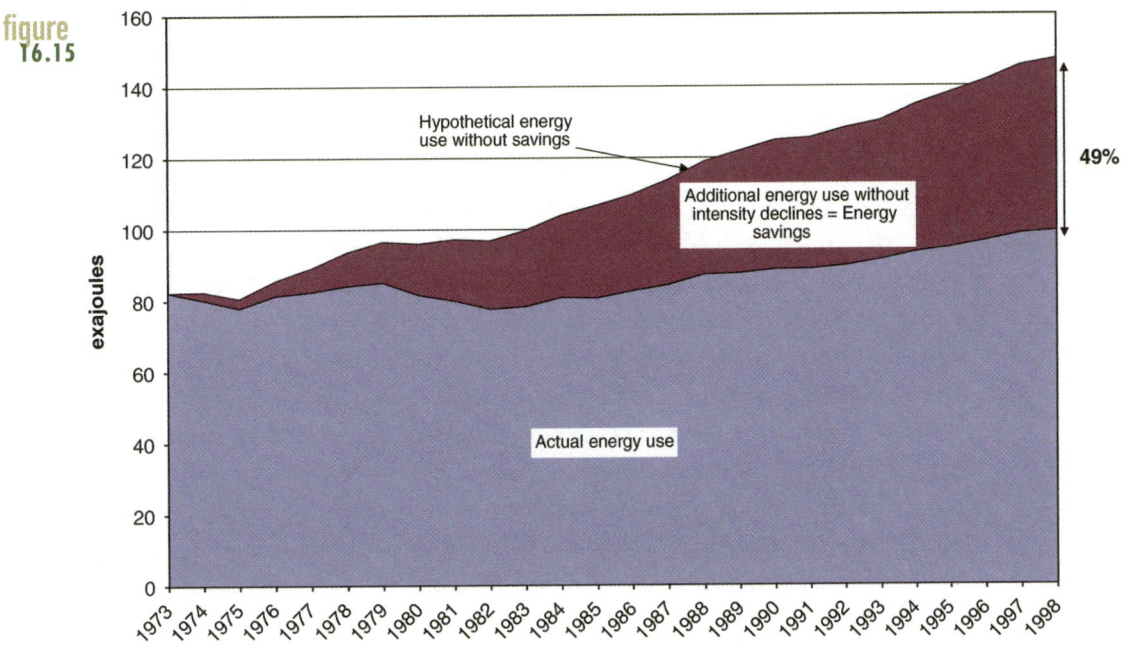

figure 16.15

Actual energy use and hypothetical energy use without energy savings in eleven OECD countries, 1973–1998.

SOURCE: IEA, 2004; Geller and Attali, 2005

Still others, like Smil (2003) and Mallon (2006), think that market imperfections and the paralysis of government policy making dictate the need for the complementary *social solutions* of civil society activism and widespread consumer choice for sustainable energy on the scale of a social movement. Such a social movement for sustainable energy would give political support to aggressive government energy market transformation policies and could lead to widespread consumer choice for both efficiency investments and conservation behavior.

16.4.1 Energy Politics: Achieving Necessary Market Transformation Policies

Development of government policy should be informed by sound technical and economic analysis, but ultimately the adoption of policy initiatives is a political process. That process is a competition of ideas, data and information, and ideologies that are somehow reconciled in legislative programs and policies described in the previous section. Energy policy initiatives are influenced by diverse stakeholders representing a wide array of financial, economic, environmental, industrial, and civil society interests in energy.

But it is rare to find common ground among political stakeholders promoting variously coal or oil and gas or nuclear or renewables and efficiency. Conflicting interests also exist between those pushing for higher efficiency standards and the manufacturers that have to respond to them. As a result policy initiatives are often plagued by political paralysis and inaction, or they try to provide something for everyone without a clear prescription for market transformation. Such appears to have been the case with the 2005 U.S. Energy Policy Act discussed in the next chapter.

Good examples of inadequate or slow responses in U.S. policy include vehicle fuel efficiency standards, research and development funding, a meaningful national strategy for GHG emissions reduction, and a national renewable portfolio standard, among others.

The political process for meaningful policy change requires converging interests of government, industry, consumers, and civil society. If public awareness and support for sustainable energy grows to the scale of a social movement, elected officials will become more responsive to public opinion, and if they do not, they will be elected out of office. Energy industries and energy-consuming product manufacturers will begin to cater to social indicators for purposes of public relations, civic responsibility, and more importantly to their bottom line, market share.

A social movement for sustainable energy can galvanize public, private, and civil society stakeholders to political action and the adoption of aggressive energy policies. This has happened in many European countries, and there are signs of a sustainable energy movement in several U.S. states and cities, as we will see in Chapter 18.

16.4.2 Consumer Values and Choice

Many analysts, and indeed much of the attention of this book, assume that we can "engineer" our way out of our energy problems. They argue that through efficiency and new technology, enabled by more favorable economics enhanced by government policy, we can have our cake and eat it too. We can maintain the increasing levels of energy services we now enjoy but with greater efficiency and a more sustainable mix of energy sources.

However, there may be three fundamental flaws with this assumption:

1. Experience shows that significant improvements in efficiency of vehicles, equipment, and buildings in the United States have been offset by greater consumption for more vehicle-miles-traveled per capita, more and bigger houses and commercial buildings per capita, more appliances and equipment per capita, all resulting in greater energy consumption not less. Despite significant improvements in vehicle, appliance, and building efficiency, U.S. energy per capita is essentially the same in 2006 as it was in 1974.

2. Because of slow adoption of sustainable energy technologies due to inadequate market signals and government policies, new technology adoption alone looks insufficient to transform markets in the time frame necessary to avoid the impacts of petroleum and carbon dependence.

3. Led by the U.S. lifestyle as a model, the world's affluent continue to expand in per capita consumption of materials and energy. There appears to be no end in sight. In fact, many argue that the world's economy requires the driving force of consumption, even over-consumption, to maintain its necessary growth. Meanwhile the poor majority of the world's people struggle to reach a subsistence level of energy use.

SOLUTION BOX 16.2

Energy Needs for U.S. or European Consumption Rates for 10 Billion People

Global annual energy consumption is about 488×10^9 GJ or 75 GJ/p-yr (2005 data). If the population grows to 10 billion as most demographers expect sometime in the second half of this century, what would the global energy consumption be at today's per capita levels? . . . at today's average European's per capita use? . . . at today's average U.S. per capita use?

Solution:

At current global per capita use: 10 billion people × 75 GJ/p-yr = 750 GJ/yr or 54% more than today

At current German per capita use: 10 billion × 185 GJ/p-yr = 1850 GJ/yr or 3.8 times today

At current U.S. per capita use: 10 billion × 359 GJ/p-yr = 3590 GJ/yr or 7.4 times today

---

Vaclav Smil (2003) and others see these flaws as the greatest challenge facing our energy future, as well as the future of our economy, environment, and global justice. Smil calculates that a subsistence level of energy use for an acceptable quality of life (based on food, water, health, education, employment, leisure, human rights) is about 50–70 gigaJoules (GJ) (47–66 million Btu) per person-year.

Coincidentally, the average world per capita consumption (2005) happens to be 75 GJ/p-yr (72 MBtu/p-yr [see Table 1.2]). But we know this is not evenly distributed. The average Bengali consumes 5 GJ/p-yr of commercial energy, the average Indian 16 GJ/p-yr, the average Chinese 54 GJ/p-yr. This compares to the U.S. average of 359 GJ/p-yr. The average for Germany, Japan, France and the UK is about 185 GJ/p-yr.

Can the world's energy support an expanding global population at U.S. or European levels of per capita consumption? This question is addressed in Solution Box 16.2. The answer is that at the U.S. level of energy consumption, a global population of 10 billion people would require more than seven times the current global energy use. Can we develop the energy capacity for this? Few think so. But which of the following do you think is more possible or likely?

- Our ability to expand global energy by more than seven times current consumption to meet a global population's demand at U.S. current per capita energy or by nearly four times to meet European per capita energy

or

- Our ability to reduce energy consumption per capita among the world's affluent without diminishing quality of life, a trend that would help to accommodate the rising per capita needs of the poor in an energy constrained world

The former prospect is plagued by current constraints of oil and carbon and the pace at which we can develop non-carbon alternatives. This is a *difficult* problem.

But the latter prospect of arresting "over-consumption" is a *wicked* problem. It assumes that technical efficiency and new sources are not enough, and that we may need to move beyond "energy efficiency" to "energy conservation." Recall from Chapter 2, energy efficiency improvements do not assume any change in the functions provided or people's behavior, and energy conservation is defined as behavioral changes to save energy by cutting back on the functions energy provides, or at least the growth of those functions.

Arresting over-consumption through energy conservation assumes that at some point, *people will voluntarily choose to be satisfied* with a level of material consumption and the energy it requires. It assumes there are limits to what, on average, each person will want in the number and size of vehicles, equipment/appliances, and living spaces; the vehicle and air miles traveled; the lumens of light, gallons of water, and food calories consumed; the list goes on.

Surely such limits exist, but will they be so high that only a precious few can attain them and the rest will be left behind within the energy constraints we face? Or will these limits be reasonable so that many can attain them, while even more are able to rise to subsistence levels of energy? Within such reasonable limits, energy use per capita would decline with greater efficiency, contrary to recent trends in the United States. We seem to not have yet found those limits in the United States, although people in other countries seem to live quite well with half the U.S. per capita energy.

As evidence of global warming becomes increasingly hard for the general public to ignore and as gas and oil prices keep rising to record levels, there is an emerging social movement for energy efficiency *and* conservation. It is well developed in Europe, and even in the United States there are signs that many are *voluntarily choosing to be satisfied* and are modifying behavior and consumption. The movement responds to dissatisfaction with some dysfunctional aspects of the fast-paced, high-energy-consumptive lifestyle: auto dependence and congestion, reduced sense of community, and wasteful practices. The movement is characterized by increasing interest in slower and simpler lifestyles (such as the "slow cities," "slow food" movements), walkable communities, transit orientation, and resource conservation.

The popular literature in 2006 and 2007 has been filled with indicators of such a movement. A mid-2006 issue of *Newsweek* carried the cover story: "The New Greening of America: From Politics to Lifestyle: Why Saving the Environment Is Suddenly Hot" with the byline: "with windmills, low-energy homes, new forms of recycling, and fuel-efficient cars, Americans are taking conservation into their own hands." Although the article presents only anecdotes of environmental and energy activism sweeping the nation, it may be indicative of a cultural "twitch," if not a genuine cultural shift. The news media continued to fixate on this social movement throughout 2007. Time will tell if it is a lasting movement or a passing fad.

If it is lasting, it could lead to widespread consumer choice for efficient vehicles, green buildings, marketed renewable power, biofuels, onsite and community distributed generation, and other sustainable technologies. Such a growing market would elicit a response by energy producers, product manufacturers, and investors. While enlightened consumers might buy sustainably, they might also limit overall material and energy consumption, voluntarily choosing to be satisfied.

Sustaining such a movement beyond a passing fad is a challenge. Successful movements of the 1960s dealing with civil rights, environmental pollution, and gender equity, ultimately became engrained in public policy and social norms (although many think there is still much work to be done). But social movements often suffer from a "social entropy," similar to the entropy facing natural and societal systems: without a constant input of "energy" (in this case leadership, hard work, and collaboration of individuals and institutions), they will tend toward disorder and disarray. This is especially true of sustainable energy, where public interest and public policy wax and wane with the volatility of energy prices.

Germany provides a useful lesson. It has had perhaps the most active "green" energy social movement in the world that led to decisions for a phaseout of nuclear power, a 21% reduction in GHG emissions from 1990 levels by 2012, and the world's most aggressive development of wind power, PVs, and biodiesel. Despite these efforts, there appears to be resistance among some stakeholders to the large incentives for renewables (coming from the nonrenewable energy industry lobby) and the siting of wind farms (coming from some community advocacy groups; Runci, 2005). (See Chapter 17.)

Some argue that the social movement for renewable energy that existed in the United States in the 1970s to 1990s met its demise when renewable technologies were taken over by big corporations such as BP, Shell, and General Electric; they argue that corporations worked more to inhibit development than to advance it. What is needed, they say, is a more community-based energy movement that tackles "contemporary society's preference for abundance over sufficiency, for waste over frugality, for replacement over repair, and for frugality over utility" (Glover, 2006, p. 263). This latter point is consistent with the need for social solutions, but necessary market transformation cannot rest alone on lifestyle changes or "backyard renewables" as Glover implies. It also needs huge learning investments, large-scale infrastructure, research and development, and growing production to slide down the Experience Curve. Private investments and corporations are critical participants in transforming energy markets. Policy and social solutions can push them in that direction.

The good news is that the context for social solutions is better today than it has been in past decades. Because of policy advances, corporate innovations, and support from civil society organizations, energy consumers are faced with a much wider range of choice for efficiency, renewables, and conservation than ever before.

In many states, they can choose renewable sources for their electricity. In some areas, they can buy or lease rooftop PV systems and run their utility meters backward with excess production. They cannot yet go to Wal-Mart or Home Depot and buy PV arrays with built-in synchronous inverters that they can simply "plug and play" these devices, but these products are not far away.

They can buy more efficient hybrid vehicles, and in a few years they can move on to flex-fuel plug-in hybrids that give them greater fuel choice (gas, electric, and/or E85 ethanol) especially when E85 becomes more available. They can replace appliances and equipment with high-efficiency models meeting improved standards or go beyond those standards with ENERGY STAR rated units.

They can buy energy-efficient "green" houses, built by certified builders following trustworthy and documented green protocols, like LEED. They have more choices to live in walkable and transit-oriented communities that are less dependent on the automobile. Better transit and light rail systems and better bikeways are giving them better choices of transportation modes.

16.5 Summary

Market transformation is necessary to transition from our current oil- and carbon-based energy patterns to sustainable energy characterized by greater efficiency of use, limited oil, and limited carbon emissions. This market transformation requires techno-economic solutions, policy solutions, and social solutions.

Previous chapters emphasized technical solutions, and this chapter looked at some concepts of market transformation including existing barriers to achieving technical potential. These barriers include imperfect market forces, market inertia, transaction costs, and social and cultural factors. Market forces are driven by the price or initial cost of a technology and its energy and dollar savings. The price of a new technology depends on its stage of development, and the Experience Curve helps track and predict price reductions as cumulative production increases. The curve can also be used to estimate learning investments necessary to achieve a certain production and price level. Government policies can help new technologies move down the Experience Curve.

In practice, even short simple payback periods do not achieve significant market penetration. Because of transaction costs and other market imperfections such as external effects of energy on the environment, there is a need for government policy to intervene into energy markets, and to accelerate the market penetration of sustainable energy through regulation, tax policy, direct funding, and planning.

But achieving meaningful energy policy is complicated by the high stakes and competitive politics of energy. Diverse interests fragment political support and many government policies fall short of the aggressive market transformation programs necessary to speed our path to sustainable energy. What may be necessary to build political support for meaningful policy is the social solution of a sustainable energy social movement. Such a movement could also effect widespread consumer choice for sustainable energy, including efficiency improvements through technical advances and conservation behavior through voluntary action.

CHAPTER 17

Energy Policy

The transition to more sustainable energy requires a rapid transformation of energy markets from current reliance on oil and carbon-based fuels to greater use of non-carbon sources and greater efficiency. Chapter 16 explained that this transformation requires technology development, market forces, consumer action, and perhaps most importantly, governmental policy and planning to accelerate emerging technology, use market forces to achieve sustainable outcomes, and encourage consumer choice for sustainable energy.

All levels of government are involved in this transformation. Energy markets and their impacts have significant global dimensions, so international agencies and agreements are necessary to drive collective government and private actions. National governments set the primary energy policies affecting national practices, although state and local governments complement and often push the policy envelope beyond federal policy, as we will see in the next chapter.

This chapter provides a summary of current energy policy approaches at the national and international level of government. First, we give some international perspectives, then we focus on the U.S. federal policies, and outline prospective policies for sustainable energy.

17.1 International Perspectives on Energy Policy

Energy is a global system. Oil, coal, and natural gas fuels have global markets. Energy technologies such as wind, nuclear, biofuels, and photovoltaics (PVs) are global industries. And energy production, transport, consumption, and fuel cycles have significant global impacts. International agencies and agreements have been developed to help manage these concerns.

For example, the International Energy Agency (IEA) is an independent body of twenty-six countries, mostly Organization for Economic Cooperation and Development (OECD) developed countries, established in 1974 to prepare for and coordinate responses to energy problems. It aims to balance energy security, economic development, and environmental protection. IEA provides a review of each member country's energy policy every two years and is the primary source of statistical data on global energy production and consumption. Check out IEA's Web site at www.iea.org.

Among United Nations agencies, the International Atomic Energy Agency (IAEA) was established in 1957 to monitor the development of the nuclear power industry. One of its principal objectives is to control the proliferation of nuclear weapons. The UN Development Programme (UNDP) and the UN Environment Programme (UNEP) have been instrumental in advancing international dialogue on energy and sustainability issues. They sponsored the 1972 UN Conference on the Human Environment in Stockholm and the subsequent 1992 UN Conference on Environment and Development in Rio de Janeiro.

Among the results of this 1992 Rio Earth Summit included the adoption of Agenda 21 for sustainable development by 178 nations and the establishment of the UN Framework Convention on Climate Change that led to the Kyoto Protocol for control of greenhouse gas (GHG) emissions in 1997.

17.1.1 International Agreements

Most often under the auspices of the United Nations, the global community has developed a number of international conventions and agreements on issues ranging from the practices of war to the protection of human rights. It has sponsored a number of conventions devoted to environmental problems that cross national boundaries or are not within the control of any one national government. For example, the 1959 Antarctica Treaty and the 1982 UN Convention on the Law of the Sea focused on protection of common environmental resources. The most far-reaching agreements to regulate environmental protection have been the 1987 Montreal Protocol and the 1997 Kyoto Protocol.

17.1.1.1 Montreal Protocol on Substances That Deplete the Ozone Layer

In the mid-1970s, scientific data began to show that anthropogenic emissions of chlorofluorocarbons (CFCs) and other substances were acting to deplete stratospheric concentrations of ozone (O_3), which helps shield the Earth's surface from damaging ultraviolet solar radiation. A hole in the ozone layer over Antarctica grew to 11 million square miles (the size of North America) and the layer thinned over most of the globe by 2003. CFCs include a family of the most efficient refrigerants ever developed, so they were a very important economic product and an outright ban was thought likely to spur severe economic and political opposition.

In 1977, the United Nations Environment Programme established a Coordinating Committee on the Ozone Layer. By the early 1980s, several Scandinavian countries spearheaded the development of a global convention, and in 1985 the Vienna Convention on the Protection of the Ozone Layer was adopted. Like many previous international agreements, however, it lacked regulatory controls and was largely ineffective. So, after two years of negotiation, twenty-four industrialized countries signed a subsequent agreement, the Montreal Protocol on Substances that Delete the Ozone Layer in 1987. By 2006, there were 189 member nations.

The original Montreal protocol called for reduction of ozone-depleting CFCs to 50% by the year 2000, but subsequent revisions to the protocol in London (1990) and Copenhagen (1992) sped up this timeline considerably, and actions by the United States and the European Community pushed their deadlines up further, with complete phaseout by 1995–1996. More recent meetings of the signatories in Vienna (1995), Montreal (1997), and Beijing in 2000 addressed other ozone-depleting substances, progress toward meeting the phaseout, and the status of ozone layer depletion.

This unprecedented international action resulted from three factors:

1. Accepted scientific theory of the mechanism of ozone layer depletion from anthropogenic ozone-depleting substances
2. Scientific evidence of the occurrence of stratospheric ozone depletion
3. Availability of suitable substitutes to ozone-depleting substances, thereby mitigating any economic impacts of their phaseout

Although the substitute refrigerants hydrochlorofluorocarbons (HCFCs) and hydrofluorocarbons (HFCs) do not have the ozone-depleting characteristics of CFCs because they decompose faster, they, like CFCs, are powerful greenhouse gases. Although they may help solve one global atmospheric problem, they contribute to another, so further alternatives may be needed.

In 2006, UNEP and the World Meteorological Organization (WMO) announced that the protocol was succeeding in closing the ozone hole, but the improvement was slower than expected. The thinning layer over most of the terrestrial world is likely to recover to pre-1980 levels by 2049, five years later than estimates made in 2002. The holes in the layer over Antarctica would take until 2065 to recover, fifteen years later than previous estimates. Thus further measures will be required.

17.1.1.2 The Intergovernmental Panel on Climate Change (IPCC) and the Kyoto Protocol

The success of the Montreal Protocol set the stage to address an even more complex atmospheric challenge, the buildup of GHGs and resulting climate change.

The Intergovernmental Panel on Climate Change (IPCC) was established by the UN and WMO in 1988. We discussed IPCC's four assessment reports in Chapter 2. Its first assessment report prompted formation of the United Nations Framework Convention on Climate Change (UNFCCC) adopted by 154 nations and the European Union (EU) at the 1992 Rio Earth Summit. It was triggered to become a binding agreement once fifty countries ratified it, which happened in 1994; 189 countries are now party to the Convention. The Convention addresses six GHG not including CFCs that were phased out by the 1987 Montreal Protocol. These GHG include carbon dioxide (82% of total GHG), methane (10%), nitrous oxide (6%), perfluorinated hydrocarbons, HFCs, and sulfur hexafluoride.

The 189 countries of the Convention are classified according to their levels of development and their commitments for GHG emission reductions and reporting. They include the following:

- Annex I Parties: Forty developed countries plus EU's fifteen states that aim to reduce emissions to 1990 levels

- Annex II Parties: The most developed countries in Annex I, which also commit to help support efforts of developing countries
- Countries with economies in transition (EITs): An Annex I subset mostly eastern and central Europe, and the former Soviet Union, which do not have Annex II obligations
- Non-Annex I Parties: All other, mostly developing countries, which have fewer obligations and should rely on external support to manage emissions

Each year the UNFCCC holds a Conference of Parties (COP). The third COP held in Japan in 1997 produced the Kyoto Protocol, which stated that by the first commitment period (2008–2012), developed countries would have to reduce combined emissions of GHG to at least 5% below 1990 levels. The Protocol would come into force when it was ratified by at least fifty-five countries, provided they comprise 55% of the CO_2 emissions of Annex I countries. This threshold was reached in November 2004, when Russia ratified the Protocol, so it became legally binding to the 128 ratifying parties 90 days hence, on February 16, 2005.

Table 17.1 gives the emissions reduction targets from 1990 required for Annex II countries by the first commitment period (2008–2012). Under these targets, many European nations must reduce their emissions by 8%, whereas Iceland can increase its emissions by 10%. To provide flexibility in meeting targets, COPs in 2000 and 2001 developed three mechanisms other than simply reducing domestic emissions, including:

- The Clean Development Mechanism (CDM) allowing developed countries to receive certified emission reduction (CER) credits by investing in "clean" projects in developing countries approved by the CDM governing board. Such projects include clean

table 17.1 **Countries Included in Annex II to the Kyoto Protocol and Their Emissions Targets**

Country	Target (1990*–2008/2012)
EU-15**, Bulgaria, Czech Republic, Estonia, Latvia, Liechtenstein, Lithuania, Monaco, Romania, Slovakia, Slovenia, Switzerland	−8%
United States***	−7%
Canada, Hungary, Japan, Poland	−6%
Croatia	−5%
New Zealand, Russian Federation, Ukraine	0
Norway	+1%
Australia	+8%
Iceland	+10%

* Some EITs have a baseline other than 1990.
** The EU's fifteen member States redistribute their targets among themselves, taking advantage of a scheme under the Protocol known as a "bubble." The EU has already reached agreement on how its targets will be redistributed.
*** The United States has indicated its intention not to ratify the Kyoto Protocol.

energy projects (nuclear not allowed) and carbon sequestration forestry projects planting new forests on land cleared prior to 1990.

- The Joint Implementation (JI) allowing essentially tradable credits for approved CDM-type projects if both countries are Annex I.
- Emissions trading allowing countries to buy and sell CER credits from or to other countries in an effort to make compliance more cost-effective. Emissions trading is discussed in Sections 17.1.2.1 and 17.2.2.4.

Some countries have made considerable progress, especially central and Eastern Europe, Germany, and the United Kingdom (Table 17.2). Although EIT country emissions dropped 40%, non-EIT Annex I increased 7.5%, and Annex I overall increased 6.7%. In addition to overall GHG emissions, another measure of progress is the GHG or carbon intensity of domestic economies of kilogram carbon per $GDP (kg C/$GDP). Both Annex I and non-Annex I countries are moving toward less-GHG-dependent economies. Between 1990 and 2005 both Annex I and non-Annex I groups decreased kg C/$GDP. But overall global GHG emissions continue to increase (see Table 17.3).

table 17.2 Changes in Greenhouse Gas Emissions by Annex I Parties, 1990–2003

Country	Change	Country	Change
Lithuania	−66.2%	European Union	−1.4%
Latvia	−58.5%	Switzerland	−0.4%
Luxembourg	−55.0%	Belgium	+1.3%
Estonia	−50.8%	Netherlands	+1.5%
Bulgaria	−50.0%	Liechtenstein	+5.3%
Ukraine	−46.2%	Denmark	+6.8%
Romania	−46.1%	Italy	+7.2%
Belarus	−44.4%	Norway	+9.3%
Russia	−38.5%	Japan	+12.8%
Poland	−34.4%	United States	+13.3%
Hungary	−31.9%	Austria	+16.5%
Slovakia	−28.3%	Finland	+21.5%
Czech Republic	−24.2%	New Zealand	+22.5%
Germany	−18.2%	Australia	+23.3%
United Kingdom	−13.0%	Canada	+24.2%
Iceland	−8.2%	Greece	+25.8%
Croatia	−6.0%	Ireland	+26.6%
Sweden	−2.3%	Portugal	+36.7%
Slovenia	−1.9%	Monaco	+37.8%
France	−1.9%	Spain	+41.7%

Source: UNFCCC, 2006

table 17.3   CO$_2$ Emissions from Fossil Fuels, 1992, 2001, 2005

Country/Region	% Contribution 1992	% Contribution 2001	% Contribution 2005
United States	23.4	23.8	21.1
Western Europe	16.4	15.6	14.2
EE/FSU	19.1	12.7	10.5
China	11.3	12.7	18.9
Japan	4.9	4.8	4.4
India	3.0	3.8	4.1
Africa	3.5	3.8	3.7
Middle East	3.8	4.8	5.1
Rest of world	14.7	18.1	18.0
Total, mil. met T	21,247	24,546	28,193

Source: U.S. EIA, 2007

Despite these efforts to increase flexibility and reduce cost of compliance under Kyoto, in 2001, President Bush announced that the United States would never ratify the Protocol because of negative impacts on the U.S. economy. With nearly one-fourth of the world's GHG emissions, this refusal by the United States put in jeopardy the potential for the Protocol to adequately address the global warming problem. Australia has also decided to not ratify the Protocol, using the U.S. decision as an excuse: "there is no clear pathway for action by developing countries, and the United States has indicated that it will not ratify. Without commitments by all major emitters, the Protocol will deliver only about a 1% reduction in global greenhouse gas emissions." In 2007, both the United Nations and the United States were looking beyond the Kyoto agreement to forge a new approach to reduce GHG emissions, but no acceptable and effective international means to mitigate emissions has emerged, although negotiations continue. Meanwhile, the IPCC AR4 Synthesis Report, unveiled at the November 2007 meeting in Valencia, Spain, issued a dire warning that to keep global temperatures from rising more than 2°C (3.6°F) above 2000 levels, GHG emissions must be stabilized by 2015 and drop quickly after that peak by 50%–85% by 2050 (IPCC, 2007).

17.1.2 Innovations in Developed Countries

17.1.2.1 European Union Energy Initiatives

Although the European Union does not have the authority of a national federal government like the United States does, it has been arguably more effective than the United States in establishing sustainable energy directives and programs for its member countries. Still, implementation of its directives is dependent on national policies and regulations. Perhaps its most well-known energy directive is the CO$_2$ emissions trading scheme, but several other EU energy directives and policy strategies have emerged since 2000.

SIDEBAR 17.1

European Union "Green Papers" on Energy

The European Commission uses "Green Papers" on select topics to generate discussion and comment to help formulate policy directives. Among the Green Papers on energy:

Green Paper on Security of Energy Supply (2000). Referred to as the "Guliver in Chains" report, this Paper projected that EU-25 energy dependency of 50% on foreign sources in 2000 would increase to 70% by 2030. Subsequent debate in more than 300 meetings in thirty countries recognized the need to address demand-side policies.

Green Paper on Energy Efficiency: Doing More with Less (2005). This June 2005 paper promoted the idea that the EU could save 20% of expected energy consumption and €60 billion of expected energy costs in 2020. In typical EU fashion, the paper posed twenty-five questions for public debate until the end of March 2006. An analysis of the debate concluded there is strong support for improved efficiency. The paper and debate contributed to an EU Energy Efficiency Action unveiled in October 2006.

Green Paper: A European Strategy for Sustainable, Competitive, and Secure Energy (2006). Yet another Green Paper was produced by the European Commission in 2006 that opened debate on a broader energy strategy for the EU. Using the themes of sustainability, competitiveness, and security, the paper outlined priority areas of interconnecting European electricity and natural gas infrastructures, assuring security of supply through a cooperative internal energy market, diversifying the energy supply mix, going beyond the CO_2 ETS to tackle climate change through efficiency improvements and increased use of renewables, encouraging technological innovation, and developing a strategic external energy policy with suppliers and other large consuming nations. One of the most controversial issues regarded national sovereignty on energy, particularly with regard to nuclear power. Europe depends on nuclear for about one-third of its electricity, and member nations have adopted far different policies on its future, ranging from France's commitment to the technology, to Germany's and Sweden's apparent rejection of it, to the United Kingdom's continuing debate on the subject.

The EU's framework for developing directives includes scientific study, the development of discussion papers (often called "Green Papers"), extensive participatory discourse and comment, and ultimately **directives** or mandates passed by the European Commission. Sidebar 17.1 gives examples of "Green Papers" developed on energy that have prompted discussion and debate. This section describes some of the more important energy directives passed by the EU, including its CO_2 cap-and-trade scheme, renewable electricity targets, biofuel targets, and building energy standards.

EU CO$_2$ Emissions Trading Scheme (2003/87/EC). Perhaps the most famous EU directive is its CO_2 emissions trading scheme. In 2005 the EU established a unique "cap-and-trade" emissions trading system (ETS) for CO_2. The scheme is designed so that EU countries will reduce their combined CO_2 emissions to 8% below 1990 levels by the end of the Kyoto Protocol's first

table 17.4   Trading Allocations—Selected Nations

EU Member State	Allocated CO$_2$ Allowances (mill. tons)	Share in EU Allowances (%)	Installations Covered	Kyoto Target (%)
Germany[1]	1497.0	22.8	1849	−21
Spain[1]	523.3	8.0	819	+15
France[1]	469.5	7.1	1172	0
Italy[1]	697.5	10.6	1240	−6.5
Poland	717.3	10.9	1166	−6
United Kingdom[1]	736.0	11.2	1078	−12.5
Total	**6572.4**	**100.0**	**11,428**	

(1) Under the Kyoto Protocol, the EU-15 has to reduce its greenhouse gas emissions by 8% below 1990 levels during 2008–2012. This target is shared among the original fifteen member states, including those marked with (1), under a 2002 legally binding burden-sharing agreement. The ten member states that joined the EU in 2004 have individual targets under the Kyoto Protocol with the exception of Cyprus and Malta, which have no targets (EC, 2005b).

commitment period of 2008–2012. Estimated cost of compliance is €$2.9–$3.7 billion per year, less than 0.1% of the EU GDP. In its first phase (2005–2007), the scheme applies only to CO$_2$ from large emitters in power and heat generation and energy intensive industrial sectors: 11,500 facilities are included that account for 45% of EU's CO$_2$ emissions.

The scheme allows companies to use credits from Kyoto's project-based multi-national approaches: JI and CDM. Each member state has a national allocation plan that must be approved by the European Commission, reflects the state's Kyoto target, allocates allowances to facilities considering their potential for emissions reduction, and allows JI and CDM. Table 17.4 gives 2005–2007 allocations for the top emitting nations and Solution Box 17.1 illustrates how the trading works. An extensive 2005 survey of companies, industry associations, governmental bodies, and NGOs involved in implementing the EU ETS revealed that the ETS is affecting corporate behavior.

- About half of the company respondents "price in" the value of CO$_2$ allowances into their operations and more than 70% intend to do so.
- For half of the companies, the ETS is one of the key issues in long-term decisions; for the other half it is among several issues.
- About half indicated the ETC has had a strong or medium impact on decisions to develop innovative technology (EC, 2005b).

Directive on Energy End-use Efficiency and Energy Services (2006/32/EC).
This directive calls for the reduction of energy use by 9% at 1% per year from 2007 to 2016. It requires each member nation to prepare by June 2007 an initial energy efficiency action plan (EEAP) to achieve this target, a second EEAP by June 2011, and a third by 2014. Each EEAP will be followed by a progress report six months after the plan.

SOLUTION BOX 17.1

EU CO$_2$ Emissions Trading in Action: A Hypothetical Example

How does CO$_2$ emissions trading work and what are its economic environmental benefits?

Solution:

Let's assume companies A and B both emit 100,000 tons of CO$_2$ per year. In their national allocation plans their governments give each of them emission allowances for 95,000 tons, leaving them to find ways to cover the shortfall of 5000 allowances. This gives them a choice between reducing their emissions by 5000 tons, purchasing 5000 allowances in the market, or taking a position somewhere in between. Before deciding which option to pursue, they compare the costs of each.

- In the market, the price of an allowance at that moment is €10 per ton of CO$_2$.
- Company A calculates that cutting its emissions will cost it €5 per ton, so it decides to do this because it is cheaper than buying the necessary allowances. Company A even decides to take the opportunity to reduce its emissions not by 5000 tons but by 10,000.
- Company B is in a different situation. Its reduction costs are €15 per ton, which is higher than the market price, so it decides to buy allowances instead of reducing emissions.
- Company A spends €50,000 on cutting its emissions by 10,000 tons at a cost of €5 per ton, but then receives €50,000 from selling the 5000 allowances it no longer needs at the market price of €10 each. This means it fully offsets its emission reduction costs by selling allowances, whereas without the emissions trading scheme it would have had a net cost of €25,000 to bear to cut 5000 tons.
- Company B spends €50,000 on buying 5000 allowances at a price of €10 each. In the absence of the flexibility provided by the ETS, it would have had to cut its emissions by 5000 tons at a cost of €75,000.

Emissions trading thus brings a total cost savings of €50,000 for the companies in this example. Because Company A chooses to cut its emissions (because this is the cheaper option in its case), the allowances that Company B buys represent a real emissions reduction even if Company B did not reduce its own emissions (EC, 2005b).

Action Plan on Energy Effciency: Realizing the Potental (COM(2006)545). Six months after directive 2006/32/EC was issued, the European Commission (EC) released this ambitious action plan calling for a 20% reduction of primary energy use by 2020. The plan cites considerable potential for savings in all sectors and calls for new policies beyond directive 2006/32/EC. The plan estimates that 2005 consumption of 1750 million tons of oil equivalent (Mtoe), or 41.3 quads, would grow to 2460 Mtoe (58 quads) by 2020 under a "business-as-usual" constant energy intensity scenario. By 2020, it assumes that continuing structural changes in the economy, autonomous national policy, and previous EU policies would likely reduce this to 1890 Mtoe (44.6 quads). But directive 2006/32/EC policies could reduce this 10% to 1700 Mtoe, and the new policies of this action plan could reduce another 10% to 1500 Mtoe (35.4 quads). These latter policies include appliance and equipment standards, building performance requirements, more efficient power generation and distribution, higher auto fuel efficiency, better financing of energy efficiency investments, more intensive investment in new member states, coherent use of taxation, greater energy efficiency awareness through education, covenant of mayors of Europe's most pioneering cities, and fostering efficiency worldwide.

Directive on Promotion of Renewable Electricity (2001/77/EC). This 2001 directive promotes renewable electricity (RES-E) by quantifying national targets (similar to renewable portfolio standards), providing support schemes, simplifying permitting procedures, and guaranteeing transmission access. If 2010 targets are met, EU total renewable electricity would be 22% of total capacity, although a 2004 assessment report indicated current policies will result in an 18%–19% renewable share. Table 17.5 gives some of the EU countries' targets. Wind power has had great success especially in Germany, Spain, and Denmark, which have 84% of EU capacity.

Directive on Promotion of Biofuels for Transport (2003/30/EC). This directive set a minimum replacement by biofuels of marketed diesel or gasoline transport fuels of 2%

table 17.5   Renewable Energy Production Targets in Selected EU Countries

EU Country	Renewable Energy Targets
Austria	78.1% of electricity output by 2010
Denmark	29% of electricity output by 2010
Finland	35% of electricity output by 2010
France	21% of electricity output by 2010
Germany	12.5% of electricity output by 2010
Greece	20.1% of electricity output by 2010
Italy	25% of electricity output by 2010
Portugal	45.6% of electricity output by 2010
Spain	29.4% of electricity output by 2010
Sweden	60% of electricity output by 2010
United Kingdom	10% of electricity output by 2010

Source: IEA, 2006

by the end of 2005 and 5.75% by the end of 2010. If this 2010 target and the 2010 target of 22% for RES-E were met, the EU would achieve a 10% renewable share of total energy; an additional 2% from heating and cooling (RES-H) would give a total share of 12%, the EU goal. But, like the RES-E target, the biofuels displacement is falling short of the target with only 0.6% biofuel replacement in 2005.

Directive on Building Energy Performance (2002/91/EC). This directive requires member states to incorporate into national legislation high-performance building energy standards by 2006 for both new construction and major renovation. European countries, especially Sweden, Denmark, Germany, and France, have developed some of the world's most stringent building energy standards over the past thirty years. In 2006, the United Kingdom upgraded its building efficiency standards to meet this directive that would improve efficiency by 40% compared to pre-2002 standards (Figure 17.1).

Voluntary Agreements. Where mandates are difficult to pass, the EU and its member nations have used voluntary agreements effectively. For example, these agreements have been effective in CO_2 emission reductions in industry and efficiency improvements in appliances and vehicles. A 1998 agreement between the European commission and automobile manufacturers aimed to achieve by 2008 a fuel economy target based on 140 g CO_2/km, or 25% below the average new car sold in Europe in 1995.

Summary. The EU has a strong commitment to sustainable energy and has been a world leader in efficient transportation and buildings, control of carbon emissions, and development

figure 17.1 Evolution of Building Standards in the United Kingdom

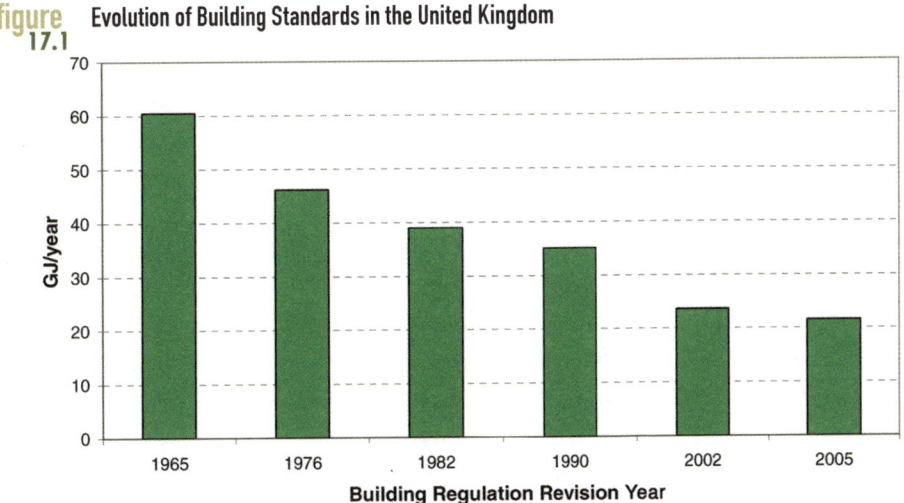

Typical space and water heating (GJ/year) under codes enacted in various years.

Source: Shorrock, 2005. Used with permission.

of renewable energy. As a union of twenty-five nations, however, the directives from the EC must be implemented at the national level, and national sovereignty issues often constrain effective implementation. Individual countries offer their own lessons and models for sustainable energy especially Germany, Spain, Sweden, Denmark, and the United Kingdom.

17.1.2.2 Germany: World Leader in Renewable Energy

By the end of 2006, Germany emerged as the global leader in wind energy (with 29% of the world's 74 GW of installed capacity), in photovoltaics (with 38% of the 6.7 GW of installed capacity; Figure 17.2), and in biodiesel production. We know from Chapter 16 that one of the secrets of market transformation is pushing down prices by fueling the Experience Curve with greater production and installation. So it is important to take note of Germany to see how it was able to accelerate the development of these emerging technologies.

How did Germany, a country smaller in size than Montana, become the global leader in renewable energy? It was not due to an abundance of wind and sun, because Germany has poor sun (Berlin is at 52°N latitude, north of Calgary, Canada) and only a small area of class 4 or better winds. Instead, it was due to a combination of political, policy, and technological commitment. With few indigenous energy resources, except for coal and nuclear power (still sources for 58% and 28% respectively of German electricity in 2002), Germany was struck especially hard by the energy events of the 1970s and 1980s. The oil crises of the 1970s, the European forest die-off due to acid rain from coal, and the Chernobyl nuclear accident spawned strong anti-nuclear and environmental movements, as well as Europe's first significant Green Party. This led to a first phase of government and private investment, experimentation, and learning in renewable energy and efficiency.

Although there was not a high rate of diffusion of new technology initially, there emerged a public vision of the future in which renewables would play a prominent role (Jacobsson, et al., 2006). This was further charged by the climate change debate, and around 1990 a series of policy initiatives and demonstration programs began to grow the market. In the 1990s, Germany's utility deregulation policies led to green power marketing used by 500,000 customers by 2003. In 1999, it passed an "eco-tax" on energy and negotiated a phaseout of nuclear power by 2025.

But it was Germany's demonstrations and feed-in rates for renewable power that have had the greatest impact on growth of renewable energy. Starting in 1989, the government seeded a 100 MW demonstration of wind power, later increasing it to 250 MW. They also called for 1000 roofs with solar photovoltaic systems, later increased to 100,000 roofs, by initially guaranteeing €0.04/kWh for electricity produced and providing low-interest loans to investors.

The 1990 **Feed-In Law** set the stage for considerable expansion of renewable power. It required utilities to connect renewable energy generators and buy power at rates about 90% of average retail rates for wind and PV, based on rationale of added benefits from lower external costs. In 2000, and again in 2004, the Renewable Energy Sources Act expanded this program, guaranteeing twenty-year rates for new systems. Those rates vary by source

figure 17.2 German Share of Installed Capacity for (a) Global PV (38% by 2006) and (b) Global Wind (29% by 2006)

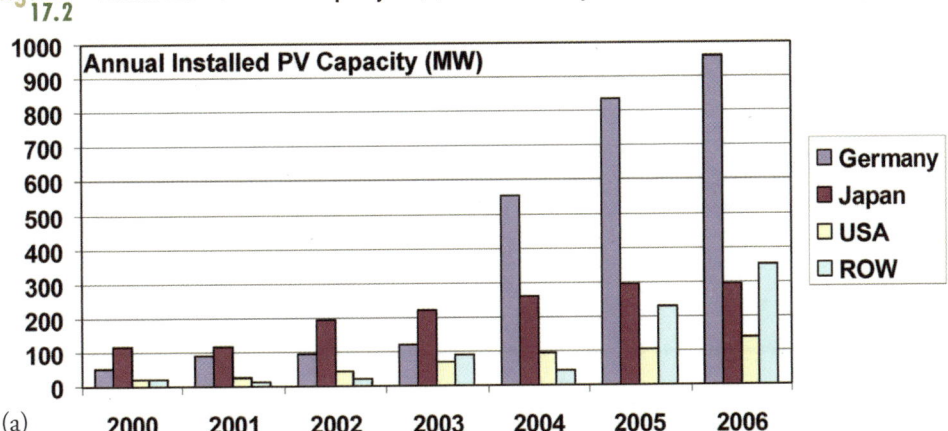

(a)

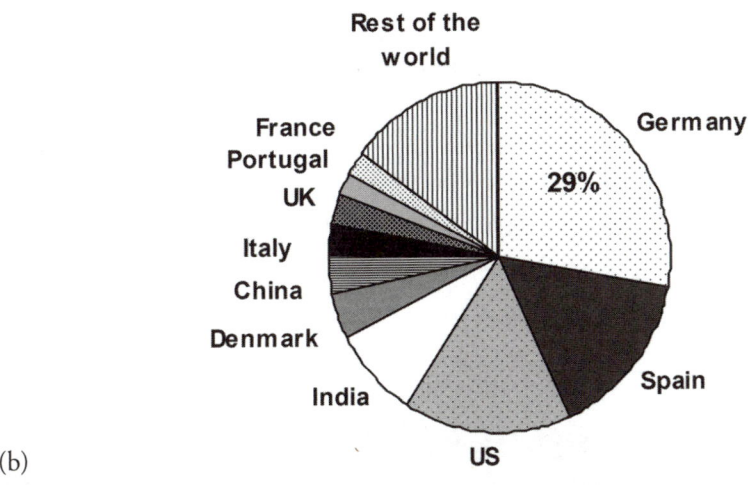

(b)

SOURCE: b: Global Wind Energy Council, 2007

(PV, wind, geothermal, biomass, hydro, landfill methane), capacity, and year of installation. Twenty-year rates decline for systems installed after a start-up period of one to five years. The 2004 Amendments increased the rate structure considerably (see Table 17.6). For example,

- On-shore wind systems installed in 2004 have twenty-year feed-in rates of 8.7 €¢/kWh. Rates for systems installed in later years decline at 2% per year of installation after 2004. Off-shore wind systems have 9.1 €¢/kWh twenty-year rates declining at 2% per year of installation after 2008.

table 17.6 **German Feed–In Rates for Renewable Electricity**

	20-yr Rate €¢/kWh	Installation Date for Full Rate	20-yr Rate Decline per Install Year after Date
Wind on-shore	8.7	2004	2%
Wind off-shore	9.1	2008	2%
Photovoltaics (< 30 kWh)	57.4	2004	5%
Photovoltaics (< 100 kWh)	54.6	2004	5%
Biomass CHP	21.5	2004	2%
Landfill Methane	11.5	2004	2%
Geothermal	15.0	2004	2%

- PV systems installed in 2004 have twenty-year rates of 57.4 €¢/kWh up to 30 kW (54.6 €¢/kWh up to 100kW) with rates for later installations declining at 5%/year.
- Twenty-year rates for other 2004-installed renewable source systems are 7.67 €¢/kWh for small hydro, 11.5 €¢/kWh for biomass (21.5 €¢/kWh for combined heat-and-power [CHP] facilities), and 15 €¢/kWh for geothermal. Like wind and PV, rates decline for later installations.

This one policy is largely responsible for Germany's incredible growth of PV and wind power, rising to more than 30% of world installed wind and PV capacity by 2006 (see Figure 17.2). Figures 17.3–17.5 give examples of German projects.

figure 17.3 **170 kW Grid-Connected PV System in Munich**

SOURCE: Joachim Berner, 2006

figure
17.4   One of WKN's Forty-Eight German Wind Farms Totaling 450 MW

SOURCE: courtesy WKN Windkraft Nord AG

figure
17.5   10 MW, 62-acre Bavaria Solarpark

Park uses 57,600 PowerLight PowerTracker panels.

SOURCE: courtesy PowerLight Corp.

Germany has also been a leader in biofuels. Spurred by its biofuel tax exemption in 2003, biodiesel production grew to 2000 ton/yr capacity in 2005 or 54% of Europe's entire capacity. Germany is expected to meet the EU biofuel target of 5.75% of fuel sales in 2006, four years ahead of the deadline.

Three factors facilitated the enactment of renewable energy policies in Germany (Wustenhagen and Bilharz, 2006): (1) a strong central government and a political culture open to government intervention, (2) a critical mass of interest groups in favor of renewables, and (3) a critical mass of elected officials with knowledge of and commitment to renewables. For other countries with these facilitating factors, the following are useful lessons for developing renewable energy policies:

- The legislature plays a critical role that can transcend industry and ministry interests.
- Interparty coalitions can cut across traditional political camps.
- Careful "burden-sharing" such as feed-in rates can spread costs over a dispersed group through utility rates with minor government budget impact.
- Power market deregulation creates a window of opportunity by establishing new rules that can dissolve traditional political camps and enable new coalitions.
- Customer choice can play a role and complement more direct policy directives and establish a long-term market for green power.
- Germany had a good deal of luck in the political process of policy development, which depended on grass-roots political support.

The incredible growth of renewable power in Germany is expected to slow as the incentives begin to decline, prime sites for facilities are developed, and the market becomes more saturated. Green power marketing may help sustain growth. And the growth has spawned a large German renewable energy industry in wind, PV, and CHP, that has benefits for the German economy as it targets international markets.

17.1.2.3 Japan: Successful Promotion of PV Industry

Japan is the world leader in photovoltaic cell production, and through 2003 Japan also led the world in PV installations until Germany's explosive growth in 2004 and 2005. Japan's PV development followed a more traditional policy model than Germany's. It included "upstream" investment in research, development, and demonstration. Whereas Germany instituted aggressive policies for deployment by setting incentive pricing for PV power, Japan has relied on industrial development policy that has driven down system cost in a competitive market.

Shum and Wanatabe (2006) maintain that Japan's emphasis on a "manufactured technology" deployment approach (focusing on grid-connected 3–5 kW systems in new residential development) is far more effective than the "information technology-like" deployment approach used in the United States. In this latter approach, diffusion is based on developing a variety of new application categories and customization, whereas the former is based on standardized package technologies. Japan's standardized grid-connected manufactured technology has driven down "balance of system" (BOS) costs (i.e., components other than the photovoltaic cells, such as inverters, net meters, controls, etc.) producing more cost-effective systems. Figure 17.6 shows PV system cost in Japan as of 2003 was about half that in the United States.

17.1.3 Innovations in Developing Countries

17.1.3.1 As China Goes, So Goes the World?

It used to be said "as the USA goes, so goes the world," but China is looking more like the determining factor for the world's energy future. Energy use in China has exploded by 58% between 2002 and 2005 with its huge investment in housing, infrastructure, and power plants. In 2005, China added 40 GW of electrical generating capacity, the equivalent of an 800 MW plant each week; in this one year it added more than the entire capacity of New York or Spain in one year! More than 80% of this capacity is coal-fired steam generation.

figure 17.6   **Historic Price of PV System in Japan and United States, 1992–2003**

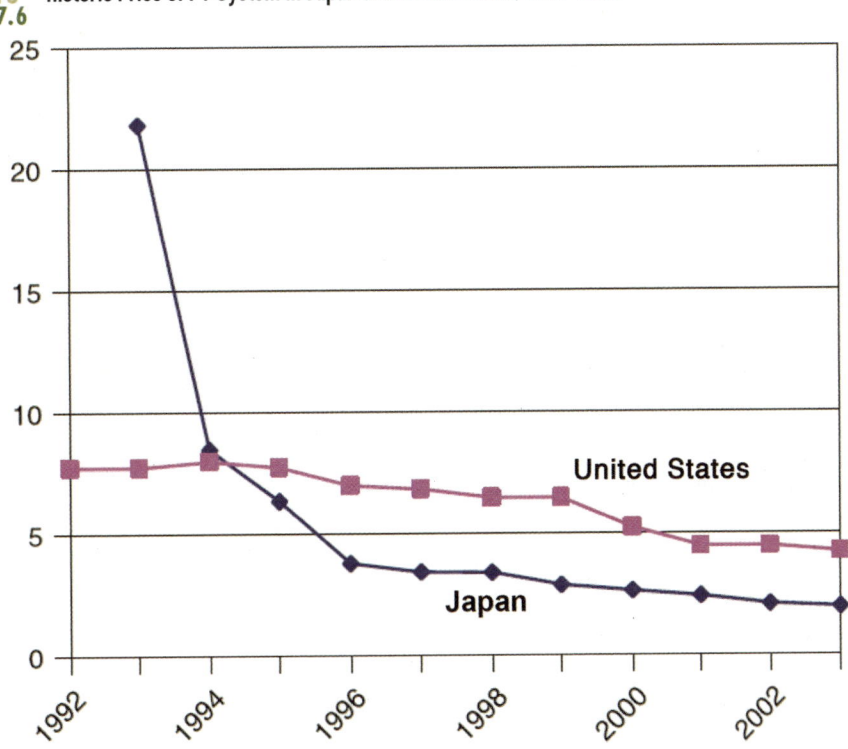

SOURCE: Shum and Wanatabe, 2006. Used with permission.

SIDEBAR 17.2

How Have Central and Eastern European Countries Reduced Energy and GHG?

One of the recent success stories in improved energy efficiency comes from the dramatic turnaround in the energy patterns of most of the former Soviet Union countries, especially those eight that have joined the European Union in 2004. Already we have seen (in Table 17.2) the significant reduction in GHG emissions of these countries.

A recent study of their transition from a centrally planned to market economy and entry into the EU showed that they have been able to break the negative legacy of their former Soviet energy economy by rewarding efficiency and discouraging waste, providing appropriate price signals, preventing oversizing of infrastructure, reducing corruption, and generally restructuring industrial economies. At the same time, these countries had certain positive legacies from their centrally planned systems, albeit not the result of user choice, including significant use of public and rail transport, concentrated land use, district heating and cogeneration, a low rate of individual consumerism, and high level of reuse and recycling.

Although most of these countries suffered significant economic decline from 1989 to 1993, they have enjoyed significant rise in GDP since 1993 by transitioning to market economies, metering and billing energy use, improving information, and conforming with stricter environmental controls. Energy intensity (energy use/GDP) has improved by 48% in Poland, 25% in Hungary, and 15% in the Czech Republic between 1989 and 2002. Still, energy intensity in the Czech Republic is 76% greater than the EU and Russia has $3\frac{1}{2}$ times the EU rate, so there is a long way to go (Urge-Vorsatz, et al., 2006).

Between 1980 and 2000, China's economy grew at an incredible rate of 9% per year, but even more impressive was that its growth of energy consumption was less than half that rate, 3.9% per year. Energy intensity (energy/GDP) dropped by 67% in this twenty years. This decoupling of energy and economic growth resulted from significant investment in efficiency using a range of policies including western-style low-interest loans and tax credits, as well as Chinese-style energy quotas (in the early years) and government funded "energy conservation service centers" to assist large users in improving efficiency. But most investment has come from business funds rather than government funds, but these sources are often blurred in China's centrally planned economy (Lin, 2006).

Although energy efficiency investment has continued to grow in China, as a percentage of investment in supply, it has dropped from 9% in 1991 to less than 4% in 2003. This percentage must increase significantly, perhaps by two or three times, if China is to meet the ambitious goal in its eleventh five-year plan to reduce energy intensity of GDP by 20% between 2005 and 2010. This may be necessary if China wishes to grow and sustain its economy within energy supply constraints, and if the world wishes to see stabilization of GHG emissions. China's CO_2 emissions are growing fast, increased by 83% from 2000 to 2005, and surpassed the United States as the world's emission leader in 2006.

SIDEBAR 17.3

Hope for Africa?

Africa poses a major challenge for development with reinforcing problems of poverty, population growth, health, economy, and environment. Energy is critical in any strategy for development. Today most African nations still depend on traditional biomass (charcoal, wood, dung) for 60%–90% of basic energy needs, with resulting indoor air pollution, health effects, and environmental degradation. More developed South Africa and northern African nations consume three-fourths of the continent's electricity, whereas the undeveloped remainder with 80% of the population consumes only one-fourth, and that only in urban areas.

On their way to modern energy, poor regions of Africa must convert to inexpensive, more efficient biofuel cookstoves (e.g., 1.5 million have been replaced in Kenya) and look to community-scale energy projects for electricity (for lighting, water pumping, communications). Biogas, PV, wind, and small hydro offer significant advantages for rural community applications, and these appropriate technologies can also help build local capacities for other development needs.

17.1.3.2 Brazil: World Leader in Biofuels

Brazil offers many lessons for the developing world. Now the ninth largest economy in the world and the fifth most populous with 185 million people, Brazil is classified as a middle-income developing country. Although Brazil has a long way to go to achieve stability and equity, it has made considerable progress, achieving energy equilibrium in 2006, exporting as much fuel as it imports. Brazil made considerable progress improving electricity efficiency beginning in 1985 through a national program run by its state-owned utility holding company. Beginning with grants, low-interest loans, and voluntary agreements with manufacturers, and culminating in new appliance efficiency standards in 2001, Brazil has achieved savings and a culture for electricity efficiency (Geller, 2003).

Brazil's biggest energy achievement, however, is its biofuels program and development of sugarcane-based ethanol production. After some fits and starts in the late 1970s and early 1980s with various subsidies for ethanol production and a production push for 100% all-ethanol vehicles, the Brazilian ethanol industry has matured. High oil prices have made domestic ethanol economical without subsidies, and the country displaces 40% of its gasoline consumption with ethanol. The all-ethanol vehicles have been replaced with flex-fuel E85 vehicles. Just three years after their introduction in Brazil, flex-fuel vehicles represent 100% of automakers' production there. Brazil has become an ethanol exporter as well, although U.S. import tariffs on ethanol make it difficult for its exports to compete in the large U.S. ethanol market. As we have seen in Chapter 14, sugarcane-based ethanol is far more efficient than corn-based, but Brazil's industry understands that the future of ethanol there resides not in sugarcane but in cellulose (Reel, 2006).

17.2 U.S. Federal Energy Policy

17.2.1 Overview and a Brief History of U.S. Federal Energy Policy

Energy has long been subject to strong market forces, but U.S. federal government intervention in energy markets also has a long history. Government energy policies prior to 1975 focused on regulation of utilities, price regulation, and incentives for production and research, especially for fossil fuels and nuclear power. Since 1975, energy policies have broadened at both federal and state levels to address both supply and demand, including efficiency of use, and to develop both conventional and alternative energy sources. Some policies created more government intervention into energy markets through regulation and tax incentives; others aimed to free energy markets and consumer choice.

The federal government has had policies affecting energy since the anti-trust laws that broke up Standard Oil Company in 1911. The Public Utility Holding Companies Act (PUHCA) of 1938 led to state regulation of utilities, and the New Deal legislation of the 1930s made the federal government a major player in energy development, especially hydropower in the Pacific Northwest, the Tennessee Valley, and other regions. Federal policies both promoted and regulated the development of nuclear power, and various subsidies and incentives helped the oil and gas industries advance. Coal was affected by federal laws for mine land reclamation and miner safety and compensation. Environmental legislation in the early 1970s affected air emissions from combustion of fuels in industry, utilities, and vehicles. Table 17.7 provides a chronology of United States federal energy policies along with some influencing events in italic.

Over the past three decades, federal energy policy has been reactive. The energy crises of 1974–1975 and 1979–1981 prompted by oil supply manipulation, the Iraq wars in 1991 and 2003, the nuclear accidents at Three Mile Island in 1978 and Chernobyl in 1986, the California electricity crisis of 2000–2001, and eastern U.S. blackout in 2003, all affected federal policy.

After the 1974–1975 energy crisis and Arab oil embargo, the first pieces of energy legislation aimed at efficiency and renewable energy were passed in 1975, 1978, and 1980. Other than appliance efficiency standards passed in 1986, little substantive energy efficiency policy legislation was passed during the low-oil-price mid-to-late 1980s. But after the Iraq wars of 1991 and 2003, and subsequent oil price hikes, comprehensive national energy policy acts were passed in 1992, 2005, and 2007.

These energy policies were driven by the desire to find new solutions to oil import dependency, but were constrained by conflicting political pressure to continue antiquated policies to serve existing energy interests.

For example, prior to 1975, federal energy policies promoted production with large subsidies for oil and gas and nuclear industries. The "oil depletion allowance" initiated in 1913 granted oil companies a lucrative 27.5% tax deduction on their income from the first ten years of production from their wells. This helped build the industry and made rich oilmen even richer, but it was a huge subsidy. Thanks to the political influence of Texan oilmen, the law stayed intact until the mid-1970s when it was retargeted to specific segments of the

table 17.7	A Brief Chronology of U.S. Federal Energy Policy *and Some Influencing Events in Italic*
1920	**Federal Water Power Act:** led the way for federal hydro energy production, ultimately the later establishment of the Bonneville Power Administration and Tennessee Valley Authority and the Federal Energy Regulatory Commission (FERC), which regulates hydro power, interstate power sales, wholesale electricity rates, and natural gas pricing
1935	**Public Utility Holding Company Act** (PUHCA): required electricity and later natural gas utilities to be regulated by state utility commissions
1946	**Atomic Energy Act:** established the Atomic Energy Commission (AEC); 1954 amendment established the civilian nuclear power program; in1957 Price-Anderson Act federal government subsidizes insurance for nuclear accident liability
1970	**Clean Air Act:** established national clean air program including controls on power plants and 90% reduction of vehicle emissions; latest amendments (1990) established first "cap-and-trade" emissions trading system for power plant SO_x
1973	*Arab Oil Embargo leads to price hikes, filling station lines, gas rationing.*
1973	President Nixon announces Project Independence with goal of achieving energy self-sufficiency by 1980 with allusions to 1940s Manhattan Project and 1960s Man-on-the-Moon
1975	**Energy Policy and Conservation Act:** extended oil price controls into 1979; established strategic oil reserve; standards for doubling auto fuel economy (Corporate Average Fleet Efficiency [CAFE]) to 27.5 mpg by 1985
1978	**National Energy Act:** included National Energy Conservation and Production Act; Power Plant and Industrial Fuel Utilization Act (limited use of oil and natural gas, later repealed); Public Utility Regulatory Policies Act (PURPA) opened the power grid to small renewable and CHP private generators, set guidelines for utility demand-side management (DSM); Energy Tax Act (income tax credits for renewables and conservation; gas-guzzler tax); Natural Gas Policy Act
1979	*Three Mile Island nuclear accident in Pennsylvania*
1979	*Iranian revolution and hostage crisis, with additional oil price hikes; prices drop by 1985*
1980	**Energy Security Act:** set of seven acts dealing with synthetic fuels, biomass and alcohol fuels, renewable and solar energy, geothermal energy, and ocean thermal energy
1981	President Reagan executive order for decontrol of petroleum prices
1982	**Nuclear Waste Policy Act:** ultimately led to choice of Yucca Mountain, Nevada, as national nuclear waste depository
1986	**Appliance Energy Conservation Act:** following California's lead, established national appliance standards, which have been expanded by subsequent amendments and administrative action
1986	*Chernobyl nuclear accident in Ukraine, USSR*
1991	*Gulf War in Kuwait, Iraq; petroleum prices increase, then drop again by 1993*
1992	**Energy Policy Act:** comprehensive act expanding access of private generators to utility grids; opportunities for state utility restructuring; directives for alternative fuel in fleets
1993	President Clinton announces U.S. goal to stabilize GHG emissions at 1990 levels by 2000; proposes a producer's energy tax of 26¢/MBtu (coal and gas) and 60¢/MBtu (oil), fails to get political support in Congress
1993–1999	Clinton administrative initiatives for improved appliance efficiency and voluntary programs, for example, ENERGY STAR, Building America, Partnership for New Generation Vehicle
2000–2001	*California electricity crisis resulting from flawed restructuring plan*
2001	President Bush proposes National Energy Policy, but it takes four years for legislation to pass.
2001	President Bush announces the United States will not sign the Kyoto Protocol for GHG emission reduction and will continue to study the matter.
2001	*September 11 terrorist attacks in New York and Washington*
2001	*Enron files for bankruptcy*
2003	*Largest U.S. electricity blackout affects 50 million in eight eastern states and Ontario*
2003	*Iraq War; petroleum prices increase and continue to rise in tight market to $100/bbl by late 2007*
2005	**Energy Policy Act:** comprehensive energy act resulting from intense four-year political debate; tax incentives for nuclear, ethanol, solar and wind, clean coal, oil and gas, and other supply options; renewable fuel standard for biofuels; repeals PUHCA, parts of PURPA; electricity reliability standards
2007	**Energy Independence and Security Act:** expands biofuel renewable fuel standard; improves light vehicle fuel economy (CAFE) to 35 mpg by 2020; expands appliance efficiency standards.

industry. It and other oil and gas subsidies remain in various forms, and some sustainable energy advocates estimate that with the addition of new $2.6 billion in tax incentives from the 2005 Energy Policy Act (EPAct), the industry gets as much as $14 billion in tax breaks at a time when it is enjoying record profits.

Nuclear power has also enjoyed substantial government support that was initially necessary to build the industry. Since the Atomic Energy Act of 1954, the federal government has supported technology development as well as direct funding of the nuclear fuel cycle, from uranium enrichment to waste disposal. In an attempt to revive the dormant industry, the 2005 EPAct renews the accident liability limits of the Price-Anderson Act (originally passed in 1957), provides "regulatory risk insurance" totaling $2 billion for the next four new plants, and has a 1.8¢/kWh tax credit for production from new plants. The *Wall Street Journal* estimates that new nuclear subsidies total $8 billion.

It is important to understand some of the political dimensions of policy development. In a democracy, everyone potentially has a voice, but in reality it is powerful interests who engage in the competition of ideas and policy options. In a high-stakes domain like energy, this debate often results in stalemate and paralysis. If compromise is reached, it generally supports the status quo rather than creating policies to transform energy markets. This is a lesson from the energy policy debates of the 1970s, and especially the 1990s and 2000s.

The 2005 EPAct is a case in point. After aggressive energy policy proposals of the Clinton administration failed in the 1990s, Congress was in no mood to take up energy until it had to. Well, it had to in 2001 as oil prices began to rise and tensions in the oil-rich Middle East grew after the 9/11 terrorist attacks. The Bush administration proposed a National Energy Policy in 2001, but it was immediately politicized because it was developed by a secret committee whose membership the administration refused to disclose. Congress began debate on an energy policy bill, and the Republican House passed its version reflecting the proposed Bush policy. The Republican Senate had some differences. The key issues of debate were drilling for oil in the Alaska National Wildlife Refuge (ANWR), increasing the CAFE vehicle efficiency standards, taking action against GHG emissions, and a national renewable portfolio standard for electricity. The debate continued until 2005, when a compromise bill was passed containing no provisions for ANWR drilling, CAFE standards, GHG emissions, or the Renewable Portfolio Standard (RPS).

The comprehensive and complex EPAct tried to give something to all of the powerful stakeholders, emphasizing incentives over regulations. It aimed to jump-start the nuclear power industry with risk insurance and production tax credit for new plants, advance biofuels with a renewable fuels standard, extend the renewable electricity production tax credit for twenty years, provide royalty relief and other incentives for frontier and unconventional oil and gas production on public lands and off-shore, and give incentives for clean coal technologies, such as IGCC and carbon capture and sequestration. Research, development, demonstration, and deployment funding saw marginal increases across the board.

Table 17.8 highlights the complicated maze of U.S. federal energy policies for six main supply and demand categories and the three principal policy groups introduced in Chapter 16: regulations, financial incentives, and information. The following sections discuss some of the specific policies, especially those for efficiency, renewables, and electricity.

17.2.2 Federal Regulations

Energy related regulations include efficiency standards, production standards, utility regulation, and environmental protection regulations.

table 17.8 Overview of U.S. Federal Energy Policies

	Efficiency	Renewables	Oil and Gas	Coal	Nuclear	Electricity
Regulations						
Efficiency standards	CAFE - vehicles Appliances/ equip					
Production standards		RFS for biofuels				
Utility regulation		Interconnection guidelines			NRC: plant licensing; streamlined by 2005 EPA	FERC: inter-state electricity; state guidance for DSM, transmission, interconnection
Environmental regulation		Various depending on type and scale	Control vehicle and power plant air emissions; regulatory relief for production	Power plant emissions; mine reclam; miner health/safety	Nuclear safety; nuclear waste management	Power plant air emissions; transmission line review
Financial						
Investment tax credits	Consumer investment tax credit: building, autos	Consumer and business investment tax credit	Business investment tax credit	Business investment tax credit	Business investment tax credit	
Production tax credits		1.8¢/kWh	Yes + royalty relief	1.8¢/kWh from new plants		
Total tax incentive for 2006–2016	$1.3 billion	$4.5 billion	$2.6 billion	$3.0 billion	$3.0 billion for electricity	$3.0 billion (includes nuclear)
Loan guarantees, liability insurance		$1 bill. ethanol loan guar., 80% non-recourse		Loan guar. for IGCC	Regulatory risk insurance; Price- Anderson	
Energy Taxes	Auto gas guzzler		Gasoline			
RD&D author./yr by 2005 EPA	$0.85 B authorized	$0.75 B auth + $0.20 B auth BF	Unconventional	Carbon capture	$0.53 B auth	$0.25 B auth
Direct Assistance	LIHEAP fuel assist Weatherization			Clean coal power	Fuel cycle; waste depository	
Energy Information	Vehicle, appliance efficiency labeling; ENERGY STAR, more	EIA	EIA	EIA	EIA	EIA Utility information programs

SOURCE: Lazarri, 2006

17.2.2.1 Efficiency Standards

As discussed in Chapter 16, national standards are the most effective means of effecting change because they are mandatory and thus can achieve a near 100% penetration rate for new efficient products. Federal standards for efficiency were first established for automobiles in 1975 and for appliances in 1986. They have had a dramatic effect in reducing energy consumption.

Vehicle efficiency standards. Federal vehicle economy standards demonstrate the effectiveness of efficiency standards, but also the need to increase the stringency of standards over time. The 1975 Energy Conservation and Policy Act mandated doubling automobile efficiency in ten years. As described in Chapter 13, the CAFE standards required the annual fleets sold by each automobile manufacturer to average 27.5 mpg for cars by 1985 and 20.7 for light trucks and SUVs by 1996. Despite several proposals since 1980 to improve these standards, they were not changed until the 2007 Energy Act, except for a 2006 upgrade of the light truck/SUV standard to 23.5 by 2010. Light trucks and SUVs now make up more than half of the passenger vehicle market. In November 2007, the federal Ninth Circuit Court of Appeals in San Francisco ruled that the new light truck/SUV standard was not sufficient.

The auto standards are credited with what improvements have been achieved in on-the-road vehicle efficiency. Technology has improved efficiency of new vehicles, but increases in average vehicle size and performance have offset most of those improvements. For 2005, the passenger car fleets of the U.S. "big three" (GM, Ford, Daimler-Chrysler) averaged 28.0 mpg, whereas those of Honda and Toyota averaged 33.5 mpg (U.S. DOT, 2005).

Figure 17.7(b) from the National Commission on Energy Policy (NCEP) shows the estimated energy savings (all oil) in 2000 of the existing standards (5.5 quads) and the additional savings that would be achieved by increasing the standards (4.5 quads for 10 mpg increase, 6.5 quads for 20 mpg, with additional savings for tractor trailer and tire standards). The bipartisan NCEP recommended that Congress direct the National Highway Traffic Safety Administration (NHTSA) to make significant increases in the CAFE standards beginning no later than 2010, considering safety, performance, jobs, and competitiveness. Senate bills included similar provisions but they were deleted from the final 2005 EPAct.

But in late 2007, Congress passed and President Bush signed the Energy Independence and Security Act (EISA) with a provision to increase the CAFE standards to 35 mpg by 2020 for all light vehicles, including SUVs and pickups. This essentially adopts the NCEP recommendation for a 10 mpg or 40% increase in fuel economy.

Appliance standards. The Appliance Energy Conservation Act of 1986 followed California's lead in creating efficiency standards for fourteen consumer appliances. The list was expanded to twenty-six appliances and equipment by the Energy Policy Act of 1992, the 2005 EPAct added another sixteen appliances, and the 2007 EISA added ten more, including

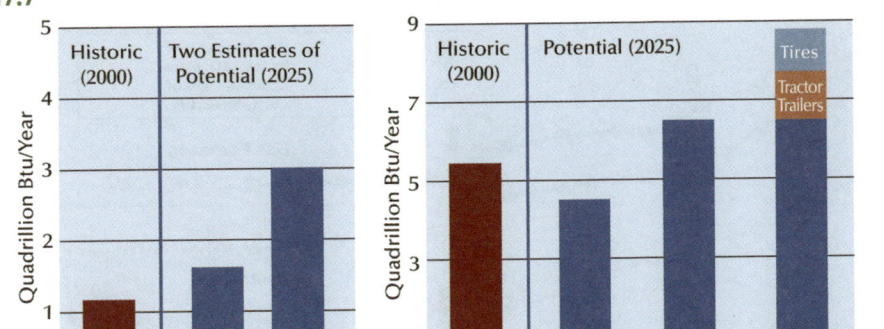

figure 17.7 Historic and Potential Savings from Appliance and CAFE Efficiency Standards

(a) Appliance standards energy savings of 1.2 quads through 2000 according to RFF (2004) and *additional* potential savings to 2025 estimated by LBL (2004) and ACEEE (2004) ranging from 1.6 to 3 quads.

(b) CAFE standards energy savings of 5.5 quads through 2000 and *additional* potential savings from increase of CAFE by 10 mpg to 37.5 mpg (4.5 quads), by 20 mpg to 47.5 (6.5 quads), and other measures.

Source: NCEP, 2004

lamps. Figure 17.7(a) from NCEP shows the Resources for the Future (RFF, 2004) estimate of 1.2 quad savings from appliance standards through 2000 and two estimates of additional potential savings through 2025 ranging from 1.6 to 3.0 quads. More recent American Council for an Energy Efficient Economy (ACEEE) estimates of the 2030 impact of federal standards adopted through 2007 include (a) savings of 577 billion kWh of electricity or 7 quads (6% of total energy); (b) reduced need for 177 GW of peak power or six hundred 300 MW power plants; (c) reduced CO_2 emissions of 45 million tons; and (d) a net economic benefit of $250 billion (Nadel, et al., 2006; ASAP/ACEEE, 2008).

Art Rosenfeld (2005), using Steve Nadel's analysis (2003), traced the improvements made in efficiency of three appliances (Figure 17.8). Annual energy usage of new gas furnaces, central air conditioners, and refrigerators sold from 1972 to 2006 dropped to 75%, 50%, and 25% respectively of 1972 values. The figure shows how appliance standards were the driving force of efficiency improvement, led first by California's state standards, then by progressively improving federal standards.

The 2006 **SEER-13 standard** for central air-conditioning (A/C) systems improved A/C efficiency by 30% compared to the previous SEER-10 standard set in 1992 (Figure 17.8). This case illustrates the effect of administrative, political, and judicial action on energy policy. After

figure 17.8 Reduction of Energy Usage by New Appliances in Year of Purchase Relative to 1972

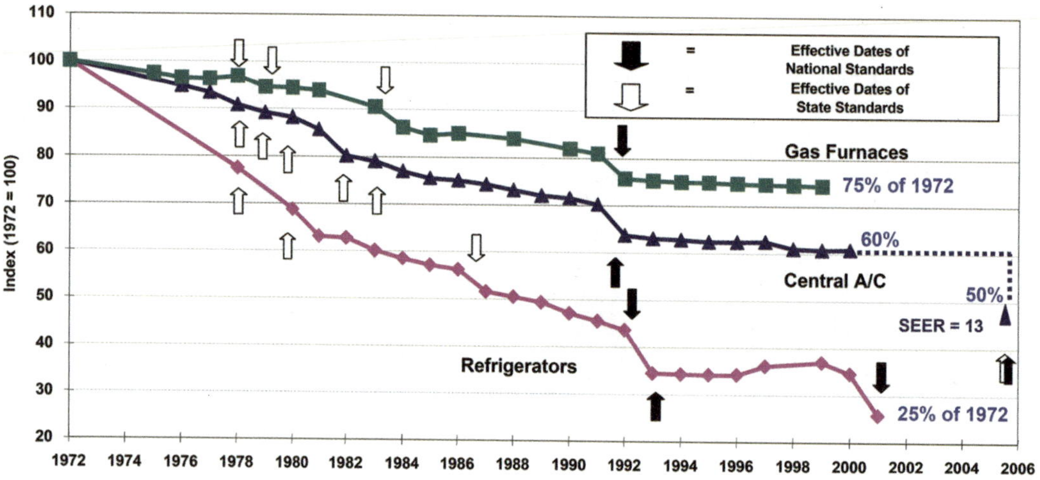

Reduction driven first by state efficiency standards, subsequently by federal standards.

SOURCE: Rosenfeld, 2005; Nadel, 2003. Used by permission.

seven years of review, this rule was promulgated in the final days of the Clinton administration in January 2001 to become effective January 2006. However, the Bush administration, under some pressure from industry, tried to roll back the standard to SEER-12. The Department of Energy (DOE) recommended this change over the objectives of the EPA, which argued that DOE underestimated the benefits and overestimated the costs of the higher standard. After three years of debate and legal action, in 2004 the Second Circuit Court of Appeals restored the SEER-13 standard for 2006.

17.2.2.2 Renewable Fuels Standard (RFS)

The 2005 EPAct included the first U.S. federal energy production standard, requiring suppliers to include 7.5 billion gallons (Bgal) of biofuels blended in the nation's transportation fuels by 2012. This is nearly double the 2005 ethanol production (see Figure 14.9), and amounts to a savings of 80,000 bbl/day of petroleum or about 5% of total vehicle fuel supply in 2005. The act also provides for a 2.5:1 trading ratio for cellulose ethanol. Each gallon of cellulose-based ethanol counts for 2.5 gallons toward the RFS until 2012. After 2012, this ratio does not apply, but the RFS must include 250 million gallons (Mgal) from cellulose.

EPA is assigned to enforce the RFS and set subsequent RFS after 2012. In late 2006, EPA issued a proposed ruling for implementation. It includes a credit trading system and requires a Renewable Identification Number (RIN) to be issued with each shipment of biofuel. Blenders would have the option of using the RIN for their own compliance or of trading it to others. Any party that produces gasoline for consumption in the United States, including refiners, importers, and blenders, would be subject to a renewable volume obligation. Each year, these parties must acquire enough RINs to demonstrate compliance with their RFS obligation. In its proposed rule, EPA indicated its expectation that 2012 biofuel production would reach 9.9 Bgal in 2012, or 32% more than the mandated RFS. The 2007 EISA increased the RFS to 35 Bgal by 2022.

The 2007 act did *not* include a House-proposed provision for an RPS that would have required that 15% of the nation's electricity come from renewables by 2020. While federal proposals for a national RPS continue, the real action in RPS is at the state level, as discussed in the next chapter on state policy.

17.2.2.3 Federal Utility Regulation Policy

Even though states regulate utilities, they must conform to federal legislation. As discussed in Chapter 9, this responsibility dates back to the PUHCA of 1935, which defined electric "utilities" as any generators or providers of power and mandated that they be subject to state regulation. The federal government has retained some regulatory authority over utility operations ever since the 1920 Federal Water Power Act. FERC has responsibility over interstate transmission, wholesale power rates, hydroelectric power facility licensing, and other issues affecting interstate commerce of electricity and natural gas.

The PURPA of 1978 established the first guidelines for state regulation of utility conservation services and interconnection of non-utility distributed generation. The 1992 and 2005 Energy Policy Acts set the stage for state restructuring and deregulation of the utility industry. These laws and their implementing regulations provided for independent and distributed power generation, flexibility in utility operations, more consumer choice for sources of electricity, and authority for states to direct utilities to DSM and restructure their utility regulation. The 1992 Energy Policy Act especially paved the way for state utility restructuring and deregulation that took off in the mid- to late 1990s.

Utility regulation from both federal and state perspectives continues to be a moving target. For example, the 1978 PURPA required utilities (a) to interconnect grid-connected "qualifying" small renewable electricity and CHP facilities and (b) to pay such generators rates equivalent to the costs the utilities avoided by buying the power. This provision was an important first step in distributed generation. However, it was repealed by the 2005 EPAct, largely because it had been superseded by better buy-back rate agreements, including net-metering, in many but not all states. If FERC determines that a state does not have sufficiently competitive markets, this interconnection requirement may still apply. The 2005 Act also repealed PUHCA, but passed a streamlined "PUHCA 2005" to improve the reliability of energy infrastructure by opening electricity and natural gas sectors to new sources of investment and giving them greater eminent domain powers for siting infrastructure. We will discuss the status of state utility deregulation in Chapter 18.

17.2.2.4 Federal Environmental Regulations Affecting Energy

Federal environmental regulations aim to reduce pollution, protect human health, and protect natural resources, including atmosphere, waters, and ecosystems. For example, the 2005 EPAct aimed to protect waters by continuing the ban on offshore oil and gas drillng in 85% of coastal waters (except central and western Gulf of Mexico and some parts of Alaska), the Great Lakes region, and the Alaska National Wildlife Refuge.

In doing so, these laws internalize environmental and social costs and create a more level economic playing field for clean energy. For example, the Clean Air Act's performance standards for coal-burning utilities increase the cost of generating electricity. Rules for coal mine land reclamation and miner safety increase the cost of coal. Uncertainties about human health radiation standards for nuclear waste management continue to cloud the future costs of nuclear power. As long as there are uncertainties, this is as it should be. Regulations intend to internalize the external environmental costs and uncertainties of energy production and use, thereby raising the price of these energy sources and making more competitive sources with fewer externalities, such as renewable energy and efficient technologies.

Clean Air Act and Power Plant Emissions. A plethora of environmental regulations affects energy, and we cannot address them all here. One example that illustrates the political process of developing and implementing environmental regulations relating to energy is the stationary source provisions of the Clean Air Act. It also illustrates implementation of emissions "cap-and-trade" programs that have become the model approach for reducing emissions, including GHG emissions, in an economically efficient manner.

In 2003, power plants emitted 69% of the nation's SO_x emissions, 40% of mercury, and 22% of NO_x. Their impact on air quality became clear when the August 2003 Northeast blackout shut down 100 mostly coal-fired power plants, and ambient levels of SO_2 and NO_x-causing ozone fell by 90% and 50% respectively.

Many of the power plants east of the Mississippi River are older coal-burning facilities that are not subject to today's strict New Source Performance Standards (NSPS) for new plants. Although many of these plants have been improved, a challenge for the Clean Air Act has been to reduce their emissions without economic hardship on utility customers. The 1990 Clean Air Act amendments (CAAA) established the first "cap-and-trade" emission allowance trading system to reduce power plant emissions of sulfur oxides (SO_x) to control acid rain. As described in Sidebar 17.4, this program has been largely successful and served as a model for other cap-and-trade programs such as the EU's CO_2 program described in 17.1.2.1. Figure 17.9 illustrates its success, showing ambient air sulfate (SO_4) concentrations resulting from SO_2 emissions in 1988 and in 2003. Although SO_x problems are not solved, they are vastly improved and the SO_x cap-and-trade program has been one of the U.S. environmental regulatory success stories.

To upgrade emission controls for older power plants further, the CAAA provided that major modifications and equipment replacements in older plants would be subject to New

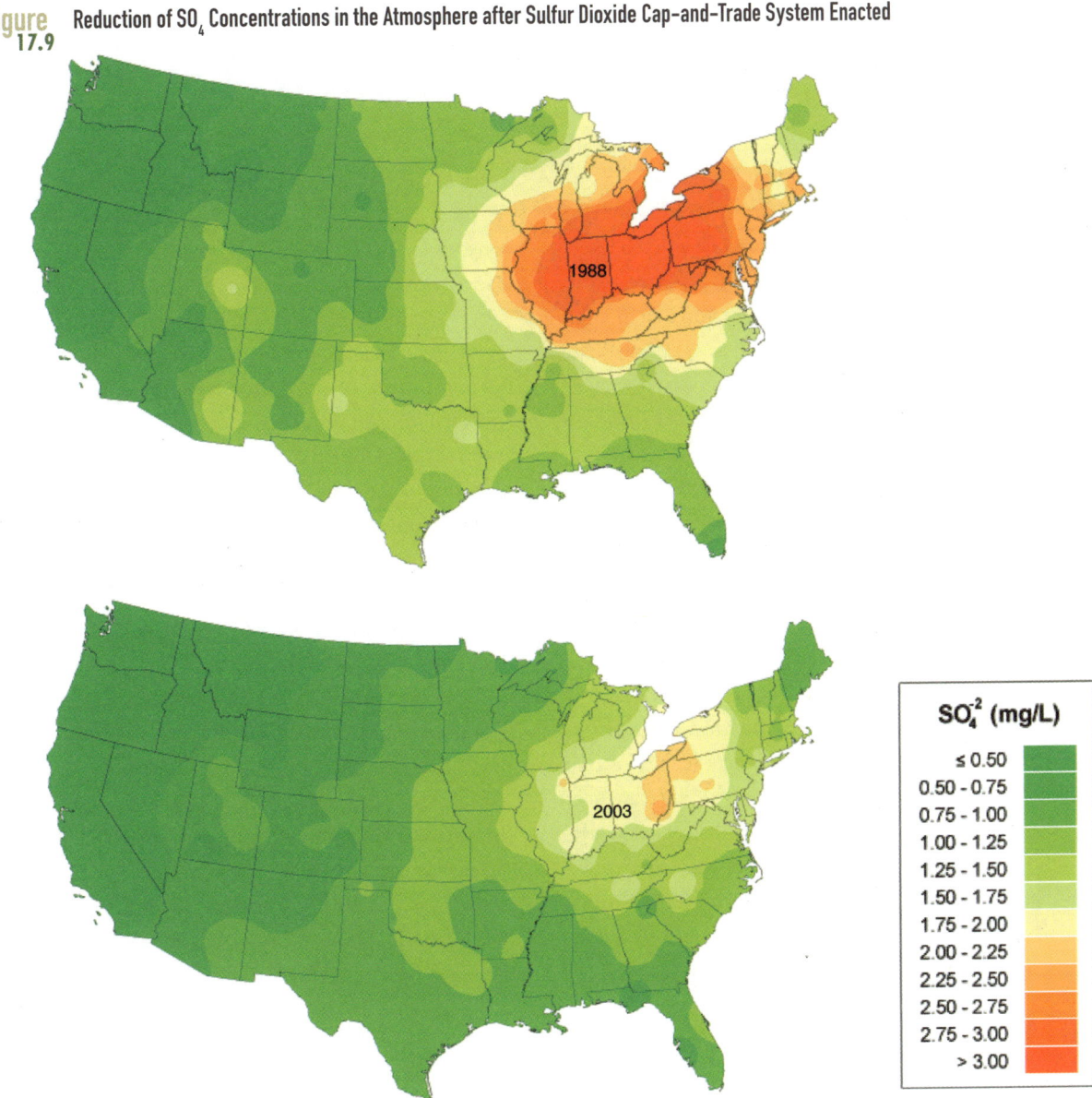

figure 17.9 Reduction of SO$_4$ Concentrations in the Atmosphere after Sulfur Dioxide Cap-and-Trade System Enacted

SO$_4^{-2}$ (mg/L)

≤ 0.50
0.50 - 0.75
0.75 - 1.00
1.00 - 1.25
1.25 - 1.50
1.50 - 1.75
1.75 - 2.00
2.00 - 2.25
2.25 - 2.50
2.50 - 2.75
2.75 - 3.00
> 3.00

Maps compare SO$_4$ concentrations in the atmosphere in 1988, before the 1990 CAAA that set up the SO$_2$ cap-and-trade program for Midwest coal-burning power plants, and in 2003, after several years of program implementation.

Source: NADP, 2007

Source Review (NSR) and the NSPS. The idea was that as these older polluting plants were renovated, they would be treated like new plants, be subject to NSR, and be brought up to modern NSPS. But utility and coal advocates and the Bush administration's EPA argued that

SIDEBAR 17.4

U.S. Emissions Cap-and-Trade Programs

1. **Acid Rain SO$_2$ program:** The 1990 CAAA set up the SO$_2$ trading system to reduce continental 48 state emissions by 10 million tons below 1980 levels. Phase I (1995–1999) covered 263 power plants larger than 100 MW with a 1985 annual emission rate more than 3.4 kg SO$_2$ per kJ heat input. Phase II began in 2000, plants larger than 25 MW were included, with a total annual cap of 8.12 million tons, equivalent to rate of 0.98 kg/kJ, or about half of the SO$_2$ emissions in the early 1980s. Tradable allowances were issued to sources based on their average heat input in 1985–1987, their emissions rate, and their plant type. Continuous emissions monitoring systems (CEMS) were required for each unit and data had to be available to the public. Those with too few allowances are subject to fines and required to acquire allowances to cover their excess emissions. Studies of the program showed that an active and efficient allowance market emerged, with significant trading and reliable pricing. Units can "bank" unused allowances from one time to use later, and this produced early emission reductions. The program has a near-perfect record of compli-

ance at costs estimated to be 57% less than without an emissions trading program, or a thirteen-year savings of $20 billion (Ellerman, et al., 2003).

2. **Northeast NO$_x$ Budget Program.** The Ozone Transport Commission, established by the 1990 CAAA, is a regional body representing northeast and mid-Atlantic states. In 1994 these member states set up the NO$_x$ Budget Program to control NO$_x$ emissions from utility power plants and large industries. After applying technology standards the program used a cap-and-trade program to limit emissions further during the May to September smog season. The unique multi-jurisdictional approach started slowly because bankable allowances were restricted, and it had price volatility in the first year. But it developed into an active market and by 2004, targeted emissions were 30% lower than 2003 levels and 50% lower than those of 2000.

3. **RECLAIM Program.** The Regional Clean Air Incentives Market was adopted by the California South Coast Air Quality Management in 1993 to reduce NO$_x$ and SO$_x$. The program was less effective than the Acid Rain and NO$_x$ budget programs

this would create a hardship for domestic energy supply, the utilities, and their customers. This has been controversial, and in 2006, the U.S. Court of Appeals for the D.C. Circuit struck down the Bush EPA rule that would have exempted most equipment replacements from NSR. The Bush administration has proposed the cap-and-trade Clear Skies initiative as an alternative (see Sidebar 17.4). In its current form, it has been characterized as a step back for the Clean Air Act and it has not gained support in Congress.

Are GHG air pollutants? Since 1970, EPA has regulated six "criteria" air pollutants under the Clean Air Act and has the authority to expand that list as necessary to protect public

because when it began emissions were below caps, additional controls were not needed, and allowance prices for later years were modest. But in 2000, prices spiked and emissions exceeded caps, largely as a result of California's electricity crisis. In a 2002 evaluation of RECLAIM, EPA cited the problems resulting from the uncertainties from electric utility restructuring, but concluded that RECLAIM demonstrated that cap-and-trade can work.

4. **Proposed Cap-and-Trade Programs:**
   a. **Clear Skies Multipollutant Proposal.** The Bush administration proposed the Clear Skies initiative that would significantly amend the CAA to establish a multipollutant cap-and-trade system for power plants, while eliminating many existing Clean Air Act regulations for those sources. Much of the debate has been over pollutants covered and stringency and timing of the caps. The administration's bill generally had more generous caps for emitters and longer time to meet them than most congressional bills. Some related congressional bills include CO_2 emissions in addition to SO_x, NO_x, and mercury.
   b. **Regional Greenhouse Gas Initiative** (RGGI): In 2006, ten northeastern states in the United States joined the first CO_2 cap-and-trade program for power plants > 25 MW. The program is described in Sidebar 18.1.
   c. **GHG Cap-and-Trade Proposals.** The 2004 National Commission on Energy Policy (NCEP) proposed a GHG cap-and-trade program with mandatory, economy-wide, tradable permits; a target (total cap) between 2010 and 2019 reflecting a 2.4% annual reduction in GHG emissions per $GDP; and increase to 2.8% beginning in 2020. In 2007, seven separate CO_2 cap-and-trade bills were introduced in Congress, including the bipartisan Lieberman-Warner America's Climate Security Act, as economy-wide hybrid act that has "upstream" provisions for transportation fuels and "downstream" provisions for utilities and large industrial sources. It calls for 2005 level emissions of GHG by 2012, 15% below 2005 by 2020, 33% below by 2030, and 70% below by 2050. For an updated review of all proposals, see Pew Center on Global Climate Change, Cap-and-Trade Proposals in the 110th Congress (http://www.pewclimate.org/).

SOURCE: NCEP, 2004; McCarthy, 2006; Pew CGCC, 2007

health and welfare (see Sidebar 17.5). Concerns about climate change and GHG emissions have prompted a debate about whether GHG emissions, especially CO_2, should be treated as air pollutants and be regulated by EPA in controls of vehicle and stationary source emissions. Several states petitioned EPA to include GHG in its control of vehicle emissions (see section 13.2.3). When the agency ruled it would not, Massachusetts and eleven other states sued EPA and the Bush administration over the ruling. Although EPA won a Circuit Court decision on the matter, in April 2007, the Supreme Court overturned that decision. Justice Stevens' opinion said GHG were air pollutants, and EPA's argument that they were not subject to the Clean Air Act and thus to vehicle emission controls was insufficient. This decision

SIDEBAR 17.5

Auto Emissions Standards: Just Do It!

In one of the most dramatic acts of Congress, the 1970 Clean Air Act mandated a 90% reduction of auto air pollutant emissions in just five years. There were no proven technologies to do it at the time and automakers said it couldn't be done, but somehow the mandate was met, albeit in 1977, after a two-year extension. One would think that Congress could be as bold in auto efficiency, GHG emission controls, or RPS. Have our elected officials lost their assertiveness to enact far-reaching policies in the past three decades? Or has the political process become too complex for them to think out of the box?

will likely have far-reaching effects not only for vehicle emissions controls but also for all GHG control programs.

17.2.3 Federal Economic and Financial Energy Policies

Federal policy has long used economic measures to employ market forces to develop and deploy beneficial technologies and influence producer and consumer behavior. These measures include price controls, taxes, tax credits and deductions, federal procurement policies, research and development (R&D) funding, and direct federal outlays. Direct regulation of prices and markets such as oil and natural gas price controls and prohibition of oil and gas use in certain electric utility plants in the mid-1970s were short-lived, and their repeal in 1981 gave way to freeing energy markets. Tax incentives, R&D funding, and direct outlays are now the predominant financial measures used to affect market transformation.

17.2.3.1 Tax Policy: Incentives for Renewables and Efficiency

It is important to understand the different types of tax incentives. They include investment tax credits (ITC), production tax credits (PTC), and investment tax deductions (ITD). The 2005 EPAct included a plethora of energy tax incentives highlighted in Table 17.9, along with their estimated eleven-year impact on federal tax revenues, totaling $14.5 billion. The table includes all energy tax incentives in the act but not some continuing tax subsidies such as the 15% oil and gas and 10% coal depletion allowance for small independent producers (about $2.2 billion/year or $24 billion over eleven years).

Many of the EPAct's tax incentives apply to renewable energy and efficiency. These are summarized below. For more detail and updates on federal tax incentives see Lazarri (2006) and DSIRE (2006).

Investment Tax Credits. These are popular tax measures that allow investors to claim a credit for a portion of the money they invest in efficiency improvements and renewable energy

table 17.9 Energy Tax Incentives of the 2005 Energy Policy Act *(with date extensions from 2007 tax legislation given in italic)*

Energy Type	Tax Incentives	Provisions	Eleven-Year Budget Impact
Energy Efficiency	Residential energy efficiency **investment** tax credit	10% for envelope, 100% for equipment up to $500 installed through 2007	$1.3 billion
	Builders' energy-efficient home **production** tax credit	Up to $2000 credit per unit exceeding IECC efficiency, through 2007	
	Corporate tax **deduction** for efficiency, new commercial property	$0.30–$1.80 per ft^2 (floor), depending on technology/reduction, through 2007	
	Manufacturer **production** tax credit, appliances > efficiency standards	$75–$175/unit, up to $75 million limit on all credits taken	
	Personal hybrid vehicle **investment** tax credit	up to $3400, qualifying vehicle, through 2010	
Renewable Energy	Residential solar and fuel cell **investment** tax credit	30% for PV and solar HW up to $2000 (30% FC up to $500/0.5kW) if installed by *2008*	$4.5 billion
	Business energy **investment** tax credit	30% for solar PV and thermal, fuel cells, geothermal, microturbines installed by *2008*, no limit on solar	
	Renewable electricity **production** tax credit	1.9¢/kWh, ten-year credit, installed by 2008: wind, geothermal, biomass, others, but NOT solar PV	
	Biofuel blender **production** credit	$0.51/gal ethanol through 2010, $1/gal agri-biodiesel through *2008*	
	Alternative refueling infrastructure **investment** tax credit	30% for E85, B-20, natural gas, H$_2$ up to $30,000 ($1000 residential) by 2009 (2014 for H$_2$)	
	Clean Renewable Energy Bonds and **investment** tax credit for bond purchaser	Interest-free bonds for non-taxable entities (e.g., power co-ops, municipal utilities) for renewable power projects, and tax credit for bond purchaser equivalent to bond interest	
Oil and Gas	Oil and gas production	Tax credit for nonconventional sources	$2.6 billion
	Oil and gas refining and distribution tax package	Accelerated depreciation, expensing provisions	
Coal	Clean coal facility tax credit	20% for IGCC, 15% for adv. coal	$3.0 billion
	Pollution control amortization rules		
Electricity	Nuclear production tax credit	1.9¢/kWh for new nuclear generation limited to first 6000 MW capacity built before 2021	$3.0 billion
	Nuclear decommissioning deduction	for contributions to fund	
	Rules on transmission depreciation	Classified as fifteen-year property	
Total			$14.5 billion

Does not include direct funding, loan, and risk guarantees.
* For more information and updates, see Database of State Incentives for Renewable Energy (DSIRE) (http://www.dsireusa.org/), managed by the North Carolina Solar Center, North Carolina State University
SOURCE: Lazzari, 2006; DSIRE, 2007

systems. They effectively reduce the capital cost of these projects, improve their cost-effectiveness, and encourage greater investment. Some of the key credits for renewables and efficiency in the 2005 EPAct are given below. Some had deadline dates of 2007 that were extended to 2008 by Congress in late 2007. These are shown in italic:

- Business ITC for 30% of cost of solar systems installed by *2008*, no monetary limit.
- Residential ITC for 30% of cost of solar systems installed by *2008*, up to $2000 credit.
- Clean Renewable Energy Bonds are a unique way to provide incentives to non-tax entities through no-interest bonds and ITC credits for bond purchasers (see Sidebar 17.6).
- Residential energy efficiency ITC for 10% of envelope improvements up to $500 credit, 100% of qualified HVAC equipment up to $500 credit, installed by 2007.
- Alternative fuel infrastructure ITC for 30% of cost of E85, B-20, H_2 refueling infrastructure installed by 2009 (2014 for H_2), up to $30,000 credit ($1000 for residences).
- Hybrid vehicle tax credit up to $3400, depending on vehicle efficiency, purchased by 2010.

Production Tax Credits. These are tax incentives provided for each unit of renewable energy or high-efficiency product produced. Whereas the ITC rewards simply investing in energy projects, the PTC rewards actual performance. For example,

- Renewable electricity PTC of 1.9¢/kWh produced for first ten years from wind, geothermal, biomass (0.9¢/kWh for small hydro, refined coal) if systems installed by *2008*. Rates are adjusted for inflation.
- Biofuel blender PTC of $0.0051/gal per 1% ethanol blend (e.g., $0.51/gal pure ethanol, $0.051/gal E10) thru 2010; $0.01/gal agri-biodiesel and $0.0050/gal waste-grease-biodiesel per 1% blend (e.g., $1/gal agri B-100)
- Manufacturer efficient appliance PTC of $75–$175/unit depending on type and savings for refrigerators, washer/dryer, dishwashers, but to $75 million total credit limit.
- Builders energy-efficient home PTC of $2000 per house built by 2007 if exceeds IECC efficiency by 50% ($1000 if exceeds by 30%); more than 20% of savings must be from house envelope.

Investment Tax Deductions. Tax deductions reduce taxable income and thus have less impact than tax credits. This is apparent in Solution Box 16.1 that described the difference between ITC and ITD for the hybrid vehicle incentive. Most deductions have been replaced by tax credits, but one remains:

- Efficient commercial property ITD of $0.30–$1.80/ft² (depending on technology used and efficiency achieved) built by 2007

Solution Box 17.2 illustrates the tax credits in action, with two examples we used in previous chapters: a residential PV system and a wind farm.

SOLUTION BOX 17.2

Effect of Federal Tax Credits on the Cost of PV and Wind Power

What is the relative effect of federal tax credits on the cost of PV and wind power?

Solution:

Let's look at two of our previous examples of economic analysis of PV and wind systems to isolate the effect of the federal tax incentives, relative to other incentives like state rebates. The incremental effect of the federal credit on electricity price is shown in bold.

- **Residential PV:** In Section 11.8, we explored the economics of a 2 kW residential PV system in Los Angeles showing the incremental effect of different incentives including the California rebate program, the federal investment tax credit, and a tax deduction on mortgage interest on the net investment. The assumptions: 2 $kW_{DC,STC}$ (1.68 $kW_{AC,PTC}$) system, $7/W installed, 6% 30-year mortgage, 28% tax bracket, California rebate $2.80/$W_{AC,PTC}$, federal tax credit 30% up to $2000.

Cost of Residential PV and Effect of Federal ITC, State Rebate, Interest Tax Deduction

	Capital Cost	Annual Savings	Resulting Rate	Incremental Change
No incentives	$14,000	–	33.8 ¢/kWh	–
California rebate	$9296	–	22.4 ¢/kWh	11.4 ¢/kWh
+ federal ITC	**$7296**	**–**	**17.6 ¢/kWh**	**4.8 ¢/kWh**
+ tax deduction savings	$7296	$122/yr	13.5 ¢/kWh	4.1 ¢/kWh

- **Wind farm:** In Section 12.6, we explored the economics of a 2 MW wind farm in an area of 7 m/s winds. We calculated the cost of energy produced assuming a capital cost of $2 million ($1000/kW) financed at 7% for twenty years. We did not calculate the effect of the ten-year 1.9¢/kWh federal PTC but we do so in the table below, and it drops the effective price per kWh by 1.9¢/kWh to 2.6¢/kWh.

Cost of Wind Farm Power and the Effect of Federal PTC

	Capital Cost	Resulting Rate	Incremental Change
No incentives	$2 million	4.5 ¢/kWh	–
+ federal PTC (10 yr)	**$2 million**	**2.6 ¢/kWh**	**1.9 ¢/kWh**

SIDEBAR 17.6

Community Renewable Energy Bonds

The problem with investment and production tax credits is that you need to pay federal taxes to take advantage of the incentives. There are certain entities that should have incentives for renewable energy, like electric cooperatives and municipal utilities, but they do not pay taxes. Community Renewable Energy Bonds (CREB) provide a unique program to get around this problem. Under the 2005 EPAct, qualified entities (electric co-ops, municipal utilities, Native American tribes, government entities) can issue bonds that are interest-free for qualified renewable energy projects. Purchasers of the bonds are not paid an interest return but can claim an equivalent tax credit. The incentive aims to be equivalent to the renewable PTC. Tax credits are limited to $800 million for 2006 and 2007.

17.2.3.2 Energy Research, Development, Demonstration, and Deployment

Federal budgets have supported energy research and development through the twelve DOE national laboratories and research grants and contracts to energy industries, research firms, and universities. As discussed in Chapter 16, Kammen and Nemet (2005) argue that the federal energy R&D budget should be increased by 5–10 times. Figures 17.10–17.11 from the American Association for the Advancement of Science (AAAS; Koizumi, 2007) and Gallagher, et al. (2007), plot the overall outlays and composition of energy research develop-

figure 17.10   DOE Energy RD&D FY1978 to FY2008 Administration Request in Million Constant 2000 Dollars

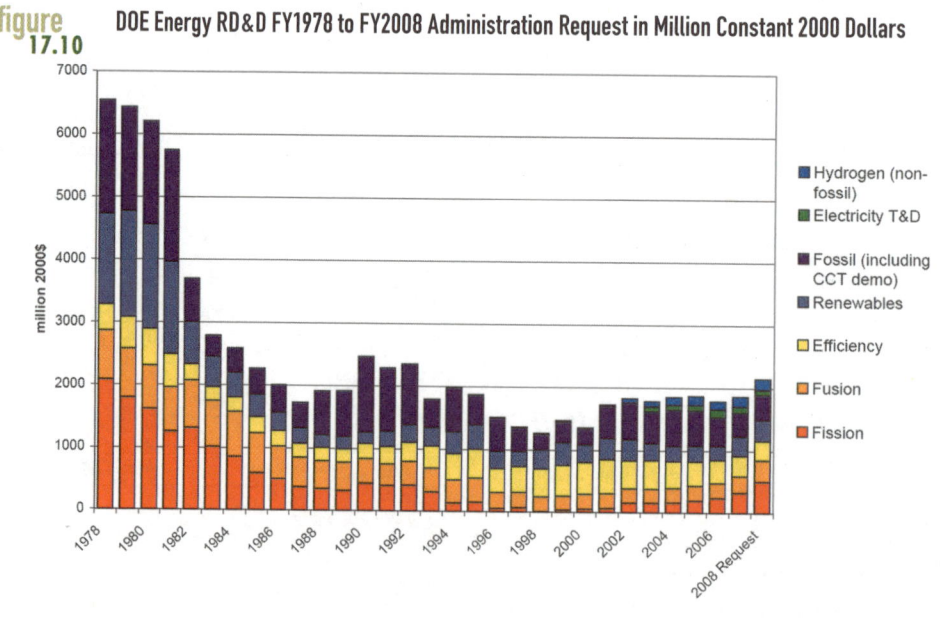

SOURCE: Koizumi, 2007, after Gallagher, et al., 2007. Used with permission.

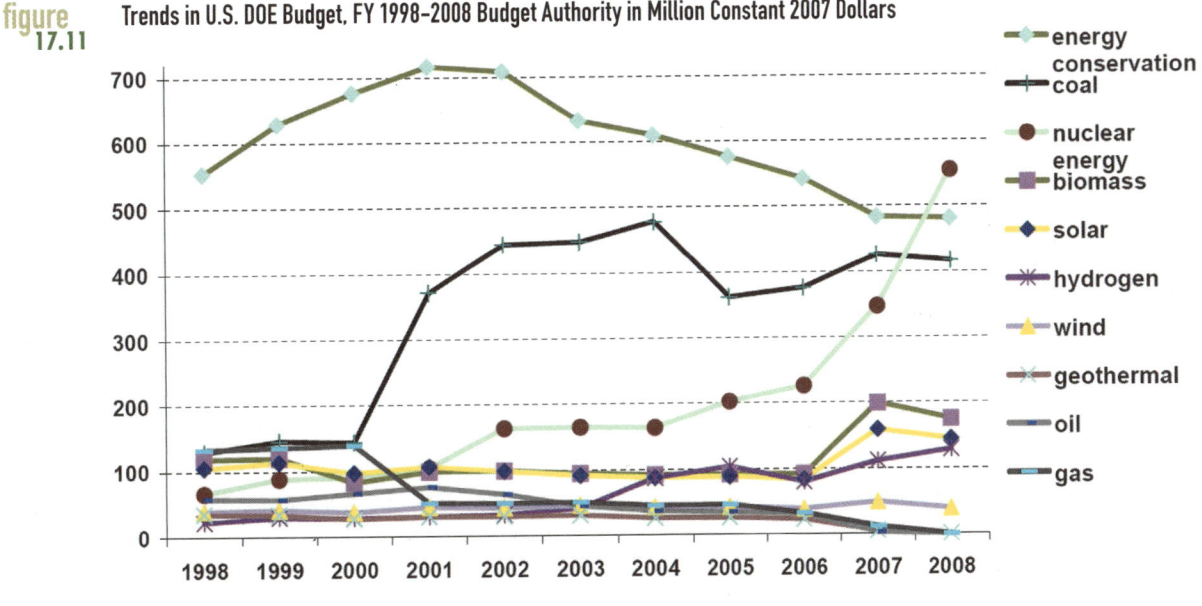

figure 17.11 Trends in U.S. DOE Budget, FY 1998–2008 Budget Authority in Million Constant 2007 Dollars

energy conservation
coal
nuclear energy
biomass
solar
hydrogen
wind
geothermal
oil
gas

SOURCE: Koizumi, 2007, AAAS. Used with permission.

ment and demonstration (RD&D) from FY1978 to FY2008, and trends for specific sources from FY1998 to FY2008.

The president's requested federal energy R&D budget for FY2008 is one-third of the 1978 budget, and it is only 1% of the total federal R&D budget. The big-ticket items for federal R&D are defense (57%), health (22%), space (8%), and environment (1.5%). Of the $2.2 billion for energy RD&D shown for FY2008 in Figure 17.10, 25% is for nuclear fission, 21% fossil energy, 16% for renewables, 14% for efficiency, 16% for fusion, 6% for hydrogen, and 3% for electricity transmission. Figure 17.10 shows that the energy efficiency R&D budget dropped by 25% from 2001 to 2008, while nuclear increased five times.

The National Commission on Energy Policy (2004) recommended a doubling of the R&D budget by 2010 and more than doubling the efficiency and renewables R&D budgets.

17.2.3.3 Loan and Risk Guarantees

Loan guarantees and risk insurance are additional financial incentives intended to lessen the risks to investors of potentially beneficial energy technologies. Funds for such assistance are generally authorized by legislation and then must be appropriated in specific budget bills. Three provisions of the 2005 EPAct illustrate this policy approach.

- Loan guarantees to cover 80% of the cost of no more than four projects, at a maximum of $250 million per project, to demonstrate commercial feasibility of converting cellulosic biomass to ethanol.

- Loan guarantees for an IGCC coal facility to be combined with renewable sources, CO_2 sequestration, and hydrogen production.
- "Regulatory risk insurance" for six new nuclear power plants to cover cost overruns arising from regulatory delays, up to $500 million per plant for the first two plants, up to $250 million for the next four, to a maximum total of $2 billion.
- Price-Anderson Act extended for new nuclear plants through 2025 to share and limit liability in case of accident.

17.2.3.4 Federal Procurement

The federal government is a big consumer of energy and energy-consuming equipment and procurement policy can help increase development of energy technologies and set an example for other consumers.

- The 1992 Energy Policy Act required federal vehicle fleets to employ alternative fuels.
- The 2005 EPAct established federal agency "Greenpower" purchasing goals of 3% for 2007–2009, 5% for 2010–2012, and 7.5% for 2013 and thereafter. The 2007 EISA increased these goals to 10% by 2010 and 15% by 2015.
- Federal energy management requirements: ENERGY STAR equipment, building performance standards, 20% reduction on federal energy use by 2015 compared to 1985–2003 baseline.

17.2.3.5 Direct Funding and Energy Assistance

Federal funds are used to assist low-income energy users and to provide grants for state and local governments to enhance their energy planning and efficiency programs. These programs have been in operation for nearly thirty years.

- LIHEAP is a $5 billion per year program to help low-income residents pay their energy bills.
- The Weatherization Assistance Program (WAP) provides energy efficiency improvements by local weatherization agencies to low-income home owners (< 150% of poverty level income). EPAct of 2005 authorized $500 million in FY2006, increasing to $700 million by FY2008.
- State Energy Program funding authorized at $100 million for FY2006 increasing to $125 million by 2008, with new 2005 EPAct requirements for state plans to achieve 25% energy efficiency improvements in 2012 over 1990 baseline.
- Community grant programs, such as the Clean Cities Program, provide small grants and technical assistance to communities wishing to develop alternative fuel programs. It has helped develop eighty Clean City Coalitions across the country, displacing over 1 Bgal of petroleum fuels. In 2005, grants to sixty community projects totaled $5.4 million.

17.2.4 Federal Energy Information and Education

Market transformation requires improved information to lower transaction costs and enhance consumer choice. The ENERGY STAR program and Appliance Energy Labeling were discussed in Chapter 8. EPA estimates that the ENERGY STAR program led to consumer savings of $10 billion in 2004 alone. The 2005 EPAct provided for an HVAC maintenance consumer education program and an energy efficiency information and public education program.

17.3 Summary and Prospects for U.S. Federal Energy Policy

In 2004, the National Energy Policy Commission (NEPC) issued its comprehensive report "Ending the Energy Stalemate: A Bipartisan strategy to meet America's Energy Challenges." The stalemate referred to the contentious and lengthy political process involved in Congress, the administration, and the wide range of political interests in agreeing to a comprehensive energy policy act, the first since 1992.

Finally, in August the 2005 EPAct was passed by the Republican-controlled Congress and signed by President Bush. The EPAct established a broad framework for federal policy including biofuel production standards, additional appliance efficiency standards, investment and production tax credits that support energy efficiency and renewables, and enhanced research and development funding for renewables. The EPAct also supports conventional energy including oil and gas subsidies, clean coal technologies, and especially a range of subsidies to create a renaissance of nuclear power.

However, many analysts thought the 2005 EPAct fell short of the following needed policies, including those recommended by the NCEP:

1. Improving the CAFE vehicle efficiency standards that had not really changed since originally enacted in 1975 for 1985 vehicles
2. Increasing significantly federal energy research, development, and demonstration funding that remains at one-third of 1978 levels
3. Providing a meaningful federal RPS for electricity generation from wind, solar, biomass, geothermal, and other sources
4. Establishing a comprehensive program to reduce GHG emissions, including a cap-and-trade or similar program for stationary sources, GHG emission standards for vehicles, or a comprehensive carbon tax.

Recognizing the limitations of the 2005 EPAct, the 110th Democrat-controlled Congress resumed energy policy deliberations and debate, and in December, 2007 passed the EISA with the following provisions, among many others. (For a summary and full text of the EISA, see http://www.govtrack.us/congress/bill.xpd?bill=h110–6.)

- Increase the CAFE standards to 35 mpg for all light vehicles by 2020
- Improve appliance and lighting efficiency standards
- Increase the RFS from EPAct's 7.5 Bgal in 2012 to 36 Bgal by 2022, 20 Bgal of which must come from cellulose
- Require half of new cars to be flex-fuel vehicles by 2015

Other proposals that were part of the debate, such as a national RPS of 15%, eliminating over ten years $28 billion in subsidies to oil companies that are currently enjoying skyrocketing profits, and using those revenue savings to fund additional tax incentives for efficiency and renewables, failed to pass because of resistance from oil and utility companies and the White House.

The prospects are uncertain for several proposed GHG emission bills in both the House and the Senate calling for a nationwide CO_2 cap-and-trade system. (For updates, see Pew Center on Global Climate Change.)

Although the federal government is critical in a nationwide transition to sustainable energy, it is not alone in public market intervention. In the absence of effective federal policy in several areas, many states have stepped forward to take leadership in energy policy. There has also been a groundswell among many communities as they enact regulations, incentives, and education programs for energy efficiency, renewable energy development, and enhanced consumer choice. In the next chapter, we discuss how states and communities can complement and exceed federal energy policies. Indeed, we will see there is much U.S. federal policy can learn from the states.

There is also much the United States can learn from other nations. Once the world leader in environmental and energy policy, the United States now lags behind many countries in developing innovative policies and new energy technologies. The United States must collaborate with the global community to forge effective agreements and strategies for sustainable energy and especially climate change. The United States must also learn from the energy policy experimentation in the European Union, Brazil, Japan, Australia, and elsewhere. And it is hoped that these experiences can also allow the developing nations to advance their economies with less dependency on carbon and oil, for we all share their fate.

CHAPTER 18

U.S. State and Community Energy Policy and Planning

Federal leadership is essential for effective policy to achieve sustainable energy. We saw in the previous chapter the effect of national standards and regulations, federal budget investments in research and development, and federal tax incentives for private investment. But we also saw that federal policy is often subject to political winds, and vested interests can slow and impede policy innovations. For example, U.S. federal energy policy has not been able to address effectively such important issues as building and land use efficiency, renewable portfolio standards, necessary funding for energy research and development, and control of greenhouse gas (GHG) emissions.

As a result of limited federal policy in these areas, many state and local governments have stepped forward to fill the void. In 2007, the hotbed of energy policy innovation resides at the state and local level in the United States. Historically the states have acted as "policy laboratories," developing models for each other and for federal action. In energy, they have functioned as the "laboratories of efficiency" and have provided the best opportunity for policy innovation for sustainable energy through utility regulation, building efficiency standards, and other measures (Prindle, et al., 2003). California is an exemplar in building and appliance standards, Minnesota in biofuels, Texas in renewable portfolio standards, and the northeastern states in GHG emission cap-and-trade, to name a few.

Local governments are also in a strategic position to initiate innovative energy policy. Although they usually have neither the financial resources nor the regulatory authority of the higher levels of government, they are closest to the point of consumption and can foster community energy action and education. In addition, they have some control over land use, transportation planning, transit development, and implementation of building efficiency standards. Some communities operate municipal utilities or electric co-ops and have the capacity to develop innovative utility programs. As of late 2007, more than 775 U.S. cities are now members of the Cool Cities program, actively seeking ways to improve energy efficiency and reduce GHG in their communities. Seattle, Portland, Austin, Sacramento, New York, Boulder, and many others provide models for local action that is necessary for sustainable energy.

This chapter presents the variety of energy policy approaches used by state and local governments in the United States, ranging from regulations to incentives to programs that enable consumer and community choice of energy.

18.1 State Energy Policy

State governments have significant authority over how energy is provided and used. They regulate utilities, establish building codes, plan transportation networks, use tax revenues and incentives for energy development, and enact environmental policies to complement national programs. States have adopted their energy policies in response to federal legislation and by their own initiative. In many cases, state policies have exceeded federal policies with innovations that have subsequently become federal policy.

States have thus provided necessary leadership in energy policy, from building codes and appliance standards in the 1970s to utility demand-side efficiency programs in the 1980s to renewable portfolio standards in the 1990s to climate change initiatives in the 2000s. Not all states have been active in energy policy; a few states have led the way.

Table 18.1 lists categories of state energy policies for renewable energy and efficiency, provides a brief description of each, and gives some states with exemplary programs. State energy plans set the stage for more specific policies. Policies for energy efficiency include building and appliance regulations, tax incentives for efficiency investments, utility regulation for demand-side management and public benefits funds, transportation and land use programs, and state building and fleet management. Policies for renewable energy include tax and financial incentives, renewable fuels standards, and a range of utility policies including renewable portfolio standards, net metering, and green power marketing.

A prerequisite for developing and implementing effective energy policy is an effective organizational structure. Nearly all states have commissions to regulate utility operations, and many states have recently established energy councils or interagency energy committees to facilitate energy planning. But few states have comprehensive energy agencies, such as the California Energy Commission (CEC), which works closely with the California Public Utility Commission (CPUC). The CEC was established in 1974 to forecast energy needs, license thermal power plants > 50 MW, promote energy efficiency through appliance and building standards, develop energy technologies and support renewable energy, and plan for and direct state response to energy emergencies. It is funded by a 0.22 mil ($0.00022) per kWh surcharge on electric utility bills, adding 12–14¢ to an average monthly bill. The CEC has a budget of more than $400 million/yr and a staff exceeding 550.

18.1.1 State Energy Planning

18.1.1.1 State Comprehensive Energy Plans

Historically, states have developed energy policies incrementally by specific legislation or administrative order. Early state energy plans focused on a certain sector, especially electricity for which they have regulatory authority. As energy concerns have increased, many states have begun to develop more comprehensive state energy plans. These plans vary. Some use an energy plan to commence development of a state energy strategy and specific policies, such as the 2006 *Virginia Energy Plan* enacted by legislative action and Michigan's 2006 *21st Century Energy Plan,* by executive order.

table 18.1 State Energy Policy Categories for Renewable Energy and Efficiency

Category	Description	Exemplars
STATE ENERGY PLANNING		
State energy plan	State plan for energy supply, demand; most utility-only, some comprehensive	CA, NY, MN, TX, NC
State climate change action plan	State inventory, registry of GHG emissions, reduction targets, action plans for utilities, transportation	28 states; exemplars: CA, MN, New England, OR, NJ, NY, CO
ENERGY EFFICIENCY		
Building efficiency standards	Energy-efficient building codes based on model codes or state design	CA, FL, TX, OR, WA
Appliance efficiency standards	Efficiency standards exceeding federal standards	CA, CT, NY, MA, WA, OR
State facilities	State energy management	many
Tax/financial incentives	Tax credits and deductions, rebates, low-interest loans for energy efficiency	CA, NY, IN, OR, MD
Utility regulation/restructuring	Range of policies affecting customer choice, rate-base funding of efficiency, rate-structure	23 states + DC pursued restructuring but vary in directives for efficiency
Demand-side management	Utility provide load reduction and energy efficiency programs using rate-base funding	CA, NY, WA
Public benefits fund	Rate-base funding of resource pool for efficiency improvement	CA, MA, CT
Incentive rates	Time-of-day, on-peak, off-peak, and inverted rates to modify demand	CA
TRANSPORTATION INITIATIVES		
Tax incentives or feebates for high-efficiency vehicles	Tax-neutral feebates: fees for low efficiency offset rebates for high-efficiency vehicles	CA, NY
State fleets	Alternative fuels and efficiency	MN, WA
Land use, smart growth, VMT reduction, transit-oriented development	Land use directives to localities to reduce vehicle miles traveled (VMT), energy, and emissions	OR, WA, MD, NJ, DE
GHG emission reduction	Mandatory and voluntary programs for vehicle GHG emission reduction	CA, 15 other states to adopt CA rules
RENEWABLE ENERGY		
Tax/financial incentives	Tax credits and deductions, rebates, low-interest loans for renewable energy	CA, NY, IN, OR, MD
Utility regulation/restructuring	Range of policies affecting power portfolios, customer choice, rate-base funding of renewables, distributed generation feed-in rates	23 states and DC have pursued restructuring but not all include directives for renewables
Renewable Portfolio Standards (RPSs)	Mandate for a certain amount of renewable power from each utility's portfolio of power sources	20 states; TX, CA, NY, ME have the highest RPS
Net metering	Allows distributed generators to feed power back into grid at retail rates	40 states
Green power choice	Customers can choose their source of power at different rates	TX, PA, NY, MA, MN, WI
Public benefits fund	Rate-base funding of resource pool for renewable power development	CA, MA, CT, NH
Community choice aggregation	Allows local governments to aggregate citizens and supply their electricity	MA, OH, CA, NJ, RI
Renewable Fuels Standard (RFS)	Mandate for transportation fuel suppliers to include a certain amount of biofuels in their supplied fuel	MN, IA, WA, AR, HI

Sources: Prindle, et al., 2003; York and Kushler, 2005; Institute for Local Self-Reliance, 2006; DSIRE, 2006; PEW CGCC, 2007

Other state plans recommend specific policies and call on legislative or executive action to enact them. For example, Texas (2005) and North Carolina (2005) have taken this approach. Table 18.2 outlines the four main components of the *Texas Energy Plan:* reorganization and energy education, traditional oil and gas industries, emerging technologies, and demand-side efficiency.

Because of the changing nature of energy markets and federal and other states policies, states should update energy plans on a regular schedule. New York updates every year its Energy Plan that was first developed in 2002. California's *Energy Action Plan* and its *Integrated Energy Policy Report* (IEPR) are updated every one to three years. The 2005 IEPR has no fewer than thirty-one policy recommendations under transportation fuels, electricity needs, demand-side resources and distributed generation, transmission, renewable energy, and natural gas (CEC, 2005). The 300-page 2007 IEPR, prepared by 160 CEC staff with the assistance of 17 consulting firms and hundreds of reviewers, is perhaps the most comprehensive state energy planning document to-date. It incorporates a long list of energy legislation recently enacted in California, developing implementation plans for such laws as AB 32 (2006) calling for reducing total GHG emissions in the state to 1990 levels by 2020 and SB 107 and subsequent directives calling for a 33% RPS by 2020. The 2007 IEPR is highlighted in Sidebar 18.3.

table 18.2   *Texas Energy Plan 2005:* **Ten Policy Recommendations under Four Categories**

1. **Better organize** the state's attention to energy matters by establishing a continuous energy policy development program, consolidating energy regulatory programs, and beginning efforts to develop an energy curriculum.

   a. **Texas Energy Planning Council Act.** Establish high level energy planning group
   b. **Texas Energy Commission Consolidation Act.** Reorganize administrative agencies
   c. **Texas Energy Education Act.** Support for public energy education

2. **Don't forget traditional oil and gas** fuels by recognizing the dependence on the traditional oil and gas industry through research, incentives, and development of LNG markets.

   a. **Resolution for Texas Leadership in Oil and Gas Research.** Position state for federal research funding
   b. **Texas Increased Rig Count and Petroleum Production Act.** Incentives for production
   c. **Resolution for Development of Texas LNG Markets.** Position Texas ports for LNG terminals.

3. **Look to emerging technologies** for the future by encouraging new gasification technology and increasing targets for renewable power sources.

   a. **Texas Gasification Technology Act.**
   b. **Texas Enhanced Renewable Portfolio Standard (RPS) Act.** Increase RPS to 5000 MW by 2015 and 10,000 MW by 2025

4. **Pay attention to demand-side management** by providing incentives and new technologies to allow utilities and consumers to realize more energy efficiency and conservation.

   a. **Texas Energy Savings Act.** Increase energy savings goal for utilities from 10 to 15% of utilities expected demand growth.
   b. **Recommendation for Advancement of Texas Smart Metering Technologies.**

18.1.1.2 State Climate Change Action Plans and Programs

Although the federal government has not addressed climate change and GHG emissions in a serious way, several states have begun to develop climate change plans and programs in their energy planning activities. As of 2007, these plans and programs include the following (Pew CGCC 2007; see www.pewclimate.org for updates):

1. **Climate Change Action Plans:** These plans identify opportunities for GHG emission reduction, but plans need inventories, targets, and controls to be effective (see Figure 18.1).

2. **GHG emission inventories and reporting registries:** The first step to control emissions is to know sources and amounts. Some states are members of regional registries.

3. **GHG emission reduction targets:** Seventeen states have targets (see Figure 18.2). For example, New England states, New York, and Oregon aim to reduce GHG emissions to 1990 levels by 2010, 10% below 1990 by 2020, and long-term reduction of 75%–85%. California AB 32 targets reductions to 2000 levels by 2010 and 1990 levels by 2020, and Governor Schwarzenegger has established a long term target of 80% below 1990 levels by 2050.

4. **Emission caps and offsets:** Five states have cap (CA, NH, MA) or offset (WA, OR, MA) programs requiring old power plants to cap their GHG emission (e.g., Massachusetts caps six older plants at 10% below 1997–1999 levels by 2006–2008) or new plants to offset anticipated emissions (e.g., OR and WA require 17%–20% offset).

figure 18.1 **States with Climate Change Action Plans, 2007**

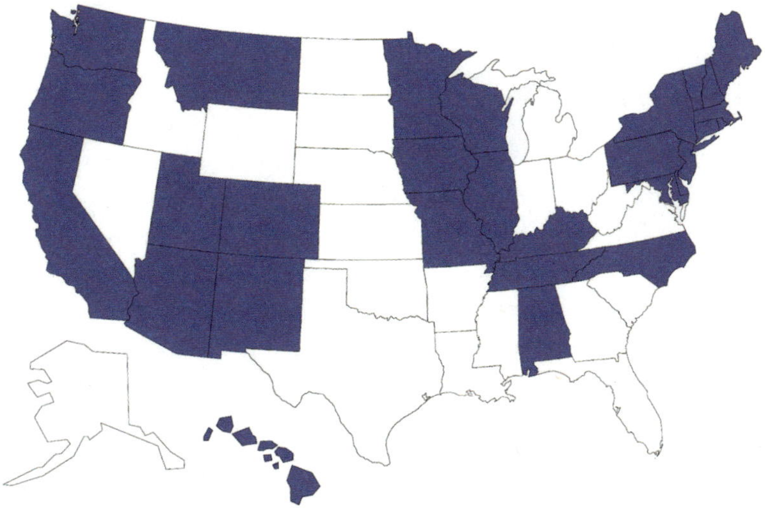

SOURCE: Pew CGCC, 2007

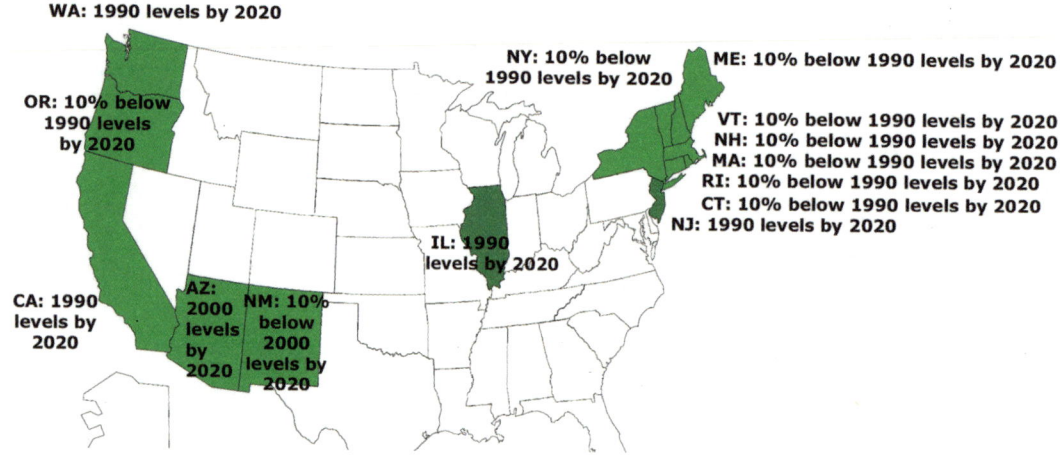

figure
18.2   States with GHG Emission Reduction Targets, 2007

SOURCE: Pew CGCC, 2007

5. **Mandates and cap-and-trade programs:** In 2006, California enacted legislation for the Air Resources Board to develop mandatory controls on GHG emissions to meet its 2020 target (reduction to 1990 levels), and the ARB is to consider a cap-and-trade program. The state's 2002 AB 1493 (Pavley) law mandates a 30% reduction in vehicle GHG emissions by 2016 (see Sidebar 18.3).

6. **Regional programs:** States have joined others in regional programs for GHG registries and inventories (Figure 18.3). Ten northeastern states (CT, DE, MA, ME, MD, NH, NJ, NY, RI, VT) formed the **Regional Greenhouse Gas Initiative** (RGGI) that developed the first mandatory GHG cap-and-trade program for CO_2, to cap emissions at 2005 levels by 2009 and reduce them by 10% by 2019 (see Sidebar 18.1).

18.1.1.3 State Land Use Planning

Land use sprawl into exurban areas dominated the patterns of development in the United States over the last half of the twentieth century and contributed to growing dependence on automobile transport, vehicle miles traveled (VMT), and transportation energy use. As a result, several states have reformed state land use planning legislation to promote "smart growth" to arrest sprawl. It aims to make it easier to invest and develop in areas of existing infrastructure and revitalize existing communities, and to make it more difficult to develop "greenfields" further and further from existing urban centers. Smart growth aims to create more livable, pedestrian- and transit-oriented neighborhoods and urban centers and to shorten travel distances. The Urban Land Institute study, *Growing Cooler*, found that by 2050 smart growth by itself could reduce VMT and transportation-related energy and GHG emissions by 7%–10% from current trends (Ewing, et al., 2007).

Figure 18.4 shows the status of states engaged in implementing or developing statewide planning legislation for smart growth. Michigan and Massachusetts also developed

SIDEBAR 18.1

Regional Greenhouse Gas Initiative

In 2006, RGGI adopted model rules for member states CO_2 cap-and-trade programs for power plants > 25 MW with allocations at 2005 levels starting in 2009 until 2015, then dropping by 10% by 2019. Utilities can trade credits exceeding their allowances and can also obtain "offsets" from nonutility sources up to 3.3% of their emissions. These offsets can be obtained from any state in the United States so long as an agency in that state oversees compliance through a memorandum of understanding. California agreed to participate in providing offsets to RGGI (Pew CGCC, 2007).

figure 18.3 Regional Climate Change Organizations, 2007

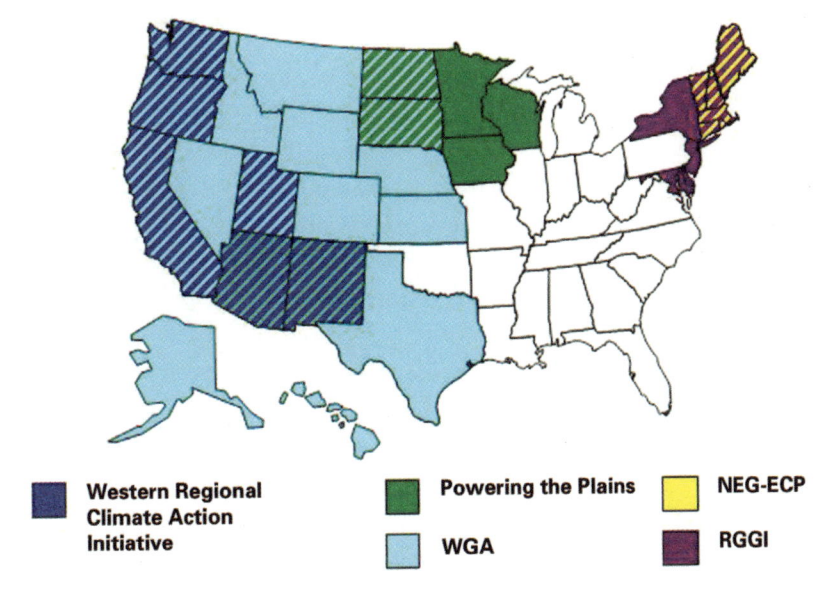

- **Western Regional Climate Action Initiative**
- **WGA**
- **Powering the Plains**
- **NEG-ECP**
- **RGGI**

Source: Pew CGCC, 2007

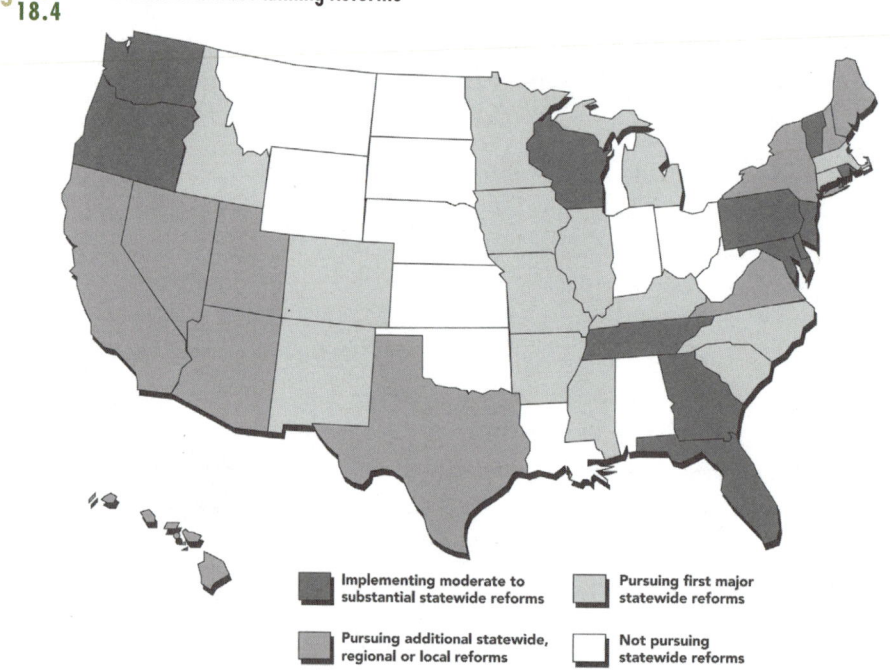

figure 18.4   **State Smart Growth Planning Reforms**

Legend:
- Implementing moderate to substantial statewide reforms
- Pursuing additional statewide, regional or local reforms
- Pursuing first major statewide reforms
- Not pursuing statewide reforms

States have promoted or mandated Smart Growth land planning at the local level to discourage sprawl development and revitalize existing communities with several public benefits including reduced VMT and enhanced opportunities for more energy-efficient transport including transit and pedestrian modes.

Source: APA, 2002

land use initiatives in 2004. The most effective statewide and local smart growth measures include the use of urban growth boundaries (UGBs). Using financial incentives or regulations, development is steered within the UGB and it is discouraged or disallowed outside the UGB. Oregon is most noted for its use of regulatory UGBs, whereas Maryland uses state funding as a means of containing growth within UGBs, or what they refer to as Priority Funding Areas. Implementing smart growth management depends on regional and local land use planning and controls, and this is where the action is. We discuss that under community planning and policy in 18.2.3.

18.1.2 State Utility Regulation

States have had the primary responsibility for regulating electric and natural gas utilities serving the public since the federal Public Utility Holding Company Act of 1935. This was necessary because most customers were captured by utility service areas, so each utility was essentially a monopoly. State regulation included approval of rates, permits and licenses for generation and transmission, and delineation of utility service areas, among other actions. As discussed in Sections 17.2.2.3 and 9.7, federal energy legislation expanded the responsibilities of state regulators to direct utilities to engage in demand-side energy services (1978 NECA), distributed generation (Public Utility Regulatory Policies Act [PURPA]), and restructuring of utility regulation (1992 Energy Policy Act). These provisions have had several effects.

18.1.2.1 Utility Restructuring and Consumer Choice

Restructuring and deregulation of utilities were pursued aggressively by several states in the mid- to late 1990s with the objective to keep prices low for consumers by creating greater competition in generation and marketing. State actions caused many utilities to reorganize, in some cases divesting generation to subsidiaries or other companies. It created a boom for energy brokers (like Enron), which bought and sold electricity in a freer market.

Figures 18.5 and 18.6 illustrate typical utility organization and regulation before and after restructuring. Figure 18.5 shows two typical **vertically integrated utilities** A and B that own and operate power plants, transmission, and distribution, and serve retail customers. This example shows interconnected transmission lines so they exchange wholesale power. The Federal Energy Regulatory Commission (FERC) regulates the wholesale electricity market and state Public Utility Commissions (PUC) regulate the in-state operations, including retail rates that provide utilities a fair rate of return (ROR).

Restructuring was designed to create competition in both retail and wholesale markets. As shown in the generalized case in Figure 18.6, restructuring encourages vertically integrated utilities to break up into, at most, four companies: **generators** (power plant owners), **transmitters** (transmission line owners), **distributors** (distribution line owners), and **marketers** (buyers and sellers of power), the latter serving as brokers for the others and for retail customers. Under this approach, FERC continues to regulate wholesale markets and states regulate retail. The goal of this arrangement is to provide retail customers the ability to purchase electricity from a range of providers, promoting competition and lowering rates.

But as Figures 18.5 and 18.6 show, the restructured environment is much more complicated than pre-restructuring. There are more parties involved, all trying to maximize their interests. A growing concern in utility restructuring, especially since California's "restructuring crisis" in 2000–2001 (see Section 9.7.6) and the Northeast blackout in 2003, is the possible reduced reliability in the power system. Instead of one integrated utility to blame for any problems, there are multiple parties in a restructured system, and they all tend to point fingers at the others if something goes wrong. In response, the 2005 EPAct authorized FERC to certify a national electric reliability organization (ERO) to enforce mandatory reliability standards for all bulk power activities.

The California experience cooled the enthusiasm of several states to restructure their utility regulation, and the deregulation movement slowed in subsequent years. Although

figure 18.5   Before Utility Restructuring

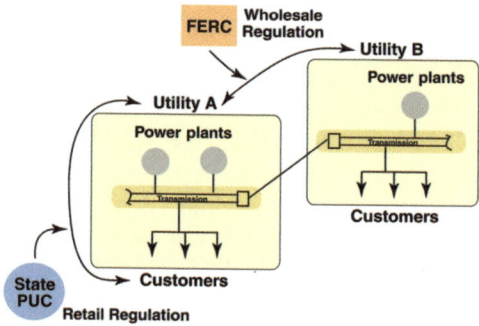

Large vertically integrated utilities did everything from building power plants to retail customer relations, responding to State regulators for retail and matters within the state and FERC for wholesale and interstate issues.

SOURCE: Abel, et al., 2005

figure 18.6   After Utility Restructuring

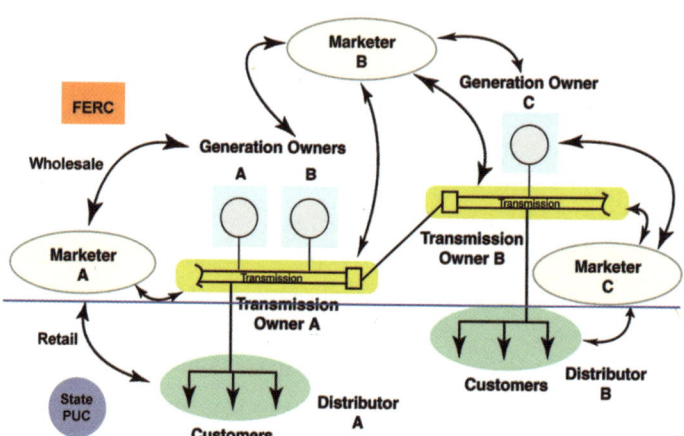

Utilities tended to break up into separate companies. Customers have greater choice of providers working through distributors or marketers.

SOURCE: Abel, et al., 2005

23 states and the District of Columbia had enacted some restructuring program by 2001 and seventeen others were considering it, by 2007 eight states followed California's example by suspending their restructuring, leaving only fourteen states and DC with active restructuring programs Figure 18.7). Virginia passed a "re-regulation" law in 2007.

figure 18.7 Status of State Utility Restructuring, 2007

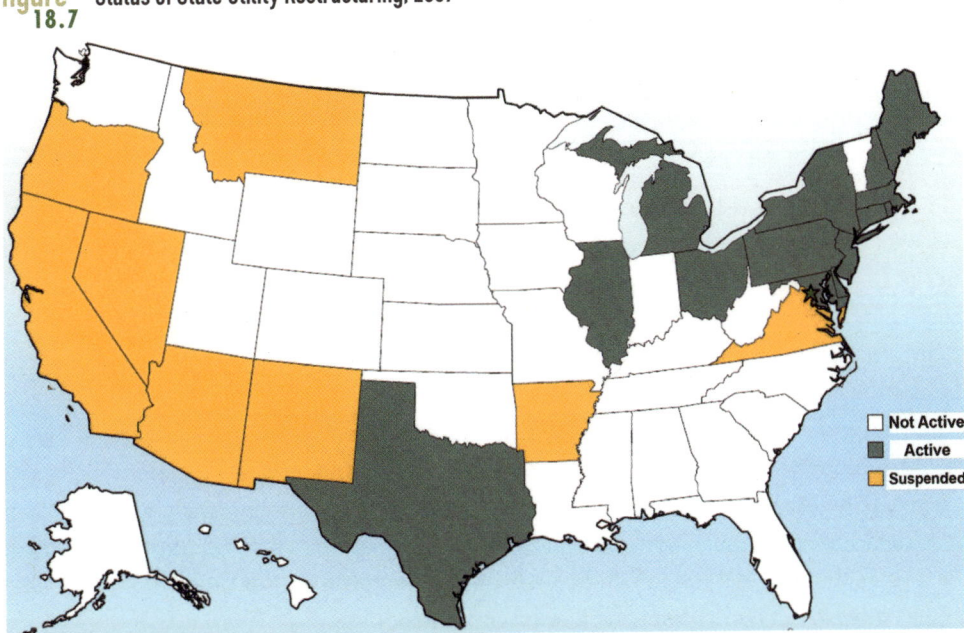

Fourteen states and the District of Columbia still have active restructuring programs. After California's power crisis in 2000–2001, it suspended its restructuring effort and seven other states with active restructuring programs followed suit.

SOURCE: U.S. EIA, 2007

Texas and Pennsylvania appear to have successful restructuring programs as measured by market competitiveness and consumer choice. They are ranked as the top two states according to the Center for Advancement of Energy Markets' Retail Energy Deregulation Index (CAEM, 2003). In Texas, as of September 2004, 1.1 million customers including 18% of residential users, had chosen a retail electricity provider (REP) other than their local utility, three times the number in 2002. (Section 18.1.2.6 describes Texas' choice program.)

18.1.2.2 Demand-Side Management

Demand-side management (DSM) programs are active efforts by utilities to modify customers' energy use patterns to achieve energy savings through efficiency and to reduce peak power demand through efficiency and load management. DSM programs include the following:

- **Incentives** such as rebates for energy-efficient appliances, HVAC systems, and building efficiency improvements
- **Direct assistance** such as on-site energy audits, commercial relamping programs, and low-income housing weatherization

- **Demand response** mechanisms including dynamic pricing (providing an incentive for less on-peak and more off-peak use), programmable communicating thermostats (enabling users to monitor and control electricity use), and automated controls of lighting and equipment (allowing utilities to remotely control users' equipment during peak demand).

DSM programs began slowly after the National Energy Conservation Act of 1978 required utilities to offer on-site energy audits to residential customers. Programs expanded in the late 1980s when state regulators gave incentives and directives to utilities for "least-cost" and "integrated resource planning." This planning approach requires utilities to give DSM services full consideration along with supply options. The costs of DSM programs can be absorbed into the rate base.

But most regulated utilities, especially investor owned utilities (IOU), did not have a strong incentive for DSM because they derived profits from sales and revenues. Their rates are generally based on a rate-of-return formula that encourages power plant construction and sales of electricity, and discourages them from providing greater efficiency that reduces sales. As long as the marginal revenue from a sale exceeds marginal cost of production, they have an incentive to sell more power and a disincentive to sell less (Eto, 1996).

In response, states created other financial incentives for DSM programs that try to dissociate earnings from sales. In some cases, utilities can earn a higher return on money spent on DSM. In others, utilities earn a bonus paid in $/kWh or $/kW based on energy or capacity saved by DSM. In the most popular approach, utilities earn a percentage of the "net resource value" or the difference between the avoided electrical system costs (i.e., fuel, new plants) from the DSM savings and the cost to implement the DSM program. This incentive promotes the most cost-effective DSM programs.

CPUC implemented such a program in 2007. It ordered a risk/reward performance-based mechanism for IOU DSM programs. IOUs must adopt goals based on 91% of achievable energy savings. If they fail to achieve less than 65% of their target savings, their shareholders are subject to penalties up to 5¢/kWh below the target. If they achieve 85% of their targets, 9% of the net benefits (resource savings minus costs) are returned to shareholders, and if 100% of targets are met, 12% of net benefits go to shareholders, with the remainder of the benefits going to ratepayers (CEC, 2007).

Utility DSM program spending grew significantly in the late 1980s and early 1990s, to $2.7 billion or about 1% of total utility spending in 1993. Studies of the effectiveness of DSM programs showed that while utilities varied in DSM performance, DSM electricity savings were achieved at an average 3.2¢/kWh, and many utilities achieved savings at less than 2¢/kWh (Eto, et al., 1995).

After 1993, utility investment in DSM declined, primarily as a result of state utility restructuring. In constant dollars, utility spending on efficiency declined by nearly 50% from 1993 to 1997. As progressive states that were active in DSM programs began to engage in utility restructuring and deregulation, it became apparent that less regulation and increasing competition for supply pushed utilities to abandon their DSM programs. To revive DSM activity and other investment in programs with public benefit, such as renewable energy, low-income programs, and public interest research and development, states established Public Benefits Funds supported by a small per-kWh charge on all utility sales.

18.1.2.3 Public Benefit Funds

Public Benefit Funds (PBF) are becoming increasingly attractive to states wishing to reinvest a small portion of utility receipts to fund projects with longer term public benefit. Funds are supported by a millage charge on each kWh sold (1 mill = $0.001). Funding rates range from 1 to 3% of utility revenue. The 22 PBF states (see Figure 18.8) are led by Connecticut (4.05 mills/kWh), California, Massachusetts, and New Hampshire (all with 3 mills/kWh).

These funds have increased overall utility funding of efficiency and renewable energy programs, turning around the utility DSM funding decline in the mid-1990s. A study of the first five years of PBF showed that energy savings are achieved at about 3¢/kWh. Annual electricity savings estimated from PBF supported programs range from 0.1% to 0.8% of electricity sales, with a mean value of 0.4%. In ten years, this annual rate would achieve cumulative savings of 4% of annual sales (Kushler, et al., 2004). This sounds small but it is a small percentage of a big number, in some cases billions of kWh.

18.1.2.4 Renewable Portfolio Standard and Energy Efficiency Portfolio Standards

Twenty-four states and DC have adopted **Renewable Portfolio Standards** (RPSs) requiring utilities to provide a certain percentage or amount of their power from renewable sources by

figure 18.8   **Public Benefit Funds, 2007 ($ millions)**

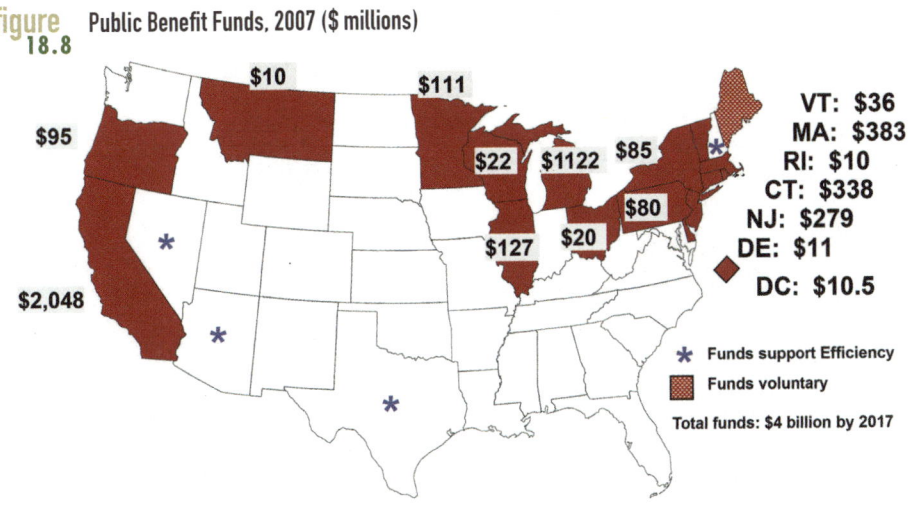

PBFs are supported by a small per-kWh charge on all electricity sales. They are used to fund energy efficiency and renewable energy programs.

Source: DSIRE, 2007

a certain date. Four other states have similar goals or voluntary programs. An RPS stimulates the development and use of renewable electricity. Other policies, such as Public Benefit Fund investments, net metering, and green power marketing, help achieve an RPS. Figure 18.9 shows the RPS states and their requirements as of September 2007. New York's 24% standard will create 3700 MW of new renewable capacity by 2013; Texas will create 2000 new MW by 2009, 5880 MW by 2015. California shortened its original timeline for its 20% RPS from 2017 to 2010 and has proposed a 33% RPS by 2020, not including the state's large hydro capacity which amounts to 19% of current total capacity. Some states like Texas allow utilities to comply with the RPS through **tradable renewable energy certificates** (REC), in which utilities that exceed their allocated requirement can sell credits to those that do not. California has chosen not to allow REC toward its RPS because it would slow renewable power's contribution to its AB 32 mandate to reduce GHG emissions to 1990 levels by 2020.

Nine states (CA, CO, CT, HI, NJ, NV, PA, TX, and VT) have adopted an **Energy Efficiency Resource Standard** (EERS). Similar to RPS, the EERS requires utilities to meet a target standard by a deadline, in this case an efficiency energy savings. The EERS is either a specific energy amount (e.g., California's 23 TWh and 5 GW peak for electric and 444 million Therms for gas utilities by 2013) or a percent of sales (e.g., Texas's 10% of forecast growth in 2004 and thereafter). All EERS states also have RPS; some combine them into one standard (Nadel, 2006).

figure 18.9   Renewable Portfolio Standards, 2007

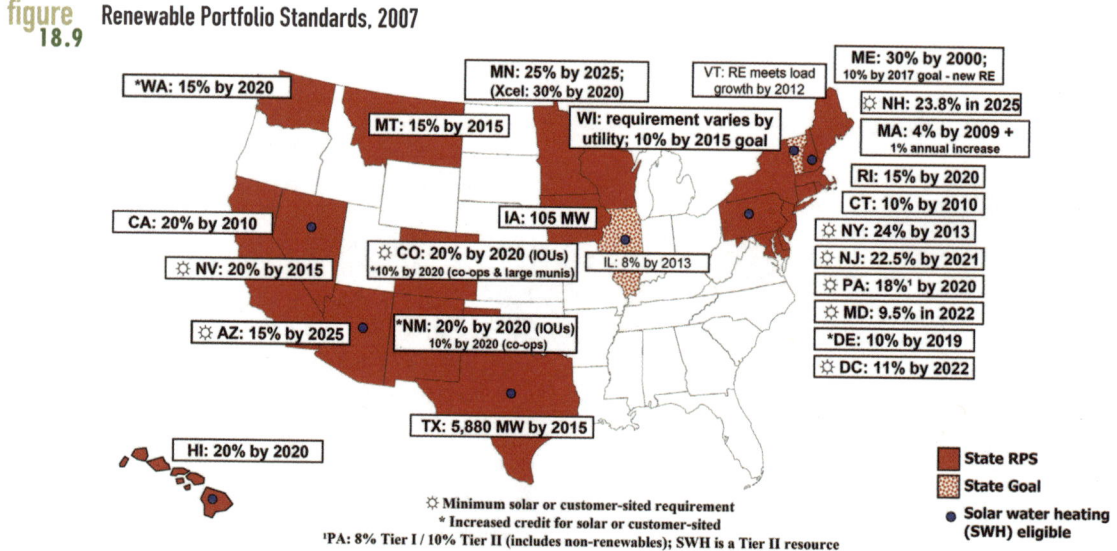

As of 2007, twenty-four states have RPSs (with goals in four others). They have been a major stimulus for renewable power in Texas, Iowa, Connecticut, and Pennsylvania.

SOURCE: DSIRE, 2007

18.1.2.5 Net Metering

In 1978, PURPA required utilities for the first time to buy excess power produced by local generators, but they only had to pay rates based on their "avoided costs." These rates were generally low and often provided little incentive for investment in local generation. But some utilities experimented with simple net metering of retail customers who had some on-site generation. As discussed in Chapters 8 and 10, net metering allows these generators to simply plug in their generating unit through a synchronous inverter on their side of the meter. Power generated on-site supplies the site as needed, but excess generation is fed back to the grid through the customer's retail meter and runs the meter backwards. Monthly meter readings display "net energy" drawn from the grid and the customer will be billed for that net energy. The on-site generator is essentially paid retail rates for the power it generates and feeds to the grid.

By August 2007, forty-two states and DC had adopted some sort of a **net metering** policy for at least some utilities in the state (Figure 18.10), providing a big incentive for on-site distributed generation (DG), especially small renewable energy systems like residential rooftop photovoltaic systems and small wind systems. The map shows that system kW limits vary among the states and in some states they are different for residential and commercial applications. New Mexico (80 MW), Colorado (2 MW), and California (1 MW) have the

figure 18.10 Net Metering in Forty–two States and DC, 2007

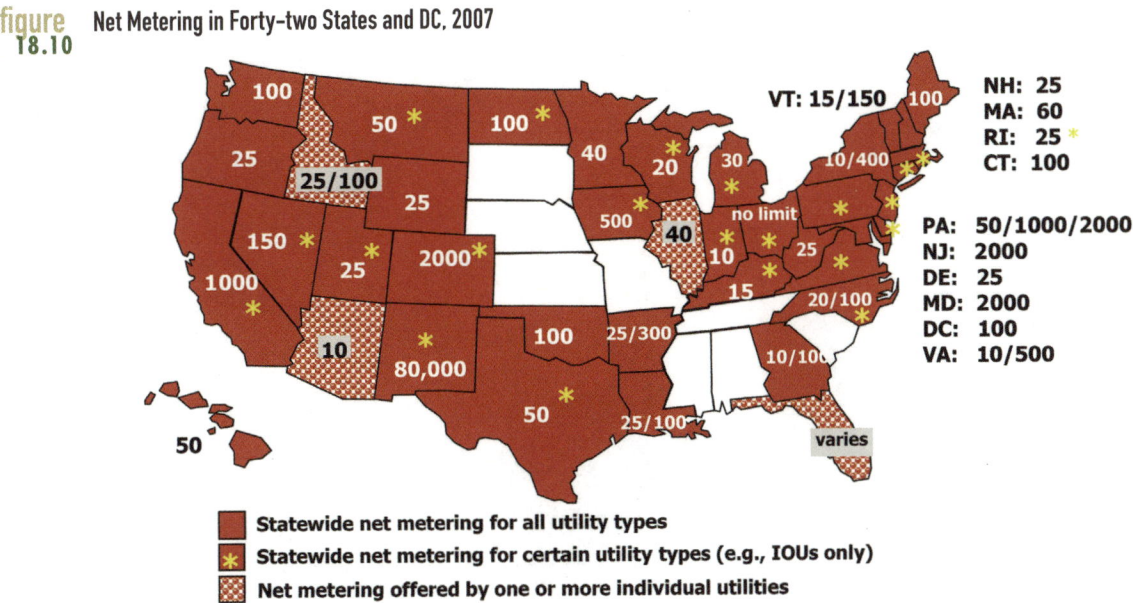

Numbers indicate limits in kW size of systems for which net metering is provided; two numbers indicate different limits for residential/commercial.

SOURCE: DSIRE, 2007

highest capacity limits for net metered systems. But specific terms of net metering agreements vary widely from state to state and from utility to utility.

18.1.2.6 Consumer Choice and Markets for Green Power

To enhance consumer choice of power source, programs have been developed by the states, by utilities, and by third-party marketers under three labels: "green marketing," "green pricing," and "Green Tags." In some restructuring states with green marketing, customers can choose the type and source of power they buy. In other states, the utilities offer customers a renewable power option with a green price premium as an alternative to their regular electricity. Regardless of location, customers can purchase "Green Tags" or REC or more recently "White Tags" or Energy Efficiency Credits (EEC) to offset their utility-provided nonrenewable power.

Green marketing offers the most formal program for consumer choice. In states with active restructuring programs, consumers are given a choice of power source in a competitive marketplace. Usually several suppliers and service offerings are available. Figure 18.11 gives the number of green power marketers offering products in the states.

figure 18.11   Green Power Marketing

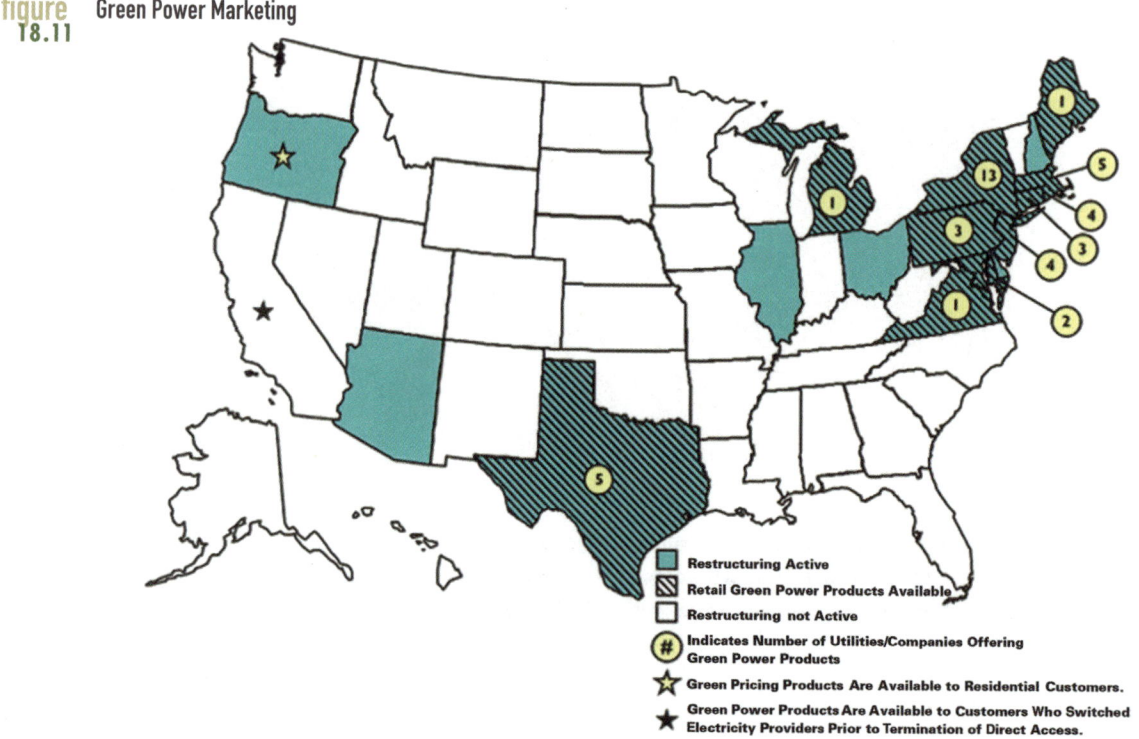

States with restructuring that offer green marketing and the number of green power marketers.

Source: Green Power Network, 2007

For example, Table 18.3 illustrates consumer choice in Texas, not only for green power but all other sources. On the utility commission Web site, consumers enter their zip codes and the array of choices is displayed. (Try it out at http://www.powertochoose.org/.) At the top of the list is the "default" or affiliate REP if no choice is made and the "price to beat" that REP charges. Table 18.3 shows just ten of the thirty-five choices available for Dallas. Five of the choices (shown in italic) are "green power." For each choice, detailed fact sheet information is provided with a click of the mouse. Table 18.4 illustrates the fact sheet information for Green Mountain Energy, a 100% renewable energy provider. Solution Box 18.1 illustrates the choice by a Dallas customer.

Green pricing offers utility customers the option of paying a price premium to cover the incremental cost of additional renewable energy. This premium is either an added cost per kWh or cost per block of kWhs. For example, the Southern Minnesota Municipal Power Authority, a wholesale power supplier in southern Minnesota, offers wind power to the retail customers of the utilities it serves at an additional $1 per 100 kWh block or a price premium of 1¢/kWh.

Figure 18.12 gives the number of utilities offering green pricing in the states where it is available. Green pricing is mandatory in Iowa, Minnesota, Montana, New Mexico, and Washington—utilities in those states have to offer it. Utilities with the largest programs include municipal utilities in California (Sacramento, Los Angeles, Palo Alto) and Texas

figure 18.12   Green Pricing

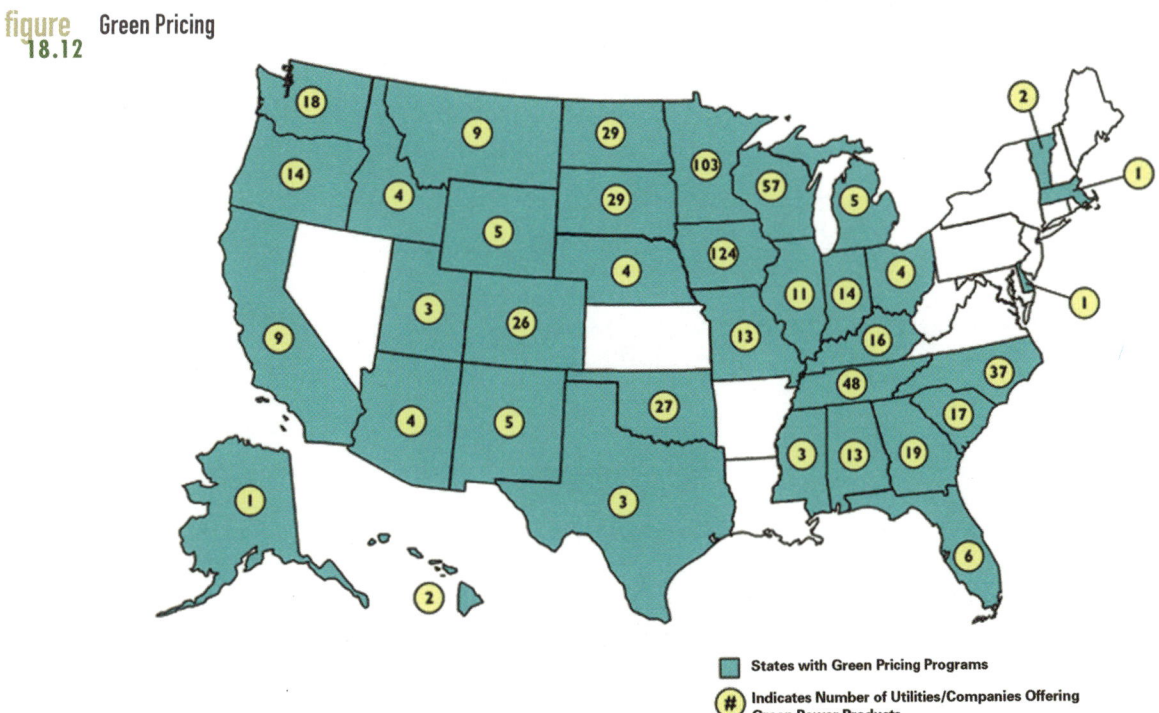

States with Green Pricing Programs

(#) Indicates Number of Utilities/Companies Offering Green Power Products

States in which green pricing for renewable power is available and the number of utilities with programs.

Source: Green Power Network, 2006

table 18.3   Sample Power Choices in Dallas, Texas

10 of 35 choices including "price to beat." Green power shown in italic.

Provider Offer	Monthly Cost (1000 kWh)	Save First Year	Min. Term (Mo.)	Env. Info	Average $ per kWh (1000 kWh)
YOUR AFFILIATE REP: TXU "Price to Beat"	$150	0	1		$0.15/kWh
Green Mountain Energy Company Pollution-Free Rate	*$145*	*3%*	*12*	X	*$0.1445/kWh*
Dynowatt	$133	11%	1		$0.133/kWh
Gexa Energy Gexa Green–100% Pollution Free	*$148*	*1%*	*12*	X	*$0.148/kWh*
Texas Power	$132	12%	1		$0.132/kWh
Amigo Energy	$126	16%	6		$0.126/kWh
Reliant Energy Renewable Plan	*$160*	*None (–7%)*	*0*	X	*$0.16/kWh*
First Choice Power, Simply Better Renewable	*$143*	*5%*	*12*	X	*$0.143/kWh*
StarTex Power	$136	9%	12		$0.136/kWh
TXU Energy EarthWise 18^{SM}	*$148*	*1%*	*18*	X	*$0.148/kWh*

table 18.4   Typical Provider Fact Sheet to Inform Consumer Choice: Green Mountain Energy

		Average Price per Kilowatt-Hour		
		Average Monthly Use		
Electricity price	Service territory	500 kWh	1,000 kWh	1,500 kWh
	TCC	14.45¢	14.45¢	14.45¢
	CenterPoint	15.42¢	15.42¢	15.42¢
	TNMP	14.11¢	14.11¢	14.11¢
	TXUED	15.80¢	15.80¢	15.80¢
	TNC	16.00¢	16.00¢	16.00¢
Contract	Minimum term: 1 Year; Penalty for early cancellation: $150			
Sources of power generation		Green Mountain Energy	Texas	
	Coal and lignite	0%	37.1%	
	Nautral gas	0%	50.1%	
	Nuclear	0%	9.9%	
	Renewable Energy (wind, water)	100%	< 1%	
	Other	0%	2.2%	
	Total	100%	100%	
Emissions and waste per kWh generated	Carbon Dixide	0		
	Nitrogen Oxides	0		
	Particulates	0		
	Sulfur Dioxide	0		
	Nuclear Waste	0		

SOLUTION BOX 18.1

Choosing Sources of Electricity in Texas's Green Marketing Program

A Dallas customer consumes 1000 kWh per month and wants to explore her choices for providers to reduce economic and environmental costs.

Solution:

From Table 18.3, she has the following choices:

Affiliate REP: TXU (default)	$150/mo
Lowest cost REP: Amigo	$126/mo, savings of $24/mo over Affiliate REP
Renewable REP: First Choice	$143/mo, savings of $7/mo over Affiliate REP and $17/mo premium over lowest cost REP

Obviously, it is worth it to make a choice rather than do nothing, and for $7/mo less than her default provider and $17/mo more than the lowest cost provider, she can be assured that her consumed electricity will result in no air pollutant or GHG emissions or nuclear wastes.

(Austin), Xcel Energy in Midwest and mountain states, Pacific Corp., and Portland (OR) General Electric.

Green Tags (also called Tradable Renewable Certificates [TRC] and RECs) are available to all consumers regardless of whether their state or utility has a formal green pricing or marketing program. Consumers buy certificates from third-party providers separately from the utility from which they buy the electricity. The proceeds are used by the third party to support renewable energy development. This lowers the cost of renewable power, making it more competitive in wholesale markets. Sixteen providers operate nationally and sell a variety of Green Tag products. The certificate cost (generally 0.5¢–2.5¢/kWh) is comparable to the premium in green pricing programs. For example, 3-Phases Energy Services offers its *Green-e* certified (see Sidebar 18.2) Green Certificates for 2¢/kWh, which supports 100% new wind power. Green tags offer consumers, especially those for whom green pricing is not available, a "proxy" for buying renewable power or at least "offsetting" their purchases of nonrenewable power. Table 5.12 describes a method for calculating the Green Tags needed for a household to become carbon neutral. "White Tags" are a more recent innovation which offsets power purchases with efficiency improvements instead of renewable energy development.

SIDEBAR 18.2

What Is Green Power? The Green-e Certification Program

If consumers are paying a price premium for "green power" under pricing, tags, or marketing schemes, they want to be assured that they are really getting or supporting renewable power. The nonprofit Center for Resource Solutions based in San Francisco developed the **Green-e** Renewable Electricity Certification Program to certify and label utility and third-party products (Center RS, 2006).

Qualifying Sources for Green-e

Eligible sources are new sources put into operation after 1997 from solar, wind, geothermal, new hydropower from non-impoundment generating capacity, fuel cells, biodiesel, and other biomass (except coated, painted, or treated wood and municipal wastes that have not first been converted to clean fuel).

Minimum Purchase for Green-e

For retail electricity, purchase must offset at least 25% of customer's electricity and be beyond the state's RPS. For block products, purchased blocks must have a minimum 100 kWh/mo.

In March 2006, Recreational Equipment, Inc. (REI), a national cooperative retailer of outdoor gear and apparel, announced that it will purchase 10 million kWh of green power, equivalent to 20% of the company's annual electricity usage. The purchase of wind, landfill gas, and solar-generated electricity will provide 100% of the power for seventeen of REI's eighty-two retail stores. REI has entered into renewable energy contracts with seven utilities and marketers in nine locales, with most of the products being **Green-e** certified (Green Power Network, 2006).

18.1.2.7 Community Choice and "Public Power"

In the United States, 43 million people in 2000 cities and towns are served by municipal-owned public utilities. The municipal utility offers significant advantages to serve the interests of its customers, which are also its owners. Municipal utilities in Seattle, Sacramento, Austin, Los Angeles, and other cities have developed some of the most innovative local energy programs in the country, including DSM and green marketing and pricing programs (see Section 18.2.6).

Utility restructuring and other state policy can provide opportunities for cities and towns without municipal utilities to play a more substantive role in community electricity. Most state deregulation plans that call for consumer choice of generators, also allow for "community aggregators" or buying groups that can make better deals for power. Municipalities can play that role and some states including Massachusetts, Ohio, and California have adopted legislation that allows the municipality to decide what the "default provider" of power will be for those customers who do not make a choice. This is the vast majority of customers. San Francisco and Chicago are two cities that are using this opportunity effectively (see Section 18.2.6.2).

18.1.3 State Renewable Fuels Standards

The 2005 and 2007 federal energy acts included a Renewable Fuels Standard (RFS) for biofuels (see Sections 14.3.1 and Sections 17.2.2.2). This provision was informed by state policies for RFS, most notably Minnesota and Iowa. Minnesota adopted a renewable fuels standard in 1991 that required all gasoline sold in the state to have 10% ethanol. In 2004, it extended this standard to overall 20% ethanol by 2013. In addition, the state and a coalition of private companies initiated an aggressive program in 1998 for advancing E85 fuel for use in flex-fuel vehicles. There are now more than 300 gas stations in the state selling E85, and a rapidly growing ethanol industry. E85 sales tripled from 2.6 million gallons (Mgal) in 2004 to 8.1 Mgal in 2005, then doubled to 17.9 Mgal in 2006. The state sees this effort as an economic development program for rural Minnesota (see Sidebar 14.1).

Figure 18.13 shows that by mid-2007 nine states have RFS and 25 others have some incentives for biofuels. These incentives include exemption from state excise taxes on biofuels and tax credits and grants promoting biofuel production.

figure 18.13 Renewable Fuels Standards and Other Biofuels Incentives, 2007

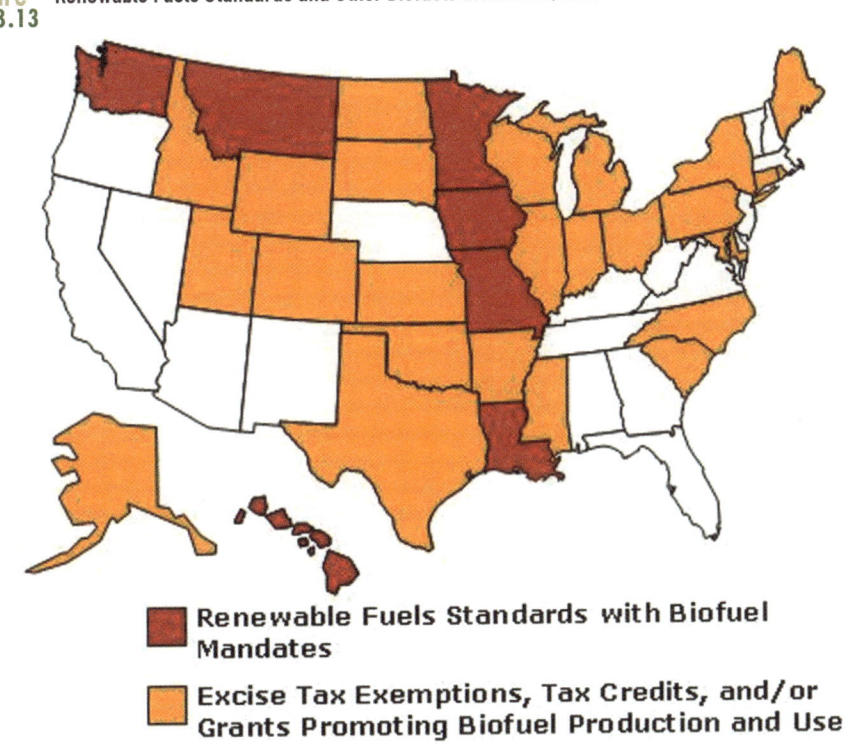

■ **Renewable Fuels Standards with Biofuel Mandates**

■ **Excise Tax Exemptions, Tax Credits, and/or Grants Promoting Biofuel Production and Use**

Most states provide some incentive for biofuels. RFS is the most aggressive approach.

Source: Pew CGCC, 2007

18.1.4 State Efficiency Standards

States have long taken authority under the constitutional police power to protect public health and safety for building codes and land use controls like zoning. Most states have delegated some of this authority to local governments, but if so, state agencies still have retained some oversight and direction. Led by California, states have also developed energy efficiency for appliances and equipment.

18.1.4.1 Building Energy Efficiency Standards

Section 8.2 discussed building energy codes and Figure 8.2 shows the 2007 status of state codes and compliance. All but nine states have a statewide building code. Under the 1992 EPAct, state codes must incorporate model energy standards equivalent to the International Energy Conservation Code (IECC). Many states have uniform building codes, which serve as both a minimum and a maximum standard to be implemented by local officials, but others allow for local codes to exceed the state minimum. Some states (e.g., California, Washington, Oregon) have developed their own codes that are equivalent to or exceed IECC standards. California's Title 24 standards are perhaps the most comprehensive in the nation and are summarized in Sidebar 8.2.

18.1.4.2 Appliance and Equipment Efficiency Standards

As shown in Figure 17.8, federal appliance and equipment standards have made a significant impact on efficiency. These national standards enacted by the 1986 Appliance Energy Conservation Act were prompted by California's successful appliance efficiency standards mandated by its 1974 Energy Resources Conservation Act. With 12% of the nation's population, California has a market big enough to set its own standards.

Although the federal standards continue to be upgraded, several states in addition to California continue to provide their own standards that go beyond the federal mandates. Figure 18.14 shows that ten far west and northeastern states have either followed California's requirements or adopted their own. The American Council for an Energy Efficient Economy (ACEEE) estimates that state and federal appliance efficiency standards in place will save 577 billion kWh or 6% of total energy by 2030 at a net cumulative savings of $250 billion to consumers (Nadel, et al., 2006; ASAP/ACEEE, 2008).

18.1.5 State Environmental Standards Affecting Energy Use

State regulations for the protection of air quality, water quality, and ecological resources and the remediation of energy related lands all affect the cost of fuels and competitiveness of clean energy sources and efficiency. States can assume responsibility for federal regulatory programs under the primary provisions of the major environmental laws. States also have their own

figure 18.14   **Appliance and Equipment Efficiency Standards**

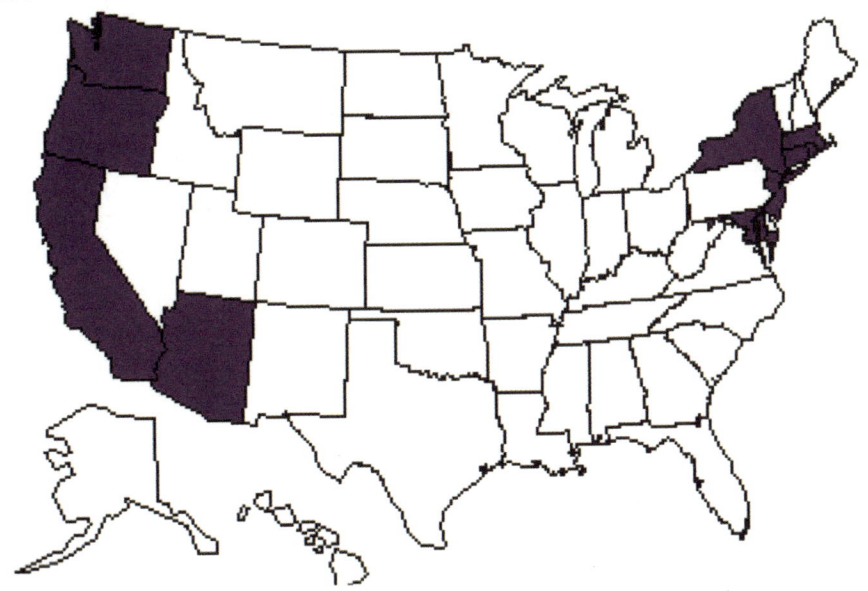

States that have adopted appliance efficiency standards that go beyond federal standards.

SOURCE: Pew CCC, 2006

laws, such as the California Environmental Quality Act (CEQA) and the Washington State Environmental Policy Act (SEPA), which require environmental assessment of local decisions. Both states are using the laws to scrutinize the energy and GHG emissions effects of local government plans and decisions.

18.1.5.1 State Implementation Plans (SIP) under the Clean Air Act

SIPs must show how clean air standards will be met in areas of nonattainment with ambient air quality standards by EPA's deadline. EPA set new SIP rules in 2004. SIPs are due in 2007–2008, and attainment deadlines are 2007–2014 depending on the area. The most important aspect of SIPs for energy is specified transportation control measures (TCM) that aim to reduce VMT to reduce emissions through transit improvements, shared ride programs, parking controls, bicycle/pedestrian measures, and land use controls. These TCM have some teeth, because transportation plans must conform to the SIP to qualify for federal transportation funding.

As we discussed in Chapter 15, the measures of land use and smart growth have the most enduring effect on emissions (and energy use), but they continue to fit awkwardly into TCM and SIPs because of their uncertain effects. But some SIPs are beginning to address land use, and the plans for Sacramento (CA) and Albany (NY) expect to achieve 15%

and 10%–12% VMT reduction respectively from land use measures. The Sacramento plan is described in Section 18.2.3.1.

18.1.5.2 Vehicle Emissions Controls and GHG

The Clean Air Act authorized California to develop its own auto emission standards because of its large market and air pollution problems. Other states have adopted California's standard. Under this authority, California passed the 2002 Pavley Act (AB 1493) calling for new emission standards that would achieve a 30% reduction in vehicle GHG emissions by 2016. Five western states (WA, OR, AZ, NM, CO) and ten eastern states (NY, PA, NJ, CT, RI, MA, VT, ME, FL, MD) have indicated they will adopt California's standard.

However, as discussed in Section 13.2.2, in December 2007, EPA denied Calfornia's request for waiver from federal preemption to implement the standard. The state sued EPA over its decision, and this case will be affected by the April 2007 U.S. Supreme Court opinion in *Massachusetts v. U.S. EPA*, supporting the case for federal vehicle CO_2 emission standards.

18.1.6 State Energy Production and Efficiency Incentives

Most states offer some type of tax or other incentives for energy production and efficiency, including tax credits, deductions, and exemptions; rebates; grants and loans; and production incentives. The state programs vary widely. Renewable energy and efficiency programs are monitored by the Database of State Incentives for Renewable Energy operated by the North Carolina Energy Center (DSIRE, 2007; http://www.dsireusa.org).

As of December 2007, there are 660 efficiency and 224 renewable energy rebate programs offered by states and utilities in thirty-nine states (Figure 18.16). There are more than 400 state and utility grant and loan programs for efficiency and renewables offered in 47 states. Individual and/or corporate tax incentives are available in thirteen states for investments in energy efficiency and in twenty-five states for renewable energy systems (Figure 18.15).

California has long relied on financial incentives along with its building and appliance regulations to implement its policies for greater efficiency and use of renewable energy. Its Emerging Renewables Program (ERP) and Self-Generation Incentive Program (SGIP) provide rebates of $2.50–$3.25/kW for renewable energy systems. These programs are part of the California Solar Initiative to grow the state's solar electric capacity to 3000 MW by 2017 with a ten-year budget of $3.3 billion (see Sidebar 18.3).

18.1.7 Energy Management in State Buildings, Fleets, and Procurement

State governments are major energy consumers in buildings, fleets, and procurement, and they not only need to manage their energy use to control costs, but their decisions can also

State Tax Credits for Renewable Energy Systems

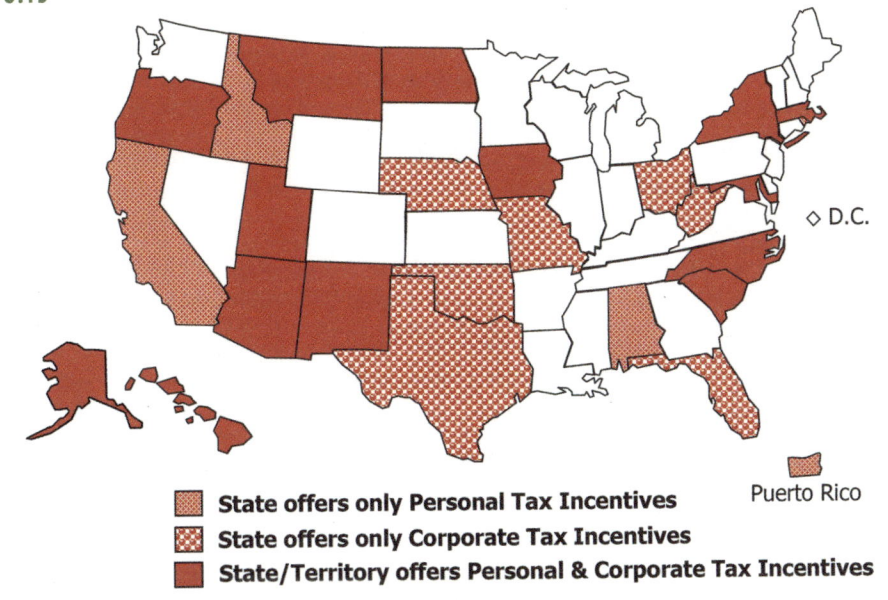

◇ D.C.

Puerto Rico

State offers only Personal Tax Incentives
State offers only Corporate Tax Incentives
State/Territory offers Personal & Corporate Tax Incentives

SOURCE: DSIRE, 2007

figure 18.16 State and Utility Rebates for Renewable Energy Systems

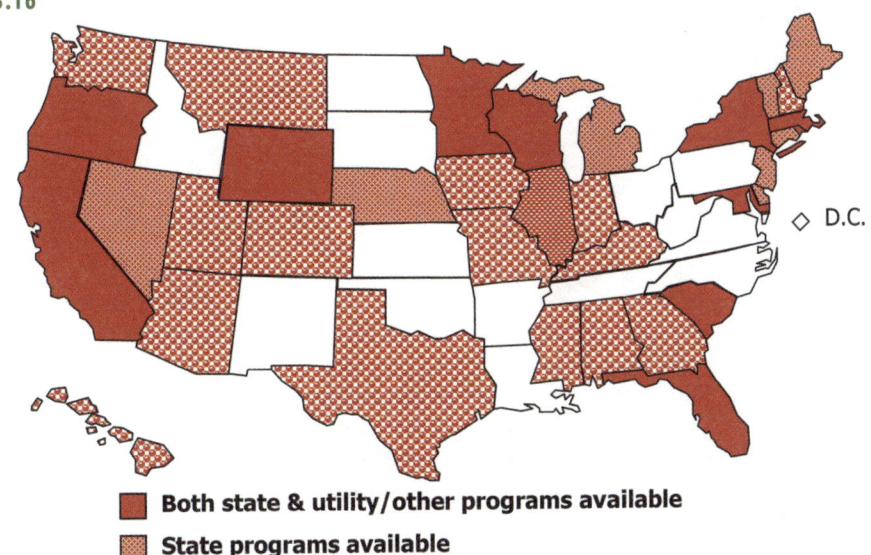

◇ D.C.

Both state & utility/other programs available
State programs available
Utility/other programs available

SOURCE: DSIRE, 2007

help stimulate sustainable energy development and serve as an example for corporate and individual consumers.

- **New Buildings:** Figure 18.17 gives the states that require or recommend that new state-funded buildings be certified green, either by LEED or Green Globes.
- **Energy Efficiency Targets:** Several states have set energy efficiency targets for state agencies and state buildings. This is a recommendation of the North Carolina Energy Plan.
- **State Vehicle Fleets:** The first market for alternative fuel vehicles is public fleets; states have mandated or encouraged flex-fuel or compressed natural gas vehicles.
- **Procurement:** State agencies and institutions are large consumers of fuel and power and some states use green pricing or green marketing programs to use state buying power to advance renewable electricity.

figure
18.17

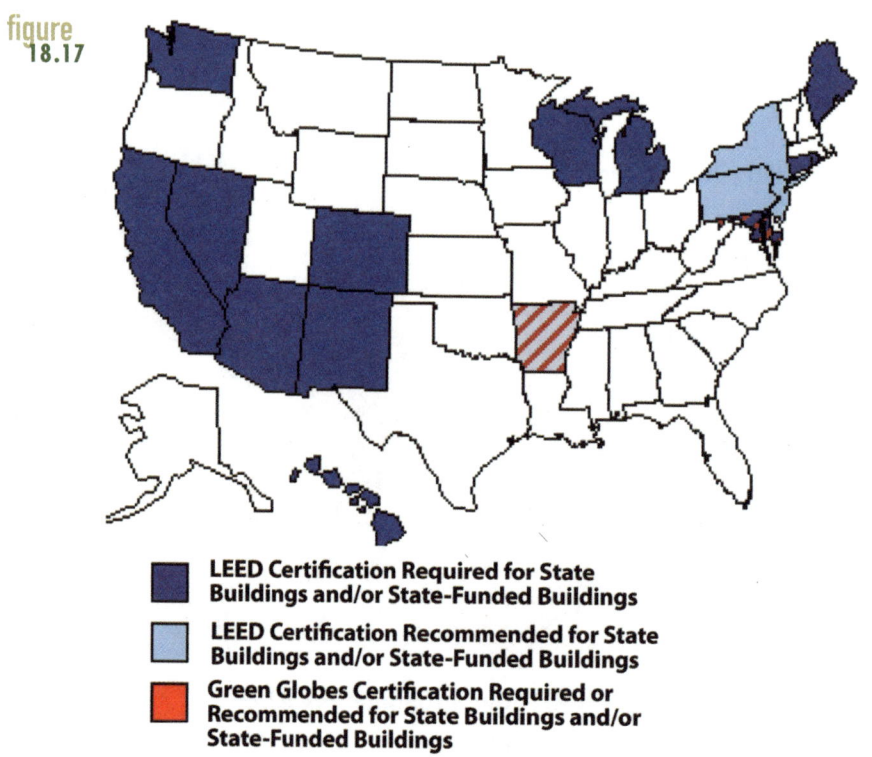

LEED Certification Required for State Buildings and/or State-Funded Buildings

LEED Certification Recommended for State Buildings and/or State-Funded Buildings

Green Globes Certification Required or Recommended for State Buildings and/or State-Funded Buildings

States requiring or recommending green building certification for new state-funded buildings.

SOURCE: Pew CCC, 2006

SIDEBAR 18.3

California: Exemplary State Policies for Sustainable Energy

California has been the most innovative of the states in developing new policies for sustainable energy. Of course, all of the state's energy policy experiments have not been successful, as witnessed by the 2000–2001 electricity crisis that resulted from its utility restructuring program. But California has led the way in several other energy areas, many of which were mentioned above and in previous chapters. Here are some highlights and emerging policies from the state's 2007 IEPR, which aims to integrate the plethora of state laws, executive orders, and administrative directives to reduce energy consumption and GHG emissions in the face of a growing population expected to exceed 44 million by 2020 and 60 million by 2050.

1. **Planning and Organizational Structure**

   California energy policies are implemented by the **California Energy Commission** (CEC) (for building and appliance standards and general policy) and the **California Public Utilities Commission** (CPUC) (for utility regulation). The two agencies work closely together. The state's energy planning uses extensive agency and stakeholder involvement in developing and updating its *Energy Action Plan* and the IERP.

2. **Energy Efficiency Policies**

   Figure 18.18 shows the cumulative effect of three programs on electricity savings:

   - **Building energy efficiency standards:** Title 24 is the strictest code in the nation and each update informs national and international model codes (see Figure 18.18[a] and Sidebar 8.2).
   - **Appliance energy efficiency standards:** The first appliance standards in the nation led to significant efficiency savings and the adoption of federal standards ten years later (see Figures 17.8, 18.18[a]).
   - **Utility DSM programs:** California's utility efficiency programs are among the most successful in the nation (see Figure 18.18[a]).

   As a result of these programs and other efforts, California has the lowest per capita electricity use in the nation, less than 60% of the U.S. average and less than half of that of Texas (see Figure 18.18[b]).

   The 2007 IEPR aims to go much further to improve electricity efficiency as a means to achieve the state's targets for GHG emission reduction. It continues to upgrade its building, appliance and lighting standards with a target to achieve zero net energy buildings by 2020 (residential) and 2030 (commercial). It continues to rely on IOUs and publicly owned utilities (POU) to provide energy efficiency services to achieve a statewide goal of 100% of economic potential. Figure 18.18(c) from the 2007 IEPR shows four possible savings targets for reductions of electricity consumption through 2016, ranging from 7% to 14% savings of California's already efficient baseline.

3. **Renewable Energy Policies**

   California has long supported development of renewable energy. It has by far the largest geothermal and photovoltaic (PV) power capacities in the nation, the largest wind capacity until taken over by Texas in 2005, and the second largest hydropower capacity. The development of solar and wind capacity resulted from favorable state policies, especially financial support. Its RPS of 20% by 2010 (accelerated from 2017) is one of the most

(continued)

figure
18.18

California's Past and Potential Future Electricity Savings

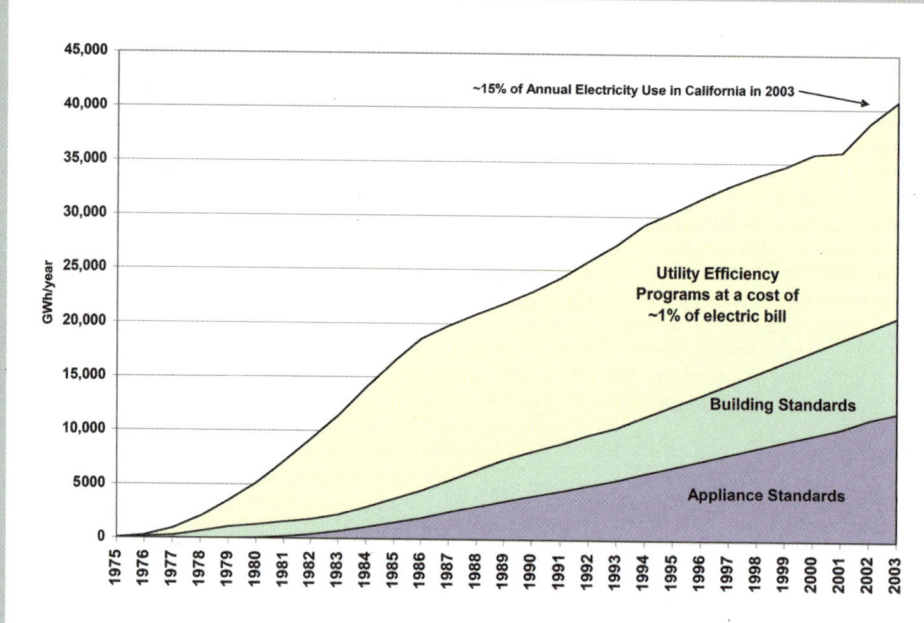

(a) Electricity savings from utility programs and building and appliance standards (15% of annual use).

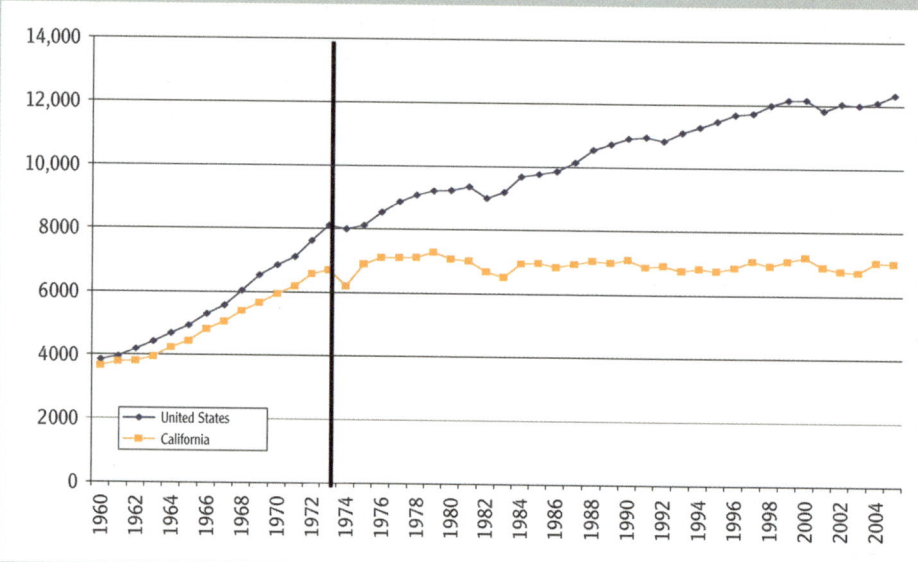

(b) California per capita electricity use is the lowest in the United States, less than 60% of the U.S. average.

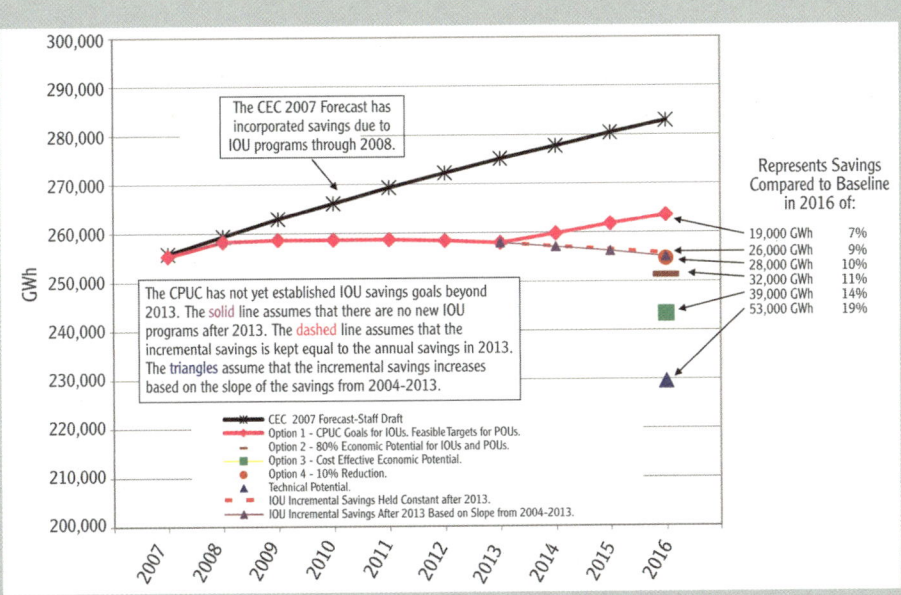

(c) Four possible savings targets of California electric utility demand-side efficiency programs.

SOURCE: CEC, 2005, 2007

aggressive in country, and the governor, CEC, and CPUC propose extending it to 33% by 2020, not including large hydro which is 19% of 2006 total capacity. Figure 18.19(b) shows these goals as well as the slow progress through 2007, although several large projects are under contract awaiting needed transmission capacity.

The CEC's 2007 IEPR assessed the feasibility of meeting the 33% goal by 2020 and found it to be feasible and desirable. However, achieving it will be challenging because of transmission constraints, project delays, a tightening market for wind and PV systems, management issues for intermittent power, and other constraints. A likely 33% portfolio of renewable electricity by 2020 (26 GW) includes 50% wind, 20% geothermal, 12% PV, 11% solar steam, and 7% biomass. Achieving these lev-

els requires innovative financial mechanisms and incentives. Utility contracts with renewable suppliers are based on a "market price referent" (MPR) baseload proxy plant. To enhance the financing of renewable energy systems CPUC is considering a "carbon-adder" to the MPR and also European-Union-like "feed-in tariffs" for RPS-eligible renewables less than 20 MW.

CPUC and CEC also provide direct rebates to renewable energy projects. The CPUC's SGIP and CEC's ERP are intended to "accelerate cost reduction and market acceptance through high volume production of renewable energy technologies." SGIP provides a $2.80/kWh rebate (reduced to $2.50/kWh in 2006 because of popularity) to larger projects and ERP gives a $2.60/kWh rebate to smaller projects. Although

SIDEBAR 18.3 (*CONTINUED*)

**figure
18.19**   California Renewable Energy, Transportation Fuels, and GHG Emissions

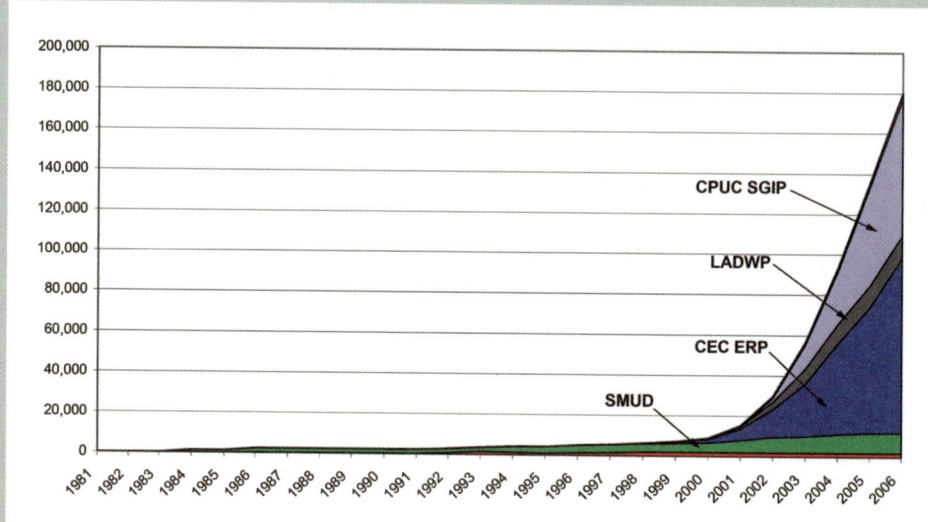

(a) Growth of residential photovoltaic systems due to CEC, CPUC, and municipal utility incentive programs.

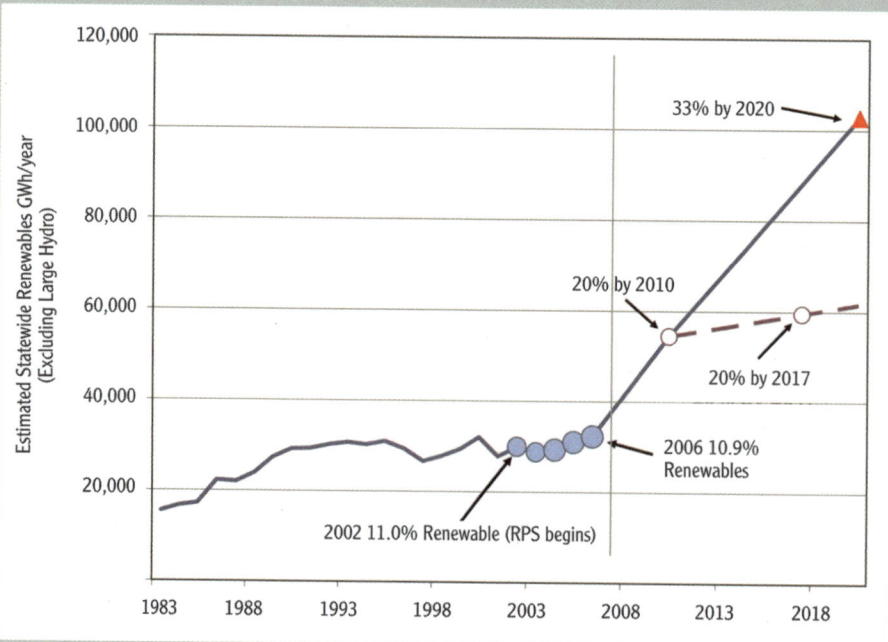

(b) California's RPS goals and progress to 2007.

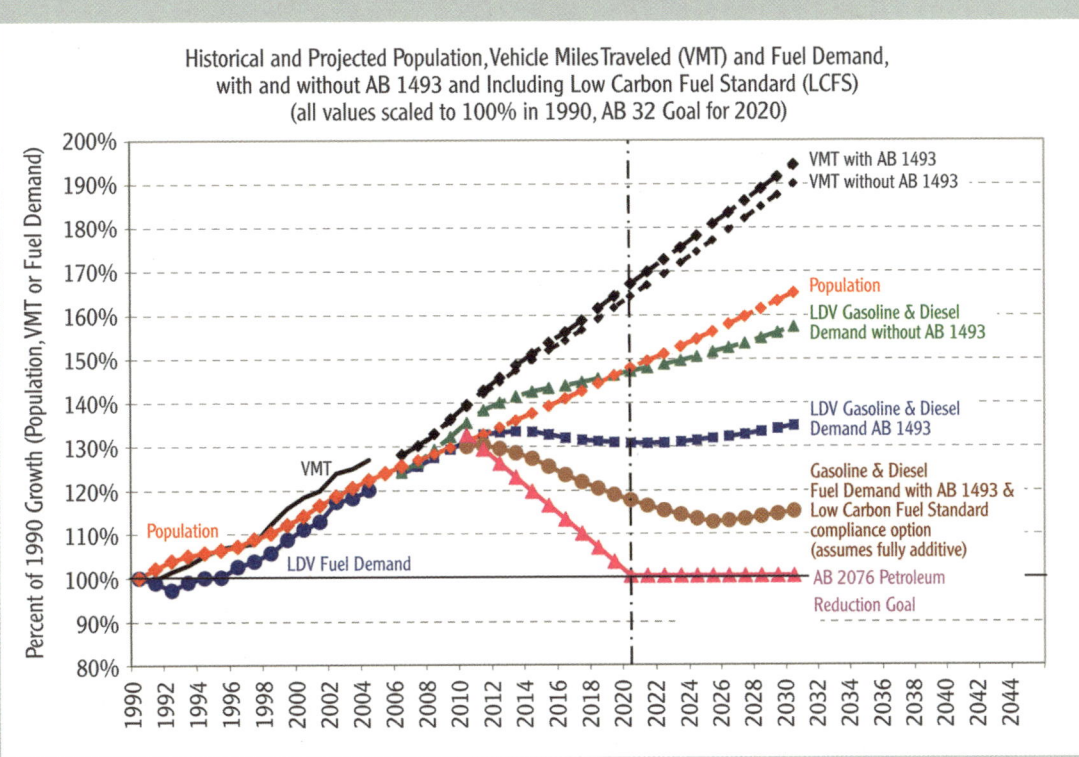

(c) Cumulative effect of three California transportation vehicle emissions and biofuels policies.

solar, wind, and other technologies qualify, most rebates have been used for solar PV. This enhances the **California Solar Initiative's** goal of adding 3000 MW of grid-connected PV by 2017 with a state subsidy of $3.3 billion. Figure 18.18(c) shows California PV capacity growth through October 2006 and the effect of SGIP and ERP, as well as municipal utility programs in Sacramento and Los Angeles.

4. **Transportation vehicle efficiency and biofuels** California is also moving aggressively to reduce transportation petroleum fuels and emissions through efficiency and renewable biofuels. The 2002 Pavley Act (AB 1493) requires reduction of vehicle GHG emissions by 30% by 2016, with phase-in beginning in 2009. As discussed in Sections 18.1.5.2 and 13.2.2, U.S. EPA denied California's request for waiver under the Clean Air Act for the Pavley standard, and the state, along with some of the 15 others poised to implement the standard, has sued EPA over its decision. AB 1007 (2005) establishes a low-carbon fuel standard (LCFS) which has been enhanced by the governor's executive order that sets a goal of 20% biofuels by 2010, 40% by 2020, and 75% by 2050. AB 2076 sets a petroleum consumption reduction goal of 1990 level of use by 2020.

The potential cumulative effect of these policies is illustrated in Figure 18.19(c) which shows

SIDEBAR 18.3 (*CONTINUED*)

historic and projected population, VMT, and light duty vehicle (LDV) fuel demand. It shows slightly elevated VMT from the assumed higher efficiency Pavley AB 1493 vehicles, but much lower fuel demand. Adding the LCFS drives demand down further but not as far as the AB 2076 reduction goal of 1990 level of consumption.

5. **GHG Emission Reduction Policies**

In addition to the Pavley Act's standard for a 30% reduction in vehicle GHG emissions by 2016, California enacted AB 32 in 2006, requiring mandatory controls to achieve 1990 levels of overall state GHG emissions by 2020. The act was passed after careful study and the support

of sixty-four prominent economists, including five Nobel Prize winners, who argued "The most expensive thing we can do is nothing!" Given the state's rate of population growth (an increase of 7 million or 19% by 2020 to 44 million) and expected economic expansion, this is a challenging goal. As shown in Figure 18.19(d), the transportation fuel policies and utility efficiency and renewable energy policies discussed above and other known actions like forest conservation are likely to achieve almost 75 percent of the needed reduction, but additional actions will be needed for the remainder.

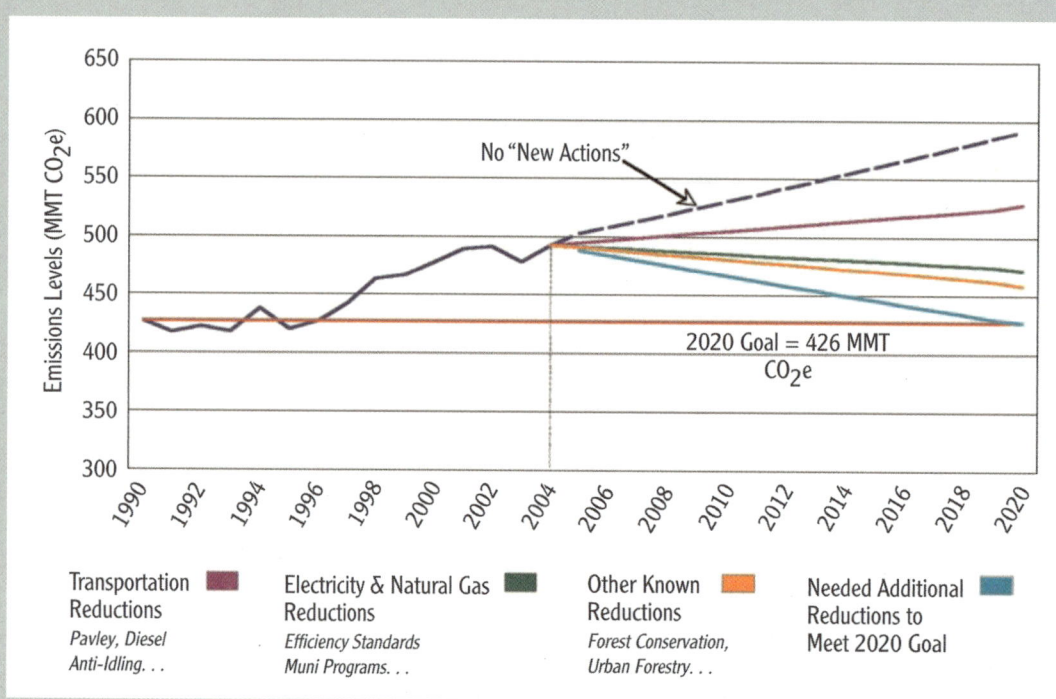

(d) GHG emission reduction goals and potential impact of existing policies.

SOURCE: CEC, 2007

18.2 Community Energy Planning and Policy

Energy planning and policy at the local level are important because it is in our communities where we consume energy in buildings and transportation. Localities have limited authority over some aspects of energy and we need broad energy policies at the federal and state level to address those, including efficiency standards for appliances and vehicles, regional air quality standards, federal and state tax incentives, energy technology research and development, and rules for utility operations, among others.

But we also need effective community planning to achieve more sustainable energy. It is local government that controls patterns of land use, permits subdivision and building construction, develops public transit, builds local roads and other infrastructure, and creates community plans for housing, land use, development, and environment. Some 2000 cities and towns own their own electric utilities, and these communities have an added stake and opportunity in community energy.

18.2.1 History

Local governments became active in their attention to energy in the late 1970s and early 1980s when energy markets were in crisis and energy was a high priority. When prices dropped, energy became less of a priority and other local issues, such as land use and urban sprawl, brownfields and redevelopment, transportation and traffic congestion, information technology infrastructure and economic development, and affordable housing, drew the attention of citizens, elected officials, and planners.

However, the early experience of some communities like Portland (OR), Seattle (WA), and Davis (CA) showed that local communities could make a difference in energy use and that their traditional authority over land use, building, transit and transportation, and in some cases utilities were means to achieve more efficient and sustainable energy use. Portland was exemplary in energy plan-making, UGB development, and light rail transit enabled by the UGB. Davis was exemplary in building codes and its Village Homes served as an early model for sustainable subdivision development. Seattle was exemplary in the use of the municipal utility to provide a range of demand-side energy services to customers (Randolph, 1981, 1988).

When energy prices dropped by the mid-1980s, energy faded as a high priority for communities relative to other issues. But after the energy price increases of the early 1990s and emerging public and consumer demands for greater livability and sustainability, several common themes began to converge in community planning that provided a foundation for addressing energy issues more substantively at the local level. We have discussed several of these themes in previous chapters, and we are seeing them increasingly manifested at the community level throughout the United States. They include the following:

- Smart growth and land use controls of urban sprawl; compact, mixed-use, and pedestrian- and transit-oriented development

- Light rail transit and other VMT reduction programs to relieve traffic congestion and improve air quality
- Green building programs and building energy codes
- Public power, community choice, and distributed resources (both efficiency and generation)
- Community education and action for GHG emission reduction, energy efficiency, and environmental protection

These community initiatives have increased exponentially in the past few years, and several of them are discussed below. There are some online database networks of community energy and climate protection activities that aim to share experiences among cities. Two good ones are the Sustainlane Government site of best practices (http://www.sustainlane.us/) and the Institute for Local Self Reliance's Democratic Energy site (http://www.newrules.org/de/). Check them out to monitor new developments in community energy.

18.2.2 Community Energy Plans and Plan-Making

Concerns over energy costs, security, and sustainability issues have caused many communities to revisit how they address energy. Developing an energy plan is an obvious first step, and several cities have drafted stand-alone energy plans. These planning exercises provide a focused effort to galvanize both local government and community action. Such planning efforts must also engage electric utilities and other energy providers. But localities can also integrate energy concerns into their core planning elements addressing land use, buildings, housing, utilities, transit, infrastructure, and environmental sustainability. Energy needs to be addressed in community comprehensive and general plans to be part of the community's vision for the future. It should be a focus of public discourse and participation. Energy improvements can be implemented through local government's regulations, incentives, and assistance policies, including land use and building regulations, capital investment plans, and partnerships with the private sector and community groups.

Communities have organized energy planning efforts in different ways. They have used city energy offices, nonprofit community groups, or municipal utilities as lead organizations. Others have integrated various aspects of energy into mission-specific offices, including land use, building, transportation, or public works agencies. However community energy planning is organized, the most effective programs have a committed and educated public constituency, a champion elected official, and a dedicated and knowledgeable staff. The sections that follow describe several core community planning areas that involve energy and provide some exemplary case examples.

Plan-making is a central part of process and focuses on a future vision of the community represented in future land use, physical development of infrastructure (including utilities, roads, and transit), and increasingly development design. This vision is usually provided in the comprehensive or general plan resulting from a **community participation process,** often involving scenario exercises, community workshops, and planning/design charrettes. About half of the states require local jurisdictions to develop community-wide comprehensive plans and update them, but most cities in other states prepare such plans without a state mandate. Energy can be addressed as a chapter of this comprehensive planning, in a separate plan, or both. Any community plan must address energy's relationship to land use, transportation, building codes, housing, utilities, environment, and economic development.

The best way to understand plan-making and the planning process is to look at case examples. We briefly describe just two below, from New York and San Francisco, but we encourage readers to review others to gain more insight. Among other better known stand-alone community energy plans are the Portland (OR) Energy Policy (1990), the Minneapolis Energy Plan (1996), the Chicago Energy Plan (2001), the New York Energy Policy: An Electricity Policy Roadmap (2004), and the growing number of community climate action plans.

San Francisco's Electricity Resource Plan (2002) and Climate Action Plan (2004).

After the California electricity crisis in 2000–2001, the city of San Francisco recognized its vulnerability to supply constraints and reliability problems caused by transmission constraints posed by its location at the end of the peninsula. In addition, air emissions from in-city power plants (Hunters Point and Potrero) were creating an environmental justice problem: nearby low-income residents were suffering high incidence of respiratory ailments. With the help of Rocky Mountain Institute and an extensive public involvement process, the city developed an electricity resource plan that allowed closure or reduced operation of the plants and a commitment to local distributed resources, including efficiency, load management, and solar, cogeneration, and other distributed generation (Figure 18.20).

Main elements of the San Francisco Electricity Resource Plan:

1. *Demand Reduction through energy efficiency and load management.* This is generally a cost-effective means of reducing electricity load. The objectives are **16 MW by 2004, 55 MW by 2008, and 107 MW by 2012.**

2. *Renewables.* Programs to harness the sun, wind, water, and other natural resources will be a high priority. The objectives for renewables are **7 MW by 2004, 28 MW by 2008, and 50 MW by 2012.**

3. *Medium-sized Generation and Cogeneration.* Mid-size plants of about 50 megawatts can provide high levels of reliability and could be built in several locations in San Francisco. This Plan assumes the capacity needed to shut down Hunters Point and Potrero Unit 3 are **150 MW by 2004, and 250 MW by 2008.**

4. *Small-scale Distributed Generation (DG).* These include fuel cells, packaged cogeneration, and micro-turbines. DG generators ranging from 10 kilowatt to 5 megawatts in size usually support single facilities. The objectives are **10 MW by 2004, 38 MW by 2008,** and **72 MW by 2012.**

5. *Transmission.* An upgrade to an existing line and a new transmission line scheduled to be built on the peninsula to service San Francisco will be necessary for long-term reliability, and should be supported by the city. At the same time the city should commit to securing a continually increasing percentage of renewable sources to feed into the transmission grid.

In 2004, the City, in working with the International Council for Local Environment Initiatives (ICLEI) cities for Climate Protection (CCP), issued a Climate Action Plan to reduce GHG emissions. In 2002, the San Francisco Board of Supervisors passed the GHG Emissions Reduction Resolution committing the City and County of San Francisco to a GHG reduction goal of 20% below 1990 levels by 2012. This target is 2.6 million tons below 2000 levels. The plan promotes the following actions to achieve this reduction:

Strategies	GHG Reduction
Transportation actions: e.g., fleet fuel efficiency, increased transit use, commuter trip reduction	963,000 tons/yr
Energy efficiency in buildings: e.g., incentives, direct retrofit, technical assistance, codes	801,000 tons/yr
Renewable energy: e.g., solar, wind, biomass projects; green power purchasing	548,000 tons/yr
Solid waste: e.g., recycling, composting	302,000 tons/yr
Total	**2,614,000 tons/yr**

The Plan states that achieving these goals requires accelerating and expanding existing programs in all four areas, developing infrastructure and resources to implement actions, and setting up mechanisms and indicators to monitor and measure progress.

PlaNYC 2030 (2007). New York City's 2007 sustainable future plan addresses ten key areas: housing, open space, brownfields, water quality, water network, transportation congestion and infrastructure, air quality, energy, and climate change. Among 14 energy initiatives are strengthening energy codes, adding 800 MW of clean distributed energy and establishing a NYC Energy Planning Board and an energy efficiency authority.

PlaNYC states that climate change is "one challenge that eclipses them all." Figure 18.21 shows the plan's goal of reducing GHG emissions by 30% (nearly 20 MMt CO_2-e) below 2005 levels (and 50% below a BAU scenario) by 2030 through avoiding sprawl by accommodating 900,000 more people in NYC (32%), clean power (22%), efficiency buildings (34%), and sustainable transportation (12%; New York City, 2007).

figure 18.20 Projected Resource Mix under San Francisco Energy Plan

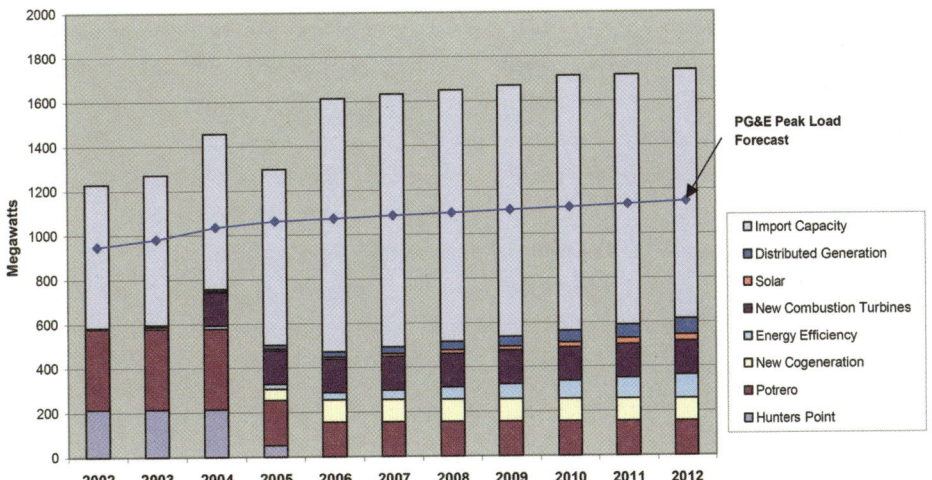

Includes retirement of power plant units and distributed resources.

SOURCE: City of San Francisco, 2002

figure 18.21 PlaNYC 2030 Goals and Strategies to reduce GHG emissions by 30% below 2005 levels

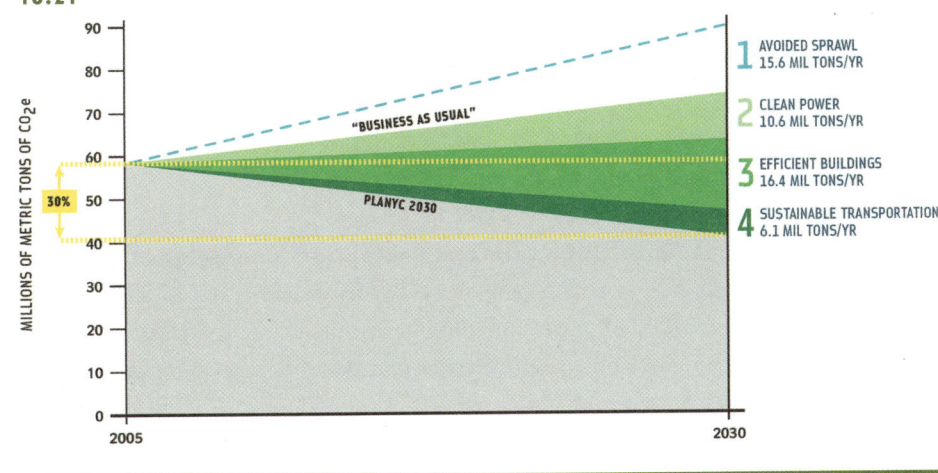

SOURCE: PlaNYC 2030 (New York City, 2007)

18.2.2.1 Community Action Initiatives

Plan-making is necessary to set goals, but plans need to be implemented with action programs to achieve goals and engage citizens. Energy-related community projects can also attract federal and state funding. For example, **Clean Cities Coalitions** have developed in eighty communities to take advantage of U.S. DOE funding for alternative fuels in local fleets. **Local weatherization** programs take advantage of federal and state funding to improve efficiency, reduce heating bills, and provide "affordable comfort" to low-income residents.

A number of cities have initiated programs to promote community sustainability. One of the first was **Sustainable Seattle,** which is a nonprofit organization established in 1991 to develop awareness, monitor progress, and foster action among diverse constituencies to enhance environmental, economic, and social sustainability in the Seattle-King County metropolitan area. **Portland's (OR) Office of Sustainable Development** is a city government office with divisions of Energy, Solid Waste and Recycling, Green Building, and Sustainable Technologies and Practices. The office monitors city progress on a number of sustainability indicators, including air quality, CO_2 emissions per capita, vehicle miles traveled per capita, central city jobs, and housing, among others.

Some other initiatives show the diversity of efforts emerging from communities in the United States:

- **Arlington County (VA)** is a model for use of alternative fuels, especially biodiesel. In response to citizen complaints about school bus emissions, the county converted all diesel vehicle and equipment to B20 starting in 2001. It uses more than 600,000 gallons of B20 (120,000 gallons of B100) per year, with no operating problems and no more citizen complaints.
- In 2005, the **City of Boulder (CO)** challenged the community to increase its renewable energy purchases through green pricing. In just two months 1150 customers signed up, far exceeding the goal of 500. Including pre-existing participants, Boulder's nearly 7000 green power customers make up 16% of total utility customers. As a result, U.S. EPA named Boulder a Green Power Community (see Boulder's Climate Action Plan on p. 741).

Cool Cities Initiative and ICLEI Climate Protection Campaign. Communities have been spurred to action by two large campaigns. We discussed the ICLEI CCP campaign in Chapter 15. More than 300 U.S. cities and 600 international cities participate in the campaign, which requires an emissions data inventory and development of a plan and action strategies (see Section 15.5.2).

The Cool Cities Initiative, or the U.S. Mayors Climate Protection Agreement, was launched in 2005 by Seattle Mayor Greg Nickels, based on the success of Sustainable Seattle, Seattle City Light energy savings (see Section 18.2.5), and other city programs. Under the agreement, participating cities commit to take the following three actions:

- Strive to meet or beat the Kyoto Protocol targets in their own communities, through actions ranging from anti-sprawl land-use policies to urban forest restoration projects to public information campaigns.

- Urge their state governments, and the federal government, to enact policies and programs to meet or beat the greenhouse gas emission reduction target suggested for the United States in the Kyoto Protocol—7% reduction from 1990 levels by 2012.
- Urge the U.S. Congress to pass the bipartisan greenhouse gas reduction legislation, which would establish a national emission trading system.

By early 2006, 200 cities had adopted the agreement, and by December 2007, the number grew to more than 775 U.S. cities from all of the states. The program has raised awareness in many cities and spawned greater attention to energy planning.

Puget Sound Roadmap for Climate Protection. Mayor Nickels's initiative was not based on wishful thinking but on sound experience and good planning. In addition to Seattle's own experience, the Puget Sound Roadmap for Climate Protection released in December 2004, gave strong evidence that carbon reduction was not only achievable but could have significant economic as well as environmental benefits.

The Roadmap identifies key actions to achieve 1990 levels of GHG emissions by 2020 and analyzed their economic effects. Figure 18.22(a) projects baseline emissions and savings to 2020 from actions directed at buildings and facilities (B&F) and electricity, transportation, and agriculture, forestry and solid waste (AFSW). Figure 18.22(b) shows the savings contribution of various actions, and Table 18.5 details the savings and net costs, including net present value (NPV). Most of the actions have negative costs, so the total net present value of the actions is an economic **benefit** of $1.4–$2.1 billion.

Boulder Climate Action Plan and the Climate Acton Plan Tax. Boulder, Colorado, was an early signatory to the Cool Cities agreement. In 2002, Boulder City Council passed Resolution 906, the so-called Kyoto Resolution, setting a goal of reducing GHG emissions to 7% below 1990 levels by 2012. The City contracted out a GHG emissions inventory that was completed in 2004, and completed the Climate Acton Plan in 2006 (City of Boulder, 2006). The plan included six strategies:

- Increase residential, commercial, and industrial energy efficiency
- Increase community-wide purchases and installations of renewable energy
- Reduce VMT, purchase more efficient vehicles, and switch to low carbon fuel
- Utilize external funding including Xcel Energy utility rebates and federal tax incentives
- Provide community education and outreach to increase awareness of energy, climate, and the emissions reduction goal and to reduce barriers to voluntary action
- Provide financial resources to implement the above strategies

The City realized that significant resources would be needed to implement the plan. Boulder has never been shy about using local taxes for public benefit. In 1967, it increased its sales tax to generate revenue to purchase public open space and greenways. In 2004, to

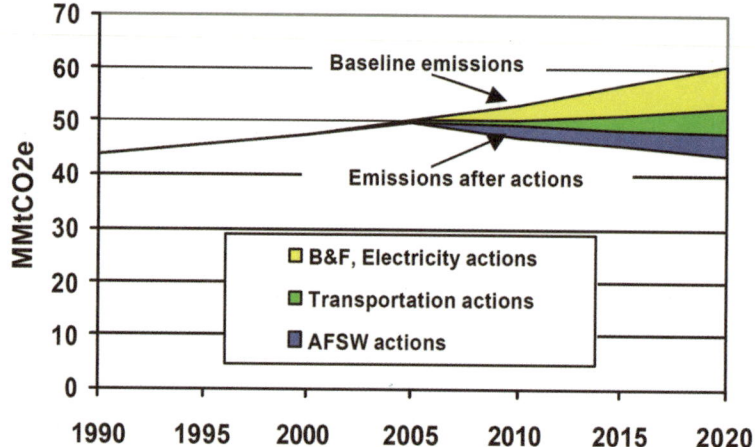

figure 18.22 (a) Puget Sound Roadmap (a) CO₂ Emissions Baseline and (b) Savings from Various Actions

(a) Baseline 2050 Development Scenario

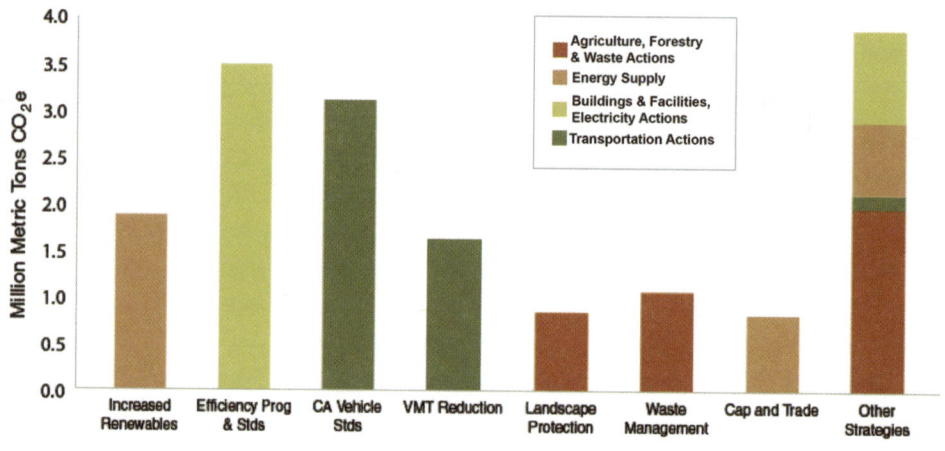

(b) Preferred 2050 Blueprint Scenario

Source: PSCAA, 2004

fund the GHG emissions inventory and planning, it enacted a two-year increase in the Trash Tax, which generated $258,000 per year for energy and GHG programs. The Climate Action Plan estimated the annual budget to implement the plan would range from $860,000 in 2007 to $1.98 million in 2012, the year when significant funding for renewable energy purchases would be necessary. The City's Climate Action Plan Committee (CAPC) studied

table
18.5   **Puget Sound Roadmap for Reducing GHG Emissions, Key Actions and Effects**

Key Actions	GHG Savings		Net Costs (million 2002$)		
	2010	2020	2010	2020	NPV 2005–2020
Develop and adopt a climate change policy framework					
Emissions trading (cap-and-trade)	0.2	0.8	16.6	4.1	18.0
Maximize energy efficiency					
Full, sustained efficiency programs, building codes, and appliance standards	1.4	3.5	($55)	($137)	($707)
Other strategies to increase efficiency improve design, reduce emissions	0.7	1.8	$17 to ($5)	$32 to ($11)	$204 to ($66)
Increase renewable energy in the region's power mix.					
Renewables portfolio standard	0.6	1.9	$16 to ($1)	$33 to ($33)	$171 to ($72)
Increase the GHG emissions of new vehicles sold					
Adopt California standards: LEV II & Pavley	0.2	3.1	($10)	($439)	($1171)
Other transportation strategies	0.1	0.1		—not estimated—	
Reduce motor vehicle miles traveled					
Location-efficient plans, transit, and demand-side measures	0.5	1.6		—not estimated—	
Protect natural landscapes and forest biomass					
Landscape protection	0.8	0.8	$6 to ($6)	$6 to ($6)	$59 to ($59)
Other AFSW Strategies	0.9	2.0	$0.1	$0.1	$1
Increase recycling and compost rates; reduce waste					
Recycling and waste reduction	0.6	1.0		—not estimated—	
Total	**6.0**	**16.6**	**($9) to ($60)**	**($501) to ($621)**	**($1425) to ($2056)**

Source: PSCAA, 2004

several options to generate revenue to support this budget, and by April 2006, it had narrowed the options to three: (1) a Climate Action Plan Tax on electricity collected on Xcel Energy power bills, (2) a building square footage fee collected on city water bills, and (3) a permanent increase in the Trash Tax. The latter two options could be adopted by the City Council, but the energy tax would require voter approval. The Committee recommended the energy tax (especially after Xcel agreed to collect the tax on their electricity bills), and in the November 2006 election, the citizens of Boulder voted to adopt Initiative 202, the Climate Action Plan Tax, the first local energy tax in the country.

Table 18.6 shows several characteristics of the three building sectors, including contributions to emissions, percent of emission reductions in the plan, percentage of public and private investment needed to achieve reductions, the initial energy tax, and maximum tax rates. The tax will generate about $1 million per year from 2007 to 2012 when it is scheduled to expire.

table 18.6   **Boulder (CO) Building Sector Emissions, Reductions, Investments, and Energy Tax Rates**

	Residential	Commercial	Industrial
Percent of Emissions (2005)	27%	53%	19%
Percent of Total Reductions	31%	41%	22%
Percent of Total Private Investment	19%	76%	5%
Percent of Total Public Investment	58%	39%	3%
2007 Energy Tax Rate	$0.0022/kWh	$0.0004/kWh	$0.0002/kWh
Maximum Energy Tax Rate	$0.0049/kWh	$0.0009/kWh	$0.0003/kWh

SOURCE: Boulder, 2006a, 2007

18.2.5 Building Energy Codes and Green Buildings

Most states adopt **statewide buildings energy codes** that comply with the IECC model code (Figure 8.2). However, it is local government that implements the code by inspecting compliance and issuing building and occupancy permits. The value of the regulations is only as good as the compliance. And many local governments are allowed to go beyond the standard code to require more stringent energy efficiency measures. Sometimes this experience effects changes in the state code.

Such was the case in Davis (CA) in the early 1980s. As a result of builders' experience in Davis (principally Michael Corbett's Village Homes subdivision), Davis developed one the country's most stringent building energy codes. Subsequently, California based its Title 24 standards on Davis's building code.

Green building programs go beyond standard building codes by using scoring systems to provide consumers with choice in building energy efficiency. Most communities' green buildings programs are voluntary and rely on consumer education and choice. For example, Austin Energy, the municipal utility in Austin (TX), operates the city's award-winning green buildings program "Green by Design." It implements its own five-star program to rate new construction, has training and information programs for both builders and consumers, manages a database of certified green builders, and prepares and posts case studies of successful projects on its web site. Check them out at http://www.austinenergy.com/, and see other Austin Energy programs in Section 18.2.6.1.

Some communities have incorporated minimum "green building" scores into their regulations. Boulder's (CO) building energy code is based on the IECC standards, but also requires new construction and remodels to add "green building" points based on energy measures, use of recycled materials, land use and water conservation, and framing techniques (Boulder, 2008). Table 18.7 gives the most recent version of "Green Points" requirements for new residential construction that became effective February 2008. Remodels and additions have similar requirements, including Green Points ranging from 10 to 45 points

table 18.7   **Boulder Green Points Program**

Mandatory Green Building Requirements: New dwelling units
1. Energy Efficiency and HERS Index Rating (by certified rater)

Type of Project	Square Footage	Required HERS Index	Energy Efficiency Above Code
Single family	up to 3000	70	30%
	3001–5000	60	50%
	5000 and up	35	75%
Multi-unit	All	70	30%

2. Construction Waste Recycling (50%) and Demolition Management (65% diverted from landfill)
3. Minimum Green Points Requirements

Project Description	Square Footage	Green Point Requirement	Green Points:
New Dwelling Single Family	1501–3000	20	• site development (up to 22 points) • building rehabilitation (up to 10 points) • waste management (up to 11 points) • energy efficiency (up to 71 points) • solar (up to 39 points) • water efficiency (up to 6 points)
	3001–5000	40	
	5001 and up	60	
New Dwelling Multi-Family	1001–2000	10	• material efficiency (up to 26 points) • sustainable products (up to 16 points) • indoor air quality (up to 19 points) • design innovation (up to 13 points)
	2001–3000	20	
	3001 and up	30	

SOURCE: Boulder, 2008

depending on size, in addition to mandatory energy audits, lighting efficiency upgrades, and a high efficiency model if the furnace is replaced. The table shows that new units must achieve (1) a Home Energy Rating System (HERS—see Section 8.5.1) index score that gets more stringent with larger unit size; (2) construction waste recycling of 50%; and (3) a minimum number of Green Points, with more points needed for larger unit size. Green Points are acquired by including measures from a long list under ten categories. Categories with the most potential points are energy efficiency, solar thermal and electricity, material efficiency, and site development. See www.BoulderGreenPoints.com for point allocation to specific measures (Boulder (2008).

18.2.3 Land Use Planning and Smart Growth Management

Land use is accepted as the domain of the local planner, and it has a significant effect on transportation patterns, vehicle miles traveled, and energy use. The trend toward urban sprawl has exacerbated auto dependence, VMT, transportation energy consumption, traffic congestion, and livability. Many communities, some with the assistance or prodding of state directives, have engaged in more aggressive growth management planning and policies to direct growth to areas of existing infrastructure and redevelopment. These efforts seem to be playing well in the market and consumer preference is growing for compact, community-oriented, mixed-use, and pedestrian- and transit-oriented development.

As discussed in Chapter 15, Smart Growth management aims to contain development within growth boundaries, increasing urban density, and enabling transit options. Infill and redevelopment in downtown and inner suburban areas have brought more residents to the urban core, avoiding the long commutes, congestion, and energy consumption of suburban life. New energy-efficient development designs foster mixed use, compactness, pedestrian- and transit-orientation, and enhanced community space. ULI's *Growing Cooler* estimates that a commitment to smart growth from now to 2050 could reduce VMT and related energy and GHG emissions by 7%–10% (Ewing, et al., 2007).

The most effective means of achieving smart growth include UGBs to contain growth, effective transit, and transit-oriented development (TOD) to create walkable, mixed-use, and attractive neighborhoods around transit stations. UGBs create densities that support transit, transit makes TODs possible, and TODs increase transit ridership making it more effective.

This pattern of development is part of the **Whole Community Energy** approach discussed in Chapter 15, and it is finding its way into green development guidelines, like LEED-ND guidelines developed for neighborhood development. The scoring system for development design includes location efficiency; compact, complete, and connected neighborhoods; transit orientation; and other factors (see Table 15.9).

These principles are beginning to be incorporated into local comprehensive and land use plans. While zoning ordinances that explicitly prescribe these patterns of development are rare, planned or negotiated developments such as Planned Unit Developments (PUDs) and rezoning negotiations (that are commonly required for nearly all new developments), are increasingly incorporating these principles for reasons of market attractiveness, politics, and/or administrative approval. But a project-by-project approach is inefficient and ineffective for smart growth, and even local plans and ordinances are not comprehensive enough to address regional growth patterns that lead to sprawl.

18.2.3.1 Regional Land Use Plans

Regional planning in the United States has been fraught with complications from competing jurisdictions and lack of regional authority, but there are a growing number of exemplary regional planning efforts that have tried to confront these constraints. The Twin Cities (MN) Metropolitan Council, the Portland (OR) Metropolitan Commission, some regional transit authorities, and other regional agencies have exhibited the authority and leadership to help guide regional development.

Portland especially demonstrates the benefits of a regional approach to land use. It employed smart growth principles before the term was coined. Following Oregon's 1973 Land Conservation and Development Act requirements, the Portland region established urban growth boundaries (UGB), encouraged development within the boundary, and restricted development outside. As a result, during the past thirty years, Portland has developed the density to support light rail transit and has become one of the most livable cities in the country (see Figures 15.12 and 15.13 on Portland's UGBs and TODs).

Other regional bodies have engaged in regional land use planning initiatives. Envision Utah (Salt Lake City Region) and Envision Central Texas (Austin Region) were two planning projects conducted by regional groups with the help of Calthorpe and Associates. They employed an extensive public involvement process to assess future growth scenarios and develop a preferred scenario for the region. The Envision Utah process led to a regional vision with far less sprawl and far more compact and transit-oriented development.

Sacramento (CA) Region Blueprint 2050 Transportation and Land Use Plan. Completed in 2004, this plan offers another good example of a regional planning approach. The Sacramento Council of Governments (SACOG) engaged a three-year planning process with thirty-seven public workshops and 5000 participants to develop different land use scenarios for the year 2050. It used a 1300-person telephone poll to help zero in on a preferred "blueprint scenario." Figure 18.23 shows the baseline scenario and the preferred blueprint. Table 18.7 gives various metrics for comparison. The preferred plan is far more compact with mixed-use "places" that serve as community centers. Although the preferred plan includes extension of commuter rail and light rail service, most of the VMT savings and air emission reductions come from shorter travel distances due to more compact and mixed-use development.

table 18.8 Sacramento Region Blueprint 2050

	Existing	Base Case Scenario	Preferred Blueprint Scenario
Housing—% rural residential	5%	5%	3%
Housing—% large-lot single family	63%	68%	45%
Housing—% small-lot single family	3%	2%	17%
Housing—% attached	29%	25%	35%
VMT, daily per capita	41.9	47.2	34.9
Transit—within 15 minute walk of jobs	5%	—	41%
Transit—within 15 minute walk of housing	2%	—	38%
Trips—% auto	92%	93.7%	83.9%
Trips—% transit	1.1%	0.8%	3.3%
Trips—% walk/bike	8.9%	5.5%	12.9%
CO_2 emissions per capita from vehicles	—	100%	85%

figure 18.23   Sacramento Region Blueprint 2050 Transportation and Land Use Scenarios

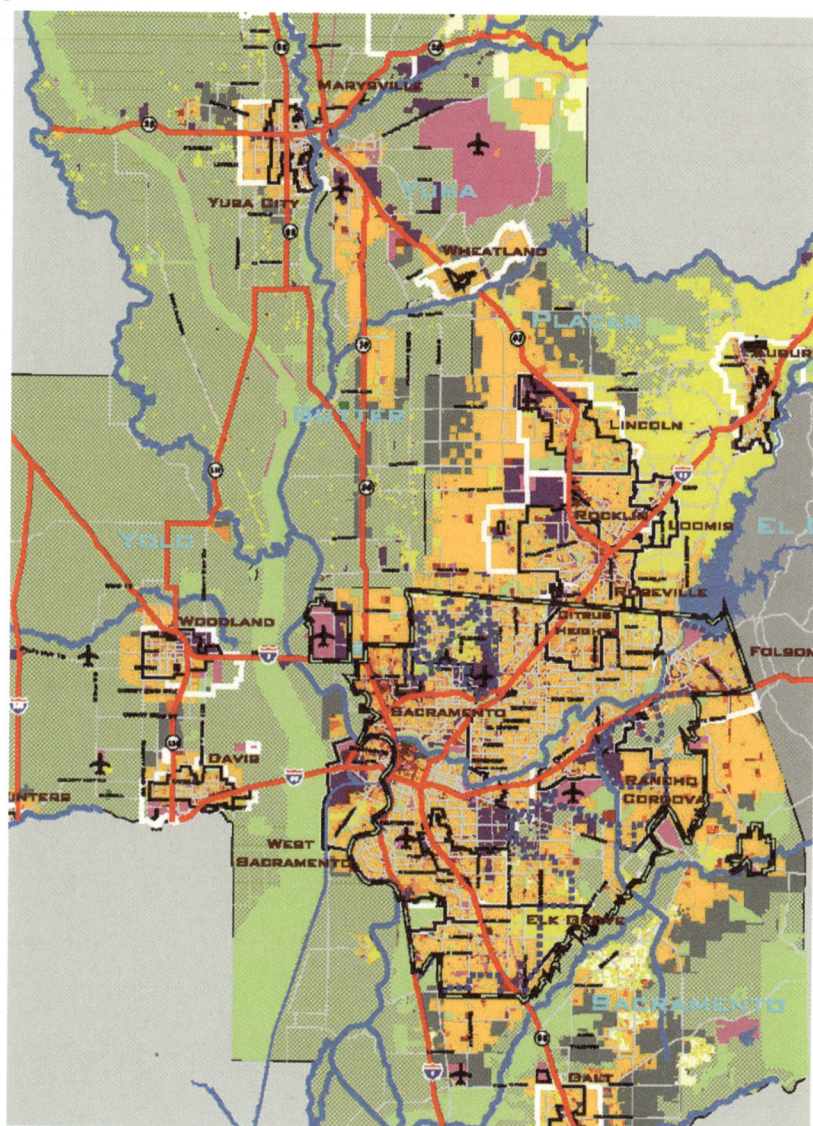

(a) Baseline 2050 Development Scenario

18.2.4 Transit and Transit-Oriented Development

Relative to the rest of the world, the United States is an auto-dependent, transit-poor nation. Except for New York and to a lesser degree Chicago, Boston, and Philadelphia, all of which have older systems and a transit culture, transit has not made a significant impact on passenger miles traveled in the United States. However, despite growing automobile VMT, transit is

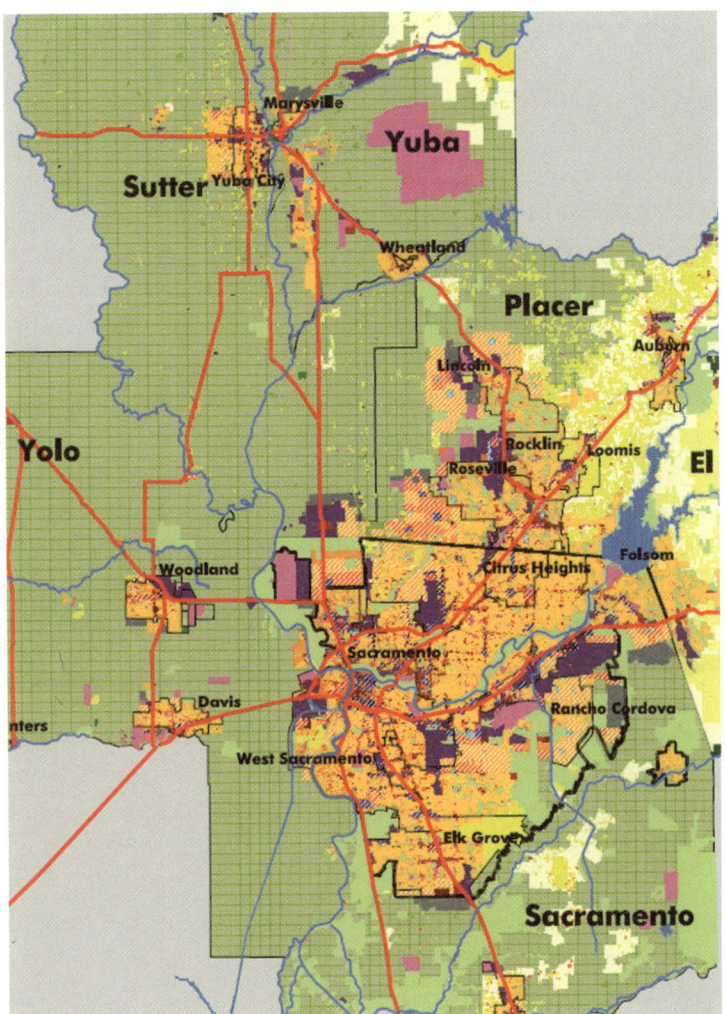

(b) Preferred 2050 Blueprint Scenario

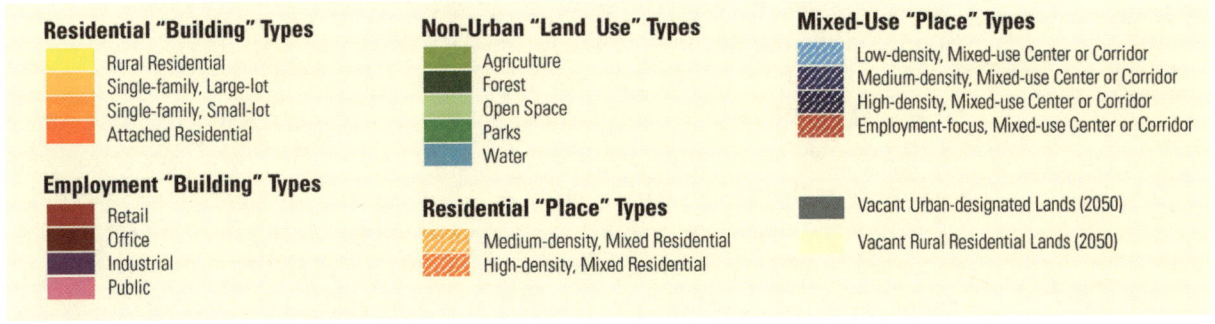

Residential "Building" Types
- Rural Residential
- Single-family, Large-lot
- Single-family, Small-lot
- Attached Residential

Employment "Building" Types
- Retail
- Office
- Industrial
- Public

Non-Urban "Land Use" Types
- Agriculture
- Forest
- Open Space
- Parks
- Water

Residential "Place" Types
- Medium-density, Mixed Residential
- High-density, Mixed Residential

Mixed-Use "Place" Types
- Low-density, Mixed-use Center or Corridor
- Medium-density, Mixed-use Center or Corridor
- High-density, Mixed-use Center or Corridor
- Employment-focus, Mixed-use Center or Corridor

- Vacant Urban-designated Lands (2050)
- Vacant Rural Residential Lands (2050)

Through compact and mixed-use development, the preferred scenario aims to reduce VMT, vehicle energy use, and air emissions.

SOURCE: SACOG, 2005

beginning to make a difference in several cities that have added rail transit in the past thirty years, including the San Francisco Bay Area, Washington (DC), Atlanta, Los Angeles, Baltimore, Portland (OR), San Diego, Denver, Miami, Minneapolis, Seattle, and Salt Lake City, among others.

An effective transit system means ridership, taking people out of single-occupancy vehicles (SOVs), and relieving traffic congestion and parking. Most of these regions and their transit systems are beginning to realize that transit and development go hand in hand, transit systems shape the region, and transit stations shape the neighborhood. Transit-oriented development really means *pedestrian-oriented* (albeit centered around a transit station). Designing a station area for *people* rather than vehicles will ultimately support healthy transit ridership. This means locating *destinations* close to the transit station, meaning residential units on the homeward end, and jobs, retail, and entertainment on the other end, all within an easy walk, say ten minutes or a half mile or less. This means that development around the transit station should be mixed use and dense. And opportunities need to be provided for connecting modes, including park-and-ride (although limited), spouse drop-off kiss-and-ride, bus transit, bicycles, and so on.

Transit agencies are playing an increasing role in TOD through **joint development** or partnerships with developers and local governments. Until 1997, federal rules provided little incentive for transit agencies to engage in joint development because federal funds used to buy land for joint development had to be reimbursed. The rules changes in 1997 and transit agencies can retain proceeds from the sale or use of land for joint development. Transit systems realize that not only does joint development provide proceeds to control costs, but if done right, it will enhance ridership.

San Francisco Bay Area Rail Transit Network. The Bay Area has a comprehensive rail system and a developing network of TODs. Figure 18.24 illustrates the transit network and the twelve TODs given in the CalTrans online searchable TOD database (CalTrans, 2006). The database also shows TODs in southern California. The figure highlights the Pleasant Hill TOD in Walnut Creek that is Phase II of development converting the large BART parking lot around the station site to a dense, mixed-use community development. (For more on BART and other California TODs, see SFBART, 2003; CalTrans, 2002, 2006.)

Washington, DC, Metro. The Washington Metro is a 103-mile heavy rail system serving DC and surrounding suburban Maryland and northern Virginia. The Washington Metropolitan Area Transit Authority (WMATA), which runs Metro has done more joint development than any other transit system. Its first twenty-seven projects through 2001 were valued at $2 billion, and earned $6 million for the Metro system. Subsequent ventures have been more lucrative, not only for WMATA but for state and local jurisdictions. One study showed that Virginia received an annual rate of return of 19% on its investment in Metro (CalTrans, 2002).

Although the first joint development projects did not incorporate all beneficial elements of TOD, essentially all projects now do. The Ballston-Rosslyn development in Arlington County (VA) was one of the first successful TODs. Figure 18.25 shows the Metro system

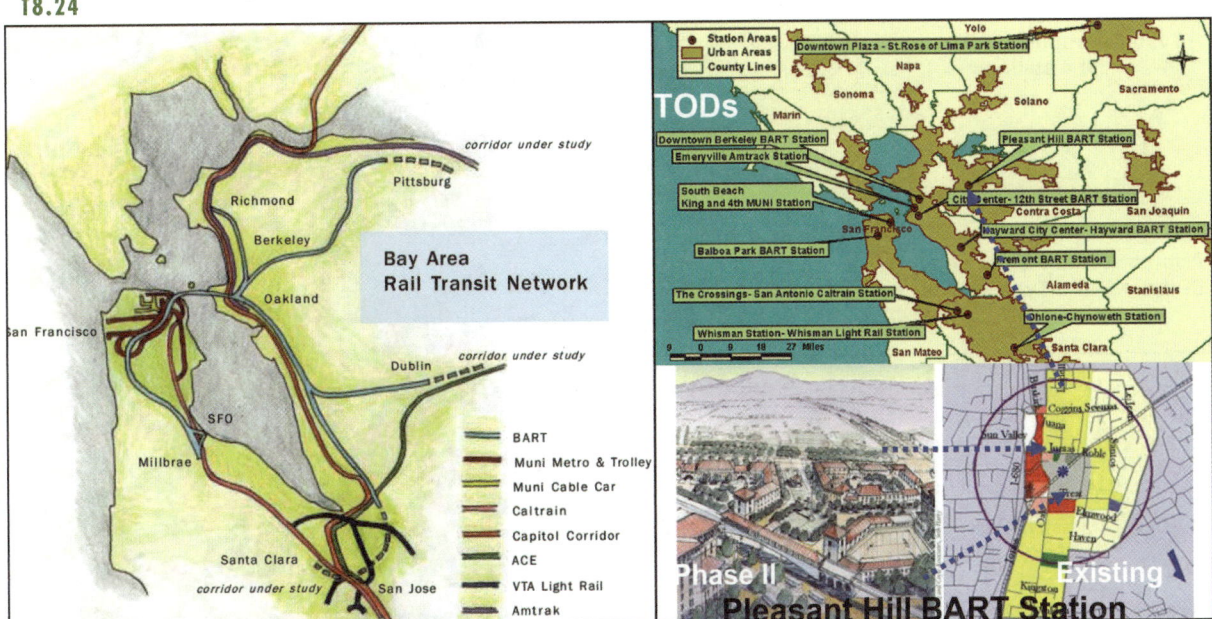

figure 18.24 San Francisco Bay Area Rail Transit Network and TODs

Pleasant Hill BART Station is highlighted: Phase II will fill in the large BART parking lot (gray) around the station expanding residential and mixed use. Existing land use shows residential in yellow, commercial in red.

Sources: BART, 2005; CalTrans, 2006

and highlights the Ballston TOD and one of the newest TODs, the approved Metro West developed by Pulte Homes at the Vienna Metro station. The figure also shows the planned extension of the Metro via the new Silver Line to Dulles Airport. Joint development TODs are planned at all eight new stations including four in Tyson's Corner.

18.2.6 Publicly Owned Utilities, Community Choice Aggregators, and Distributed Energy

Communities with POUs, including municipal utilities and electric co-ops or authorities have additional opportunities for energy planning and development of distributed energy. But communities without POUs also have choice about sources of power under rules established in many states supporting Community Choice Aggregators (CCAs).

18.2.6.1 Communities with Publicly Owned Utilities

In the United States, 2000 cities and towns own their own electric utilities and serve 43 million people or 14% of the nation's consumers. These communities have special opportunities

figure
18.25   Washington (DC) Metro Transit System

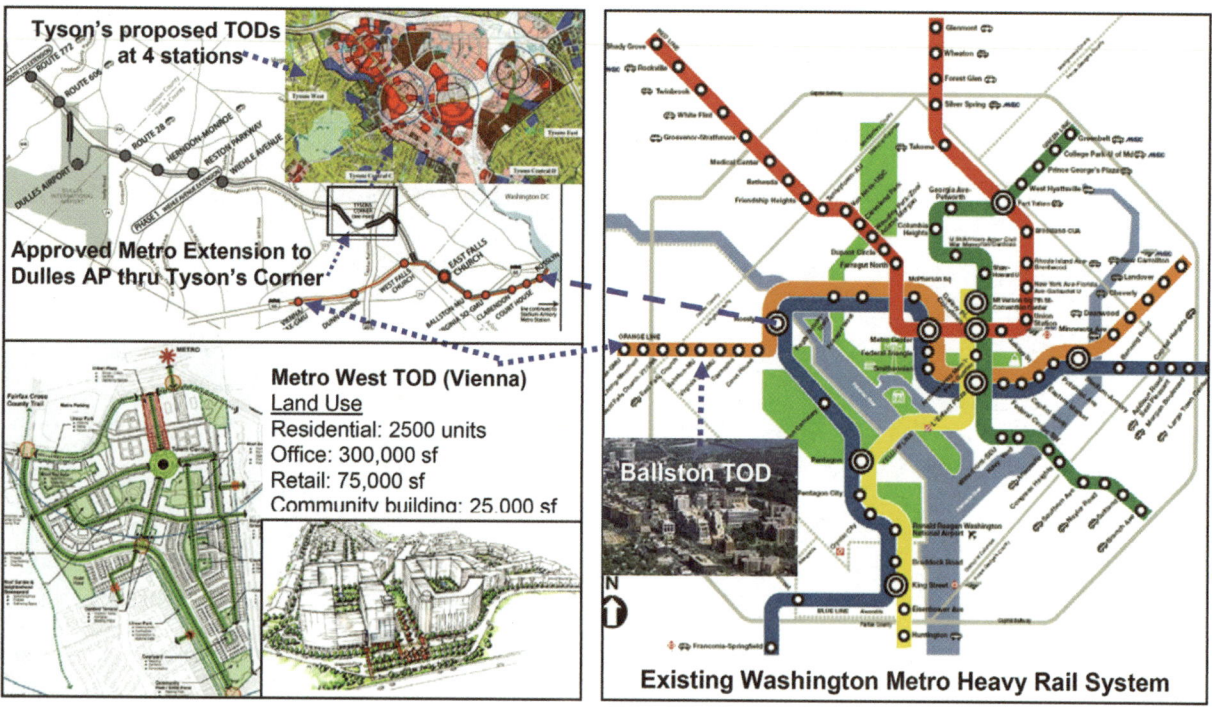

Metro uses joint development projects for TOD. Ballston in Arlington County (VA) is one of Metro's most successful TODs. Metro West is a large approved TOD at the furthest west Vienna Metro Station. The approved Metro extension to Dulles Airport through Tyson's Corner has been accompanied by TODs at eight stations.

to provide energy services to customers. The customers are also the citizens that elect the officials who manage the utilities, so they can use the political and planning process to guide the utility's decisions about sources of power, local generation or distributed power, and demand-side efficiency and load management.

Seattle City Light. Seattle City Light has been active in providing demand-side management services to its customers since 1977. An evaluation conducted in 2004 showed that efficiency measures had reduced electric system load by 11%. Figure 18.26 shows that electrical energy saved reached nearly 1 GWh by 2004. Program participants saved over $430 million on electric bills between 1977 and 2004 (Seattle City Light, 2004).

Sacramento Municipal Utility District. In 1980 SMUD abandoned its operating Rancho Seco nuclear power plant as a result of a local voter referendum opposed to nuclear power. The municipal utility installed a 2 MW solar photovoltaic power plant on the Rancho Seco site not only to offset some of the power lost from the closed plant but also as a symbol of the new age of the utility.

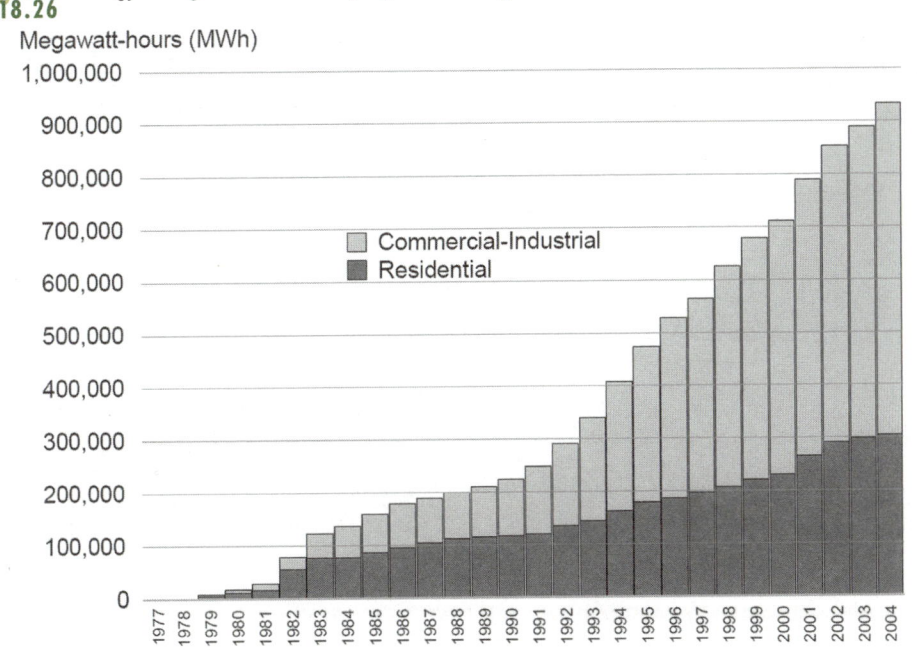

figure 18.26 Energy Savings from Seattle City Light's DSM Programs, 1977–2004

Megawatt-hours (MWh)

☐ Commercial-Industrial
■ Residential

SOURCE: Seattle City Light, 2005

The utility embarked on an aggressive program to develop renewable sources of electricity, including solar photovoltaics and wind, with a goal that 20% of its customers' electricity needs would be met by non-hydro renewable energy by 2011. In 1993, SMUD initiated its **PV Pioneer** program, which first installed utility-owned PV arrays on residential rooftops, then converted the program to a $2.80/W$_{pAC}$ rebate program for resident-owned systems. By 2006, SMUD had more than 10 MW of installed PV capacity in its service area (see Figure 18.19[a] for the role SMUD's program plays in California's PV growth). Figure 18.27(a) shows an example solar PV subdivision in SMUD's service area, and 18.27(b) gives results of an evaluation of the program benefits. In 2007, SMUD announced a partnership with Lennar Homes to collaborate on 1254 solar-powered homes in eleven Sacramento-area communities. Los Angeles Water and Power and Long Island Public Authority (NY) have similar PV programs.

Among some of SMUD's other incentive programs (listed in Table 18.9) are rebates for solar water heaters, duct sealing, and energy-efficient appliance and HVAC systems. SMUD will buy old refrigerators for $35. Its customers can buy compact fluorescent lightbulbs for 99¢ and 100% green power for a flat $6 per month. Each of these programs has an annual budget and once the allocation is depleted the program ends until the next year, as long as evaluation shows it benefits the utility, the customers, and the community.

Austin Energy. Austin's (TX) municipal utility, Austin Energy, has taken the lead on several of the city's energy initiatives including green buildings (see Section 18.2.5), DSM, and

figure 18.27 Sacramento Municipal Utility District PV Pioneer Program

(a) SMUD PV Pioneer program has spawned residential rooftop PV systems across its service area, including this PV subdivision.

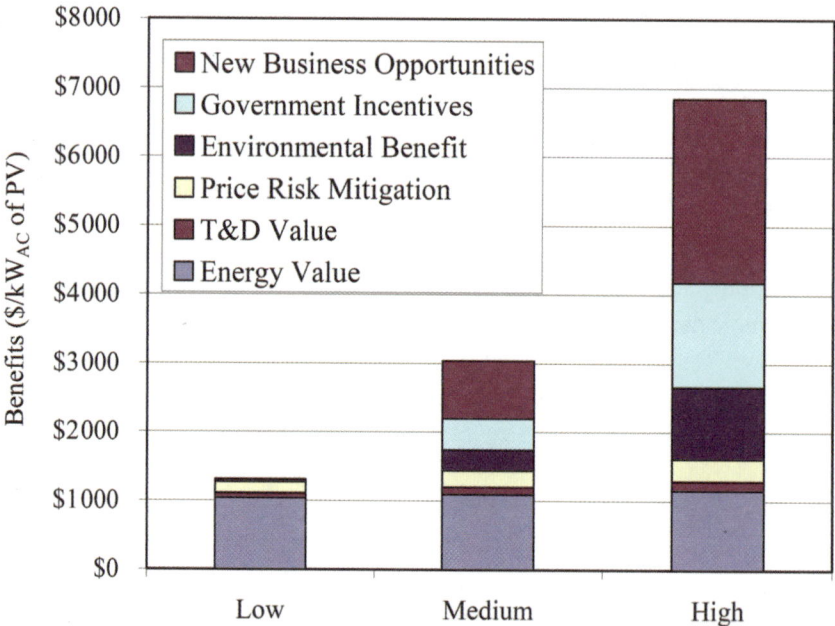

(b) An evaluation of the SMUD PV Pioneer program estimated a range of benefits provided the utility, its customers, and the community.

"Low benefits" assume only benefits to the utility and customers, "Medium" and "High" benefits assume benefits from environmental quality, government investment, and business opportunities with different assumptions.

Sources: CEC, 2007b; Hoff, 2002

table 18.9 **SMUD Programs for Efficiency and Renewables**

	Rebate/Incentive	Consumer Cost
Greenergy pricing for renewable power		$6/mo for 100%, $3/mo for 50%
Solar Pioneers	$2.80/kW	
Rebates/Incentives		
Solar water heater	$1500	
CFL lightbulbs		99¢/bulb
Old refrigerator buy-back	$35	
Duct cleaning	$300	
Ceiling fan with CFL light	$20	
Cool roofs	20¢/ft^2	
Central AC/heat pump	$550–$750	
Whole house fan	$100	
Pools and spas pumps, etc.	$200	
ES Clothes washer	$75–$125	
ES Dishwasher	$30	
ES Refrigerator	$50	
ES Room AC	$50	

renewable power. Its GreenChoice® Renewable Energy Program offers subscribing customers the option of renewable power from wind, landfill gas, and solar. With 665 million kWh in subscriptions, they claim this to be the most successful utility green-power program in the country. A new batch of renewable power subscriptions is expected in early 2008.

Austin Energy also provides rebates and loans for efficiency improvements and solar systems for both residential and commercial customers. Its solar rebates are among the best in the country:

- Solar photovoltaic system rebate: $4.50/W$_{P,DC}$ ($5.60/W$_{P,DC}$ if > 60% of PV manufacture and assembly done in Austin) for residential customers (up to $13,500 per year) and commercial customers (up to $100,000 per year). For example, a 2-kW PV system typically costs $12,000–$20,000, and the rebate would cover $9000 of that cost (2000W × $4.50/W).
- Solar hot water rebate: $1500–$2000 rebate for residential, commercial and municipal customers' systems.

Through its GreenChoice®, Austin Energy is contracting an increasing amount power from Texas' growing wind energy capacity. But we know that wind power cannot always be dispatched to meet peak demand. In 2007, the utility and Austin's Mayor Will Wynn are promoting the development of more plug-in vehicles to make Austin a demonstration of a vehicle-to-grid (V2G) system, such as that described in Section 10.3.2.

18.2.6.2 Communities without Publicly Owned Utilities: Community Choice Aggregators and Distributed Resources

Communities without municipal utilities or other POUs do not have the full range of opportunities for power options, but under emerging policies they may have choice over their source of power. For example, to implement its 2001 Energy Plan, Chicago negotiated with its IOU Commonwealth Edison to provide 25% of new power from renewable sources by 2010.

In some states with deregulation rules or specific legislation, municipalities can aggregate their citizens' demand into a unit and bid the community's electricity service to competing suppliers. Local elected officials have control over what kind of energy the community will purchase depending on the community's desired mix of resources. Although options for "green" power choices by consumers have expanded, most customers settle for the "default" power provider because they choose not to choose. In a Community Choice Aggregator (CCA), localities have the option to become the default provider and can serve their own citizens.

Massachusetts passed the first Community Choice law in 1999 and several states followed suit including Ohio, California, Rhode Island, and New Jersey. In Massachusetts, the Cape Light Compact is a regional aggregation of twenty-one towns and negotiates rates on the part of 200,000 customers. The Rhode Island Energy Aggregation Program includes thirty-six communities and purchases low-cost power for municipal facilities.

In Ohio, 90% of residential and commercial customers who switched from their local IOU joined a community choice program. The Northeast Ohio Public Energy Council (NOPEC) is a Community Choice aggregation of 118 member cities with 600,000 customers across eight Ohio counties. Through select electricity purchasing, NOPEC is the largest clean power purchaser in the country. Purchased electricity is 70% less polluting than what other Ohio customers use and the contracts provide this power at rates 4–6% less than IOU rates.

During California's period of utility restructuring (1997–2001) a plethora of new power suppliers entered the state and many customers switched from their IOU. After restructuring was suspended in 2001, customers could no longer choose. To remove this new barrier for choice for communities, the state passed the Community Choice law (AB 117) in 2002. It provides a process for a city or a consortium of jurisdictions to solicit a new electric-service provider, set higher goals for green power, and use bonds to revenue bonds for distributed generation and efficiency, financed through customer rates. The CPUC issued implementation rules for CCA in 2005, including cost recovery surcharges or payments to IOUs to protect them and their remaining customers for CCAs' exit from their service.

San Francisco Community Choice and Distributed Generation. San Francisco is served by IOU PG&E, and starting in 1999, it lobbied for the 2002 state law so it could serve as the community aggregator for the city's residents. In addition, in 2001 three-fourths of the voters approved Proposition B, a $100 million bond initiative to pay for solar systems on municipal buildings, and Proposition H that gave the city authority to issue additional bonds for renewables and efficiency without voter approval. The city's signature Proposition B project is the 60,000 square-foot, 675 kW PV system on the Moscone Center.

In 2004, the city supervisors unanimously passed the Energy Independence Ordinance to prepare a CCA implementation plan and use Proposition H bonds to build 360 MW of new distributed resources (solar PV, other distributed generation, and energy efficiency). This 360 MW is 42% of the city's 850 MW peak load and 55% of its 650 MW nighttime load and it exceeds the Energy Plan adopted in 2002 (see Figure 18.20).

Following San Francisco's lead, at least fourteen cities and counties and one consortium of twelve cities in California are in various stages of forming a CCA. These communities have a goal of achieving 40% renewable energy portfolio.

18.3 Summary

These last two chapters have highlighted the significant level of energy policy activity, focusing on the United States, but also drawing lessons from other countries. In the United States, despite national energy policy legislation in 2005 and 2007, most agree that energy policy at the federal level has been insufficient to achieve the rapid transformation of energy markets to more sustainable patterns needed to arrest oil dependence and fossil-fuel induced climate change. Fortunately much is occurring at the state and local level in the United States to fill the void left by policy limitations at the federal level.

States have shown innovation and experimented with new policies. Although some policies have failed, many have succeeded and served as models for other states and for federal policies. States that have been aggressive in energy policy believe their primary benefit will come from their state's increased economic development and competitiveness in the coming carbon-rich and oil-poor world. These state and local policy makers are not eco-centric activists—they are smart entrepreneurs.

The transformation to sustainable energy will require significant progress in three areas:

1. **Sustainable Energy Technologies.** This book has highlighted developments in the wide range of technologies for renewable energy and efficiency for buildings, transportation, and electricity. Additional technical potential exists, new technologies need be developed, and much is needed to diffuse these technologies quickly into the market.

2. **Consumer and Community Choice.** Many believe that technology alone will not solve our energy dilemma in a timely way. To accelerate market penetration of sustainable technologies, consumers must consciously choose to invest in sustainable energy and understand that simple behavioral change can greatly speed the transition to more efficient energy use. Acting collectively through the planning and political process, citizens can help shape their communities' energy use through land use planning, transit development, green building programs, and Community Choice Aggregators.

3. **Energy and Environmental Policies.** Most believe that while the market is the primary driver of technology and consumer choice, the market alone cannot lead to the rapid transformation needed. Government intervention is needed to develop technologies, to increase market penetration through investment and regulation, and to affect consumer choice through incentives and education.

Among the needed policies to assist technology development and consumer choice for sustainable energy, the list is long. They include

- Policies for **technology research,** demonstration, and especially deployment are needed to develop sustainable technologies and bring them to market.
- Policies for **mandated standards** of efficiency and production, such as product efficiency standards (e.g., vehicles, buildings, equipment), portfolio standards (e.g., renewable power and efficiency, renewable fuels), and land development controls (e.g., zoning, urban growth boundaries) are the most effective for rapid change.
- Policies for **environmental protection** (e.g., air pollutant and GHG emission controls, nuclear waste and decommissioning management) can internalize externalities of fossil and nuclear energy, develop cleaner fossil and nuclear energy, and make efficiency and renewable energy options more competitive.
- Policies for **institutional structures and rules** (e.g., utility regulation, community choice aggregators, facility siting) can remove barriers and provide opportunities for innovative distributed electricity resources and renewable fuel production.
- Policies for **investment and production financial incentives** for sustainable efficiency and renewable energy can increase competitiveness and accelerate investment and development.
- Policies for **public investment** in sustainable energy research, transit, public power, public facilities and fleets, public procurement, and community programs provide a sustainable public infrastructure.
- Policies for **community action and education** raise public awareness and affect consumer and community choice for sustainable energy.

Through effective energy policy, smart energy planning, market penetration of emerging renewable and efficient energy technologies, and informed social activism and choice, we can arrest our dependence on oil and carbon fuels, further dissociate growth of energy demand from economic growth, and achieve sustainable energy.

References and Further Reading

Chapter 1

Bell, D. 1973. *The coming of the post industrial society*. New York: Harper Colophon Books.

California Energy Commission. 2004. Energy time machine. Retrieved from http://www.energyquest.ca.gov/time_machine/

Organisation for Economic Cooperation and Development (OECD). 2001. Maddison, A. The world economy: A millennial perspective.

Rosenfeld, A. 2006. "Fermi Award Talks," June 21–22, 2006. Retrieved from http://www.energy.ca.gov/commission/commissioners/rosenfeld.html

Smalley, R. 2005. Our energy challenge. Illinois Science and Humanities Symposium.

U.S. Energy Information Administration. 2006. Annual Energy Review 2005.

U.S. Energy Information Administration. 2007a. Annual Energy Review 2006.

U.S. Energy Information Administration. 2007b. International Energy Annual 2005.

U.S. Energy Information Administration. 2007c. Monthly Energy Review.

Chapter 2

Arctic Climate Impact Assessment. 2004. Impacts of a Warming Arctic. Retrieved from http://www.amap.no/acia/

Bourzac, K. 2005. "A Photo Essay: Dirty Oil". *Technology Review*. December.

Campbell, C. J. 2003. Industry urged to watch for regular oil production peaks, depletion signals. *Oil & Gas Journal* (July 14): 38.

Campbell, C. J., and J. H. Laherrere. 1998. The end of cheap oil. *Scientific American* 278, no. 3 (March): 78.

Clean Air Initiative-Asia. 2006. Air quality of major Asian cities. Retrieved from http://www.cleanairnet.org/caiasia/1412/channel.html

Hubbert, M. K. 1969. Energy resources. In National Academy of Sciences, *Resources and Man*. San Francisco: W.H. Freeman.

Hubbert, M.K. 1971. The energy resources of the earth. *Scientific American* (September).

Intergovernmental Panel on Climate Change. 1990. IPPC First Assessment Report (FAR). Retrieved from http://www.ipcc.ch/

Intergovernmental Panel on Climate Change. 1996. IPPC Second Assessment Report (SAR). Retrieved from http://www.ipcc.ch/

Intergovernmental Panel on Climate Change. 2001. Climate change 2001: Summary for policy makers. IPPC Third Assessment Report (TAR). Retrieved from http://www.ipcc.ch/

Intergovernmental Panel on Climate Change. 2007. Climate change 2007: Summary for policy makers. IPPC Fourth Assessment Report (AR4). Retrieved from http://www.ipcc.ch/

Keeling, C. D., and T. P. Whorf. 2005. Atmospheric CO_2 records from sites in the SIO air sampling network. *Trends: A Compendium of Data on Global Change*. Carbon Dioxide Information Analysis Center, Oak Ridge National Laboratory, U.S. Department of Energy, Oak Ridge, TN.

Linden, H. 1998. Flaws seen in resource models behind crisis forecasts for oil supply, price. *Oil & Gas Journal* (December 28).

Lovins, A., E. Kyle Datta, O.-E. Bustnes, J. Koomey, and N. Glasgow. 2004. *Winning the oil endgame: American innovation for profits, jobs, and security.* Snowmass, CO: Rocky Mountain Institute.

Lynch, M. 2003. Petroleum resources pessimism debunked in Hubbert model and Hubbert modelers' assessment. *Oil & Gas Journal* (July 14): 38.

SolarBuzz™. 2007. MarketBuzz™ 2007: Annual World Solar Photovoltaic Industry Report. Retrieved from http://www.solarbuzz.com/

Massachusetts Institute of Technology. 2003. The future of nuclear power. Retrieved from http://web.mit.edu/nuclearpower/

Maugeri, L. 2004. Oil: Never cry wolf—Why the petroleum age is far from over. Science 304:1114.

National Atmospheric and Space Administration. 2007. Global annual mean surface air temperature change. http://data.giss.nasa.gov/gistemp/graphs/

National Biodiesel Board. 2007. http://www.biodiesel.org/

National Climate Data Center. 2007. Climate 2006 in historical perspective. Retrieved from http://www.ncdc.noaa.gov/oa/climate/research/2006/ann/ann06.html

Oil & Gas Journal. 2004. Hubbert revisited: Six part series revisiting the debate over Hubbert's peak.

Oil & Gas Journal. 2006. World Oil and Natural Gas Reserves and Production. December. v.104.47

Oreskes, N. 2004. The scientific consensus on climate change. *Science* 306 (December 3): 1686.

Renewable Fuels Association. 2007. Building new horizons. *Ethanol Industry Outlook 2007.* Washington, DC. http://www.ethanolrfa.org/

Simmons, M. 2006. *Twilight in the Desert: The Coming Saudi Oil Shock and the World Economy.* New York: John Wiley & Sons.

Solar Energy Industries Association. 2004. Our solar power future: The solar industries roadmap through 2030 and beyond (September). Washington, DC.

United Nations Framework Convention on Climate Change (UNFCCC). 2004. The first ten years. Bonn, Germany. Retrieved from http://www.unfccc.int

U.S. Department of Energy. 2004. *Carbon sequestration technology roadmap and program plan—Developing the technology base and infrastructure to enable sequestration as a greenhouse gas mitigation option.* Office of Fossil Energy.

U.S. Energy Information Administration. 2003. Long term world oil supply (A resource base/production path analysis). J. Wood, G. Long, D. Morehouse.

U.S. Energy Information Administration. 2005. Oil Market Basics. http://www.eia.doe.gov/

U.S. Energy Information Administration. 2006a. *Annual Energy Review* 2005.

U.S. Energy Information Administration. 2006b. *International Energy Annual* 2004.

U.S. Energy Information Administration. 2007a. *Annual Energy Review* 2006.

U.S. Energy Information Administration. 2007b. *International Energy Annual* 2005.

U.S. Energy Information Administration. 2007c. *Monthly Energy Review.*

U.S. Energy Information Administration. 2007d. *International Petroleum Monthly.*

U.S. Environmental Protection Agency. 2006. Ambient air concentrations trends. Retrieved from http://www.epa.gov/airtrends/ozone.html

U.S. Environmental Protection Agency. 2007. Air emissions trends. http://www.epa.gov/airtrends/

U.S. Federal Highway Administration. 2006. Air quality non-attainment area maps. Retrieved from www.fhwa.dot.gov/environment/conformity/naaqs.htm

U.S. Geological Survey. 2000. World oil and natural gas resources. Retrieved from http://pubs.usgs.gov/dds/dds-0-060/

Chapter 3

Baka, J., H. Ling, and D. Kammen. 2006. Towards energy independence in 2025. UC-Berkeley, Renewable and Appropriate Energy Laboratory.

Bernstein, M., J. Griffin, and R. Lempert. 2006. Impacts on U.S. energy expenditures of increasing renewable energy use. RAND Corporation.

Brown, L. R. 2003. *Plan B: Rescuing a planet under stress and a civilization in trouble.* New York: W.W. Norton.

Craig, P., A. Gadgil, and J. Koomey. 2002. What can history teach us? A retrospective examination of long-term energy forecasts for the United States. *Annual Review of Energy and the Environment 2002.* 27: 83–118

Electric Power Research Institute (EPRI). 2005. Program on technology innovation: Electric power industry technology scenarios.

English, B., D. De La Torre Ugarte, K. Jensen, C. Hellwinckel, J. Menard, B. Wilson, R. Roberts, and M. Walsh. 2006. 25% renewable energy for the United States by 2025: Agricultural and economic impacts. Department of Agricultural Economics, The University of Tennessee.

"Four Futires for China, Inc." Global Business Network. Illustrations by Ben Shannon. *Business* 2.0. August 2006.

Hanson, D., I. Mintzer, J. Laitner, and J. A. Leonard. 2004. Engines of growth: Energy challenges, opportunities, and uncertainties in the 21st century. Decision and Information Sciences Division, Argonne National Laboratory. Retrieved from http://amiga.dis.anl.gov/Engines_of_GrowthJan 17-0–04_rev161.pdf

Hirsch, R. L., R. Bezdek, and R. Wendling. 2005. Peaking of world oil production: Impacts, mitigation, and risk management. Report prepared for U.S. DOE, National Energy Technology Laboratory.

Hefner, R. 2002, 2007. The age of energy gases in the new millennium. Oklahoma City, OK: The GHK Company. Retrieved from http://www.ghkco.com/html/age.htm

Intergovernmental Panel on Climate Change. 2002. Climate change 2001: Summary for policy makers. IPPC Third Assessment Report (TAR). Retrieved from http://www.ipcc.ch/

International Energy Agency. 2006. World energy outlook. Retrieved from http://www.worldenergyoutlook.org/

Laitner, J. "Skip," S. Decanio, J. Koomey, and A. Sanstad. 2003. Room for improvement: Increasing the value of energy modeling for policy analysis. *Utilities Policy* 11:87–94. (also LBNL–50627)

Laitner J. and D. Hanson. 2004. Adapting for uncertainty: A scenario analysis of technology energy futures. 24th USAEE/IAEE North American Conference. July 9. Washington, DC. Retrieved from http://www.iaee.org/documents/washington/Skip_Laitner.pdf

Lovins, A. 1976. Energy strategy: The road not taken. *Foreign Affairs* (October).

Lovins, A., E. Kyle Datta, O.-E. Bustnes, J. Koomey, and N. Glasgow. 2004. *Winning the oil endgame: American innovation for profits, jobs, and security.* Snowmass, CO: Rocky Mountain Institute.

Massachusetts Institute of Technology. 2003. The future of nuclear power. Retrieved from http://web.mit.edu/nuclearpower/

Nordhaus, W. D. 2004. The outlook for energy three decades after the energy crisis. June 2004. Retrieved from http://www.iiasa.ac.at/Research/ECS/IEW2004/docs/2004P_Nordhaus.pdf

O'Neill, B. C., and M. Desai. 2005. The historical accuracy of projections of U.S. energy consumption. *Energy Policy* 33(8):979–993.

Pacala, S., and R. Socolow. 2004. Stabilization wedges: Solving the climate problem for the next 50 years with current technologies. *Science* 305 (August 13): 968.

Population Reference Bureau. 2004. World Population Growth through history. Retrieved from http://www.prb.org/Publications/GraphicsBank/PopulationTrends.aspx

Smil, V. 2003. *Energy at the crossroads: Global perspectives and uncertainties.* Cambridge, MA: MIT Press.

U.S. Department of Energy. 2002. Nuclear Energy Research Advisory Committee and the Generation IV International Forum. A Technology Roadmap for Generation IV Nuclear Energy Systems http://gif .inel.gov/roadmap/pdfs/gen_iv_roadmap.pdf

U.S. Department of Energy.2002. National Hydrogen Energy Roadmap. Retrieved from http://www1.eere
.energy.gov/hydrogenandfuelcells/pdfs/national_h2_roadmap.pdf

U.S. Energy Information Administration. 2007. Annual Energy Outlook. Washington, DC. Retrieved from
http://www.eia.gov

U.S. Energy Information Administration. 2006. International Energy Outlook. Washington, DC. Retrieved
from http://www.eia.gov

Wald, M. L. 2004. Questions about a hydrogen economy. *Scientific American* 290(5): 66.

Chapter 4

THERE ARE NO REFERENCES FOR CHAPTER 4

Chapter 5

American Automobile Association (AAA) Fuel Gauge Report. 2007. Retrieved from www.aaa.com

Cleveland, C. J. 2005. Net energy from the extraction of oil and gas in the United States. *Energy* 30:769–782.

Consortium for Research on Renewable Industrial Materials (CORRIM). 2005. Life cycle environmental performance of renewable building materials in the context of residential construction. Phase I Research Report. http://www.corrim.org/

Farrell, A., R. Plevin, B. Turner, A. Jones, M. O'Hare, and D. Kammen. 2006. Ethanol can contribute to energy and environmental goals. *Science:* 311, 506.

Gagnon, L., C. Belanger, and Y. Uchiyama. 2002. Life-cycle assessment of electricity generation options: The status of research in year 2001. *Energy Policy* 30:1267–1278.

Golove, W. H., and J. H. Eto. 1996. Market barriers to energy efficiency: A critical reappraisal of the rationale for public policies to promote energy efficiency. Berkeley, CA: Lawrence Berkeley National Laboratory, LBL-38059.

Knapp, K., and T. Jester. 2001. Empirical investigation of the energy payback time for photovoltaic modules. *Solar Energy* 71(3):165–172.

Lovins, A. 1985. Saving gigabucks with negawatts. *Public Utilities Fortnightly* (March 21).

Meier, A., J. Wright, and A. H. Rosenfeld. 1983. *Supplying energy through greater efficiency.* Berkeley, CA: University of California Press.

Pimentel, D., and T. W. Patzek. 2005. Ethanol production using corn, switchgrass, and wood; Biodiesel production using soybean and sunflower. *Natural Resources Research* 14 (1): 65–76.

Rosenfeld, A. 1999. The art of energy efficiency: Protecting the environment with better technology. *Ann. Rev. Energy Environ.* 24:33–82.

Rosenfeld, A. H., C. Atkinson, J. G. Koomey, A. Meier, R. Mowris, and L. Price. 1993. Conserved energy supply curves. *Contemporary Policy Issues* XI (1): 45–68.

Rufo, M., and F. Coito. 2002. *California's secret energy surplus: The potential for energy efficiency.* San Francisco, CA: The Energy Foundation and The Hewlett Foundation. http://www.ef.org/documents/Secret_Surplus.pdf

Rydh, C. J., and B. A. Sanden. 2005. Energy analysis of batteries in photovoltaic systems. Part II: Energy return factors and overall battery efficiencies. *Energy Conversion and Management* 46:1980–2000.

Shapouri, H., J. A. Duffield, and M. Wang. 2002. The energy balance of corn ethanol: An update. USDA. *Agricultural Economic Report* 814.

Shapouri, H., J. A. Duffield, and M. Wang. 2004. The 2001 net energy balance of corn ethanol. USDA. Office of the Chief Economist, Office of Energy Policy and New Uses.

Stoft, S. 1995. The economics of conserved-energy "supply" curves. PWP-028, Program on Workable Energy Regulation (POWER). University of California Energy Institute.

U.S. Department of Energy, National Renewable Energy Laboratory (NREL). 2007. Life Cycle Inventory Database. Retrieved from http://www.nrel.gov/lci/

U.S. Department of Energy, USDA, NREL. 1998. Life Cycle Inventory of Biodiesel and Petroleum Diesel for Use in an Urban Bus. (Sheehan, J, Camobreco, V, Duffield, J, Graboski, M, Shapouri, H.)

Wang, M. 2005. Update on energy and greenhouse emission impacts of fuel ethanol. Presentation at 10th Annual National Ethanol Conference, Scottsdale, Arizona (February 8).

Chapter 6

ASHRAE (American Society of Heating, Refrigerating and Air Conditioning Engineers). 1988. Standard 119-1988. Air leakage performance for detached single-family residential buildings.

ASHRAE.1995. 90.1 Code compliance manual. New York: American Society of Heating, Refrigerating and Air Conditioning Engineers.

ASHRAE. 2003. Standard 62-2-2003. Ventilation standard for acceptable indoor air quality for residences.

ASHRAE. 2004. Standard 55-2004. Thermal environmental conditions for indoor occupancy.

ASHRAE. 2005. *Handbook of fundamentals*.

ASHRAE. 2005. Standard 62-1-2004. Ventilation standard for acceptable indoor air quality.

Building Science Corporation. 2001. HVAC equipment sizing strategies: Taking advantage of high-performance buildings. EEBA Excellence in Building Conference. Online http://www.buildingscience.com

Habitat for Humanity of Metro Denver. 2000. An advanced systems engineering approach to affordable single family homes. Energy Efficient Building Association, Building Science Consortium.

Kerrigan, P., and J. Larsen. 2003. Windows and thermal comfort. Building Science Consortium.

Kolle, J. 1999. Taking a look at windows. In *Energy-Efficient Building*. Newtown, CN: The Taunton Press.

Pettit, B. 2005. Building science for architects: Thermal and air leakage control; Moisture control; Choosing heating and cooling systems; Well-ventilated buildings and ventilation systems; Design specifications for high performance buildings; Slabs, crawlspaces, basements and below-grade spaces; Wall system options. Building Science Corporation. Online http://www.buildingscience.com.

Randolph, J., K. Geeley, and W. Hill. 1991. *Evaluation of Virginia weatherization program*. Blacksburg, VA: Virginia Center for Coal and Energy Research.

Sherman, M. 1998. The Use of Blower-Door Data. LBL #35173. Berkeley: Lawrence Berkeley Laboratory.

Sterling, E. 2005. Update on Status of ASHRAE Standards 62 and 55. Vancouver, BC: Theodor Sterling Associates Ltd. Online: www.sterlingiaq.com.

U.S. Department of Energy. 2006. 2006 Buildings Energy Data Book. Office of Energy Efficiency and Renewable Energy. See http://buildingsdatabook.eere.energy.gov/.

Chapter 7

Balcomb, D., R. Jones, C. Kosiewicz, G. Lazarus, R. McFarland. and W. Wray. 1983. *Passive solar design handbook* (3). New York: American Solar Energy Society.

Department of Business, Economic Development & Tourism (DBEDT) and the American Institute of Architects. 2001. *Field guide for energy performance, comfort, and value in Hawaii homes*. Honolulu, HI: State of Hawaii.

Home Energy. 1993. *Discovering ducts* (no author cited), September/October.

International Energy Agency. 2000. *Daylight in buildings: A source book on daylighting systems and components*.

Lawrence Berkeley National Labs. *Tips for daylighting with windows* http://windows.lbl.gov/daylighting/designguide/browse.htm

Masters, G. 2004. *Renewable and efficient electric power systems*. Hoboken, NJ: Wiley Interscience.

National Renewable Energy Laboratory. 1994. *Solar radiation data manual for flat-plate and concentrating collectors*. http://rredc.nrel.gov/solar/pubs/redbook/.

The solar radiation data manual for buildings, which is also downloadable from the web http://rredc.nrel.gov/solar/old_data/nsrdb/bluebook/.

Stein, B., and J. Reynolds. 1992. *Mechanical and electrical equipment for buildings*, 8th ed. New York: John Wiley & Sons.

Swisher, J.N. 1985. Measured performance of passive solar buildings. *Annual Review of Energy* 10 (November): 201–216.

Chapter 8

Arthur D. Little, Inc. 2001. Energy savings potential of solid state lighting in general lighting applications. U.S. Department of Energy.

California Energy Commission. 2005. Title 24, Part 6, of the California Code of Regulations: California's Energy Efficiency Standards for Residential and Nonresidential Buildings. http://www.energy.ca.gov/title24/

Consortium for Research on Renewable Industrial Materials (CORRIM). 2005. Life cycle environmental performance of renewable building materials in the context of residential construction. Phase I Research Report. http://www.corrim.org/

Keoleian, G., S. Blanchard, and P. Reppe. 2001. Life cycle energy, costs, and strategies for improving a single family house. *Journal of Industrial Ecology* 4 (2): 135–157.

Krames, M. 2003. LumiLEDs: Lighting, progress and future direction of LED lighting. SSL Workshop, 13 November 2003, Arlington, VA

Larsen, J. 2003. 20 years of energy efficient buildings and windows. Cardinal Glass Industries, Inc. 2003 national workshop on state building energy codes, Atlanta, GA.

McDonough, W., and M. Braungart. 2002. *Cradle To Cradle: Remaking the Way We Make Things*. North Point Press.

Mumma, T. 1995. Reducing the embodied energy of buildings. *Home Energy*. January/February.

Residential Energy Services Network (RESNET). 2005. ENERGY STAR Homes Accreditation. Retrieved from http://www.resnet.us/standards/default.htm

Rosenfeld, A. 2006. Summing up energy symposium: The "Rosenfeld effect." California Energy Commission. April 28, 2006. http://www.energy.ca.gov/commission/commissioners/rosenfeld.html

Scheuer, C. W., and G. A. Keoleian. 2002. Evaluation of LEED using life cycle assessment methods. Building and Fire Research Laboratory, National Institute of Standards and Technology. Gaithersburg, MD.

U.S. Department of Energy. 2002. Volume 1: National Lighting Inventory and Energy Consumption Estimate.

U.S. Department of Energy. 2005. Appliance and Commercail Equipment Standards. Retrieved from http://www.eere.energy.gov/buildings/appliance_standards/

U.S. Department of Energy. 2005. Affordable high-performance homes: The 2002 NREL Denver Habitat for Humanity house, A cold-climate case study. Building America Program.

U.S. Department of Energy. 2006. Building America Research is Leading the Way to Zero Energy Homes. Energy Efficiency and Renewable Energy. Retrieved from http://www.eere.energy.gov/buildings/building_america/

U.S. Department of Energy. 2006. Residential compliance using REScheck. Energy Efficiency and Renewable Energy. http://www.energycodes.gov/rescheck/

U.S. Department of Energy. 2006. 2006 Buildings Energy Data Book. Office of Energy Efficiency and Renewable Energy. See http://buildingsdatabook.eere.energy.gov/

U.S. Environmental Protection Agency. 2006. ENERGY STAR Qualified Homes National Performance Path Requirements.

U.S. Environmental Protection Agency. 2007. Energy Star program factsheets. Office of Air and Radiation. http://www.energystar.gov/index.cfm?c=home.index

U.S. Green Building Council. 2005. LEED-EB: Green building rating system for existing buildings: Upgrades operations and maintenance. Version 2 (July)

U.S. Green Building Council. 2005. LEED for Homes: Rating system for pilot demonstration of LEED for homes program. Version 1.72

U.S. Green Building Council. 2007. LEED for New Construction: Rating system Version 2.1 (September 8)

U.S. Green Building Council. 2007. Project Profiles. Retrieved from http://www.usgbc.org/DisplayPage .aspx?CMSPageID=1721

Chapter 9

Bachrach, D., M. Ardema, and A. Leupp. 2003. *Energy efficiency leadership in California: Preventing the next crisis*. Natural Resources Defense Council, Silicon Valley Manufacturing Group, April.

Energy Information Administration (EIA). 2001. *The changing structure of the electric power industry 2000: An update*. DOE/EIA-0562(00). Washington, DC.

Massachusetts Institute of Technology. 2003. *The future of nuclear power: An interdisciplinary MIT study*. Cambridge, MA: MIT. ISBN 0-615-12420-8

Masters, G. M. 1998. *Introduction to environmental engineering and science*, 2nd ed. Upper Saddle River, NJ: Prentice Hall.

Masters, G. M. 2004. *Renewable and efficient electric power systems*. Hoboken, NJ: John Wiley & Sons.

Nakicenovic, N., 1996. Energy primer. In *Climate Change 1995, Impacts, Adaptations and Mitigation of Climate Change: Scientific-Technical Analyses*, Intergovernmental Panel on Climate Change. Cambridge, UK: Cambridge University Press.

U.S.-Canada Power System Outage Task Force. 2004. *Final report on the August 14, 2003. Blackout in the United States and Canada: Causes and Recommendations*. https://reports.energy.gov/ BlackoutFinal-Web.pdf

Chapter 10

Aisin SRI/USEPA-GHG-VR-37. Environmental Technology Verification Program, September 2005. http://www.epa.gov/etv/pdfs/vrvs/03_vs_aisin.pdf

Ceder, G. 2005. *Materials design with first principles computations*. The Reynolds Lecture, Stanford University.

Kempton, W., and J. Tomic. 2005. Vehicle-to-grid power fundamentals: Calculating capacity and net revenue. *J. of Power Sources*.

Kintner-Meyer, M. 2007. *Regional PHEV demonstration: A grid perspective*. Pacific Northwest National Laboratory, PNNL-SA-55212, November.

Masters, G. 2004. *Renewable and efficient electric power systems*. Hoboken, NJ: Wiley Interscience.

Swisher, J. N. 2002. *Cleaner energy, greener profits: Fuel cells as cost-effective distributed energy resources*. Snowmass, CO: Rocky Mountain Institute.

Yoshida, Y., N. Hisatome, and K. Takenobu. 2003. Development of SOFC for products, Mitsubishi heavy industries. *Technical Review* 40 (4) (August).

Chapter 11

Chang, A. 2000. *Residential systems summary report: Analysis of TEAM-UP residential installations*. Washington, DC: Utility Photovoltaic Group.

Chaudhari, M., L. Frantzis, and T. Hoff. 2004. *PV grid connected market potential under a cost breakthrough scenario.* Navigant Consulting report for the Energy Foundation, San Francisco, September.

ERDA/NASA. 1977. *Terrestrial photovoltaic measurement procedures.* ERDA/NASA/1022-77/16, NASA TM 73702. Cleveland, Ohio.

Masters, G. M. 2004. *Renewable and efficient electric power systems.* Hoboken, NJ: Wiley Interscience.

National Renewable Energy Laboratory. 1994. *Solar radiation data manual for flat-plate and concentrating collectors.* Golden, CO: NREL/TP-463-5607.

Shockley, W., and H. J. Queisser. 1961. *J. Applied Physics* 32:510.

Swanson, R. 2004. *A vision for crystalline silicon solar cells.* Sunnyvale, CA: SunPower Corporation.

Wiser, R., M. Bolinger, P. Cappers, and R. Margolis. 2006. *Letting the sun shine on solar costs: An empirical investigation of photovoltaic cost trends in California.* Lawrence-Berkeley National Laboratory, January.

Chapter 12

Archer, C. L., and M. Z. Jacobson. 2005. Evaluation of global wind power. *J. Geophys. Res.*: 110.

Erickson, W. P., G. Johnson, D. Strickland, and D. Young. 2002. *Avian collisions with wind turbines: A summary of existing studies and comparisons to other sources of avian collision mortality in the United States.* 2002 American Wind Energy Association Conference. Portland.

Global Wind Energy Council. 2006. Record year for wind energy: Global wind power market increased by 43% in 2005. http://www.gwec.net. March 2006.

Jacobson, M. Z., and G. M. Masters. 2001. Exploiting wind versus coal. *Science* 293 (August): 1438.

Masters, G. M. 2004. *Renewable and efficient electric power systems.* New Jersey: Wiley Interscience.

National Renewable Energy Laboratory. 1987. *Wind energy resource atlas of the United States.* Golden, CO.

Pasqualetti, M. J. 2004. Wind power: Obstacles and opportunities. *Environment* 46 (7) (September): 22–38.

U. S. Government Accounting Office. 2005. *Wind power impacts on wildlife and government responsibilities for regulating development and protecting wildlife.* GAO-050906, September 2005.

Wiser, R., and M. Bolinger. 2007. *Annual report on U.S. wind power installation, cost, and performance trends: 2006.* U.S. Department of Energy, Energy Efficiency and Renewable Energy, Washington, DC (May).

Chapter 13

Argonne National Laboratory. 2007. Argonne to Lead DOE's Effort to Evaluate Plug-in Hybrid Technology. Retrieved from http://www.anl.gov/Media_Center/News/2006/ES061201.html

Brinkman, N., M. Wang, T. Weber, and T. Darlington. 2005. Well-to-wheels analysis of advanced fuel/vehicle systems—A North American study of energy use, greenhouse gas emissions, and criteria pollutant emissions.

California Energy Commission. 2007. Integrated Energy Policy Report. Sacramento. Retrieved from http://www.energy.ca.gov/2007publications/CEC-100-2007-008/CEC-100-2007-008-CMF.PDF

Demirdöven, N. and J. Deutch. 2004. Hybrid Cars Now, Fuel Cell Cars Later. *Science.* 305 (5686): 974–976.

Dierkers, G. 2005. Reducing Frieght GHGs" What are the Possibilities? Freight Solutions Dialogue. Center for Clean Air Policy.

Eberhard, M. and M. Tarpenning. 2006. The 21st Century Car. Tesla Motors.

Greene, N. 2004. Growing energy: How biofuels can help end America's oil dependence. Natural Resources Defense Council. http://www.nrdc.org/air/energy/biofuels/biofuels.pdf

Kenworthy, J. R. 2003. Transport energy use and greenhouse gases in urban passenger transport systems: A study of 84 global cities. Third Conference of the Regional Government Network for Sustainable Development. September 17–19, 2003.

Khosla, V. 2006. A near term energy solution. A white paper by Vinod Khosla (April). http://www
.khoslaventures.com/resources.html

Kierkers, G. 2005. Reducing freight GHGs: What are the possibilities? Center for Clean Air Policy, 2005.

Lovins, A., E. Kyle Datta, O.-E. Bustnes, J. Koomey, and N. Glasgow. 2004. *Winning the oil endgame: American innovation for profits, jobs, and security.* Snowmass, CO: Rocky Mountain Institute.

Lovins, A., and D. Cramer. 2004. Hypercars®, hydrogen, and the automotive transition. International Journal of Vehicle Design. 35 (1/2): 50–85.

Pew Center on Global Climate Change. 2004. Comparison of Passenger Vehicle Fuel Economy and Greenhouse Gas Emission Standards around the World. F. An and A. Sauer. Retrieved from http://www.pewclimate.org/docUploads/Fuel%20Economy%20and%20GHG%20Standards _010605_110719.pdf

Rosenfeld, A. 2006. Fermi Award Talks. June 21–22, 2006. California Energy Commission. Retrieved from http://www.energy.ca.gov/commission/commissioners/rosenfeld.html

Sanna, L. 2005. Driving the solution: The plug-in hybrid vehicle. *EPRI Journal* (Fall).

Union of Concerned Scientists. Clean Vehicles. Retrieved from http://www.ucsusa.org/clean_vehicles/

U.S. Department of Energy. 2006. Transportation energy data book. Davis, S., and S. Diegel. Oak Ridge National Laboratory. Edition 25.

U.S. Department of Energy. 2007. *Transportation energy data* book. Davis, S., and S. Diegel. Oak Ridge National Laboratory. Edition 26.

U.S. Energy Information Administration. 2006. International Energy Outlook. Washington, DC. Retrieved from http://www.eia.gov

U.S. Energy Information Administration. 2007. International Petroleum Prices. Washington, DC. Retrieved from http://www.eia.doe.gov/emeu/international/oilprice.html

U.S. Department of Transportation, NHTSA. 2004. Summary of Fuel Economy Performance. Washington, DC.

U.S. Environmental Protection Agency. 1994. Automobile Emissions: An Overview. Office Of Mobile Sources. EPA 400-F-92-007

U.S. Environmental Protection Agency. 2005. Tier 2 Vehicle and Gasoline Sulfur Program. Retrieved from http://www.epa.gov/tier2/basicinfo.htm

Wang, M. 2005. Well-to-wheels results of advanced vehicle systems with new transportation fuels. Center for transportation research, Argonne National Laboratory, Advanced Transportation Workshop, Global Climate and Energy Project, Stanford University, CA, October 10–11, 2005

Weiss, M., J. Heywood, A. Schafer, E. Drake, F. AuYeung. 2000. On the road in 2020: A life cycle analysis of new automobile technologies. MIT Energy Lab Report EL 00-003. http://lfee.mit.edu/public/ el00-003.pdf

Chapter 14

Anderson, M., and D. Gardiner. 2006. Climate risk and energy in the auto sector: Guidance for investors and analysts on key off-balance sheet drivers. Commissioned by Ceres.

Badger, P. Ethanol from cellulose: A general review. From J. Janick and A. Whipkey (eds). *Trends in New Crops and New Uses.* Alexandria, VA: ASHS Press.

California Energy Commission. 2005. Alternative fuels commercialization in support of the 2005 integrated energy policy report, Technical recommendations for B20 fleet use based on existing data. B20 Fleet Evaluation Team.

Chen, C., and N. Greene. 2003. Is landfill gas green energy? Natural Resources Defense Council.

Farrell, A., R. Plevin, B. Turner, A. Jones, M. O'Hare, and D. Kammen. 2006. Ethanol can contribute to energy and environmental goals. *Science* 311 (27 January 2006): 506.

Gerlach, T. 2005. Minnesota's transition to ethanol and biodiesel driving blends fuel blends. Webcast (16 March 2005). Outdoor Air Programs, Twin Cities Clean Cities Coalition, American Lung Association of Minnesota, http://www.Cleanairchoice.Org

Greene, N. 2004. Growing energy: How biofuels can help end America's oil dependence. Washington: Natural Resources Defense Council. Retrieved from http://www.nrdc.org/air/energy/biofuels/contents.asp

Greene, N., and Y. Mugica. 2005. Bringing biofuels to the pump: An aggressive plan for ending America's oil dependence. Washington: Natural Resources Defense Council.

Greer, D. 2005. Creating cellulosic ethanol: Spinning straw into fuel. eNews Bulletin.

Harrison. W. Office of the Secretary of Defense (OSD) Assured Fuels Initiative. The Drivers for Alternative Aviation Fuels. U.S. Department of Defense.

Heaton, E., J. Clifton-Brown, T. Voigt, M. Jones, and S. Long. 2004. *Miscanthus* for renewable energy generation: European Union experience and projections for Illinois. *Mitigation and Adaptation Strategies for Global Change* 9:433–451.

International Energy Agency. 2004. Biofuels for Transport, An International Perspective.

Jacobson, M. 2007. Effects of ethanol (E85) versus gasoline vehicles on cancer and mortality in the United States. *Environmental Science and Technology*. April 18 online.

Karekezi, S., K. Lata, S. Teixeira. 2004. Traditional biomass energy: Improving its use and moving to modern energy use. Secretariat of the International Conference for Renewable Energies, Bonn.

Khosla, V. 2006. A near term energy solution. A white paper by Vinod Khosla (April). http://www.khoslaventures.com/resources.html

Kohl, S. 2003. 2004. Ethanol 101. Ten part series on ethanol production processes. *Ethanol Today* (July 2003–June 2004).

Milbrandt, A. 2005. A geographic perspective on the current biomass resource availability in the United States. Golden, CO: National Renewable Energy Laboratory.

Minnesota, State of, Department of Commerce 2007. Energy Info Center. E-85 Fuel Use Data.

National Biodiesel Board. 2007. Commercial Biodiesel Production Plants . . . and under Construction and Expansion. Retrieved from http://www.biodiesel.org/buyingbiodiesel/producers_marketers/ProducersMap-Existing.pdf

National Renewable Energy Laboratory. 1998. J. Sheehan, T. Dunahay, J. Benemann, P. Roessler. A Look Back at the U.S. Department of Energy's Aquatic Species Program—Biodiesel from Algae. NREL/TP-580-24190

National Renewable Energy Laboratory. 2005. A. Milbrandt, A Geographic Perspective on the Current Biomass Resource Availability in the United States. Golden, CO. Retrieved from http://www.nrel.gov/docs/fy06osti/39181.pdf

Natural Resources Defense Council and Climate Solutions. 2006. Ethanol: Energy well spent. A survey of studies published since 1990.

Passmore, J. 2006. Cellulose is ready to go. Presentation to Governor's Ethanol Coalition and US EPA Environmental Meeting "Ethanol and the Environment." Executive Vice President, Iogen Corporation.

Perlack, R., L. Wright, A. Turhollow, R. Graham, B. Stokes, and D. Erbach. 2005. Biomass as feedstock for a bioenergy and bioproducts industry: The technical feasibility of a billion-ton annual supply. DOE/GO-102005-2135, ORNL/TM-2005/66. Oak Ridge, TN: Oak Ridge National Lab.

Ragauskas, A., C. Williams, B. Davison, G. Britovsek, J. Cairney, C. Eckert, W. Frederick Jr., J. Hallett, D. Leak, C. Liotta, J. Mielenz, R. Murphy, R. Templer, and T. Tschaplinski. 2006. The path forward for biofuels and biomaterials. *Science* 311 (27 January 2006): 484.

Renewable Fuels Association. 2007. Ethanol Industry Outlook 2007. Building New Horizons. Retrieved from http://www.ethanolrfa.org/

Renewable Fuels Association. 2007. Industry Statistics. Retrieved from http://www.ethanolrfa.org/

Sheehan, J., T. Dunahay, J. Benemann, and P. Roessler. 1998. A look back at the U.S. Department of Energy's aquatic species program—biodiesel from algae. National Renewable Energy Laboratory.

Sieger, R., and P. Brady. 2004. High performance anaerobic digestion. White paper. Water Environment Federation, Residuals and Biosolids Committee, Bioenergy Technology Subcommittee.

Smith, S., M. Wise, G. Stokes, J. Edmonds. 2004. Near-term US biomass potential: Economics, land-use, and research opportunities. Battelle Memorial Institute. Joint Global Change Research Institute.

U.S. Department of Agriculture. 2007. Economic Research Service. Feed Grain Database. Retrieved from http://www.ers.usda.gov/Data/FeedGrains/

U.S. Department of Energy. 2002. Roadmap for biomass technologies in the United States. Biomass Research and Development Technical Advisory Committee.

U.S. Department of Energy. 2003. Roadmap for agriculture biomass feedstock supply in the United States. Office of Energy Efficiency and Renewable Energy, Biomass Program.

U.S. Department of Energy. 2007. Alternative Fuel Data Canter. Retrieved from http://www.eere.energy .gov/afdc/data/fuels.html

U.S. Department of Energy. 2007. Alternative Fuel Price Report. Retrieved from http://www.eere.energy .gov/afdc/price_report.html

U.S. EPA. 2002. A Comprehensive Analysis of Biodiesel Impacts on Exhaust Emissions. EPA 420-P-02-00. Retrieved from http://www.epa.gov/OMS/models/analysis/biodsl/p02001.pdf

U.S. EPA. 2005. Municipal Solid Waste Generation, Recycling, and Disposal in the United States. Facts and Figures. Retrieved from http://www.epa.gov/msw/msw99.htm

U.S. EPA Landfill Methane Outreach Project. 2006. Energy Projects and Candidate Landfills. Retrieved from http://www.epa.gov/lmop/proj/index.htm

Wright, L., B. Boundy, R. Perlack, S. Davis, and B. Saulsbury. 2006. Biomass energy data book. U.S. Department of Energy. Oak Ridge National laboratory.

Wu, M., Y. Wu, and M. Wang. 2005. Mobility chains analysis of technologies for passenger cars and light-duty vehicles fueled with biofuels: Application of the GREET model to the role of biomass in America's energy future (RBAEF) project. Center for Transportation Research, Argonne National Laboratory.

Yacobucci, B. 2006. Fuel ethanol: Background and public policy issues. CRS Report for Congress. RL33290. Congressional Research Service, Library of Congress. Washington, DC.

Chapter 15

American Public Transportation Association. 2006. *Public transportation fact book*. 57th ed. Washington, DC.

American Public Transportation Association. 2007. *Public transportation fact book*. 57th ed. Washington, DC.

Bailey, L. 2007. Public transportation and petroleum savings in the U.S.: Reducing dependence on oil, ICF International, prepared for APTA.

Bureau of Transportation Statistics. 2006. National Transportation Statistics.

Burroughs, T., G. Fitzgerald, and B. Lee. 2006. Clean Air and Climate Protection Software Online Training Session. International Council for Local Environmental Initiatives (ICLEI).

California, Department of Transportation. 2005. The Transit-Oriented Development Compendium.

California Energy Commission. 2001. Shining PLACE³S: Sacramento and national examples of smart growth. Retrieved from http://www.energy.ca.gov/places/index.html

California Energy Commission. 2001.

California Environmental Protection Agency, Air Resources Board. 1995. Transportation related land use strategies to minimize motor vehicle emissions: An indirect source research study.

Calthrope, P. 1993. *The next American metropolis: Ecology, community and the American dream*. Princeton Architectural Press.

Calthrope, P. 2002. *The Urban Network: A New Framework for Growth*. Calthorpe Associates.

Calthorpe, P. and W. Fulton. 2001. *The Regional City. Washington*: Island Press.

Center for Clean Air Policy. Transportation emissions guidebook. http://www.ccap.org/guidebook

Corbett, M. 1981. *A better place to live*. Rodale Press.

Frece, J. 2000. Smart Growth in Maryland. Presentation in Blacksburg, VA.

International Council for Local Environmental Initiatives (ICLEI). 2003. Clean air and climate protection software users' guide.

Jaffe, M., and D. Erley. 1980. Protecting solar access for residential development. A guidebook for planning officials. American Planning Association. U.S. Department of Housing and Urban development, U.S. Department of Energy. Washington: U.S. Government Printing Office.

Jonathan Rose Companies. 2004. Environmentally responsible development. *Developing Times*. Newsletter 6.

Kenworthy, J. R. 2003. Transport energy use and greenhouse gases in urban passenger transport systems: A study of 84 global cities. Third Conference of the Regional Government Network for Sustainable Development, September 17–19.

Lang, R. E. and A. C. Nelson. 2007. America 2040: The rise of the megapolitans. *Planning*. American Planning Association. January.

Oregon Department of Energy, Washington State Energy Office, California Energy Commission. 1996. *The energy yardstick: Using PLACE3S to create more sustainable communities*. Prepared for the Center for Excellence for Sustainable Development, U.S. Department of Energy. Retrieved from http://www.energy.ca.gov/places/index.html

Randolph, J. 2004. *Environmental land use planning and management*. Covelo, CA: Island Press.

Southworth, F. 2001. On the potential impacts of land use change policies on automobile vehicle miles of travel. *Energy Policy* 29 (14): 1271–1283.

San Joaquin Valley Air Pollution Control District. 2006. Rule 9510 – Indirect Source Review. Fresno, CA.

Stone, B. 2006. Urban sprawl and air quality in large U.S. cities. Presentation at Association of Collegiate Schools of Planning.

Stone, B. 2007. Urban sprawl and air quality in large U.S. cities. *Journal of Environmental Management*, in press.

Stone, B., and M. Rodgers. 2001. Urban form and thermal efficiency: How the design of cities influences the urban heat island effect. *Journal of the American Planning Association* 67 (2): 186–198.

U.S. Department of Energy. 2007. Transportation energy data book. Davis, S., and S. Diegel. Oak Ridge National Laboratory. Edition 26.

U.S. Environmental Protection Agency. 1992. Cooling Our Communities—A Guidebook on Tree Planting and Light-Colored Surfacing. EPA 22P-2001.

U.S. Green Building Council. 2005. LEED-ND: LEED for neighborhood developments rating system—preliminary draft.

Van der Ryn, S., and P. Calthorpe. 1984. *Sustainable communities*. San Francisco: Sierra Club Books.

Chapter 16

Arthur D. Little, Inc. 2001. Energy savings potential of solid state lighting in general lighting applications. U.S. Department of Energy.

Beurskens, L. 2003. Experience curve analysis: Concerns and pitfalls in data use. PHOTEX Workshop—June 2003 ECN Policy Studies.

Byrne, J., N. Toly, and L. Glover (ed.). 2006. Transforming power: Energy, environment, and society in conflict. London: Transaction Publishers.

Database of State Incentives for Renewable Energy (DSIRE). 2006. North Carolina Renewable Energy Center. http://www.dsireusa.org/

Duke, R., and D. Kammen. 1999. The economics of energy market transformation programs. *The Energy Journal* 20 (4): 15–64.

Federal Reserve Bank of Dallas. 1996. Annual Report. The economy at Light Speed: Technology and Growth in the Information Age—and Beyond. Dallas, TX.

Geller, H. 2003. Energy revolution: Policies for a sustainable future. Washington: Island Press.

Geller, H., and S. Attali. 2005. The experience with energy efficiency policies and programmes in IEA countries: Learning from the critics. International Energy Agency.

Glover, L. 2006. From love-ins to logos: Charting the demise of renewable energy as a social movement. In Byrne, J., N. Toly, and L. Glover (ed.). 2006. *Transforming Power: Energy, Environment, and Society in Conflict*. London: Transaction Publishers.

Holt, M., and C. Glover. 2006. Energy policy act of 2005: Summary and analysis of enacted provisions. CRS Report for Congress. RL33302. Congressional Research Service, Library of Congress. Washington, DC.

Institute for Local Self-Reliance. 2005. Democratic energy: Communities and government working on our energy future. Web site: http://www.newrules.org/de/aboutde.html

Jenkins, N., L. Campoy, E. Becker, J. Livingston. 2004. Emerging Technologies, energy Efficiency, Roles and Linkages. American Council for an Energy Efficient Economy (ACEEE). San Francisco.

Kammen, D. 2005. An energy policy for the 21st century. *Policy Matters:* 14–19.

Kammen, D., and G. Nemet. 2005. Reversing the incredible shrinking energy R&D budget. *Issues in Science and Technology* (Fall 2005):84–88.

Kulakowski, S. 1999. Large organizations' investments in energy-efficient building retrofits. Energy Analysis Department, Environmental Energy Technologies Division, Ernest Orlando Lawrence Berkeley National Laboratory. Retrieved from http://enduse.lbl.gov/projects/mktimperfect.html

LaCommare, K., J. Edwards, E. Gumerman, and C. Marnay. 2005. Distributed generation potential of the U.S. commercial sector. Environmental Energy Technologies Division. May 2005. Ernest Orlando Lawrence Berkeley National Laboratory. Retrieved from http://eetd.lbl.gov/ea/EMS/EMS_pubs.html

Lazzari, S. 2006. Energy tax policy. CRS Report for Congress. IB10054. Congressional Research Service, Library of Congress. Washington, DC.

Lovins, A., E. Kyle Datta, O.-E. Bustnes, J. Koomey, and N. Glasgow. 2004. *Winning the oil endgame: American innovation for profits, jobs, and security*. Snowmass, CO: Rocky Mountain Institute.

Mallon, K. (ed.). 2006. *Renewable Energy Policy and Politics*. London: Earthscan.

Margolis, R. 2003. Photovoltaic technology experience curves and markets. Presentation at NCPV and Solar Program Review Meeting. Denver, Colorado (March 24).

Margolis, R., and D. Kammen. 1999. Underinvestment: The energy technology and R&D policy challenge. *Science* 285:690–692.

Nadel, S., A. deLaski, M. Eldridge, and J. Kleisch. 2006. Leading the way: Continued opportunities for new state appliance and equipment efficiency standards. Report # ASAP-6/ACEEE-A062. American Council for an Energy-Efficient Economy, Appliance Standards Awareness Project.

Organisation for Economic Development and Cooperation and International Energy Agency. 2000. Experience Curves for Energy Technology Policy. Paris.

Rufo, M. and Coito, F. 2002. *California's Secret Energy Surplus: The Potential for Energy Efficiency*. San Francisco, CA: The Energy Foundation and The Hewlett Foundation. http://www.ef.org/documents/Secret_Surplus.pdf

Runci, P. 2005. Renewable energy policy in Germany: An overview and assessment. The Joint Global Change Research Institute. Retrieved from http://www.globalchange.umd.edu/?energytrends&page=germany.

Sathaye, J., S. Murtishaw. 2004. Market failures, consumer preferences, and transaction costs in energy efficiency purchase decisions. California Climate Change Center, Report Series Number 2005-2010. Lawrence Berkeley National Laboratory CEC-500-2005-020

Smil, V. 2003. *Energy at the crossroads*. Cambridge, MA: The MIT Press.

U.S. Department of Energy. 2005. Basic Research Needs for Solar Energy Utilization. Retrieved from http://www.sc.doe.gov/bes/reports/files/SEU_rpt.pdf

Chapter 17

Berner, J. 2006. German PV-industry: Market leader and technology trendsetter. Federal Ministry of Economics and Technology, Renewable Energy. Retrieved from http://www.german-renewable-energy.com/Renewables/Redaktion/PDF/de/de-RE-Asia-2006-Photovoltaics-Berner,property=pdf,bereich=renewables,sprache=de,rwb=true.pdf

Database of State Incentives for Renewable Energy (DSIRE). 2006. North Carolina Renewable Energy Center. http://www.dsireusa.org/

Ellerman, A. D., P. Joskow, and D. Harrison. 2003. Emissions trading in the U.S.: Experience, lessons and considerations for greenhouse gases. Pew Center for Global Climate Change.

Eto, J. 1996. The past, present, and future of U.S. utility demand-side management programs. Berkeley, CA: EOLBNL.

European Commission. 2005a. Green paper on energy efficiency. Directorate-General for Energy and Transport.

European Commission. 2005b. EU action on climate change: EU emissions trading—an open scheme to promote global innovation.

European Commission. 2006. Green paper: A European Strategy for sustainable, Competitive, and Secure Energy.

Gallagher, K., A. Sagar. 2007. Federal energy technology spending, 1978–2007. Belfer Center for Science and International Affairs, Harvard University. Retrieved from http://www.hcs.harvard.edu/~hejc/papers/FederalEnergyTechSpending1978-2007.xls

Geller, H. 2003. *Energy revolution: Policies for a sustainable future*. Washington: Island Press.

Geller, H., P. Harrington, A. Rosenfeld, S. Tanishima, and F. Unander. 2006. Policies for increasing energy efficiency: Thirty years of experience in OECD countries. International Energy Agency.

Global Wind Energy Council. 2007. Global Wind 2006 Report. Retrieved from http://www.gwec.net/fileadmin/documents/Publications/gwec-2006_final_01.pdf

Holt, M. 2006. Nuclear power issues. CRS Report for Congress. IB88090. Congressional Research Service, Library of Congress. Washington, DC.

Holt, M., and C. Glover. 2006. Energy policy act of 2005: Summary and analysis of enacted provisions. CRS Report for Congress. RL33302. Congressional Research Service, Library of Congress. Washington, DC.

Institute for Local Self-Reliance. 2005. Democratic energy: Communities and government working on our energy future. Web site: http://www.newrules.org/de/aboutde.html

Intergovernmental Panel on Climate Change. 2007. Climate change 2007: Summary for policy makers. IPPC Fourth Assessment Report (4AR). Retrieved from http://www.ipcc.ch/

International Energy Agency. 2006a. Global Energy Indicators Database and Maps. http://www.iea.org/textbase/subjectqueries/maps/world/tpes.htm

International Energy Agency. 2006b. Dealing with climate change. Policies and measures database http://www.iea.org/textbase/envissu/pamsdb/index.html

International Energy Agency. 2006c. Global Renewable Energy Policies and Measures Database. http://www.iea.org/textbase/pamsdb/grindex.aspx

Jacobsson, S., & Lauber, V. (2006). The politics and policy of energy system transformation—Explaining the German diffusion of renewable energy technology. *Energy Policy, 34*(3):256–276.

Kammen, D., and G. Nemet. 2005. Reversing the incredible shrinking energy R&D budget. *Issues in Science and Technology* (Fall 2005): 84–88.

Koizumi, K. 2007. Federal Funding for Clean and Renewable Technology. CTSI Meeting. AAAS R&D Budget and Policy Program. Retrieved from http://www.aaas.org/spp/rd

Lazarri, S. 2006. Energy tax policy. CRS Report for Congress. IB10054. Congressional Research Service, Library of Congress. Washington, DC.

Lin, J. 2007. Energy conservation investments: A comparison between China and the US. Energy Policy 35 (2).

Margolis, R., and D. Kammen. 1999. Underinvestment: The energy technology and R&D policy challenge. *Science* 285:690–692.

McCarthy, J. 2006. Clean Air Act Issues in the 109th Congress. CRS Report for Congress. Congressional Research Service Report IB10137.

Nadel, S., A. deLaski, M. Eldridge, and J. Kleisch. 2006. Leading the way: Continued opportunities for new state appliance and equipment efficiency standards. Report # ASAP-6/ACEEE-A062. American Council for an Energy-Efficient Economy, Appliance Standards Awareness Project.

National Atmospheric Deposition Program. 2007. Isopeth maps. Retrieved from http://nadp.sws.uiuc.edu/amaps2/

National Commission on Energy Policy. 2004. Ending the energy stalemate: A bipartisan strategy to meet America's energy challenges.

Pew Center on Global Climate Change. 2007. Senate Greenhouse Gas Cap-And-Trade Proposals in the 110th Congress.

Reel, M. 2006. Brazil's road to energy independence. *Washington Post* (20 August 2006).

Rosenfeld, A. 2005. Extreme efficiency, lessons from California. Presentation to AAAS Conference, Washington, DC (February 21). Retrieved from http://www.energy.ca.gov/commission/commissioners/rosenfeld.html

Runci, P. 2005. Renewable energy policy in Germany: An overview and assessment. The Joint Global Change Research Institute. Retrieved from http://www.globalchange.umd.edu/?energytrends&page=germany.

Shorrock, L. 2005. Assessing the effects of energy efficiency policies applied to the UK housing stock. *Proceedings of the 2005 ECEEE Summer Study on Energy Efficiency*. Paris: European Council for an Energy-Efficient Economy: 933–945.

Shum, K., and C. Watanabe. 2007. Photovoltaic deployment strategy in Japan and the USA—an institutional appraisal. *Energy Policy* 35(2).

United Nations Environment Programme. 2000. The Montreal Protocol on Substances that Deplete the Ozone Layer. UNEP Ozone Secretariat

United Nations Framework Convention on Climate Change (UNFCCC). 2006. Annex I GHG Emission Data. Retrieved from http://unfccc.int/ghg_emissions_data/items/3800.php

Urge-Vorsatz, D. G. Miladinova, and L. Paizs. 2006. Energy in transition: From the iron curtain to the European Union. *Energy Policy* 34(15).

U.S. Department of Energy, Alternative Fuels Data Center. Energy efficiency and renewable energy. Federal and State Incentives and Laws for Alternative Fuels http://www.eere.energy.gov/afdc/laws/incen_laws.html

U.S. Energy Information Administration. 2007b. International Energy Annual 2005.

U.S. Energy Information Administration. 2007. Monthly Energy Review.

U.S. Energy Information Administration. 2006. Annual Energy Review, 2005.

U.S. Environmental Protection Agency. 2005. Cap and trade: Acid rain program results. Retrieved from www.epa.gov/airmarkets/cap-trade/docs/ctresults.pdf

Wustenhagen, R., and M. Bilharz. 2006. Green energy market development in Germany: effective public policy and emerging customer demand. *Energy Policy* 34(13).

Yacobucci, B. 2006. Fuel ethanol: Background and public policy issues. CRS Report for Congress. RL33290. Congressional Research Service, Library of Congress. Washington, DC.

Chapter 18

American Planning Association. 2002. Planning for Smart Growth: Stat of the States. Chicago

Austin energy. 2007. GreenChoice program. Retrieved from http://www.austinenergy.com/Energy%20Efficiency/ Programs/Green%20Choice/index.htm

Boulder, City of. 2003. Residential building guide: Green points program guidelines. Planning and Development Services. Retrieved from http://www.ci.boulder.co.us/cao/brc/10-7.html

Boulder, City of. 2006. Climate Action Plan. Retrieved from http://www.bouldercolorado.gov/files/ Environmental%20Affairs/climate%20and%20energy/cap_final_14aug06.pdf

Boulder, City of. 2008. Residential Building Guide. Green Building and Green Points Guideline Booklet. Retrieved from http://www.bouldercolorado.gov/files/PDS/boards/Planning%20Board/1_17 _08GreenPoints/greenpoints_booklet.pdf

City of San Francisco. http://www.local.org/sfccaord.pdf

California, State of. 2005. *Energy Action Plan II: Implementation Roadmap for Energy Policies.* Energy Commission and Public Utility Commission. Sacramento (September).

California Energy Commission. 2005. Emerging Renewables Program.

California Local Government Commission. 2006. Community Choice Aggregation.

California Energy Commission. 2007. 2007 Integrated Energy Policy Report. Approved December 2007. Retrieved from http://www.energy.ca.gov/2007_energypolicy/index.html

California Energy Commission. 2007b. New Solar Homes Partnership Guidebook. Retrieved from http://www.gosolarcalifornia.ca.gov/documents/CEC-300-2007-008-CMF.PDF

California Public Utilities Commission. 2005. Final Report on California Public Utilities Commission Process to Implement Community Choice Aggregation (Task 2).

Caltrans. California Department of Transportation. 2002. Statewide Transit-Oriented Development Study: Factors for Success in California. Sacramento.

Caltrans. California Department of Transportation. 2006. TOD database http://transitorienteddevelopment .dot.ca.gov/miscellaneous/NewHome.jsp

Center for Advancement of Energy Markets. 2003. Electricity retail energy deregulation index.

Chicago, City of. 2001. Energy plan. Department of Environment.

Congressional Research Service. 2005. State Policies. Retrieved from http://www.eere.energy.gov/greenpower/ markets/states.shtml

Database of State Incentives for Renewable Energy (DSIRE). 2007. North Carolina Renewable Energy Center. http://www.dsireusa.org/

Eto, J. 1996. *The past, present, and future of U.S. utility demand-side management programs.* Berkeley, CA: EOLBNL.

Geller, H. 2003. *Energy revolution: Policies for a sustainable future.* Washington: Island Press.

Green Power Network. 2007. State Policies. Retrieved from http://www.eere.energy.gov/greenpower/ markets/states.shtml

Hoff, T. 2002. Final Results Report with a Determination of Stacked Benefits of Both Utility-Owned and Customer-Owned PV Systems. Sacramento Municipal Utility District.

Institute for Local Self-Reliance. 2005. Democratic Energy: Communities and government working on our energy future. Web site: http://www.newrules.org/de/aboutde.html

Kron, N. F., and J. Randolph. 1983. *Problems in implementing energy programs in selected United States communities.* ANL/CNSV-TM-139. Argonne National Laboratory. Argonne, IL. 41 pp.

Kushler, M., D. York, and P. Witte. 2004. Five years in: An examination of the first half-decade of public benefits energy efficiency policies. RN U041. Washington, DC: ACEEE.

Nadel, S. 2006. *Energy Efficiency Resource Standards: Experience and Recommendations.* ACEEE report. E063. Retrieved from http://www.aceee.org/pubs/e063.pdf

Nadel, S., A. deLaski, M. Eldridge, and J. Kleisch. 2006. Leading the way: Continued opportunities for new state appliance and equipment efficiency standards. Report # ASAP-6/ACEEE-A062. American Council for an Energy-Efficient Economy, Appliance Standards Awareness Project.

National Commission on Energy Policy. 2004. Ending the energy stalemate: A bipartisan strategy to meet America's energy challenges. .

New York City Energy Policy Task Force. 2004. New York City energy policy: An electricity resource roadmap.

North Carolina State Energy Office. 2005. North Carolina State energy Plan, June 2003 (revised January 2005).

Oregon Department of Energy, Washington State Energy Office, California Energy Commission. 1996. *The energy yardstick: Using PLACE3S to create more sustainable communities.* Prepared for the Center for Excellence for Sustainable Development, U.S. Department of Energy. Retrieved from http://www .energy.ca.gov/places/index.html

Pew Center on Global Climate Change. 2007. What's being done . . . in the States. Retrieved from http://www.pewclimate.org/what_s_being_done/in_the_states

Portland, City of. 2000. 1990 Energy Policy: Impacts and Achievements. Retrieved from http://www .portlandonline.com/shared/cfm/image.cfm?id=111740

Prindle, W. N. Dietsch, R. N. Elliot, M. Kushler, T. Langer, and S. Nadel. 2003. Energy efficiency's next generation: Innovation at the state level. RN E031. Washington, DC: ACEEE

Public Utilities Commission of Ohio (PUCO). Energy government aggregation: Local community buying power. http://www.puco.ohio.gov/PUCO/Consumer/information.cfm?doc_id=102

Puget Sound Clean Air Agency, Climate protection Committee. 2004. Retrieved from http://www .pscleanair.org/programs/climate/rptfin.pdf

Puget Sound Regional Council. 1999. Creating transit station communities in the central Puget Sound region – A transit-oriented development workbook.

Randolph, J. 1981. The local energy future: A compendium of community programs. *Solar Law Reporter* 3 (2) (July/August): 253–282.

Randolph, J. 1984. Energy conservation programs: A review of state initiatives in the USA. *Energy Policy* 12 (4) (December): 425–438.

Randolph, J. 1988. The limits of local energy programs: The experience of U.S. communities in the 1980s. Proceedings of the International Symposium: Energy Options for the Year 2000. Wilmington, DE: 41–50.

Rocky Mountain Institute. 2002. An energy resource investment strategy (ERIS) for the city and county of San Francisco. Retrieved from http://www.rmi.org/images/other/Energy/E02-16_AnalysisForSF.pdf. October 2005.

Sacramento Area Council of Governments. 2005. Sacramento region blueprint transportation-land use study: Preferred blueprint alternative. Retrieved from http://www.sacregionblueprint.org/ sacregionblueprint/the_project/2005-06-11-wrapup.cfm

Sacramento Municipal Utilities District (SMUD). Promotions, Rebates and Financing. Retrieved from http://www.smud.org/rebates/index.html

San Francisco Bay Area Rapid Transit. 2003. BART transit-oriented development guidelines.

San Francisco Public Utilities Commission, San Francisco Department of the Environment. 2002. Choosing San Francisco's Energy Future: The Electricity Resource Plan.

San Francisco Public Utilities Commission, San Francisco Department of the Environment 2004. Climate Action Plan for San Francisco. Retrieved from http://www.sfenvironment.org/downloads/library/ climateactionplan.pdf

Seattle City Light. 2004. Energy conservation accomplishments: 1977–2003. Evaluation Unit, Energy Management Services Division. Retrieved from http://www.seattle.gov/mayor/climate/

Texas Energy Planning Council. 2005. Texas energy plan 2005. Energy security for a Bright Tomorrow.

U.S. DOE Alternative Fuels Data Center. Energy efficiency and renewable energy. Federal and State Incentives and Laws for Alternative Fuels. Retrieved from http://www.eere.energy.gov/afdc/laws/ incen_laws.html

York, D., and M. Kushler. 2003. America's best: Profiles of America's leading energy efficiency programs. RN U032. Washington, DC: ACEEE.

Index

About the Authors

John Randolph is professor of Urban Affairs & Planning and director of the School of Public & International Affairs at Virginia Tech where he teaches environmental and energy planning. He is author of the popular Island Press textbook *Environmental Land Use Planning & Management*.

Gilbert Masters is professor emeritus of Civil & Environmental Engineering at Stanford University where he still teaches popular classes in energy systems. He is author of the textbooks *Renewable and Efficient Electric Power Systems* (John Wiley & Sons) and *Introduction to Environmental Science & Engineering* (Prentice Hall, 3rd edition).